Methods in Enzymology

Volume 193
MASS SPECTROMETRY

METHODS IN ENZYMOLOGY

EDITORS-IN-CHIEF

John N. Abelson   Melvin I. Simon

DIVISION OF BIOLOGY
CALIFORNIA INSTITUTE OF TECHNOLOGY
PASADENA, CALIFORNIA

FOUNDING EDITORS

Sidney P. Colowick and Nathan O. Kaplan

Methods in Enzymology

Volume 193

Mass Spectrometry

EDITED BY

James A. McCloskey

DEPARTMENT OF MEDICINAL CHEMISTRY
UNIVERSITY OF UTAH
SALT LAKE CITY, UTAH

ACADEMIC PRESS, INC.

Harcourt Brace Jovanovich, Publishers

San Diego   New York   Boston
London   Sydney   Tokyo   Toronto

QP601
C733
vol. 193

Academic Press, Inc.
San Diego, California 92101

United Kingdom Edition published by
Academic Press Limited
24-28 Oval Road, London NW1 7DX

Library of Congress Catalog Card Number: 54-9110

ISBN   0-12-182094-7   (alk. paper)

Printed in the United States of America
90 91 92 93   9 8 7 6 5 4 3 2 1

Table of Contents

Section I. General Techniques

JUL 22 1991

Section IV. Nucleic Acid Constituents

Appendixes

Contributors to Volume 193

Article numbers are in parentheses following the names of contributors.
Affiliations listed are current.

ANNE-SOPHIE ANGEL (32), *BioCarb Technology AB, S-22370 Lund, Sweden*

KALYAN R. ANUMULA (27), *Department of Analytical Chemistry, SmithKline Beecham Pharmaceuticals, King of Prussia, Pennsylvania 19406*

JOHN R. BARR (27), *Department of Physical and Structural Chemistry, SmithKline Beecham Pharmaceuticals, King of Prussia, Pennsylvania 19406*

KLAUS BIEMANN (13, 18, 25, A.5, A.6), *Department of Chemistry, Massachusetts Institute of Technology, Cambridge, Massachusetts 02139*

ROBERT K. BOYD (7), *Institute for Marine Biosciences, National Research Council, Halifax, Nova Scotia B3H 3Z1, Canada*

A. L. BURLINGAME (36, 37), *Department of Pharmaceutical Chemistry, School of Pharmacy, University of California-San Francisco, San Francisco, California 94143*

RICHARD M. CAPRIOLI (9), *Analytical Chemistry Center, and the Department of Biochemistry and Molecular Biology, University of Texas Medical School at Houston, Houston, Texas 77030*

STEVEN A. CARR (27), *Department of Physical and Structural Chemistry, SmithKline Beecham Pharmaceuticals, King of Prussia, Pennsylvania 19406*

KEITH L. CLAY (17), *Department of Pediatrics, National Jewish Center for Immunology and Respiratory Medicine, Denver, Colorado 80206*

PHILIP COHEN (26), *Department of Biochemistry, University of Dundee, Dundee DD1 4HN, Scotland*

CATHERINE E. COSTELLO (40, A.3), *Mass Spectrometry Facility, Department of Chemistry, Massachusetts Institute of Technology, Cambridge, Massachusetts 02139*

ROBERT J. COTTER (1), *Department of Pharmacology, Middle Atlantic Mass Spectrometry Facility, The Johns Hopkins University School of Medicine, Baltimore, Maryland 21205*

PAMELA F. CRAIN (42, 47), *Department of Medicinal Chemistry, University of Utah, Salt Lake City, Utah 84112*

ANNE DELL (35), *Department of Biochemistry, Imperial College of Science, Technology, and Medicine, London SW7 2AZ, England*

EDWIN DE PAUW (8), *Department of Chemistry, Liege University, B-4000 Liege, Belgium*

MIRAL DIZDAROGLU (46), *Center for Chemical Technology, National Institute of Standards and Technology, Gaithersburg, Maryland 20899*

GREGORY G. DOLNIKOWSKI (2), *USDA Human Nutrition Research Center at Tufts University, Boston, Massachusetts 02111*

BRUNO DOMON (33), *Physics Department, Ciba-Geigy Ltd., CH-4002 Basel, Switzerland*

CHARLES G. EDMONDS (22), *Chemical Sciences Department, Pacific Northwest Laboratory, Richland, Washington 99352*

HEINZ EGGE (38), *Institut für Physiologische Chemie, Universität Bonn, D-5300 Bonn 1, Federal Republic of Germany*

JAMES I. ELLIOTT (21), *Department of Molecular Biophysics and Biochemistry, Yale University, New Haven, Connecticut 06510*

SYD EVANS (3), *Kratos Analytical Instruments, Urmston, Manchester M31 2LD, England*

BRADFORD W. GIBSON (26), *Department of Pharmaceutical Chemistry, School of Pharmacy, University of California at San Francisco, San Francisco, California 94143*

BETH L. GILLECE-CASTRO (37), *Department of Pharmaceutical Chemistry, School of Pharmacy, University of California at San Francisco, San Francisco, California 94143*

GARY R. GRAY (31), *Department of Chemistry, University of Minnesota, Minneapolis, Minnesota 55455*

MICHAEL L. GROSS (6, 10), *Midwest Center for Mass Spectrometry, Department of Chemistry, University of Nebraska, Lincoln, Nebraska 68588*

GUNNAR C. HANSSON (39), *Department of Medical Biochemistry, Göteborg University, S-400 33 Göteborg, Sweden*

ALEX G. HARRISON (1), *Department of Chemistry, University of Toronto, Toronto, Ontario M5S 1A1, Canada*

ROGER N. HAYES (10), *Department of Chemistry, Midwest Center for Mass Spectrometry, University of Nebraska, Lincoln, Nebraska 68588*

CARL G. HELLERQVIST (30), *Department of Biochemistry, Vanderbilt University School of Medicine, Nashville, Tennessee 37232*

FRANZ HILLENKAMP (12), *Institut für Medizinische Physik, Universität Münster, D-4400 Münster, Federal Republic of Germany*

IAN JARDINE (24), *Analytical Biochemistry Division, Finnigan MAT, San Jose, California 95134*

KEITH R. JENNINGS (2), *Department of Chemistry, University of Warwick, Coventry CV4 7AL, England*

MICHAEL KARAS (12), *Institut für Medizinische Physik, Universität Münster, D-4400 Münster, Federal Republic of Germany*

HASSE KARLSSON (39), *Department of Medical Biochemistry, Göteborg University, S-400 33 Göteborg, Sweden*

KARL-ANDERS KARLSSON (34), *Department of Medical Biochemistry, University of Göteborg, S-400 33 Göteborg, Sweden*

DANIEL R. KNAPP (15), *Department of Cell and Molecular Pharmacology and Experimental Therapeutics, Medical University of South Carolina, Charleston, South Carolina 29425*

ROGER A. LAINE (29), *Departments of Biochemistry and Chemistry, Louisiana State University and The LSU Agricultural Center, Baton Rouge, Louisiana 70803*

TERRY D. LEE (19), *Division of Immunology, Beckman Research Institute of the City of Hope, Duarte, California 91010*

RONALD D. MACFARLANE (11), *Department of Chemistry, Texas A&M University, College Station, Texas 77843*

STEPHEN A. MARTIN (28), *Genetics Institute, Andover, Massachusetts 01810*

JAMES A. MCCLOSKEY (16, 41, 44, 45, A.1, A.4), *Department of Medicinal Chemistry, University of Utah, Salt Lake City, Utah 84112*

WALTER MCMURRAY (21), *Comprehensive Cancer Center, Yale University, New Haven, Connecticut 06510*

RICHARD M. MILBERG (14, A.2), *School of Chemical Sciences, University of Illinois, Urbana, Illinois 61801*

WILLIAM T. MOORE (9), *Analytical Chemistry Center, and the Department of Biochemistry and Molecular Biology, University of Texas Medical School at Houston, Houston, Texas 77030*

DIETER R. MÜLLER (33), *Physics Department, Ciba-Geigy Ltd., CH-4002 Basel, Switzerland*

ROBERT C. MURPHY (17), *Department of Pediatrics, National Jewish Center for Immunology and Respiratory Medicine, Denver, Colorado 80206*

BO NILSSON (32), *BioCarb Technology AB, S-22370 Lund, Sweden*

JASNA PETER-KATALINÍC (38), *Institut für Physiologische Chemie, Universität Bonn, D-5300 Bonn 1, Federal Republic of Germany*

GLENN PETERSON (21), *Adirondack Environmental Services, Albany, New York 12207*

WESTON PIMLOTT (34), *Department of Medical Biochemistry, University of Göteborg, S-400 33 Göteborg, Sweden*

STEVEN C. POMERANTZ (44), *Department of Medicinal Chemistry, University of Utah, Salt Lake City, Utah 84112*

LINDA POULTER (36), *Biotechnology Department, ICI Pharmaceuticals, Cheshire SK10 4TG, England*

WILHELM J. RICHTER (33), *Physics Department, Ciba-Geigy Ltd., CH-4002 Basel, Switzerland*

GERALD D. ROBERTS (27), *Department of Physical and Structural Chemistry, SmithKline Beecham Pharmaceuticals, King of Prussia, Pennsylvania 19406*

PETER ROEPSTORFF (23), *Department of Molecular Biology, Odense University, DK 5230 Odense M, Denmark*

BO E. SAMUELSSON (34), *Department of Medical Biochemistry, University of Göteborg, S-400 33 Göteborg, Sweden*

KARL H. SCHRAM (43), *Department of Pharmaceutical Sciences, College of Pharmacy, University of Arizona, Tucson, Arizona 85721*

HUBERT A. SCOBLE (28), *Genetics Institute, Andover, Massachusetts 01810*

JOHN E. SHIVELY (19), *Division of Immunology, Beckman Research Institute of the City of Hope, Duarte, California 91010*

DAVID L. SMITH (20), *Department of Medicinal Chemistry and Pharmacognosy, Purdue University, West Lafayette, Indiana 47907*

RICHARD D. SMITH (22), *Chemical Sciences Department, Battelle, Pacific Northwest Laboratory, Richland, Washington 99352*

KATHRYN L. STONE (21), *Department of Molecular Biophysics and Biochemistry, Yale University, New Haven, Connecticut 06510*

PAUL B. TAYLOR (27), *Department of Macromolecular Sciences, SmithKline Beecham Pharmaceuticals, King of Prussia, Pennsylvania 19406*

JAMES E. VATH (40), *Genetics Institute, Andover, Massachusetts 01810*

MARVIN L. VESTAL (5), *Vestec Corporation, Houston, Texas 77054*

J. THROCK WATSON (4), *Departments of Biochemistry and of Chemistry, Michigan State University, East Lansing, Michigan 48824*

KENNETH R. WILLIAMS (21), *Department of Molecular Biophysics and Biochemistry, Yale University, New Haven, Connecticut 06510*

RICHARD A. YOST (7), *Department of Chemistry, University of Florida, Gainesville, Florida 32611*

ZHONGRUI ZHOU (20), *Department of Medicinal Chemistry and Pharmacognosy, Purdue University, West Lafayette, Indiana 47907*

Preface

Throughout the past decade, mass spectrometry has undergone a dramatic expansion in experimental methodology and in the structural range of molecules of biological importance to which it can be applied. As a consequence, new approaches and protocols have been added to numerous existing mass spectrometry-based methods which have found routine use in the biological sciences. In addition, there are many procedures found in both the mass spectrometry and biochemical literature—some of which have undergone subtle but important refinements with time—which have never been available in one source. Many new workers wishing to assess the capabilities of mass spectrometry or to integrate these methods into the design of biological experiments require a concise and up-to-date treatment of experimental principles and detailed descriptions of specific methods.

To address these issues, this volume of *Methods in Enzymology* is devoted exclusively to mass spectrometry. The first section covers a range of general techniques and topics of contemporary importance in the applications of mass spectrometry to the biological sciences in general. The remaining three sections emphasize specific applications in what is broadly termed structural biology, and are divided among three major classes of biological molecules (peptides and proteins, glycoconjugates, and nucleic acid constituents). Each of these three sections opens with a critical overview of the applications of mass spectrometry to that area, written by a specialist in the field.

Coverage of the topics is oriented very strongly, with several exceptions, toward techniques which have demonstrated value in problem-solving. Some existing newer methods have been excluded because they are still under development and have not yet reached a stage of routine application, although they may ultimately find their way into the repertoire of routine methods. Also not covered are protocols whose current impact lies primarily in other areas to which mass spectrometry has made major contributions, such as toxicology or environmental science.

I am grateful to many of the authors and their colleagues who offered advice on the coverage of specific topics, including the interrelatedness of certain chapters. Particular appreciation is expressed to Klaus Biemann, Roger A. Laine, and Robert C. Murphy for their critical advice on the organization of the volume and on the contents of specific chapters.

JAMES A. MCCLOSKEY

METHODS IN ENZYMOLOGY

VOLUME 81. Biomembranes (Part H: Visual Pigments and Purple Membranes, I)
Edited by LESTER PACKER

VOLUME 82. Structural and Contractile Proteins (Part A: Extracellular Matrix)
Edited by LEON W. CUNNINGHAM AND DIXIE W. FREDERIKSEN

VOLUME 83. Complex Carbohydrates (Part D)
Edited by VICTOR GINSBURG

VOLUME 84. Immunochemical Techniques (Part D: Selected Immunoassays)
Edited by JOHN J. LANGONE AND HELEN VAN VUNAKIS

VOLUME 85. Structural and Contractile Proteins (Part B: The Contractile Apparatus and the Cytoskeleton)
Edited by DIXIE W. FREDERIKSEN AND LEON W. CUNNINGHAM

VOLUME 86. Prostaglandins and Arachidonate Metabolites
Edited by WILLIAM E. M. LANDS AND WILLIAM L. SMITH

VOLUME 87. Enzyme Kinetics and Mechanism (Part C: Intermediates, Stereochemistry, and Rate Studies)
Edited by DANIEL L. PURICH

VOLUME 88. Biomembranes (Part I: Visual Pigments and Purple Membranes, II)
Edited by LESTER PACKER

VOLUME 89. Carbohydrate Metabolism (Part D)
Edited by WILLIS A. WOOD

VOLUME 90. Carbohydrate Metabolism (Part E)
Edited by WILLIS A. WOOD

VOLUME 91. Enzyme Structure (Part I)
Edited by C. H. W. HIRS AND SERGE N. TIMASHEFF

VOLUME 92. Immunochemical Techniques (Part E: Monoclonal Antibodies and General Immunoassay Methods)
Edited by JOHN J. LANGONE AND HELEN VAN VUNAKIS

Section I

General Techniques

[1] Methods of Ionization

By ALEX G. HARRISON and ROBERT J. COTTER

Introduction

Historically, in the development and application of mass spectrometry to the analysis of organic molecules, during the 1950s and early 1960s, electron ionization (EI) was the only practical ionization method available. The development of chemical ionization (CI) in the late 1960s provided a complementary method for ionizing gaseous molecules. A disadvantage of both methods is that the sample of interest must be present in the gas phase at a pressure of 10^{-5} to 10^{-4} torr. With the extension of mass spectrometry to large, involatile, and, often, thermally fragile biomolecules, research during the 1970s and 1980s has been directed toward the development of ionization methods capable of ionizing such molecules directly from the solid or solution state; this work has led to the development of a variety of desorption ionization techniques. This chapter reviews these various methods of ionizing molecules in the gas phase and directly from the liquid or solid state.

Electron Ionization

Basic Principles

When gaseous polyatomic molecules are ionized by interaction with a beam of electrons the production of ions can be classified, phenomenologically, under the following schemes:

Ionization

$$ABC + e \rightarrow ABC^+ + 2e \qquad (1)$$

Dissociative Ionization

$$ABC + e \rightarrow AB^+ + C + 2e \qquad (2a)$$
$$\rightarrow AC^+ + B + 2e \qquad (2b)$$

Ion-Pair Formation

$$ABC + e \rightarrow AB^+ + C^- + e \qquad (3)$$

Electron Capture

$$ABC + e \rightarrow ABC^- \qquad (4)$$

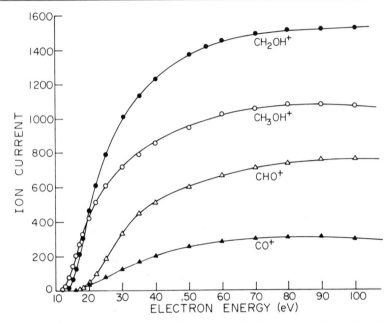

FIG. 1. Ionization efficiency curves for major ions in the mass spectrum of methanol. (From Ref. 4, with permission.)

Dissociative Electron Capture

$$ABC + e \rightarrow AB^- + C \tag{5}$$

Under normal operating conditions (20–70 eV electron energy) reactions (1) and (2a,b) account for the major part of the ionization with, in many cases, a minor contribution from reaction (3). In the electron-capture reactions (4) and (5) there are no product electrons to carry away the excess energy; hence, these reactions are resonance processes which have significant cross sections only over a very narrow range of electron energies, usually in the range 0–10 eV, i.e., normally below the threshold for processes (1) and (2). As a result, the literature of electron ionization is primarily concerned with positive ion mass spectra; for a discussion of electron-capture negative ion mass spectrometry, see later in this chapter.

Reaction (1) occurs when the energy of the bombarding electrons exceeds the ionization energy of the molecule. Typically a greater electron energy is required for dissociative ionization to take place because of the energy required to break chemical bonds either in simple bond rupture [reaction (2a)] or in bond rupture accompanied by rearrangement [reaction (2b)]. Typical ionization efficiency curves are shown in Fig. 1 for the major

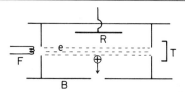

FIG. 2. Electron ionization source. F, Filament; B, source block; T, trap; R, repeller.

ions in the mass spectrum of the simple molecule CH_3OH. Of particular interest is the broad plateau in the ion yields between about 50 and 100 eV ionizing electron energy. In this region both the total ion yield and the relative ion abundances are relatively insensitive to the ionizing electron energy employed. Consequently, reproducible mass spectra and a constant sensitivity can be achieved by operating in this plateau region. By contrast, operating below 20–30 eV electron energy results in mass spectra and sensitivities that are strongly dependent on the ionizing electron energy.

In the plateau region, the total positive ion current, I_+, produced in the ion source is given by

$$I_+ = Q_i l [N] I_e \tag{6}$$

where Q_i is the total ionization cross section, l is the ionizing path length, [N] is the concentration of neutral molecules per cm^3, and I_e is the ionizing electron current. Measured ionization cross sections lie in the range 0.42 $\times$ 10^{-16} cm^2 for helium to 30×10^{-16} cm^2 for n-decane[1]; for more complex molecules higher cross sections can be expected since molecular cross sections are approximately an additive function of atomic ionization cross sections[2] or, alternatively, increase roughly linearly with increasing molecular polarizabilities.[3]

Instrumentation

A schematic diagram of a typical electron ionization source is shown in Fig. 2. Electrons, emitted from an electrically heated filament, are accelerated through the necessary potential (usually 70 V) and introduced into the ionization region where a small fraction interact with the gaseous sample molecules; the remainder impinge on the trap electrode. Usually a constant ionizing electron current is maintained by a feedback circuit to the filament power supply to maintain a constant trap current. A positive

[1] A. G. Harrison, E. G. Jones, S. K. Gupta, and G. P. Nagy, *Can. J. Chem.* **44**, 1967 (1966).
[2] J. W. Otvos and D. P. Stevenson, *J. Am. Chem. Soc.* **78**, 549 (1956).
[3] F. W. Lampe, J. L. Franklin, and F. H. Field, *J. Am. Chem. Soc.* **79**, 6129 (1957).

voltage, with respect to the cage, applied to the repeller electrode helps remove the ions for mass analysis; there often are other electrodes (not shown) to extract, shape, and guide the ion beam before entry into the mass analyzer. This type of ion source can be employed with any of the mass analyzers discussed in [2] of this volume, although with the time-of-flight (TOF) analyzer it is necessary to pulse the ionizing electron beam and/or the extraction lenses.

The sample to be ionized must be in the gas phase at a pressure in the ion source of less than 10^{-4} torr; higher pressures will lead to ion/molecule reactions between the primary ions and the neutral molecules. Such reactions can distort the mass spectrum by producing new species such as the protonated molecule, MH^+. This problem is more severe for combined EI/CI sources which, even when operated in the EI mode, usually are more gas-tight than sources designed for EI operation only, thus producing a higher ion source pressure for the same pressure reading on a remote pressure gauge. The sample may be introduced as a gas from a heated inlet system, by evaporation from a direct insertion probe, or as the effluent from a gas chromatograph.

Using Eq. (6) it can be derived that a pressure of 10^{-5} torr of a sample with $Q_i = 100 \times 10^{-16}$ cm^2 will produce a total source ion current of approximately 10^{-7} A for a typical ionizing electron current of 100 μA and a source temperature of 150°. Since this ion current is spread over many m/z values and it may be desired to detect a minor component in a gas mixture, the need for sensitive ion detection is evident. This is particularly true for sector instruments where 1% or less of the ions formed in the ion source reach the detector. In this respect, quadrupole analyzers have the advantage of a much higher percentage ion transmission.

Origin of Electron Ionization Mass Spectra

To reach some understanding of the factors which determine the final EI mass spectrum observed it is necessary to consider briefly and qualitatively the mechanisms by which molecular and fragment ions are formed. A more comprehensive discussion can be found elsewhere.[4–6]

For polyatomic molecules it is well-established that the ionization step is separate in time from the decomposition reactions which give rise to the

[4] A. G. Harrison and C. W. Tsang, in "Biochemical Applications of Mass Spectrometry" (G. R. Waller, ed.), pp. 135–156. Wiley, New York, 1972.

[5] M. E. Rose and R. A. W. Johnstone, "Mass Spectrometry for Chemists and Biochemists." Cambridge Univ. Press, New York, 1982.

[6] I. Howe, D. H. Williams, and R. D. Bowen, "Mass Spectrometry, Principles and Applications." McGraw-Hill, New York, 1981.

fragment ions. The electron/neutral interaction occurs in the order of 10^{-16} sec and leads to formation of molecular ions with various amounts of excitation energy (electronic and vibrational) with respect to the ground state of the molecular ion. The range of excitation energies is determined by the accessible states and the transition probabilities to these states and leads to a distribution of internal energies of the molecular ions which may extend to 10–20 eV. The electronic states formed are nonrepulsive and have a significant lifetime during which the excess internal energy is randomly redistributed among the vibrational degrees of freedom of the ground electronic state. Since ion/neutral collisions are negligible these energy randomization processes occur intramolecularly, primarily by radiationless transitions at the numerous crossings of potential energy surfaces.

Decomposition of a molecular ion occurs whenever sufficient energy accumulates in the appropriate vibrational mode or modes to cause bond rupture. The fragment ions formed may have sufficient internal energy to fragment further and rearrangement of the molecular framework may occur at any time. The reactions leading to the formation of a mass spectrum, thus, are a series of competing and consecutive unimolecular decomposition reactions originating from the molecular ion, the rate of each reaction being dependent on the internal energy. In principle, the rate constant for each step can be calculated by applying an appropriate form of the absolute reaction rate theory.

In their exact formulation and quantitative application to the origin of EI mass spectra the theory of rate processes, whether the quasi-equilibrium theory (QET)[7] or the similar Rice, Ramsperger, Kassel, and Marcus (RRKM) theory,[8] is highly physical and mathematical in approach. A less precise approach which leads to at least a qualitative understanding of the factors which determine the appearance of mass spectra leads[4-6] to Eq. (7) for the rate constant for an ionic fragmentation reaction

$$k(E) = \nu[(E - E_0)/E]^S \qquad (7)$$

where ν is a frequency factor, E the internal energy of the ion, E_0 the activation energy (or critical reaction energy), and S is the effective number of oscillators [often taken as one-half to one-third the total number of oscillators (3N-6 for an N-atom molecule)]. The frequency factor ν is effectively a measure of the entropy of activation. For a simple bond rupture reaction it usually is taken as the frequency of a bond vibration;

[7] H. M. Rosenstock, M. B. Wallenstein, A. L. Wahrhaftig, and H. Eyring, *Proc. Natl. Acad. Sci. U.S.A.* **38,** 667 (1952).
[8] R. A. Marcus, *J. Chem. Phys.* **20,** 359 (1952).

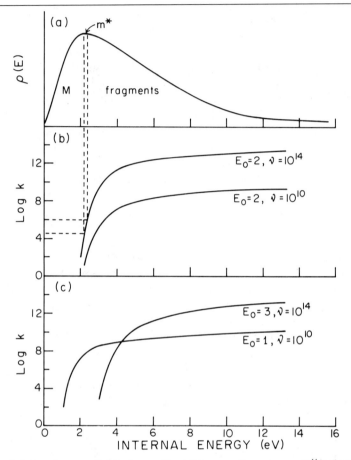

FIG. 3. Internal energy distribution and log $k(E)$ versus E curves. $\nu = 10^{14}$ is characteristic of a simple cleavage, while $\nu = 10^{10}$ is characteristic of a rearrangement.

however, for a rearrangement reaction involving a particular spatial arrangement of atoms the entropy of activation will be less favorable and this is reflected in a lower frequency factor.

Equation (7) leads to rate constants which increase rapidly with increasing internal energy to a limiting value determined by the value of ν. Two curves are shown in Fig. 3b where E_0 is the same but different frequency factors are invoked corresponding qualitatively to a simple bond rupture ($\nu = 10^{14}$ sec^{-1}) and a rearrangement reaction ($\nu = 10^{10}$ sec^{-1}). Figure 3a shows a hypothetical, but plausible, distribution of internal energies deposited in the molecular ions in the initial ionization step.

A consequence of the rate formulation is that an assembly of molecular ions will have a distribution of lifetimes which depends on the $k(E)$ function for the lowest energy fragmentation process and on the distribution of internal energies. This has a direct influence on the mass spectrum observed. In a typical sector mass spectrometer ions will spend $1-10$ μsec in the ion source and $10-30$ μsec in transit to the detector. It follows that molecular ions with internal energies sufficient to give $k(E)$ greater than about 10^6 sec^{-1} will fragment in the ion source and will be analyzed as fragment ions. Molecular ions with internal energies appropriate to give $k(E)$ in the approximate range $10^{4.5}-10^6$ sec^{-1} will fragment during transit and will either be lost in the background or, if the fragmentation occurs in an appropriate field-free region, can be detected as so-called metastable ions (for a comprehensive discussion of metastable ions see Ref. 9). Molecular ions with internal energies appropriate to $k(E) \leq 10^{4.5}$ sec^{-1} will survive until after mass analysis and will be detected as molecular ions. This breakdown into the different categories of ions is illustrated in Fig. 3a. Similar arguments apply to quadrupole and TOF instruments although the time scales may be slightly different and metastable ions are not observed. However, it should be noted that decompositions which occur during the acceleration region in TOF mass spectrometers will lead to tailing of the fragment ion signals to the high mass side.

It will be noted that the abundance of the molecular ion in the mass spectrum is determined largely by the activation energy, E_0, for the lowest energy fragmentation reaction and by the fraction of molecular ions with internal energies below E_0. Two situations can lead to the absence of the molecular ion in the mass spectrum: (a) the activation energy of the lowest energy fragmentation reaction is zero and/or (b) there are no states of the molecular ion with internal energies between 0 and E_0; the former is much more frequent.

Two further kinetic aspects deserve mention. As shown by the two $k(E)$ curves in Fig. 3b, a rearrangement fragmentation reaction cannot compete effectively with a simple bond rupture fragmentation reaction if both have the same activation energy. It follows that, if a fragment ion formed by rearrangement is observed in the mass spectrum, the activation energy for this reaction must be lower than the activation energy for any simple bond rupture reaction. This is shown schematically in Fig. 3c, where, between 1.5 and 4.5 eV internal energy, the molecular ions will fragment largely by the rearrangement reaction (low ν); at higher internal energies the simple bond fission reaction becomes dominant. A corollary of this statement is that, as one lowers the ionizing electron energy (i.e.,

[9] R. G. Cooks, J. H. Beynon, R. M. Caprioli, and G. R. Lester, "Metastable Ions." Elsevier, New York, 1973.

shifts the internal energy distribution to lower energies), the abundance of rearrangement fragment ions will increase relative to fragment ions formed by simple bond rupture. Frequently it is easier to establish molecular structures through fragment ions arising by simple bond rupture. A further point which has been ignored so far is the effect of sequential fragmentation reactions. Clearly, if the activation energy for further fragmentation of a primary fragment ion is low, the primary fragment ion will be of low abundance in the mass spectrum even though the reaction to form this ion may be favorable.

Finally, it should be noted that it is common in organic mass spectrometry to rationalize mass spectra in terms of charge localization and the rupture of adjacent bonds to give stable fragmentation products,[10,11] particularly the ionic species. Essentially, this approach rationalizes the rates of competing fragmentation reactions in terms of relative activation energies for fragmentation of the charge-localized species. Consecutive fragmentation reactions are considered in terms of relative activation energies as well. The rationale is that the activated complex will resemble the final products so that the relative energies of activated complexes can be related to the relative energies of the final products. Charge localization is most likely to occur in π-bonded systems such as carbonyls or on heteroatoms such as nitrogen and oxygen and leads to fragmentation alpha to the charge localization site. Thus one observes cleavage alpha to carbonyl bonds and alpha to the heteroatom in compounds such as ethers, alcohols, and amines. This is illustrated in Fig. 4a where the main primary fragmentation reactions in the EI mass spectra of leucine correspond to cleavage alpha to the amine function leading to $[(CH_3)_2CHCH_2CHNH_2]^+$ (m/z 86) and to $[HO_2CCHNH_2]^+$ (m/z 74).[12-14] Note the absence of a significant molecular ion in the EI mass spectrum. [The spectra shown in the NIH–EPA collection do show a low-intensity (<1% of base peak) molecular ion.]

Chemical Ionization

Introduction

The essential reactions in chemical ionization are presented in general form in Eqs. (8)–(10). The initial ionization is usually by electron impact

[10] H. Budzikiewicz, C. Djerassi, and D. H. Williams, "Mass Spectrometry of Organic Compounds." Holden-Day, San Francisco, 1967.
[11] F. W. McLafferty, "Interpretation of Mass Spectra," 3rd Ed. Univ. Sci., Mill Valley, CA, 1980.
[12] G. Junk and H. Svec, *J. Am. Chem. Soc.* **85,** 839 (1963).
[13] C. W. Tsang and A. G. Harrison, *J. Am. Chem. Soc.* **98,** 1301 (1976).
[14] M. Meot-Ner and F. H. Field, *J. Am. Chem. Soc.* **95,** 7207 (1973).

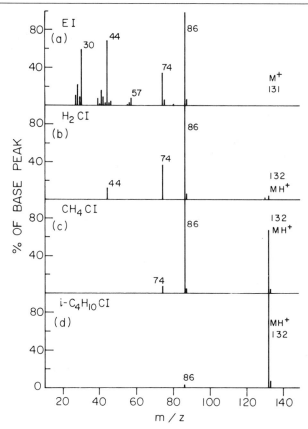

FIG. 4. Mass spectra of leucine: (a) EI, (b) H_2 CI, (c) CH_4 CI, and (d) i-C_4H_{10} CI. EI data from Ref. 12, H_2 and CH_4 CI data from Ref. 13, i-C_4H_{10} CI data from Ref. 14. Reagent ions not shown in CI mass spectra.

on the reagent gas which is present in large excess over the sample molecules of interest, A. Ionization of the reagent usually is followed by ion/molecule reactions involving the primary ions and the reagent gas neutrals and produces the chemical ionization reagent ion or reagent ion array [reaction (8)] as the stable end products. Collision of the positively or negatively charged reagent ion(s), $R^{+/-}$, with the additive, usually present

$$e + R \rightarrow R^{+/-} \qquad (8)$$
$$R^{+/-} + A \rightarrow A_1{}^{+/-} + N_1 \qquad (9)$$
$$A_1{}^{+/-} \rightarrow A_2{}^{+/-} + N_2$$
$$\rightarrow A_3{}^{+/-} + N_3 \qquad (10)$$
$$\rightarrow A_i{}^{+/-} + N_i$$

at less than 0.1% of the reagent gas pressure, produces [reaction (9)], an ion characteristic of the additive. This additive ion may fragment by one or more pathways or, less frequently, may react with the reagent gas. The final array of ions, $A_1^{+/-}$ to $A_i^{+/-}$, when mass analyzed, constitutes the chemical ionization mass spectrum of the additive A as effected by the reagent gas R. (See Fig. 4 for examples.)

In the initial report of chemical ionization,[15] methane was used as the reagent gas in the positive ion mode, with the major ions reacting as Brønsted acids. Since that time, a variety of reagent gases have been used in both the positive ion and negative ion mode. The major species used will be discussed in the following.

Instrumentation

The yield of additive ions, and thus the sensitivity of chemical ionization, depends on the magnitude of the ion/molecule reaction rate constant, on the concentrations (or partial pressures) of the reactants R and A, and on the time available for reaction between these two species. As discussed in detail elsewhere[16] the reaction rate constant frequently is close to the collision rate constant determined by ion/induced dipole and ion/dipole interactions. The partial pressures of reactants and the reaction time are interrelated quantities which depend on the type of instrument used.

The majority of CI studies have been carried out using slightly modified EI sources on quadrupole or sector mass spectrometers. In such sources diffusional loss of ions to the walls and removal of ions by the extraction fields applied limit ion source residence times to approximately 10 μsec. With this reaction time it is necessary to operate the source with reagent gas pressures in the range 0.2–2.0 torr and to use sample pressures in the range 10^{-3}–10^{-4} torr. However, the mass analysis region must be maintained at 10^{-6} torr or lower to prevent ion scattering, while the region surrounding the ion source must be maintained at 10^{-4} torr or lower to prevent ion/molecule collisions in the region between the source exit slit and the point of entry into the analyzer. To achieve these conditions, the ion source must be made as gas-tight as possible, the pumping speed at the ion source housing must be as high as practicable, and the analyzer region normally is differentially pumped. The CI ion source normally is an EI source (Fig. 2) with reduced dimensions for the electron beam entrance aperture and the ion exit aperture. Such sources can be operated under EI conditions, often employing a larger replaceable ion exit aperture or replaceable ion source volumes, but the quality of the spectra obtained

[15] M. S. B. Munson and F. H. Field, *J. Am. Chem. Soc.* **88**, 2621 (1966).
[16] A. G. Harrison, "Chemical Ionization Mass Spectrometry." CRC Press, Boca Raton, FL, 1983.

may not be optimum because of the occurrence of ion/molecule reactions at the higher pressures in the ion source due to its tighter construction.

With a source pressure of 0.2 to 2.0 torr and electron energies as high as 400 eV, the electron beam will be completely attenuated before it reaches the trap electrode. Consequently, the filament is regulated to give a constant emission current to the source block rather than a constant trap current as employed in the EI mode. With most commonly used reagent gases, the increased pressure does not adversely affect filament lifetimes. However, if oxidizing gases, such as nitric oxide or oxygen, are used filament lifetimes are severely reduced. For these cases, ion sources have been developed in which ionization results from the action of an electrical discharge.[17,18]

Less commonly used are instruments in which the chemical ionization occurs at atmospheric pressure.[19,20] In such atmospheric pressure ionization (API) instruments, positive and negative ions are formed in a flowing gas at one atmosphere pressure either by a corona discharge or by β-particles from a ^{63}Ni source. The ions present in the gas are sampled through a small aperture and focused into a quadrupole mass filter for analysis. Samples are introduced into the flowing gas from a direct probe, as the effluent from a chromatograph or by injection in a solvent; if the carrier gas is ambient air, trace impurities can be detected directly. When air is used as the flowing gas, the major positive and negative reagent ions are $H^+(H_2O)_n$ and $O_2^-(H_2O)_n$, respectively. In API, the concentration and reaction time parameters are such that the ionization usually ends up residing in the most stable positive and negative ion species, which may represent only trace quantities in the flowing gas. Since many species of interest in atmospheric, biological, and environmental studies have either large electron affinities or large proton affinities, API can be an ultrasensitive analyzer for trace amounts of these components.

Although the majority of CI studies have been carried out in medium pressure sources, much lower pressures could be used if the reaction time could be made correspondingly longer. This is possible using instruments which trap the ions in some containing field. Although chemical ionization has been reported at low pressures using a quadrupole ion trap[21] and

[17] D. F. Hunt, C. N. McEwen, and T. M. Harvey, *Anal. Chem.* **47**, 1730 (1975).
[18] B. Schneider, M. Breuer, H. Hartmann, and H. Budzikiewicz, *Org. Mass Spectrom.* **24**, 216 (1989).
[19] E. C. Horning, M. G. Horning, D. I. Carroll, I. Dzidic, and R. N. Stillwell, *Anal. Chem.* **45**, 936 (1973).
[20] D. A. Lane, B. A. Thomson, A. M. Lovett, and N. M. Reid, *Adv. Mass Spectrom.* **8**, 1480 (1980).
[21] J. S. Brodbelt, J. N. Louris, and R. G. Cooks, *Anal. Chem.* **59**, 1278 (1987).

TABLE I
BRØNSTED ACID CHEMICAL IONIZATION REAGENTS

Reagent gas	B	BH^+	$PA(B)^a$ (kcal/mol)
H_2	H_2	H_3^+	101
CH_4	CH_4	CH_5^+	132
	C_2H_4	$C_2H_5^+$	163
H_2O	H_2O	$H_3O^+(H_2O)_n^{b}$	167
CH_3OH	CH_3OH	$CH_3OH_2^+$	182
$i-C_4H_{10}$	$i-C_4H_8$	$C_4H_9^+$	196
NH_3	NH_3	$NH_4^+(NH_3)_n^{b}$	204

[a] From Ref. [23].
[b] Degree of solvation depends on partial pressure of reagent gas, proton affinity of higher solvated species greater than for BH^+.

Fourier transform ICR,[22] neither method has, as yet, seen extensive use in practical applications.

Brønsted Acid Chemical Ionization

By far the most widely used reagent gas systems have been those which yield Brønsted acid reagent ions (BH^+). The major reaction of these acids is proton transfer to give initially the protonated molecule MH^+.

$$BH^+ + M \rightarrow MH^+ + B \tag{11}$$

This reaction usually occurs with high efficiency if the proton affinity (PA) of M is greater than that of B. The proton affinity is defined as the exothermicity of the reaction

$$B + H^+ \rightarrow BH^+, \quad PA(B) = -\Delta H^0 \tag{12}$$

Table I lists the most widely used Brønsted acid reagents, the BH^+ reagent ions formed, and the proton affinities of the respective conjugate bases B.[23] One of the major advantages of Brønsted acid CI is that, by choosing the appropriate reagent ion(s), one can readily vary the exothermicity of the protonation reaction (11). Gentle protonation will lead predominantly to formation of MH^+; as the protonation reaction becomes more exothermic, more extensive fragmentation is observed. This fragmentation usually involves elimination of stable even-electron molecules,

[22] C. L. Wilkins, A. K. Chowdhury, L. M. Nuwaysir, and M. L. Coates, *Mass Spectrom. Rev.* **8,** 67 (1989).
[23] S. G. Lias, J. E. Bartmess, J. F. Liebman, J. L. Holmes, R. D. Levin, and W. G. Mallard, *J. Phys. Chem. Ref. Data* **17,** Suppl. 1 (1988).

XH, incorporating functional groups, X, present in the molecule. Thus, OH groups are readily lost as H_2O, RNH groups as RNH_2, etc. (For a more complete discussion of the fragmentation of the MH^+ ions of simple molecules see Ref. 16.) Fragmentation reactions are, of course, essential in establishing the structure of the molecule M while the MH^+ ions are necessary for establishing molecular weights. The effect of varying the protonation exothermicity is illustrated in Fig. 4 which presents the H_2, CH_4, and i-C_4H_{10} CI mass spectra of leucine. In general, CH_4 CI provides the best compromise in providing both MH^+ ions for molecular weight determination and fragment ions for structure elucidation. The strongly exothermic protonation by H_3^+ is useful only where MH^+ is particularly stable and fragment ions are desired for structure elucidation. Although i-C_4H_{10} CI is useful in giving molecular weight information, some caution is required since the reaction of $C_4H_9^+$ with molecules having PAs less than 196 kcal mol^{-1} will be endothermic and, consequently, will be relatively inefficient. (For an extensive listing of proton affinities see Ref. 24.)

Ammonia CI deserves special mention since NH_4^+ and its solvated forms are very weak Brønsted acids and will react by proton transfer only with compounds having proton affinities greater than 205 kcal mol^{-1}; in effect, this means compounds having a nitrogen function. With compounds of lower PA one frequently observes formation of an adduct ion, [M + $NH_4]^+$, although even this cluster ion is not observed if PA(M) is less than 188 kcal mol^{-1}. The use of ammonia as a CI reagent gas recently has been reviewed.[25]

The formation of cluster ions is much less prevalent with hydrocarbon reagents, although mention should be made of the low-intensity [M + $C_2H_5]^+$ and [M + $C_3H_5]^+$ ions which are sometimes observed in CH_4 CI. Similarly, [M + $C_3H_3]^+$ and [M + $C_4H_9]^+$ ions sometimes may be observed with low intensities in isobutane CI mass spectra. It also should be noted that, in H_2 and CH_4 CI mass spectra, [M − H]$^+$ ions may be observed as the result of hydride abstraction; these are particularly prominent in H_2 CI mass spectra and, in the absence of abundant MH^+ ion signals, may be mistakenly identified.

The question may remain as to whether good CI conditions have been established since many commercial instruments make no provision for measuring the source pressure. For methane and isobutane CI, a frequently used test compound is one of the xylenes.[26] In CH_4 CI ratios of

[24] D. H. Aue and M. T. Bowers, in "Gas Phase Ion Chemistry" (M. T. Bowers, ed.), Vol. 2. Academic Press, New York, 1979.

[25] J. B. Westmore and M. M. Alauddin, *Mass Spectrom. Rev.* **5,** 381 (1986).

[26] J. Robertson, VG Canada, personal communication (1989).

TABLE II
CHARGE-EXCHANGE REAGENT SYSTEMS

Reagent	Reagent ion	$RE(R^{\ddot{+}})(eV)$
C_6H_6	$C_6H_6^{\ddot{+}}$	9.3
N_2/CS_2	$CS_2^{\ddot{+}}$	10.2
CO/COS	$COS^{\ddot{+}}$	11.2
Xe	$Xe^{\ddot{+}}$	12.1, 13.6
CO_2	$CO_2^{\ddot{+}}$	13.8
CO	$CO^{\ddot{+}}$	14.0
N_2	$N_2^{\ddot{+}}$	15.3
Ar	$Ar^{\ddot{+}}$	15.8, 15.9

m/z 107 : m/z 106 ≥ 10 : 1 and m/z 106 : m/z 91 ≥ 10 : 1 are taken as indicative of good CI conditions. For i-C_4H_{10} CI ratios of m/z 107 : m/z 106 ≥ 3 : 1 and m/z 106 : m/z 91 ≥ 10 : 1 are taken as indicative of good CI conditions. No similar test conditions have been determined for other reagent systems.

Charge-Exchange Chemical Ionization

An alternative to proton transfer in positive ion CI is charge-exchange ionization. In this case a reagent gas is chosen which produces an odd-electron species, $R^{\ddot{+}}$, which reacts with the additive, M, by charge exchange, as illustrated in Eqs. (13) and (14).

$$R^{\ddot{+}} + M \rightarrow M^{\ddot{+}} + R \qquad (13)$$
$$M^{\ddot{+}} \rightarrow \text{fragments} \qquad (14)$$

Clearly, since the same odd-electron molecular ion is formed as in electron ionization, the fragment ions observed will be the same as those observed in electron ionization. The difference is that the $M^{\ddot{+}}$ ions produced by electron ionization have a wide range of internal energies (see Section II,C) while the $M^{\ddot{+}}$ ions produced in reaction (13) have an internal energy given approximately by

$$E(M^{\ddot{+}}) = RE(R^{\ddot{+}}) - IE(M) \qquad (15)$$

where IE(M) is the ionization energy of M and $RE(R^{\ddot{+}})$ is defined by

$$R^{\ddot{+}} + e \rightarrow R, \qquad -\Delta H^0 = RE(R^{\ddot{+}}) \qquad (16)$$

Charge-exchange reagent ions with recombination energies covering the range 9.3–15 eV have been developed (Table II). The major uses of charge-exchange CI are to achieve selective ionization of components of mixtures and to enhance differences, relative to EI, in the mass spectra of isomeric molecules. One example of each use will suffice. In the former area, Sieck has used both $C_6H_{12}^+$ (cyclohexane) and $C_6H_5Cl^+$ to achieve selective ionization of the aromatic compounds present in petroleum products.[27] In the latter area, Keough et al.[28] have shown that the CS_2^+ CE mass spectra of isomeric monoepoxides of arachidonyl acetate are significantly different although the 70 eV EI mass spectra are identical.

Miscellaneous Positive Ion Reagents

A wide variety of other positive ion reagents have seen limited use. Many of these have been developed for specific applications such as the determination of double bond position, differentiation of geometrical isomers, or analysis of specific classes of compounds. These specific reagent ions have been the subject of a recent review.[29]

Electron-Capture Chemical Ionization (ECCI)

The interest in negative ion CI has grown immensely over recent years,[30] lured, in part, by the prospect of an increase in sensitivity compared to EI or positive ion CI.

Electron capture is not, correctly speaking, a chemical ionization process, but it usually is included as such because a high pressure of a moderating or reagent gas is used. This moderating gas serves to "thermalize" the electrons in the ion source by inelastic scattering and dissociative ionization processes. The resulting quasithermal electrons are captured by the substrate molecules in resonant electron capture or dissociative electron-capture processes [reactions (3) and (4)]. Ideally, the moderating gas should not form stable anions and moderating gases which have been used include H_2, He, CH_4, NH_3, N_2, Ar, CO_2, and i-C_4H_{10}.[30] Of these, CH_4, NH_3, and i-C_4H_{10} have seen the most use, probably because they are commonly used in Brønsted acid CI.

A major attraction of ECCI is the possibility of enhanced sensitivity compared to EI or other forms of CI. This arises because the rate constants for electron capture can be as high as 10^{-7} cm^3 molecule^{-1} sec^{-1} compared to maximum rate constants for particle transfer ion/molecule reactions[16] of 2 to 3 × 10^{-9} cm^3 molecule^{-1} sec^{-1}; the high rate constants for electron capture are the result of the low mass and high mobility of the electron. Thus, in favorable cases, a sensitivity enhancement of a factor of 100 can be obtained. However, electron-capture rate constants as low as 10^{-13} cm^3

[27] L. W. Sieck, *Anal. Chem.* **51**, 128 (1979); *ibid.,* **55**, 38 (1983).

[28] T. Keough, E. D. Mihelich, and D. A. Eickhoff, *Anal. Chem.* **56**, 1849 (1984).

[29] M. Vairamani, U. A. Mirza, and R. Srinivas, *Mass Spectrom. Rev.* **9**, 235 (1990).

[30] H. Budzikiewicz, *Mass Spectrom. Rev.* **5**, 345 (1986).

molecule^{-1} sec^{-1} have been measured and ECCI does show variable and, sometimes, very low sensitivity. The present data indicate that only molecules with a substantial electron affinity (probably greater than 0.5 eV) will show reasonable responses in ECCI. The presence of halogen or nitro groups (especially when attached to π-bonded systems), highly conjugated carbon systems, or a conjugated carbonyl system lead to a reasonably high electron affinity. Another approach commonly taken is to derivatize the molecule to increase the electron capture probability. Perfluoroacyl, perfluorobenzoyl, pentafluorobenzyl, or nitrobenzoyl derivatives have been successfully employed for this purpose as they have been in electron capture gas chromatography. However, in contrast to gas chromatography, it is not sufficient that electron capture be facile; one also requires that ions characteristic of the analyte be formed since ions characteristic of the derivatizing group only leave the identity of the analyte in doubt. For example, the trifluoroacetyl and pentafluoropropionyl derivatives of phenols show predominantly the respective carboxylate anions in ECCI.[31] Thus, it probably is necessary that a variety of derivatives be examined for each new class of compounds studied.

Apart from variable sensitivity, a frequent problem which plagues ECCI is the nonreproducibility of the spectra obtained, making the establishment of reliable data bases difficult. Not only do the relative intensities of ions anticipated from the analyte vary but also artifact peaks often are observed. Extensive studies, primarily of environmental contaminants, have shown that ECCI mass spectra are strongly dependent on experimental conditions such as source temperature, identity and pressure of the moderating gas, impurities in the moderating gas, the instrument used, instrument focusing conditions, and the amount of analyte introduced. (For a more detailed discussion and leading references see Ref. 32.) Clearly it is necessary to establish standard conditions on a given instrument if reproducible spectra and sensitivities are to be obtained.

Artifact peaks, or those not explainable on the basis of electron capture by the analyte, also may be present. In some cases, these may be explainable on the basis of ion/molecule reactions, for example, reaction of M$^-$ with trace levels of oxygen present in the source. In other cases, more complex origins are involved. Spears and co-workers[33] have considered this topic in detail and have shown that four pathways leading to artifact peaks can be identified: (1) neutralization of the molecular anion at the

[31] T. M. Trainor and P. Vouros, *Anal. Chem.* **59,** 601 (1987).

[32] A. G. Harrison, *Adv. Mass Spectrom.* **11,** 582 (1989).

[33] L. J. Spears, J. A. Campbell, and E. P. Grimsrud, *Biomed. Environ. Mass Spectrom.* **14,** 401 (1987).

TABLE III
BRØNSTED BASE REAGENT IONS

Reagent ion(B⁻)	PA(B⁻)[a] (kcal/mol)	EA(B)[a] (kcal/mol)
OH^-	391	42.2
$O^{\overline{\cdot}}$	382	33.7
CH_3O^-	380	37.4
F^-	371	78.4
$O_2^{\overline{\cdot}}$	353	10.1
Cl^-	333	83.4

[a] From Ref. [23].

walls or by ion/ion recombination, (2) neutralization of positive ions either by recombination or at the walls, (3) reaction of free radicals derived from the moderating gas with the analyte, and (4) reaction of the analyte on the walls of the source (or chromatograph?). In each case, if the pathway leads to neutral species with higher electron-capture probabilities than the original analyte significant artifact peaks will be observed. The moral seems to be that if the analyte is a poor candidate for ECCI artifact peaks may well be observed.

Brønsted Base Chemical Ionization

In negative ion CI Brønsted bases play a role similar to that played by Brønsted acids in positive ion CI. Brønsted bases react primarily by the reaction

$$B^- + M \rightarrow [M - H]^- + BH \tag{17}$$

Reaction (17) will be exothermic and will occur efficiently for simple bases B^- if the proton affinity of B^- is greater than the proton affinity of $[M - H]^-$. The proton affinities of gaseous anions often are expressed in terms of the gas-phase acidities of the conjugate acid, i.e.,

$$BH \rightarrow B^- + H^+, \quad PA(B^-) = \Delta H^0_{acid}(BH) = \Delta H^0 \tag{18}$$

For a recent compilation of gas-phase acidities see Ref. 34.

Table III records the proton and electron affinities of those bases that have been used as Brønsted base reagent ions. The OH^- ion is readily prepared by electron impact on an N_2O/CH_4 mixture, while $O^{\overline{\cdot}}$ is prepared by electron impact on a N_2O/N_2 mixture. F^- usually is prepared by dissociative electron capture by NF_3 while Cl^- is produced by dissociative

34 J. E. Bartmess, *Mass Spectrom. Rev.* **8**, 297 (1989).

electron capture by almost any chlorine-containing compound. Methyl nitrite in admixture with any moderating gas produces good yields of CH_3O^-. The reagent ion O_2^- is used primarily in atmospheric pressure CI where it is produced by electron capture by O_2.

Of the reagent ions listed in Table III, OH^- has seen, by far, the greatest use. It usually reacts [reaction (17)] to produce $[M - H]^-$ ions which rarely undergo further fragmentation; thus Brønsted base CI is particularly useful for molecular weight determination. The CH_3O^- ion is used less frequently since its reactions are similar to those of OH^-. The halide ions, particularly Cl^-, are relatively weak bases and have seen limited use, although Cl^-, which frequently reacts by clustering to form $[M + Cl]^-$, has been employed in molecular weight determination.[30] The O^- species is a radical anion which reacts by a variety of routes, including H^+ and H_2^+ abstraction, and H atom and alkyl radical displacement.[16] In addition, it reacts, in part, by H atom abstraction to form OH^- which reacts further to produce $[M - H]^-$; thus O^- CI mass spectra are dependent on the partial pressure of analyte in the ion source. O_2^- is also a radical anion which undergoes a variety of reactions including electron transfer, since it has a relatively low electron affinity; it is primarily used in atmospheric pressure CI.

Desorption Ionization

Introduction and Definition

The transfer of a proton to a sample molecule in chemical ionization makes it possible to produce intact protonated molecular ions of compounds for which EI results in extensive fragmentation. Since the determination of molecular weights is critical for structural analysis, the development of chemical ionization has greatly increased the power and scope of mass spectrometry for synthetic organic chemistry, natural products chemistry, and biochemistry. However, a major limitation of both EI and CI is the requirement that the sample be volatile. Since volatility is reduced by the presence of polar functional groups, formation of the methyl or ethyl esters of carboxylic acids, permethylation and peracetylation of hydroxy groups, and trimethylsilyl derivatization have all been employed to increase sample volatility. However, this increases the molecular ion mass, and raises uncertainties about the molecular weight of the parent compound when the number of derivatization sites is not known or when derivatization is incomplete. Derivatization generally increases sample preparation and handling. In addition, samples which contain a quaternary ammonium center (e.g., phosphatidylcholine) remain intractable. Fortunately, the last several years have seen intense development of methods capable of directly analyzing involatile samples.

Field Desorption. The field desorption (FD) technique[35] was the first viable (and commercially available) method for the ionization of nonvolatile molecules in the condensed state. The ion source is a thin (1–5 μm) wire at high voltage (6–10 kV), positioned close to a grounded extraction electrode. The wire is activated by the growth of thin whiskers or dendrites[36] on its surface to increase the electrical field strength. Samples in solution are deposited on the wire, and ions are extracted by the high electrical field. Heating the emitter wire enhances desorption, and several investigators[37] have described the importance of establishing the best emitter temperature (BET) for different classes of compounds. FD is an outgrowth of the field ionization technique[38] for which quantum electron tunneling mechanisms were generally invoked to explain the production of $M^{\ddot{+}}$ ions from volatile samples. In contrast, FD mass spectra are characterized by the formation of even-electron $[M + H]^+$ or $[M + Na]^+$ ions from nonvolatile samples, accompanied by minimum fragmentation. While inherent experimental difficulties associated with the use of thin wires in high electrical fields have limited its routine use to a few laboratories, FD has continued to be the method of choice for the measurement of the molecular weight distributions of industrial polymers[39] and other high-molecular weight (and less polar) samples.

High-Energy Beams. The desorption techniques most commonly used today can be classified as beam techniques. Samples are deposited on a surface and gas phase ions are produced by bombarding the surface with high-energy particle or photon beams. Particle beams include ions (Ar^+, Xe^+, Cs^+) or neutral atoms (Ar, Xe) with energies of 5 to 10 keV, as well as ions with energies up to 100 MeV, generated from spontaneous nuclear reactions or from a tandem accelerator. Photon beams are generated by pulsed lasers from the far-IR to the near-UV with power densities in excess of 10^6 W/cm^2. The use of such large energies would seem to be incompatible with the formation of intact molecular ion species; however, in desorption techniques primary energy is not transferred directly to the sample molecules. Rather, the transfer of energy is mediated by absorption, excitation, and relaxation processes that are summarized in Fig. 5. Primary particles or photons penetrate the surface on which the sample is adsorbed, and

[35] H. D. Beckey and D. Schuelte, *Z. Instrum.* **68,** 302 (1960).
[36] S. J. Reddy, F. W. Rollgen, A. Maas, and H. D. Beckey, *Int. J. Mass Spectrom. Ion Phys.* **25,** 147 (1977).
[37] J. F. Holland, B. Soltmann, and C. C. Sweeley, *Biomed. Mass Spectrom.* **3,** 340 (1976).
[38] H. D. Beckey, *J. Sci. Instrum.* **12,** 72 (1979).
[39] R. P. Lattimer and G. E. Hansen, *Macromolecules* **14,** 1776 (1981).

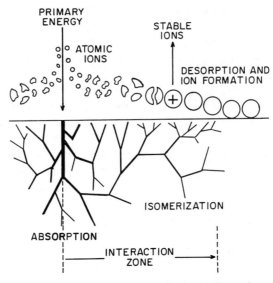

FIG. 5. Schematic of the events occurring in the desorption of ions.

their energies are absorbed as vibrational or electronic excitation of the solid lattice (or liquid) which supports the sample. Sample molecules close to the point of impact are sputtered from the surface as atoms or atomic ions with high kinetic energy. At the same time the primary energy is dissipated and converted, or isomerized, to primarily vibrational excitation. At increasingly greater distances from the point of impact this results in the desorption of radical ions, small stable ionic fragments, and, finally, intact molecular ions and neutral molecules near the fringes of the interaction zone.[40] Although the resulting mass spectra are generally dominated by atomic ions, nonspecific fragment ions, and [in the case of fast atom bombardment (FAB)] matrix ions in the low mass region, desorption spectra provide structurally informative molecular and fragment ions in the high mass region.

Mechanisms. A interesting feature of desorption methods is that a wide array of instrumentally different approaches result in mass spectra which are surprisingly similar and generally resemble CI spectra. While there has been a prolific array of mechanistic explanations, it is generally accepted that absorption of energy from particles in the megaelectron volt (MeV) range proceeds via electronic excitation, while absorption of kiloelectron volt (keV) particle energy proceeds by vibrational excita-

---

[40] R. D. Macfarlane, *Acc. Chem. Res.* **15**, 286 (1982).

tion.[41,42] The means by which energy isomerization leads to the formation and desorption of ions are less clear. For the direct sputtering of small ions from a solid matrix using keV particles, the collision cascade model has been the accepted mechanism.[43] Somewhat more complex coupled anharmonic oscillator models have been proposed to explain the disequilibrium between the internal energy (<0.1 eV/degree) and the kinetic energy (>1 eV) of larger molecules and ions leaving the surface.[44,45] The liquid matrix used in FAB mass spectrometry[46] (see below) complicates the mechanistic picture; however, in this case, solvent shedding[47] may account for the low internal energy (and stability) of molecular ions. The question of whether molecular ions are desorbed directly from the surface or formed in the gas phase after desorption is heavily debated. While it appears that certain stable ions (e.g., quaternary ammonium ions) may be preformed,[48] the origin of protonated species is not as clear. In FAB mass spectra, the relative abundances of multiply protonated molecular ions of large peptides are a function of the pH of the liquid matrix, and, therefore, reflect solution equilibria.[49] Conversely, it has been suggested that molecular ions are formed by CI processes in the gas phase.[50] While it is likely that both types of processes occur, it is well understood that positive, protonated molecular ions of peptides are best produced when the solution pH is below the isoelectric point.

Despite this less than complete understanding, desorption can be defined (in general) as the direct formation of stable molecular and fragment ions from nonvolatile samples in the condensed phase or, alternatively, as methods in which vaporization and ionization appear as a single event.

Particle Beam Techniques

The sample ions produced by particle beam techniques are, in essence, secondary processes. Conventional EI sources, gas discharges, or nuclear disintegrations are used to prepare a beam of primary ions which bombard

[41] C. J. McNeal, *Anal. Chem.* **54,** 43A (1982).
[42] C. Fenselau and R. J. Cotter, *Chem. Rev.* **87,** 501 (1987).
[43] P. Sigmund, *Phys. Rev.* **184,** 383 (1969).
[44] P. Williams and B. U. R. Sundqvist, *Phys. Rev. Lett.* **58,** 1931 (1982).
[45] R. J. Luchhese, *J. Chem. Phys.* **86,** 443 (1987).
[46] M. Barber, R. S. Bordoli, R. D. Sedgwick, and A. N. Tyler, *J. Chem. Soc. Chem. Commun.* p. 325 (1981).
[47] R. G. Cooks and K. L. Busch, *Int. J. Mass Spectrom. Ion Phys.* **53,** 111 (1983).
[48] C. Fenselau and R. J. Cotter, *in* "IUPAC, Frontiers of Chemistry" (K. J. Laidler, ed.), p. 207. Pergamon, New York, 1982.
[49] L. R. Schronk and R. J. Cotter, *Biomed. Environ. Mass Spectrom.* **13,** 395 (1986).
[50] J. A. Sunner, R. Kulatunga, and P. Kebarle, *Anal. Chem.* **58,** 1312 (1986).

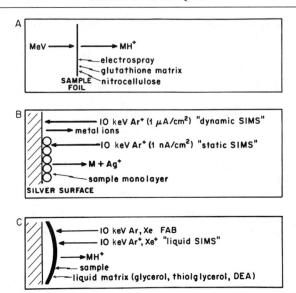

FIG. 6. An overview of particle beam techniques. (A) Plasma desorption mass spectrometry (PDMS); (B) secondary ion mass spectrometry (SIMS); (C) fast atom bombardment (FAB).

a surface and release sample, or secondary ions. However, the term secondary ion mass spectrometry (SIMS) has been used primarily to designate mass spectrometric methods in which the primary ion beam has energies in the kiloelectron volt range. Initially, such methods were used for the sputtering of metal ions from surfaces, and not for the desorption of intact molecular ion species.

Plasma Desorption Mass Spectrometry. Ionization of heavy molecules using multiply charged megaelectron volt ions is known as plasma desorption mass spectrometry (PDMS).[51] In this technique, energetic particles are generated from the spontaneous fission events occurring in a 10 μCi sample of californium-252, and pass through a thin aluminum foil on which the sample is deposited (Fig. 6). Sample ions are extracted by an electrical field and mass analyzed by a TOF mass spectrometer. In addition to its high mass range and high ion transmission, the TOF analyzer is the most practical, since the low ionization rates essential to this method preclude mass-scanning techniques.

Particles in the same 1 MeV/dalton range have also been generated

[51] R. D. Macfarlane, R. P. Skowronski, and D. F. Torgerson, *Biochem. Biophys. Res. Commun.* **60,** 616 (1974).

from tandem accelerators[52] and produce similar results. However, the instrumental simplicity of the californium source (and the TOF analyzer) has been an advantage in the development of low-cost instruments for routine analysis.[53] Sample and surface preparation have been the key to the success of this technique in measuring molecular masses up to 35 kDa.[54] Because of the extraction and focusing properties of the TOF source, it is essential that the sample be present as a thin, uniform layer. This can be achieved by electrospraying the sample onto the aluminum or aluminized Mylar sample foil.[55] In addition, dissolving peptide samples in solutions containing the tripeptide glutathione (prior to electrospraying) results in improvement in the molecular ion signal, due (presumably) to preservation of the folded form of the peptide during the electrospray process.[56] Enhancement of molecular ion yield can also be achieved by adsorption of the sample to a nitrocellulose-coated sample foil.[57] Such foils are prepared by electrospraying a nitrocellulose/acetone solution onto the foil. Adsorption of peptides to this hydrophobic surface permits the removal (by washing with deionized water or acid solutions) of inorganic cations (primarily Na^+) which interfere with the desorption process. In many cases, it is advantageous to use both the glutathione matrix and the nitrocellulose surface.

Although PDMS has been used primarily for measuring the molecular weights of peptides, Fig. 7 shows an example of its use in surveying the carbohydrate heterogeneity in the high mannose glycopeptide from the variant surface glycoprotein (VSG) from trypanosomes.[58] The mass resolution is low; however, the measured molecular ion masses correspond well to the calculated masses for a glycopeptide: Phe-Asn(GlcNAc$_2$Man$_n$)-Glu-Thr-Lys, containing from 5 to 9 mannose units.

Molecular Secondary Ion Mass Spectrometry (SIMS). An analogous method, using kiloelectron volt primary ions for the desorption of large molecules, is static or molecular SIMS[59] (Fig. 6). Primary ion current densities of about 1 nA/cm^2 distinguished this method from the 1 μA/cm^2

[52] P. Hakansson, I. Kamensky, B. Sundqvist, J. Fohlman, P. Petersen, C. J. McNeal, and R. D. Macfarlane, *J. Am. Chem. Soc.* **104**, 2948 (1982).

[53] R. J. Cotter, *Anal. Chem.* **60**, 781A (1988).

[54] A. G. Craig, A. Engstrom, H. Bennich, and I. Kamensky, *Proc. 35th ASMS Conf. Mass Spectrom. Allied Topics, Denver, CO* p. 528 (1987).

[55] C. J. McNeal, R. D. Macfarlane, and E. L. Thurston, *Anal. Chem.* **51**, 2036 (1979).

[56] M. Alai, P. Demirev, C. Fenselau, and R. J. Cotter, *Anal. Chem.* **58**, 1303 (1986).

[57] G. Jonsson, A. Hedin, P. Hakansson, B. U. R. Sundqvist, G. Sawe, P. F. Nielsen, P. Roepstorff, K. E. Johansson, I. Kamensky, and M. Lindberg, *Anal. Chem.* **58**, 1084 (1986).

[58] M. F. Bean, J. D. Bangs, T. L. Doering, P. T. Englund, G. W. Hart, C. Fenselau, and R. J. Cotter, *Anal. Chem.* **61**, 2686 (1989).

[59] A. Benninghoven, D. Jaspers, and W. Sichtermann, *Appl. Phys.* **11**, 35 (1976).

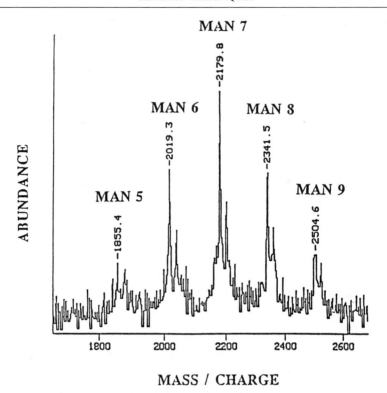

FIG. 7. Plasma desorption mass spectrum of the gp432 glycopeptide from the variant surface glycoprotein (VSG) from *Trypanosoma brucei*.[58]

(or higher) current densities used in classical sputtering (atomic or dynamic SIMS). In this case, samples are deposited as a monolayer on a silver surface so that the predominant molecular ion species are [M + Ag]$^+$ ions. While early experiments using quadrupole mass analyzers limited the mass range of this technique, both sector instruments[60] and TOF analyzers[61,62] have been utilized for high mass analysis.

Fast Atom Bombardment. Fast atom bombardment (FAB)[46] employs a neutral 10 keV Ar (or Xe) beam to sputter ions from samples dissolved

[60] T. M. Barlak, J. R. Wyatt, R. J. Colton, J. J. DeCorpo, and J. E. Campana, *J. Am. Chem. Soc.* **104**, 1212 (1982).
[61] E. Niehuis, T. Heller, H. Feld, and A. Benninghoven, *in* "Secondary Ion Mass Spectrometry, SIMSV" (A. Benninghoven, R. J. Colton, D. S. Simons, and H. W. Werner, eds.), p. 188. Springer-Verlag, Berlin, 1986.
[62] X. Tang, R. Beavis, W. Ens, F. Lafortune, B. Schueler, and K. G. Standing, *Int. J. Mass Spectrom. Ion Proc.* **85**, 43 (1988).

in a liquid matrix. It appeared initially that the use of neutral primary atoms was the major advantage for injection of high-energy particles into the high-voltage sources of double-focusing mass spectrometers. Almost immediately, however, the important role of the liquid matrix in reducing sample damage under high-flux primary particle bombardment was appreciated. This enables the production of high secondary ion currents which is convenient for scanning instruments. Thus, it has been described as a method which employs the primary ion current densities of dynamic SIMS but achieves the results of static SIMS.[63] Because FAB sources could be easily retrofitted to existing sector or quadrupole instruments, this technique enjoyed almost immediate popularity, although the sector instruments took the best advantage of the high mass capabilities of the FAB technique. In addition to neutral atom sources, cesium ion guns[64] can also be used for fast atom bombardment, and underscores the fact that the charge on the primary particle is less important than the use of the liquid matrix. Cesium ion sources can be made to produce more focused primary beams (and more efficient ionization) than neutral atom sources, and result in lower detection limits. Thus, many laboratories have replaced their neutral atom sources with cesium ion guns, now offered by most manufacturers of mass spectrometers. In addition, since scanning over wide mass ranges reduces ultimate sensitivity, an interest has developed in array detectors,[65] for simultaneous recording of a portion of the mass range.

The liquid matrix reduces sample damage by the diffusion of fresh sample to the surface from the bulk of the solution. When mixtures are dissolved in a glycerol matrix and deposited on the sample probe, the most surface-active[66,67] or hydrophobic[68] components will populate the surface and suppress ionization of the remaining components. Lowering the solution pH or the use of thioglycerol as a matrix can increase protonation and improve the $[M + H]^+$ ion signal in positive ion mass spectra. Alternatively, diethanolamine (DEA) is an appropriate matrix for recording negative ion spectra of less basic or anionic samples. The advantage of recording mass spectra compatible with the natural charge sign of the analyte is illustrated in Fig. 8. In this case, a negative ion FAB mass spectrum was

[63] R. J. Colton, *J. Vac. Sci. Technol.* **18,** 731 (1981).
[64] W. Aberth and A. L. Burlingame, *Anal. Chem.* **56,** 2915 (1984).
[65] J. S. Cottrell and S. Evans, *Anal. Chem.* **59,** 1990 (1987).
[66] S. S. Wong and F. W. Rollgen, *Nucl. Instrum. Methods Phys. Res.* **B14,** 436 (1986).
[67] M. Barber, R. S. Bordoli, G. S. Elliot, R. D. Sedwick, and A. N. Tyler, *J. Chem. Soc. Faraday Trans.* **679,** 1249 (1983).
[68] S. Naylor, G. Moneti, and S. Guyan, *Biomed. Environ. Mass Spectrom.* **17,** 393 (1988).

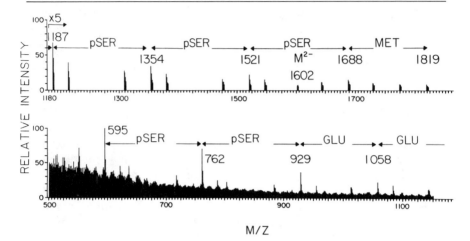

FIG. 8. A portion of the negative ion fast atom bombardment mass spectrum of the phosphate region of the phosphorylated riboflavin-binding protein from hen egg white. The sequence of that region was determined to be:

Ser-Glu-pSer-pSer-Glu-Glu-pSer-pSer-pSer-Met-pSer-pSer-pSer-Glu-Glu

where pSer corresponds to a phosphoserine residue.[69]

recorded to reveal the location of phosphate groups in the sequence of a tryptic phosphopeptide from riboflavin-binding protein.[69]

Laser Beam Techniques

Laser Microprobes. One of the earliest combinations of laser ionization and mass spectrometry to be offered commercially was the laser microprobe mass analyzer.[70] These instruments used a frequency-doubled (530 nm) or frequency-tripled (353 nm) neodymium/yttrium–aluminum–garnet (Nd/YAG) laser, and were designed primarily for elemental analysis of thin samples with high spatial resolution. However, such instruments have also been used for the laser desorption of high-molecular weight, nonvolatile samples.[71]

Infrared Laser Desorption. For the past several years many of the instruments used for the laser desorption of intractable molecules have used infrared lasers. For the most part, these instruments have been

[69] C. Fenselau, D. N. Heller, M. S. Miller, and H. B. White III, *Anal. Biochem.* **150,** 309 (1985).
[70] F. Hillenkamp, E. Unsold, R. Kaufmann, and R. Nitsche, *Nature (London)* **256,** 119 (1975).
[71] S. W. Graham, P. Dowd, and D. M. Hercules, *Anal. Chem.* **54,** 649 (1982).

constructed in the laboratories of individual investigators and represent a variety of instrumental configurations, including a Nd/YAG laser (1.06 μm) with a sector instrument and an array detector,[72] and a TOF system using a pulsed CO_2 laser (10.6 μm).[73] For such infrared systems, mechanisms have been proposed[74,75] in which molecular ion species are produced by gas phase attachment of alkali ions to codesorbed neutral molecules. This mode of ionization appears to be particularly advantageous for the analysis of large, but relatively neutral, molecules such as polyethylene and polypropylene glycols,[76,77] oligosaccharides,[78] and glycolipids,[79] and is therefore considered by many to be an alternative to field desorption. Infrared CO_2 laser desorption has also been developed on Fourier transform mass spectrometers (FTMS)[80] and sold commercially. This instrumental combination has enabled some spectacular high-resolution, high-mass measurements of industrial polymers.[77]

Although CO_2 laser desorption is not as widely used as either plasma desorption or fast atom bombardment, it has an important advantage for the structural analysis of carbohydrates, since the predominant fragmentation occurs in the sugar rings themselves, and not at the glycosidic bond.[78] This mode of fragmentation has been used as a means of locating the positions of fatty acid attachment to the diglucosamine backbone in the anchor of the lipopolysaccharides from gram-negative bacteria.[80]

Matrix-Assisted Laser Desorption. During the past year, spectacular high mass results have been achieved by matrix-assisted laser desorption. In the method of Tanaka *et al.*,[81] the sample is mixed with a slurry of finely divided platinum powder in glycerol and a volatile solvent (Fig. 9). The mixture is deposited on a surface and both the volatile solvent and glycerol are allowed to evaporate. Generally the analog signals from 200 to 500

[72] M. A. Posthumus, P. G. Kistemaker, H. L. C. Meuselaar, and M. C. Ten Noever de Brauw, *Anal. Chem.* **50,** 985 (1978).

[73] R. B. Van Breemen, M. Snow, and R. J. Cotter, *Int. J. Mass Spectrom. Ion Phys.* **49,** 35 (1983).

[74] G. J. Q. van der Peyl, K. Isa, J. Haverkamp, and P. G. Kistemaker, *Org. Mass Spectrom.* **16,** 416 (1981).

[75] J.-C. Tabet and R. J. Cotter, *Anal. Chem.* **56,** 1662 (1984).

[76] R. J. Cotter, J. P. Honovich, J. K. Olthoff, and R. P. Lattimer, *Macromolecules* **19,** 2996 (1986).

[77] R. S. Brown, D. A. Weil, and C. L. Wilkins, *Macromolecules* **19,** 1255 (1986).

[78] M. L. Coates and C. L. Wilkins, *Biomed. Mass Spectrom.* **12,** 424 (1985).

[79] K. Takayama, N. Qureshi, K. Hyver, J. Honovich, R. J. Cotter, P. Mascagni, and H. Schneider, *J. Biol. Chem.* **261,** 10624 (1986).

[80] D. A. McCreary, E. B. Ledford, Jr., and M. L. Gross, *Anal. Chem.* **54,** 1437 (1982).

[81] K. Tanaka, H. Waki, Y. Ido, S. Akita, Y. Yoshida, and T. Yoshida, *Rapid Commun. Mass Spectrom.* **2,** 151 (1988).

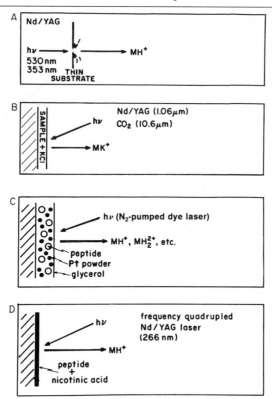

FIG. 9. An overview of laser desorption techniques. (A) Laser microprobes; (B) infrared laser desorption; (C) matrix-assisted laser desorption; (D) UV-absorbing matrix laser desorption.

laser shots from a N_2-pumped dye laser are accumulated to produce a mass spectrum. A spectrum of chicken egg white lysozyme reported by this group[81] reveals both multiply charged ions and molecular clusters up to 100,872 Da.

The method of Karas and Hillenkamp[82] employs a pulsed beam from a frequency-quadrupled Nd/YAG laser at 256 nm and a UV-absorbing matrix, nicotinic acid, mixed with the sample. Singly charged molecular ions dominate the mass spectrum, and have been observed above 200 kDa. In one case, they recorded the molecular ion of porcine trypsin (measured at 23,450 Da) with a single laser shot. Their method was developed on a laser microprobe, but has been duplicated on other instruments in several other laboratories.

[82] M. Karas and F. Hillenkamp, *Anal. Chem.* **60,** 2299 (1988).

While matrix-assisted UV laser desorption is a new technique, it can be expected to enjoy wide use in the near future.

Similar Spectra from Different Techniques

With the exception of some organometallic compounds and quinones, the molecular and fragment ions observed in desorption mass spectra are even-electron ions. Since this is a property they share with chemical ionization mass spectra, we have noted that it is tempting to conclude that ion formation always proceeds by gas-phase processes following desorption, and that desorption spectra are really CI spectra. Organic salts such as tetramethylammonium chloride, $(CH_3)_4NCl$, provide a notable exception. These involatile and thermally labile salts decompose to the tertiary amine when heated on a direct probe, and subsequent chemical ionization produces the protonated species, $(CH_3)_3NH^+$ at m/z 60. Desorption, however, results in the direct formation of the molecular cation, $(CH_3)_4N^+$, at m/z 74.[83] While this is clearly a case in which chemical ionization and desorption produce different results, it should be noted that both techniques produce stable even-electron ions.

Heteroatoms and Electrons. The sites for chemical bond cleavage in electron ionization have generally been predicted by assuming that ionization occurs at heteroatoms (those atoms which have nonbonding electrons) and that subsequent rearrangement of electrons directs the fragmentation. For example, electron ionization of the ethyl ester of methionine can result in the removal of a nonbonding electron from the sulfur atom and formation of the radical ion:

$$CH_3\overset{}{-}\dot{S}^+\overset{}{-}CH_2CH_2\underset{\underset{NH_2}{|}}{CH}\overset{\overset{O}{\|}}{C}OCH_2CH_3$$

The remaining unpaired electron and an electron from an adjacent C—C bond results in the formation of the fragment ion: $^+CH_3S{=}CH_2$. Alternatively, an electron may be removed from the nitrogen atom,

$$CH_3SCH_2CH_2\underset{\underset{^+NH_2}{|}}{CH}\overset{\overset{O}{\|}}{C}OCH_2CH_3$$

resulting in cleavage of the $CH{-}COOC_2H_5$ bond and the fragment ion:

[83] R. J. Cotter and A. L. Yergey, *J. Am. Chem. Soc.* **103**, 1596 (1981).

$$CH_3-\overset{\cdot}{S}{}^+-CH_2CH_2CH{=}NH_2$$

Heteroatoms, Even-Electron Ions, and Hydrogen Transfers. It is possible to develop similar rational predictions of fragmentation in desorption mass spectra by assuming (analogously) that proton attachment also occurs preferentially at heteroatoms, and that hydrogen transfers will occur to ensure the formation of stable even-electron ions. For a peptide, for example, a proton may be attached to the amide nitrogen to form an even-electron $[M + H]^+$ ion.

$$\begin{array}{c} \quad R \quad\; H^+\, R' \\ \quad | \quad\;\; |\;\; | \\ \text{---NHCHCNHCHC---} \\ \quad\quad\; \| \\ \quad\quad\; O \end{array}$$

Homolytic cleavage of the C—N bond and transfer of a neutral hydrogen atom results in retention of the charge on the C-terminal portion.

$$\begin{array}{c} \quad\;\; R' \\ \quad\;\; | \\ {}^+NH_3CHC\text{---} \\ \quad\;\; \| \\ \quad\;\; O \end{array}$$

Alternatively, heterolytic cleavage of the same bond transfers both electrons to the N atom, leaving the positive charge on the N terminus.

$$\begin{array}{c} \quad\;\; R \\ \quad\;\; | \\ \text{---NHCHC}{\equiv}O^+ \end{array}$$

Similar predictions can be made for the fragmentation of oligosaccharides by considering that a proton becomes attached to a nonbonding electron pair on the glycosidic oxygen. This then directs fragmentation to either side of the glycosidic oxygen and, with appropriate hydrogen transfers, results in a series of ions which enable sequencing from either the reducing or nonreducing terminus.

It is difficult to overemphasize the importance of drawing even-electron structures, which generally differ by a single mass unit from simple cleavages, when attempting to predict or explain fragmentation. A single mass unit can distinguish between the presence of Asp/Asn and Glu/Gln in peptides or glucose/glucosamine in carbohydrates. Specific details of the fragmentation of both peptides and oligosaccharides are presented in subsequent chapters in this volume.

Notable Differences in Spectra and Selectivity

Because stable, even-electron molecular ions are produced it is possible to establish general rules for interpreting desorption mass spectra, somewhat independent of the technique. At the same time, the desorption techniques are not all the same. There are some interesting differences in capabilities and some opportunities for obtaining complementary information. These differences can be generalized as differences in mass range, ionization efficiency for specific classes of compounds (selectivity), and fragmentation.

High-Mass Range. PDMS has been the most effective, routine method for obtaining molecular masses of peptides in the 5–20 kDa range. Used in conjunction with TOF mass analyzers, the resulting low-resolution mass spectra can nevertheless provide accurate mass measurements,[57] since they are generally uncomplicated by fragmentation. Molecular mass measurements by FAB mass spectrometry have been obtained for several insulins (>5 kDa) and proinsulins (>8 kDa), but the technique is more widely used for the sequence analysis of tryptic peptides in the 500–3500 Da range. Compatibility with high-resolution, double-focusing instruments is an advantage for producing high-quality sequence ion spectra. Recently, however, impressive high mass results have been achieved by the matrix-assisted laser desorption.[81,82] Equally impressive are the results which have recently been obtained by electrospraying techniques (described below), where large molecular masses have been determined on a quadrupole mass spectrometer from a series of multiply charged ions. Both of these approaches can be expected to gain in importance in the near future.

Selectivity and Suppression in Mixtures. For both FAB and PDMS the natural *charge sign* at neutral pH has generally dictated the recording of either a positive or negative ion mass spectrum. Peptides that contain one or more basic residues are most easily analyzed by recording their positive ion mass spectra. In the FAB technique, thioglycerol, thioglycerol/glycerol mixtures, and other acidified matrices encourage protonation by lowering the pH with respect to the isoelectric point. Conversely, oligonucleotides, sulfatides, phosphorylated peptides (see Fig. 8), gangliosides, and glycopeptides containing sialic acid are analyzed by recording their negative ion mass spectra, and the triethanolamine matrix is used (in FAB) to enhance deprotonation. Thus, in a mixture, the positive and negative ion mass spectra may reveal selective desorption of different components.

There is an additional aspect of selectivity in FAB mass spectra which can be attributed to the use of the liquid matrix. In a mixture of analytes, the most surface-active components[42,66,67] will be selectively ionized by FAB mass spectrometry, since they more easily diffuse to the surface of the liquid droplet. Since the other analytes might well be successfully analyzed in their pure states, these effects are generally described as suppression effects. When extended to the analysis of unfractionated tryptic digests, the FAB technique favors the desorption of the most hydrophobic peptides.[68,84] In PDMS, a liquid matrix is not used, so that analysis of tryptic mixtures is primarily sensitive to the charge sign at neutral pH.[85] However, the ionization efficiencies of the components in the tryptic peptide mixture continue to be sensitive to their relative hydrophobicities, particularly when the sample has been adsorbed to a nitrocellulose surface.[86] These effects demand some caution when using FAB or PDMS for mapping tryptic digests.

Complementary Fragmentation. FAB mass spectra of neutral oligosaccharides often provide very low molecular ion signals and little sequence information from fragmentation.[87,88] The molecular ion yield and fragmentation can both be enhanced by analyzing the permethylated or peracetylated derivatives.[87] Alternatively, collision-induced dissociation (CID) can be used to increase the amount of fragmentation.[88] In this area CO_2 laser desorption may have some advantages. The infrared laser desorption mass spectrum of underivatized maltotetraose (Fig. 10) shows several fragment ion series, resulting predominantly from two-bond ring cleavages rather than fragmentation at the glycosidic linkage.[88] Such cleavages are sensitive to the presence of amino or N-acetylated sugars, nonreducing termini (e.g., trehalose), and may be useful (in some cases) for distinguishing linkage positions.[89]

Fields and Sprays

Recently there has been a resurgence of interest in techniques using a combination of high electric fields, resistive heating, and nebulization. The earliest technique, an outgrowth of field desorption, was electrohydrodynamic (EHD) ionization.[90,91] This is not a spray technique, but employs direct field evaporation of ions from a liquid droplet. In contrast, the thermospray[92] technique does not employ an electrical field, but achieves nebulization by resistive heating. As the solvent evaporates, ions are ejected from the charged droplet when the charge density exceeds the

[84] S. Naylor, A. F. Findeis, B. W. Gibson, and D. H. Williams, *J. Am. Chem. Soc.* **108**, 6339 (1986).

[85] P. F. Nielsen and P. Roepstorff, *Biomed. Environ. Mass Spectrom.* **18**, 131 (1989).

[86] L. Chen, R. J. Cotter, and J. T. Stults, *Anal. Biochem.* **183**, 190 (1989).

[87] A. Dell and P. R. Tiller, *Biochem. Biophys. Res. Commun.* **135**, 1126 (1986).

[88] S. A. Carr, V. N. Reinhold, B. N. Green, and J. R. Hass, *Biomed. Mass Spectrom.* **12**, 288 (1985).

[89] W. B. Martin, L. Silly, C. M. Murphy, T. J. Raley, R. J. Cotter, and M. F. Bean, *Int. J. Mass Spectrom. Ion Proc.* **92**, 243 (1989).

[90] B. N. Colby and C. A. Evans, *Anal. Chem.* **45**, 1884 (1973).

[91] K. W. S. Chan and K. D. Cook, *Macromolecules* **16**, 1736 (1983).

[92] C. R. Blakley and M. L. Vestal, *Anal. Chem.* **55**, 750 (1983).

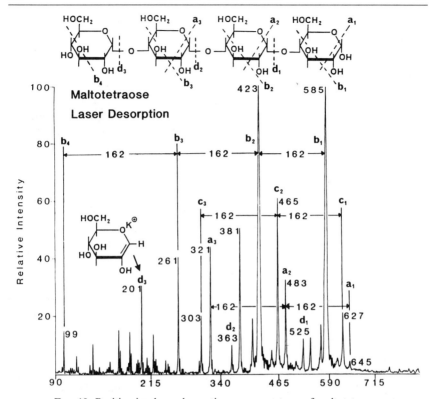

FIG. 10. Positive ion laser desorption mass spectrum of maltotetraose.

Rayleigh limit.[92] Because this ionization technique can tolerate relatively high liquid flow rates, it has been used primarily as an interface between reversed-phase high-performance liquid chromatography (HPLC) and a mass spectrometer.

An important new technique for the analysis of large biomolecules is electrospray[93] and its several variations using atmospheric pressure ionization (API) mass spectrometers.[94,95] These techniques produce a distribution of highly charged ions of macromolecules with a maximum mass-to-charge ratio (fortuitously) within the mass range of a quadrupole ana-

[93] C. M. Whitehouse, R. N. Dreyer, M. Yamashita, and J. B. Fenn, *Anal. Chem.* **57,** 675 (1985).

[94] T. R. Covey, R. F. Bonner, B. I. Shushan, and J. Henion, *Rapid Commun. Mass Spectrom.* **2,** 249 (1988).

[95] R. D. Smith, C. J. Berinaga, and H. R. Udseth, *Anal. Chem.* **60,** 1948 (1988).

lyzer. With this technique, molecular masses in excess of 100 kDa have been recorded.

Future Directions

With current, rapid explosion of new ionization techniques it is difficult to predict the future and the impact that mass spectrometry will have on structural analysis of large, more complex, biomolecules. The fact that high mass molecular ions are observed by such widely disparate techniques as UV laser desorption (for which singly charged ions predominate) and electrospray (which produces multiply charged ions) attests to the unusual stability of biological macromolecules in the gas phase as well as in solution. That fact, combined with what we have learned thus far about extracting ions from solution, the importance of matrices and surfaces, and the coupling of primary energy to the substrate, can be expected to produce new methods which increase both mass range and sensitivity.

The majority of instruments used in analytical laboratories today are equipped with interfaces to either packed-column or capillary gas chromatographs. Combined GC/MS analysis has become routine and is generally carried out with either electron ionization or chemical ionization sources, since these are appropriate for the introduction of volatile compounds. We are currently witnessing a similar explosion in the development of methods which combine HPLC and electrophoretic separation techniques with mass spectrometry. To date, reversed-phase HPLC, high-performance anion-exchange (HPAE) chromatography, microbore LC, supercritical fluid chromatography (SFC), capillary zone electrophoresis (CZE), isotachyphoresis, and thin-layer chromatography (TLC) have all been used as on-line mass spectrometric techniques. Most of these have depended upon the development of appropriate desorption techniques. Thermospray has been the most appropriate method for handling the large flow rates (1–5 ml/min) from reversed-phase HPLC, while continuous-flow FAB[96] has been developed as a technique compatible with microbore LC. Capillary zone electrophoresis has been used successfully with both electrospray[95] and continuous-flow FAB ionization. The combination of separation techniques with mass spectrometry is one of the most important directions for mass spectrometry, particularly for the analysis of complex biological mixtures. Thus, its development goes hand-in-hand with the development of appropriate desorption techniques.

[96] R. M. Caprioli, T. Fan, and J. S. Cottrell, *Anal. Chem.* **58,** 2949 (1986).

In the chapters which follow, these techniques are described in more detail and their importance to chemistry, biochemistry, and biology illustrated by their applications to the structural analysis of peptides, carbohydrates, glycoconjugates, and other macromolecules.

[2] Mass Analyzers

By KEITH R. JENNINGS and GREGORY G. DOLNIKOWSKI

Introduction

All mass spectrometers consist of three basic components: the ion source, the mass analyzer, and the detector. Ions are produced from the sample in the ion source (see [1] in this volume). The mass analyzer separates ions according to their mass-to-charge ratio, m/z. Each type of ion strikes the detector (see [3] in this volume) and produces a signal proportional to its relative abundance. A plot of the relative abundances of each ion against m/z yields a mass spectrum.

Mass analyzers separate ions by making use of appropriate electric fields, sometimes in combination with magnetic fields. The purpose of this chapter is to review the advantages and disadvantages of the most widely used mass analyzers including those employed in single- and double-focusing magnetic deflection instruments, time-of-flight (TOF) instruments, and quadrupole instruments. In addition, the different methods of scanning the analyzers to extend their range of applicability will be discussed.

The choice of a mass analyzer depends on a number of interrelated factors, such as (i) mass range (maximum m/z detectable); (ii) resolving power (ability to separate ions of closely similar m/z); (iii) accuracy of mass measurement (useful only if ions are completely resolved from interfering ions); (iv) ion transmission, sensitivity, and Q value ("figure of merit" for a mass analyzer); (v) scanning speed [usually quoted as time taken to scan a decade in mass (e.g., m/z 1000 → m/z 100)]; and (vi) ease of use with ancillary equipment (such as chromatographic equipment). No one mass analyzer is suitable for all applications and the choice of instrument is determined by the type of problem under investigation.[1]

[1] C. Brunee, *Int. J. Mass Spectrom. Ion Proc.* **76,** 1 (1987).

Elementary Considerations

Singly charged ions of mass m_1, subjected to an accelerating voltage V_1 acquire a translational energy as shown in Eq. (1)

$$V_1 e = m_1 v_1^2/2 = (m_1 v_1)^2/2m_1 \tag{1}$$

where e is the electronic charge and v_1 is the velocity of the ion m_1^+. Ions of different masses have the same kinetic energy, $V_1 e$. However, the velocity of an ion of a given mass is proportional to the reciprocal of the square root of its mass, whereas the momentum increases with the square root of its mass [Eq. (2)]:

$$m_1 v_1 = (2 V_1 e m_1)^{1/2} \tag{2}$$

At a fixed accelerating voltage, therefore, it follows that a separation of ions based on their momentum or their velocity is equivalent to a separation based on mass. The momentum separation is the basis of a magnetic deflection instrument; velocity separation is the basis of a TOF instrument. A quadrupole instrument, on the other hand, uses dc and rf voltages and separates ions directly according to m/z. Multiply charged ions behave as singly charged ones of apparent mass m/z, where z is the number of charges on the ion.

Electric Sector

A conventional electric sector consists of two parallel cylindrical plates across which an electric field E is applied (Fig. 1). Ions pass between these plates and follow an approximately circular path of radius r. The sector acts both as a focusing device for diverging ion beams and as a device for dispersing ions according to their kinetic energies.[2] It is first-order direction focusing in that all ions of the same energy emanating from the source slit with a small angular divergence are brought to focus and follow a path of approximate radius r providing that

$$Ee = m_1 v_1^2/r = 2Ve/r = 2(\text{ion energy})/r \tag{3}$$

where

$$r = 2V/E \tag{4}$$

It is energy dispersing in that ions having greater or lower energies follow paths of greater or lower radii and are brought to focus at different positions. Since most ion sources produce beams of ions with a significant

[2] R. G. Cooks, J. H. Beynon, R. M. Caprioli, and G. R. Lester, "Metastable Ions." Elsevier, Amsterdam, 1973.

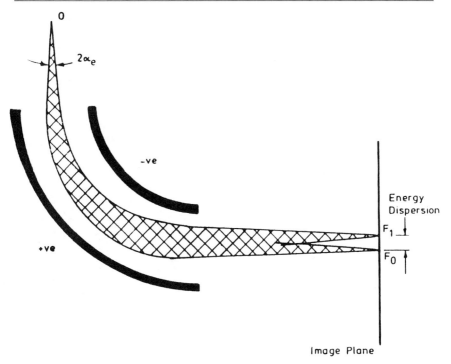

FIG. 1. Principle of operation of an electic sector. O, Object point for source of ions with mass m_0 and m_1, energies E_0 and E_1, with angular divergence of $2\alpha_e$; F_0, focus point for ions with energy E_0, mass m_0 and m_1, and with angular divergence of $2\alpha_e$; and F_1, focus point for ions with energy E_1, mass m_0 and m_1, and with angular divergence of $2\alpha_e$.

energy spread, an electric sector can be used to reduce the energy spread in the transmitted ion beam at the cost of a decrease in ion abundance.

Consequently, if all ions entering an electric sector have the same energy, there is no separation of ions according to their m/z ratio. If, on the other hand, an ion m_1^+ fragments to give an ion m_2^+ in the field-free region before the electric sector, the energy of the product ion will be $(m_2/m_1)Ve$ and different product ions may be distinguished by scanning E to produce an energy spectrum that is equivalent to a mass spectrum of the products.[3]

[3] R. W. Kiser, "Introduction to Mass Spectrometry and Its Applications." Prentice-Hall, New York, 1965.

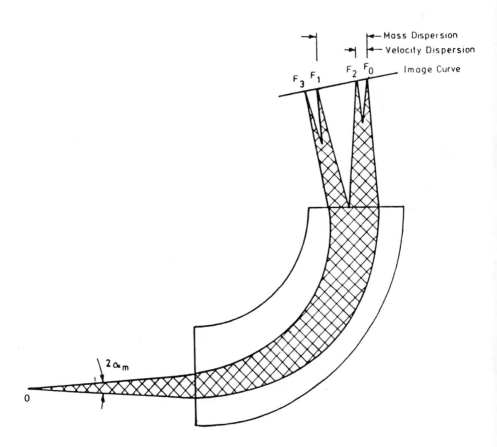

FIG. 2. Principle of operation of a magnetic sector. O, Points source for ions of mass m_0 and m_1 and velocity v_0 and v_2 with angular spread of $2\alpha_m$; F_0, direction focus point for ions of mass m_0 and velocity v_0; F_2, direction focus point for ions of mass m_0 and velocity v_1; F_1, direction focus point for ions of mass m_1 and velocity v_0; and F_3, direction focus point for ions of mass m_1 and velocity v_1.

Magnetic Sector

When accelerated $m_1{}^+$ ions enter a magnetic field of strength B_1, the ions follow a circular path of radius R, perpendicular to the direction of the field (Fig. 2), where R is given by

$$R = m_1 v_1 / B_1 e \qquad \text{or} \qquad m_1 v_1 = R B_1 e \qquad (5)$$

If B is scanned at a fixed value of R, therefore, ions of different momenta, and hence of different mass, can be made to pass through a collector slit

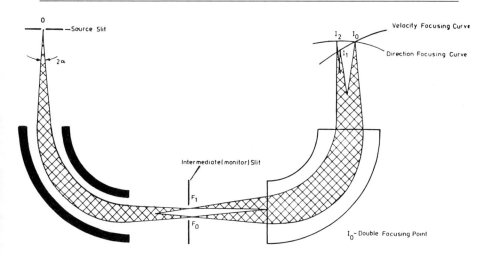

FIG. 3. Principle of operation of a double-focusing mass spectrometer. The point I_0 at which the velocity focusing image I_1 and the direction focusing image I_2 coincide is the double focusing point.

to give a mass spectrum.[2] Alternatively, ions that follow paths of different radii at a fixed magnetic field strength may be collected simultaneously using either a photographic plate or, as used more recently, an array detector (see [3] in this volume). Combining the above equations gives the standard expression for the separation of ions by a magnetic sector [Eqs. (6)–(8)]:

$$m_1/e = R_2 B_1^2/2V_1 \qquad (6)$$
$$= KB_1^2 \text{ (at fixed } V) \qquad (7)$$
$$= K'/V_1 \text{ (at fixed } B) \qquad (8)$$

If m is expressed in atomic mass units (daltons), R in centimeters, B in tesla (T), and V in volts, for singly charged ions, Eq. (6) reduces to

$$m = (4.8 \times 10^3)(R^2 B^2/V)$$

Hence, at a constant accelerating voltage, a scan linear in m is proportional to B^2 whereas a scan linear in B is proportional to $m^{1/2}$. Similarly, at a fixed magnetic field strength, a scan linear in V is proportional to $1/m$, or mV is a constant which varies with B^2.

By choosing appropriate combinations of electric and magnetic sectors, it is possible to focus an ion beam which is both divergent and inhomogeneous in energy, thereby achieving both direction and velocity first-order focusing. Such instruments are called "double-focusing" mass spectrometers (Fig. 3) and their performance as mass analyzers is deter-

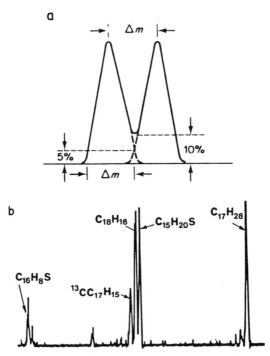

FIG. 4. Resolving power of a magnetic sector instrument: (a) 10% valley definition; (b) m/z 232 region of an oil sample recorded at a resolving power of 70,000 (Kratos Analytical, Manchester, England).

mined primarily by the size of aberrations arising from second-order terms which are due to such factors as field inhomgeneities and fringing fields.[4–6] These aberrations are normally reduced by using electrostatic lenses so as to optimize the performance of an instrument of a particular size and geometry. Instruments in which the electric sector precedes the magnetic sector are known as forward geometry or *EB* instruments; when the magnetic sector precedes the electric sector, they are known as reverse geometry or *BE* instruments.

The resolving power of a magnetic sector mass analyzer is usually defined as the value of $m/\Delta m$ for which the height of the valley between peaks of equal height at m and $m + \Delta m$ is 10% of the height of the peaks (Fig. 4). For a typical magnet of radius 30 cm, the resolving power is in

[4] A. O. Nier and T. R. Roberts, *Phys. Rev.* **81,** 507 (1951).
[5] W. H. Johnson and A. O. Nier, *Phys. Rev.* **105,** 1014 (1957).
[6] H. Hintenberger and L. A. Konig, *Adv. Mass Spectrom.* **1,** 16 (1959).

the region of 5000, but when combined with a suitable electric sector in a double-focusing instrument, resolving powers in excess of 100,000 may be obtained with a mass range of up to 2000 and an accuracy of mass measurement of better than 1 ppm.

In order to increase the mass range of a sector mass spectrometer, it can be seen from Eq. (6) that this can be achieved by decreasing V or increasing B or R. Most instruments allow the user to decrease V but the increase in mass range is obtained only at the cost of decreased sensitivity. For the majority of sources, the efficiency of ion extraction is designed to be greatest at full accelerating potential so that the number of ions leaving the source falls as V decreases. Even if the source design minimizes this effect, the fraction of ions transmitted by a mass analyzer having a fixed acceptance angle is approximately proportional to V when ions are formed with a large energy spread as in a desorption source.[7] Consequently, in order to maintain sensitivity, it is desirable to work with the full accelerating potential whenever possible.

Although increasing the strength of the magnetic flux B is attractive,[8] practical limitations on magnetic flux density arise from the materials used in constructing the magnet yoke. Very pure iron and low carbon steel are relatively inexpensive but begin to saturate at fluxes above about 1.8 tesla so that the ability to control the shape of the magnetic field and hence the ion optical performance is reduced. Vanadium Permendur (49% Co, 49% Fe, 2% V) gives very good results up to 2.35 tesla but is extremely expensive so that the bulk of the yoke is constructed of laminated low carbon steel to which tapered Permendur pole pieces are attached. For fixed values of V and R, this increases the mass range by a factor of $(2.35/1.8)^2 = 1.7$, so that for a 30-cm radius magnet working at 8 kV, the practical limit of the mass range is about 3000.[7]

The increases of R should, in principle, allow one to increase the mass range very considerably, but there are again practical limitations of size and rigidity of the instrument. If the angle of a simple magnetic sector is reduced and the object distance (distance between the object slit and the front poleface boundary of the magnetic sector) is increased, the mass range of the instrument may be increased but the minimum path length of the ions rises very rapidly reaching 7–8 m for m/z 10,000 under typical operating conditions. The use of one or more additional focusing elements to reduce the object and image distances can, however, allow one to reduce the overall size of an instrument while still providing a high mass

[7] J. S. Cottrell and R. J. Greathead, *Mass Spectrom. Rev.* **5**, 215 (1986).
[8] A. Dell and G. W. Taylor, *Mass Spectrom. Rev.* **3**, 357 (1984).

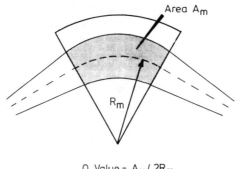

Q Value = $A_m / 2R_m$

Fig. 5. Illustration of the Q value of a magnetic sector for a given slit height.[2]

range. This also has implications for the Q value[9] of an instrument which is a figure of merit (Fig. 5) defined by

$$Q = C \times (\text{source slit width}) \times (\text{full analyzer}$$
$$\text{acceptance angle}) \times (\text{resolving power}) \quad (9)$$
$$= (\text{sensitivity}) \times (\text{resolving power}) \quad (10)$$
$$= (\text{volume swept out by beam in magnetic sector})/\text{radius} \quad (11)$$

In Eq. (9), the constant C depends on factors such as the slit length and it thereby influences the volume swept out by the beam within the magnetic sector. Using Q to characterize the overall performance of an instrument, one sees that if the radius is increased but the size of the magnet is kept approximately constant, the overall performance falls. If the analyzer acceptance angle is increased to increase Q, aberrations increase and the final design is a compromise which requires one to consider many factors such as Q value, mass range, size, and cost.[1]

Quadrupole or hexapole lenses produce a beam which is convergent on one plane but divergent in the perpendicular plane. If two such lenses are combined, overall focusing of the beam in the two planes can be effected, giving a significant reduction in the ion path length for a given magnetic sector.[10-12] A similar focusing action is obtained if the angle

[9] H. Wollnik, *Nucl. Instrum. Methods* **95**, 453 (1971).
[10] C. D. E. Ouwerkerk, A. J. H. Boerboom, T. Matsuo, and T. Sakurai, *Int. J. Mass Spectrom. Ion Proc.* **65**, 23 (1985).
[11] H. Matsuda, M. Naito, and M. Takeuchi, *Adv. Mass Spectrom.* **6**, 407 (1973).
[12] A. J. H. Boerboom, D. B. Stauffer, and F. W. McLafferty, *Int. J. Mass Spectrom. Ion Proc.* **63**, 17 (1985).

between an ion beam and the poleface boundary is not 90°, thereby increasing the differences between the path lengths of ions following trajectories of different radii.[9,13] Another method of reducing the object and image lengths of a sector magnet is to make use of an inhomogeneous magnetic field. A small radial field gradient is produced by the use of tapered polefaces so that, for example, ions which follow paths of larger radii experience a slightly greater magnetic flux than those following paths of lower radii, or vice versa. As before, this produces convergence in one plane and divergence in the other but by using a magnetic sector made up of several segments which focus alternately in one plane and then the other, an overall focusing action is achieved and a high mass range obtained with little increase in the overall size of the instrument.[9,14,15] Since aberrations occur in all focusing devices, it is usual to use a combination of two of the above methods so that corrections for first- and second-order aberrations are made. Commercial instruments using quadrupole or hexapole lenses in combination with inhomogeneous magnetic fields or nonnormal poleface boundaries are available from several manufacturers which give mass ranges up to about m/z 15,000 (Fig. 6).

The use of magnets which have laminated yokes has led to scan speeds in the region of 0.1 sec per decade in mass. Even faster scan speeds are obtained by using an air–coil magnet which is completely free from hysteresis effects.[16] If, however, a high mass range is required, the relatively large radius of the sector leads to a lower Q value than is obtained for comparable iron-cored magnetic sector instruments.

The major advantages of a magnetic sector mass analyzer, when combined with an electric sector, are a high mass range, high resolution, and accurate mass determination, together with good scanning speed and dynamic range. The major disadvantages are cost and size and the need to operate the ion source at 8–10 kV above ground potential, which complicates its use with ancillary equipment such as chromatographs. When used with a single electron multiplier, the efficiency of ion collection is necessarily low since, for much of a typical scan, no ions are collected. Nevertheless, the overall sensitivity is good and can be increased by

[13] H. Matsuda and T. Matsuo, *Proc. 10th Int. Mass Spectrom. Conf., Swansea*, p. 931. (1985).

[14] C. Brunee, G. Jung, R. Pesch, and U. Rapp, *Abstr. Pittsburgh Conf. Anal. Chem. Appl. Spectros.*, p. 317 (1953).

[15] P. A. Lyon, S. L. Hunt, S. Evans, and H. Tudge, *Proc. 34th Annu. Conf. Mass Spectrom. Allied Topics (Cincinnati)*, p. 162 (1986).

[16] R. H. Bateman, P. Burns, R. Owen, and V. C. Parr, *Proc. 10th Int. Mass Spectrom. Conf. Swansea*, p. 863 (1985).

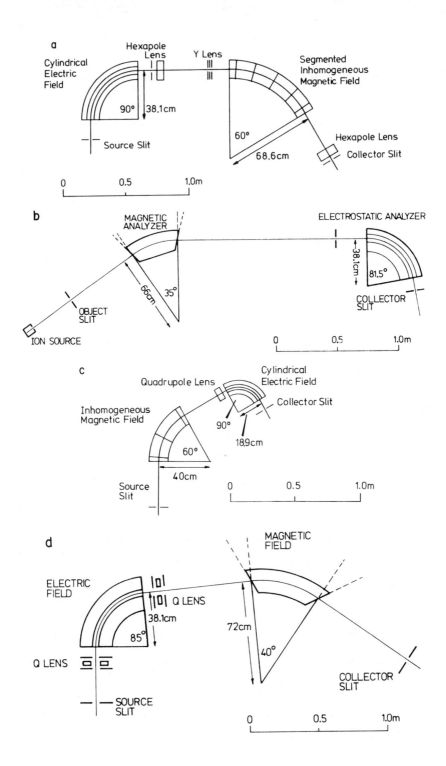

a

Cylindrical
Electric
Field

Hexapole
Lens

Y Lens

Segmented
Inhomogeneous
Magnetic Field

90°   38.1cm

Source Slit

60°

68.6cm

Hexapole Lens

Collector Slit

0   0.5   1.0m

b

MAGNETIC
ANALYZER

ELECTROSTATIC ANALYZER

66cm   35°

38.1cm

81.5°

COLLECTOR
SLIT

OBJECT
SLIT

ION SOURCE

0   0.5   1.0m

c

Quadrupole Lens

Cylindrical
Electric
Field

Collector Slit

Inhomogeneous
Magnetic Field

90°

18.9cm

60°

40cm

Source
Slit

0   0.5   1.0m

d

ELECTRIC
FIELD

Q LENS

MAGNETIC
FIELD

38.1cm

85°

72cm

40°

Q LENS

SOURCE
SLIT

COLLECTOR
SLIT

0   0.5   1.0m

approximately two orders of magnitude by using an array detector to collect all ions simultaneously over a limited mass range.

Use of Magnetic Sector Instruments for Study of Collision-Induced Decomposition of Ions

The control of three parameters, namely, the accelerating voltage V, the electric sector field strength E, and the magnetic sector flux B determines the type of mass spectrum recorded by a double-focusing magnetic deflection instrument. In normal operation, V and E are held constant, their ratio being determined by the size and geometry of the instrument, and B is scanned. In an EB instrument, if an ion m_1^+ fragments to give an ion m_2^+ in the field-free region between the two sectors, a low-intensity, diffuse peak, known as a metastable peak,[17] may be observed at an apparent mass m^*, where $m^* = m_2^2/m_1$. These peaks are not seen in the spectrum recorded using a BE instrument. The B scan of an EB instrument therefore gives a spectrum almost entirely of ions formed in the source together with low-intensity peaks arising from decomposition products formed between the two sectors and whose masses are often difficult to assign. Other methods of scanning V, E, or B, or two of them simultaneously, have been developed which discriminate against ions formed in the source and allow one to collect only those ions formed in unimolecular or collision-induced fragmentation reactions in the first or second field-free regions of instruments of either geometry.[18] These methods are outlined below, with emphasis placed on those scans that are most widely used.

Scans with Constant Magnetic Flux B

Of the three scans of this type, the two for use with EB instruments are now rarely employed since they require the accelerating voltage V to be scanned. This usually leads to detuning of the source and irreproducible relative intensities of peaks. These scans (V scan at constant E, V and E scan such that V/E^2 is constant) will not be discussed further.

In a BE instrument, suppose that precursor ions m_1^+ are collected

[17] J. A. Hipple, R. E. Fox, and E. U. Condon, *Phys. Rev.* **69,** 347 (1946).
[18] K. R. Jennings and R. S. Mason, *in* "Tandem Mass Spectrometry" (F. W. McLafferty, ed.), Ch. 9. Wiley, New York, 1983.

FIG. 6. Schematics of the ion optics of commercial double-focusing mass spectrometers[2]: (a) MS50RF (Kratos Analytical, Manchester, England); (b) ZAB-E (VG Analytical, Manchester, England); (c) MAT-8500 (Finnigan Corp., San Jose, CA); and (d) HX/110 (JEOL, Tokyo, Japan).

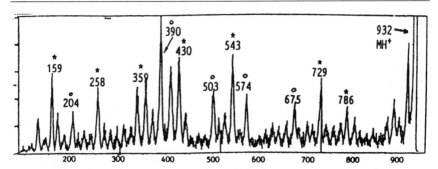

FIG. 7. Mass-analyzed ion kinetic energy (MIKES) spectrum of the MH$^+$ ion, m/z 932 of the βT2 peptide [D. Prome and J. C. Prome, *Spectros. Int. J.* **5**, 157 (1987)].

under normal operating conditions (V_1, E_1, B_1). Any m_2^+ ions formed by unimolecular decomposition of m_1^+ ions between the two sectors, or by collision-induced decomposition in a collision cell placed at the focal point between the two sectors, will possess an energy $(m_2/m_1)Ve$ and will be transmitted by the electric sector only if the field strength is reduced to E_2 where

$$E_1/E_2 = m_1/m_2 \qquad \text{or} \qquad E_2 = (m_2/m_1)E_1 \qquad (12)$$

Hence a spectrum of all decomposition products from a chosen precursor ion may be obtained by scanning E downward from E_1 and the masses of product ions deduced from Eq. (12). This technique is sometimes referred to as MIKES (mass-analyzed ion kinetic energy spectrum).[3] Since the sensitivity of detection decreases approximately linearly as E is reduced, this imposes a practical limit on the lowest useful value of E. If translational energy is released during the fragmentation, the product ion peak is broadened so that the resolving power of the product ion spectrum is usually less than 100. Nevertheless, the scan can be carried out simply and rapidly and is often useful despite the low resolution (Fig. 7).

Linked Scans with Constant Accelerating Voltage V

Four scans have been described, usually referred to as linked scans, in which B and E are scanned simultaneously at a constant value of V such that a particular relationship between B and E is maintained throughout the scan. Of these, that used with a BE instrument in which B^2E remains constant has not found widespread application and will not be considered further. The other three linked scans may be used with EB or BE instruments but will be described as used with the former. The scans are now normally carried out under computer control in which prior mass calibra-

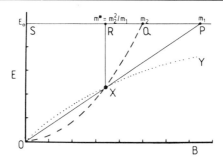

FIG. 8. Diagram of B,E plane. Loci of linked-scan lines are represented by PXO (——) for B/E = constant, QXO (---) for B^2/E = constant, and YXO ($\cdots$) for $B(1 - E')^{1/2}/E'$ = constant. PQRS represents a conventional B scan at $E = E_0$.[30]

tion is used to produce accurate mass-to-time correlation and the appropriate scan is generated (Fig. 8).

The most widely used linked scan is that in which the ratio B/E remains constant and which gives a spectrum consisting of product ions formed from a chosen precursor ion in a collision cell placed close to the source-defining slit.[19,20] If m_1^+ ions fragment to give m_2^+ ions, let E_1 and E_2 be the respective electric sector field strengths required to transmit these ions through the electric sector and B_1 and B_2 be the respective magnetic fluxes required to transmit the ions through the magnetic sector. If one assumes that the velocities of the precursor and product ions are the same, it follows from Eqs. (3) and (5) that

$$m_1/m_2 = E_1/E_2 = B_1/B_2 \qquad \text{or} \qquad B/E = \text{constant} \qquad (13)$$

The value of the constant is fixed by the choice of m_1^+ which determines the value of B_1. The resolving power for m_2^+ obtainable using this scan is somewhat dependent on the translational energy released during fragmentation but is typically in the region of 500 to 1000 with mass assignment of the product ion accurate to the nearest integer (Fig. 9). The resolving power with which m_1^+ can be selected is only in the region of 300 to 400 so that for high mass organic ions, product ions formed by the fragmentation of precursor ions containing one or more ^{13}C isotopes will be collected when the magnet is set to transmit the precursor ion containing only ^{12}C isotopes.[21,22]

[19] D. S. Millington and J. A. Smith, *Org. Mass Spectrom.* **12**, 264 (1977).
[20] A. P. Bruins, K. R. Jennings, R. S. Stradling, and S. Evans, *Int. J. Mass Spectrom. Ion Phys.* **26**, 395 (1978).
[21] C. J. Porter, A. G. Brenton, and J. H. Beynon, *Int. J. Mass Spectrom. Ion Phys.* **36**, 69 (1980).
[22] R. S. Stradling, K. R. Jennings, and S. Evans, *Org. Mass Spectrom.* **13**, 429 (1978).

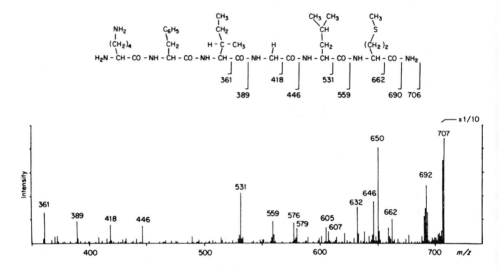

FIG. 9. A linked scan spectrum at constant B/E showing products formed by the decomposition of the MH$^+$ ion, m/z 707, of a peptide amide [T. Matsuo, H. Matsuda, I. Katakuse, Y. Shimonishi, Y. Maruyama, T. Higuchi, and E. Kubota, *Anal. Chem.* **53**, 416 (1981)].

The resolving powers for both $m_1{}^+$ and $m_2{}^+$ are very much improved when this scan is used with a four-sector tandem mass spectrometer. The first two sectors are used to select the precursor ion and it is usual to select those ions containing only ^{12}C isotopes. Collision-induced decomposition occurs in a cell between the two mass spectrometers and the electric and magnetic sectors of the second instrument are scanned such that B/E is constant to provide a product ion spectrum of unit mass resolution and a mass assignment accuracy of about 0.3 u. Usually, the collision cell is floated at 2 to 5 kV to improve the collection efficiency of the product ions; this requires a modification to the scan law as discussed more fully in [6] in this volume.

The second linked scan, which gives a spectrum of all precursor ions that form a chosen product ion by collision-induced decomposition in a collision cell placed near the source-defining slit, is that in which the ratio B^2/E is kept constant throughout the scan.[23] Suppose that $m_1{}^+$ ions having a velocity v_1 fragment to give $m_2{}^+$ ions which are transmitted at field

[23] E. A. I. M. Evers, A. J. Noest, and O. S. Akkerman, *Org. Mass Spectrom.* **12**, 419 (1977).

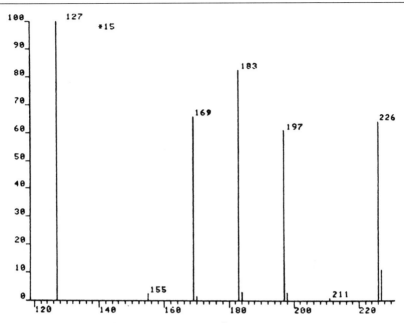

FIG. 10. A linked-scan spectrum at constant B^2/E showing precursors of the m/z 127 ion in the mass spectrum of hexadecane (Kratos Analytical).

strengths of E_2 and B_2. If one assumes that the product ions have the same velocity as the prercursor ions, then from Eqs. (3) and (5),

$$RB_2e = m_2v_1 \qquad (14)$$
$$E^2e = m_2v_1^2/r \qquad (15)$$
$$m_2/e = R^2(B_2^2/E_2)/r \quad \text{or} \quad B_2^2/E_2 = m_2(r/R^2e) \qquad (16)$$

From Eq. (16), one sees that for a given instrument, B_2^2/E_2 is a constant determined by the choice of m_2. Peaks are broadened due to the release of translational energy in fragmentations. Precursor ions are usually identified with sufficient accuracy, however, for this to be a useful technique for demonstrating the presence of related compounds, such as homologs, in mixtures (Fig. 10).

The two linked scans described above allow one to obtain spectra of all product ions formed from a chosen precursor ion and all precursors formed from a chosen product ion. The constant neutral loss scan, on the other hand, produces a spectrum of all product ions formed in a collision cell close to the source-defining slit by the loss of a chosen neutral species from precursor ions. This is again useful in identifying structurally similar compounds in mixtures.[24]

Suppose that $m_1{}^+$ ions formed in the source are transmitted by the electric and magnetic sectors under conditions specified by V_1, E_1, and B_1. If a neutral species of mass m_n is lost in forming an ion $m_1{}^+$, this ion will be transmitted by the electric sector only if Eq. (3) is satisfied:

$$m_2/m_1 = E_2/E_1 = E' = 1 - m_n/m_1 \tag{17}$$

These ions are transmitted by the magnetic sector, set at B_2, only when Eq. (5) is satisfied for ions of hypothetical mass m^*, where m^* is given by

$$m^* = m_2{}^2/m_1 = m_2 E' = (m_1 - m_n)E' \tag{18}$$

Since

$$m_1 = m_n/(1 - E')$$

this gives

$$m^* = (m_n/(1 - E') - m_n)E' = m_n E'^2/(1 - E') \tag{19}$$

so that from Eq. (6),

$$m^*/e = m_n E'^2/(1 - E')e = R^2 B_2{}^2/2V_1 \tag{20}$$

which, on rearrangement, gives

$$B_2{}^2(1 - E')/E'^2 = (2V_1/R^2 e)m_n \tag{21}$$

The right-hand side of this equation is a constant determined by the choice of m_n and ions formed by the loss of this neutral are transmitted by the two sectors only when the expression on the left-hand side of the equation is held constant throughout the scan. A resolution of approximately 800 has been reported for this type of scan.[25]

All linked scans may contain low-intensity artifact peaks which arise from the collection of product ions formed during acceleration or during passage through the electric sector. In addition, $m_2{}^+$ ions formed with the release of translational energy may give rise to such peaks. These peaks can usually be recognized with a little experience, however, and rarely cause problems.[18]

As may be seen from Fig. 8, the various scans described above for an *EB* instrument represent different methods of investigating ion intensity distribution within the *E,B* plane. Consequently, a complete set of scans at constant *B/E* would contain all the data contained in complete sets of

[24] W. F. Haddon, *Anal. Chem.* **51**, 987 (1979).
[25] W. F. Haddon, *Org. Mass Spectrom* **15**, 539 (1980).

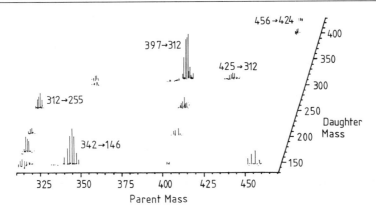

FIG. 11. Part of a two-dimensional mass spectrum produced by investigating the ion intensity within the B,E plane due to decomposition products of the peptide Ac-Val-Ile-Gly-Leu-OMe [M. J. Farncombe, K. R. Jennings, R. S. Mason, and U. P. Schlunegger, *Org. Mass Spectrom.* **18**, 612 (1983)].

the other two types of linked scan. Several methods have been developed for obtaining and displaying data contained within the B,E plane,[26–30] one of which will be described to illustrate the general principles.

The ion intensity within the plane can be obtained by repetitive scanning of B with E being reduced stepwise between scans. This allows information to be built up in a manner similar to that used to produce an image on a television screen. This information is stored in the data system and may be transformed into m_1, m_2 intensity data, prior to its being displayed in one of several ways. By choosing appropriate values of m_1, m_2 or m_n, one can use the data system to simulate any of the linked scans described above or one can obtain a three-dimensional display of ion intensities within the plane, examples of which are illustrated in Fig. 11.

Time-of-Flight Mass Analyzers

The time-of-flight (TOF) mass analyzer is conceptually the simplest of the mass analyzers in common use. It is based on the fact that if ions of different masses leave the source simultaneously, they will take different

[26] R. W. Kiser, R. E. Sullivan, and M. S. Lupin, *Anal. Chem.* **41**, 1958 (1969).

[27] J. C. Coutant and F. W. McLafferty, *Int. J. Mass Spectrom. Ion Phys.* **8**, 323 (1972).

[28] M. J. Lacey and C. G. McDonald, *Org. Mass Spectrom.* **12**, 587 (1977).

[29] G. R. Warburton, R. S. Stradling, R. S. Mason, and M. J. Farncombe, *Org. Mass Spectrom.* **16**, 507 (1981).

[30] R. S. Mason, M. J. Farncombe, K. R. Jennings, and J. H. Scrivens, *Int. J. Mass Spectrom. Ion Phys.* **44**, 91 (1982).

times to reach the detector.[3] It is, therefore, used with sources which can provide pulses of ions and requires the use of a collector with very fast time resolution. For singly charged ions of masses m_1 and m_2 having velocities v_1 and v_2

$$m_1 v_1^2/2 = m_2 v_2^2/2 \quad \text{or} \quad m_1/m_2 = (v_2/v_1)^2 \qquad (22)$$

If the distance to be travelled is d so that the time taken, t, is equal to d/v,

$$t_1/t_2 = v_2/v_1 = (m_1/m_2)^{1/2} \qquad (23)$$

so that a mass spectrum may be obtained by measuring the times taken for different ions to travel this distance, and there is no theoretical limit to the upper mass range. From Eq. (2), it is clear that $m = at^2$, where a is a constant of proportionality, so that $dm = 2at^2$ and the resolving power is given by

$$m/dm = t/(2dt) \qquad (24)$$

If ions of mass m (daltons) and charge ze are subjected to an accelerating potential of V volts, the velocity of the ion in m sec^{-1} is given by

$$v = 1.39 \times 10^4 (zV/m)^{1/2} \qquad (25)$$

If $V = 5000$ V and $m = 2500$ D, $v = 2 \times 10^4$ m sec^{-1} approximately. If the path length between source and collector is 2 m, $t = 1 \times 10^{-4}$ sec. If the smallest time interval that can be accurately measured at the collector is 5 nsec, the resolving power predicted from Eq. (24) is 10,000. In principle, the resolving power can be increased by increasing t, either by increasing the length of the flight path or by reducing the velocity, but practical considerations, primarily concerned with the ion source, reduce the performance of TOF instruments.

The two requirements of the ion source are (i) all ions should be formed in the same plane and (ii) all ions should have the same energy. In the original Wiley–McLaren instrument,[31] a multifield ion source together with a time delay between the ionizing and extraction pulses produced space focusing and energy focusing, respectively (Fig. 12), but with an electron ionization source, typical resolving powers were only in the region of 400. If ionization takes place at the surface of a solid, as in the case of SIMS,[32] plasma desorption[33] and, to a good approximation, laser desorp-

[31] W. C. Wiley and I. H. McLaren, *Rev. Sci. Instrum.* **26,** 1150 (1955).
[32] A. Benninghoven, E. Niehuis, T. Friese, D. Griefendorf, and P. Steffens, *Org. Mass Spectrom.* **19,** 346 (1984).
[33] R. D. Mcfarlane, *Anal. Chem.* **55,** 1247A (1983).

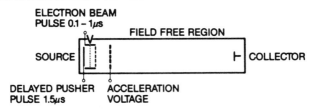

ELECTRON BEAM
PULSE 0.1 – 1μs   FIELD FREE REGION

SOURCE   ⊢ | COLLECTOR

DELAYED PUSHER   ACCELERATION
PULSE 1.5μs   VOLTAGE

FIG. 12. Schematic of the Wiley-McLaren TOF mass spectrometer.[1]

tion followed by multiphoton ionization,[34] the ionization plane is well defined, and much higher resolving power can be obtained. The thermal energy of sample molecules formed by laser desorption has also been reduced by using a supersonic jet before generating ions by multiphoton ionization.[35] These techniques are discussed further in [1] in this volume.

The main requirement, therefore, is to compensate for the energy spread within the ion beam. Several instruments have now been described that incorporate a reflecting electrostatic lens at the end of the flight tube (Fig. 13), which has been termed a "reflectron."[36,37] Ions of higher energy arrive first and penetrate more deeply into the lens and so describe a longer path length than those of lower energy so that ions of similar mass reach the collector over a much reduced time span.

As ionization methods improve, the upper usable mass range of the TOF mass analyzer rises; the detection of ions of m/z in excess of 100,000 has recently been reported.[38] The accuracy of mass measurement is about 1 D in the region of m/z 5000 and although the standard deviation rises significantly as m/z rises, it is the only analyzer capable of measuring m/z to ±100 D at m/z in the region of 50,000. Resolving powers of the larger modern instruments are in the region of 5000 to 10,000.

In addition to the very high mass range, the major advantages of the TOF instrument are its sensitivity, scanning speed, and simplicity. Typically, 10,000 spectra per second are recorded and if the sample is consumed only during the ionization pulse, essentially all ions produced

[34] M. W. Williams, D. W. Beekman, J. B. Swan, and E. T. Arakawa, *Anal. Chem.* **56,** 1348 (1984).

[35] J. Grotemeyer, U. Boesl, K. Walter, and E. W. Schlag, *Z. Naturforsch.* **40A,** 1349 (1985).

[36] W. Gohl, R. Kutscher, H. J. Laue, and H. Wollnik, *Int. J. Mass Spectrom. Ion Phys.* **48,** 411 (1983).

[37] E. Niehuis, T. Heller, H. Feld, and A. Benninghoven, *in* "Ion Formation from Organic Solids" (A. Benninghoven, ed.), p. 198. Springer, Berlin, 1986.

[38] M. Karas and F. Hillenkamp, *Anal. Chem.* **60,** 2299 (1988).

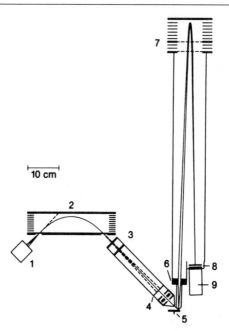

10 cm

Fig. 13. Schematic of Benninghoven's SIMS-TOF instrument comprising: (1) primary ion source; (2) 90° deflector; (3) bunching system; (4) mass separation slit; (5) Arget carousel; (6) Einzel lens; (7) two-stage reflector; (8) channel plate and scintillator; (9) photomultiplier.[37]

are collected, giving very high sensitivity.[39] Modern TOF instruments fall into two general classes (i) fast-scanning, low-mass, low-resolution GC/MS instruments[40] and (ii) higher-resolution, high-mass desorption ionization instruments which have recently renewed interest in the technique.[41]

Quadrupole Mass Analyzer

The quadrupole mass analyzer, originally developed by Paul and co-workers,[42] forms the basis of what is probably the most widely used mass spectrometer. Its popularity is based on the following (i) it requires no magnet and thus is very compact, (ii) the source operates at a potential very close to ground potential, (iii) it may be scanned very rapidly, and

[39] R. B. Opsal, K. G. Owens, and J. P. Reilley, *Anal. Chem.* **57**, 1884 (1985).
[40] J. F. Holland, C. G. Enke, J. Allison, J. T. Salts, J. D. Pinkson, B. Newcome, and J. T. Watson, *Anal. Chem.* **55**, 997A (1983).
[41] R. D. Macfarlane, J. C. Hill, D. L. Jacobs, and R. G. Phelps, *in* "Mass Spectrometry in the Analysis of Large Molecules" (C. J. McNeal, ed.), p. 1. Wiley, New York, 1986.
[42] W. Paul, H. P. Reinhard, and U. von Zahn, *Z. Phys.* **152**, 143 (1958).

(*iv*) it requires no slits and hence, in principle, has high transmission. When compared with a magnetic mass analyzer, however, its mass range, accuracy of mass measurement, and resolving power are relatively modest.

A quadrupole electric field is created by electrical potentials applied to four parallel rods of hyperbolic cross section, although properly spaced rods of circular cross section give results which are not dissimilar (Fig. 14). In mass spectrometry, such fields are widely used to filter ions according to their m/z ratios but are also important in trapping ions and in focusing ion beams.[43]

At sufficiently high translational energies, all ions will pass through a quadrupole mass filter, but at energies below about 100 eV, mass selectivity can be obtained using rods of 20 to 30 cm. length. Under such conditions, the mass filter acts as a path stability device if opposite pairs of rods are connected electrically and a dc voltage U and an rf voltage $V_0 \cos (wt)$ are applied to the rods. For particular fields, ions having a small range of m/z values have stable paths through the filter and all other ions are not transmitted. The stability and paths of singly charged ions within an ideal quadrupole mass filter in which the separation of opposite rods is $2r_0$ can be calculated from the Mathieu equations. These are equations of motion of an ion of mass-to-charge ratio m/e along the x and y axes, perpendicular to the direction of motion of the ions in the z direction:

$$d^2x/d(wt/2)^2 + (a + 2q \cos 2(wt/2)x = 0 \qquad (26)$$
$$d^2y/d(wt/2)^2 - (a + 2q \cos 2(wt/2)y = 0 \qquad (27)$$

where

$$a = 8eU/mr_0^2w^2 \qquad (28)$$

and

$$q = 4eV_0/mr_0^2w^2 \qquad (29)$$

so that

$$a/q = 2U/V_0 \qquad (30)$$

In these expressions, x and y are distances from the center of the field (Fig. 14). For ions of a given mass, certain values of a and q lead to

[43] P. H. Dawson (ed.), "Quadrupole Mass Spectrometry and Its Applications." Elsevier, New York, 1976.

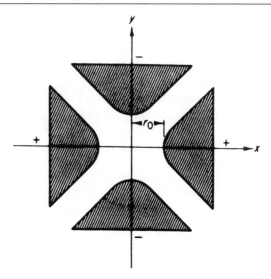

FIG. 14. Schematic of quadrupole geometry.[3]

stable oscillations in the x and y directions (Fig. 15). For other ions, the oscillations are unstable in either the x or y direction so that the ions are lost by striking the rods or by escaping between them. For a fixed frequency, m is directly proportional to U and V_0 so that if these are scanned such that their ratio, and hence a/q, remains constant, ions of different masses will, in turn, follow stable paths and a mass spectrum, linear in mass, will be obtained. The optimum resolution is obtained by arranging that the scan line intersects the tip of the stability region and this requires that $U/V_0 = 0.167$. Theoretically, the resolving power can be increased by lengthening the rods but it is usually limited to unit mass resolution by imperfections in the quadrupole field. The resolving power of the mass filter, unlike that of a magnetic deflection or TOF analyzer, increases with mass. This is due to the fact that the higher mass ions have lower velocities and so spend a longer time in the analyzer. They therefore experience more cycles of the rf field, increasing resolving power and giving unit mass separation throughout the mass range.

In principle, the mass range can be increased by increasing the maximum values of U and V_0. The above simplified analysis assumes, however, that the ions entering the mass filter have translational energies only along the z axis. In practice, components along the x and y axes, arising from fringing field effects, increase ion losses and since the time spent within the fringing fields is proportional to $m^{1/2}$, losses increase as m increases

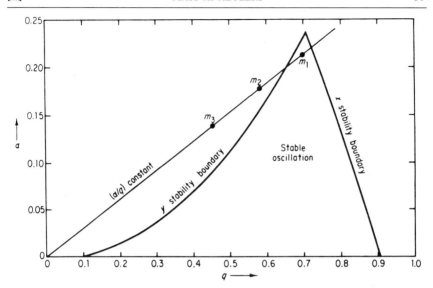

FIG. 15. Quadrupole stability diagram illustrating conditions for stable and unstable oscillations.[3]

and become increasingly important above m/z 800–1000. These effects can be reduced by the use of a circular aperture to reduce radial divergence at the entrance to the analyzer, thereby defining more closely the trajectories of the incoming ions, and by the use of pre- and postfilters that are short (2–3 cm) rf-only quadrupole fields which are used to steer the ions through the fringing fields.[44] The recent introduction of quadrupole rods with truly hyperbolic cross sections in commercial mass filters has also increased the resolution attainable.

An indication of the practical limitations in extending the mass range and resolving power can be seen as follows.[1] If $U/V_0 = 0.167$, ions of mass-to-charge ratio m/z are transmitted when

$$m/z = 0.136 V_0/r_0^2 f^2 \tag{31}$$

where V_0 is in volts, r_0 is in centimeters, and f is in megahertz (MHz) ($f = w/2\pi$). Typically, $r_0 = 0.4$ cm and $f = 1$ MHz, so that for $m/z = 10000$, $V_0 = 11.76$ kV which is difficult to work with. If f is reduced, V_0 decreases but the resolving power, RP, also falls since

$$RP \propto (\text{length of quadrupole field/radius})^2(V_0/E_{tr}) \tag{32}$$

[44] N. M. Brubaker, *Adv. Mass Spectrom.* **4**, 293 (1968).

where E_{tr} is the ion injection energy. If this is reduced in order to try to increase RP, the transmission is reduced because of ion losses arising from radial translational energy.

Because the quadrupole mass filter measures mass directly, minor variations in the translational energy of the ion and deflections due to collisions with gas molecules in the analyzer have little effect on the performance of the instrument. They may, therefore, be operated at pressures up to 10^{-4} torr which allows them to be used with a variety of ion sources having high pressure requirements or intrinsically large kinetic energy spreads.[45,46]

Quadrupole mass filters have frequently been used for GC/MS since the gas chromatograph and ion source are simply interfaced as the latter is operated at close to ground potential. Because there are no slits in a quadrupole mass filter, transmission of ions is very high, especially at lower masses, and when used in the selected-ion monitoring mode, detection limits for environmental and biological samples are frequently in the picogram or femtogram region. In most instruments, however, the sensitivity decreases markedly for ions of m/z above 800–1000 and depends on the sample and ionization method employed.[47,48] Because quadrupole mass filters have line-of-sight geometry from the ion source to the detector, photons, fast atoms, molecules, and electrons contribute to background noise, reducing detection limits. This has largely been overcome using off-axis detection systems into which the slower moving mass filtered ions are deflected.

Modern, high-performance quadrupole mass filters having rods of hyperbolic cross section with pre- and postfilters have mass ranges of up to 4000 at unit mass resolution and scan speeds of up to 4000 u per second and for masses up to 800–1000, the detection limits are in the picogram range. The instruments can be used with a wide variety of ion sources and inlet systems and the absence of a magnet simplifies computer control of all aspects of the instrument.

Conclusion

This chapter has summarized the main features of magnetic deflection, TOF, and quadrupole mass analyzers. It should be apparent that no one

[45] R. S. Honk, *Anal. Chem.* **58**, 97A (1986).
[46] E. C. Horning, M. G. Horning, D. I. Carroll, I. Dzidic, and R. N. Stillwell, *Anal. Chem.* **45**, 936 (1973).
[47] A. J. Alexander, E. W. Dyer, and R. K. Boyd, *Rapid Commun. Mass Spectrom.* **3**, 364 (1989).
[48] A. J. Alexander and R. K. Boyd, *Int. J. Mass Spectrom. Ion Phys.* **90**, 211 (1989).

TABLE I
CHARACTERISTICS OF MASS ANALYZERS

Characteristic	Magnetic deflection	TOF	Quadrupole
Mass range	Good	Excellent	Fair
Resolving power	Good	Fair	Fair
Exact mass determination	Good	Poor	Poor
Sensitivity	Good[a]	Excellent	Good[b]
Scanning speed	Good	Excellent	Good
Dynamic range	Excellent	Poor	Good
MS/MS potential	Excellent[c]	Poor	Excellent[d]
Ease of operation	Fair	Good	Good

[a] Excellent with an array detector.
[b] Decreases above m/z 1000.
[c] When used as the first stage of a hybrid or four-sector instrument.
[d] When used in a triple quadrupole or hybrid instrument.

analyzer is ideal for all purposes. Table I lists the major advantages and disadvantages of each type of analyzer.

For work with relatively low-mass compounds at unit mass resolution, the quadrupole is most widely used. For work at higher masses requiring high resolving power and accurate mass measurement, the magnetic deflection instrument is the best general-purpose instrument. If a very high mass range is the major requirement in which pulsed ionization techniques are used to maximize sensitivity, the TOF analyzer is the most suitable.

[3] Detectors

By SYD EVANS

Introduction

Detectors for mass spectrometers fall into two main categories: direct measurement detectors, which detect the charge arriving at the detector, and multiplier detectors, which use electron multiplication to give a signal of sufficient intensity to enable single-ion detection and rapid scanning. The choice of detector depends on the application. Direct measurement methods are essential for high-accuracy isotope ratio measurements, as the gain of an electron multiplier varies not only with mass but also with the structure of the ion.

Electron multipliers are used universally as the workhorse detector for all general-purpose instruments, but if negative ion analysis is required, particularly on low-voltage instruments, the multiplier may require the addition of a conversion dynode.

Since the introduction of fast atom bombardment (FAB) in 1981, the rapidly increasing requirement for instruments capable of the analysis of high-molecular weight samples has led to the development of postacceleration detectors. Improved ion source technology has increased the accessible mass range well beyond the presently available capability of existing detectors. Although very high mass ions can be detected, the efficiency of detection even using high-voltage postacceleration is poor.

The development of multisector instruments has more recently led to the development of array detection systems. Although still in their infancy, array detectors have been used successfully on several instruments, and it is likely that new instrument geometries will be developed over the next few years to enable the use of arrays with increased mass range and resolution.

Direct Measurement Methods

Plate Detector

The simplest way to detect an ion beam is to place a metal plate in its path and measure the charge falling on the plate by connecting it to an amplifier using a large feedback resistor. However, when an ion strikes a surface with the sort of energy typically used in a sector-field mass spectrometer, secondary electrons are produced. The charge associated with an electron leaving a surface has the same magnitude and polarity as a positive ion arriving at the surface, and typically several electrons will be produced for each incident ion. Unless these secondary electrons are suppressed, the apparent ion current is likely to be in error by several hundred percent. The energy of the secondary electrons is very low, the peak of the energy distribution being only 5–10 eV, so a suppressor plate carrying a slit through which the ion beam can pass only needs a small negative potential applied to it to return all but a few of the secondaries to the plate from which they originated. Secondary electrons will also be produced at the edges of the resolving slit, which defines the width of the ion beam. The suppressor, mounted between the resolving slit and the collector plate, will suppress these secondaries efficiently, but it is important that the ion beam cannot strike the suppressor slit edges. Secondary electrons originating from the suppressor would be accelerated to the collector plate, and would cause peak distortion (Fig. 1a). The voltage applied to the suppressor must be high enough to ensure that the potential

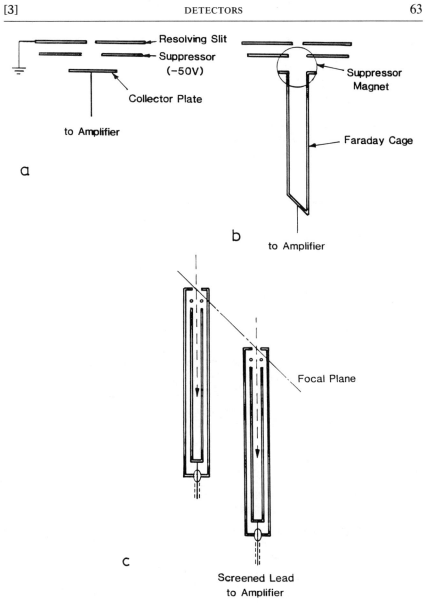

FIG. 1. (a) Plate detector; (b) Faraday cage detector; and (c) multiple detector.

in the center of the suppressor slit, which is affected by field penetration from the resolution and collector plates, is adequate to give efficient secondary suppression. Usually, about -50 V would be sufficient.

Faraday Cage Detector

In principle, the Faraday cage is very similar to the plate (Fig. 1b). It is a long, thin rectangular box arranged such that the incoming ion beam hits the bottom of the box, and if it strikes the side walls at all, does so at glancing incidence. The beam enters via a narrow rectangular aperture, and the depth of the box is usually at least ten times the narrow dimension of the entrance aperture. Thus the solid angle presented by the aperture to electrons originating at the bottom of the box is very small. Other features to prevent or suppress secondaries are sometimes included, such as internally coating the box with carbon, or using a small magnetic field to cause the secondary electrons to spiral back to the surface at which they are released.

The Faraday cage is used when the best possible quantitative measurements are required. It is particularly useful when two or more detectors are required in close proximity, for example, on thermal ionization or isotope instruments. Such detectors may require mounting no more than 2 mm apart, or may need to be adjustable to enable the peaks being monitored to be changed (Fig. 1c). Each collector box is enclosed within a grounded box, which usually includes the resolving slit for the detector. This box provides electrical screening of the collector to minimize noise and prevents secondaries arising from impact of the ion beam elsewhere in the detector region from reaching the detector.

Signal Detection and Amplification

The smallest signal which can be routinely and reliably measured is 10^{-15} A, for a signal-to-noise (S/N) ratio of about 2 : 1. The amplifier uses a maximum feedback resistance of $10^{11}\Omega$, giving an output of 100 μV for 10^{-15} A. With extreme care, the minimum signal can be reduced to an absolute limit of 2×10^{-16} A, but at this level an amplifier time constant of at least 5 sec is required. This means that scanning is impractical, and applications are therefore limited to quantitative analysis of stable beams using either multiple detectors or magnetic peak switching techniques. Larger signals can be detected by reducing the feedback resistance, or if a minimum current of 10^{-14} A is acceptable, the bandwidth can be increased. This is useful for applications such as type analysis of hydrocarbons introduced via a batch inlet system.

For the vast majority of analytical applications, speed and single-ion detection capability are of overriding importance and a multiplier detector is obligatory.

Multiplier Detectors

Electron Multiplier Detectors

Electron multipliers can be divided into two principal types, discrete and continuous dynode multipliers. Discrete dynode multipliers consist of a series of separate shaped plates or assemblies of plates connected together via a chain of resistors, usually of equal value. A high voltage applied across the resistor chain will therefore apply an equal voltage difference between each dynode, and provided that the first dynode (cathode) is at a negative voltage with respect to the final dynode, electrons will be accelerated from one dynode to the next. A screened collector plate (anode) positioned after the last dynode collects the multiplied electron stream and is connected to an amplifier, usually at ground potential.

An ion striking the first dynode will produce secondary electrons which are accelerated to the second dynode, and so on through the multiplier. Multiplier gain is varied by changing the applied voltage usually over the range 1000–3000 V. The number of stages varies between 16 and 20 and an interstage voltage of more than about 200 V is likely to cause field emission and hence a high noise level, giving a maximum practical voltage of 4000 V for a 16-stage multiplier.

Dynode material is usually beryllium/copper, but aluminum is beginning to be used by some manufacturers. The secondary emission properties of beryllium/copper depend on the distribution of the two principal oxides of beryllium on the surface of the material. To produce this surface condition, a specialized heat treatment process is required. Aluminum dynodes require a surface of aluminum oxide, which forms naturally in atmosphere. Demountable aluminum multipliers, which can be cleaned abrasively, are an interesting development.

The most common discrete dynode multiplier is the "venetian blind" type (Fig. 2a). Each dynode consists of a series of slats arranged like a venetian blind. On the input side of the blind is a fine grid, to give a uniform field which penetrates between the slats of the preceding stage. Gain of the first stage, which performs the ion–electron conversion, is dependent on the impact position of the ion beam on the dynode surface. If the ion beam is traversed across the slats of this dynode, the output signal will show a sawtooth variation reproducing the shape of the dynode. It is thus usual to position the multiplier far enough away from the resolving slit to allow the beam to spread before impact, and to orient the multiplier in a

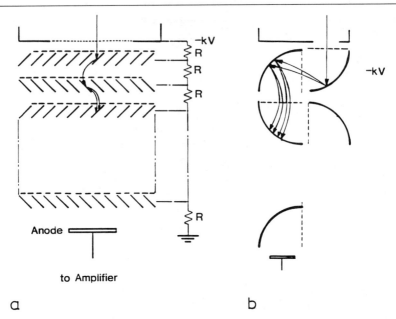

FIG. 2. (a) Venetian blind electron multiplier; (b) box and grid electron multiplier.

sector instrument such that the long dimension of the slit is at right angles to the slats.

The other common type is the box and grid (Fig. 2b) where each dynode is quadrant shaped. Typically these multipliers have a much smaller entrance aperture, and for maximum gain need to be positioned fairly accurately as gain will still be dependent on the impact position on the first dynode.

Multiplier gains of up to 10^7, more than adequate for analog detection, can be obtained from 17-stage multipliers, but for ion counting applications it is usual to fit 20-stage multipliers capable of over 10^8 gain to give a high enough signal from a single ion for unambiguous identification by a discriminator amplifier. The gain of a multiplier decreases continuously with use, and is dependent on the vacuum conditions and the total charge collected by the last few stages. It is therefore desirable to keep the output current low by using a high value of amplifier feedback resistor to generate the required output voltage.

The photomultiplier is used in scintillation detectors and is identical to an electron multiplier, but enclosed in a sealed glass vacuum envelope with a photocathode coated on the internal surface of the glass in front of the first dynode. Photons hitting the photocathode cause electrons to be emitted, which are accelerated to the first dynode. Photomultipliers have extremely stable gain with time, a direct consequence of the vacuum encapsulation, and, in some instances, gain curves plotted after up to 5 years continuous use have shown good agreement with the original gain data. Photocathodes have an inherently high noise level, so to obtain single-ion detection it is necessary to ensure that each incoming ion generates multiple photons.

Channel Electron Multipliers. Channel electron multipliers (CEMs) are tubes made from a lead doped glass which, when suitably processed, has good secondary emission properties and is electrically resistive. A voltage applied between the ends of the tube will therefore be dropped uniformly along its length. An ion striking the inner surface will release secondary electrons which are then accelerated by the field within the tube to strike the wall, causing more secondaries to be released.[1] The output signal is detected on a separate collector plate at the end of the tube.

Straight CEM's become unstable at gains in excess of 10^4, due to positive ion feedback, caused by ionization of residual gas molecules within the tube. These ions can occasionally travel back through the tube to strike the inner wall near the input end, producing secondary electrons which multiply through the device to produce spurious pulses. Channeltrons are therefore curved, to prevent positive ions from moving far within the tube, so reducing any resulting secondary pulse to much smaller intensity than a "real" pulse (Fig. 3a).

Gain is a function of the length-to-diameter ratio of the tube, so the inner diameter is kept fairly small to keep the overall size of the multiplier small. Single channeltrons usually have a conical input stage, to give a larger target area for the incoming ion beam. As with discrete dynode multipliers, gain varies considerably with the position at which the ion beam strikes the surface of the cone. Ions entering the tubular section should be at an angle to the axis, or they will penetrate too far before striking the wall.

Performance of channeltrons is very similar to that of the discrete dynode types, and are extensively used in quadrupoles, and, more rarely, in sector instruments. In the author's experience, single CEM's have a shorter life than the discrete dynode types, and with direct ion input gain decreases more rapidly with mass. For applications where physical size is not a problem, as in sector instruments, discrete dynode multiplers are preferred.

Because gain is a function of the length-to-diameter ratio, channel multipliers can be made extremely small, using fiber optic technology (Fig.

[1] E. A. Kurz, *Am. Lab*. March, 1979.

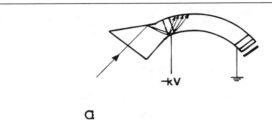

a

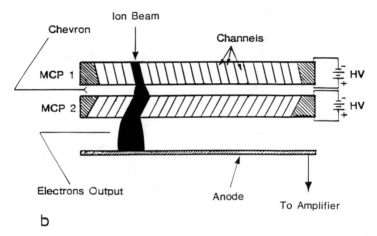

b

FIG. 3. (a) Channeltron electron multiplier; (b) microchannel plate multiplier.

3b). Microchannel plates are now available with a tube diameter as small as 10 μm in the form of thin flat plates containing millions of individual channels. The plates have a bias angle of about 7° and to prevent positive ion feedback, two plates arranged in chevron are used. These devices are used in array detection (Section IV) and frequently in time-of-flight (TOF) instruments where the large surface areas and flat uniform face offer ideal characteristics.

Scintillation Detectors

In the scintillation detector, ions are first converted to photons at a scintillator, and the photons released are detected by a photomultiplier. If the ion beam is allowed to strike the scintillator directly, the normally insulating surface would charge up until no further ions could reach it. Therefore, it is usual to coat the surface of the scintillator with a very thin

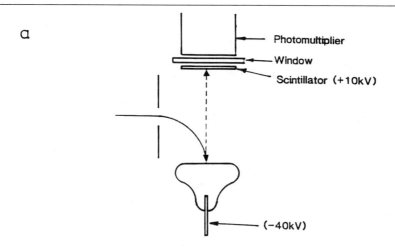

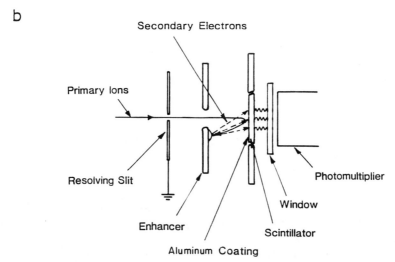

FIG. 4. (a) Daly detector; (b) Daly metastable detector.

layer of aluminum to make this surface conducting. An ion beam which strikes the aluminum will initially not penetrate to the scintillator, but will sputter the aluminum layer away. In the form of detector used by Daly,[2] (Fig. 4a), the ion beam is accelerated and deflected, after passing through the collector slit, to a polished aluminum or stainless steel button held at

[2] N. R. Daly, *Rev. Sci. Instrum.* **34**, 1116 (1963).

up to -40 kV. This produces a high yield of secondary electrons, which are accelerated to the aluminum-coated scintillator, which is at a high positive potential. The electrons readily pass through the aluminum coating, without causing any significant damage, and enter the scintillator. The photons produced are detected by an externally mounted photomultiplier. The extra stages of gain at the secondary electron surface and the scintillator ensure that the spike produced by a single ion is far greater than the photocathode noise background.

An interesting variant of this detector is the Daly metastable detector (Fig. 4b), in which the ion beam is accelerated through a slit in an enhancer plate toward a mounted scintillator held at a few volts greater than beam energy. The ion beam is therefore repelled by the scintillator plate and accelerates to strike the back of the enhancer. Secondary electrons are produced which accelerate to the scintillator. If the scintillator potential is reduced to just less than beam energy, all ions originating in the ion source will drift very gently into the scintillator with insufficient energy to release photons, but ions which have lost energy by unimolecular decomposition or collision will be repelled to the enhancer, produce secondary electrons, and be detected. By scanning the scintillator potential, a metastable spectrum can thus be generated. Photons can be emitted in any direction, and the aluminum coating acts as a reflector, to direct photons toward the photocathode.

Velocity Threshold

It has been shown that the yield of secondary electrons from a surface is proportional to the velocity, rather than the energy, of the impinging ion. Schram et al.[3] showed that the velocity at which the secondary yield drops to zero, determined by extrapolation of experimental data for all the noble gas ions, is approximately 5.5×10^4 m/sec (Fig. 5). In practice, the secondary electron yield for any given energy has a wide statistical variation, and even at velocities well above threshold, occasional ions may fail to yield any secondaries. Similarly, at velocities below threshold, some ions will produce secondaries. The velocity threshold theory is not yet universally accepted, but experimental evidence supporting it is accumulating.

At high molecular weights, the measurement of detector efficiency, defined as the percentage of impinging ions producing secondaries, is difficult. The signal levels obtained are usually too low to be measured by a plate collector, and as an ion which fails to produce any secondaries is

[3] B. L. Schram, A. J. H. Boerboom, W. Kleine, and J. Kistemaker, *Physica* **32,** 749 (1966).

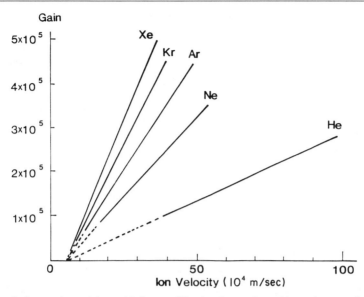

FIG. 5. Comparison of the multiplier amplification factors for noble gas ions as a function of their velocity.[3]

not detected there is no way of knowing the total number of ions arriving. Measurement of multiplier gain does not help, as those ions which do produce even a single secondary electron will yield a normal output pulse by electron multiplication. The measurement, therefore, has to be made by counting the number of single ion pulses produced from a steady beam as impact velocity is changed. Hedin *et al.*[4] has made such measurements on a number of ions, including bovine insulin at a mass of 5733 Da. He found that the velocity threshold for insulin was 1.8×10^4 m/sec, and increasing the velocity to 4×10^4 showed an increase of an order of magnitude in the number of ion pulses produced, and still increasing at this velocity. For a lower mass ion, the peptide luteinizing hormone-releasing hormone (LHRH), of 1182 Da, the number of pulses saturated at a velocity of 5×10^4 m/sec, indicating that all the ions were being detected (Fig. 6).

The velocity of an ion on impact with the multiplier surface is given by Eq. (1):

$$v = 1.39 \times 10^4 \left[(V_a - V_t)/M \right]^{1/2} \text{ m/sec} \tag{1}$$

[4] A. Hedin, P. Hakansson, and B. U. R. Sundquist, *Int. J. Mass Spectrom. Ion Proc.* **75**, 275 (1987).

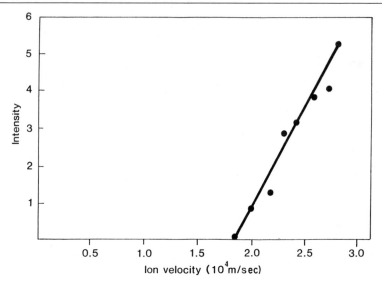

Fig. 6. The yield of bovine insulin molecular ions as a function of ion velocity (after Hedin *et al.*[4]).

where V_a is the source accelerating voltage; V_t is the voltage on the target surface; and M is the mass of the ion in daltons.

Thus, for an ion of mass 1000 Da, impacting with 10 keV energy, the velocity will be 4.4×10^4 m/sec. These are fairly routine conditions for large-sector mass spectrometers, and where detection by electron multiplier is efficient. At high molecular weight, for example, insulin, under the same conditions, impact velocity drops to 1.9×10^4 m/sec and detection efficiency decreases dramatically. It is therefore necessary to increase the velocity of the ion to obtain efficient detection. These considerations led to the development of the postacceleration detector (PAD).

Postacceleration Detectors

The PAD is now used, in a variety of forms, by all commercial high mass instruments and in its simplest form is very similar to the Daly detector.

The ion beam, after passing through the resolving slit of the instrument, is deflected and accelerated to an off-axis target, which is usually made of polished aluminum. Electrons from the target are then accelerated to an electron multiplier (Fig. 7). The deflection which an ion experiences is governed by the potential gradient between the multiplier and the target. Thus if the target voltage or the ion energy is changed, the position of

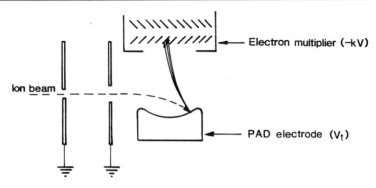

Fig. 7. Postacceleration detector.

impact on the target will change. With a flat target surface, the field above the target is convex, and electrons from ions impacting away from the center of the target will tend to curve away from the multiplier axis. To overcome this, the version of detector shown has a concave target which focuses the electrons into the multiplier for a wide range of impact positions. The target is usually operated at a potential of between -8 and -30 kV. Thus, for an insulin ion with $V_a = 8$ kV, $V_t = -8$ kV, impact energy is 16 keV, and velocity on impact is 2.4×10^4 m/sec. In the author's experience, collector efficiency, although less than 100%, is better than would be expected from the data of Sundqvist, so presumably threshold is affected by factors such as target work function and impact angle. For efficient detection of ions of higher molecular weight than this, a higher target potential is desirable.

Postacceleration is also used on TOF instruments, which normally operate with a source voltage of about 3 kV. Postacceleration in a TOF must be uniform over the full cross section of the ion beam to avoid causing differences in transit time, so the off-axis arrangement is impractical. Cotter[5] has used postacceleration of -20 kV to a venetian blind dynode mounted perpendicular to the beam axis, with electron extraction from the back of the dynode to an electron multiplier. Using this arrangement, he has reported detection of ions in excess of 100,000 Da. Hillenkamp[6] has also used 20 kV postacceleration to detect good signals approaching 250,000 Da. In this case, impact velocity is less than 5×10^3 m/sec, well below the threshold velocity determined for insulin. It is possible that the very high sensitivities for massive ions reported for the electrospray source

[5] R. Cotter, private communication, 1989; see also A. G. Harrison and R. J. Cotter, this volume [1].
[6] F. Hillenkamp and M. Karas, this volume [12].

are related to the higher impact velocities possible with the multicharged ions produced by this technique. For singly charged ions, as velocity is proportional to the square root of $V_a - V_t$, efficient detection of ions in excess of 10,000 Da is difficult to achieve. With $V_a = 10$ kV, $V_t = -30$ kV, a mass 10,000 ion would only have a velocity of 2.8×10^4 m/sec. Clearly there are opportunities here for the development of a totally new detection system.

Negative Ion Detection

Plate and Faraday cage detectors can be used to detect negative ions as readily as positive ions, but this is not necessarily so with multipliers. The normal mode of operation for a multiplier is to maintain the output stage, connected to the amplifier, at ground potential. This means that, in order to accelerate electrons through the multiplier, the first dynode must be at a negative potential. Negative ions approaching the first dynode are therefore decelerated before impact, giving reduced secondary electron yield. This is not normally a problem for low mass range, i.e., <1000 Da, sector field instruments, which operate at high accelerating voltage (e.g., 8 kV) unless voltage scanning or peak switching is used, or accelerating voltage is reduced to extend the mass range. Quadrupole instruments, in which the ion beam energy is typically less than 20 eV, would be unable to detect negative ions using the conventional arrangement, so the whole electron multiplier must be floated to a high positive potential such that the ions are accelerated sufficiently to give a good secondary electron yield. This means that the anode and preamplifier must also be floated to a high potential. Amplifiers incorporating optical isolation capable of floating to as high as 10 kV are readily available, but because stray electrons transmitted through the quadrupole are not distinguished from negative ions, a high noise background is observed.

In 1977, Stafford et al.,[7] described a detector for quadrupoles in which the negative ion beam is accelerated to a surface held at a high positive potential to yield secondary positive ions, which are then accelerated to the cathode of an electron multiplier, operated in the grounded anode mode. Secondary electrons are, of course, formed at the target surface, but cannot reach the electron multiplier. The positive ions are believed to be produced by sputtering of the metal surface, fragmentation of the polyatomic negative ion, and charge stripping. Sputtering of adsorbed gases from the metal surface also contribute, and efficiency of positive ion production can be increased by ensuring that the target surface is not clean!

[7] G. Stafford, J. Reeker, R. Smith, and M. Storey, *in* "Dynamic Mass Spectrometry" (D. Price and J. F. J. Todd, eds.), Vol. 5, p. 55. Heyden, London, 1977.

This detector is a form of the PAD, developed independently as a high-mass positive ion detector for sector machines. To operate a PAD for negative ion detection, it is only necessary to reverse the voltage on the postacceleration electrode to give secondary positive ions. Gain in the negative ion mode is typically about a factor 3 less than for positive ions, but depends on factors such as the cleanliness of the target.

Gain has been shown to increase with mass, certainly over the mass range from 28 to 600. This is ascribed to fragmentation of the polyatomic ion on impact. Above this mass, no detailed information is available, but comparison of cesium iodide spectra for positive and negative ions carried out in the author's laboratory suggest that the velocity threshold limitations apply equally to negative ions (see p. 81).

Scintillator detectors as described previously will not detect negative ions, as their operation depends on the production and acceleration of electrons to the scintillator using the same field used to deflect the incoming positive ions. The Daly metastable detector has been reconfigured by replacing the enhancer plate with a single venetian blind dynode at ground potential, accelerating the electrons produced to the scintillator plate. Postacceleration could be achieved by floating this dynode to a high voltage and maintaining the scintillator at a few thousand volts positive with respect to the dynode. An alternative arrangement[8] is shown in Fig. 8. In this case, positive ions are deflected to a conventional PAD electrode, and the secondary electrons travel through a short stainless steel cylindrical electrode to the scintillator surface. In negative ion mode, negative ions are deflected into the cylindrical electrode and electrons extracted by the penetrating field from the scintillator.

As with all high-voltage (>20 kV) detectors, extreme care must be taken to radius and polish all electrodes, and to avoid using thin connecting wires to minimize field emission effects. This is particularly true for negative ion scintillation detectors, as field emission of electrons can occur from the grounded housing which therefore also has to be suitably polished. Vacuum also must be good to minimize the production of electrons by ionization of residual gas molecules.

Selecting Multiplier Gain

To make full use of the available sensitivity of a mass spectrometer, particularly under scanning conditions, the multipler gain should be set such that the smallest peak which can be reliably detected contains the minimum number of ions to give the data quality required. Resolution and scan speed enable the amplifier bandwidth to be selected. From these

[8] R. H. Bateman and P. A. Butt, *Proc. 35th ASMS Conf. Mass Spectrom. Allied Topics,* p. 253 (1987).

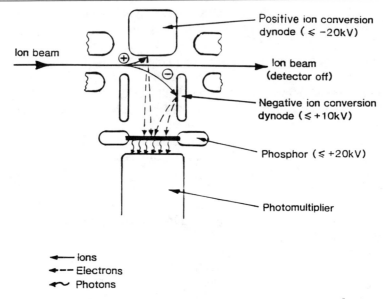

FIG. 8. Positive/negative ion postacceleration detector (Bateman[8]).

factors, the most appropriate maximum gain can be determined. At this gain, single ions should be detectable. Higher gain will make the single ion signals bigger but will not provide any additional data, and will reduce the dynamic range obtainable in the scan. How these factors are interrelated are discussed below.

Peak Width. The preferred scan law for most applications using magnetic sector instruments is an exponential scan of either field or current, scanning from high to low field. This form of scan is used because it has the unique property that the peak width measured in time (*tp*) is constant throughout the scan. This means that all peaks will be observed for the same length of time, thus eliminating one source of error in determining intensity ratios. More importantly, it means that the amplifier bandwidth can be constant throughout the scan and can be selected to minimize peak distortion. The width of the peak is calculated from the equation:

$$tp = \frac{t_{10}}{\ln 10R} = \frac{t_{10}}{2.3R} \tag{2}$$

where t_{10} is the time to cover a decade in mass (seconds) and R is resolution. Thus, for a scan speed of 10 sec/decade at a resolution of 10,000, *tp* is 435 μsec.

Bandwidth. The amplifier bandwidth should be selected to smooth out the intensity variations occurring across the width of a peak, which are due to statistical variations in the arrival rate of ions and in the pulse height produced by the multiplier for each individual ion. A symmetrical peak is essential for accurate mass measurement and, if bandwidth is too low, peaks will be reduced in height and distorted. Low bandwidth will also increase the width of the single ion peaks relative to that of genuine multiple ion peaks, making it more difficult for the data system to discriminate. Too high bandwidth can lead to peaks being detected as multiplets.

Banner[9] developed the relationship between peak width and bandwidth. From this work it can be deduced that an amplifier time constant, τ, of 0.08 tp will give less than 5% loss of peak height and a smaller increase in peak width. Applying this to the 10,000 resolution, 10 sec/decade case, we get a value for τ of 35 μsec. Amplifier bandwidth, f, is related to time constant by:

$$f = \frac{1}{2\pi\tau} \tag{3}$$

In this case, the minimum acceptable frequency is approximately 5 kHz. Using this criterion, the width of a single-ion pulse would be about 0.2 tp, which is readily discriminated by a data acquisition system.

Mass Measurement Accuracy. In the early 1960's, Campbell and Halliday[10] showed that the standard deviation of mass measurement (σ parts per million) is related to resolution and to the number of ions in the peak being measured (N) by the relationship [Eq. (4)]:

$$N = \frac{1}{24}\left(\frac{10^6}{R\sigma}\right)^2 \tag{4}$$

assuming that all other sources of error are small compared to the ion statistical component. This means that for a mass measurement accuracy of 5 ppm at 10,000 resolution, only 17 ions are needed; at 20,000 resolution, 4 ions are sufficient. At low resolution, only a few ions are needed to give nominal mass accuracy but normally 10 ions would be considered necessary to ascertain the presence of a genuine signal.

Optimum Multiplier Gain. The height (V) of the smallest peak to be detected should be at least ten times the peak-to-peak noise on the amplifier baseline at the bandwidth selected. The output current from the multiplier to give this voltage is given by:

[9] A. E. Banner, *J. Sci. Inst.* **43** (March), 138 (1966).
[10] A. J. Campbell and J. S. Halliday, *in* "13th Annual Symposium on Mass Spectrometry" (R. M. Elliott, ed.), Vol. 2. Pergamon (Oxford), Oxford, 1962.

$$i = \frac{V}{R_f} \tag{5}$$

where R_f is the feedback resistance of the amplifier system.

The total charge q in Coulombs contained within the peak is thus, assuming for simplicity, a triangular peak:

$$q = \frac{i(tp)}{2} \tag{6}$$

To find the gain G to give this peak height, it is only necessary to divide q by the number of ions required in the smallest peak times the electronic charge:

$$G = \frac{q}{Ne} \tag{7}$$

The value of feedback resistance is usually kept as high as possible commensurate with the bandwidth required, since, for a discrete dynode multiplier, gain decays principally as a function of the current flowing through the last few stages. Unpublished data obtained in the author's laboratories showed clearly that reducing the output current by a factor of 10 increased multiplier life by at least a factor of 50. A higher value of R_f therefore means that a lower output current is required from the multiplier for a given indicated voltage. Typically, with R_f equal to $10^8 \, \Omega$, the maximum bandwidth attainable is about 5 kHz. For higher bandwidths, it is necessary to reduce the resistance to $10^7 \, \Omega$, with a maximum bandwidth of about 40 kHz.

For the 10,000 resolution, 10 sec/decade example, R_f would usually be $10^7 \, \Omega$, and for a peak-to-peak noise of say 2 mV, we would require a 20 ion peak to be 20 mV high. This gives a gain requirement of only 1.35 $\times 10^5$. A 20 ion peak will be 20 times the area of a single ion peak, so because bandwidth has been selected to give a single ion peak width of 0.2 tp, it follows that, on average, a 20 ion peak will be four times the height of a single ion peak under these conditions.

Array Detectors

Introduction

The most efficient detector for a sector field mass spectrometer is one which records all of the ions all of the time. Indeed, the first commercial double-focusing mass spectrometer produced, the AEI MS7, was a Mattauch–Herzog instrument with a flat focal plane and a photoplate detector.

Other photoplate instruments were manufactured by CEC in the United States and until quite recently by MAT (now Finnigan-MAT), many of which are still in use. The photoplate has severe disadvantages as a universally useful detector, because of the limited number of spectra which can be recorded on a single plate, and the lack of immediate information about the data recorded. Sensitivity is also poor, requiring 300 to 2000 ions to define a single line.

Photodiode Arrays

In the mid-1970's, Griffin et al.,[11] at the Jet Propulsion Laboratory, made a 10-inch array detector based on the use of microchannel plate (MCP) electron multiplier and reticon solid state detectors. Others in the United States and in Europe, for example, Tuithof et al.[12] at the FOM Institute in Amsterdam, made smaller 1-inch long devices based on the same technology. However, although these systems worked well and gave considerably improved detection efficiency (defined in this context as the fraction of the total analysis time spent on any one peak), the detection limit did not decrease sufficiently to justify the high cost. Detection limit in a two-sector instrument is usually a function of signal intensity relative to chemical background rather than of sensitivity.

In 1986 the first commercial array detector was introduced.[13] It was intended primarily for use on high-molecular weight fast atom bombardment (FAB)/liquid secondary ion mass spectrometry (LSIMS) instruments and on the second stage (MS-2) of a four-sector MS/MS instrument. Similar systems have been described since by Hill et al.[14] and by Gross.[15] The array has an advantage for two-sector operation with FAB/LSIMS as the extra detection efficiency provided enables lower primary beam intensities to be used, which enhances the signal-to-background ratio by disproportionately reducing the intensity of matrix-related adduct ions. In the four-sector case, the spectrum produced by MS-2 is almost entirely free of chemical background as all the ions transmitted through MS-2 originate from an ion beam preselected by MS-1.

[11] C. E. Griffin, H. G. Boettger, and D. D. Norris, *Int. J. Mass Spectrom. Ion Phys.* **15,** 437 (1974).

[12] H. H. Tuithof, A. J. H. Boerboom, and H. L. C. Meuzelaar, *Int. J. Mass Spectrom. Ion Phys.* **17,** 299 (1975).

[13] J. S. Cottrell and S. Evans, *Anal. Chem.* **59,** 1990 (1987).

[14] J. A. Hill, J. E. Biller, S. A. Martin, K. Biemann, K. Yoshidome, and K. Sato, *Int. J. Mass Spectrom Ion Proc.*, **92,** 211 (1989).

[15] M. L. Gross, this volume [6].

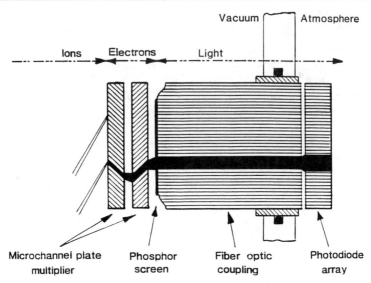

FIG. 9. Schematic diagram of the electrooptical array detector.

The array described by Cottrell[13] is shown schematically in Fig. 9 and consists of a pair of MCP's, a phosphor screen, fiber optic coupling, and a 1024-channel photodiode array. The focal plane of the instrument lies at about 30° to the direction of the incoming ion beam, and is flat over at least the 1-inch length of the array.

The two channel plates are arranged in chevron, with the adjacent faces touching, such that a single voltage of up to 1.2 kV can be applied across the pair. The maximum gain of such an arrangement is about 10^7, but in practice a gain of between 10^3 and 10^4 is sufficient for single-ion detection. The MCP's are mounted such that the input face lies along the focal plane of the instrument, which means that ions enter the channels at 30° to the surface, and the input face is held at ground potential. With a conventional detector, the resolution is defined by a grounded slit. Once an ion beam has passed through the slit, ion trajectory changes caused by the detector fields are irrelevant provided all the ions are collected. With an array, collecting over a wide mass range, no slit can be fitted, hence even a small voltage on the input face is sufficient to deflect the beam, causing a mass scale shift and loss of resolution. Postacceleration is, therefore, not possible with an array, but evidence currently available implies that this is not a serious problem for four-sector applications since the accessible mass range for efficient production of fragment ions by gas collision or indeed by any currently available method is well below 5000

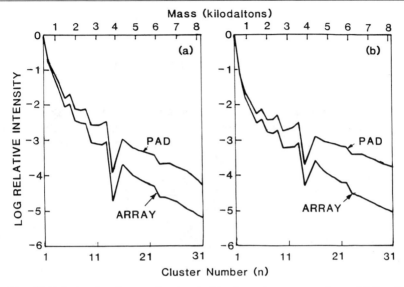

FIG. 10. Relative abundances of cesium iodide cluster ions measured on a PAD and on an array detector. All values are normalized to the intensity of the $n = 1$ cluster ion.[3] (a) Positive ion mode; (b) negative ion mode.

Da. It is not possible to count individual ions with this type of array, due to the integrating function of the photodiodes, making it difficult to measure collection efficiency. Comparison of the relative intensity of cesium iodide cluster ions for a PAD with 16 keV impact energy and an array with 8 keV impact energy is shown in Fig. 10. At first sight, this indicates that velocity threshold effects may be affecting the array result, but other explanations for the deviation between the two curves may be possible (Cottrell).

The electrons emerging from the output of the channel plates are accelerated to an aluminum-coated phosphor screen on the face of a fiber optic bundle. The gap between the channel plates and the screen is kept as small as possible, typically less than 0.5 mm, and the electron accelerating voltage as high as possible, typically about 3 kV, to minimize divergence of the electrons. Photons produced in the phosphor are transmitted through the fiber optic to a 1024-channel charge or plasma-coupled device (CCD or PCD) mounted outside the vacuum. Each channel or pixel of the CCD is 25 μm wide by 2.5 mm long, and converts the photons to charge, which is integrated and stored. The CCD is cooled by a peltier device to approximately $-20°$ to minimize noise, and when the MCP gain is set to detect a single ion with an average signal to noise of about 5 : 1, each channel can accumulate about 100 ions before saturating. The CCD is read at a repetition rate of between 10 and 40 msec, depending on the type of CCD used, each pixel taking a few microseconds to read. Reading implies measuring the charge necessary to neutralize the charge accumulated in each pixel, and transferring the data to memory. An "exposure" of 1 sec thus consists of say 25 integrations of 40 msec, giving a saturation level of 25×100 ions/second/pixel. If the gain is reduced, or the integration time decreased, clearly the maximum number of ions per second will increase.

The gain of the overall system varies slightly with position so it may be necessary to produce a "shading" function. A signal of constant intensity is swept slowly across the channel plate, and the resulting output function is stored to give a pixel-by-pixel intensity correction which can be applied automatically to subsequent data. Gain of channel plates will decrease with time, as a function of the total charge taken from each channel. It is therefore desirable to avoid prolonged exposure to intense beams in one spot on the array, which, over many mouths, can lead to a localized dip in the gain characteristic.

Resolution

Resolution can be limited by the characteristics of the array system. Even a narrow ion beam is likely to enter at least two adjacent channels in the first MCP and electrons will spread through more channels of the second plate. Further beam spread occurs at the phosphor, and still more at the CCD. Thus a single ion will be visible on at least four channels of the CCD, and an ion beam will cover five to six channels, with low-level tails. The resolution this corresponds to depends on the characteristics of the instrument. The Kratos array, for example, covers 4% of the mass range on 1024 channels, or approximately 40 ppm per channel. This gives a minimum beam width of about 200 ppm or 5000 resolution, FWHM. Minimum obtainable resolution, with maximum available beam width, is about 1000 resolution, FWHM. Biemann's array covers approximately 6.6% over a 2048 channel length, obtaining similar resolution. Biemann has added quadrupole lenses to rotate the focal plane, which has the effect of decreasing mass dispersion along the array and hence increasing mass range. The quadrupoles can also be used to compress the mass spectrum still further, to give approximately 25% maximum mass range, but with severe mass discrimination caused by the difficulty of extracting ions with a wide range of energies efficiently from the collision cell, and electrostatic analyzer transmission characteristics, and as this mode gives 125 ppm per channel, resolution is considerably reduced. This resolution is adequate to give unit mass resolution at low mass, and the quadrupole settings and array angle can be changed during a scan to progressively reduce the mass range and hence increase resolution as the center mass on the array is increased.

Sensitivity

If an instrument is scanned exponentially at a given resolution, over a decade in mass, then the time spent on any individual peak is given from Eq. (2) by:

$$\frac{tp}{t_{10}} = \frac{1}{\ln 10R} \tag{8}$$

For a resolution of 2000, this becomes 1 : 4600. As the beam is being swept across a slit, the number of ions collected in this peak is equivalent to observing the beam steadily for only 0.5 tp, so that the number of ions collected to the number of ions of that species produced per second is 1 : 9200.

With only a 4% array, a decade in mass can be covered by stepping the magnet by 4% fifty-six times, which, allowing for magnet switching times and other overheads, gives a genuine ion collection efficiency increase of more than a hundredfold. This increase should be realizable as a genuine decrease in detection limit. Increasing the array width to say 20% gives a further linear increase, that is, a further factor of 5, or perhaps more importantly enables the same collection efficiency to be obtained for a shorter "scan" time. The benefit of this is perhaps doubtful if it is at the expense of resolution.

Mass Calibration and Accuracy

The mass dispersion across the array must be calibrated by moving a known ion beam across the array, and calculating a mass-to-channel curve. Providing that the double-focusing characteristics of the instrument are good, this curve should hold for ions originating from an ion source, and therefore of constant energy, or from a collision cell and hence of varying energy. This relationship can then be used to assign mass corresponding to any channel position relative to a known mass, defined from magnet calibration, at a nominated channel. A complete spectrum is obtained by acquiring a series of frames, stepping the magnet to change the center mass of the array by the precise array width between each frame. The accuracy of magnet positioning should be no worse than one channel (40 ppm for a 4% array). The individual frames are then assembled to give a complete spectrum, but mass conversion is still carried out on a frame-by-frame basis. Mass assignments to within 0.1 Da can usually be obtained.

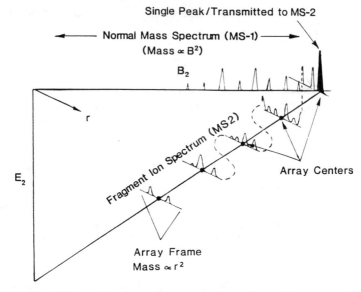

FIG. 11. The B/E plane for a stepped array scan. Array "frames" are centered on a B/E line, with mass dispersion proportional to r^2. See in this volume [2].

The center mass of the array, when used to detect fragment ion spectra, must lie on the B/E line from the nominated parent, and so both B and E[16] must be stepped between each frame. The mass distribution across the array, however, is not affected, as this is obtained under static conditions, the variable being r, the radius of deflection within the magnetic sector (Fig. 11). Further details of the performance of this detector and its applications in four-sector tandem mass spectrometry are given by Walls *et al.*[17]

Position and Time-Resolved Arrays

Another type of array detector which detects the electron cloud produced by the MCP directly was developed at the FOM Institute in Amsterdam, and a derivation of this was described in 1981 by Stultz *et al.*[18] This

---

[16] For definition of terms, see K. R. Jennings and G. G. Dolnikowski, this volume [2].

[17] F. C. Walls, M. A. Baldwin, A. M. Falick, B. W. Gibson, S. Kaur, D. A. Maltby, B. L. Gillece-Castro, K. F. Medzihradszky, S. Evans, and A. L. Burlingame, *in* "Biological Mass Spectrometry" (A. L. Burlingame and J. A. McClosky, eds.), p. 197. Elsevier, Amsterdam, 1990.

[18] J. T. Stultz, C. G. Enke, and J. F. Holland, *Anal. Chem.* **55,** 1323 (1983).

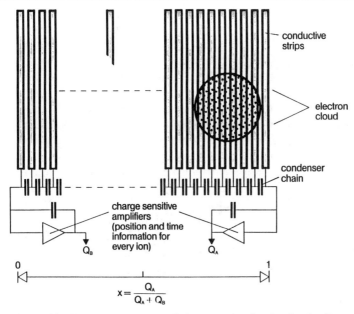

FIG. 12. Schematic diagram of the PATRIC detector, showing details of collector plate.

detector has since been developed commercially; the principle of operation is shown in Fig. 12.[19]

The ion initially strikes the first MCP as in the photodiode array system, but in this case the channel plate pair is operated at a gain of at least 10^6. The electron cloud produced by the ion is accelerated to a collector plate. This plate consists of 54 conductive strips, connected by capacitors, mounted far enough away from the MCP to allow the electron cloud to expand to a circular spot covering several of the strips. Charge-sensitive amplifiers located at each end of the collector plate detect the current flowing along the capacitor chain, and by comparing the way in which the current is shared between the two amplifiers, the center of charge of the electron cloud can be determined, to an accuracy of typically about one-tenth of a strip width.

Only one event can be detected at a time, and as it takes approximately 5 μsec for the charge pulse to be read and to decay, the theoretical maximum count rate across the whole array, which in this case covers 8% of the mass range, is 200,000 per second. At this rate, there is a high

[19] R. Pesch, G. Jung, K. Rost, and K.-H. Tietje, *Proc 37th ASMS Conf. Mass Spectrom. Allied Topics*, p. 1079 (1989).

statistical probability that ions will frequently arise less than 5 μsec apart causing overlap between pulses. The second pulse is therefore not counted, but is detected and allowed for by fast signal processing and any resultant errors in intensity distribution corrected for, up to a maximum count rate of 400,000 per second. High-intensity beams can be switched in less than 0.1 sec to a PAD detector mounted alongside the array.

Since ions are detected individually, the arrival time of an ion can also be measured, which makes it possible to use the array while the instrument is scanning. By relating the arrival time to the instantaneous value of the magnetic field, the mass corresponding to any position on the array can be calculated, and ions of a particular mass can be tracked and integrated as they cross the array, at scan speeds of up to 1.5 sec/decade.

Performance of the detector is very similar to that of the photodiode array in terms of resolution and sensitivity. The best resolution so far achieved is 7500 FWHM, which corresponds to a spatial resolution of 0.1 mm. This is equivalent to the four channels minimum of the photodiode array. So far, this detector has only been applied to two-sector instruments.

[4] Selected-Ion Measurements

By J. Throck Watson

Introduction

There are two fundamental methods by which a mass spectrometer can be used to detect ion currents. The most common method is that of scanning a parameter that controls the mass axis and, thus, that allows ion current at a given mass value to be focused on the detector at any given time. The process of scanning a mass spectrometer is relatively inefficient in that only ion current at a given mass value along the mass axis is recorded or detected at any given moment and, at that moment, all ion currents at all other mass values along the mass axis are being ignored by the data acquisition system. An alternative to this wasteful process is the use of some form of array detection which records the ion current at all mass values of the mass axis simultaneously. Array detection in mass spectrometry is a developing and promising technology (see [3], this volume).

Selected-ion monitoring (SIM) is a specialized mode of operation of the mass spectrometer in which some of the attributes of array detection

METHODS IN ENZYMOLOGY, VOL. 193

are achieved. During SIM the single detector of the typical mass spectrometer is time-shared among ion currents at selected mass values along the mass axis. In this way, most of the ion current at the selected mass values is recorded rather than being wasted during the analysis of the compound of interest.

In order to use SIM effectively, it is necessary to first characterize the compound of interest by obtaining its full mass spectrum under standard conditions. The analyst then identifies characteristic peaks in the mass spectrum which correspond to the analyte and chooses appropriate mass values along the mass axis from which to monitor ion current during analysis of samples which are thought to contain the analyte. In this way, much more efficient use is made of the ion current that is generated from the analyte during the analytical process.

Two distinct limits of detection can be achieved by a given mass spectrometer depending on whether the technique of scanning or the technique of SIM is employed. The limit of detection established by SIM is approximately 1000 times lower than that established during repetitive scanning. This disparity exists because the single detector must be time-shared among all of the mass values along the mass axis should the technique of scanning be used, but among only selected (relevant) mass values in the SIM technique. In the typical mass spectrum there may be between 1000 and 10,000 mass values along the mass axis depending on the resolving power achieved by the mass spectrometer during the process of complete scanning. During SIM, the ion current corresponding to characteristic peaks in the spectrum may be recorded at only one mass value for a given mass-to-charge ratio. Thus, the number of mass values examined during each analysis is responsible for the disparate ratio of limits of detection between the two techniques. Only if all the ion current is recorded along all mass values of the mass axis all the time can the limit of detection for recording a complete spectrum be achieved at the same low level as otherwise achieved by SIM. This ambitious goal can only be achieved by the technology called array detection.

Array detection will be summarized briefly because of the promising potential of this developing technology. However, the main emphasis in this chapter will be on the SIM technique which is applicable to most of the existing instruments already in use in laboratories. Array detection was effectively initiated with the advent of the mass spectrograph in which the complete mass spectrum is dispersed in space, yet brought into focus along the plane of a photographic emulsion. Because of the relatively low inherent sensitivity of the photographic emulsion, this particular arrangement for array detection is not advantageous over that of using a single electron multiplier detector which must be time-shared among all of the mass values of the complete mass spectrum. The first significant break-through in array detection was achieved by Boettger *et al.*[1] through their pioneering efforts of using an "electronic photoplate" with a magnetic mass spectrometer. They used a microchannel electron multiplier coupled with a phosphorous screen and an array of photodiodes to record all the ion current all the time in a Mattauch–Herzog instrument. Work in this area was continued by Hedfjall and Ryhage[2] using a single-focusing mass spectrometer with a microchannel electron multiplier plate of limited dynamic range which recorded only a segment of the complete mass spectrum. More recently, the use of array detectors based on the microchannel plate detectors has been quite successful in the application to fast atom bombardment (FAB) mass spectrometry[3] as described in [3] in this volume.

The use of an ion cyclotron resonance (ICR) mass spectrometer in the Fourier transform (FT)-MS mode of operation serves effectively as an array detector. In the pulsed mode of operation of the ICR instrument, ions of all masses are excited simultaneously and, thus, all resonant frequencies are detected simultaneously which are then deconvoluted through the FT process. Whereas this technology has the attribute of recording all the ion current all the time, overall it is not yet competitive with the more common analyzers based on magnetic dispersion or quadrupole mass filter approaches to mass spectrometry for purposes of high sensitivity, especially when applied to GC/MS.

Time array detection is a promising new technology which is being developed in an area of time-of-flight (TOF) mass spectrometry.[4,5] In this case, the mass spectrum is dispersed in time and an integrating transient recorder is used to record ion currents at all mass values in the spectrum after each extraction pulse of the ion source in the TOF operational cycle. Array detection is not amenable to the popular quadrupole mass spectrometer because it acts as a mass filter thereby allowing only ions of a given mass-to-charge to be detected under a given set of electrical conditions. The only mass spectrometric methods that are amenable to array detection are those that employ some means of dispersing the various ion currents such that they are all available to detection simultaneously as in the magnetic instrument (spatial dispersion), the ICR instrument (frequency dispersion), or the TOF mass spectrometer (time resolution).

[1] H. G. Boettger, C. E. Giffen, and D. D. Norris, *in* "Multichannel Image Detectors" (Y. Talmi, ed.), p. 291. American Chemical Society, Washington, D.C., 1979.

[2] B. Hedfjall and R. Ryhage, *Anal. Chem.* **53,** 1641 (1981).

[3] J. Cottrell and S. Evans, *Anal. Chem.* **59,** 1990 (1987).

[4] J. F. Holland, C. G. Enke, J. T. Stults, J. D. Pinkston, B. Newcome, J. Allison, and J. T. Watson, *Anal. Chem.* **55,** 997A (1983).

[5] J. Allison, J. F. Holland, C. G. Enke, and J. T. Watson, *Anal. Instrum.* **16,** 207 (1987).

Selected-Ion Monitoring

SIM is the technique in which a mass spectrometer is dedicated to monitoring ion current at only a few selected mass values throughout the complete mass spectrum as a function of time. It is important to realize that the output from the mass spectrometer detector will be available as a function of time and, thus, will provide a record of ion current that corresponds to designated m/z values throughout the mass spectrum as chosen or selected by the analyst. The profile of ion current at selected mass values will depend on the dynamics of the sample in the ion source. For example, if the sample is introduced into an electron ionization (EI) ion source via a gas chromatographic inlet system, the profile produced during SIM will have the shape of a typical gas chromatogram. If the sample is introduced via the direct-probe inlet or via a FAB probe inlet, the profile will be much more broad and slowly changing as the sample slowly distills or desorbs into the ion source. Some of the key features of SIM as compared to those of the technique of repetitive scanning are summarized in Fig. 1. The history of the development of SIM by Sweeley, Ryhage, Holmstedt, and other pioneers has been reviewed.[6,7]

Nomenclature

The term, selected-ion monitoring, is recommended for the following reasons.[8] The term selected is appropriate because it implies both choice and specificity; furthermore, it imposes no restriction as to the number of ions involved. The word ion accurately describes the species being monitored. The term monitoring is preferred because it connotes a temporal component in this specialized technique, which records profiles of ion currents as a function of time.

The data records are referred to as selected-ion current profiles at a given mass-to-charge value. This name may be considered cumbersome, but it has the advantage of being explicit and descriptive.

Qualitative Example of SIM

An example, designed to assist the reader in making the transition between comprehending mass spectral information in the format of a mass spectrum to that of appreciating the data format available from SIM, is provided by Figs. 2 and 3. The mass spectral data in Fig. 2A are in the format of a conventional bar graph mass spectrum of the pentafluoropro-

[6] F. C. Falkner, *Biomed. Mass Spectrom.* **4**, 66 (1977).

[7] J. T. Watson, "Introduction to Mass Spectrometry," 2nd Ed. Raven, New York, 1985.

[8] J. T. Watson, F. C. Falkner, and B. J. Sweetman, *Biomed. Mass Spectrom.* **1**, 156 (1974).

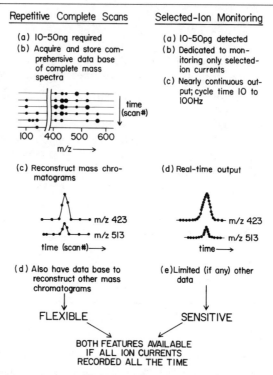

FIG. 1. Summary of operational aspects and mutually exclusive advantages of instrumental techniques leading to generation of reconstructed mass chromatograms versus selected-ion current profiles. (Reprinted from J. T. Watson, "Introduction to Mass Spectrometry," 1985, p. 325, with permission from Raven Press.)

pionyl (PFP) derivative of normetanephrine (NMN) in which the ion current throughout the mass spectrum is represented as discrete peaks at corresponding mass-to-charge values. From inspection of Fig. 2A, it is apparent that ion currents at m/z 445 and at m/z 458 would be good candidates for serving as characteristic indicators of the presence of NMN-PFP in a given sample as analyzed by mass spectrometry.[7,9]

If one adjusts a given mass spectrometer to monitor only ion current at m/z 445 and 458 as a function of time, then a response will be produced at the detector of the mass spectrometer when a material which can produce ion current at m/z 445 and/or 458 enters the ion source of the mass spectrometer. The duration of the signal at the detector will depend on the lifetime of the compound in the ion source of the mass spectrometer.

[9] D. Robertson, E. Heath, F. Falkner, R. Hill, G. Brilis, and J. T. Watson, *Biomed. Mass Spectrom.* **5,** 704 (1978).

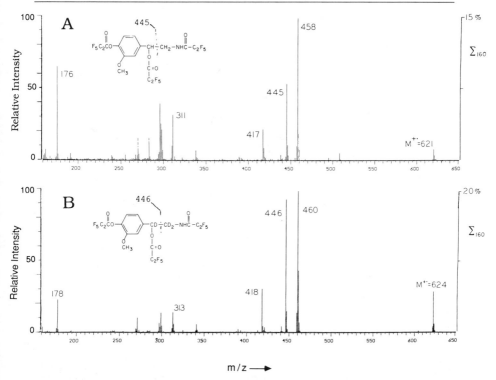

FIG. 2. (A) Mass spectrum of [²H₀]normetanephrine-PFP derivative ([²H₀]NMN-PFP). (B) Mass spectrum of [²H₃]normetanephrine-PFP derivative ([²H₃]NMN-PFP). (Adapted from J. T. Watson, "Introduction to Mass Spectrometry," 1985, p. 61, with permission from Raven Press.)

Figure 3 is an example of the SIM output from a GC/MS instrument adjusted to monitor ion current at m/z 445 and 458. This selected ion current profile at m/z 445 and 458 takes the form of a gas chromatogram because the sample was introduced into the mass spectrometer through a gas chromatographic inlet system.

There are several peaks in the selected ion current profile in Fig. 3, but it is important to realize that only one of these peaks has all the characteristics that are appropriate to relate it to NMN-PFP. For example, notice the peak that occurs at 2 min along the retention time axis. This peak corresponds to a response at m/z 458 with no response at m/z 445. Thus, the material represented by this peak surely is not NMN-PFP, which would generate both ion currents at the same time. The peak at 3.7 min is more appropriate for NMN-PFP because it consists of a response at both m/z 445 and at 458, and the ratio of these two ion currents is in the ratio

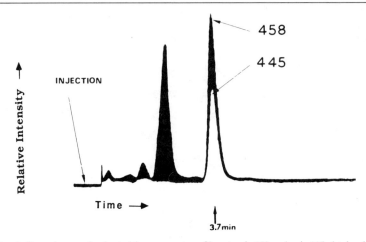

FIG. 3. Superimposed selected-ion current profiles at m/z 458 and m/z 445 obtained during the analysis of an aliquot of derivatized urine extract by GC/MS with SIM. This output from a light beam oscillograph produces the selected-ion profile at m/z 458 in black; that at m/z 445 in white. The peak at 3.7 min corresponds to [$^2H_0$]NMN-PFP (40 ng injected on-column). (Adapted from J. T. Watson, "Introduction to Mass Spectrometry," 1985, p. 28, with permission from Raven Press.)

of 1 : 2 as expected from the mass spectrum shown in Fig. 2A. The ratio of the ion current at m/z 458 to that at m/z 445 in both Figs. 2A and 3 is fixed by the fragmentation pattern of NMN-PFP. This ratio of certain ion currents recorded during the process of SIM is one of the qualitative characteristics that helps to strengthen the basis for selectivity by this technique. Of course, the major selectivity is brought about by detecting ion current at only selected mass values in the first place. It is worthy of note that when the biological sample (an extract from urine) represented in Fig. 3 was analyzed by conventional gas chromatography under these same conditions, there was no discernible peak from the flame ionization detector (FID) at a retention time of 3.7 min. However, from the results of SIM, as is apparent in Fig. 3, one recognizes a readily discernible and well-resolved peak at 3.7 min on the retention time axis.

Quantitative Use of SIM

The sensitivity also is greatly improved when using GC with SIM over that of GC with FID; for example, 50 pg of NMN-PFP is readily detectable using SIM. Although the data in Fig. 3 could be used for quantitative analysis, the quantitation would have to be judged either on an absolute basis or on the basis of using a "bracketing" technique whereby standard aliquots of a known sample would be analyzed alternately with aliquots of this unknown mixture. The responses at the retention time of

3.7 min for the unknown and the standard would then be compared to quantitate NMN-PFP in the unknown. The accuracy and precision of such a procedure would likely benefit from use of an "injection standard"[10] to account for variable amounts of sample remaining in the sampling syringe following injection of an aliquot into the GC/MS instrument.

Ideally, for quantitative purposes, an internal standard is added to the unknown sample at an early stage in the analysis to facilitate quantitative estimation that is independent of problems with isolation techniques and variations in instrument performance. With mass spectrometry, it is possible to use a stable isotope-labeled analog of the analyte as an internal standard. In this way, corrections for losses due to chemical and physical characteristics of the analyte are automatically compensated. This is so because the stable isotope-labeled analog has virtually the same chemical and physical characteristics as the analyte and, thus, will suffer the same losses as those of the analyte. If the stable isotope-labeled internal standard is added early in the assay process, losses due to these chemical and physical parameters will be proportionally the same for the internal standard and the analyte. Thus, the *ratio* of the stable isotope-labeled analog to the analyte should be the same in the final step of analysis as it was in the first step of the analytical procedure, although the absolute levels of both materials may be lower in the final step.

An example of a stable isotope-labeled analog as a candidate for an internal standard is shown in Fig. 2B. In this case, the analog of NMN contains three deuterium atoms: two on the α-carbon and one on the β-carbon. As illustrated in Fig. 2B, cleavage of the bond between the α- and β-carbons leads to charge retention on the phenyl-substituted portion of the molecule which now contains one of the three deuterium atoms. Thus, this ion, which in the unlabeled case corresponds to mass 445, now corresponds to mass 446 because of retention of the deuterium atom. Similarly, the ion current at m/z 458, in the case of the unlabeled material in Fig. 2A, now shifts to m/z 460 (in Fig. 2B) because the rearrangement process involved in forming this ion expels one of the three deuterons[7,9] and, thus, with retention of two deuterons, the mass of the ion shifts from 458 to 460 Da. Now, if one has a mixture of unlabeled and labeled materials in a given sample, one can monitor the ion current at m/z 458 for the unlabeled material and the ion current at m/z 460 as an index of the response to the labeled analog. The ion current profiles in Fig. 4 correspond to the analysis by SIM of a standard mixture of labeled and unlabeled species. Thus, in this case, the ratio of the peak areas under the ion current profile at m/z 458 to that at m/z 460 has a quantitative aspect. This quantitative aspect must be calibrated by analyzing, with the mass spectrometer, a series of

[10] H. Valente and K. M. Aldous, *Anal. Chem.* **60,** 1478 (1988).

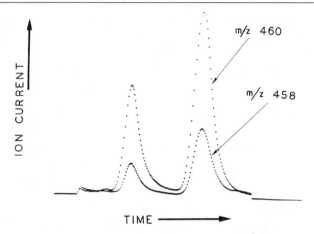

FIG. 4. Selected-ion current profiles at m/z 460 representing [^{2}H$_3$]NMN-PFP and at m/z 458 representing [^{2}H$_0$]NMN-PFP; earlier unlabeled peak represents response at the same two mass values, but for metanephrine-PFP. (Reprinted from J. T. Watson, "Introduction to Mass Spectrometry," 1985, p. 62, with permission from Raven Press.)

standard samples which contain known ratios of labeled or unlabeled material, and plot the response of mass spectrometric responses at appropriate mass values as a function of the known ratios of the standards. This, then, provides a calibration plot (see Fig. 5) which can be used for determining the quantity of unlabeled material in a given sample by virtue of the ratio of responses at m/z 445 and m/z 458.[9]

The important feature in Fig. 5 is that the ratios of ion currents are linear functions of the ratios of quantities of material used to generate the ion currents. The fact that the two plots have different slopes is related to an isotope effect on the fragmentation behavior described earlier. These differences in slope have no bearing on the final quantitative result so long as the relationship of the ratio of ion currents to the quantity ratio of materials is linear. One can use the calibration plot in Fig. 5 to determine the ratio of an unknown quantity of analyte to the standard quantity of internal standard added at the beginning of the assay. With this ratio and knowledge of the quantity of labeled material (namely, the internal standard) that was added to a known quantity of biological sample in the first place, one can mathematically calculate the quantity of unlabeled material or analyte that must have been in the sample at the outset (an example of this calculation is given later).

Finally, it is worth reviewing the selected-ion current profile information in Figs. 3 and 4. In Fig. 3, the ratio of ion current at m/z 458 to that at m/z 445 for the peak at retention time at 3.7 min is a *qualitative* ratio established by the fragmentation pattern of NMN-PFP. In Fig. 4, the ratio of ion currents

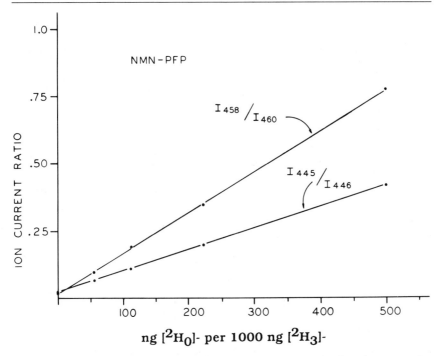

FIG. 5. Calibration plot for indicated ratios of ion currents monitored versus ratios of amounts of [$^2H_0$]NMN-PFP (I_{458} or I_{445}) and [$^2H_3$]NMN-PFP (I_{460} or I_{446}) injected into the GC/MS instrument with SIM operation. (Reprinted from J. T. Watson, "Introduction to Mass Spectrometry," 1985, p. 63, with permission from Raven Press.)

at m/z 460 to that at m/z 458 has a *quantitative* aspect because each profile originates from a different compound; that at m/z 460 comes from the stable isotope-labeled analog of NMN-PFP and that at m/z 458 comes from the unlabeled species (the analyte). Thus, the ratio of ion current at m/z 458 to that m/z 460 can be related through a calibration plot to the relative amounts of these two materials present in a given sample.

Selection of m/z Values

The detection limits available with SIM are in large measure due to the careful selection of peaks in the mass spectrum chosen to characterize the compound of interest. Of course, maximum selectivity can be ensured by choosing an ion that is unique among all the ions generated and detected during the analysis of a complex mixture. Whereas, uniqueness cannot be guaranteed, selection of ions at high mass helps to meet the goal of uniqueness as much as possible and also helps to minimize background signals, an important consideration in high sensitivity applications. Selecting an ion at relatively low mass works the opposite; selection of an ion of mass 43, for example, would not be good because many different compounds could fragment to form an ion at this low mass value thereby reducing the selectivity of the overall technique. Monitoring an ion at high mass improves the overall selectivity because fewer compounds are likely to generate such a high mass ion by a fragmentation process, thus the response to an ion which represents a larger proportion of a given molecule helps to ensure greater selectivity in detecting that molecule.

A second important characteristic to consider when selecting a given ion current is the intensity of the peak in the mass spectrum of the compound of interest. The greater the fraction of total ion current represented by a given peak, the lower will be the limit of detection that can be established for that particular compound. Thus, in summary, there is a trade-off between selectivity and sensitivity in choosing the ion currents to be monitored during SIM.

Number of Ion Currents Monitored

The detection limits that can be achieved by SIM are related to the efficiency with which the ion current at any given m/z value can be acquired during the analysis of a given sample. Time-sharing the detector with only a *few* ion currents, rather than time-sharing it among several hundred ion currents throughout the mass spectrum during the process of scanning, gives SIM its great advantage of lower detection limits over those achieved by complete scanning.

The very lowest detection limit for a given compound during SIM can be achieved by continuously monitoring the ion current at only a single m/z value. As one time-shares the detector with other ion currents in the process of detecting more than one ion current in a contemporaneous fashion, the ion current accumulated at any given m/z value will drop in a reciprocal function of (at least) the number of different ion currents being multiplexed. Thus, under circumstances where one wishes to time-share a detector between a dozen or more ions during a given analysis, the detection limit advantage of SIM over scanning becomes greatly diminished. In fact, if one wishes to time-share the detector among more than a dozen ion currents during a given analysis, it may not be worth giving up the flexibility of a comprehensive data base, as is available from complete scanning, for the limited advantage of somewhat lower detection limits from SIM under *those* circumstances. In summary, if one anticipates monitoring more than a dozen ion currents in a contemporaneous fashion, one may as well consider scanning the mass spectrometer.

Considering the number of ion currents to be chosen, however, from the standpoint of the analytical objective, it is important to monitor at least two ion currents from the analyte so that one will have qualitative information "built in" to a set of selected-ion current profiles. That is, the ratio of two ion currents from the same compound will provide confirmatory proof that the compound is likely to be present because the ratio of the two ion currents would be established by the characteristic fragmentation pattern of the analyte. This kind of characteristic information builds confidence into the methodology by providing confirmatory evidence for the presence of the material in addition to the characteristic information provided by the retention time (if using GC/MS). On the other hand, it is a "good insurance policy" to monitor at least three ion currents from a given analyte so that in case there is interference in the detection of one of the ion currents, at least there would remain two ion currents which would give a characteristic ratio for qualitative identification of the compound of interest.

In the case of the internal standard, which may be an isotopically labeled analog of the compound of interest or a structurally similar compound, it may be possible to successfully complete the analysis by monitoring ion current at only one m/z value for the internal standard. This is possible because the analyst may be able to choose an internal standard which will elute during the chromatographic analysis at a retention time known to be fairly "quiet" from the standpoint of response to any compound. Thus, by selecting a material which elutes during this "quiet" interval, the likelihood that it is a material other than the internal standard is reduced or is remote. In addition, because the analyst can control the amount of internal standard used, it may be possible to add an excess of internal standard which may be two or three times the amount of the analyte; in this case, the internal standard would be more easily detected and, thus, less likely to be confused with some exogenous material. Of course, this procedure has the disparate feature that much better ion statistics would be achieved for detection of the internal standard than of the analyte if the same amount of time were spent time-sharing the detector for ion currents of each species. One way to handle this situation would be to monitor the ion currents from the analyte for greater periods of time than for the ion of the internal standard so that comparable ion statistics would be available from detection of both species. Again, from the insurance policy point of view, it would be desirable to monitor at least two ion currents from the internal standard so that characteristic information would be available to verify that the internal standard is being detected (i.e., agreement of ion current ratios with the ratio expected from the fragmentation pattern of the internal standard).

Following the strategy outlined above, it is clear that if one is interested in the detection of more than two or three analytes and has more than one internal standard, that there could easily be as many as a dozen or so ion currents that should be monitored during the analysis of a given sample. This could force one into an analytical plan that would diminish the advantage of low detection limits offered by SIM. One way that this can be avoided is to determine whether there are groups of ions that could be monitored during different time periods in the overall analysis. For example, if some of the analytes are well separated chromatographically from others, it may be possible to program the data system to monitor one group of ion currents after another so that during any given segment of the overall analysis, no more than five to six individual ion currents would be time-shared with the detector at any given time interval. In this way, the overall detection limits are maintained at a reasonably low level while accommodating the detection of a relatively large number of different analytes.

Selection of Dwell Times

Dwell time is the period of time that the detector is allowed to integrate the ion current at a given mass value before it is time-shared with ion currents at other mass values. The selection of a dwell time, typically 50 to 100 msec, involves consideration of trade-offs between precision (and accuracy) of the detector output as related to ion counting statistics and the degree of accuracy desired in representing a chromatographic profile (or other rapidly changing function of the analyte concentration). Precision of the detector output improves with the square root of the number of ions (which is equal to the integral of ion current over time at a given mass value) detected, and thus, for a given ion current, the precision will improve with the length of the dwell time. On the other hand, the longer the dwell time, the smaller the number of data points available to represent discrete points along the chromatographic profile.

The greater the desired accuracy in representing a given chromatographic profile, the greater the number of discrete data points required. For example, Chesler and Cram[11] have estimated that as many as ten data points per sigma (σ) are required to accurately characterize a chromatographic profile, where sigma is the standard deviation of the chromatographic peak profile. As a rule of thumb, a minimum of ten data points are required to provide a reasonable representation of a chromatographic peak; if accurate detail on inflection points is desired, many more data points are required.[11] The reader can appreciate the difficulty in providing

[11] S. N. Chesler and S. P. Cram, *Anal. Chem.* **43**, 1922 (1971).

a sufficient number of data points in this context of SIM by considering the following analytical problem. If one works with a capillary column that produces peaks that are 2 sec wide at the base, and monitors two different ion currents with a dwell time of 100 msec for each ion current, this procedure will produce five data points per second for each of the two ion currents; this results in the minimum number of data points to adequately represent the specified chromatographic peak in either selected-ion current profile. If, in this example, the operator had monitored four ion currents with a dwell time of 100 msec on each, there would have been only five data points (2.5 data points per second) available to plot a chromatographic profile in any one of the four selected-ion current profiles; this is because, under the stated conditions, for a given selected-ion current profile, discrete data points are produced only once every 400 msec. Clearly, the operator must make some decisions on selecting an optimum dwell time based on the desired accuracy of profile representation and the number of ion currents being monitored at any given time. Some compromises could include the choice of a short dwell time, e.g., 25 msec, for an ion current that is expected to be large, and a longer dwell time, e.g., 100 msec, for an ion current that is expected to be smaller, thereby achieving comparable ion statistics at each point.

The demands[4,12] on mass spectrometry as described in the previous paragraph are becoming more critical as improvements[13,14] in high-resolution gas chromatography produce more narrow peak widths on the order of tens of milliseconds.

Mechanics of SIM Operation

Magnetic Mass Spectrometer

The most popular means of using a magnetic mass spectrometer for purposes of SIM involves adjusting the magnetic field to a value that corresponds to the lowest mass of the group of ions of interest for monitoring. During the analysis, the magnetic field strength is maintained at a constant value and the selected mass value in the spectrum is focused onto the detector by attenuating the magnitude of the accelerating voltage. This technique of acceleration voltage alternation has the advantage of accurate and rapid adjustment of the accelerating voltage between different values

[12] P. A. Leclereq and C. A. Cramers, *J. High Resolut. Chromatogr. Chromatogr. Commun.* **11,** 845 (1988).
[13] L. A. Lanning, R. D. Sacks, R. F. Mouradian, S. P. Levine, and J. A. Foulke, *Anal. Chem.* **60,** 194 (1988).
[14] Z. Linarcland and J. B. Phillips, *J. Microcolumn Sep.* **1,** 159 (1989).

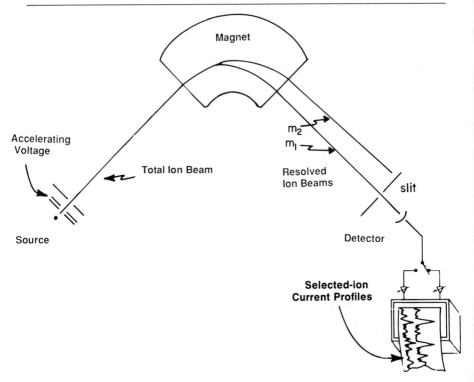

FIG. 6. Schematic diagram of magnetic mass spectrometer showing arrangement for contemporaneous recording of two selected mass-resolved ion currents at a single detector. Alternation of appropriate values of accelerating voltage brings selected-ion currents at m_1 or m_2 alternately onto the detector.

which bring the selected-ion currents of interest into focus alternately onto the detector for given periods of time. Figure 6 is a schematic diagram of a magnetic sector instrument illustrating the different flight paths of two ion currents of m/z m_1 and m_2 (where $m_2 > m_1$). At full accelerating voltage, ion current corresponding to m_1 is focused on the detector; the accelerating voltage can be attenuated abruptly to bring ion current for m_2 onto the detector. Corresponding switching of the detector output to a different channel of the data acquisition system effectively allows both ion current profiles to be recorded contemporaneously.

The technique of accelerating voltage alternation as described above has the disadvantage that as the magnitude of the accelerating voltage is decreased, the efficiency of transmission of a given ion current is also decreased. Because of this feature, there frequently will be a limit to the range of mass values that can be monitored using the technique of accelerating voltage alternation. For example, this limitation may range anywhere from 10 to 50% of the mass range of interest (which corresponds to an attenuation of 10 to 50% of the accelerating voltage) depending on the manufacturer, so that the performance of the instrument is not compromised prohibitively.

Another concern during SIM with a magnetic instrument is the possibility of a "drift" in the mass scale calibration. Note from the above paragraph, that the magnetic field is set to a value which transmits the ion of lowest mass onto the detector. If this mass value were at a relatively high mass such as mass 400, for example, a fairly high level of current will be driven through the magnet for the duration of the analysis. Should the analysis continue for several hours, thermal effects could influence the resistivity of magnetic current circuit which, in turn, would influence the magnet current (and magnetic field) causing the ion beam to shift slightly. The principal concern here is that the ion current monitored during SIM be sampled at the center of the mass spectrometric peak. As is often the case in SIM, there is no attempt to ascertain the peak shape of a given mass spectrometric peak to ensure that the center of the peak is focused onto the detector. Of the many data systems described in the literature, only one provides for automatic focus[15]; unfortunately, as of this writing, no commercial version of this original work[15] is available, in spite of the heavy involvement of computers in modern mass spectrometry. Of course, use of modern laminated magnets, which also have the facility for thermal control, greatly minimizes the problem of drift in the mass calibration.

It is also possible to conduct SIM on a magnetic instrument by modulating the magnetic field. In this circumstance, the mass range is adjusted from one mass value to another by controlling the magnetic field.[16] Whereas this technique has the advantage of monitoring a group of ions which vary over the complete mass range, there is always the inherent disadvantage of hysteresis in the magnetic field, making it difficult to reestablish exactly the same value of magnetic field strength on a repetitive basis.

Quadrupole Mass Spectrometer

The quadrupole mass spectrometer is ideally suited for SIM because of the ease with which ion currents at alternately selected mass values can be transmitted to the detector. First, the mass axis is completely controlled by electrical parameters so that there is no hysteresis problem involved.

---

[15] J. F. Holland, C. C. Sweeley, R. E. Thrush, R. E. Teets, and M. A. Bieber, *Anal. Chem.* **45,** 308 (1973).
[16] N. D. Young, J. F. Holland, J. N. Gerber, and C. C. Sweeley, *Anal. Chem.* **47,** 2373 (1975).

Second, because only electric potentials are involved in selecting a given mass value, the different ion currents can be alternately switched to the detector very promptly. Third, because of the linear relationship between the magnitude of the mass axis and the electrical potentials, there is no limitation as to the range of mass-to-charge values that can be monitored during any given analysis.

Time-of-Flight Mass Spectrometer

The mechanics of data acquisition from the time-of-flight (TOF) mass spectrometer are ideally suited for SIM operation. Because the m/z value of a given ion is determined by its arrival time at the detector, several ion currents from any part of the mass spectrum can be monitored simultaneously[4] without affecting optimum operating conditions of the instrument. The analog-to-digital converter can be synchronized to sample the electron multiplier during the arrival interval (10 nsec) of the selected ions. The computer can be programmed to accumulate data from several consecutive pulses of the ion source; TOF instruments routinely scan the complete spectrum 10,000 times per second, each scan requiring 100 μsec. An integrating transient recorder (ITR) can record all the mass spectral information from each pulse of the ion source of a TOF instrument.[5] The ITR provides time array detection, which permits full-spectrum sensitivity to be achieved at a level presently possible only by SIM.

Accuracy and Precision

The accuracy of SIM assays relies heavily on the degree of recovery of the analyte throughout a given procedure and on proper calibration of the mass spectrometer. Use of an internal standard as described above addresses requirements of both assessing recovery and providing calibration. Colby and McCaman[17] examined some of the limiting approximations of the original sample and internal standard isotope ratios; they found that certain data reduction procedures allow determinations over wider ranges of mole ratios and with internal standards that contain a greater amount of unlabeled impurity than others. Matthews and Hayes[18] studied the sources of systematic errors in isotope ratio measurements during GC/MS and found that the mass-cycling error is minimized during a unidirectional scanning pattern (ABCABC...) rather than in the bidirectional pattern (ABCCBA...). Schoeller[19] has examined the assumptions implicit in isotope dilution calculations and has reviewed the statistical considerations

[17] B. N. Colby and M. W. McCaman, *Biomed. Mass Spectrom.* **6,** 225 (1979).
[18] D. E. Matthews and J. M. Hayes, *Anal. Chem.* **48,** 1375 (1976).
[19] D. A. Schoeller, *Biomed. Mass Spectrom.* **3,** 265 (1976).

involved in treatment of the calibration data. Pickup and McPherson[20] described a model for studying isotope dilution assays based on the use of binomial probability theory to evaluate the effects of changing isotope abundance during a dynamic analysis.

The accuracy of SIM also depends in large measure on the selectivity achieved by the SIM process and by the chromatographic system or other sample introduction system to preclude interference from extraneous materials. Low et al.[21] have described distortions in quantitative measurements due to ion current corresponding to $(M - H)^+$ and $(M - 2H)^+$ when relying on ion abundance measurements from a stable isotope-labeled analog. In these cases, increasing the resolving power to 10,000, which improves discrimination against interference,[22,23] does not solve the problem. The best strategy is to avoid the problem, that is, should interference from $(M - 2H)^+$ be possible, use a labeled analog which contains three or more deuterium atoms or other heavy isotopes.

The precision of SIM analyses is inversely related to the quantities of material being analyzed because it is a function of the ion statistics available for quantitation. For an evaluation of precision to be meaningful, the conditions under which it is determined should be clearly stated. For example, the precision of analyzing a sample that corresponds to a quantity of material at the midpoint of a calibration curve will be quite different from that obtained with a quantity near the detection limit. Furthermore, analyses of extracts from biological samples usually show a higher relative standard deviation (RSD) than do analyses of a pure standard of the same concentration.

In most cases, an RSD of ±3% or less can be expected for repeated analyses of an amount of material corresponding to the midpoint of the calibration curve. The RSD increases as quantities approaching the detection limit are analyzed. For example, replicate analyses of 8 pmol (2 ng) of N-(2-trifluoroacetamidoethyl)perhydroazocine were accomplished with an RSD of ±2.9%. In contrast, for replicate analyses of 0.4 pmol (100 pg) the RSD increased to ±12%.[24] Other detailed analyses of variance for SIM have been described in the context of biogenic amines[25] and prostaglandins.[26]

[20] J. F. Pickup and K. McPherson, Anal. Chem. 48, 1885 (1976).
[21] I. A. Low, R. H. Liu, S. A. Barker, F. Fish, R. L. Settine, E. G. Piotrowski, W. C. Damert, and J. Y. Liu, Biomed. Mass Spectrom. 12, 633 (1985).
[22] D. Millington, D. A. Jenner, T. Jones, and K. Griffiths, Biochem. J. 139, 473 (1974).
[23] G. C. Thorne, S. J. Gaskell, and P. A. Payne, Biomed. Mass Spectrom. 11, 415 (1984).
[24] F. C. Falkner, B. J. Sweetman, and J. T. Watson, Appl. Spectrosc. Rev. 10, 51 (1975).
[25] M. Claeys, S. P. Markey, and W. Maenhaut, Biomed. Mass Spectrom. 4, 122 (1977).
[26] K. Green, E. Granstrom, B. Samuelsson, and U. Axen, Anal. Biochem. 54, 434 (1974).

Liu *et al.*[27] emphasize the importance of instrumental parameters[21] such as resolution, dwell time, etc., as well as availability of comparable ion currents for both the analyte and internal standard in providing good precision. Thorne *et al.*[28] tried to optimize peak height measurements to gain an improved measure of ion abundance ratios. They found that a regression analysis approach[28] to the determination of response ratios was better than a peak-smoothing approach,[29] which gave poorer precision as well as a bias in the results.

Quantitative Computation from SIM

The following example illustrates the use of 15-[$^2H_4$]keto-13,14-dihydroprostaglandin E_2 ([$^2H_4$]15K-H_2-PGE$_2$) as an internal standard during the quantitative analysis of [$^2H_0$]15K-H_2-PGE$_2$.[30] Because the analysis was performed with a packed column which offered the possibility of adsorptive losses of the analyte, a large excess (>200-fold) of the internal standard was added to the sample so that the labeled compound would serve as a carrier for the analyte. In this case, the mass spectral response at m/z 375, corresponding to M^+-31-90, represents the analyte, and that at m/z 379 corresponds to the same ionic species for the internal standard ([$^2H_4$]15K-H_2-PGE$_2$). Standard quantities of the [$^2H_0$]- and [$^2H_4$]15K-H_2-E_2 were introduced into the mass spectrometer, and ion currents at m/z 375 and 379 were measured to provide data for the calibration plot in Fig. 7. The nonzero intercept indicates that the $^2H_4$ internal standard contains 0.26% of the unlabeled material.

Because the internal standard and the analyte have nearly identical chemical and physical properties, they elute from the gas chromatograph at nearly the same time; the presence of deuterium shortens the retention time slightly by an interval that is structure dependent. For this reason, the unlabeled and deuterium-labeled forms of 15K-H_2-PGE$_2$ as the methyl ester (ME), methoxime (MO), trimethylsilyl (TMS) ether derivative have nearly the same retention time in Fig. 8, which resulted from analysis of a plasma extract (Fig. 8C and D) and an aliquot of the pure internal standard [$^2H_4$]15K-H_2-PGE$_2$-ME-MO-TMS (Fig. 8A and B). Figure 8A and B shows the same data from analysis of an aliquot of "pure" internal standard. The presentation in Fig. 8B is magnified 64 times over that in Fig. 8A; the selected-ion current profile at m/z 379 goes off scale in this

[27] R. H. Liu, F. P. Smith, I. A. Low, E. G. Piotrowski, W. C. Damert, J. G. Phillips, and J. Y. Liu, *Biomed. Mass Spectrom.* **12,** 638 (1985).

[28] G. C. Thorne, S. J. Gaskell, and P. A. Payne, *Biomed. Mass Spectrom.* **11,** 415 (1984).

[29] G. C. Thorne and S. J. Gaskell, *Biomed. Mass Spectrom.* **13,** 605 (1986).

[30] W. C. Hubbard and J. T. Watson, *Prostaglandins* **12,** 21 (1976).

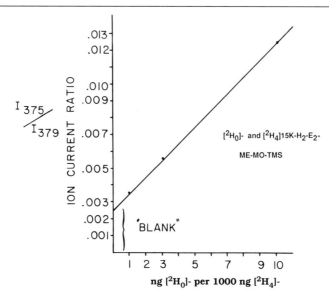

FIG. 7. Calibration plot relating ratio of ion currents at m/z 375 and m/z 379 to ratios of quantities of $[^2H_0]$- and $[^2H_4]$15K-H$_2$-PGE$_2$ derivatives injected into GC/MS instrument. The slope of the line is unity indicating that the $[^2H_0]$- and $[^2H_4]$- analogs have the same ionization and fragmentation efficiency. The intercept is 0.0026 indicating that "pure" $[^2H_4]$15K-H$_2$-PGE$_2$ contains 0.26% $[^2H_0]$15K-H$_2$-PGE$_2$. (Reprinted from J. T. Watson, "Introduction to Mass Spectrometry," 1985, p. 312, with permission from Raven Press.)

case. The area under the small peak (arrow) in Fig. 8B represents signal from intrinsic unlabeled material in the internal standard. Figure 8C and D represent data from analysis of an aliquot of derivatized extract from plasma. Again, the data in Fig. 8D are magnified 64 times over the display in Fig. 8C. The somewhat larger peak (arrow) in Fig. 8D represents the signal from a combination of the biological material plus intrinsic unlabeled material in the internal standard.

The objective is to determine the net ratio of $[^2H_0]$15K-H$_2$-E$_2$ to $[^2H_4]$15K-H$_2$-E$_2$. This is done by obtaining the gross ratio of I_{375}/I_{379} from Fig. 8C or D and subtracting the "blank" ratio I_{375}/I_{379} from Fig. 8A or B. For panel D, the ratio I_{375}/I_{379} is 0.0036. For panel B, the ratio I_{375}/I_{379} is 0.0026. Therefore, the net ratio is 0.0010.

To the original sample of 20 ml plasma was added 1500 ng $[^2H_4]$15K-H$_2$-E$_2$; thus, the original sample had the following concentration of metabolite:

I_{375}/I_{379} × amount of internal standard/volume of sample
   = 0.0010 × 1500 ng/20 ml
   = 75 pg/ml of $[^2H_0]$15K-H$_2$-E$_2$

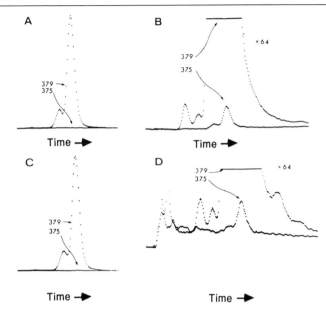

FIG. 8. Selected-ion current profiles (acquired and displayed by computer) at m/z 375 and 379 obtained during analysis of plasma extract for 15K-H_2-E_2-ME-MO-TMS. A and B are selected-ion current profiles from analysis of an aliquot of "pure" $^2H_4$ internal standard; C and D represent analysis of a biological sample. (Reprinted from J. T. Watson, "Introduction to Mass Spectrometry," 1985, p. 313, with permission from Raven Press.)

This example illustrates how the use of a high level of internal standard can limit the effective detection limit of the technique. In this case, the blank or the response due to the intrinsic level of unlabeled material in the internal standard was comparable to that from the biological experiment. Better overall statistics can be achieved when the internal standard and the analyte produce approximately the same response, as emphasized by Liu et al.[27] Numerous other applications based on the use of SIM can be found in the biennial general reviews on mass spectrometry[31,32] and in reviews emphasizing the SIM technique.[33-35]

[31] A. L. Burlingame, D. Maltby, D. H. Russell, and P. T. Holland, *Anal. Chem.* **60**, 294R (1988).
[32] A. L. Burlingame, D. S. Millington, D. L. Norwood, and D. H. Russell, *Anal. Chem.* **62**, 268R (1990).
[33] W. J. A. VandenHeuvel and J. M. Liesch, *in* "Ultratrace Analysis of Pharmaceuticals and Other Compounds of Interest" (S. Ahuja, ed.), p. 91. Wiley, New York, 1986.
[34] W. J. A. VandenHeuvel, *Drug Metab. Rev.* **17**, 67 (1986).
[35] G. N. Thompson, P. J. Pacey, G. C. Ford, and D. Halliday, *Biomed. Environ. Mass Spectrom.* **18**, 321 (1989).

[5] Liquid Chromatography–Mass Spectrometry

By MARVIN L. VESTAL

Introduction

Combined gas chromatography–mass spectrometry (GC/MS) is a very well-established, powerful technique with many applications. Unfortunately, a great many samples of biological interest are either insufficiently volatile or too labile to be amenable to GC/MS analysis. In some cases this limitation can be overcome by chemical derivatization.[1] High-performance liquid chromatography (HPLC) using reversed-phase columns has emerged over the last 15 years as a preferred method for separating biological mixtures, but has been somewhat limited by the properties of available detectors. Concurrent development of liquid chromatography–mass spectrometry (LC/MS) techniques have overcome some of the limitations imposed by the detector, but until recently mass spectrometry was of limited utility for very large molecules, such as biopolymers. Recent developments in mass spectrometry such as continuous-flow fast atom bombardment (FAB),[2] electrospray,[3] plasma desorption,[4] and UV laser desorption[5] provide much higher sensitivity for high-molecular weight compounds than was previously possible. Continuous-flow FAB and electrospray have been successfully interfaced with micro-LC and with capillary electrophoresis, but interfaces with conventional HPLC are not yet available.

The emphasis in this chapter is on combined liquid chromatograph–mass spectrometer interfaces which can be used with a variety of separation techniques including standard 4.6-mm reversed-phase columns employing aqueous buffers at flows in excess of 1 ml/min. Early approaches to LC/MS, such as direct liquid introduction (DLI), and moving belt transport interfaces were useful for many applications, but never achieved widespread acceptance at least partly because of difficulties in dealing with this most common form of liquid chromatography.[6]

---

[1] D. R. Knapp, this volume [15].
[2] R. M. Caprioli and W. T. Moore, this volume [9].
[3] C. G. Edmonds and R. D. Smith, this volume [22].
[4] R. D. Macfarlane, this volume [11].
[5] F. Hillenkamp and M. Karas, this volume [12].
[6] For a recent comprehensive review of LC/MS techniques and applications including a complete bibliography through 1988, see A. Yergey, C. Edmonds, I. Lewis, and M. Vestal, "Liquid Chromatography/Mass Spectrometry, Techniques and Applications." Plenum, New York, 1989.

METHODS IN ENZYMOLOGY, VOL. 193

Thermospray

The technique known as thermospray has emerged as a practical LC/MS technique applicable to nonvolatile samples in aqueous effluents at conventional flow rates.[7] It is now widely recognized as an efficient method for interfacing these two powerful techniques, and is applicable to a wide range of samples including volatile and nonvolatile, polar and nonpolar, labile and stable, under a variety of chromatographic conditions.[8] The term thermospray was derived by combining *thermo* with *spray* to provide a descriptive term for "production of a jet of fine liquid or solid particles by heating." Briefly, thermospray is a method for directly coupling liquid chromatography (LC) with mass spectrometry (MS) which involves controlled, partial vaporization of the LC effluent before it enters the ion source of the mass spectrometer. This is accomplished by heating the capillary tube connecting the two devices. It is vital that the heat input be properly controlled so that premature complete vaporization does not occur inside the capillary. As a result of heating, the liquid is nebulized and partially vaporized; any unvaporized solvent and sample are carried into the ion source as microdroplets or particles in a supersonic jet of vapor. Nonvolatile samples may be ionized by direct ion evaporation from highly charged droplets or particles, and more volatile samples may be ionized directly or by ion–molecule reactions in the gas phase.

Apparatus for Thermospray

A schematic diagram of a recent version of the thermospray LC/MS interface is shown in Fig. 1. This device employs direct electrical heating of the capillary and electrical heating of the ion source to convert a liquid stream into gas-phase ions for introduction into a conventional mass analyzer. It appears that any LC/MS interface must accomplish nebulization and vaporization of the liquid, ionization of the sample, removal of the excess solvent vapor, and extraction of the ions into the mass analyzer, although different techniques may involve variations in the methods by which these steps are accomplished.

In the thermospray system depicted in Fig. 1, the chromatographic effluent is partially vaporized and nebulized in the directly heated vaporizer probe to produce a supersonic jet of vapor containing a mist of fine droplets or particles. As the droplets travel at high velocity through the heated ion source, they continue to vaporize due to rapid heat input from

[7] C. R. Blakley and M. L. Vestal, *Anal. Chem.* **55,** 750 (1983); M. L. Vestal and G. J. Fergusson, *Anal. Chem.* **57,** 2372 (1985).
[8] M. L. Vestal, *Science* **221,** 275 (1984).

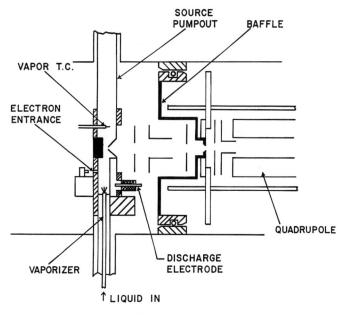

FIG. 1. Schematic diagram of thermospray ion source.

the surrounding hot vapor. A portion of the vapor and ions produced in the ion source escape into the vacuum system through the sampling cone and the remainder of the excess vapor is pumped away by a mechanical vacuum pump. A description of the major components of the thermospray LC/MS system and their primary functions are given below.

Vaporizer Probe. The vaporizer is the heart of the thermospray system. In most presently used systems it consists of a stainless steel tube 0.1 to 0.15 mm id by 0.8 to 1.6 mm od. The heated length does not appear to be a very critical parameter and is usually chosen for mechanical convenience or to match the characteristics of the heater power supply. Heated lengths between 15 and 50 cm are common. The vaporizer is heated by passing either an ac or dc current through the heated length, and the power input is controlled by a thermocouple attached near the input end. A second temperature sensor is usually attached near the exit from the vaporizer. The latter is very useful for sensing the state of the fluid exiting the vaporizer and determining that complete vaporization or "take-off" has not occurred, but this temperature depends on several parameters and is generally not useful for direct feedback control. The vaporizer may be mounted in a probe for insertion through a vacuum lock, as indicated in Fig. 1, or it may be coiled up inside the vacuum system.

Thermospray Ion Source. Vapor produced by vaporizing 1 ml/min of solvent is about one hundred times the amount that can be accommodated by a typical mass spectrometer equipped for chemical ionization (CI). The thermospray ion source allows high mass flows to be vaporized by using a very tight ion chamber similar to that used in CI but with a mechanical vacuum pump (ca. 300 liter/min) attached directly to the source chamber through a port opposite the vaporizer. Orifices for ion exit and electron entrance are typically about 0.5 mm in diameter and only about 1% of the total vapor exits through these to the source housing vacuum manifold. Ions may be produced by direct ion evaporation from charged droplets or by CI, initiated either by an electron beam or a low current discharge. The source block is strongly heated by high-capacity electrical cartridge heaters (ca. 50–100 W), and a thermocouple is normally positioned to monitor the vapor temperature just downstream of the ion sampling orifice in addition to a temperature sensor to monitor the source block. In a recent version of the thermospray ion source shown in Fig. 1, a separately heated block surrounds the vaporizer tip. This allows the region cooled by the rapid adiabatic expansion to be properly heated without overheating the portions of the ion source further downstream. Thermospray ion sources are often equipped with repeller electrodes which, under some circumstances, can be used to either enhance high mass sensitivity or to produce increased fragmentation for qualitative identification.[9]

Electron Gun. The electron gun serves the same function as in a conventional CI source, but some minor changes are generally required. Thoriated iridium filaments, such as those used in nonburnout ionization gauges, are used in place of tungsten or rhenium to achieve satisfactory operating life in the presence of high pressures of water and methanol vapors. Also the pressures inside the ion source tend to be somewhat higher than in conventional CI (ca. 10 torr at 1 ml/min); thus electron energies of 1 kV or more are required to efficiently penetrate into the ion source. The high velocity of the vapor through the ion source requires that the electron beam entrance be displaced upstream from the ion sampling cone, otherwise ions are swept downstream and few diffuse to the ion sampling cone and into the mass analyzer.

Discharge Electrode. Even with the thoriated iridium filaments it is difficult to maintain satisfactory emission at high flows of water; the discharge electrode overcomes this limitation. The lifetime of the discharge electrode is determined by the time required for it to be fouled by conductive deposits to the point that electrical leakage paths prevent the discharge

[9] R. H. Robbins and F. W. Crow, *Rapid Commun. Mass Spectrom.* **2**, (1988); J. Yinon, T. Jones, and L. Betowski, *Rapid Commun. Mass Spectrom.* **3**, 38 (1989).

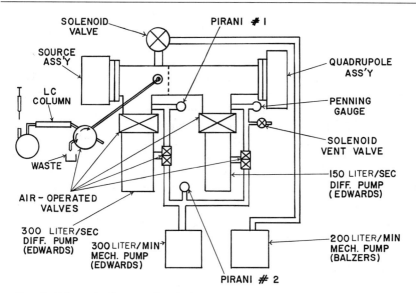

FIG. 2. Schematic diagram of typical vacuum system required for thermospray on a quadrupole mass spectrometer.

from striking. With water, fouling is not a problem, and the lifetime is virtually infinite. Running the discharge in the presence of organic solvents produces carbonaceous deposits which will eventually cause fouling. Running for a few hours in water vapor can sometimes clean a partially fouled discharge electrode. The electron beam and the discharge generally give very similar results in thermospray.

Vacuum System. A typical vacuum system for a thermospray LC/MS system is shown schematically in Fig. 2. Except for the added mechanical pump connected directly to the ion source, this system is similar to those normally provided with commercial mass spectrometers equipped for CI. The pumping speeds indicated in Fig. 2 are sufficient for liquid flows of at least 1.5 ml/min. By increasing the capacity of the diffusion pump on the source housing and its associated mechanical pump, higher flow rates can be accommodated. Since commercial MS systems rarely are designed to operate at maximum output, the mechanical pumps backing the diffusion pumps are often the limiting factor, and replacing a mechanical backing pump with one of larger capacity is sometimes required. Earlier versions of thermospray systems used a refrigerated cold trap between the ion source and the source pump. To be effective this trap must be cooled to liquid nitrogen temperature, since otherwise the organic solvents are not efficiently trapped and the lifetime of conventional mechanical vacuum pumps may be seriously shortened. The new "hot" pumps which operate with oil reservoirs at ca. 80°, provided they are always used with full gas ballast, appear to perform satisfactorily without use of a cold trap.

Ion Optics and Mass Analyzer. Thermospray LC/MS can, in principle, use any type of mass analyzer. Most of the work to date has employed quadrupoles, but thermospray is now commercially available on magnetic deflection instruments as well. The high voltage employed on the ion source of magnetic instruments requires that the thermospray ion source and its associated control electronics also be at high potential. The liquid chromatograph can be electrically isolated from the vaporizer probe by using 1 or 2 m of fused silica in the flow path,[10] and several satisfactory solutions to other problems associated with operating thermospray at high voltage have been developed. In both magnetic and quadrupole instruments, it is desirable to employ ion lenses between the source and analyzer which are as open as possible to allow efficient pumping of this region.

Thermospray Control System. Precise temperature control, particularly of the vaporizer, is required for the best performance. The vaporizer control system should normally be capable of maintaining the control temperature to within 1°. The source block and tip heater must also be controlled for reproducible results, but the control requirements are not as stringent. In order to reproduce conditions it is necessary to monitor both the vaporizer exit temperature and the vapor temperature, since these temperatures which are not directly controlled are the best indicators of system performance.

A triac-controlled ac supply may be used for supplying the power to the directly heated thermospray vaporizer. A block diagram of this controller is shown in Fig. 3. The power output to the capillary is controlled by feedback so as to maintain the temperature, T_1 constant, as indicated by a thermocouple attached to the capillary near the inlet end where no vaporization occurs.

By using this control point, the indicated temperature, T_1, is linearly related to the fraction vaporized, f, and independent of flow rate. Thus by controlling the power input so as to maintain a constant T_1, the fraction vaporized may be maintained at a constant selected value even though the flow rate may change. However, the set point required to maintain a certain fractional vaporization depends on the composition of the fluid. As the fluid composition changes, the power required to achieve a particular degree of vaporization also changes. If the set point T_1 is changed accordingly, then the fraction vaporized may be maintained constant even though the solvent composition may vary from pure water, for example, to pure

[10] M. L. Vestal, *Anal. Chem.* **56**, 2590 (1984).

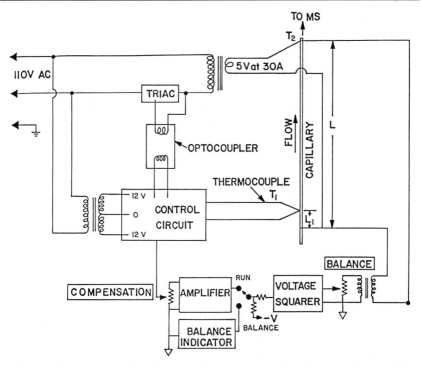

FIG. 3. Block diagram of the triac-controlled ac supply used for direct heating of capillary vaporizers shown with added feedback system for automatic gradient compensation.

methanol. This compensation can be accomplished automatically by sensing the change in heat capacity which accompanies a change in composition at constant flow rate.

The minimum temperature for complete vaporization of a fluid flowing through a capillary tube depends on the liquid velocity and the thermal properties of the fluid. Calculated minimum temperatures for complete vaporization for several common solvents are given as functions of liquid velocity in the capillary in Fig. 4. If the heat supplied to the liquid is more than that required to reach this temperature (for a given flow rate) vaporization will occur prematurely and superheated, dry vapor will emerge from the capillary. If the heat supplied is slightly less than the critical value, then a portion of the liquid is not vaporized and will emerge along with the vapor jet as small entrained droplets. For most applications of thermospray it appears that the best operating point corresponds to a fluid temperature at which partial but nearly complete vaporization occurs. In this range the residual droplets tend to be relatively small and are

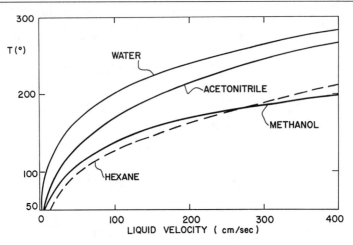

FIG. 4. Maximum exit temperature (at which complete vaporization occurs) as a function of liquid velocity for some common HPLC solvents. For a typical capillary (ca. 0.15 cm id) a flow rate of 1 ml/min corresponds to a velocity of about 100 cm/sec.

accelerated to high velocities by the expanding vapor. Since the vapor pressure is a very steep function of the temperature, it is essential to have very precise control of the temperature if a stable fraction vaporized is to be maintained at nearly complete vaporization.

A typical plot of vaporizer exit temperature as a function of control point temperature is shown in Fig. 5 for a vaporizer with an internal diameter of 0.15 mm. When liquid is forced at high velocity through an unheated capillary tube, a solid jet issues from the tube and breaks up into regular droplets. When the tube is heated gently, the properties of the jet are modified slightly by the drop in surface tension accompanying the increase in temperature, but very little change is observed visually. This corresponds in Fig. 5 to the initial linear rise in exit temperature with increasing control temperature up to ca. 60°. In this region the liquid is being heated but no vaporization occurs.

When enough heat is applied to produce significant vaporization inside the capillary, the appearance of the jet changes drastically and a very fine, high-velocity spray is produced. In this region the exit temperature increases more slowly with control temperature as the extra heat input goes primarily into vaporizing the liquid. In this plateau region the fraction vaporized varies linearly with control temperature as indicated in Fig. 5. Further increase in the applied heat reduces the visibility of the jet as the number and size of the droplets decreases until complete vaporization occurs within the capillary and the jet becomes invisible. The onset of

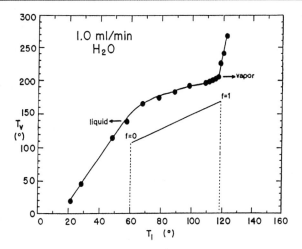

FIG. 5. Exit temperature and fraction vaporized as a function of control temperature for a 0.15 mm id capillary at 1 ml/min of water.

complete vaporization corresponds to the sharp upward break in the curve shown in Fig. 5. Beyond this "take-off" point, additional power input goes to increase the temperature of the vapor. Since the heat capacity of the vapor is only about one-half that of the liquid, the slope of this linear portion is approximately twice that of the low temperature linear portion where liquid is being heated. In this case the exit vapor temperature at complete vaporization ($f = 1$) is about 200°, in agreement with theoretical expectations based on the data shown in Fig. 4.

Automatic Compensation for Changes in Solvent Composition. As discussed above the control temperature must be programmed with changes in solvent composition such as occur in gradient elution if proper thermospray operation is to be maintained. This can be accomplished automatically by a fairly simple modification to the original thermospray controller[11] as shown in Fig. 3. The additional circuitry added for gradient compensation allows the specific heat of the solvent actually present in the thermospray vaporizer to be determined by measuring the power input required to maintain the preset control temperature. As there is a direct relationship between the specific heat of the solvent mixture and the optimum control temperature, a signal to adjust the control temperature set point can be derived from the specific heat measurement and fed back to the temperature control circuit.

[11] C. H. Vestal, G. J. Fergusson, and M. L. Vestal, *Int. J. Mass Spectrom. Ion Proc.* **70**, 185 (1986).

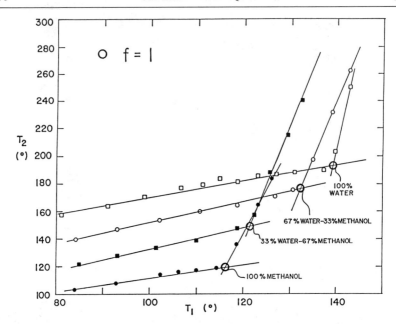

FIG. 6. Plots of exit temperature as a function of control temperature for water/methanol mixtures. The large circles indicate the experimentally determined "take-off" points at which complete vaporization occurs just at the vaporizer exit.

To set up the gradient controller, the set point temperatures for the desired degree of vaporization of two solvent compositions at the given flow rate must first be determined. These usually may be taken as 100% solvent A (for example, water) and 100% solvent B (for example, methanol or acetonitrile). The proper set point temperatures can be determined experimentally by measuring the vaporizer exit temperature, T_v, as a function of control temperature, T_1, to determine, as illustrated in Fig. 6, the point at which complete vaporization occurs ($f = 1$). As can be seen from the figure, the dependence on composition is fairly substantial, corresponding to about 20° difference in control temperature, T_1, and to about 70° difference in exit temperature between water and methanol at 1.0 ml/min flow rate.

The desired operating temperature (f slightly less than 1) can be determined directly from Fig. 6. Generally, operating at a set point, T_1, about 5% below that corresponding to $f = 1$ gives satisfactory results. With solvent A flowing through the vaporizer and the controller set at the desired operating point for that solvent, the balance control is adjusted until the balance meter is zeroed. At balance, the reference level is set so that the input to the compensation amplifier is zero. The liquid flow is that switched to solvent B (e.g., methanol) and the temperature, T_1, for the desired f is determined as described above. The controller is then switched to the gradient mode and the gain of the compensation is adjusted until the correct T_1 for solvent B is obtained. The system may now be operated with any mixture of A and B and the vaporizer controller will automatically track to maintain a constant fraction vaporized.

Optimizing Thermospray Liquid Chromatography–Mass Spectrometry for Particular Applications

In setting up to perform an analysis by thermospray LC/MS, several choices must be made correctly for best results to be obtained. These include the following: (1) liquid chromatographic conditions, (2) ionization technique, and (3) vaporizer and ion source temperatures. These choices depend on the properties of both the chromatographic mobile phase and those of the sample, as well as on the kind of analytical result that is sought. While it is not yet possible to give a precise prescription that is applicable to all cases, the present level of understanding of the processes involved, as summarized in a recent monograph,[6] provides considerable guidance in narrowing the choices.

Selection and Modification of Liquid Chromatography Conditions. Most HPLC procedures using standard size columns can be used without modification, but in some cases changes will be required. At present, thermospray works best at flow rates between 0.5 and 1.5 ml/min. Somewhat higher and lower flow rates can be accommodated, but some loss in sensitivity generally occurs outside this range. If the separation normally uses a nonvolatile buffer or other additive, such as an ion-pairing reagent, it is generally necessary to switch to a volatile alternative. Phosphates and alkali salts as major components of the mobile phase must be avoided since they cause very rapid contamination of the mass spectrometer ion source and ion optics. More volatile salts, such as ammonium acetate, ammonium formate, ammonium alkylsulfonates, trifluoroacetic acid, and tetrabutylammonium hydroxide, are no problem, and can often be substituted without severe loss in chromatographic performance, particularly if the more important parameters, such as pH, are not changed.

Selection of Ionization Technique. Most commercially available thermospray systems provide three alternate modes of ionization including: (1) direct ion evaporation, sometimes referred to as "thermospray ionization"; (2) chemical ionization (CI) initiated by an electron beam, sometimes called "filament on" operation; and (3) chemical ionization initiated by a low-current Townsend discharge, sometimes called "discharge ionization."

The mass spectrometer is also normally equipped with the capability to analyze and detect both positive and negative ions. Both the properties of the sample and those of the mobile phase must be considered in choosing which of the six possible operating modes is likely to be best for a particular analysis. In general, for positive ion detection, samples must be more basic than the mobile phase (to form MH^+) or be sufficiently polar to form stable adducts (e.g., $M + NH_4^+$). For negative ion detection samples must be more acidic than the mobile phase [to form $(M - H)^-$] or have a higher electron affinity (to form M^-). Use of either the filament or discharge is required to form M^-, since no free electrons are produced by the direct ionization mechanism.

Direct ion evaporation. This technique involves direct evaporation into the gas phase of ions present in solution. Since it does not require vaporization of the sample to produce a neutral vapor, it is applicable to totally nonvolatile samples. If the sample is ionized in water or methanol, this technique can be used for these cases without the addition of a buffer, but the sensitivity is generally poor and the response is nonlinear, varying approximately with the square root of sample concentration. Addition of a volatile buffer, such as ammonium acetate at a concentration in the 0.01–0.1 M range, generally enhances the sensitivity for these kinds of samples rather dramatically and yields a linear response over several orders of magnitude of sample concentration. In a few cases, e.g., polysulfonated azo dyes, addition of buffer decreases response. Sensitivity is best when the mobile phase is predominantly aqueous. This technique can be used with essentially 100% methanol but the sensitivity is often marginal if the water fraction is less than ~20%. If a buffer is used, the sample need not be ionized in solution since often the buffer ionization can be transferred to the sample, either in the vaporizing solution or in the gas phase, to efficiently ionize the sample.

Molecules containing both acidic and basic functional groups (such as peptides) can sometimes be detected efficiently as negative ions even though they may exist in the mobile phase as positively charged species.

This mode of ionization is generally favored in the following cases: (1) the sample is ionic, polar, nonvolatile and (2) the preferred mobile phase is water with either methanol or acetonitrile and a volatile buffer.

Filament on operation. The use of the filament to initiate CI is most effective when the mobile phase contains a large organic fraction. It can be used with pure aqueous phases but it may be difficult to maintain full emission and the filament lifetime may be shortened. The use of the filament is almost essential for normal-phase chromatography and it can significantly enhance the sensitivity for reversed-phase separations, particularly when the samples are at least slightly volatile. Use of the filament with truly nonvolatile samples is generally not recommended, since it does not enhance sensitivity, and may, in many cases, cause a decrease in sensitivity for nonvolatile samples of interest and an increase in the solvent-related chemical noise.

Discharge ionization. The discharge electrode provides an alternative technique for initiating chemical ionization. It is most useful when the mobile phase contains high water fractions. It can be used in the presence of organic mobile phases, but it is not recommended for extended use with organic fractions greater than about 60% because carbon deposits on the electrode build up and short out the discharge. Carbon deposits are indicated by erratic behavior and higher than normal discharge currents. This condition can sometimes be reversed by running the discharge for 15 to 30 min with pure water as the mobile phase.

Positive ion vs. negative ion detection. All three of these ionization techniques produce approximately equal amounts of positive and negative ions from most common solvents and for many relatively neutral analytes. The general rule is that basic compounds give higher sensitivity in positive ion mode and acidic compounds in negative ions, but unfortunately, the proton affinities of many molecules are unknown. This general principle is a useful guide in deciding which mobile phases may be most suitable for analyzing particular classes of samples in either ionization mode, but in many cases the most effective method will need to be determined empirically.

Operating Temperatures. For optimum performance of thermospray LC/MS two deceptively simple criteria must be met. For the best sensitivity, sufficient heat must be supplied to completely vaporize the sample, since an unvaporized sample cannot contribute ions to the mass spectrum. At the same time, the sample must not be heated so much as to cause pyrolysis or other uncontrolled chemical modification. Nonvolatile samples do not produce intact molecules in the gas phase by vaporization, but they may produce ions in the gas phase by direct ion evaporation. In such cases virtually all of the sample that is vaporized is ionized; thus, it is possible to obtain acceptable sensitivity even though the fraction of the sample vaporized may be relatively small.

Samples may be divided roughly into the following categories: (1) at least slightly volatile [can be analyzed by introduction on a direct chemical ionization (DCI) probe]; (2) slightly volatile, but very labile (e.g., glucuronides); (3) nonvolatile, ionic; and (4) nonvolatile, neutral. The first category encompasses a fairly large fraction of low-molecular weight compounds. Most of these can be analyzed satisfactorily with a single set of standard operating conditions. Typical operating conditions consist of setting the vaporizer control temperature about 2° to 5° below "take-off" (see Fig. 6) and the source block at ca. 250°–300°. Under these conditions the vapor temperature in the ion source may be 25°–50° below the block temperature. Optimum conditions may vary somewhat depending on details of the ion source design, but these kinds of compounds generally give excellent results over a wide range of operating conditions.

Labile compounds. The above conditions may also be suitable for labile compounds, but molecular ions may be weak. Molecular ions can often be increased (and fragment intensities decreased) by lowering source block temperature. Block temperatures as low as 200° with vapor temperatures of 150° are sometimes useful for modestly volatile but thermally labile compounds. For less volatile neutral compounds, sensitivity can often be improved by increasing source block temperature, but this may cause additional pyrolysis and fragmentation if the samples are thermally labile.

Nonvolatile, ionic compounds. The best performance for nonvolatile, ionic compounds is generally obtained under rather different conditions. To obtain intense molecular ions for these samples, it is often necessary to set the vaporizer control temperature 20°–50° below the "take-off" point, and drastically increase the block temperature to as much as 450°. Vapor temperature under these conditions may be in the 250°–300° range.

Nonvolatile, neutral compounds. The most difficult samples for thermospray are nonvolatile neutrals, particularly those that are not water soluble. In some cases, these can be ionized by attachment of an ammonium or sodium ion which presumably occurs prior to vaporization. Thermal conditions are generally similar to those for nonvolatile, ionic compounds when ion attachment can be made to work. Some success has been obtained using 0.1 *M* ammonium acetate in neat methanol as a mobile phase for this class of compounds.

Vaporizer tip temperature. The vaporizer tip temperature is an important variable to monitor since it closely correlates with thermospray performance, but it is not directly controllable since it depends on the vaporizer control temperature, the flow rate, the mobile phase composition, the diameter of the vaporizer nozzle, and, weakly, on the temperature of the tip heater. It is important to monitor the tip temperature since if all of the above are constant, then it also should be constant. For a given set of operating conditions changes in tip temperature normally indicate a change in the diameter of the vaporizer nozzle, although sudden changes can indicate a malfunction in the chromatographic system. It is not unusual for the tip temperature, for a given set of operating conditions, to change by a few percent over the course of a day of running. An increase generally means that the tip nozzle is reduced in area which may result from deposition of materials leached from the column packing. Variations in tip temperature of as much as 20% can generally be tolerated without substantially affecting performance. If the tip temperature increases more, it may be necessary to replace the vaporizer insert or to clean it using a procedure such as that described by Hsu and Edmonds.[12]

Present Status of Thermospray

Thermospray is now established as a practical technique for LC/MS interfacing. Its utility for a large number of applications is indicated in other sections of this volume, and in more detail in a recent monograph.[6] For small molecules such as amino acids, nucleosides, and many drugs and drug metabolites, sensitivities are sufficient to allow reliable detection at the low picogram level using selected-ion monitoring (SIM) and for obtaining useful full-scan spectra in the low nanogram range. For larger, more ionic compounds, such as peptides and nucleotides, sensitivities are not as high, and typically nanomole sample quantities are required for full-scan spectra. Perhaps the most favorable attribute of thermospray is its relative simplicity. It is now developed to the point that it can be used routinely for a variety of quantitative and qualitative applications.

Despite this obvious success, there remain a number of valid criticisms. Some of the deficiencies which have been noted include the following listed below:

1. Results very often allow unambiguous determination of molecular weight, but fragmentation is either absent, insufficient, or of insufficient reproducibility to allow definitive identification of known compounds or much in the way of structure elucidation of unknown compounds.

2. Sensitivity, particularly for large, nonvolatile compounds such as peptides, nucleotides, oligosaccharides, and complex lipids, is disappointingly low.

3. The technique is limited to a fairly narrow range of chromatographic conditions.

These deficiencies primarily result from the fact that the solvent vapor is introduced into the ion source along with the sample which limits the range of operating parameters and ionization techniques which can be employed.

---

[12] F. F. Hsu and C. G. Edmonds, *Vestec Thermospray Newsl.* **1**, 4 (1985).

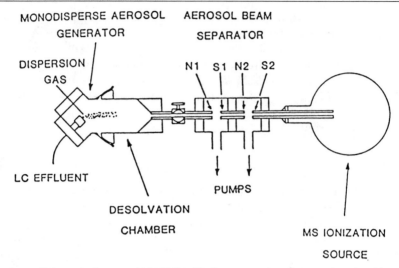

FIG. 7. Schematic diagram of MAGIC-LC/MS. N1, Nozzle 1; N2, nozzle 2; S1, skimmer 1; S2, skimmer 2. (From Ref. 14, with permission.)

Particle Beam Interfaces

The major disadvantage of thermospray and other direct coupling techniques is that the ionization occurs in a bath of the solvent vapor at a relatively high source pressure of typically 1 torr or more. This effectively precludes the use of electron ionization (EI) and also limits the choice of reagents in CI. Attempts to overcome this limitation have generally focused on various transport devices designed to allow removal of the solvent while transporting the sample to the ion source. The most successful of these have involved moving wires or belts,[13] but this approach has not achieved widespread acceptance.

An alternative transport system involving no moving parts was first described by Willoughby and Browner.[14] This approach was given the acronym MAGIC which stands for monodisperse aerosol generation interface for chromatography. A schematic diagram of the MAGIC interface is shown in Fig. 7. In this device the chromatographic effluent is forced under pressure through a small orifice (typically 5–10 μm in diameter) and

[13] R. P. W. Scott, C. G. Scott, M. Munroe, and J. Hess, Jr., *J. Chromatogr.* **99**, 395 (1974); W. H. McFadden, H. L. Schwartz, and D. C. Bradford, *J. Chromatogr.* **122**, 389 (1976).

[14] R. C. Willoughby and R. F. Browner, *Anal. Chem.* **56**, 2626; P. C. Winler, D. B. Perkins, W. K. Williams, and R. F. Browner, *Anal. Chem.* **60**, 489 (1988).

as a result of the Rayleigh instability the liquid jet breaks up into a stream of relatively uniform droplets whose initial diameter is approximately 1.9 times the nozzle diameter. A short distance downstream the stream of particles is intersected at 90° by a high-velocity gas stream (usually helium) to disperse the particles and prevent coagulation. The dispersed droplets fly at relatively high velocity through the desolvation chamber where vaporization occurs at atmospheric pressure and near ambient temperature. Heating is provided to the desolvation chamber not to raise the aerosol temperature above ambient but to replace the latent heat of vaporization necessary for solvent evaporation. Ideally all of the solvent is vaporized and the sample remains as a solid particle or less volatile liquid droplet.

The MAGIC interface is designed to handle 0.1 to 0.5 ml/min of liquid and up to 1 liter/min of dispersion gas is required. To reach the low ion source pressures required for EI, an efficient method for separating gas from the sample particles is required. As indicated in Fig. 7, a two-stage momentum separator is used to form a particle beam from the sample and pump away the vapor and dispersion gas. During the expansion in the first capillary nozzle the particles are accelerated to a velocity approaching that of the gas. The high-momentum particles tend to remain on the axis of the separator while the light molecules diffuse away. The first aerosol–beam separator chamber is pumped with a 300 liter/min mechanical pump which maintains the pressure at between 2 and 10 torr. The second chamber is pumped by a 150 liter/min pump which keeps the pressure in this chamber at between 0.1 and 1 torr. A commercial version of a MAGIC or particle beam interface has recently been presented by Hewlett-Packard.

Willoughby and co-workers from Extrel Corp. have described a modified version of a particle beam interface which employs a thermospray vaporizer as a nebulizer. While no details of this thermobeam apparatus have yet been published, it appears that this approach may have some significant advantages. In particular, it appears that the thermospray nebulizer produces smaller initial droplets at higher temperatures thus substantially facilitating desolvation. Furthermore, it does not require such small nebulizer orifice diameters thus providing more immunity to plugging.

Both versions of the particle beam interface have been demonstrated to give EI mass spectra in good agreement with library spectra using sample injections of 100 ng or more. Generally, these spectra do not include the low mass region where serious solvent interference may be expected. These particle beam interfaces show considerable promise for future LC/MS applications, particularly for samples with at least modest volatility. However, these interfaces do not meet the criteria established for inclusion in this chapter in that they have difficulty handling the higher flow rates of aqueous media encountered with reversed-phase chromatography using standard 4.6-mm columns. The system described below was developed in an attempt to overcome the limitations of these earlier EI interfaces.

Universal Interface

The new "universal interface" was developed to allow HPLC using standard 4.6-mm columns to be coupled to conventional EI and CI mass spectrometry.[15] A block diagram of the new interface (Vestec Corp., Houston, TX) is shown in Fig. 8. The chromatographic effluent is directly coupled to a thermospray vaporizer in which most, but not all, of the solvent is vaporized and the remaining unvaporized material is carried along as an aerosol in the high-velocity vapor jet that is produced. The thermospray jet is introduced into a spray chamber which is heated sufficiently to complete the vaporization process. Helium is added through a gas inlet in sufficient quantity to maintain the desired pressure and flow rate. The fraction of the solvent vaporized in the thermospray vaporizer and the temperature of the desolvation chamber is adjusted so that essentially all of the solvent is vaporized within the desolvation region.

After exiting the spray chamber, the aerosol, consisting of unvaporized sample particles, solvent vapor, and inert carrier gas, passes through a condenser and a countercurrent membrane separator where most of the solvent vapor is removed. The resulting dry aerosol is then transmitted to the mass spectrometer using a momentum separator to increase the concentration of particles relative to that of the solvent vapor and carrier gas. The coupling between the gas diffusion cell and the momentum separator employs a length of Teflon tubing, typically 4 mm id and as much as several meters long. The length of this connection is noncritical since the dry aerosol is transmitted with no detectable loss in sample and a negligible loss in chromatographic fidelity.

The apparatus in its simplest form is shown schematically in Fig. 9. The thermospray vaporizer is installed in the heated desolvation chamber

[15] M. L. Vestal, D. H. Winn, C. H. Vestal, and M. L. Vestal, in "LC–MS: New Developments and Application to Pesticide, Pharmaceutical and Environmental Analysis" (M. A. Brown, ed.), ACS Symp. Ser., pp. 215–231. American Chemical Society, Washington, D.C., 1990.

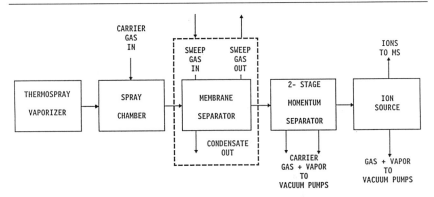

FIG. 8. Block diagram of new "universal interface" between HPLC and EI mass spectrometry.

through a gas-tight fitting. The helium is introduced through a second fitting and flows around the thermospray vaporizer and entrains the droplets produced in the thermospray jet. In general, the carrier gas flow required is at least equal to the vapor flow produced by complete vaporization of the liquid input. The heated zone within the spray chamber should be sufficiently long to allow the particles to approach thermal equilibrium with the vapor phase. The minimum temperature for complete vaporization of the solvent is that at which the vapor pressure of the solvent is just greater than the partial pressure of the completely vaporized solvent at the particular flow of liquid employed. Ideally, the effluent from the desolvation region consists of dry particles of unvaporized sample, solvent vapor at a partial pressure somewhat less than one-half of the total pressure, and the balance is the carrier gas.

Flow velocities through the system are not critical, but must be high enough to efficiently carry the aerosol, but not so high as to cause extensive turbulence. Under the correct flow conditions in which essentially laminar flow is maintained, the aerosol is carried preferentially by the higher velocity gas stream near the center of the tube, and the aerosol particles can be transported for large distances with negligible losses. Because of the very large mass of these particles relative to the gas molecules, the diffusion coefficients for the particles is sufficiently small that diffusion of aerosol to the walls is very slow, and the parabolic velocity profile provides an aerodynamic restoring force which continually pushes particles toward the center of the tube where the gas velocity is highest. On the other hand, diffusion of solvent molecules in the carrier gas is relatively rapid.

As effluent passes from the heated zone of the spray chamber to the

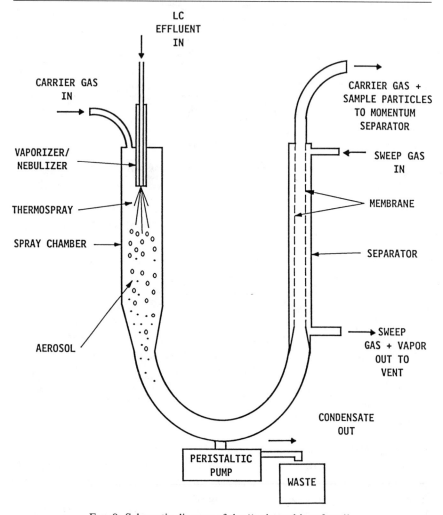

FIG. 9. Schematic diagram of the "universal interface."

unheated zone of the condenser the vapor may become supersaturated and begin to condense on the walls. The condenser and transition region are arranged as shown in Fig. 9 so that the liquid condensate flows under the effect of gravity to the drain where it is pumped away to waste. A small positive displacement pump, such as a peristaltic tubing pump, is used to pump away the liquid without allowing a significant amount of gas or vapor to escape.

Membrane Separator

The unique part of the universal interface is the membrane separator or gas diffusion cell which allows the solvent vapor to be efficiently removed with essentially no loss of sample contained in the aerosol particles. In this device the aerosol is transported through a central channel bounded on the sides by a gas diffusion membrane or filter medium which is in contact with a countercurrent flow of a sweep gas. For EI mass spectrometry, helium is used for both the carrier and sweep gas. The membrane must be sufficiently permeable to allow free diffusion of carrier gas and solvent vapor while dividing the macroscopic carrier flow from the oppositely directed sweep flow. In this cell the concentration of vapor in the central channel decreases exponentially, with the value of the exponent depending on the geometry of the cell, the effective diffusion velocity, and the relative velocities of the carrier and sweep flows. A consequence of this exponential dependence is that the solvent vapor transmission can be reduced as low as required for any application merely by increasing the length of the cell or by increasing the sweep flow. If a certain solvent vapor removal is achieved under one set of conditions, the square of this value can be achieved by either doubling the length of the cell, or by doubling the sweep gas flow.

Momentum Separator

The two-stage momentum separator used in this interface is conceptually similar to those used in other MAGIC or particle beam interfaces. However, since most of the solvent vapor is removed in the gas diffusion cell, this separator is required primarily to remove sufficient helium to allow the standard mass spectrometric pumping system to achieve the good vacuum required for EI operation. The performance of this device can be optimized much more readily for separating helium from macroscopic particles than when copious quantities of condensible vapors are present as in the more conventional particle beam systems.

Ion Source

The sample is transmitted to the ion source as a very narrow and well-defined, high-velocity beam of macroscopic particles of solid sample. Any standard electron ionization ion source can be used so long as it has an opening into the source at least 2 or 3 mm in diameter, which can be aligned with the particle beam, and the opposite side of the source must be closed to provide a heated surface for the particle beam to strike.

Recently, this interface has also been used to successfully couple liquid chromatography to CI and FAB ion sources, and with further work it appears that virtually any type of liquid chromatograph can be coupled to any ionization technique using this type of interface.

Vacuum System

The momentum separator normally requires two mechanical vacuum pumps with capacities in the 200–300 liter/min range. For most research-grade mass spectrometers no other changes in the vacuum system are required since the total gas load transmitted along with the particle beam is roughly equivalent to that involved in capillary gas chromatography using the larger bore columns. The gas load may be somewhat more than can be tolerated by some of the benchtop GC/MS instruments. If necessary a third stage can be added to the momentum separator to further reduce the gas transmission. If the correct orifice sizes are chosen in the separator, this reduction can be accomplished with essentially no additional sample loss.

Mass Spectrometer

Since the momentum separator is not directly connected to the ion source, this interface can be used with equal facility on both quadrupole and magnetic mass spectrometers.

Control System

When the thermospray vaporizer is used as a nebulizer, the basic control system is the same as that described above for directly coupled thermospray; however, the system performs best with a much smaller fraction of the liquid vaporized within the capillary and control is much less critical. The spray chamber is also heated and temperature controlled to complete the vaporization, and for most applications all that is required is to provide enough heat to vaporize the least volatile solvent, usually water. The only other controls required are the flow rates of the carrier and sweep gases. The carrier gas flow required is set by the conductance of the nozzle in the momentum separator. With the nozzles generally used, a flow of 1.5 to 2.5 liter/min is required to maintain the carrier channel at atmospheric pressure. The sweep gas flow is a free parameter which determines the degree of solvent vapor removal that is achieved. In recent work,[15] the solvent transmission was reduced to about 0.3 ppm using a sweep flow of 10 liter/min and to about 5 ppm for a sweep flow of 5 liter/min. The latter is more than adequate for most LC/MS applications, but these results show the very high degree of solvent removal that is possible at the expense of increased gas consumption.

Optimizing Universal Interface for Particular Applications

One of the strengths of this approach is that very little optimization is required. Sample transmission through the interface is very nearly independent of both solvent flow rate and composition.[15] Also, since nearly complete removal of solvent vapor is possible, the liquid chromatograph is quite efficiently decoupled from the mass spectrometer ion source. Thus, in contrast to directly coupled thermospray, the choice of ionization technique can be made on the basis of the information required from the sample analysis; it is not dictated by the chromatographic technique or choice of mobile phase.

Liquid Chromatographic Conditions. Essentially all liquid chromatographic procedures can be used except those requiring nonvolatile buffers. The interface has no difficulty with nonvolatile materials; it very efficiently transmits the nonvolatile materials to the ion source of the mass spectrometer. For some types of specialized mass spectrometer ion sources such as inductively coupled plasma (ICP)/MS or FAB this is not a problem, but for EI mass spectrometry it is disastrous. Contamination by phosphates and alkali salts is much more serious than in the case of the thermospray source, but switching to volatile buffers overcomes this limitation in both cases.

Ionization Techniques. Essentially any ionization technique can be used with this interface. Most of the work to date has focused on EI mass spectrometry, but results have recently been obtained for CI and FAB. The major appeal of EI is that it provides an extensive and reproducible fragmentation pattern which can be used for unambiguous compound identification by comparison with the extensive EI data bases that exist. Electron ionization in conjunction with liquid chromatographic separation is still quite new, and fewer than 1% of the compounds in the data base have been studied using the universal interface, but for those compounds for which spectra have been obtained the agreement with the library is excellent. Some minor difficulties might be expected for nonvolatile and labile compounds because of differences in the way the sample is presented to the ion source. In GC/MS and conventional solids probe introduction the sample is vaporized before it enters the ion source, while the universal interface injects a high-velocity beam of macroscopic sample particles into the ion source. These particles are presumably vaporized by impact on the heated surfaces in the ion source and the vapor ionized in the usual way. For relatively nonvolatile samples it is generally necessary to operate the ion source somewhat hotter than usual (ca. 300°–350°) and it would not be surprising if some differences in fragmentation patterns might result from such high temperatures either on the ion source or on the direct insertion probe.

Critical Adjustments. As discussed above there are few critical adjustments on this interface. Once the set of temperatures and gas flow rates have been established, there is little need for reoptimization. One exception applies to samples of relatively high volatility. The high degree of solvent vapor removal implies that any portion of the sample which is vaporized will also be removed. Since the samples are transmitted as fine particles, quite modest vapor pressure can lead to significant sample loss. Generally, samples with vapor pressures less than 10^{-7} torr at room temperature are transmitted efficiently independent of operating conditions employed. The transmission of more volatile samples depends on the temperature of the vaporizer and spray chamber, and nonlinear response may be observed in these cases since the small particles produced at lower sample concentrations vaporize more rapidly than larger particles. Samples with vapor pressures less than 10^{-4} torr at room temperature are unsuitable for use with this technique since they tend to be completely vaporized and removed along with the solvent vapor.

Conclusion

Particle–beam interfaces between liquid chromatography and mass spectrometry are not yet fully established for applications to nonvolatile biological samples, but with further development may provide the ultimate solution to this challenging and important problem. The combination of the gas diffusion cell for removing solvent vapor with the momentum separator for removing both carrier gas and solvent vapor allows the sample to be almost completely separated from the mobile phase and efficiently transported to the mass spectrometer. This allows, at least in principle, all kinds of chromatography to be coupled to all types of mass analyzers using the ionization technique of choice for a particular application. Considerable work remains before this "Utopia" approaches reality; nevertheless, it now appears feasible that a truly "universal" solution to the liquid chromatography detector problem is on the horizon. On the other hand, direct coupling of liquid chromatography with mass spectrometry via the thermospray technique is now a practical, routine technique with many established applications, although limitations are imposed by the presence of the solvent vapor along with the sample in the ion source of the mass spectrometer.

[6] Tandem Mass Spectrometry: Multisector Magnetic Instruments

By MICHAEL L. GROSS

Introduction

Tandem magnetic sector mass spectrometry was originally developed for investigations of ion chemistry and ion–molecule reaction dynamics. In the mid 1970s, its application to chemical analysis was realized particularly for complex mixtures.[1] However, as tandem mass spectrometry evolved, it became clear that its major application was in determining the structure of biomolecules and that truly complex mixture analysis was better accomplished by incorporating some kind of chromatography in the analytical protocol. Its utility for mixture analysis is not to be negated, since many isolates of biomolecules, even after chromatographic separation, are mixtures. Thus, the structure determination or identification of the analyte requires both separation and analysis, both of which can be accomplished with this method.

Ionization of biomolecules usually requires desorption methods [e.g., fast atom bombardment (FAB), liquid secondary ion (LSIMS)]. These methods employ a liquid matrix from which desorption occurs, with the consequent formation of ions from the matrix. Tandem mass spectrometry can be employed to separate these matrix ions from those of the analyte.

Further limitation of desorption ionization, especially when the analyte has a large molecular weight (i.e., >1000), is that the desorption is accompanied by minimal fragmentation. Tandem mass spectrometry incorporates some method of activation (usually collisional activation[2]) that causes much more extensive fragmentation than that which occurs during desorption alone. The limitations of matrix and other component interference and limited fragmentation serve as the major motivations for using tandem mass spectrometry in bioanalytical chemistry. Most of the experiments involve product (daughter) ion scans, although constant neutral loss scans and precursor (parent) ion scans are possible and should not be overlooked in the early stages of the analysis when the ion signals of interest are unknown. The capability for product ion scans is emphasized in this

---

[1] R. W. Kondrat and R. G. Cooks, *Anal. Chem.* **50**, 81A (1978); see also *ibid.*, *in* "Tandem Mass Spectrometry" (F. W. McLafferty, ed.). Wiley, New York, 1983.

[2] R. N. Hayes and M. L. Gross, this volume [10].

METHODS IN ENZYMOLOGY, VOL. 193

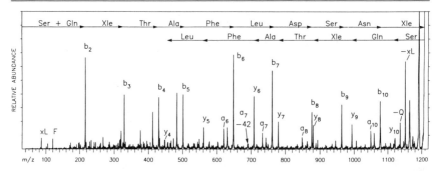

FIG. 1. Tandem CAD mass spectrum of a tryptic peptide $(M + H)^+$ of m/z 1208 from thioredoxin. (Reprinted from Ref. 6 with permission.)

chapter because the vast majority of applications in structure determination makes use of product ion spectra.

Tandem mass spectrometry has gained with acceptance in determining peptide sequences, especially because of the high structural diversity of peptides, their importance in biochemistry, and the limitations of other more classical methods (e.g., Edman degradation). These subjects are discussed elsewhere in this volume.[3-5] One advantage of using tandem mass spectrometry is illustrated by the collisionally activated decomposition (CAD, also termed collision-induced dissociation or CID) spectrum of a tryptic peptide from the protein thioredoxin from *Chromatium vinosum* (Fig. 1).[6] The spectrum was obtained by Johnson and Biemann[6] using a four-sector tandem mass spectrometer. The reader should note that the entire sequence is revealed by using a series of ions that contain the N terminus (*b*-ions; see Appendix 5 at end of volume). Moreover, the sequence of seven of twelve amino acids is confirmed by reading a series of ions that contain the C terminus (*y*-ions).

Other applications of tandem mass spectrometry (MS/MS) are less well-developed. Its use in sequencing carbohydrates[7] is an important application because there are few other structural methods. Fragmentation

[3] K. Biemann, this volume [25].
[4] B. W. Gibson and P. Cohen, this volume [26].
[5] S. A. Carr, J. R. Barr, G. D. Roberts, K. R. Anumula, and P. B. Taylor, this volume [27].
[6] R. S. Johnson and K. Biemann, *Biochemistry* **26,** 1209 (1987).
[7] L. Poulter, J. P. Earnest, R. M. Stroud, and A. L. Burlingame, *Proc. Natl. Acad. Sci. U.S.A.* **86,** 6645 (1989).

of FAB-desorbed nucleosides,[8] oligonucleotides,[9,10] modified nucleosides and nucleotides,[11] and fatty acids/complex lipids[12-15] have been investigated. Two examples of structure determination by FAB and MS/MS can be found in recent literature.[16,17] Its application to glycolipids is discussed elsewhere in this volume.[18]

Two- and Three-Sector Instruments

Two-Sector Instruments

Double-focusing mass spectrometers were developed principally as instruments to make accurate mass measurements.[19] As these two-sector instruments were used more extensively, it became apparent that they could be scanned in special ways so that metastable ion decompositions could be observed without interference of source-produced precursor and product ions. The development of these special scan methods[20] was largely focused on obtaining a spectrum of product ions from a given precursor ion so that fragmentation mechanisms could be understood. A common double-focusing instrument of the 1960s and 1970s was the so-called forward-geometry, Nier–Johnson mass spectrometer (often denoted as an *EB* instrument because the electric sector *E* precedes the magnetic sector *B*). The most expeditious means of obtaining a "pure" product ion mass spectrum turned out to be, after considerable research, the *B/E*-linked scan in which both *B* and *E* were scanned together such that the ratio of the magnet field strength and the electric field strength of the electrostatic

[8] K. B. Tomer, F. W. Crow, M. L. Gross, J. A. McCloskey, and D. E. Bergstrom, *Anal. Biochem.* **139**, 243 (1984).

[9] R. L.Cerny, M. L. Gross, and L. Grotjahn, *Anal. Biochem.* **156**, 424 (1986).

[10] R. L. Cerny, K. B. Tomer, M. L. Gross, and L. Grotjahn, *Anal. Biochem.* **165**, 175 (1987).

[11] K. B. Tomer, M. L. Gross, and M. L. Deinzer, *Anal. Chem.* **58**, 2527 (1986).

[12] N. J. Jensen and M. L. Gross, *Mass Spectrom. Rev.* **6**, 497 (1987).

[13] N. J. Jensen, K. B. Tomer, and M. L. Gross, *Lipids* **21**, 580 (1986).

[14] N. J. Jensen, K. B. Tomer, and M. L. Gross, *Lipids* **22**, 480 (1987).

[15] B. Domon and C. E. Costello, *Biochemistry* **27**, 1534 (1988).

[16] E. G. Rogan, E. L. Cavalieri, S. R. Tibbels, P. Cremonesi, C. D. Warner, D. L. Nagel, K. B. Tomer, R. L. Cerny, and M. L. Gross, *J. Am. Chem. Soc.* **110**, 4023 (1988).

[17] M. Ubukata, K. Isono, K. Kimura, C. C. Nelson, and J. A. McCloskey, *J. Am. Chem. Soc.* **110**, 4416 (1988).

[18] C. E. Costello and J. E. Vath, this volume [40].

[19] K. Biemann, this volume [13].

[20] K. R. Jennings and G. G. Dolnikowski, this volume [2].

analyzer (ESA) were held constant.[21-23] Linked B/E scans give reasonable product ion resolution, but poor precursor ion selection and are sometimes subject to artifacts. The linked scan certainly provides MS/MS information, but forward-geometry double-focusing instruments are usually *not* considered tandem instruments because the first sector cannot serve as a mass analyzer in this instrument configuration.

The first "true" tandem mass spectrometer for chemical analysis was introduced by Beynon and Cooks[24,25] in 1971. The instrument consisted of a magnetic sector followed by an ESA (i.e., a BE configuration). Although the original spectrometer design was not double focusing, it did allow a precursor ion to be selected by setting the magnetic sector and the product ion spectrum to be acquired by scanning the ESA field. In early applications, only spontaneous (metastable ion) decompositions were observed. Later, the addition of a gas cell allowed the acquisition of CAD spectra. The advantage of the design was that the precursor ion of m/z <1000 would be selected with unit mass resolution, but the product ion spectrum would be at low resolution (~100) because the inevitable kinetic energy release accompanying fragmentations would broaden product ion peaks in a scan of the field of the electric sector. Wachs, Bente, and McLafferty[26] also built a reverse-geometry instrument by modifying a commercial Hitachi Perkin-Elmer RMU-7 EB mass spectrometer.

Later, a reverse-geometry tandem mass spectrometer called the ZAB-2F was built by VG Organic Ltd. (now VG Analytical) at the urging of John Beynon.[27] This instrument is not only a tandem but is also a double-focusing mass spectrometer.

Three-Sector Instruments

In an effort to improve on the reverse-geometry instrument, two different three-sector instruments were designed and built. The first, an EBE

[21] R. K. Boyd and J. H. Beynon, *Org. Mass Spectrom.* **12**, 1643 (1977).

[22] D. S. Millington and J. A. Smith, *Org. Mass Spectrom.* **12**, 264 (1977).

[23] A. P. Bruins, K. R. Jennings, and S. Evans, *Int. J. Mass Spectrom. Ion Phys.* **26**, 395 (1978).

[24] J. H. Beynon and R. G. Cooks, *Res. Dev.* **22**, 26 (1971).

[25] J. H. Beynon, R. G. Cooks, J. W. Amy, W. E. Baitinger, and T. Y. Ridley, *Anal. Chem.* **45**, 1023A (1973).

[26] T. Wachs, P. F. Bente III, and F. W. McLafferty, *Int. J. Mass Spectrom. Ion Phys.* **9**, 333 (1972).

[27] R. P. Morgan, J. H. Beynon, R. H. Bateman, and B. N. Green, *Int. J. Mass Spectrom. Ion Phys.* **28**, 171 (1978).

design, was built by Kratos for the author of this chapter.[28] This instrument permitted high-resolution selection of the precursor ion; a resolution of 100,000 was demonstrated in the original report.[28] Although ultrahigh resolution is usually not needed for investigations of biomolecules, this capability is important in studies of ion chemistry and in investigations of energy materials (complex mixtures of hydrocarbons). The mass resolving power of the second stage (an ESA only) is still limited. The *EBE* instrument has been used for a decade to solve a number of ion chemistry, analytical, and bioanalytical problems in the author's laboratory despite the low resolving power of the second stage.[29]

Another three-sector instrument was built by VG Analytical which utilized a reverse-geometry (*BE*) Nier–Johnson instrument as the first stage and a magnetic sector as the second stage, with an overall configuration of *BEB*. By using *BE* as the first stage, high resolution can be employed in the precursor ion selection. The resolution of the second stage, now a magnetic sector, is improved with respect to an ESA, but only by a factor of two or three. An advantage of this design is that B_1 can be used for mass selection and EB_2 (instead of B_2 only) for double-focusing product analysis.

Tandem Nier–Johnson Mass Spectrometers (Four-Sector Instruments)

The immediate *raison d'être* for a four-sector instrument emerges from experiences with two- and three-sector instruments. The means of improving the mass resolving power of the second stage while still retaining high resolving power in the first stage is to use a tandem double-focusing or four-sector instrument. An additional motivation is linked to retaining the advantages of high-energy compared to low-energy collisional activation.[2,30]

A variety of four-sector instruments have been built over the past decade. The basic instrument types, their designs, and capabilities will be the emphasis of the remainder of this chapter. Many mass spectrometrists consider four-sector tandem instruments as the cutting edge in tandem mass spectrometry; two and three sectors were necessary experiments in the four-sector evolution.

[28] M. L. Gross, E. K. Chess, P. A. Lyon, F. W. Crow, S. Evans, and H. Tudge, *Int. J. Mass Spectrom. Ion Phys.* **42**, 243 (1982).
[29] M. L. Gross, *Mass Spectrom. Rev.* **8**, 165 (1989).
[30] R. A. Yost and R. K. Boyd, this volume [7].

Cornell Four-Sector Instrument

The first four-sector instrument for analytical applications was built by McLafferty and co-workers[31] in the late 1970s and was first described in 1980. Futrell and Miller[32] had assembled a four-sector instrument in the mid-1960s, but this instrument was designed for ion chemistry and ion physics and not for analytical chemistry. The first stage of the McLafferty instrument is a Nier–Johnson MS-1, which is a modified Hitachi RMH-2 mass spectrometer. An inhomogeneous magnetic field for MS-1 was incorporated in the design to give a mass range of 10,000 u (atomic mass units) at 10 kV of acceleration although the instrument has not been used for high mass studies. The second-stage mass spectrometer was custom-built and also employed an inhomogeneous magnetic field for high-mass applications. This instrument was among the first—if not the first—in analytical mass spectrometry to make use of inhomogeneous magnetic fields. All high-mass Kratos mass spectrometers now take advantage of this technology.

Other features of the design include a very high-energy collisional activation (at 30 keV) which was achieved by crossing the ion beam with a uniquely designed helium molecular beam. Although this system has not been incorporated into commercial designs, it did set a benchmark for high-efficiency collisional activation that had not been achieved in the custom-built instruments. Most contemporary commercial designs incorporate improved pumping and ion beam guiding so that efficiencies similar to that reported by McLafferty et al.[31] are now achieved with simple collision cells.

The Cornell instrument was intended to be used for both fundamental ion chemistry studies and analysis of large molecules. The former application has dominated its use because of unexpected problems with the mass resolving power of MS-2 and transmission of high-mass ions. In fact, the instrument was recently modified for fundamental studies involving neutralization/reionization.[33] Nevertheless, this instrument, along with the reverse-geometry Nier–Johnson and the three-sector instruments, set the stage for the development of analytical four-sector tandem mass spectrometers. The appearance of the first commercial four-sector instrument followed quickly on the heels of the Cornell instrument.

[31] F. W. McLafferty, P. J. Todd, D. C. McGilvery, and M. A. Baldwin, *J. Am. Chem. Soc.* **102,** 3360 (1980).

[32] J. H. Futrell and C. D. Miller, *Rev. Sci. Instrum.* **37,** 1521 (1966).

[33] R. Feng, C. Wesdemiotis, M. A. Baldwin, and F. W. McLafferty, *Int. J. Mass Spectrom. Ion Proc.* **86,** 95 (1988).

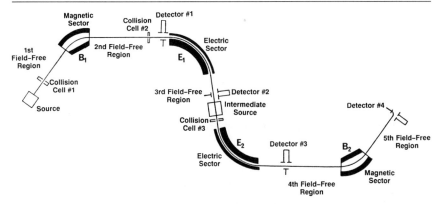

FIG. 2. Diagram of the VG Analytical ZAB-4F tandem double-focusing mass spectrometer. The original instrument did not have an intermediate ion source for calibrating MS-2. (Adapted from Ref. 37 with permission.)

VG ZAB-4F Mass Spectrometer

The first commercial four-sector mass spectrometer (VG ZAB-4F) was built for Dr. J. R. Hass of the National Institute of Environmental Health Science (NIEHS) by VG Analytical of Manchester, U.K.[34] The instrument was installed in 1984. Its design, shown in Fig. 2, consists of linking together two VG ZAB double-focusing mass spectrometers of reverse-geometry, Nier–Johnson design. Collision cells are located in the first, second, and third field-free regions. This design allows one not only to use the first stage as a stand-alone two-sector tandem instrument for various linked scanning experiments, but also to utilize the entire system for higher order MS/MS experiments, as was originally demonstrated with a three-sector instrument.[35]

To illustrate higher-order MS/MS, consider using B_1 to separate a precursor ion, m_1^+, for activation in collision cell No. 2. A particular product ion (e.g., m_2^+), is then separated by E_2 and, in turn, activated in collision cell No. 3. Second-generation product ions (m_3^+, m_4^+, . . .) are then mass analyzed by making use of the second-stage E_2B_2 mass analyzer. This experiment is often called MS/MS/MS or MS^3 and is used when one wishes to examine the spectrum of a product ion produced in one specific reaction, an important capability in ion chemistry investigations. Another application occurs when m_1^+ gives only one facile and dominant, but not

[34] J. R. Hass, B. N. Green, P. A. Bott, and R. H. Bateman, *Proc. 32nd ASMS Conf. Mass Spectrom. Allied Topics, San Antonio, TX,* p. 380 (1984).

[35] D. J. Burinsky, R. G. Cooks, E. K. Chess, and M. L. Gross, *Anal. Chem.* **54,** 295 (1982).

particularly illuminating, fragmentation to give $m_2{}^+$ (e.g., loss of water). A second activation step may then provide several, more illuminating fragmentation reactions. Tomer and co-workers[36] have demonstrated MS[5] experiments with this particular instrument. Although their demonstration is hardly routine and not broadly applicable because of loss of sensitivity at each stage of MS/MS, it does provide a challenging benchmark for MS^n. Moreover, its success points to more routine utility of lower order MS^n (e.g., MS^3), an experiment that is currently underutilized on tandem sector instruments.

An early and interesting application of tandem mass spectrometry with this four-sector was the direct analysis of stereoisomeric steroid conjugates by Cole, Guenat, and Gaskell,[37] who used both MS/MS and MS^3 experiments. The CAD spectra were found to vary as a function of concentration of the analyte. The explanation for the variation is either different sites for protonation and, as a result, different ion structures at different concentrations. Another possibility is an internal energy effect causing changes in low-energy reaction channels. In any event, one should be wary when using tandem mass spectrometry for distinguishing subtle structural features even if the activation is by high-energy collisions. Control experiments that focus on effects of concentration may be necessary.

Deterding et al.[38] are using a novel form of continuous-flow FAB and capillary high-performance liquid chromatography (HPLC) with the ZAB-4F four sector and are obtaining tandem mass spectra on low picogram quantities of model peptides. An ultimate limitation of tandem mass spectrometry, when coupled with desorption ionization, is the reduced detection limit because of interferences from matrix ions (the so-called "peak-at-every-mass" phenomenon). The interferences become particularly pronounced as the concentration of the analyte is reduced such that detection of the analyte becomes impossible in the overwhelming milieu of matrix ions. This problem obviously limits desorption ionization coupled with MS and MS/MS. One solution is to use simple matrices (e.g., water) that do not produce a peak-at-every-mass. Because these simple solvents are volatile, they must be introduced by flow methods (i.e., continuous-flow FAB[39]). It is expected that more applications of continuous-flow FAB will be seen with four-sector tandem mass spectrometers as the demands for lower and lower detection limits are pressed.

[36] K. B. Tomer, C. R. Guenat, and L. J. Deterding, Anal. Chem. 60, 2232 (1988).
[37] R. B. Cole, C. R. Guenat, and S. J. Gaskell, Anal. Chem. 59, 1139 (1987).
[38] L. J. Deterding, A. Moseley, K. B. Tomer, and J. W. Jorgenson, Anal. Chem. 61, 2504 (1989).
[39] R. M. Caprioli and W. T. Moore, this volume [9].

VG ZAB SE-4F Tandem Mass Spectrometer

A second four-sector instrument was built by VG Analytical for Dr. S. Carr of Smith Kline & French (SKF). Similar in design to that of the ZAB-4F (see Fig. 2), it is of the B_1E_1/E_2B_2 design and is designated as a VG ZAB SE-4F tandem double-focusing mass spectrometer.[40,41] The major developments accompanying this instrument are the use of high-mass, nonnormal entry magnetic sectors and the incorporation of an intermediate ion source for calibration of MS-2 (to be described in the next section). The magnets increase the mass range from 3000 u at 8 kV of acceleration (NIEHS) to 12,000 u at 10 kV (15,000 u at 8 kV). Although it is not possible to activate and fragment singly charged biomolecule ions of $m/z > 3000$, many mass spectrometrists need both high mass and MS/MS capabilities on a single instrument. Moreover, the upper mass limit for collisional activation had to be established experimentally, and the ZAB SE-4F and HX110/HX110 (discussed next) instruments played the key roles in determining that mass limit. This instrument is among the first of the second-generation tandem four-sector mass spectrometers.

Instruments of this generation are often operated with an electrically floated collision cell. The magnitude of the voltage applied to the cell is usually less than 50% of the accelerating voltage when product ion spectra are scanned across a single detector and may be 90% when array detectors are used (as will be discussed later). Thus, the ion beam is decelerated as it approaches the cell; however, the extent of deceleration is still sufficient to ensure that the activation is at high energy. A major advantage is that the lowest energy of the fragment ions is now bounded; that is, if the accelerating voltage is 8 kV and the cell is at 4 kV, then the minimum energy for an infinitesimally small mass product ion is 4 kV. If the cell is at ground potential, the minimum energy is 0 kV at the low mass limit. Because most detectors respond to ion kinetic energy, low mass ions are detected with greater efficiency when the collision cell is floated. The reader should be alert to the experimental condition of the collision cell (both potential and collision gas) when comparing spectra of the same compound taken at different times or with different instruments.[2]

The ZAB SE-4F tandem instrument has been used extensively for peptide and carbohydrate structural analysis in a pharmaceutical chemistry research. The interested reader is directed to Refs. 40 and 41 for more details.

[40] S. A. Carr, G. D. Roberts, and M. E. Hemling, in "Mass Spectrometry of Biological Materials" (C. N. McEwen and B. S. Larsen, eds.), p. 89. Dekker, New York, 1990.

[41] W. E. DeWolf, Jr., S. A. Carr, A. Varrichio, P. J. Goodhart, M. A. Mentzer, G. D. Roberts, C. Southan, R. E. Dolle, and L. I. Kruse, Biochemistry 27, 9093 (1988).

Before considering the MIT instrument, it should be noted that VG Analytical now produces a medium-mass range tandem mass spectrometer (the so-called VG 70-SE4F),[42] which is based on the recognition that MS/MS is not tenable today for molecule ions with masses above 3000 u (there are a few exceptions to this rule). Moreover, the two ZAB four-sector instruments are S configuration tail-to-tail designs, an arrangement that was determined to be less than optimum.[43] Thus, this instrument, which has, in theory, an upper mass limit of 3000 at 8 kV of acceleration, is of a E_1B_1/E_2B_2 C configuration. A practical advantage is that the size of this four sector is comparable to that of a stand-alone two-sector high-performance instrument, and thus the instrument is quite manageable.

JEOL HX110/HX110 Four-Sector Mass Spectrometer

The first of the second-generation analytical four-sector tandem mass spectrometers was installed at MIT in late 1985 in the laboratory of Professor K. Biemann. This instrument (see Fig. 3) is much better characterized than the ZAB-4F and ZAB SE-4F instruments,[44] in part, because instrument development and evaluation is compatible with the academic nature of the laboratory.

The instrument consists of two JEOL HX110 mass spectrometers. The ion optics of the HX110/HX110 were developed by Matsuda *et al.*[45] The major development is the inclusion of quadrupole lenses before and after the electric sector (Q_1 lens and Q_2 lens in Fig. 3) to limit ion beam dispersion in the z axis (perpendicular to the plane of the instrument) so that a narrow gap magnet can be incorporated. Such magnets have enhanced field strengths giving, in this case, an upper mass limit of 14,500 u at 10 kV acceleration potential. Moreover, the transmission is improved because ion loss due to z dispersion is minimized. In contrast to the ZAB-4F and SKF designs, this spectrometer is of E_1B_1/E_2B_2 design, and the overall layout is a more manageable C, rather than S, configuration.

Both the HX110/HX110 and the ZAB SE instruments can be used as three-sector instruments; that is, the product ion spectrum is obtained by scanning E_2, an advantage if one wishes to study peak shapes, energy shifts, and kinetic energy release. This is an important capability if dynamics of ion fragmentation are to be investigated.

The HX110/HX110 tandem spectrometer contains an intermediate ion

[42] R. M. Milberg, this volume [14].
[43] T. Matsuo and H. Matsuda, *Int. J. Mass Spectrom. Ion Proc.* **91**, 27 (1989).
[44] K. Sato, T. Asada, M. Ishihara, F. Kunihire, Y. Kammei, E. Kubota, C. E. Costello, S. A. Martin, H. A. Scoble, and K. Biemann, *Anal. Chem.* **59**, 1652 (1987).
[45] H. Matsuda, T. Matsuo, Y. Fujita, and H. Wollnik, *Mass Spectrom. (Jpn.)* **24**, 9 (1976).

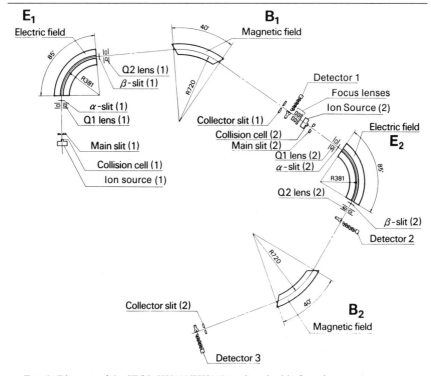

FIG. 3. Diagram of the JEOL HX110/HX110 tandem double-focusing mass spectrometer. (Reprinted from Ref. 44 with permission.)

source that is used to provide "standard" mass ions for calibrating the scan of MS-2.[44] The calibration procedure is considerably simpler and more precise and accurate than the original method used on the ZAB-4F instrument.[46] A calibrant is introduced into the intermediate source (usually a mixture of LiI, NaI, KI, RbI, and CsI so that numerous reference ions are produced), and a mass spectrum is taken by scanning B_2 while keeping E_2 constant. The resulting data set is used to calculate the B_2/E_2-linked scan function for any mass precursor ion that is selected by using MS-1. It is well known that a scan of an electric field can be done with high accuracy whereas magnetic fields are more difficult to control because of hysteresis and nonlinearities in the scan. For the HX110/HX110 instrument, the two fields are scanned linearly in a linked fashion. In addition, the magnetic field scan is corrected, on the basis of the calibration, so

[46] R. K. Boyd, P. A. Bott, D. J. Harvan, and J. R. Hass, *Int. J. Mass Spectrom. Ion Proc.* **69,** 251 (1986).

that its scan is linear (linearized). Thus, the appropriate B_2/E_2 ratio is maintained. Scans of product ions from precursors of up to m/z 3500 have been successfully calibrated, and mass assignment accuracy is maintained at least ±0.3 u over a period of a few days. The interested reader is directed to the reference 44 for details.

This instrument, as well as other high-performance four-sector mass spectrometers, is set to at least unit resolution in the static mode of MS-1. This permits precursor ions to be selected with no significant coselection of an $A + 1$ ion (e.g., an ion containing a ^{13}C isotope). The resolving power of MS-2 is typically 500–1000. The capability to do MS/MS at these resolving powers with good sensitivity is an advantage of multisector magnetic instruments.

Tandem four-sector instruments have been used extensively for peptide and protein sequencing and for determining the nature and location of posttranslational modifications (e.g., N-acylation, O-phosphorylation, O-sulfation, O- and N-glycosylation).[3,4,40,47] The latter problem is difficult by conventional biochemical methods and is an important application for mass spectrometry. In the area of protein sequencing, at least high picomole to low nanomole quantities are needed, although the detection limit is in the mid picomole range in favorable cases. Future improvements are possible with array detector-based instruments (*vide infra*) and with a better understanding of desorption ionization. The sequencing is limited to peptides containing 20–25 amino acids, as was determined with the MIT instrument.[47] Fortunately, this size limit corresponds closely to the largest peptides liberated in the tryptic digest of a protein. The instrument has also been actively used for understanding the mechanisms of fragmentation of peptides[48] and for establishing the capabilities of MS/MS. The reader is referred to an interesting paper in which the FAB mass spectra (without CAD) and tandem mass spectra (with CAD) obtained with this instrument are compared,[49] establishing clearly the advantages of tandem mass spectrometry.

Another use of four-sector mass spectrometers was developed theoretically and then demonstrated experimentally by using an HX110/HX110 instrument.[50] Because there are two double-focusing analyzers, the resolving power of the tandem should be better than that of either of the individual mass spectrometer stages. A similar enhancement occurs in optical

---

[47] K. Biemann and H. A. Scoble, *Science* **237,** 992 (1987).

[48] R. S. Johnson, S. A. Martin, and K. Biemann, *Int. J. Mass Spectrom. Ion Proc.* **86,** 137 (1988).

[49] S. A. Martin and K. Biemann, *Int. J. Mass Spectrom. Ion Proc.* **78,** 213 (1987).

[50] T. Matsuo, M. Ishihara, S. A. Martin, and K. Biemann, *Int. J. Mass Spectrom. Ion Proc.* **86,** 83 (1988).

spectroscopy, especially Raman, by using a double monochromator. The predicted resolution improvement is 2.49, and the achieved factor is 2.37.

JEOL also builds a smaller version of the HX110/HX110, the SX102/SX102. The instrument is similar in design and intent to the VG 70-SE4F, which was mentioned earlier.

Kratos IIHH Four-Sector Mass Spectrometer

Professor A. L. Burlingame (University of California at San Francisco) contracted the design and construction of a four-sector instrument based on the newer design of the double-focusing Kratos Concept instrument. The Kratos Concept instrument has ion optics that are similar to the older Kratos MS-50, but the analyzers are arranged in a horizontal rather than vertical plane, giving a much more flexible layout (see Fig. 4). Moreover, the first instrument has been equipped with an array detector.

The Kratos Concept IIHH is a four-sector E_1B_1/E_2B_2 configured instrument.[51] Each instrument is of the well-known Nier–Johnson geometry with 90° electric sectors (381 mm radius) and 60° sector magnets (600 mm radius) of inhomogeneous field. The upper mass limit of each analyzer of the tandem is 10,000 u for an acceleration potential of 8 kV.

The collision cell region is of a new design that allows for focusing and decelerating the beam before introduction into the gas cell and for refocusing the beam onto the object slit of MS-2. The collision cell region also contains a second ion source for producing ions for calibrating the scan of MS-2. The design and approach to calibration are similar to those of the HX110/HX110 and ZAB SE-4F instruments.

Although this instrument has been used less than other instruments, its operating specifications seem comparable to ZAB SE-4F and HX110/HX110. To date, fewer examples of higher mass (>1500 u) peptides, however, have been sequenced with this instrument.

Tandem Nier–Johnson Instruments with Array Detectors

Evolution and Advantage

Experiences with tandem mass spectrometry—even with the simpler two- and three-sector instruments—led to the conclusion that sensitivity improvements are necessary and can be made by employing simultaneous

[51] F. C. Walls, M. A. Baldwin, A. M. Falick, B. W. Gibson, S. Kaur, D. A. Maltby, B. L. Gillece-Castro, K. F. Medsihradszky, S. Evans, and A. L. Burlingame, in "Biological Mass Spectrometry" (A. L. Burlingame and J. A. McCloskey, eds.). Elsevier, Amsterdam, in press (1990).

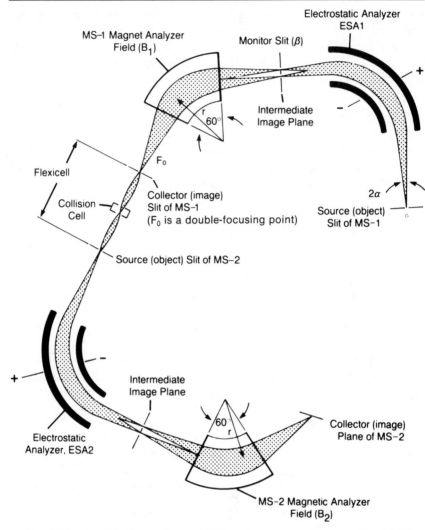

FIG. 4. Diagram of the Kratos Concept IIHH tandem mass spectrometer. (Reprinted from Ref. 51 with permission.)

detection with a channel electron multiplier array. The principles of array detectors and their application in double-focusing or two-sector mass spectrometry are discussed elsewhere,[52] and so only the application to four-sector instruments will be briefly discussed here. Proposals to build tandem mass spectrometers with array detectors were made in the mid-1980s by a few investigators including the author of this chapter. The first reports of success were by Biemann and Burlingame and their co-workers in side-by-side posters at the 1988 ASMS meeting in San Francisco.

Scanning mass spectrometers of the Nier–Johnson design will always be limited in sensitivity because they detect one ion at a time. The same is true of scanning or continuous-wave optical spectrometers. However, sensitivity need not be compromised in a static optical spectroscopy experiment. Only the time of data acquisition is extended with respect to an integrating or multichannel instrument. Mass spectrometers, unlike optical spectrometers consume sample and, as a result, scanning is not only time-consuming but also wasteful of sample.

Integrating instruments, such as, the photoplate or Mattauch–Herzog double-focusing mass spectrometer, the time-of-flight (TOF) mass spectrometer, and the more recent Fourier transform (FT) mass spectrometer have been used for many years. Thus, the goal to modify four-sector instruments for simultaneous recording seems, in retrospect, quite natural and evolutionary rather than revolutionary. In fact, in the early 1980s, A. J. H. Boerboom of the FOM Institute in Amsterdam built tandem mass spectrometers with array detection and demonstrated spectacular detection limits for MS/MS experiments.[53,54] Unfortunately, these instruments do not meet the requirements of tandem mass spectrometry of middle mass biomolecules. Furthermore, a two-sector Nier–Johnson mass spectrometry equipped with a 25-mm array was built and evaluated at Kratos in the mid-1980s.[55]

The Nier–Johnson instruments can only accommodate a small range of simultaneous detection, especially when the collision cell is at ground potential and the laboratory energy of the colliding beam is maximum. This is because the range of constant velocity product ion masses that can be transmitted by E_2 is small ($\sim$5%). Furthermore, Nier–Johnson instruments are second-order double focusing at a single point, not over a line such as would be occupied by the array detector. The double-focusing line for direction focusing (to compensate for angular divergence)

[52] S. Evans, this volume [3].
[53] A. J. H. Boerboom, in "Tandem Mass Spectrometry" (F. W. McLafferty, ed.), p. 239. Wiley, New York, 1983.
[54] R. Weber, K. Levsen, G. J. Louter, and A. J. H. Boerboom, Anal. Chem. 54, 1458 (1982).
[55] J. S. Cottrell and S. Evans, Anal. Chem. 59, 1990 (1987).

does not coincide with the double-focusing line for energy focusing in a standard Nier–Johnson instrument, although on some instruments the lines are nearly overlapping. Therefore, a reasonably good focusing line, which is a compromise for energy and direction focusing, can be located for an array. This has been done for both the HX110/HX110 and the Kratos Concept IIHH instruments.

Despite the small mass range of such integrating detectors on the MIT and UCSF spectrometers, a significant advantage can still be achieved with respect to a single-point detector. Consider obtaining a mass spectrum by an exponential mass scan of the magnetic sector. The width of a single peak in a 10 sec/decade of mass scan at a resolution of 1000 is approximately 4 msec. If one decade of mass could be monitored with an array detector for the same time period of 10 sec, the increase in signal would be 10 sec/0.004 sec or 2500. Even if 100 different exposures with a small array were required to obtain one decade of mass, the signal gain is still a factor of 25, which is not insignificant. To gain a factor of 10- to 100-fold increase in signal intensity (a range that encompasses a factor of 25), array detectors are incorporated in tandem Nier–Johnson mass spectrometers. The improvement in signal-to-noise is the square root of the improvement factor, but this is true only if random noise is important in tandem mass spectrometry experiments. Because random noise is *not* important, the factor of 10 to 100 is more applicable.

Array Detection with JEOL HX110/HX110 Instrument

A JEOL HX110/HX110 mass spectrometer can accommodate a focal plane that will have in reasonable focus a 1.1 : 1.0 mass range over a distance of 78 mm. This pertains if the ions have constant energy and are transmitted equally through the ESA. The product ions emerging from a grounded collision cell into MS-2 have constant velocity, and the ESA now discriminates in terms of mass. This discrimination is minimized and high-energy collisional activation is still maintained by floating the collision cell at some intermediate voltage.

It was found that by using a 50-mm array (containing 2048 pixels or separate detectors), the collision cell at 5 kV and an accelerating potential of 10 kV, that 6.6% mass range of product ions can be simultaneously detected.[56] Although the transmission of product ions with a 6.6% mass difference is not equal across the array, this problem can be corrected in the data processing step.

A full-product ion spectrum must be taken in steps. The sequence of

[56] J. A. Hill, J. E. Biller, S. A. Martin, K. Biemann, K. Yoshidome, and K. Sato, *Int. J. Mass Spectrom. Ion Proc.* **92,** 211 (1989).

events involves taking a 6.6% "exposure" and simultaneously jumping B_2 and E_2 to higher field strengths such that there is some overlap with the first "exposure." A second "exposure" is then taken, and the sequence is repeated until a suitably complete product spectrum has been acquired. Each "exposure" or integration takes between 0.1 and 1.0 sec, and usually 10 consecutive integrations are made before B_2 and E_2 are jumped to new values. For a 6.6% mass range, 35 separate exposures are required to record product ions over one decade (factor of 10) of mass.

On the basis of sample requirements needed to give comparable results for the array and the single-point detector, a factor of 50 improvement in "practical sensitivity" was achieved for the array installed on the MIT instrument operating in the tandem mode. The test sample was a peptide with an $(M + H)^+$ of m/z 1759. Actually better resolution and peak shape, as well as improved sensitivity, were obtained with the array compared to the single-point detector, because a beam having the width of a slit appears twice as broad in a scanning instrument. In the array mode, where there is no detector slit, the natural width is observed. Good accuracy in mass measurement (± 0.25 u) was also obtained, and the array spectrum showed a comparable pattern to that obtained with the single-point detector.

Array Detection with Kratos Concept IIHH Instrument

The Kratos Concept IIHH instrument can be equipped with an 25-mm array.[51] This length of an array is located along a compromise focal line, as is the case with the HX110/HX110 instrument. A mass resolution of 5000 was reported to be achievable. Because the array is only one-half the length of the MIT array, a smaller mass range (4%) is detected simultaneously. This means that 59 "exposures" are required to obtain product ion spectrum over one decade of mass (compare with 36 exposures with the 50-mm array and 6.6% range on the MIT instrument). As was pointed out earlier, a significant improvement in sensitivity is still expected even though a rather large number of "exposures" are required to obtain a reasonably complete product ion mass spectrum. Although no comparison between results obtained with the array and with the single-point detector has been published as yet, the UCSF instrument has been used to obtain a product ion spectrum of substance P (mol wt = 1346) at the 10 pmol level. With 1 pmol, it is still possible to identify 50% of the fragments from substance P, but the spectrum also shows considerable chemical noise from fragmentation of coselected matrix ions. The chemical noise is a problem of ionization, not detection.

The pioneering evaluations of array detectors on four-sector tandem

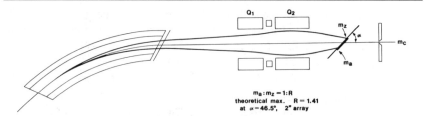

FIG. 5. Diagram of the ion optics starting with B_2 of the JEOL HX110/HX110 modified for variable dispersion and extended-array detection. (Reprinted from Ref. 57 with permission.)

mass spectrometers are certainly encouraging. Additional improvements are expected with more extended arrays (i.e., arrays that cover more than 4 or 6.6% of the mass range). Two developments that show potential for extended arrays will be discussed in the next section.

Tandem Four-Sector Instruments with Extended Arrays

An extended-array instrument is defined as one that can detect simultaneously mass ranges greater than a few percent. The MIT four-sector instrument has been recently modified to permit simultaneous recording of product ion spectra over a larger range than 6.6%. VG Analytical has built tandem four-sector mass spectrometers with a reverse-geometry Mattauch–Herzog-type mass spectrometer as MS-2. MS-2 can be equipped with a 152-mm array detector that can simultaneously record a mass range of 1.4 : 1.0 (grounded collision cell) and 1.6 : 1.0 (collision cell floated at 50% of the acceleration potential). A third extended-array instrument was recently proposed. These will be discussed in turn.

Extended MIT Array

Despite the limitations of Nier–Johnson instruments for simultaneous detection over wide mass ranges, Biemann and co-workers[57] have extended the simultaneous detecting capability of MS-2 of the HX110/HX110 at MIT from 1.06 to 1.3 : 1.0 *without changing the size of the array*. They accomplished this by adding two quadrupole lenses after the second-stage mass analyzer (B_2) (see Fig. 5). By changing the potentials of these quadrupole lenses, the focal plane can be moved forward from the single-point detector, which in the first modification incorporating the array was

[57] K. Biemann, *in* "Biological Mass Spectrometry" (A. L. Burlingame and J. A. McCloskey, eds.). Elsevier, Amsterdam, in press (1990).

located along a focal line containing the single-point detector. Thus, a focal line can be inserted between the B_2 and the final, single-point detector (see Fig. 5).

The detected mass range can be increased or decreased by simultaneously varying the quadrupole potentials and the angle (α) of the focal plane. The theoretical limitation is the range or dispersion of masses that be transmitted through B_2 and the flight tube. For the HX110/HX110, the range is 1.4 : 1.0, but this requires that the energy of the ions to be nearly constant over the mass range. In an MS/MS experiment in which the array detector is used for simultaneous detection of product ions, the collision cell would need to be floated at nearly the value of the accelerating voltage. Thus, the increase in mass range is achieved at the expense of collision energy. It remains to be seen whether the good precision of tandem sector instruments will be compromised by working at reduced collision energy.

The Biemann group[58] reported a nearly complete mass spectrum (from m/z 60–2500) of a peptide having an $(M + H)^+$ of m/z 2465.2 by taking 15 "exposures" over a mass range of 1.3 : 1.0 for each exposure. This is a fourfold reduction in the number of "exposures" that would be required with 6.6% array, giving, in principle, a fourfold improvement in sensitivity or reducing the time of the experiment by four times. The larger range of 1.4 : 1.0 could have been achieved but the reduction in "exposures" is not significant (12 vs. 15 for the 1.3 : 1.0 array). Moreover, the number of channels (pixels) per mass must be decreased as the mass range per "exposure" is widened, limiting the mass resolving power and peak definition.

A CAD spectrum of the same peptide was obtained by 13 "exposures" at a mass range of 1.3 : 1.0 each and a detector angle of 40° from m/z 50–1500 and then 5 "exposures" to cover the m/z 1430–2450 range (detector angle at 28°). The collision cell was floated at 8 kV (accelerating potential was 10 kV) for this experiment. This change in exposure range permits the peaks at high mass to be better defined. For example, at an exposure range of 1.3 : 1.0, the mass range of 1896 to 2465 u is defined by 2048 points (i.e., the number of channels on the array). At an exposure range of 1.12 : 1.0, the same number of channels is used to record a mass range of 2201 to 2465 u. This example shows that by varying the exposure range, one can minimize the integration time with respect to a narrow array and still obtain reasonable definition for the peaks.

[58] J. A. Hill, J. E. Biller, S. A. Martin, K. Biemann, and M. Ishihara, *Proc. 37th ASMS Conf. Mass Spectrom. Allied Topics, Miami Beach, FL* p. 990 (1989).

Nier–Johnson/Mattauch–Herzog-Type Four-Sector Instrument

A different approach to wide mass range simultaneous detection is to use a focal plane instrument as MS-2. These designs, called Mattauch–Herzog instruments, are well known in mass spectrometry and were employed with photographic detection for chemical analysis in the 1960s and 1970s. Although initially the design of Mattauch–Herzog instruments utilized forward geometry (EB), the instrument described herein is constructed differently.

The difficulty with a Mattauch–Herzog forward-geometry instrument is that changes in the focal plane (i.e., location, flatness, and angle) are determined by the ion beam-deflection properties of the second analyzer, which is a magnetic sector. These properties are extremely difficult to vary on a routine basis because they would involve changes in the shape of the field, which is determined by the shape of the pole faces. Reverse-geometry designs use changes in electric field which are easier to make.

A versatile Mattauch–Herzog-type design needs to accommodate not only changes in the focal plane, as discussed above, but also variations in the nature of the ion beam (i.e., are the ions of constant energy, constant velocity, or of some compromise of energy and velocity?). The goal is to maintain the double-focusing properties of MS-2 under conditions of either a grounded collision cell or a collision cell floated at any potential between ground and the acceleration potential. The instrument should also be double focusing as a stand-alone spectrometer (i.e., for ions at constant energy).

A second-stage mass spectrometer of a design that fulfills all the above requirements was designed by R. Bateman of VG Analytical to meet the specifications of our laboratory (see Fig. 6).[59] The design is a reverse-geometry Mattauch–Herzog type (i.e., BE). The magnetic sector has an upper mass limit of 15,000 at 8 kV of acceleration. Although this is a mass range that far outstrips the present capabilities of MS/MS, the second stage is also designed to be a stand-alone high-mass instrument. A fully optimized ion source will be available for the intermediate region (along with a collision cell and a calibration ion source).

The second electric sector of MS-2 is of a novel design, employing an inhomogeneous electric field. The sector is comprised of a series of shim electrodes (three rows of 21 pairs extending across the width of the ESA). The ion beam travels between the electrode pairs. By applying different voltages to the electrodes, the electric field can be modified so that it

---

[59] R. Bateman and M. L. Gross, manuscript in preparation.

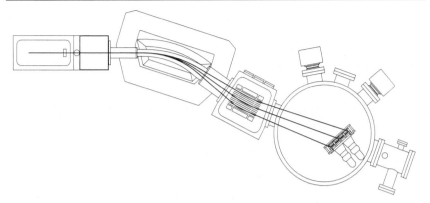

FIG. 6. Diagram of the MS-2 of the VG ZAB-T four-sector tandem mass spectrometer, giving the ESA as a schematic representation. The three ion beams are intended to represent three different mass product ions that are formed in a collision cell located at the origin of the trajectory on the left and detected at an array on the right.

serves as more than only a ion deflection device. The electric field for this ESA is described by a series of terms as shown in Eq. (1).

$$E = E_0 + E_1 y + E_2 y^2 + E_3 y^3 \tag{1}$$

where E_0 is the constant electric field in any normal double-focusing mass spectrometer and $E_1 y$ is the field as a function of y (the dimension across the ESA) that controls focal length. By simply rotating the focal plane, one can make use of the full array detector to look either at a wide mass range at low mass resolution or at a narrow mass range at higher mass resolution. To maintain double focusing, a third term, $E_2 y^2$ is added to the full description of the electric field. Finally, the focal plane should be flat because contemporary array detectors are planar. Flatness is achieved by modifying the electric field by adding a fourth term describing the electric field, $E_3 y^3$.

The full tandem mass spectrometer consists of a ZAB-SE as MS-1 (like the SKF instrument) and the new design as MS-2. The detector is a 152-mm array, the first to be built of this length. The array, which contains at least 6144 channels, is interfaced to a 4096 channel photodiode array. With the collision cell at ground potential, product ions over a 1.4 : 1.0 mass range can be observed simultaneously. With ions produced at an intermediate source and having constant energy (or with the collision cell floated at the acceleration potential), the mass range is 2.0 : 1.0 for simultaneous detection. A 1.6 : 1.0 mass range of product ions was found when the collision cell was floated 50% of the acceleration potential (i.e., at 4 kV). Narrower mass ranges can be simultaneously observed over the entire array by rotating the array away from the normal of the ion beam trajectory and adjusting the E_2 electric field.

Mass resolution, detection limit, and routine operability of this instrument remain to be established. It does appear, however, that the resolving power of MS-2 with the array is comparable to those of the HX110/HX110 and the Concept IIHH instruments.

Wienfilter/Magnetic Sector MS-2

An alternative approach to extended-array instruments was recently proposed by Matsuda and Wollnik.[60] Given that conventional double-focusing instruments employ an ESA to energy-analyze source-produced ion beams of nearly constant energy but of different masses, an MS-2 should employ an analyzer that similarly passes those product ions of nearly constant velocity but of different masses that emerge from a grounded collision cell. That is, the analyzer should compensate for velocity dispersion just as the ESA compensates for energy dispersion in conventional double-focusing mass spectrometers. This is true for conditions of high-energy CAD. A Wienfilter was chosen because it is a velocity analyzer. Such a system was built in the early 1940s,[61] but the mass analyzer exhibited a large variation of focusing properties as a function of ion mass. The recently proposed design incorporates a reduction in the overall focusing power.

Six different double-focusing Wienfilter/magnetic sector instruments (*WB* configuration) were presented in the proposal.[60] All involve joining the *W* and *B* so that the same magnet flux was shared by both. This design is compact and eliminates tracking problems that will pertain if the two magnet analyzers are separated. One design was found to be particularly suitable as an MS-2 of a tandem mass spectrometer. The predicted mass range is 1.4 : 1.0, and the expected mass resolving power is 2000.

Conclusion

Four-sector instruments represent many aspects of ultimate performance in tandem mass spectrometry. They permit mass selection of precursor ions at unit resolution or significantly higher and mass analysis at

[60] H. Matsuda and H. Wollnik, *Int. J. Mass Spectrom. Ion Proc.* **86,** 53 (1988). The combination of a Wienfilter and magnet sector for MS/MS applications was first considered by H. H. Tuithof, "Simultaneous Ion Detection in a Mass Spectrometer with Variable Mass Dispersion," Ph.D. Thesis, F. O. M. Institute, Amsterdam, 1977.

[61] E. B. Jordan, *Phys. Rev.* **60,** 710 (1941).

resolving powers of at least 500–1000. They take advantage of the good reproducibility of high-energy collisional activation. These are significant advantages and, for some applications, justify the additional cost with respect to hybrids and triple-stage quadrupole instruments. Nevertheless, there are many more two- than three- and four-sector instruments, and readers are reminded that adequate MS/MS experiments can often be done with such instruments.

Unlike multiple quadrupole and contemporary hybrid instruments,[30] the four-sector instrument can accommodate simultaneous (array) detection. Simultaneous detection over narrow mass ranges was demonstrated (4 and 6.6% on the Concept IIHH and the HX110/HX110 instruments, respectively). The next generation instruments have simultaneous detection over mass ranges from 1.4 : 1.0 to 2.0 : 1.0. Additional improvements may be expected.

Four-sector instruments may also accommodate opportunities for overcoming the limitations of collisional activation.[2] Three possibilities are photo-, surface-, and electron activation. Preliminary photoactivation experiments with the HX110/HX110 instrument point to improved efficiency of activation of angiotensin III with respect to collisional activation.[62] Electron activation on sector instruments is under investigation at the University of California–San Francisco.[63]

Acknowledgment

The author thanks K. Biemann, S. Carr, K. Tomer, and R. Bateman for helpful discussions. Support from the U.S. NSF is gratefully acknowledged (Grant CHE 8620177).

[62] S. A. Martin, J. A. Hill, C. Kittrell, and K. Biemann, *J. Am. Soc. Mass Spectrom.* **1,** 107 (1990).
[63] W. Aberth and A. L. Burlingame, *in* "Biological Mass Spectrometry" (A. L. Burlingame and J. A. McCloskey, eds.). Elsevier, Amsterdam, in press (1990).

[7] Tandem Mass Spectrometry: Quadrupole and Hybrid Instruments

By Richard A. Yost and Robert K. Boyd*

Introduction

The general concept of tandem mass spectrometry (MS/MS) as a tool which delineates connectivity relationships between precursor and product ions[1] will be reviewed briefly below. The present chapter is mainly concerned with instruments, designed specifically for experiments of this type, which use quadrupole mass filters[2] as analyzers for the product ion(s) finally detected in the experiment. The precursor ion analyzer is most frequently another quadrupole (the triple quadrupole instruments), but an increasing number of tandem instruments incorporate a sector analyzer as the first stage; such "hybrid" instruments combine the high-mass/high-resolution advantages of the modern sector instrument with the flexibility of the quadrupole.

In practically every case, the instruments discussed here have also used a quadrupole with only radio frequency (rf) potentials (no dc) applied to pairs of opposite rods, an efficient device for the containment and transmission of ions, as the reaction cell. Indeed, the rf-only quadrupole cell is the only device involved whose principles of operation have not been discussed in earlier chapters. Accordingly, an outline of the operational principles of this and related devices will be presented. (Recall[2] that a quadrupole mass filter employs superimposed rf and dc fields; the reaction cell device uses the rf field only.) Inherent in the selection of an rf-only quadrupole as the reaction cell is the use of lower energy collisions (up to a maximum of a few hundred electron volts, but most often a few tens of electron volts).

It is convenient here to define some of the nomenclature to be used in this chapter. The symbols Q, q, B, and E denote a quadrupole mass filter, an rf-only quadrupole, a magnetic sector analyzer, and an electric sector analyzer, respectively.[2] Using this nomenclature, a triple quadrupole tandem mass spectrometer is conveniently denoted as Q_1qQ_2, while $BEqQ$ denotes a hybrid instrument comprising a double-focusing (BE) analyzer of reversed configuration[2] plus a qQ combination, and so on. The notation XqQ is sometimes used to denote all instruments covered in the present

---

* National Research Council of Canada.
[1] F. W. McLafferty (ed.), "Tandem Mass Spectrometry." Wiley, New York, 1983.
[2] K. R. Jennings and G. G. Dolnikowski, this volume [2].

METHODS IN ENZYMOLOGY, VOL. 193

chapter, where X can represent either another quadrupole mass filter or a suitable sector analyzer. Confusion can sometimes arise over the distinction between the analyzers themselves (denoted B, E, Q) and the field strengths acting on the ion beams within them [written in boldface characters $\mathbf{B}$, $\mathbf{E}$, $\mathbf{Q}$, respectively, where in the case of the mass filter function $\mathbf{Q}$ there are actually *two* fields involved, U (dc) and V_0 (rf amplitude[2])]. Another potential source of confusion is the use of the letter q to denote both a physical device (rf-only quadrupole) and also conventionally one of the Mathieu parameters[2] which characterize the behavior of all rf-quadrupoles; accordingly the symbol q_u will be used here for the Mathieu parameter, where u can signify either or both of the transverse directions x and y (the main axis of the quadrupole is designated as the z axis, by convention). Finally the translational energy of an ion at any point within the apparatus will be denoted E_{lab}, to distinguish it from E and $\mathbf{E}$ defined above and to emphasize that this energy is measured relative to a reference frame fixed with respect to the laboratory.

One characteristic of any analyzer which is of crucial importance in the context of tandem mass spectrometry is the degree of linearity of its mass–scan law, i.e., the relationship between the applied electromagnetic field strength(s) and the m/z values of the ions thereby transmitted. Quadrupole mass filters Q display a high degree of linearity, which greatly facilitates their use in tandem instruments; most magnetic sector field instruments (e.g., double-focusing analyzers), on the other hand, are highly nonlinear in the relationship of magnetic field strength to transmitted mass, and this represents an operational disadvantage for hybrids relative to triple quadrupoles. This point is referred to below. Of course the same point is even more pertinent with respect to multisector tandem instruments.[3] It must be added, however, that recent development[4] of hysteresis-free air-cored magnets for mass spectrometers has paved the way for medium resolving-power capability (up to 7×10^3, 10% valley definition) combined with linearity and ease of control comparable with those characteristic of quadrupoles.

The present chapter contains four main sections. The first two of these cover material (general concepts of tandem mass spectrometry and operational properties of rf-only quadrupoles) pertinent to both kinds of XqQ instruments. Each of the two remaining sections is devoted to one of these two types, their special advantages and problems, and is illustrated, where appropriate, by examples from the recent literature. The chapter concludes with a summary and discussion of future directions.

[3] M. L. Gross, this volume [6].
[4] R. H. Bateman, P. Burns, R. Owen, and V. C. Parr, *Adv. Mass Spectrom.* **10**, 863 (1985).

Tandem Mass Spectrometry: General Concepts

In their most general form, the concepts of tandem mass spectrometry may be discussed in terms of ion–molecule reactions of the type shown in Eq. (1); for convenience only singly charged positive ions are considered, but the generalization to other ion types (negative and/or multiply charged ions) is straightforward:

$$\underline{R}^+ + T \rightarrow \underline{P}^+ + \underline{N}_1 + N_2 + \dots \tag{1}$$

where the mass-selected precursor ion $\underline{R}^+$ possesses a highly nonthermal translational energy which transports the species through the instrument (including the reaction cell), and is thus signified by underlining the symbol for that species. As a consequence of energetic collisions with a thermal target gas (T), $\underline{R}^+$ ions mav be converted into product ions $\underline{P}^+$, accompanied by one or more neutral species $\underline{N}$ which are not directly detectable. The most commonly studied reactions of this type are fragmentations [Eq. (1a)], in which the target species T (typically an inert gas such as argon) is unaltered chemically (though it does necessarily acquire some additional translational energy in the collision):

$$\underline{R}^+ + T \rightarrow \underline{P}^+ + \underline{N} + T + \dots \tag{1a}$$

In the special case of collision-induced dissociation [CID, Eq. (1a)] the mass balance relationship (Eq. 2) becomes particularly simple (Eq. 2a):

$$m_R + m_T = m_P + m_{N1} + m_{N2} + \dots \tag{2}$$
$$m_R = m_P + m_N \tag{2a}$$

where clearly $m_P < m_R$ in Eq. (2a). However, a characteristic of rf-only quadrupole cells (q) is that the translational energy of the precursor ions can be kept sufficiently low (down to 1 eV or so) that fragmentation becomes inefficient relative to chemical reaction between R^+ and some suitably chosen target gas T; in such cases it is possible that $m_P > m_R$. Such reactive collisions are, in general, inaccessible in multisector instruments unless special facilities are provided to permit low-energy collisions. Examples of exploitation of such intermolecular chemistry in tandem mass spectrometry are included below.

Another advantage of low-energy reaction cells is their ability to facilitate studies of the energy (E_{lab}) dependence of the collision-induced dissociation (CID) process in an energy regime where interesting effects are frequently observed. It is important to emphasize, however, that results obtained via exploitation of rf-only quadrupole reaction cells for this purpose must be interpreted with care, in view of the complex dependence of ion transmission efficiency upon several parameters including E_{lab} (see below).

For an overall process (Eq. 1) or (Eq. 1a) it is possible to operate any tandem mass spectrometer in one of three basic MS/MS modes, depending on whether the mass of the precursor ion, or of the product ion, or of the neutral fragment, is fixed by the experimental parameters. For example, if the precursor ion analyzer (Q_1 in a triple quadrupole, BE in a $BEqQ$ hybrid) is set so that only precursor ions restricted to some narrow range of values of m/z are transmitted to the reaction region (q in this case), scanning the final quadrupole mass filter will provide a product ion spectrum (often referred to as a "daughter-ion spectrum") of the mass-selected precursor (or "parent") ion. Such a "fixed-precursor ion" experiment is the operational mode most frequently used in tandem mass spectrometry. It is used primarily to provide *structural* information on the chosen precursor ion, even though the latter may arise from ionization of a sample which is a mixture; in such a case, the precursor ion analyzer may also be considered to be acting as a chemical clean-up stage, which is potentially capable of much higher discrimination, in the case of a hybrid instrument (high-mass–high-resolution double-focusing BE or EB analyzer), than in that of a triple quadrupole.

Applications of tandem mass spectrometry to *mixture analysis* often employ experiments which fix the mass of either the product ion P^+ or of the neutral fragment N [the latter in the case of fragmentation reactions, Eq. (1a)]. As a simple example, protonated amino acids fragment to give a characteristic ion at m/z 74, usually assigned the structure $H_2N^+ = CHCOOH$; by setting the product ion analyzer (final Q in the present context) to transmit only ions with this m/z value to the final detector, and scanning the precursor ion analyzer, a mass spectrum will be obtained which contains peaks corresponding only to those precursor ions which entered the reaction cell (q) and fragmented there to yield product ions of the selected mass (m/z 74). Such a precursor ion spectrum (often called a "parent-ion spectrum") could provide the basis for a rapid means of identifying which amino acids were present in an unseparated mixture, supplied as sample to an ionization source[5] which produces protonated molecules.

A related experiment which specifies the mass of the neutral fragment [N in Eq. (1a)] can achieve a similar result. For example, precursor ions containing O-acetyl groups readily expel a neutral fragment of molecular mass 60 Da (acetic acid); if the precursor and product ion analyzers can be scanned synchronously in a linked-scan relationship, experimentally defined by a fixed mass difference [Eq. (2a), $m_N = (m_R - m_P) = 60$ Da, in this example], the resulting spectrum will correspond to all precursor

---

[5] A. G. Harrison and R. J. Cotter, this volume [1].

ions which underwent loss of neutrals with mass 60 Da within the reaction region. Such a scan for fixed neutral loss is simply achieved in a triple quadrupole instrument, because of the excellent linearity of the mass filters Q_1 and Q_2; it is somewhat more difficult to achieve such a linked scan in most hybrid instruments, due in part to the highly nonlinear nature of the mass–scan law for magnetic sector analyzers with iron cores.

It should be emphasized that no detailed mechanistic implications are intended by writing chemical equations such as Eqs. (1) and (1a); in particular, it is *not* implied that the reactions thus written necessarily proceed in a single mechanistic step. Rather, the implications are entirely phenomenological, namely, the reactions proceed to observable extents within the appropriate instrumental time scale. [Characteristic available reaction times, defined as flight times for the $\underline{R}^+$ ions, are 1–10 μsec for a field-free region in a sector analyzer, and tens to hundreds of microseconds for an rf-only quadrupole cell (q).] It is particularly important to keep in mind this lack of detailed mechanistic implications when considering two-step experiments for which the chemistry is written as:

$$\underline{R}^+ + T_1 \longrightarrow \underline{I}^+ (+ \underline{N}_1 + T_1)$$
$$+ T_2 \big|$$
$$\qquad\qquad \longrightarrow \underline{P}^+ + \underline{N}_2 + T_2 \tag{3}$$

where the intermediate $\underline{I}^+$ must have a lifetime sufficient that it survives the flight time between the first reaction cell (containing target gas T_1) and the second (containing target gas T_2). Either or both of the reactions appearing consecutively in Eq. (3) may, in fact, proceed via more than one mechanistic step; the only implications in writing Eq. (3) in this way are purely phenomenological, as discussed above for Eqs. (1) and (1a). Such two-step reactions [Eq. (3)] may be studied using hybrid instruments, which have suitable reaction regions in the sector analyzer and in the rf-only quadrupole q. Similar studies in tandem instruments incorporating only quadrupoles require extension to, for example, a pentaquadrupole $Q_1q_1Q_2q_2Q_3$, but such instruments are rare[6–8]; in part, this is due to the fact that the available sensitivity in such two-stage experiments [Eq. (3)] is too low for many practical applications. The popularity of hybrid instruments is due primarily to their other advantages (availability of a high-performance analyzer), with the facility for two-step experiments offering an added bonus.

[6] J. D. Morrison, D. A. Stanney and J. Tedder, *Proc. 34th ASMS Conf. Mass Spectrom. Allied Topics, Cincinnati, OH* p. 222 (1986).

[7] C. Beaugrand, G. Devant, D. Jaouen, H. Mestdagh, N. Morin, and C. Rolando, *Adv. Mass Spectrom.* **11**, 256 (1989).

[8] J. C. Schwartz, K. L. Schey, and R. G. Cooks, *Int. J. Mass Spectrom. Ion Proc.*, in press.

It is worthwhile to mention briefly here that the tandem instruments discussed in this and the preceding chapter are "tandem-in-space," in the sense that the ion reactions studied are isolated from possible interfering reactions by spatial dispersion of the appropriate ions; study of sequential reactions in such instruments thus requires a separate well-defined spatial region for each reaction step. This is in contrast to "tandem-in-time" instruments, in which sequential steps are studied in the same physical space, but separated from one another via an appropriate timing sequence of excitation and detection; such devices are exemplified by ion traps, involving either magnetic fields [as in Fourier transform (FT) ion cyclotron resonance spectrometers[9]] or a combination of ac and rf electric fields (as in the quadrupole ion trap, or "Quistor"[10]). Both of these devices show considerable promise, but at present have not yet proved capable of providing information on real-world biological samples on a routine basis.

Experimental realization of experiments based on Eq. (3) implies use of three stages of mass analysis (tandem in space or time), one for each of the precursor, intermediate, and product ions. The number of conceivable experiments is now quite large, since it is possible to specify fixed mass values for two independent mass variables (via two of the spatially distinct analyzers in the case of the "tandem-in-space" instruments discussed here), with the third analyzer being scanned to provide the spectrum. A simple example would be to fix the first and third analyzers to specify the m/z values for $\underline{R}^+$ and $\underline{P}^+$, and to scan the intermediate analyzer; the resulting spectrum contains peaks arising from mass-selected $\underline{P}^+$ ions (the only species detected directly), formed ultimately from the mass-selected precursors $\underline{R}^+$ via a series of intermediates $\underline{I}^+$ whose m/z values are indicated on the m/z scale of the intermediate analyzer. Systematization of the many related possibilities, and experimental realization of most of these, have been achieved by Cooks and collaborators[11]; examples of practical applications of this approach are included below in the section on hybrid instruments.

rf-Only Quadrupole Reaction Cells: General Principles

The rf-only quadrupole reaction cell (q) is common to the triple quadrupole and to the hybrid tandem mass spectrometers. Although this device can provide efficient containment of product ions, it also exhibits complex dependences on mass, E_{lab}, and other operating parameters. Below we

[9] A. G. Marshall, *Adv. Mass Spectrom.* **11,** 651 (1989).

[10] J. F. J. Todd, *Adv. Mass Spectrom.* **10,** 35 (1985).

[11] J. N. Louris, L. G. Wright, R. G. Cooks, and A. E. Schoen, *Anal. Chem.* **57,** 2918 (1985); R. G. Cooks, J. W. Amy, M. E. Bier, J. C. Schwartz, and K. L. Schey, *Adv. Mass Spectrom.* **11,** 33 (1988).

discuss the complexity of these apparently simple devices, and present simple results of the detailed theory (together with related experimental evidence) in order to provide the nonspecialist user a working understanding of some of the peculiarities of XqQ tandem mass spectrometers.

The first published account (in 1965) of the use of an rf-only quadrupole in the present context was by von Zahn and Tatarczyk,[12] who constructed a hybrid instrument ($EBqQ$) in order to study ion lifetimes; both unimolecular and collision-induced dissociations were thus observed. Subsequently, triple quadrupole instruments (Q_1qQ_2) were used by Vestal and Futrell[13] and by McGilvery and Morrison[14] in photodissociation spectroscopy experiments. This early work[12-14] was motivated by an interest in the physical chemistry of gaseous ions. Yost and Enke,[15,16] initially through the assistance[17] of one of the physical chemistry groups, were the first to realize that such techniques were potentially powerful tools in analytical chemistry. Since then, the exploitation of the ion containment and transmission properties of the rf-only quadrupole in tandem mass spectrometry has developed dramatically.

This section contains a brief account of some of the more important features of this device as applied to reactions [most often dissociations, Eq. (1a)] of ions. That the device is not simple can be readily appreciated from the results of a round robin study organized by Dawson and Sun,[18] in which CID spectra obtained in Q_1qQ_2 instruments under nominally identical experimental conditions (collision energy and collision gas pressure within q, etc.) were found to show variations in relative peak intensities by factors of up to hundreds! More recently, attempts to better understand and control the experimental parameters have been made by Martinez,[19,20] with a view to eventual establishment of MS/MS spectral libraries. This situation is not so serious in the context of structure elucidation of wholly unknown compounds, for which spectral libraries are of limited usefulness. Nonetheless effective use of an XqQ tandem instrument requires some appreciation of the complexity of the device; this section describes aspects of the trajectory stability and directional focusing

[12] U. von Zahn and H. Tatarczyk, *Phys. Lett.* **12,** 190 (1964); *ibid., Z. Naturforsch.* **20A,** 1708 (1965).

[13] M. L. Vestal and J. H. Futrell, *Chem. Phys. Lett.* **28,** 559 (1974).

[14] D. C. McGilvery and J. D. Morrison, *Int. J. Mass Spectrom. Ion Phys.* **25,** 81 (1978).

[15] R. A. Yost and C. G. Enke, *J. Am. Chem. Soc.* **100,** 2274 (1978).

[16] R. A. Yost and C. G. Enke, *Anal. Chem.* **51,** 1251A (1979).

[17] R. A. Yost, C. G. Enke, D. C. McGilvery, D. Smith, and J. D. Morrison, *Int. J. Mass Spectrom. Ion Phys.* **30,** 127 (1979).

[18] P. H. Dawson and W.-F. Sun, *Int. J. Mass Spectrom. Ion Proc.* **55,** 155 (1983/84).

[19] R. I. Martinez, *Rapid Commun. Mass Spectrom.* **2,** 8 (1988).

[20] R. I. Martinez, *Rapid Commun. Mass Spectrom.* **3,** 127 (1989).

properties of q. In addition, the implications of the variable translational energy of product ions formed within q, for operation of the product ion analyzer Q, are also discussed below.

Trajectory Stability

The question of stability of the trajectories of ions in an rf-only quadru-pole (q) can be discussed via an extension of the treatment[2] for rf/dc quadrupole mass filters (Q). Since the dc potential difference U is zero in the present case, the Mathieu parameter a is also zero[2] and the region of stable trajectories in the two-dimensional stability diagram[2] now collapses to a segment along the q_u axis (denoted[2] as the q axis). It is important to realize that not all trajectories within such a device are stable, despite the frequent use of phrases such as "ion-funnel mode" or "total ion mode" to describe operation of a quadrupole with U = zero (i.e., rf-only). In particular, as indicated previously,[2] there is a *maximum* value for q_u of 0.908 for stable trajectories [Eq. (4)]:

$$q_u = 4V_0/[(m/ze)(2\pi f r_0)^2] = 9.775 \times 10^6 \, (zV_0/Mf^2r_0^2)$$
$$< 0.908, \text{ for stable trajectories} \quad (4)$$

where V_0 is the zero-to-peak rf amplitude (in volts) measured between one pair of rods and V_q, the mean potential of the device q; V_q is often referred to as the rod offset of q in triple quadrupoles, or as the float potential of q in XqQ hybrids. The rf frequency f is in $\sec^{-1}$, the inscribed radius r_0 of q is in meters, and the ion mass M is in daltons (m in kg); z is the number of elementary charges on the ion, most usually one (plus or minus). The inscribed radius r_0 is a fixed parameter for any given device, and so also is f in the great majority of cases; thus the only tuning parameter available [Eq. (4)], to ensure trajectory stability for ions of a specified value of m/z, is the rf amplitude V_0. This parameter must be controlled with some care, as discussed further below.

The implications of Eq. (4), for product ions P^+ formed within q, are a little more complicated. When an ion R^+ reacts within q as per Eq. (1), the value of q_u [Eq. (4)] for the product ion P^+ changes from that for the precursor by the ratio (M_R/M_P). If the reaction involves intermolecular chemistry such that $M_P > M_R$, the value of q_u drops and the trajectory stability is ensured. However, in the much more common case of fragmen-tation reactions [Eq. (1a)], $M_P < M_R$ so the value of q_u instantaneously increases, thus endangering the trajectory stability of product ions whose mass is a small fraction of that of the precursor. One solution to this problem is to ensure that the starting value of q_u for R^+ is sufficiently low, via low values for the rf amplitude V_0, that the q_u value for the product ions does not increase to the stability limit [0.908, Eq. (4)] until M_P has decreased to some uninterestingly low value. This approach works, but the low V_0 value simultaneously introduces limitations in operation of q for high precursor masses and higher collision energies (see discussion of directional focusing effects below). Alternatively, V_0 can be scanned in a linked fashion with the mass function **Q** of the product ion mass filter (final quadrupole Q in XqQ instruments), so as to keep q_u for the product ions approximately constant. This is, in principle, the preferable option, but in practice requires user tuning of slope and intercept controls in the **Q**/V_0 linked-scan circuit or else blind acceptance of preset adjustments of these controls.

None of these options is ideal, but the situation is somewhat alleviated by the fact that product ions are formed throughout the length of q; if q_u > 0.908, all such ions are lost [Eq. (4)] *except* for those formed sufficiently close to the exit of q that the number of rf cycles experienced is insufficient for the inherent instability to lead to actual loss of the ions.[21] This is a rare example of nonideal behavior actually benefiting the experimentalist!

Directional Focusing

dc Quadrupole assemblies are widely used as ion optical lenses, so it is not surprising that their rf counterparts also possess directional focusing properties. Indeed, a remarkable device has been described[22] which exploits these properties to permit simultaneous mass and energy selection of ions. The effects of directional focusing within q upon performance of XqQ tandem mass spectrometers have been the subject of several studies,[17,21,23–31] and become particularly important[21] for situations typical

[21] A. J. Alexander, E. W. Dyer, and R. K. Boyd, *Rapid Commun. Mass Spectrom.* **3**, 364 (1989).

[22] E. Teloy and D. Gerlich, *Chem. Phys.* **4**, 417 (1974).

[23] D. M. Mintz, C. A. Bitnott, and U. Steiner, *Proc. 29th ASMS Conf. Mass Spectrom. Allied Topics, Minneapolis, MN* p. 168 (1981).

[24] P. H. Dawson and J. H. Fulford, *Int. J. Mass Spectrom. Ion Phys.* **42**, 195 (1982).

[25] J. D. Ciupek, J. W. Amy, R. G. Cooks, and A. E. Schoen, *Int. J. Mass Spectrom. Ion Proc.* **65**, 141 (1985).

[26] C. Hagg and I. Szabo, *Int. J. Mass Spectrom. Ion Proc.* **73**, 295 (1986).

[27] P. E. Miller and M. B. Denton, *Int. J. Mass Spectrom. Ion Proc.* **72**, 223 (1986).

[28] A. E. Schoen and J. E. P. Syka, *Proc. 34th ASMS Conf. Mass Spectrom. Allied Topics, Cincinnati, OH* p. 722 (1986).

[29] J. E. P. Syka and I. Szabo, *Proc. 36th ASMS Conf. Mass Spectrom. Allied Topics, San Francisco, CA* p. 1328 (1988).

[30] J. J. Monaghan and B. Wright, *Proc. 36th ASMS Conf. Mass Spectrom. Allied Topics, San Francisco, CA* p. 819 (1988).

[31] A. J. Alexander and R. K. Boyd, *Int. J. Mass Spectrom. Ion Proc.* **90**, 211 (1989).

of biochemical applications, namely, high mass precursor ions requiring high collision energies to effect observable extents of CID.

These focusing properties are best described in terms of the secular motion of an ion through q (where the term is used with a particular mathematical connotation). Briefly, as a consequence of the exponential form of the Mathieu equations which describe the motion,[2] the mathematical solutions corresponding to stable trajectories are periodic functions with secular frequency f_{sec} given to a good approximation[22,32] by [Eq. (5)]:

$$f_{sec} = f[q_u/2(2)^{1/2}] \tag{5}$$

It is clear from Eq. (5) that the secular motion is different from the low-amplitude motions in which the ions follow the rf oscillations (frequency f) directly; the latter are superimposed upon the much larger amplitude of the slower secular motion. Ions which enter the q cell close to the central axis will, as a result of this secular periodicity, return close to the axis during their flight through q at regular invertals corresponding to one-half of the secular wavelength. The number of such crossings of the axis is then simply the number (n_{sec}) of half-wavelengths of this secular wavelength occurring within the cell length.

The significance of the quantity n_{sec} is that an *integral* number of half-wavelengths implies that ions focused on to the entrance aperture of q will thereby also be focused on to the exit aperture, thus maximizing transmission efficiency. If changes occur in ion mass (via reaction within q) or in velocity [magnitude $(2E_{lab}/m)^{1/2}$ and/or direction, caused by reaction and/or collisional scattering], Eqs. (4) and (5) imply that the ion trajectory now corresponds to a nonintegral value of n_{sec} so that the ions will arrive at the exit aperture of q with large radial displacements. Such loss of the direction-focusing condition corresponds to poor transmission efficiency. The use of small exit and entrance apertures for q, in order to effectively contain the collision gas, also exacerbates these directional-focusing effects which impose periodicities upon the transmission efficiency as any one of E_{lab}, f, or V_0 is varied. These effects are also more extreme if the incoming ion beam is tightly focused on to the entrance aperture and has a narrow spread in E_{lab}; these conditions are characteristic of the ion beam transferred from a sector analyzer into q, and much less so of that exiting from a quadrupole mass filter. There is, indeed, evidence to support the implication that these periodicities should be noticeably less marked for triple quadrupole than for hybrid instruments. The smearing effect of collisions, alluded to above, also tends to reduce

[32] R. E. March and R. J. Hughes, "Quadrupole Storage Mass Spectrometry." Wiley (Interscience), New York, 1989.

the periodic effects under the multiple-collision conditions often employed in analytical studies.

For a precursor ion of biochemical interest, e.g., $M = 1500-2000$ Da, which may require a large collision energy E_{lab} (e.g., 400–500 eV) to induce sufficient fragmentation, the result of these effects may be particularly severe. If low values of V_0 are used to ensure trajectory stability of low-mass product ions (see above), for high-mass precursors at high values of E_{lab} the value of n_{sec} will be less than unity and thus inherently noninte-gral.[21] The poor transmission efficiency, resulting from this lack of direc-tional focusing is believed[21] to be a principal reason for the current lack of success in operating some XqQ instruments at adequate sensitivity at collision energies E_{lab} in the range of a few hundreds electron volts, for high-mass precursor ions. Use of longer collision cells or higher values of the rf amplitude V_0 increases the value of n_{sec}, thus helping to ensure that the value of this parameter does not drop below unity.

Examples of some of these effects can be seen in Fig. 1. These data are for transmission of $Cs_4I_3{}^+$ precursor ions (m/z 912) and their dissociation products through the qQ section of a $BEqQ$ hybrid instrument, modified to provide higher values for V_0; the original work[21] demonstrated that these transmission characteristics were dominated by those of q rather than Q. The observed positions of the transmission maxima corresponded closely to integral values of n_{sec} evaluated from the theory, as indicated on the curves in Fig. 1.

More detailed discussion of this periodic response can be found else-where.[21] The curves for the product ions comprise more complex structure superimposed on a relatively more important general rise in transmission efficiency, which in some cases subsequently falls to low values at higher V_0 (Fig. 1). The onset of this last effect occurs at values of V_0 corresponding to the stability limit $q_u = 0.908$, as discussed above [Eq. (4)]. The local maxima in the product ion transmission curves are of two kinds, namely, those corresponding to integral values of n_{sec} calculated for the *precursor* ion, plus those calculated for the *product* ion actually observed. In general, the fragmentation reaction occurs at random throughout the length of q, and no structure is thus expected in the transmission curves. However, those product ions formed close to the entrance of q must traverse the full length, and this fraction of the total should thus display their own characteristic periodicity. Similarly, those product ions formed near the exit of q were derived from precursor ions which had already traversed almost the entire length, and will thus display the periodicity characteristic of the *precursor*, as demonstrated experimentally.[21] It is important to note (Fig. 1) that the periodic components are relatively much less significant

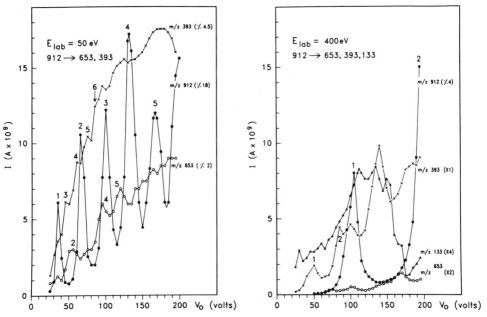

FIG. 1. Transmission of $Cs_4I_3^+$ precursor ions, and their CID fragment ions, through the deceleration lens/qQ assembly of a $BEqQ$ hybrid instrument. Data normalized to the beam current of precursor ions measured at the point of double focus of the BE stage, and plotted as a function of V_0 (the rf amplitude in q) for different values of E_{lab} (the axial translational energy of the precursor ions within q). Argon was used as collision gas, at a pressure corresponding to 50% attenuation of the precursor beam intensity. For all points plotted, the transmission of the mass filter Q was optimized [see text, Eq. (7)]. The integers labeling the maxima are the values of n_{sec}. [Adapted from *Rapid Commun. Mass Spectrom.* **3**, 364 (1989).]

for the product ions that for the precursors; the information content of an MS/MS spectrum is contained largely in the product ion signals rather than in the surviving precursor signal, so the practical consequences of the periodicities are somewhat less than might appear.

The periodicities (Fig. 1) are susceptible to a satisfactory interpretation. The general rise in transmission efficiency with V_0, which underlies these periodicities, is not so well understood but is clearly of great practical significance in the context of overall sensitivity. It probably corresponds[21] to an increased efficiency in containment of ions with large transverse velocity components arising from collisional scattering, etc., since a larger V_0 implies larger transverse containment forces. This intuitive suggestion has recently been supported by results of computer simulations of trajectories of precursor and product ions.[33]

From the point of view of this section, it is best to operate q at the largest available value of V_0 consistent with a transmission maximum (integral value for n_{sec} for the precursor), at the selected value of collision energy E_{lab}. However, such a choice also leads to a parallel increase in the low-mass limit for trajectory stability [$q_u = 0.908$, Eq. (4)], i.e., to loss in sensitivity for low-mass product ions (see above). Suitable compromises must be sought in practice, sometimes via suitable linkage of V_0 to the mass-filter function of Q; the details of how this linkage relationship is arranged will determine the relative transmission efficiencies for the product ions, i.e., will strongly affect the relative intensities in the product ion spectrum. Selection of an rf-only cell (q) with a long path length not only yields an intrinsic increase in n_{sec}, as discussed above, but also permits use of lower steady-state pressures of collision gas within q to achieve a given average number of collisions; in turn, this permits the use of larger entrance and exit apertures for q since the collision gas need not now be contained so stringently.

The present discussion has centered on optimizing experimental conditions within q to obtain usuable product ion spectra for difficult cases, without being concerned about reproducibility of relative intensities. However, Fig. 1 is illuminating in this regard, since such a spectrum is represented by relative intensities at a fixed value of V_0, or else at intersections along a curve determined by the ramping of V_0 with the mass filter function of the final quadrupole Q (see above). In either case it is clear that the relative intensities are functions of the detailed settings or linked-scan settings for V_0; it is thus not surprising that such divergent results were obtained in the round robin[18] in which the value of V_0, or its variation relative to the mass–filter function of Q, was not controlled. The relative intensities of the precursor ions were particularly irreproducible, reflecting the dominance of the periodic component in the response curves for these ions (Fig. 1).

Translational Energy Considerations

The discussion thus far has concentrated upon the transmission properties of the rf-only quadrupole reaction cell q, since these dominate the overall sensitivity of tandem mass spectrometry experiments in XqQ instruments as well as the relative intensities of different product ions.[21]

---

[33] S. C. Davies and B. Wright, *Rapid Commun. Mass Spectrom.* **4**, 186 (1990).

However, the final Q mass filter does affect the outcome in ways which are additional to those characteristic of stand-alone mass filters, e.g., the well-known "quadrupole roll-off" of transmission efficiency at higher m/z values.[2]

The most important additional effect concerns the translational energy of the product ions, formed within q and subsequently presented to Q for mass analysis. To a first approximation, *assuming negligible fractional loss of translational energy in the activating collisions within q* and ignoring effects of translational energy release in the dissociation itself,[34] the translational energy E_P of P^+ within q is [Eq. (6)]:

$$E_P = E_{lab}(M_P/M_R) \qquad (6)$$

where E_{lab} is the translational energy of R^+ within q (the collision energy). Unlike the case of a stand-alone mass filter supplied with ions formed within an ionization source, the translational energies of the product ions are now strongly dependent upon their masses [Eq. (6)]. For efficient operation of Q in an XqQ instrument, some compensation for this dependence is essential; this is done by linking V_Q (the float potential of Q) to its mass filter function **Q**, i.e., to the mass M_P of the product ion currently transmitted. One possible functional relationship used[31] to optimize this aspect, based upon the theoretical relationship Eq. (6), is [Eq. (7)]:

$$V_Q = V_s + [(M_P/M_R) - 1]E_{lab} - [kM_P + C] \qquad (7)$$

The collision energy E_{lab} is defined experimentally [Eq. (6)] as $ze(V_s - V_q)$, where V_s and V_q are the float potentials applied to ion source and to q, respectively. The second term in square brackets in Eq. (7) represents a linear ramp (operator-tuned slope k and low-mass intercept C) intended to optimize the transmission of ions of higher mass; such a ramp is normally applied even in the case of stand-alone mass filters [corresponding to $M_P = M_R$ in Eq. (7), so that the first bracket is zero, in this case]. The first bracket is a correction applied automatically[31] for the partitioning [Eq. (6)] of E_{lab} between P^+ and the neutral fragment; the value of M_R in Eq. (7) is readily derived experimentally from the (linear) mass filter function **Q₁** in a triple quadrupole, and less easily in the case of a hybrid from the combination of magnetic field strength **B** (Hall effect probe signal) and source potential V_s used to provide the mass indicator reading for the magnetic sector analyzer.

That this relationship [Eq. (7)] is not universally valid was shown[35] by direct experimental measurements of E_P in one case; the simple linear

---

[34] R. N. Hayes and M. L. Gross, this volume [10].

[35] B. Shushan, D. J. Douglas, W. R. Davidson, and S. Nacson, *Int. J. Mass Spectrom. Ion Phys.* **46**, 71 (1983).

relationship [Eq. (6)], upon which Eq. (7) was based, was shown to be inadequate in the case thus investigated. Presumably this was due in part[35] to significant transformations of translational energy into internal energy, as a prerequisite for fragmentation, which varied with the reaction channel and thus with the mass of P^+. To some extent, the superimposed linear ramp [last term in Eq. (7)] can compensate[31] for such inadequacies.

In any event, Eqs. (6) and (7) are expected to be valid for unimolecular dissociation reactions (zero loss of translational energy in activating collisions), and approximately so[35] under conditions such that the majority of precursor ions undergo only a single collision during their passage through the reaction cell. (Such "single-collision conditions" can be characterized via a low fractional suppression of the intensity of the precursor ion beam.[34]) This low-collisional suppression extreme of operating conditions is exemplified by experiments in which the rleative intensities in the product ion spectra are required to be reproducible in some sense. Examples of such experiments include acquisition of product ion spectra for interlaboratory comparisons; the implications for the transmission of ions from q into Q are that the product ions will possess considerable residual translational energy [up to the ideal values predicted by Eq. (6)], and can thus be efficiently transported[31] from q to Q by suitable manipulation [Eq. (7)] of V_Q.

At the other extreme, of very high pressures of collision gas, corresponding to multiple-collision conditions where the ions may have very little (or even zero) translational energy within q, Eqs. (6) and (7) are wholly invalid and the best method of extraction of the ions from q and transmission into Q is now quite different. Simple application of a large difference $(V_q - V_Q)$ in rod offset potentials does indeed suck the "stopped" ions out of q, but results in a large value for E_{lab} in Q and thus serious deterioration in peak-shape. This problem is very important in the present context of biochemical applications, since extreme multiple-collision conditions offer one solution to the problem of inducing a useful degree of fragmentation of large organic ions at the lower values of E_{lab} characteristic of XqQ instruments. A better solution to the problem of efficient extraction of "stopped" ions, commonly practiced in triple quadrupole instruments, is to insert an extraction lens between q and Q thus permitting independent optimization of V_Q; there is no problem in principle in applying this solution to hybrid instruments, but in practice it would require yet another independently controlled high-voltage power supply to provide the float potential for the extraction lens (see Fig. 8 and related discussion below).

One final question will be dealt with briefly in this section. It is common to operate an XqQ instrument with values of $(V_s - V_q)$ of 20 to 50 V, and an rf amplitude V_0 (for q) of 40 to 100 V or even higher. It thus appears that the collision energy associated with the transverse motion (proportional to V_0) should dominate that derived from the axial motion of R^+ [E_{lab} = $ze(V_s - V_q)$]. That the transverse motion does *not* make a significant contribution was first experimentally demonstrated in a qualitative way by comparison of CID spectra[17] obtained by using different values of V_0; more recent experimental investigations,[31,36] via studies of competing fragmentation reactions for which some knowledge of the energy dependence is available, have confirmed this general conclusion in a more quantitative way. The inefficiency of the rf field in inducing ion dissociation is due to the high rf frequency f, which is too fast for the ions to follow. Thus studies on energy dependence of collision-induced dissociation are indeed justified in considering only the axial energy $E_{lab} = ze(V_s - V_q)$, except near the reaction threshold where the contribution of the transverse motion to the collision energy may become an appreciable fraction of the total.

Related Devices

Hexapoles and octapoles are similar to quadrupoles, but incorporate six or eight uniformly spaced rods, respectively, instead of four. Theory[26] and experiment[29,30] agree that, in the rf-only mode, these higher-order fields should not give rise to the marked periodicities evident in the transmission curves for their quadrupole counterparts (Fig. 1). This striking difference between rf-only quadrupoles and their higher-order counterparts is related, at least in part, to the insensitivity of the *phase* of the secular motion through a quadrupole to the position and angle of entry of an ion into the device[33]; this leads to a high degree of coherence of the motions of precursor ions with a range of initial conditions at the entrance to the quadrupole. In contrast, the higher-order multipole rf fields are characterized[33] by a strong dependence of the phase of the (more complex) secular motion upon the initial entry conditions; since the latter are distributed over a range of values, the observable result corresponds to an integration over phases, from which essentially all periodic features have been averaged out. At the very least, this property offers the possibility of less demanding tuning procedures required of the user, and commercial manufacturers now offer Q_1hQ_2 and Q_1oQ_2 instruments, where h and o denote rf-only hexapole and octapole, respectively. What is not clear, at present, is whether or not a properly tuned rf-only quadrupole (see above) would provide greater transmission efficiency for product ions, thus justifying the additional effort required in the tuning procedure. Considerations

[36] J. E. Fulford and B. Shushan, *Proc. 31st ASMS Conf. Mass Spectrom. Allied Topics,* Boston, MA p. 322 (1983).

of interlaboratory reproducibility of CID spectra are, in any case, pertinent only at or near the single-collision limit[19,20]; in the present context of structure elucidation of large organic ions typical of biochemical applications, the principal problem is to obtain sufficient fragmentation to provide worthwhile information.

Finally, in this section it seems appropriate to acknowledge the detailed theoretical foundations for an understanding of rf quadrupoles, provided by several pioneering authors and which underly the above brief summary. Notable among these are the contributions of Dawson[37-39]; a recent monograph[32] contains an excellent updated account of this detailed theory.

Triple Quadrupole Tandem Mass Spectrometers

The triple quadrupole implementation of MS/MS is a particularly important one, since its introduction, especially as commercial computer-controlled instruments, transformed MS/MS from a research technique for physics and physical chemistry into a widespread, popular technique for analysis. Indeed, a majority of the instruments currently in use for analytical MS/MS are of this Q_1qQ_2 configuration.

The first tandem quadrupole instruments were designed for studies of ion–molecule reactions.[40,41] These were double quadrupole (Q_1Q_2) instruments, with a field-free collision region situated between the two mass filters. Triple quadrupole (Q_1qQ_2) instruments soon followed, employing an rf-only quadrupole as a photodissociation region.[13,14] Even at pressures in q of 10^{-8} torr, however, the ions produced by CID exceeded those from photodissociation,[14] suggesting the high efficiency that could be achieved.

The first computer-controlled triple quadrupole system designed for analytical MS/MS was developed by Yost and Enke[15,16]; their early studies were performed on one of the early photodissociation systems,[14] modified to perform CID.[17] The earliest applications of this instrument were for direct mixture analysis[16] and structure elucidation.[42] The commercial introduction of computer-controlled Q_1qQ_2 instruments soon followed; a few hundred instruments have now been delivered by several manufacturers. Modern triple quadrupole instruments generally offer mass range to 4000 Da, collision energies to 200 eV, complete computer control and automa-

[37] P. H. Dawson (ed.), "Quadrupole Mass Spectrometry and Its Applications." Elsevier, Amsterdam, 1976.
[38] P. H. Dawson, Adv. Electron. Electron Phys. 53, 153 (1980).
[39] P. H. Dawson, Adv. Electron. Electron Phys. Suppl. 13B, 173 (1980).
[40] T.-Y. Yu, M. H. Cheng, V. Kempter, and F. W. Lampe, J. Phys. Chem. 76, 3321 (1972).
[41] C. R. Iden, R. Lairdon, and W. S. Koski, J. Chem. Phys. 21, 349 (1976).
[42] R. A. Yost and C. G. Enke, Am. Lab. 88 (June) (1981).

tion of all instrumental parameters, and sophisticated instrument control and data systems.

Comparison to Other MS/MS Instruments

The major advantages of the triple quadrupole approach to MS/MS include relatively low cost, instrumental simplicity, and ease of computer control. Disadvantages include limited mass range, inability to separate precursor ions with high resolution or to provide accurate mass measurement (to within a few parts per million) of ions formed in the ion source, and restriction to low-energy CID. Some of these features merit further discussion.

The Q_1qQ_2 offers a degree of instrumental simplicity not found with sector tandem mass spectrometers. Since quadrupoles optimally operate on ions with energies below 100 eV, there is no need to float the ion source (see discussion of Figs. 7 and 8 below); this provides for easy interfacing with a variety of sample inlets. Furthermore, it eliminates arcing problems observed due to contamination of sources floated at high potential. Indeed, the tolerance of ion source contamination is accentuated by the common practice of interchanging ion volumes within the ion source without breaking vacuum. The linear scan function of the quadrupole mass filter greatly simplifies the implementation of the variety of scan modes useful for analytical MS/MS (e.g., product, precursor, and neutral loss scans). In addition to the rapid speeds achievable for these scans, the ability to quickly switch between masses makes multiple selected-reaction monitoring (SRM) experiments practical, even on the chromatographic time scale. The fact that quadrupoles alone are true mass analyzers (i.e., their separation of ions is a function of only m/z, and is independent of the velocity of the ions) greatly simplifies their control in MS/MS, in which the conversion of precursor ions to product ions involves not only a change in m/z, but also a change in (and generally a range of) ion energy. This instrumental simplicity, combined with the ease of computer control and the large number of instruments delivered, has led to an early maturity in computer control and data processing software which is only now being approached by other MS/MS systems. Finally, the independence on ion polarity of the quadrupole's mass separation, combined with the use of conversion dynodes and the need to switch only low voltages, makes it routine to switch between positive and negative ions on a millisecond time scale.

The CID process is extremely efficient at low collision energies (<100 eV),[17] at least for ions <1000 Da or so. This is largely due to the strong focusing nature of the rf-only field of the collision quadrupole, which is able to focus ions scattered by collisions through large angles and thus transmit them on toward the second mass analyzer. This also results in the ability to employ relatively high-collision gas pressures (a few millitorr) to effect multiple collisions. Those ions scattered through the largest angles are generally those to which the most internal energy has been imparted; it is thus particularly important to collect these ions. In contrast, in the field-free collision region of a sector MS/MS instrument, scattering of the ions is in direct competition with collisions which produce detectable product ions (i.e., those which remain on course through ensuing slits). The high efficiency of the low-energy CID process is reflected in the frequent reporting of subpicogram detection limits by triple quadrupole MS/MS.[43]

A further advantage that accrues from the use of low-energy collisions is a versatility not available with high-energy collisions. This versatility includes the straightforward acquisition of energy-resolved CID spectra and ready observation of associative ion–molecule reactions on triple quadrupole instruments.[44] Detailed energy-resolved CID "breakdown curves" can routinely and rapidly (and even automatically) be obtained on modern triple quadrupole instruments[44] as an aid in optimizing collision conditions or for differentiating isomeric ion structures; this is, in principle, a feature of all XqQ instruments, not only triple quadrupoles. The acquisition of similar data by angle-resolved experiments on all-sector MS/MS instruments is far from routine.[34] Associative ion–molecule reactions[44] offer the opportunity to add chemical selectivity to the collision process (e.g., to differentiate isomeric ion structures[45,46]) and are easily implemented on triple quadrupole and other XqQ instruments. A recent application of biochemical interest is the modeling of *in vivo* reactions involved in carcinogenesis by gas-phase ion–molecule reactions.[47] In this study, reactions between DNA nucleosides or bases (or model bases) and electrophiles (introduced via capillary gas chromatography) were studied in either the ion source or the center quadrupole of the triple quadrupole. The results suggest that this method may permit screening of complex mixtures for potential carcinogenic components, as well as predicting their carcinogenic potential.

One major limitation of current triple quadrupole mass spectrometers is their limited mass range (maximum of 4000 Da). It is informative, however, that some of the highest-mass mass spectrometry ever per-

[43] J. V. Johnson and R. A. Yost, *Anal. Chem.* **57**, 758A (1985).

[44] D. D. Fetterolf and R. A. Yost, *Int. J. Mass Spectrom. Ion Proc.* **62**, 33 (1984).

[45] D. D. Fetterolf, R. A. Yost, and J. R. Eyler, *Org. Mass Spectrom.* **19**, 104 (1984).

[46] R. Kostiainen and S. Auriola, *Rapid Commun. Mass Spectrom.* **2**, 135 (1988).

[47] J. A. Freeman, J. V. Johnson, M. E. Hail, R. A. Yost, and D. W. Kuehl, *J. Am. Soc. Mass Spectrom.* **1**, 110 (1990).

formed (for 2-propanol clusters up to 60,000 Da) was performed with a quadrupole mass filter.[48] As is common even today in mass spectrometry above 30,000 Da, no attempt was made to achieve unit mass resolution in these studies. The real limitation for high-mass MS/MS with triple quadrupole instruments, then, is the same as for other MS/MS instruments, the low efficiency of CID at high mass.[32] An alternate approach to the analysis of high-mass molecules is to produce from them multiply charged ions such that the m/z (as opposed to the mass) is within the range of common triple quadrupole instruments. This is achieved in an elegant way with electrospray ionization (see below).

Instrumental Considerations

There are two instrumental issues which bear special consideration in the discussion of triple quadrupole instruments: potential problems with the line-of-sight geometry and the possible need for an intermediate detector following Q_1. These are detailed below.

The line-of-sight geometry afforded by the quadrupole contributes to the mechanical simplicity of the Q_1qQ_2 instrument. However, it also makes these instruments less resistant than sector instruments to "neutral noise," in which fast neutrals and photons produced in the ion source reach the detector. The common use of detectors which have their entrances off-axis prevents this source of noise from being significant in most cases. However, in the case of fast atom bombardment (FAB), the large number of fast atoms deflected off the probe tip down the axis of the quadrupole may, upon striking surfaces near the detector, produce charged particles which may be detected as noise. Most off-axis detectors still have on-axis surfaces which lead to neutral noise. One method to significantly reduce this noise is to move the detector farther away from the ion source, as is accomplished with triple quadrupole instruments, especially given the restrictive apertures often employed to aid in pressurizing the collision cell. A more successful approach is to bend the ion beam axis of the instrument somewhere between the source and Q_2. Any charged particles produced by collisions will be strongly discriminated against by the mass filtering action of Q_2. This can be efficiently accomplished via the use of a bent quadrupole (or octapole) collision cell.[49] This approach reduces the neutral noise observed in FAB–MS/MS experiments by two orders of magnitude; it appears to impose no particle restrictions upon the MS/MS process, aside from the need to maintain enough rf amplitude in the bent

[48] R. J. Beuhler and L. Friedman, *Nucl. Instrum. Methods* **170**, 309 (1980).
[49] J. E. P. Syka and A. E. Schoen, *Int. J. Mass Spectrom. Ion Proc.*, in press.

rf-only quadrupole (or octapole) device to successfully steer ions of high mass or high kinetic energy efficiently around the bend.

A second instrumental consideration is whether an intermediate detector (after the first mass filter) is useful, in analogy to the intermediate detectors employed in sector and hybrid tandem mass spectrometers. The ability to switch either mass filter into rf-only (total-ion) mode makes it possible to nearly instantaneously switch a Q_1qQ_2 instrument between MS/MS and MS modes of operation, using only a single detector after the third quadrupole. Indeed, it is not even necessary to remove the collision gas from the collision cell; assuming that the collision cell pressure is low enough to provide reasonable collection efficiencies (i.e., small scattering losses), the mass spectrum obtained by scanning Q_1 will be little affected by CID occurring in q, since Q_2 will pass nearly all m/z ions with similar efficiency.

Examples of Biochemical Applications of Triple Quadrupole Instruments

As a result of the large number of triple quadrupole tandem mass spectrometers in use, there is an enormous number of examples of biochemical applications in the literature. Below are described a few select examples which illustrate the capabilities of these systems.

FAB–MS/MS Analysis of Peptides. The combination of MS/MS with FAB for the sequencing of peptides is a promising one. Whereas the FAB mass spectra of low-molecular weight peptides often contain useful sequence ions, higher mass peptides often fail to show adequate fragmentation to useful sequence ions in the FAB mass spectrum itself. Further complicating the problem is the "chemical noise" throughout the mass spectrum produced from the FAB matrix, making it difficult to pick sequence ions (especially for unknown peptides) out of the "grass." Tandem mass spectrometry is well suited to the task of reducing this chemical noise problem by mass selecting the FAB-produced peptide ions and further fragmenting them.

The first report of MS/MS employing FAB and CID for the sequencing of oligopeptides utilized the triple quadrupole instrument in Hunt's laboratory.[50] Previous MS/MS studies of peptide sequencing by triple quadrupole MS/MS[51] predated the FAB technique, and required the derivatization of the peptides to enhance volatility. Low-energy CID produces product ions

[50] D. F. Hunt, W. M. Bone, J. Shabanowitz, J. Rhodes, and J. M. Ballard, *Anal. Chem.* **53,** 1704 (1981).
[51] D. F. Hunt, A. M. Buko, J. M. Ballard, J. Shabanowitz, and A. B. Giordani, *Biomed. Mass Spectrom.* **8,** 397 (1981).

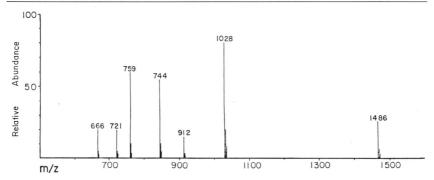

FIG. 2. The normal FAB mass spectrum of one HPLC fraction of the soluble peptides produced by trypsin digestion of a cyanogen bromide-treated apolipoprotein B (molecular mass ~500,000 Da).

characteristic of both the N and C termini of the peptide (types a, b, and c, as well as x, y, and z). Figure 2 shows the normal FAB mass spectrum of one high-performance liquid chromatography (HPLC) fraction of the soluble peptides produced by trypsin digestion of a cyanogen bromide-treated apolipoprotein B (molecular mass ~500,000 Da).[52] Figure 3 shows the product ion spectrum of m/z 912 from Fig. 2, the molecular ion of one of the tryptic peptides in the fraction, obtained under CID conditions of 20 eV collision energy and 5 mtorr of argon. The utility of this approach for sequencing proteins via MS/MS is clear. In an elegant extension to this work,[51] the derivatization of the N terminus of each peptide in the digest mixture to form mixed [$^2H_0/^2H_3$]acetyl derivatives permits "labeling" of the N terminus. Detuning the first mass filter to pass a window of a few m/z values (uniquely possible on the triple quadrupole instrument) provides a product ion spectrum containing both labeled and unlabeled N terminus product ions (series A) as doublets separated by 3 Da.

A unique hybrid quadrupole/FT tandem mass spectrometer has been developed in Hunt's laboratory specifically for such protein sequencing experiments.[53] In this instrument, ions produced by FAB are mass-selected (as either a single m/z or a range of m/z values) by a quadrupole mass filter, and transmitted via an rf-only quadrupole into a Fourier transform ion cyclotron resonance cell (FT-ICR), where they are stored, mass-selected, subjected to low-energy CID, and the product ion spectra recorded. More recent studies have employed laser photodissociation of

[52] D. F. Hunt, J. R. Yates, III, J. Shabanowitz, S. Winston, and C. R. Hauer, *Proc. Natl. Acad. Sci. U.S.A.* **83**, 6233 (1986).

[53] D. F. Hunt, J. Shabanowitz, J. R. Yates III, R. T. McIver, Jr., R. L. Hunter, J. E. P. Syka, and J. Amy, *Anal. Chem.* **57**, 2728 (1985).

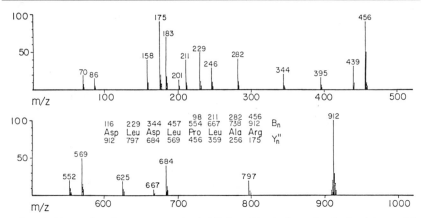

FIG. 3. The product ion spectrum of m/z 912 from Fig. 2, the molecular ion of one of the tryptic peptides in the fraction, obtained under CID conditions of 20 eV collision energy and 5 mtorr of argon; triple quadrupole instrument, E_{lab} 20 eV, argon collision gas.

peptide ions stored in the ICR cell.[54] The product ion spectrum of the m/z 1772.9 $[M + H]^+$ ion of 10 pmol of a 15-residue tryptic peptide from digestion of beef spleen purple acid phosphatase, a 38,000 Da glycoprotein of unknown structure, is shown in Fig. 4.

Two related developments in the sequencing of peptides by FAB and tandem mass spectrometry employing low-energy CID deserve brief mention here. One is the use of a pentaquadrupole $(Q_1qQ_2qQ_3)$ instrument to obtain reaction-intermediate spectra [i.e., an MS/MS/MS scan in which the precursor and final product ion m/z values are fixed, while the intermediate mass analyzer (Q_2) is scanned] (see later, Ref. 104). This provides a spectrum of those sequence ions produced by CID of the precursor ion selected by Q_1 (and hence of a selected-peptide molecular ion) which then further fragments via a second stage of CID to produce a characteristic product ion (e.g., a C or N terminus ion) selected by Q_3. This MS/MS/MS approach is discussed in more detail below in the section on applications of hybrid instruments, where it is compared to the analogous experiment conducted (see later, Ref. 104) using a $BEqQ$ hybrid. It is important to point out here, however, that the pentaquadrupole approach eliminates artifact peaks observed with the particular technique (linked scan at constant B/E) used (see later, Ref. 104) with the hybrid instrument, and provides abundant fragmentation over the entire mass range through effi-

[54] D. F. Hunt, J. Shabanowitz, and J. R. Yates III, *J. Chem. Soc. Chem. Commun.* p. 548 (1987).

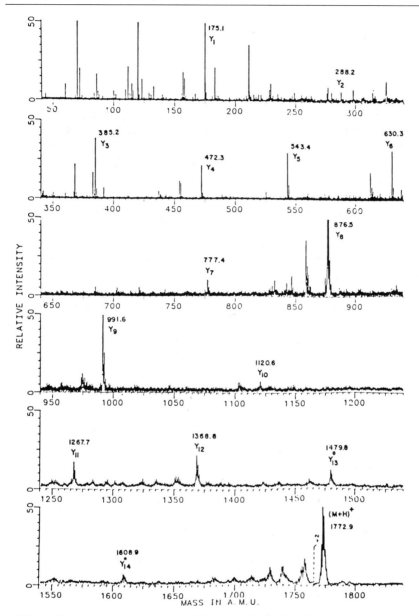

FIG. 4. The product ion spectrum of the m/z 1772.9 $[M + H]^+$ ion of 10 pmol of a 15-residue tryptic peptide from digestion of beef spleen purple acid phosphatase, a 38,000 Da glycoprotein of unknown structure; laser photodissociation in the FT-ICR cell of a Qq(FT-ICR) hybrid instrument.

cient low-energy CID under multiple collision conditions (see later, Ref. 104). Preliminary experiments employing MS/MS with FAB and low-energy CID on a quadrupole ion trap mass spectrometer[55] have demonstrated the capability to obtain peptide sequence information with incredibly high sensitivity (product ion mass spectra on subfemtomole levels of gramicidin S, 1142 Da), as well as capability for MS^n.

Electrospray–MS/MS Analysis of Peptides. A number of researchers have recently demonstrated the utility of electrospray ionization for the analysis of biological molecules.[56–60] Since the electrospray mechanism involves the desorption of ions from solution into the gas phase without fragmentation, multiply charged (usually multiply protonated) ions are produced with a range of charges. For proteins and peptides, the range of charges that are observed depends upon the number of basic residues (arginine, lysine, or histidine) in the molecule. This multiple charging phenomenon has permitted direct molecular mass analysis of proteins up to 133,000 Da on conventional quadrupole mass spectrometers, with precisions and accuracies approaching 0.01%.[57] In addition, triple quadrupole MS/MS of multiply charged peptide ions produced by electrospray suggests that this approach may be particularly useful for sequence determination at high sensitivity.[59]

In the case of tryptic peptides, predominantly doubly charged ions are produced. Figure 5 shows a typical electrospray mass spectrum of a tryptic peptide obtained on a triple quadrupole instrument, demonstrating the production of the intense doubly protonated molecular ion at m/z 786 and the singly protonated molecular ion (m/z 1571) at much lower intensity.[60] Presumably, the peptide protonates on the free amino group at the N terminus (Glu) and on the arginine at the C terminus. A calculated 1.7 pmol of peptide was consumed for this analysis, providing ample intensity and signal-to-noise.

The doubly charged tryptic peptide ions produced by electrospray are well suited for MS/MS analysis. One advantage is that CID of the doubly charged precursor ion produces predominantly singly charged product

[55] R. E. Kaiser, R. G. Cooks, J. E. P. Syka, and G. C. Stafford, *J. Am. Chem. Soc.,* in press.

[56] C. M. Whitehouse, R. N. Dryer, M. Yamashita, and J. B. Fenn, *Anal. Chem.* **57,** 675 (1985).

[57] T. R. Covey, R. F. Bonner, B. I. Shushan, and J. D. Henion, *Rapid Commun. Mass Spectrom.* **2,** 249 (1988).

[58] J. A. Loo, H. R. Udseth, and R. D. Smith, *Anal. Biochem.* **179,** 404 (1989).

[59] E. D. Lee, J. D. Henion, and T. R. Covey, *J. Microcolumn Sep.* **1,** 14 (1988).

[60] M. E. Hail, S. Lewis, J. Zhou, J. Schwartz, I. Jardine, and C. Whitehouse, *in* "Current Research in Protein Chemistry" (J. J. Villafranc, ed.), in press. Academic Press, San Diego, 1990.

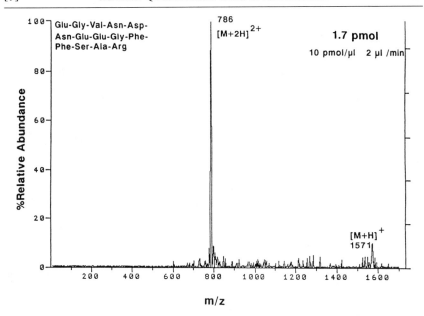

FIG. 5. The electrospray mass spectrum of 1.7 pmol of a tryptic peptide (Glu-fibrinopeptide B) of molecular mass 1570 Da; triple quadrupole instrument (Q_1qQ_2 operated in single MS mode (i.e., qqQ).

ions.[60] This may result from localization of the charge at both ends of the peptide. In addition, initial experiments[60] indicate a remarkably high fragmentation efficiency upon CID of the doubly charged molecular ions, significantly higher than that obtained upon CID of singly charged molecular ions. This should increase the mass range of peptides that can be efficiently fragmented and sequenced via low-energy CID. For example, Fig. 6 shows that CID of the doubly charged molecular ion from Fig. 5 produces abundant product (sequence) ions under relatively modest CID conditions (collision energy and collision gas pressure); in contrast, very little fragmentation of the singly charged molecular ion was observed under the same collision energy and pressure. It should be noted that the singly charged molecular ion could be fragmented, but only at two times higher collision gas pressures.

Although only a limited number of peptides have been analyzed to date by these methods,[60] these initial studies indicate that the MS/MS spectra from electrospray-produced doubly charged molecular ions may be more readily interpreted than MS/MS spectra from FAB-produced singly charged molecular ions. For example, it appears that predominantly y ions are produced from CID of doubly charged molecular ions. One disadvan-

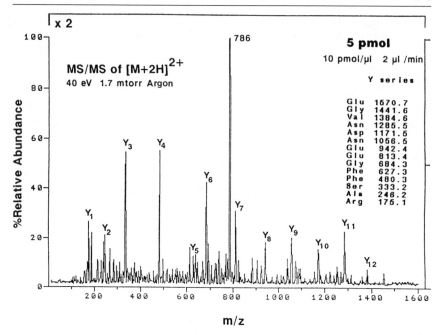

FIG. 6. The product ion spectrum of the doubly protonated molecule $(M + 2H)^{2+}$ from Fig. 5; triple quadrupole instrument.

tage is that insufficient overlapping sequence information is sometimes obtained to completely define the sequence; for example, complementary *b* ions are often absent from the electrospray–MS/MS data. The extent of the utility of this approach for protein sequencing remains to be fully evaluated; nevertheless, it is clear that electrospray in conjunction with triple quadrupole MS/MS is promising for peptide sequencing at the sub-10 pmol level.

Hybrid Sector-Quadrupole Mass Spectrometer

Although the first published account[12] (in 1965) of a tandem mass spectrometer incorporating an rf-only quadrupole involved an *EBqQ* hybrid, for several years thereafter triple quadrupole instruments dominated this area of analytical instrumentation. Although other hybrid instruments have been constructed subsequently for specialized applications in physical chemistry,[61-67] the more recent popularity of hybrid instruments in analytical chemistry originated with work of Glish[68] and Cooks in the late 1970s on design and construction of a BqQ instrument. As elaborated further below, the design problems are far from trivial, and the principles thus established[68] were the basis of subsequent developments by Cooks and collaborators[11,25,69-71] and by several other groups.[72-78] Notable among these investigations was construction and evaluation of a simple EQ combination[73,74] (no q cell), which permitted experimental evaluation of concepts which had been delineated previously[72]; this work[72] was important in identifying the $BEqQ$ configuration as that most likely to provide the greatest flexibility in operation of a hybrid instrument as a tandem mass spectrometer.[11,25,71,77] One of the most recent developments[78] has been the construction of a QEB instrument; this design has some intrinsic advantages for tandem mass spectrometry, including its potential to provide medium-to-high resolution spectra of both precursor and product

[61] J. J. Leventhal and L. Friedman, *J. Chem. Phys.* **49**, 1974.(1968).

[62] P. C. Cosby and T. F. Moran, *J. Chem. Phys.* **52**, 6157 (1970).

[63] J.-P. l'Hote, J. Ch. Abbe, J. M. Paulus, and R. Ingersheim, *Int. J. Mass Spectrom. Ion Proc.* **7**, 309 (1971).

[64] A. Giardini-Guidoni, R. Platania, and F. Zocchi, *Int. J. Mass Spectrom. Ion Phys.* **13**, 453 (1974).

[65] T. F. Thomas, F. Dale, and J. F. Paulson, *J. Chem. Phys.* **67**, 793 (1977).

[66] S. S. Medley, *Rev. Sci. Instrum.* **49**, 698 (1978).

[67] P. B. Armentrout and J. L. Beauchamp, *J. Am. Chem. Soc.* **102**, 1736 (1980).

[68] G. L. Glish, Ph.D. Thesis, Purdue University, West Lafayette, IN, 1980.

[69] S. A. McLuckey, G. L. Glish, and R. G. Cooks, *Int. J. Mass Spectrom. Ion Phys.* **39**, 219 (1981).

[70] G. L. Glish, S. A. McLuckey, T. Y. Ridley, and R. G. Cooks, *Int. J. Mass Spectrom. Ion Phys.* **41**, 157 (1982).

[71] A. E. Schoen, J. W. Amy, J. D. Ciupek, R. G. Cooks, P. Dobberstein, and G. Jung, *Int. J. Mass Spectrom. Ion Proc.* **65**, 125 (1985).

[72] J. H. Beynon, F. M. Harris, B. N. Green, and R. H. Bateman, *Org. Mass Spectrom.* **17**, 55 (1982).

[73] F. M. Harris, G. A. Keenan, P. D. Bolton, S. B. Davies, S. Singh, and J. H. Beynon, *Int. J. Mass Spectrom. Ion Proc.* **58**, 273 (1984).

[74] C. W. Trott, F. M. Harris, M. S. Thacker, S. Singh, and J. H. Beynon, *Comput. Enhanced Spectrosc.* **2**, 43 (1985).

[75] R. H. Bateman, B. N. Green, M. H. Tummers, and D. C. Smith, *Fresenius Z. Anal. Chem.* **316**, 217 (1983).

[76] L. C. E. Taylor, R. S. Stradling, K. L. Busch, and W. C. Qualls, *Proc. 32nd ASMS Conf. Mass Spectrom. Allied Topics, San Antonio, TX* p. 384 (1984).

[77] A. G. Harrison, R. S. Mercer, E. J. Reiner, A. B. Young, R. K. Boyd, R. E. March, and C. J. Porter, *Int. J. Mass Spectrom. Ion Proc.* **74**, 13 (1986).

[78] G. L. Glish, S. A. McLuckey, E. H. McBay, and L. K. Bertram, *Int. J. Mass Spectrom. Ion Proc.* **70**, 321 (1986).

ions. At present, there does not appear to be much activity in further development of this concept, at least in part due to uncertainty as to the degree to which the initial Q analyzer compromises the high-performance characteristics of the double-focusing stage. A possible application of such instruments would involve operation in a qEB configuration (i.e., the quadrupole in rf-only mode) to permit easier interfacing of an electrospray ionization source[56-60] (operating at atmospheric pressure) to a double-focusing analyzer.

An earlier review of hybrid instruments,[79] and a recent monograph,[80] contain a great deal of information concerning configurations for hybrid tandem mass spectrometers, their scan modes for realizing the various experiments discussed briefly above, and published applications. In what follows below, the emphasis will be placed upon further descriptions of design and operating principles to assist the nonspecialist in appreciating the strengths and limitations of this approach, together with recent examples of interest to the biochemist wishing to solve real problems.

Interfacing Sector Analyzers to rf Quadrupoles

There are two principal problems associated with transferring an ion beam from a sector analyzer to an rf quadrupole. The first of these concerns the mismatch between the pertinent translational energies, that through a sector being typically several kiloelectron volts and that through an rf quadrupole two orders of magnitude smaller. The only available sink for translational energy of an ion in a high-vacuum enclosure is electromagnetic potential energy, and this is almost invariably provided by floating the entire qQ assembly (*and* its associated dc and rf power supplies) to a potential close to that of the ionization source. The second problem arises from the shapes of the ion beam cross sections in the two cases; a sector analyzer is a dispersive device,[2] the mass selection being provided by optical slits which can be closed down to extremely narrow widths in the dispersive direction. In order to achieve useful beam intensity, the extension of the slits in the nondispersive direction is made as large as possible. Thus, the beam emerging from the image slit of a sector analyzer is typically 10 mm long in the nondispersive dimension and perhaps 0.05 mm wide. This long narrow ribbon must be transformed into a roughly circular shape, suitable for transmission through a quadrupole; this has to be achieved simultaneously with deceleration of the beam by running it

---

[79] G. L. Glish and S. A. McLuckey, *Anal. Instrum.* **15**, 1 (1986).

[80] K. L. Busch, G. L. Glish, and S. A. McLuckey, "Mass Spectrometry/Mass Spectrometry: Techniques and Applications of Tandem Mass Spectrometry." VCH Publ., New York, 1988.

up an "electrostatic hill," and all with minimum intensity loss! The sequence of ion optical elements which accomplishes this dual purpose will, for convenience, be referred to below as the deceleration lens.

These two problems are peculiar to hybrid instruments; the energy mismatch can be visualized from Figs. 7 and 8. The diagrams representing Q_1qQ_2, $BEqQ$, and $B_1E_1E_2B_2$ tandem mass spectrometers in Fig. 7 serve as aids in understanding the corresponding graphs (Fig. 8) of electrical potential experienced by the ions as they traverse the various instruments. No element of the triple quadrupole instrument is at a potential more than a few tens of volts from ground, apart from the conversion dynode and the detector.[81] In the case of the four-sector instrument,[3] the ion source is necessarily at several kilovolts from ground, but no other element *need* be, although it is common (but not universal) practice to use the collision cell at a potential about 50% of that of the ion source to improve the transmission properties of the E_2B_2 fragment analyzer for low-mass product ions.[82] Moreover, in both of these cases (Q_1qQ_2 and $B_1E_1B_2E_2$), the beam cross sections remain the same shapes throughout (approximately circular and long, narrow ribbon, respectively).

However, in the case of the hybrid designs there is no escape from the requirements to float the qQ section and its power supplies, and to change the basic shape of the beam cross section, as described above. It should be recalled that the float potential V_Q applied to Q is usually linked to the mass filter function $\mathbf{Q}$ [e.g., Eq. (7)]; thus, in any experiment which requires that $\mathbf{Q}$ be scanned, V_Q is not static as might be inferred from Fig. 8b. The change of beam shape has a nonobvious effect arising from the particularly strong focusing required, in the beam length direction, to image the narrow ribbon into the circular entrance aperture of q. This requirement implies large angles of approach to the main axis for ions redirected from the extremities of the ribbon, and thus large subsequent excursions away from the main axis during the secular motion of the ion through q. In turn, this implies that the direction-focusing effects within q, discussed above (Fig. 1), are likely to be more extreme for a hybrid than for a triple quadrupole, and the available evidence suggests that this is indeed so.

In view of these comments it seems appropriate to assess the inherent advantages and disadvantages of hybrid instruments, relative to those of the two competing types of tandem mass spectrometers considered here. In addition to the complications involved in drastically changing beam energies and cross section shapes, hybrids suffer from the disadvantages relative to

[81] S. Evans, this volume [3].
[82] R. K. Boyd, *Int. J. Mass Spectrom. Ion Proc.* **75**, 243 (1987).

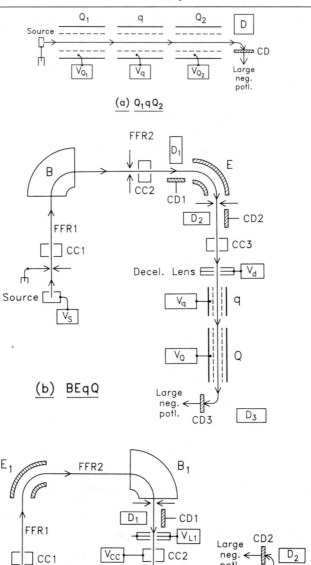

(a) $Q_1 q Q_2$

(b) BEqQ

(c) $E_1 B_1 E_2 B_2$

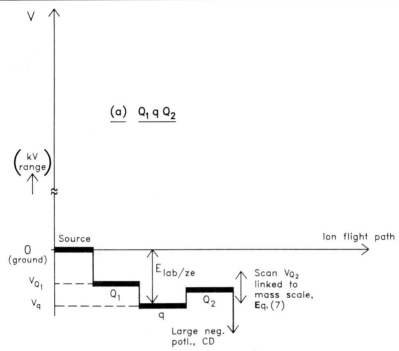

FIG. 8. Variation of electric potential V experienced by (assumed positively charged) ions traversing the tandem mass spectrometers described in the diagrams of Fig. 7. The initial translational energy is zeV_s, and that at any point at potential V' is $ze(V_s - V')$. It was assumed that the collision cell CC2 in FFR3 of the tandem double-focusing instrument (c) is to be operated at a potential (V_{CC}) 50% of that (V_s) of the ion source. In all cases, note the break in the V axis separating potentials in the kilovolt range from those within a few hundreds of volts from ground potential.

FIG. 7. Diagrams of three different types of tandem mass spectrometers: (a) Triple quadrupole Q_1qQ_2; (b) $BEqQ$ hybrid instrument; (c) $E_1B_1E_2B_2$ tandem double-focusing instrument. Collision cells (CC) in field-free regions (FFR) between sector analyzers are generally operated at ground potential except in the case of the cell (CC2) in FFR3 of the four-sector tandem double-focusing design (c). Double arrow symbol → ← denotes an optical slit, D denotes an off-axis detector, and CD its conversion dynode. The two lenses in FFR3 of (c) provide transfer optics linking the two double-focusing combinations via the collision cell CC2.

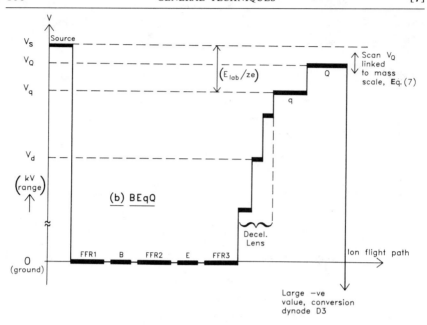

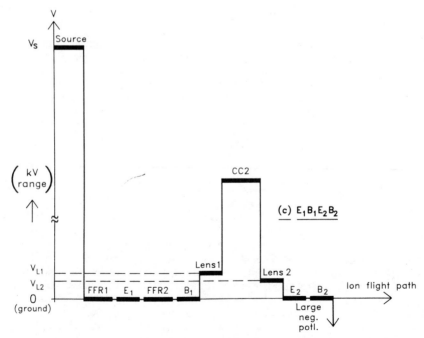

FIG. 8. (*continued*)

triple quadrupoles of having one highly nonlinear analyzer (B) and of an ion source operated at potentials far from ground, and thus appreciably less tolerant of contamination; offsetting these disadvantages is the availability of a high-performance mass spectrometer stage, the possibility of performing two-step experiments [Eq. (3)], and the availability of collisional activation of ions at energies from very low values up to the kiloelectron volt range. (This last point is sufficiently important, in the context of applications to biochemistry, that it is discussed separately below.)

When the same comparison is made between hybrids and multisector instruments the beam-shape/beam-energy transformations still represent disadvantages of hybrid designs; in addition, use of a double-focusing analyzer for the product ions greatly facilitates study of high-energy CID of larger ions typical of biochemical applications.[83,84] Another intrinsic disadvantage of any XqQ instrument is a consequence of the nature of Q as a mass filter, implying that it *must* be scanned. It is possible, on the other hand, to design the E_2B_2 product ion analyzer to have a flat focal plane, thus permitting use of a photographic plate or an array detector in simultaneous detection of ions covering a range of masses without scanning.[3,81] The only conceivable method whereby an XqQ design might be able to recover a comparable gain in sensitivity would involve replacement of q by an rf ion storage trap such as a Quistor[10,32]; some progress along these lines has been reported,[85] but this work is still in its preliminary stages. On the other hand, advantages of a hybrid over a multisector instrument include the linearity and ease of control of the Q mass filter, which make performance of scans for fixed product ion or for fixed neutral mass loss quite feasible. Such scans on a four-sector instrument, using the highly efficient collision cell between the two double-focusing stages, would be extremely difficult since it requires one highly nonlinear device (B_2) to be scanned in a fixed relationship to a similarly nonlinear device (B_1) as well as linked to E_2. Finally, it should be mentioned that there is a clearcut gradation in cost, independent of particular manufacturers, in the order: (triple quadrupoles) < (hybrid instruments) < (multisector instruments).

High-Energy Collisional Activation in Hybrid Instruments

The need for a section on this topic is illustrated by Fig. 9, which compares spectra of product ions, formed within the q cell of a hybrid

[83] L. Poulter and L. C. E. Taylor, *Int. J. Mass Spectrom. Ion Proc.* **91**, 183 (1989).
[84] A. J. Alexander, P. Thibault, R. K. Boyd, J. M. Curtis, and K. L. Rinehart, *Int. J. Mass Spectrom. Ion Proc.*, in press.
[85] M. F. Sutter, H. Gfeller, and U. P. Schlunnegger, *Adv. Mass Spectrom.* **11**, 284 (1989); *ibid., Rapid Commun. Mass Spectrom.* **3**, 89 (1989).

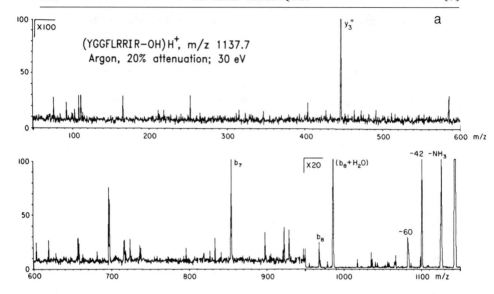

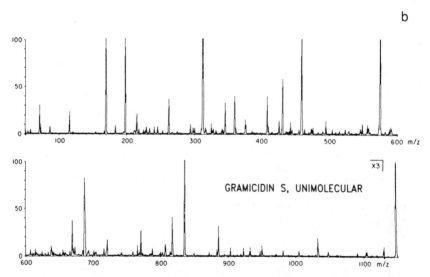

FIG. 9. Product (fragment) ion spectra of two protonated peptides, obtained in the same *q* cell of a *BEqQ* hybrid instrument. (a) Tyr-Gly-Gly-Phe-Leu-Arg-Arg-Ile-Arg-OH (dynorphin fragment 1–9), m/z of protonated form = 1137.7. Collision energy E_{lab} = 30 eV; argon collision gas, to attenuate precursor beam intensity by 50%. (b) cyclic-[-Phe-Pro-Val-Orn-Leu-]$_2$ (gramicidin S), m/z of protonated form = 1141.5. Experimental conditions were identical to those used to obtain (a) (experiments were performed within a few minutes of one another), *except* that no collision gas was used in the case of (b).

instrument, from protonated peptides of very similar masses (1137 and 1141 Da). The detailed interpretation of these spectra is not the point here. Rather, the contrast between the almost negligible information content in Fig. 9a and the rich spectrum observed in Fig. 9b is even more striking when it is noted that these spectra were obtained under identical experimental conditions, within a few minutes of one another, except that the rich spectrum was obtained with no collision gas in the q cell! Clearly, under the low-energy collision conditions used ($E_{lab} = 30$ eV), the precollision internal energy imparted to these medium-sized peptide precursor ions in the ionization process itself dominates the additional energy deposited by the collisions. A highly informative spectrum was obtained[84] by using kiloelectron volts collisions from the same peptide sample as was used to obtain the spectrum Fig. 9a. The collisional activation process itself is discussed elsewhere[34] for both low- and high-energy regimes.

The obvious solution, for cases in which the precollision internal energy and the collisional energy deposition are too low for their combined effect to give rise to significant levels of dissociation, is to significantly increase the collision energy E_{lab}. This raises the question as to the best way to achieve this end using a hybrid instrument.

The first approach is to use the techniques appropriate[2] to a standalone double-focusing mass spectrometer, i.e., to ignore the qQ section. The techniques of this kind which are in most common use are the linked-scan at constant **B/E** and the mass-analyzed ion kinetic energy (MIKES) technique (the latter possible in only a *BE* instrument). Useful analytical information can be obtained by using either of these techniques. For example, Dass and Desiderio[86] have exploited the linked-scan method in studies of simple peptides, while the MIKES technique has provided useful information on structures of small cyclic peptides[87] and of cyclic nucleotides[88]; a useful review of biochemical applications of tandem mass spectrometry, including these linked-scan and MIKES techniques, has been published by Tomer.[89] However, both of these techniques suffer from disadvantages which limit their usefulness. The low translational energies of product ions with low masses [expressed as fractions of the precursor mass; see Eq. (6)] lead to poor transmission efficiencies for such ions through analyzers designed for optimum performance at an energy close to that of the precursor ions; this problem can be alleviated by electrically floating the collision cell,[82] although this strategy has problems

[86] C. Dass and D. M. Desiderio, *Anal. Biochem.* **163,** 52 (1987).

[87] K. B. Tomer, F. W. Crow, M. L. Gross, and K. D. Kopple, *Anal. Chem.* **56,** 880 (1984).

[88] R. P. Newton, T. J. Walton, S. A. Basaif, A. M. Jenkins, A. G. Brenton, D. Ghosh, and F. M. Harris, *Org. Mass Spectrom.* **24,** 679 (1989).

[89] K. B. Tomer, *Mass Spectrom. Rev.* **8,** 445 (1989).

of its own when applied to MIKES or to linked scanning in a stand-alone double-focusing instrument. In addition, the effects of translational energy release limit the resolving power for product ions to perhaps 200 in MIKES, and to similar values $(300-400)^{90}$ for the *precursor* ion in linked scans at constant $\mathbf{B}/\mathbf{E}$; the latter problem is actually the more serious since the effect shows up as artifact peaks in the product ion spectrum, and identification of artifacts requires either construction of a complete ion-intensity map[2] (expensive in terms of both time and sample) or else exploitation of phase relationships in a field modulation experiment.[90] Neither solution is realistic in many biological applications where sample availability is limited.

The second approach to high-energy CID in a hybrid instrument is to obtain unit mass-resolved MIKE spectra by linked scanning of the electric sector field $\mathbf{E}$ and the mass filter function $\mathbf{Q}$. The reactions thus studied occur in the collision cell in the second field-free region of the $BEqQ$ instrument (Fig. 7b), with the q cell operated without collision gas as an ion transmission device. This technique was first proposed by Beynon *et al.*,[72] and is usually referred to as a linked scan at constant $\mathbf{Q}/\mathbf{E}$; however, one of the reasons why this technique is of limited practical use is connected with the fact that this apparently trivial specification is, in fact, a gross oversimplification. Since the product ions from the fragmentations are of lower translational energies E_P [Eq. (6)], not only $\mathbf{Q}$ must be scan-linked to the sector field strength $\mathbf{E}$, but also the float potentials applied to the deceleration lens, to q, and to Q. For a precursor of mass 1000 Da, for example, whose product ions down to mass 50 Da are of interest, this experiment[72] requires that five independent fields ($\mathbf{E}$, $\mathbf{V_d}$, $\mathbf{V_q}$, $\mathbf{V_Q}$, and $\mathbf{Q}$; see Figs. 7b and 8b) must be scanned simultaneously, in a fixed proportion to one another to within a tolerance of about 1 in 10^4, over a range of a factor of 20 (the mass range). This is an extremely demanding requirement which usually implies that this technique is restricted to a limited useful scan range. More recently,[91,92] a method has been developed which greatly alleviates this particular problem, but unfortunately it was concluded[92] that intrinsic sensitivity limitations seemed likely to restrict any version of this technique to cases where large amounts of sample are available.

A third possibility is to conduct the CID reactions in a collision cell located in the third field-free region of a hybrid instrument (either $EBqQ$ or

[90] R. K. Boyd, *Spectrosc. Int. J.* **1,** 169 (1982).
[91] R. Guevremont and R. K. Boyd, *Int. J. Mass Spectrom. Ion Proc.* **84,** 47 (1988).
[92] R. K. Boyd, E. W. Dyer, and R. Guevremont, *Int. J. Mass Spectrom. Ion Proc.* **88,** 147 (1989).

$BEqQ$), after the slit at the point of double focus but before the deceleration lens.[93] This technique is similar to that described in the preceding paragraph, except that the product ions P^+ do not now traverse the electric sector. This distinction has both advantages and disadvantages. Transmission losses in the electric sector (as in MIKES) are thereby avoided, but the energy filter properties of E are also lost; this latter point is a disadvantage because the ensemble of ions emerging from such a collision cell includes *all* ions of all masses and, therefore [Eq. (6)], of all translational energies up to that of the precursor R^+. When the mass filter function $\mathbf{Q}$ is set to a mass 0.5 M_R, for example, the three float potentials will all be set to a few volts below half of that of the ion source to ensure efficient transmission. This implies that all ions with masses in the range 0.5 M_R to M_R will also be transmitted unless some additional energy filter is inserted[93,94] between the collision cell and the final detector. Further, all fast neutral products will be transmitted whereas these would have been screened out via the circular flight path through the electric sector; however, use of off-axis detectors[81] alleviates this problem. Very little information is available concerning this approach; published spectra[94] are for only small organic precursor ions ($M <$ 150 Da), for which low-energy CID in the q cell is known to be highly efficient, so the utility of this approach in the present context of a *requirement* for high-energy CID of larger ions is largely unknown at present.

The final possibility is to operate the q cell at much higher values of E_{lab} than the usual range (few tens of electron volts) employed in XqQ instruments. This is easier to achieve in hybrid instruments than in triple quadrupoles, since the qQ section is already floated to high potentials with respect to ground in any case (Fig. 8b). To obtain higher values of E_{lab} it is necessary only to drop V_q, the float potential of q, to lower values thus increasing the potential difference ($V_s - V_q$); since appreciable pressures of collision gas will be present in the vicinity of what will now be sizable potential gradients on either side of q, there will be some danger of electrical discharges and thus practical limits on this strategy. It is also necessary[21] to operate q with proportionally higher values of the rf amplitude V_0, as discussed at length above. An example of the kind of improvement in spectral quality which can be achieved[21] in this way is shown in Fig. 10. This approach has not yet been evaluated as a realistic tool for real-world situations, but appears to offer considerable promise.

[93] R. G. Cooks, M. S. Story, G. Jung, and P. Dobberstein, UK Patent Application GB 2129607A (1984).
[94] Finnigan MAT GmbH, Tech. Information, H-SQ 30 hybrid mass spectrometer.

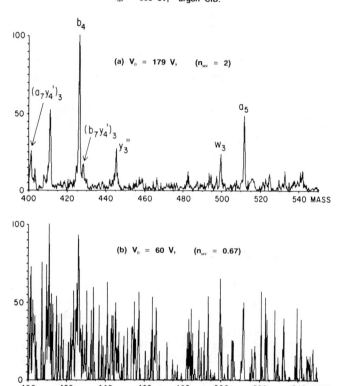

FIG. 10. Partial product (fragment) ion spectra of protonated dynorphin fragment 1–8 (Tyr-Gly-Gly-Phe-Leu-Arg-Arg-Ile-OH, m/z of protonated form = 981.6), in the q cell of a *BEqQ* hybrid instrument. Argon collision gas to give 80% attenuation of precursor beam intensity; collision energy E_{lab} = 300 eV. Spectra adapted from *Rapid Commun. Mass Spectrom.* **3,** 364 (1989). (a) V_0 = 179 V, corresponding to n_{sec} = 2 (i.e., direction-focusing condition satisfied). All other experimental conditions were identical to those used to obtain Fig. 10a. (b) V_0 (rf amplitude in q) = 60 V (maximum available in original configuration of instrument); this corresponds to n_{sec} = 0.67 for the instrument used, i.e., a nonintegral value corresponding to no directional focusing in q.

Examples of Biochemical Applications of Hybrid Instruments

The somewhat arbitrary choice of examples, used here to illustrate the power of hybrid tandem mass spectrometers, was guided by the principle that each example should emphasize advantages of hybrids over either triple quadrupoles or multisector instruments. Of course, these examples reflect the working experience of the contributing author of this section, and alternative sets of examples could undoubtedly have been chosen from the literature.

Advantages of Accurate Mass Measurement: Alamethicins. The structures of these peptide antibiotics were proposed by Pandey, Cook, and Rinehart,[95] who used high-resolution mass spectrometry [both field desorption (FD) and electron ionization (EI) of intact peptides and their hydrolysis products, both derivatized and underivatized], [^{13}C] NMR spectroscopy, and gas chromatographic analyses on a chiral stationary phase. This work[94] provides a beautiful example of the interplay of different techniques for structure elucidation, and suggested that the structures of alamethicins I and II were those shown below, where for present purposes the optical configurations of the amino acids have been omitted [note that Aib denotes 2-methylalanine]:

$$Ac\text{-}Aib_1\text{-}Pro_2\text{-}Aib_3\text{-}Ala_4\text{-}Aib_5\text{-}[Ala(I);Aib(II)]_6\text{-}Gln_7\text{-}Aib_8\text{-}Val_9\text{-}$$
$$Aib_{10}\text{-}Gly_{11}\text{-}Leu_{12}\text{-}Aib_{13}\text{-}Pro_{14}\text{-}Val_{15}\text{-}Aib_{16}\text{-}Aib_{17}\text{-}Glu_{18}\text{-}Gln_{19}\text{-}Phol_{20}$$

Features of this primary structure which present difficulties for classical strategies for peptide analysis include the blocked N terminus, its insensitivity to peptidases, and the presence at the C terminus of the phenylalaninol residue (Phol, equivalent to Phe with the carboxyl function reduced to a primary alcohol). It was suggested[95] that alamethicins I and II were simple homologs, the homology arising at residue 6 via replacement of Ala in I by its methylated homolog Aib in II.

At the time at which this work[95] was completed (1977), the modern advances in ionization techniques (particularly FAB) and in tandem mass spectrometry, were as yet unavailable. Independent work by Taylor and collaborators (Brewer *et al.*[96]) resulted in isolation from the same source of a complex mixture of peptides characterized by C-terminal phenylalaninol residues and a high proportion of Aib residues. Mass spectrometric techniques were applied by P. Thibault[97] to help with this structural problem. The FAB mass spectrum[97] of this sample showed intense peaks at m/z values corresponding to those predicted[95] for the protonated alamethicins. [These predicted m/z values are 1964.1 and 1978.1 for protonated I and II, respectively, calculated from the molecular formulas using accurate masses for the most abundant isotopes (^{12}C, ^{1}H, ^{14}N, and ^{16}O), although mass *numbers* calculated from integral mass numbers for the atoms are

[95] R. C. Pandey, J. C. Cook, Jr., and K. L. Rinehart, Jr., *J. Am. Chem. Soc.* **99**, 8469 (1977).
[96] D. Brewer, A. W. Hanson, I. M. Shaw, A. Taylor, and G. A. Jones, *Experientia* **35**, 294 (1979).
[97] P. Thibault, unpublished work (1989).

1963 and 1977, respectively; these discrepancies emphasize the importance of measuring m/z values to within ± 0.2 Da or better.] Attempts[97] to obtain useful product ion spectra for these ions, using a hybrid instrument with E_{lab} in the range 30–100 eV, failed; in the context of structural applications, this is a realistic example of the limitations of low-energy CID of large precursor ions, discussed above.

However, it was possible[97] to obtain useful CID spectra for several fragment ions prominent in the FAB mass spectrum. Provided these fragment ions are chosen with care as precursor ions for tandem mass spectrometry, it is possible to obtain much useful structural information about the intact analytes. For example, in the present case of the suspected alamethicins, there were three prominent ions in the FAB mass spectrum at m/z 1203.7, 1189.7, and 775.4; these are consistent with the b_{13} ions of alamethicins I and II and the y_7'' ion from both, respectively, and complement one another in the sense that the entire masses of the protonated precursors are thus accounted for. (The sum of masses of complementary b and y'' ions must equal that of the protonated linear peptide plus one additional hydrogen atom.[98]) The CID spectra of the two putative b_{13} ions are shown in Fig. 11. The important points are that these spectra are of excellent quality, although the corresponding spectra obtained in absence of collision gas were also intense (see discussion of Fig. 9, above), and are entirely consistent with the alamethicin I and II structures (note the 14 Da displacements); the spectra below m/z 500 were identical in the two cases, as predicted from Fig. 9, where the homology occurs at residue 6, so these portions are not shown here in the interests of clarity. This result (Fig. 11) is encouraging, but peaks corresponding to the b_7 fragments are missing; this was also true of the FAB mass spectra. Accordingly, CID spectra of the two homologous b_8 fragment ions were obtained; unfortunately, these were not as clear-cut as those shown in Fig. 11, in that each of the predicted series of b sequence ions (including the desired b_7 ions) was accompanied by a parallel series at 42 Da lower. While it is possible to provide a reasonable interpretation of this feature in terms of expulsion of an additional stable C_2H_2O neutral fragment (probably ketene originating in the N terminus acetyl group), it was felt that additional confirmation was necessary.

This was obtained[97] by accurate mass measurement of relevant ions in the FAB mass spectra, which contained peaks at m/z values corresponding to those predicted (Fig. 9) for the two homologous pairs of b_8 and $(b_7-C_2H_2O)$ ions. The measured masses, and deviations from those predicted from Fig. 9, were: $b_8(\text{II})$, 764.4294 (-2 ppm); $b_8(\text{I})$, 750.4117

---

[98] This volume, Appendix 5.

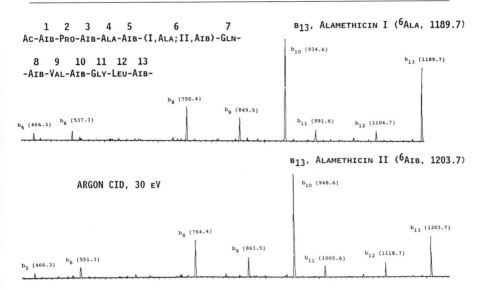

```
 1  2   3   4   5    6       7
Ac-Aib-Pro-Aib-Ala-Aib-(I,Ala;II,Aib)-Gln-

 8  9  10  11  12  13
-Aib-Val-Aib-Gly-Leu-Aib-
```

B₁₃, ALAMETHICIN I (⁶ALA, 1189.7)

b₁₀ (934.6)

b₁₃ (1189.7)

b₈ (750.4)

b₉ (849.5)

b₁₁ (991.6)   b₁₂ (1104.7)

b₅ (466.3)   b₆ (537.3)

B₁₃, ALAMETHICIN II (⁶AIB, 1203.7)

ARGON CID, 30 EV

b₁₀ (948.6)

b₈ (764.4)

b₉ (863.5)

b₁₃ (1203.7)

b₁₁ (1005.6)   b₁₂ (1118.7)

b₅ (466.3)   b₆ (551.3)

FIG. 11. CID product ion spectra (*BEqQ* hybrid instrument) of the b_{13} fragment ions formed by FAB ionization of alamethicins I (m/z of b_{13} ion = 1189.7) and II (m/z of b_{13} = 1203.7, note homology at residue position 6). Collision energy E_{lab} = 40 eV; argon collision gas added to give 50% suppression of precursor ion intensity.

$(-4$ ppm$)$; $[b_7$-$C_2H_2O](II)$, 637.3513 $(-25$ ppm$)$; $[b_7$-$C_2H_2O](I)$, 623.3464 $(-8$ ppm$)$. With one exception, the measured masses agree with those predicted to within better than 10 ppm, and thus provide confirmatory evidence for the suggested structures.

The more general conclusions to be drawn from this work concern the applicability of hybrid tandem mass spectrometers to problems of this kind. The failure of low-energy CID for the protonated molecules themselves was entirely consistent with the generalizations outlined above; successful CID (Fig. 11) of precursor ions with $M_R > 1000$ Da was obtained when these ions (b_{13}) showed intense fragmentation in the absence of collision gas (not shown in this case). High-energy CID (four-sector instruments) would undoubtedly have provided useful information directly for the protonated molecules. The availability of a high-resolution double-focusing stage in the hybrid permitted mass measurement to an accuracy and precision unattainable from a triple quadrupole instrument, and this information was invaluable in the example under discussion.[97] Thus, the middle position of hybrids in the price ranking (see above) is reflected in a similar position in performance ratings.

Use of High Resolution for Precursor Ion Selection. The double-focusing stage of a hybrid instrument offers the potential for mass selection of the precursor ion at high resolving power; indeed, examples of this feature have been published.[25,75,77] However, it is notable that all such examples have involved precursor ions of low mass (M_R < 200 Da). The reasons for this restriction are important in the present context of applicability to biochemistry.

For purposes of illustration one of the published examples[75] will be considered. The two precursor ions whose separate CID spectra were required were both of nominal m/z 182, but had different atomic compositions and thus different accurate masses, namely, $C_7H_6N_2O_4^+$ (M = 182.033) and $C_{13}H_{26}^+$ (M = 182.203). The mass difference ΔM_R of 0.170 Da requires a resolving power $(M/\Delta M)_R$ of 1070 for M_R = 182; the resolving power of sector instruments is conventionally defined in terms of a 10% valley definition, implying that use of a resolving power of 1070, thus defined, would not result in a completely clean separation of the CID spectra of the two incompletely resolved precursor ions. In fact, a resolving power of 2000 (10% valley definition) was found[75] to be necessary to achieve CID spectra with negligible cross-talk between the two precursor ions. Such a resolving power is readily achievable, with a transmission reduction of only 10–20% compared with that at a resolving power of 1000, using any modern double-focusing analyzer.

However, if the same mass difference separated two precursor ions at nominal mass 1000 Da, for example, the corresponding value of $(M/\Delta M)_R$ is 5900. Thus, by analogy with the low-mass example,[75] a resolving power of about 10,000 would be required on the 10% valley definition to provide clean separation of the two spectra. The best modern double-focusing analyzers give 10,000 resolving power at 10% transmission, relative to that at 1000 resolving power; as with most such figures of merit, such performance requires that the instrument be maintained in perfect operating condition (including cleanliness of ion source and analyzer), a requirement not always easy to meet in a laboratory. In other words, for precursor ions of mass 1000 Da or so, increasing the resolving power to the required values will be accompanied by loss in transmission by a factor of 10 to 20. Such losses usually preclude the possibility of obtaining useful product ion spectra, particularly when the difficulties associated with CID of larger precursor ions (see above), together with the possibility of limited sample quantities, are also taken into consideration.

The conclusion to be drawn from this simplified analysis is that, for applications to larger precursor ions characteristic of biochemical analytes, mass selection at high resolving power is not generally useful. An example of alternate strategies employed in a real-world situation is provided by the work of Gaskell (Mathews *et al.*[99]), in which impurities in a synthetic renin substrate decapeptide were characterized by a combination of hybrid tandem mass spectrometry and well-designed derivatization chemistry. In this particular case,[99] it turned out that the two impurities, shown to occur at the same nominal mass (1236 Da), had the same atomic constitution and thus identical accurate masses and, therefore, could not have been resolved from one another in any case; of course, this was not known during the course of the investigation, and the decision to *not* attempt separation of the precursors by high resolving power was based[99] on considerations of the kind summarized above.

The general conclusion to be reached in this regard is that, for most biochemical applications, the chief importance of the double-focusing analyzer in a hybrid instrument lies in its ability to provide accurate mass measurements.[100] The high mass range of modern magnetic sectors can also be exploited[101] to provide molecular mass information (to within a few daltons) on small proteins (up to 15 kDa or so). The practical benefits, likely to accrue from the potential for precursor ion selection at high mass resolving power, are limited to masses of a few hundred daltons.

Investigations of Two-Step Fragmentations. As discussed above with reference to Eq. (3), hybrid mass spectrometers lend themselves to investigations of fragmentation reactions proceeding via intermediate ions, provided the latter have lifetimes at least as long as the flight time between collision cells. While the number of possible experiments based on Eq. (3) is quite large,[11] most practical applications have employed a scan of the intermediate analyzer, with the masses of both the precursor ion and the final product ion fixed experimentally. An example of such an application is provided by the work of Thorne and Gaskell[102] on elucidation of mechanisms of two important fragmentation modes of protonated peptides, namely, internal fragments (loss of both N and C terminus residues) and loss of the entire C terminus residue alone. An example of application of a different scan mode to mechanistic investigations of sequential reactions is that of Guevremont and Wright,[103] who elucidated the isomer-specific fragmentations of ions derived from peracetylated hexoses; in this case, the first and second analyzers (*B* and *E* of a *BEqQ* hybrid) were set to

[99] W. R. Mathews, T. A. Runge, P. E. Haroldsen, and S. J. Gaskell, *Rapid Commun. Mass Spectrom.* **3**, 314 (1989).

[100] K. Biemann, this volume [13].

[101] M. Barber and B. N. Green, *Rapid Commun. Mass Spectrom.* **1**, 80 (1987).

[102] G. C. Thorne and S. J. Gaskell, *Rapid Commun. Mass Spectrom.* **3**, 217 (1989).

[103] R. Guevremont and J. L. C. Wright, *Rapid Commun. Mass Spectrom.* **1**, 12 (1987); *ibid.*, **2**, 47 (1988).

specify the precursor and intermediate ion, respectively [Eq. (3)], and the third analyzer (Q) scanned to provide spectra of final product ions formed via the specified route.

As mentioned above, the sensitivity is rather low for such sequential reaction experiments, and it is thus not surprising that most published applications involve mechanistic studies rather than analytical investigations per se. Recently,[104] however, Cooks et al. have proposed use of the intermediate ion scan in simplifying sequence analysis of linear peptides by tandem mass spectrometry (this work was also mentioned above in the section on triple quadrupoles). Briefly, the final product ion is specified to be a fragment derived from the N terminus residue (which must, therefore, be known, via one Edman cycle, for example); the precursor ion is also fixed experimentally at the appropriate m/z value observed in the FAB mass spectrum. A scan of the intermediate analyzer will now give a spectrum of all intermediate fragment ions which contain the N terminus residue; the process can be repeated with the final product ion specified in terms of the C terminus residue, if it is known. This experiment was conducted[104] using both a pentaquadrupole ($Q_1q_1Q_2q_2Q_3$) and a $BEqQ$ hybrid; in the latter case, a linked-scan at constant $\mathbf{B/E}$ was employed, a technique subject[90] to both artifacts (interferences from precursors close in mass to that selected) and to loss of signal at the lower end of the product ion mass range due to difficulties in correctly tracking $\mathbf{E}$ with $\mathbf{B}$. There is no question that such a strategy would greatly facilitate deduction of primary sequence from mass spectrometric data[105] by identifying those product ions, observed in the simple one-step experiment, which are N terminus fragments (a, b, c'', etc.[98]). The strategy[104] also possesses advantages for analysis of peptide mixtures. However, as emphasized by the authors,[104] the detection limits must be lowered appreciably if this approach is to be generally useful. Thus an intermediate ion spectrum with a signal-to-noise ratio of $10:1$ was obtained by expending 9 nmol of a simple pentapeptide (Leu-enkephalin), known[31] to give intense beams of protonated molecules under FAB conditions and also to exhibit extensive fragmentation in the absence of collision gas and thus excellent efficiency in low-energy CID (see above); that is to say, this choice represented almost a "best case" from the point of view of sensitivity. Nonetheless, this approach[104] holds considerable promise of simplifying interpretation of fragment ion spectra of peptides, and thus increasing confidence in the resulting sequence assignment.

[104] K. L. Schey, J. C. Schwartz, and R. G. Cooks, Rapid Commun. Mass Spectrom. 3, 305 (1989).
[105] K. Biemann, this volume [18].

Quantitative Analysis: Platelet-Activating Factor. One potential advantage of tandem mass spectrometry, in general, is that it lends itself to minimizing fractionation and clean-up of complex matrices which are to be quantitatively analyzed for some specific analyte(s). This advantage is, of course, a result of the two-stage analysis intrinsic to the technique, and can be realized in practice provided that careful consideration is given to the appropriate degree and type of fractionation used prior to the mass spectrometry. A good example of a successful procedure of this type is the development by Haroldsen and Gaskell[106] of an analytical method for platelet-activating factor (PAF) in cells.

The chemical nature of PAF is one or more 1-*O*-alkyl-2-acetyl-3-glycerophosphocholines, where the alkyl substituent can be 16:0, 17:0, 18:0, 18:1, etc. Positive ion FAB mass spectra of these compounds give a base peak at m/z 184, due to choline phosphate [$^+N(CH_3)_3]C_2H_4[-O-PO_3H_2]$], as well as intense peaks corresponding to intact molecular ions (charge localized on the quaternary nitrogen). Low-energy CID of the molecular ions gives only the product ion at m/z 184, in about 20% yield.[106] Formation of the choline phosphate requires transfer of a hydrogen atom from some other portion of the molecule; a PAF species incorporating a 2-[$^2H_3$]acetyl group gave m/z 185 as the only product ion, indicating that the additional hydrogen atom is derived entirely from the acetyl substituent.

These observations were exploited[106] as the basis for an analysis for PAF by tandem mass spectrometry, using the [$^2H_3$]acetyl compound as internal standard. A conventional extraction method for phospholipids was used, followed by separation into phospholipid classes by a simple thin-layer chromatography procedure; no further clean-up was found to be necessary for the method to give reliable results. The tandem mass spectrometry involved fixing the *Q* mass filter of a *BEqQ* hybrid at m/z 184, scanning the double-focusing analyzer over a limited mass range appropriate to the PAF molecular ions, and recording the intensities at appropriate m/z values. In alternate scans, the *Q* analyzer was set to m/z 185, instead of 184, in order to record the intensity of the internal standard. In this regard, the linearity and ease of control of the quadrupole provided a significant advantage over the same experiment had it been attempted using a four-sector instrument, even though modern laminated and air-core magnets[4] are much easier to control than their predecessors of even 10 years ago; the experimental procedure[106] could have been conducted just as readily using a triple quadrupole instrument. Use of the internal standard permitted quantitative detection of PAF down to the low picogram level, with a linear response curve up to 300 ng. While the method

[106] P. E. Haroldsen and S. J. Gaskell, *Biomed. Environ. Mass Spectrom.* **18**, 439 (1989).

is sufficiently susceptible to interferences that it cannot yet be regarded as a reference method, its combination of reliability, speed, and relative simplicity offers a real advantage.

Conclusions and Future Directions

Tandem mass spectrometry is a powerful approach to solving analytical problems in biochemistry. It offers the sensitivity, selectivity, speed, and versatility to solve a wide range of important problems. The triple quadrupole and hybrid tandem mass spectrometers described in this chapter are particularly important implementations of MS/MS. They are practical, analytically powerful systems which, due in part to their relatively low cost compared to other MS/MS implementations, are in widespread use.

Future directions in quadrupole and hybrid tandem mass spectrometers will certainly include new generations of smaller, lower-cost, more fully automated instruments. Equally important, ongoing studies will strengthen the fundamental foundations upon which these MS/MS approaches are supported, and lead to new and better analytical methods. New instrumental approaches which bear careful watching include tandem-in-time mass spectrometry in ion traps (both quadrupole ion traps and FT-ICR spectrometers). These systems offer the potential for high-order MS^n, for high-efficiency CID or photo-induced dissociation, and for high mass. The lower cost range of sector-quadrupole hybrids, based upon air-cored magnets, seem likely to make further inroads due to their ease of control (comparable with that of all-quadrupole instruments), provision of access to accurate mass measurement at medium resolving power (up to 7×10^3, 10% valley definition), and availability of CID at both low and high energies.[107]

[107] M. R. Clench, R. Owen, V. C. Parr, and D. Wood, *Proc. 36th ASMS Conf. Mass Spectrom. Allied Topics, San Francisco, CA* p. 719 (1988); *ibid., Adv. Mass Spectrom.* **11,** 248 (1989).

[8] Matrix Selection for Liquid Secondary Ion and Fast Atom Bombardment Mass Spectrometry

By EDWIN DE PAUW

Introduction[1-2]

The need for structural and quantitative information from biological samples led to the development of ionization methods able to handle fragile, nonvolatile, high-mass molecules, avoiding as much as possible the delicate step of derivatization. Various methods, described in [1] in this volume, have been established for that prospect. These methods have in common the ability to bring directly into the gas phase analytes supported by or embedded in a solid or a liquid matrix.

Two general strategies have been adopted. The first consists of applying a high density of energy to the sample for a short time, which leads to the emission of intact ions. This can be achieved by discrete impacts of primary particles, which have been accelerated to a kinetic energy in the kiloelectron volt (keV) range, using pulsed laser irradiation or fission product irradiation. The second consists of direct introduction of the sample in a solution through a nebulizer into the source of the mass spectrometer. Ions can either result from this process or are formed in a subsequent ionization step. Both approaches create an unusual situation since the use of matrices introduces many new parameters not otherwise relevant when working under low-pressure, gas-phase conditions. In fact, a judicious matrix choice often appears as the key requirement in the overall analytical procedure.

In this chapter, we describe the critical parameters necessary in the selection of an appropriate matrix for liquid secondary ion mass spectrometry (LSIMS) and fast atom bombardment (FAB) mass spectrometry. Requirements for acquisition of full mass spectra will be considered. Dedicated matrix properties for applications of tandem mass spectrometry will also be evaluated.

Why LSIMS or FAB?

Most of the so-called "soft" ionization methods yield molecular weight as well as structural information. Semiquantitative analysis can be per-

---

[1] H. R. Morris (ed.), "Soft Ionization Biological Mass Spectrometry." Wiley, New York, 1982.

[2] E. de Pauw, *Mass Spectrom Rev.* **5**, 191 (1986).

METHODS IN ENZYMOLOGY, VOL. 193

formed once the relation between the signal from the target compound and its concentration is not modified by matrix effects. As a consequence, except for specific applications, the choice of an ionization method for fragile molecules analysis will be guided mainly by its ease of implementation and operation on multipurpose mass spectrometers.

The bombardment of the sample, introduced as a solution in a low-volatility matrix, by an intense (1 μA/cm^2) kiloelectron volt primary particle beam (sputtering ion source) has been widely adopted as a standard soft ionization method. FAB and LSIMS use primary neutrals or ions as the projectiles. With little sample preparation and handling the method gives high-intensity, long-lasting signals.

The complexity of many biological or environmental samples necessitates the separation of their components prior to analysis. In addition, a time concentration effect (as in chromatography) allows large gains in sensitivity for analytes present in trace amounts. Off-line analysis of chromatographic fractions is possible, but tedious. Great efforts have been made toward the direct coupling of liquid chromatography and mass spectrometry, which is experimentally more difficult as compared to gas chromatography coupling. The nebulization technique mentioned above is theoretically the best suited technique for this purpose (see [5] in this volume). The direct coupling of LSIMS or FAB with microbore high-performance liquid chromatography (HPLC) was first introduced as the method termed continuous-flow FAB.[3] However, it resulted in relatively poor chromatographic resolution. The recent coupling of FAB with the sample introduction by nebulization appears promising but still needs to be completely evaluated as far as the use of matrices is concerned.

Matrix Effect in LSIMS and FAB

We will not fully describe here the mechanisms of ion formation, which is still under debate; instead, we will summarize its main aspects as directly related to choice of the matrix.

The complete process giving rise to the emission of ions has been described as the combination of ejection (sputtering) and ionization. This separation of events, even if artificial, provides a good scheme to rationalize the experimental observations. A third part of the discussion will be devoted to the artifacts induced by presence of the matrix.

[3] R. M. Caprioli, T. Fan, and J. S. Cottrell, *Anal. Chem.* **58**, 2949 (1986).

Sputtering Process[4-6]

Solubility of Analyte. The major and advantageous effect of the matrix is to allow long-lasting, high-intensity signals without apparent accumulation of damage, in contrast to direct solid bombardment, which produces transient mass spectra and which shows radiation damage products. This effect is attributed to the buffering of the impact energy by the surrounding solvent molecules. Long-lasting signals are required for structure determination when various types of scans are recorded and for exact mass determinations for which instrument tuning can be time consuming.

A good dispersion of the analyte in the matrix is required to avoid the problems encountered when bombarding solids. It implies good solubility of the analyte, which ensures not only suitable dispersion but also a good source of supply of analyte molecules (surface renewal). Practical consequences of these requirements can be summarized as follows: (1) The matrix must be a good solvent, or additives must be added to solubilize the analyte; (2) solubilization may be a slow process because of matrix viscosity. Solutions often give better signals some hours after their preparation. Solubilization can be performed more rapidly by first dissolving the sample in a good volatile solvent before mixing with the matrix. However, one must verify that precipitation of the sample has not occurred after evaporation of the volatile solvent; (3) chemical modification of the analyte (derivatization) can be used to improve the solubility of the analyte. As in classical solution chemistry, a good match between the functional groups of the analyte and of the solvent will allow the required solubility.

Surface Effects. Origin. The sensitivity of the desorption ionization method is determined more by the ion signal of the analyte relative to that of the matrix rather than from its absolute value. This ratio is, in turn, related to the analyte concentration rather than to its total amount, and also to the spectral "transparency" of the matrix.

The sputtering process has been shown to produce ions mostly from the uppermost layers of the matrix solution. All the properties leading to concentration changes between the bulk solution and the surface of the liquid will thus strongly affect the sensitivity of the method.

Consequences of matrix choice. The consequences of surface effects are numerous, and can be beneficial or unwanted. Even in the case of small molecules such as amino acids, small peptides, or organic salts, the

[4] S. S. Wong and F. W. Rollgen, *Nucl. Instr. Meth. Phys. Res.* **B(14),** 436 (1986).

[5] W. V. Ligon and S. B. Dorn, *Int. J. Mass Spectrom. Ion Proc.* **78,** 99 (1987).

[6] J. Sunner, A. Morales, and P. Kebarle, *Int. J. Mass Spectrom. Ion Proc.* **86,** 169 (1988).

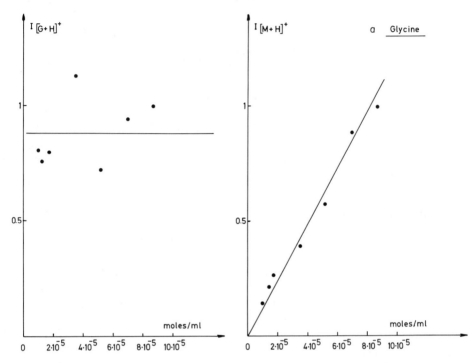

FIG. 1. Effect of the surface activity on ion currents from glycerol (G) matrix and analyte as a function of concentration: (a) hydrophilic amino acid, glycine; (b) hydrophobic amino acid, valine.

surface concentration of the most surface-active species is increased. The concentration dependence of the signal of small amino acids is illustrated in Fig. 1. Preferential sputtering occurs, producing a time dependence of the sample surface composition.

This surface concentration effect is useful to lower the detection limit of a selected target compound. However, it must be avoided when dealing with semiquantitative determinations in mixtures. In such cases, analytes may be in part or completely suppressed. This occurs, for example, in some cases when performing peptide mapping after enzymatic digestion of proteins.

The surface activity of the analyte in the chosen matrix is a very important parameter. It can be adjusted by modification of either the matrix or the analyte. Modification of the analyte is achieved by derivatization. In the case of hydroxylic matrices, the addition of a large alkyl group to the analyte will increase its surface activity, as was discussed, for

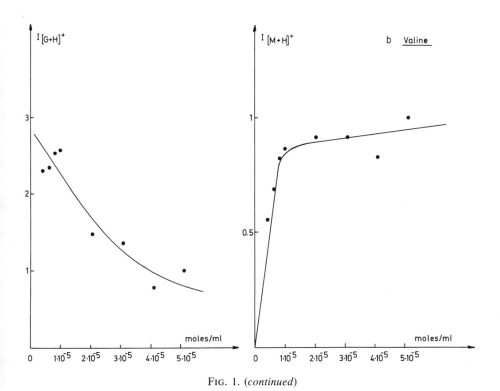

FIG. 1. (*continued*)

example, this volume [37]. It will also equalize the surface activity of components of mixtures and, as a consequence, lower the discrimination due to surface activity, as in the case of modified ribonucleosides (Fig. 2). The same effect can be obtained for ionic samples by the addition of a surfactant counterion that will keep the analytes in the sampled area by their strong Coulombic interactions. The use of positively as well as negatively charged additives have been proposed, such as, camphorsulfonic acid, quaternary ammonium salts, and perfluorocarboxylic acids. The matrix itself can be tailored to give to the analyte the appropriate surface activity. This can be realized by changing the ratio between hydrophilic (polar) and hydrophobic (hydrocarbon) groups. Polyol ethers including specific alkyl groups have been reported as well as long-chain aliphatic alcohols, amino alcohols, thioalcohols, etc. Availability, safety, and cost considerations may, however, limit the use of such "custom-made" matrices.

Sample Renewal. In order to obtain long-lasting, high-intensity signals,

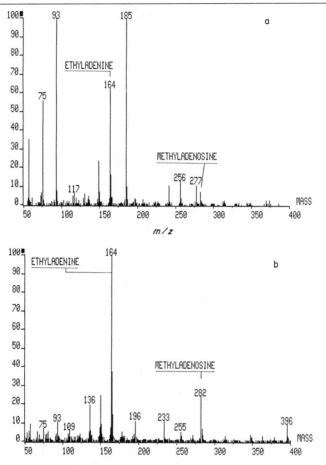

FIG. 2. Effect of the addition of a surface activity-promoting agent on the FAB mass spectrum of a mixture of ribonucleosides. (a) Surfactant acid, neat glycerol; (b) camphorsulfonic acid.

the surface of the sample submitted to bombardment must be continuously refreshed. Several different processes have been claimed to contribute to this renewal. The rate of removal by the primary beam itself could be sufficient to explain the apparent "self-cleaning" of the sample. All of the damage area is vaporized during the impact of the primary particle. This can, however, not explain the enrichment of the surface in surfactant species.

The study of the secondary ion signal intensity for matrices of different molecular weight, at temperatures at which they have same viscosity,

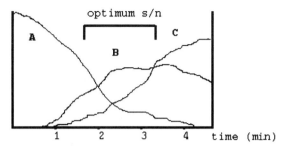

FIG. 3. Time dependence of a sulfolane–glycerol matrix spectrum. (A) m/z 121, sulfolane $\cdot$ H$^+$; (B) unknown analyte; (C) m/z 93, glycerol $\cdot$ H$^+$.

permits the exclusion of diffusion as a major contributor. The analysis of the time dependence of secondary ion signals using a pulsed primary beam confirms this observation. Mass transport by surface tension-driven convection is also a likely contributor to surface renewal.

These results point to the importance of the evaporation rate of the solvent. However, a suitable vapor pressure of the matrix alone is not sufficient. If evaporation is quenched by the formation of an nonvolatile polymeric skin at the surface of the droplet, the sample behaves as a solid and the signal quickly vanishes. The same observation has been made for low-temperature FAB spectra. Supplements can be formulated should the need for a "clean" decomposition of the matrix be required; only volatile compounds will thus be produced under the bombardment by energetic primary particles. Low-molecular weight compounds, such as alcohols, polyols, amino alcohols, and small aromatics, will usually satisfy this requirement.

Secondary Properties. A relatively high viscosity of the matrix is usually useful for physical stability of matrix on the probe tip. It will probably also lower the rate of recombination of reactive species, but this has never been clearly demonstrated.

Mixtures of compounds may provide appropriate matrices. In particular, the fusion point of room temperature melting chemicals can be lowered upon formation of an eutectic mixture. The dithioerythritol–dithiothreitol (DDTE) mixture has been used effectively for various analytes.

A better surface concentration effect may also be attained by mixing two solvents of different volatility, as, for example, glycerol and thioglycerol or glycerol and sulfolane (Fig. 3), or glycerol and pyridine.

Summary. As far as the desorption process is involved, the criteria for matrix selection are the following: the matrix must solubilize the analytes; the analytes are better detected if they present a constant excess in the surface region of the matrix. This condition is influenced by adequate surface activity of the analyte, and by favorable physical (vapor pressure, viscosity) and chemical (radiation chemistry) properties of the matrix.

Ionization Process[7-9]

To be analyzed by mass spectrometers, the particles emitted from the matrix must be charged. The second important factor that determines the overall sensitivity of LSI–MS–FAB is the ionization efficiency (the ratio between the emitted ions and neutrals) and the survival of ions during their path through the spectrometer.

Most work to date has been carried out using polar hydroxylic matrices. The following relative sensitivity scale has usually been observed in acidic matrices such as glycerol (G) or thioglycerol (TG): organic salts (mostly ammonium and phosphonium ions) > protonated species > cationized species > radical ions. When using basic matrices such as triethanolamine (TEA), diethanolamine (DEA), or 3-amino-1,2-propanediol, only quaternary salts are correctly detected as positive ions. Negative ions corresponding to the deprotonated or to the anionized species (usually resulting from chlorine attachment) are readily detected. When using aprotic matrices such as sulfolane, ethylene carbonate, or dimethyl sulfoxide, radical ions are often detected as base peaks.

In view of the above observations, one can assume, even if the ion formation process is not clearly understood in terms of gas phase or solution processes, that the ion composition in the gas phase largely reflects the ionic composition of the solution. The detection of ions formed exclusively by solution reactions supports this assumption. Modification of the surface composition and recombination of oppositely charged ions after their ejection may modulate this similarity. Ions can also be formed by chemical ionization reactions involving the abundant matrix-derived ionic species.

The key to sensitive detection is thus to preform ions in solution and to ensure their survival in the gas phase. As a consequence, the matrix must be able to separate and solvate the ion pairs to avoid recombination.

The following recommendations follow from the above conclusions. Most of the samples amenable to LSIMS–FAB are polar amphoteric molecules. Simple acid–base reactions will strongly increase the ion yield. The required protons can be those of the acidic matrix (e.g., thioglycerol), or can be provided by additives (e.g., surface-active acids). For negative

[7] M. J. Connoly and R. G. Orth, *Anal. Chem.* **79**, 903 (1987).
[8] T. Keough, *Anal. Chem.* **57**, 2027 (1985).
[9] E. de Pauw, *Anal. Chem.* **55**, 2196 (1983).

ion detection, proton abstraction can be promoted by the matrix itself or by added surface-active bases. Complexation by alkalic cations or halogen anions is used for polar compounds not amenable to acid–base reactions (such as sugars, ethers).

Derivatization reactions have also been employed, even though the initial tendency during early studies was to avoid them. They confer the required charge and surface activity to the target compound. The classical organic reactions which produce ions in solution can also be used; the following are taken from the recent literature. Girard's reagents easily condense on carbonyl functions. This reagent has been modified by replacement of a methyl by a *tert*-butyl group. Quaternization of heteroatoms by butyl iodide instead of the methyl analog produces a tenfold increase of the resulting ion signal, as, for example, in the case of telluronium salts. Specific addition of a charged moiety on the C- or N-terminal positions of peptides not only enhances the ion signal but also induces specific fragmentation pathways useful for structural determinations. Nonpolar compounds, if LSI–MS or FAB is required for their analysis, can be measured using oxidizing or reducing additives in nonprotic or redox matrices such as aromatic nitro compounds. Charge transfer complexation as well as electrochemical reactions may be useful for specific applications.[10]

Drawbacks in Use of Liquid Matrices[11–15]

Although many advantages have been pointed out resulting from the use of liquid matrices in SIMS and FAB, significant drawbacks may also be encountered.

Spectroscopic Artifacts. The most obvious artifact is the presence of interfering peaks in the mass spectrum produced by the matrix. This problem is usually easy to solve when dealing with pure compounds. A simple change of matrix is often sufficient. Addition of a compound that suppresses the matrix peaks is another alternative. When dealing with mixtures and with biological extracts contaminated by alkali salts, a complete suppression of the analyte signal may be observed, leaving only peaks resulting from the cation attachment to the matrix. Desalting of the

[10] J. E. Bartmess and L. R. Phillips, *Anal. Chem.* **59**, 2014 (1987).
[11] W. Kausler, K. Schneider, and G. Spiteller, *Biomed. Environ. Mass Spectrom.* **17**, 15 (1988).
[12] G. Pelzer, E. de Pauw, D. V. Dung, and J. Marien, *J. Phys. Chem.* **88**, 5065 (1984).
[13] T. Keough, *Int. J. Mass Spectrom. Ion Proc.* **86**, 155 (1988).
[14] C. Dass and D. Desiderio, *Anal. Chem.* **60**, 2723 (1988).
[15] D. C. Moon and J. A. Kelly, *Biomed. Environ. Mass Spectrom.* **17**, 229 (1988).

sample may then be required. Matrices with a low affinity for cations, such as, 3-nitrobenzyl alcohol, are then recommended.

Besides mass spectral interferences, the use of liquid matrices also gives rise to discrimination according to the surface activity of the analytes. Some matrices are claimed to be nondiscriminatory, as for example *n*-dodecanol. In fact, this property is obtained at the expense of sensitivity, as the analytes have little or no surface activity within that solvent. The same situation pertains for sugars in glycerol, due to their chemical similarity. A more attractive approach consists in increasing substantially the surface activity of all the analytes, as mentioned above, making their initial differences negligible.

Chemical Modification of Analyte.[16] Up to now, matrices were considered inert solvents, but this is by far not the case. Many chemical reactions, either spontaneous or induced by particle bombardment, have been reported.

Reduction reactions have been reported for organic as well as for inorganic compounds. It has now been found to occur for many classes of compounds. Reduction is likely to proceed through the release of hydrogen during the initial steps of the impact energy dissipation. The extent of reduction leading to the $(M + 2)^+$, $(M + 3)^+$ ions can be predicted on the basis of electrochemical data in such a way that incorrect assignment of molecular weights can be avoided. However, in the case of strong reduction, fragment ions originating selectively from the various redox species are mixed in the full-scan spectra, and MS/MS at unit mass resolution may be required in order to establish fragmentation pathways before structure can be reliably assigned. This is shown in the daughter ion spectra of various redox forms of crystal violet (Fig. 4). Various addition reactions have also been described. They include CH_2 insertion, formaldehyde addition, dehalogenation, ligand exchange, ring openings for lactones, sulfur–sulfur bond cleavage, and hydrolysis. The condensation of various hydroxylic solvents with loss of water from aldehydes, ketones, and amides has been reported and studied in detail. The partial exchange of oxygen by sulfur is observed in the spectra of aldehydes taken in thioglycerol. The corresponding exchange of sulfur by oxygen has been observed for the thioaldehyde in glycerol. This exchange occurs through the formation of a cyclic acetal, for example, as shown by the scheme depicted in Fig. 5.

The energy involved in the desorption ionization process may lead to a complex and sometimes unpredictable chemistry. A critical examination

---

[16] L. D. Detter, O. W. Hand, R. G. Cooks, and R. A. Walton, *Mass Spectrom. Rev.* **7**, 465 (1988).

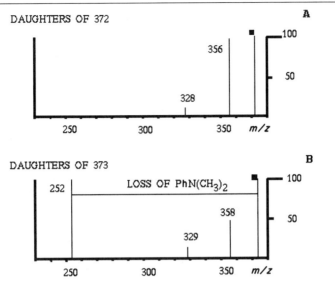

FIG. 4. Unit mass resolution MS/MS of redox forms of crystal violet showing the specificity of fragmentation channels. (A) Oxidized form C^+; (B) semiquinonic form $(C + 1)^+$.

of the data, therefore, is required before final compound identification or structure determination. A careful experiment should include the analysis of isotopic distributions, MS/MS verification of the origin of key fragments, and the design of a realistic fragmentation scheme, excluding even-electron to odd-electron pathways. Interpretation of the spectra can, in any case, be performed keeping in mind that in FAB any type of fragmentation may be possible.

Matrix Effects for MS/MS[17,18]

The choice of matrix for LSIMS and FAB–MS is guided mainly by the required quality of the full-scan mass spectra. Additional requirements may, however, be added to increase the information content obtained when MS/MS is used, with respect to both sensitivity and selectivity. One may then either use the inherent matrix effects or design specific ones. Let us first consider the potential use of generally encountered matrix effects.

[17] E. de Pauw, K. Nasar, and G. Pelzer, *Biomed. Environ. Mes Spectrom.* **15,** 577 (1988).
[18] K. L. Busch, G. L. Glish, and S. A. McLuckey, "Mass Spectrometry/Mass Spectrometry, Techniques and Applications of Tandem Mass Spectrometry." VCH Publ., New York, 1988.

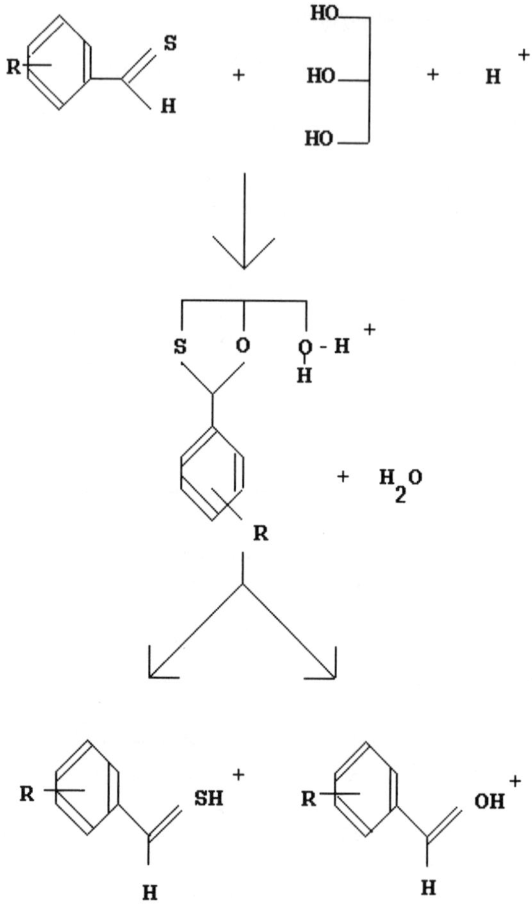

FIG. 5. Mechanism of oxygen–sulfur exchange following the solvent condensation induced by the primary particle bombardment.

All of the effects improving the general sensitivity of the method will also improve its MS/MS aspect (such as surface activity, preformation of ions) in terms of signal intensity. This is, however, not the most interesting approach. The reactions leading to the formation of new species (addition reactions, cationization) provide new and effective means to obtain data from MS/MS experiments.

Fragmentation of clusters formed by sugars and matrix (glycerol, diethanolamine) has been analyzed by mass-analyzed ion kinetic energy spectroscopy (MIKES). It shows a reversible loss of solvent, which can be used to selectively detect solvent–analyte adducts, even at the chemical noise background level in the full-scan spectra.

The daughter-ion spectra of cationized species have been shown to depend upon the nature of the cation. The reduced species has been reported to give different fragments as compared to the original form of the molecule, giving rise to odd-electron-type fragmentation patterns, similar to the one obtained under electron impact.

Selected neutral losses are induced by specific matrix additives, as, for example, the loss of the lithium oxide from the C-terminal end of peptides, in the presence of Li^+. The loss of CO in peptide spectra, observed as a_n–b_n ion pairs (see this volume, Appendix 5), is promoted by the formation of peptide–copper adducts.

Acidification of the matrix strongly promotes the condensation of hydroxylic matrices on aldehydes giving an adduct which reversibly decomposes. Aldehydes in complex mixtures can be selectively detected by a single constant neutral loss scan by monitoring the reverse reaction upon collision.

As mentioned earlier, the limitation of LSIMS–FAB sensitivity is mainly due to the presence of a high chemical noise level. These interferences are not fully eliminated by the use of MS/MS scans. The constant neutral loss of water can be cited as an example in peptide analysis. Weak signals are effectively extracted from the matrix background but intense losses of water also occur from matrix ions. In such cases, the addition of a second selection step, the loss of CO, leading to MS/MS/MS scans, could provide an unambiguous detection of peptide ions.

Conclusions

A flow chart can be constructed from the above considerations leading to a matrix choice in agreement with the planned experiment. A recent survey of the physical properties of common matrices has appeared in the literature.[19]

Before acquiring mass spectra with LSIMS or FAB, one should perform the following steps: (a) Check and optimize the solubility of the analyte. In the case of trace amounts, the solubility can only be evaluated on the basis of experience with similar compounds. Do not use all the sample in the first test matrix, but dissolve with a known good solvent. (b) Adjust the surface activity of the analyte. Hydrophobicity indexes as well as rough estimations of the known or assumed molecular structure may be helpful. (c) Check the acid–base properties of the compounds to ensure

[19] K. D. Cook, P. J. Todd, and D. H. Friar, *Biomed. Environ. Mass Spectrom.* **18,** 492 (1989).

maximum ion preformation in the matrix. For positive ion detection, a matrix with a low pK_a is best suited; for negative ion detection use a high pK_a. (d) Evaluate the possibilities of other ion-promotion reactions, such as complexation, or redox reactions. (e) Check for interferences due to matrix components. Desalting of the sample may then be necessary. (f) On the basis of solution chemistry, examine the possibility of artifact occurrence. (g) Keep in mind that the sensitivity will be increased for more volatile matrices (thioglycerol versus glycerol) but at the expense of the lifetime of the sample. Starting with usual matrices, such as glycerol, thioglycerol, or 3-nitrobenzyl alcohol for positive ion detection, or diethanolamine or aminopropanediol for negative ion analysis, will often be a good compromise, as more than 95% of reported measurements are performed with these matrices.

[9] Continuous-Flow Fast Atom Bombardment Mass Spectrometry

By RICHARD M. CAPRIOLI and WILLIAM T. MOORE

Introduction

Fast atom bombardment (FAB) mass spectrometry has been used extensively in the past for the analysis of compounds of biological origin because it provides a means for ionizing molecules that are polar and/or charged without prior derivatization.[1] The fact that this is done with a liquid sample droplet provides an ease of use which has greatly contributed to its widespread acceptance. In a typical analysis, 1 μl of a sample solution is mixed with 3–5 μl of glycerol and 2 μl of this mixture subjected to atom bombardment. Such an analysis has been termed standard (std)- or static-FAB. The high concentration of glycerol gives a highly viscous nature to the liquid sample droplet that serves important functions: (1) it keeps the sample a liquid during its introduction into the high vacuum ion source of the spectrometer, and (2) provides the sample with a surface which can be constantly renewed as the bombardment process occurs. Several drawbacks of the use of high glycerol concentrations include the concomitant dilution of the sample on addition of glycerol, a significant increase in

---

[1] M. Barber, R. S. Bordoli, G. Elliott, R. D. Sedgwick, and A. N. Tyler, *Anal. Chem.* **54**, 645A (1982).

METHODS IN ENZYMOLOGY, VOL. 193

"background" chemical signals in the sample because of cluster and fragmentation ions derived from glycerol, and the formation of surface distribution effects, e.g., the tendency of molecules to distribute themselves in the droplet either into or away from the liquid/vacuum interface.

Continuous-flow FAB (CF-FAB) mass spectrometry was devised to eliminate or diminish some of the major difficulties with the std-FAB technique while retaining the essential advantages of this bombardment ionization process.[2] Basically, the technique entails the use of a sample introduction probe that provides a continuous flow of liquid into the mass spectrometer ion source and onto the sample stage where atom bombardment takes place. This dramatically reduces the requirement for a viscous carrier and allows for the use of volatile solvents such as water, methanol, and acetonitrile. The rate of flow of liquid is generally in the range of about 1 to 20 μl/min for commercial instruments. Samples can be analyzed by flow-injection techniques where typically 0.5–1.0 μl samples are injected into a liquid carrier stream at a flow rate of about 5 μl/min. For peptide analyses, as for most other analyses of biologically derived samples, water is the carrier solvent of choice. In this case, 5% glycerol is added to the carrier solution to help maintain stable operating conditions. CF-FAB has significant advantages in several areas relative to std-FAB, most noteworthy of which are a significantly lower limit of detection[3] and decreased ion suppression effect.[4] It also provides greater accuracy for sample delivery for quantitative analyses and provides a fast and easy technique for the analysis of large numbers of samples. These advantages will be illustrated with specific examples in the following sections.

For biochemists, the utility of CF-FAB is that it effectively allows the direct analysis of compounds dissolved in water by mass spectrometry.[5] This is accomplished while retaining the specific utility of FAB for producing mass spectra from relatively large and polar molecules. This direct analysis of aqueous solutions eliminates the very time-consuming preparation methods formerly employed where the compound of interest had to be extracted or otherwise isolated and put through derivatization steps to render it amenable to mass spectrometric analysis. Often these preanalysis procedures introduced additional problems of poor and variable extraction and derivatization yields, added impurities and side products, and sample loss due to adsorption on surfaces during these sample handling procedures. The effective use of water as a medium for the analysis of biological

[2] R. M. Caprioli, T. Fan, and J. S. Cottrell, *Anal. Chem.* **58**, 2949 (1986).
[3] R. M. Caprioli and T. Fan, *Biochem. Biophys. Res. Commun.* **141**, 1058 (1986).
[4] R. M. Caprioli, W. T. Moore, and T. Fan, *Rapid Commun. Mass Spectrom.* **1**, 15 (1987).
[5] R. M. Caprioli, *Biochemistry* **27**, 513 (1988).

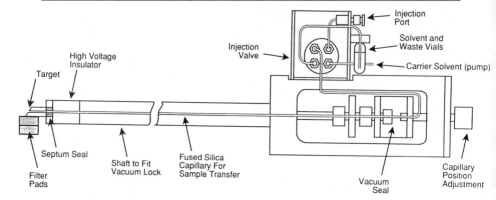

FIG. 1. A typical CF-FAB sample introduction probe designed for a high-field magnetic mass spectrometer. This particular design incorporates filter pads to adsorb excess liquid and has a capillary position-adjustment mechanism. It is built for the Finnigan MAT 90 high-field instrument.

samples is a significant asset and opens up many possibilities for on-line process monitoring in a variety of areas including batch reaction monitoring, specific enzyme reaction analysis, on-line pharmacokinetic studies in live subjects, etc. In addition, direct combination of aqueous-based separations techniques with mass spectrometry can be accomplished, e.g., liquid chromatography/MS, capillary electrophoresis/MS, microdialysis/MS, as well as others. Some other mass spectrometric methods which are amenable to coupling with liquid chromatography (LC) are described elsewhere in this volume.[6]

Operational Considerations

CF-FAB Probe Design

The CF-FAB probe consists of a capillary bore tube inside of a stainless steel shaft, the latter fitting the introduction lock of the particular spectrometer used.[2] A drawing of one such probe for a high-field magnetic mass spectrometer is shown in Fig. 1. The capillary tube is made of fused silica (75 μm id, 280 μm od) to insulate the target tip, which is at high voltage during normal operation, from the handle which is at ground potential. Although a high vacuum seal is made at the handle end of the shaft, a septum is placed as close to the target tip as possible to prevent liquid on the surface from running down the outside of the capillary inside the probe

[6] M. L. Vestal, this volume [5].

tip. Should such leakage occur, the ion current produced by bombardment of the target will be unstable as a result of solvent vapor bursting from inside the probe tip, disrupting the liquid sample on the target. The probe design shown in Fig. 1 incorporates a mechanism for moving the capillary in and out of the plane of the target surface when the probe is in operation.

Generally, stable operation is achieved when the amount of liquid on the surface is in a steady state, i.e., when the rate of evaporation is equal to the rate of flow onto the surface, and where the target surface contains only a thin film of liquid. Commercially available probes use several methods to achieve these conditions. The probe depicted in Fig. 1 for the Finnigan MAT 90 utilizes a filter paper pad to remove excess liquid from the surface and to prevent a droplet buildup. The volatile solvents then evaporate from this pad in a steady manner, eliminating disruptive bursting, and permitting the probe to remain in place and in stable operation for 4–5 hr or more. Other commercial CF-FAB probes rely on direct evaporation from the target tip and careful flow control to achieve this stability, such as those offered by Kratos and VG Instruments. In an associated technique called frit-FAB, JEOL employs a fine mesh frit to disperse the solvent.[7] Whatever the mechanism, it is clear that optimal performance is obtained when a thin film is maintained on the surface of the target rather than a liquid droplet.

Carrier Solvent Composition

Many different types of solvents have been used as carrier liquids for CF-FAB analyses and range from completely aqueous systems to organic liquids, such as toluene, chloroform, methanol, and acetonitrile.[8] However, much of the use of FAB involves the analysis of compounds derived from biological origin and, therefore, compounds that are generally water-soluble. For totally aqueous samples, a small amount (5–10%) of glycerol or other viscous solvent is included in the carrier solution to aid the FAB process and coat the sample stage to permit smooth and even flow of liquid over the surface. Generally, the higher the percentage of glycerol used, the easier it is to achieve stable operation, but the greater the intensity of interfering background ions. Many laboratories have started CF-FAB operation with 15 or 20% glycerol in the carrier solution and subsequently dropped this to about 5% as their experience and skill with the technique improved.

[7] Y. Ito, T. Takeuchi, D. Ishi, and M. Goto, *J. Chromatogr.* **346,** 161 (1985).
[8] M. Barber, L. W. Tetler, D. Bell, A. E. Ashcroft, R. S. Brown, and C. Moore, *Org. Mass Spectrom.* **22,** 647 (1987).

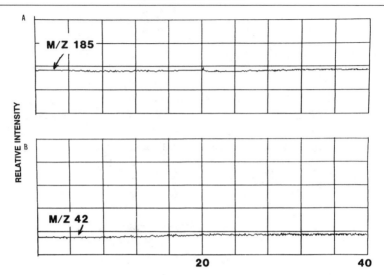

Fɪɢ. 2. Selected ion traces taken by monitoring the carrier solvent using the CF-FAB for sample introduction for (A) the protonated diglycerol cluster and (B) the protonated molecular ion of acetonitrile. The solvent was 95% water, 5% glycerol, and 3% acetonitrile (v/v/v).

Stability

Stable operation of the CF-FAB interface is generally characterized by a stable baseline for the total ion or specific ion monitor mode of operation of the mass spectrometer. For example, Fig. 2 shows the specific ion traces for m/z 42, the $(M + H)^+$ ion for acetonitrile, and m/z 185, the $(M + H)^+$ ion for the glycerol dimer, after approximately 40 min of monitoring the carrier solution which contained 92% water, 5% glycerol, and 3% acetonitrile. This level of performance is considered quite stable and is suitable for quantitative analysis. Such stability can generally be achieved in about 15 min, following the insertion of the probe.

The degree of stability required is dependent on the analytical task to be performed and this may differ from user to user for a given application. For qualitative analyses or where only semiquantitative estimates are necessary for compounds of the same approximate molecular weight range in a single sample, the requirement for a stable baseline is not stringent and one could do well with even moderate baseline variation. On the other hand, if the intensity of one or more compounds is being measured in a series of samples where comparisons of peak areas are important, then a high degree of stability should be maintained where short-term baseline shifts are not present.

For the bulk of the samples that are analyzed by CF-FAB, the following criteria for stability should be met: (1) The peak areas for three or more consecutive injections of a standard compound should not vary more than ±10% from the mean. (2) The ion current recorded for the total ion or a specific ion chromatogram should not vary more than ±10% over a period of 15 min. (3) The mass spectrometer ion source pressure gauge, which typically operates at about 2×10^{-4} torr when using a flow of 5 μl/min for an aqueous carrier solvent, should be steady. At a practical level, criterion (3) is the parameter that is most often used to evaluate stability over operating periods of many hours.

Sensitivity

Peptides of molecular weight of up to about 3000 have been routinely analyzed by CF-FAB at the 1 pmol level over a mass range of m/z 400–3000 using a high sensitivity magnetic instrument. At higher molecular weights, larger amounts of sample are required. The mass spectra produced using CF-FAB are characterized by a significantly lower background ion current than that obtained with std-FAB.[3] This is the major factor in the increased limit of detection of CF-FAB over that of std-FAB. At the low picomole level, this increase in signal-to-background noise can be as great as 150-fold or more. For example, Fig. 3 shows a portion of the molecular ion region from mass spectra taken for the analysis of 0.5 pmol of the peptide angiotensin II. The spectrum taken with std-FAB, shown in the top portion of the figure, contains intense background ions (most of it derived from glycerol) which mask the $(M + H)^+$ ion at m/z 1046. However, this ion is clearly observed in the spectrum taken with CF-FAB for the same amount of this sample. Some of this effect, in this case, may also be a consequence of the decreased "ion-suppression effect," discussed in the next section.

The high apparent sensitivity of the technique is a primary consequence of the reduced glycerol concentration in the sample. When bombarded by energetic atoms, glycerol contributes two types of ions to the background spectrum. The first is the result of radiation-damage ions produced by direct impact, which give rise to a signal at every mass throughout the spectrum. The second is derived from desorbed clusters of glycerol molecules that are protonated and appear at m/z $(92n + 1)$, where n is the number of glycerol molecules in the cluster, and fragmentation products showing successive losses of H_2O and CH_3O from these clusters. Since CF-FAB uses about a twenty-fold decrease in the glycerol concentration, ions derived from this compound are significantly decreased.

Ion-Suppression Effects

FAB is a surface analysis technique for liquid samples. For std-FAB, if the molecule of interest distributes itself in the interior of the droplet

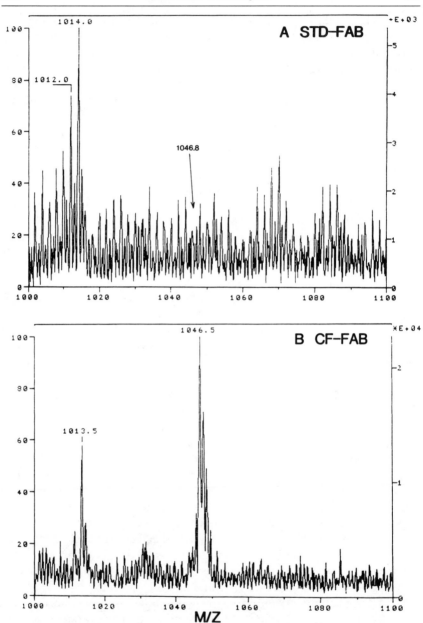

FIG. 3. The FAB mass spectra of the molecular ion $(M + H)^+$ regions from the analysis of 500 fmol of the peptide angiotensin II taken using (A) std-FAB and (B) CF-FAB.

TABLE I

COMPARISON OF ION SUPPRESSION EFFECTS FOR CF-FAB AND STD-FAB
WITH HYDROPHILIC PEPTIDES

Com-pound	Sequence	$(M + H)^+$	Hydro-philic index[a]	Charge at pH 1	Relative ion intensity[b]	
					CF-FAB	std-FAB
I	Ala-Phe-Lys-Lys-Ile-Asn-Gly	777.4	37	+3	59	4
II	Ala-Phe-Asp-Asp-Ile-Asn-Gly	751.3	80	+1	64	35
III	Ala-Phe-Lys-Ala-Lys-Asn-Gly	735.4	331	+3	31	7
IV	Ala-Phe-Lys-Ala-Asp-Asn-Gly	722.3	353	+2	18	4
V	Ala-Phe-Lys-Ala-Ile-Asn-Gly	720.4	59	+2	100	28
VI	Ala-Phe-Asp-Ala-Ile-Asn-Gly	707.3	80	+1	100	70
VII	Ala-Phe-Ala-Ala-Ile-Asn-Gly	663.3	80	+1	100	100

[a] Calculated by the method of Naylor *et al.*[10] from the data of Bull and Breese, *Arch. Biochem. Biophys.* **161,** 665 (1974).

[b] Measured from an average of five mass spectra, with reproducibility of repetitive analyses less than ±10%.

rather than in the surface layers, it will be missed in the analysis. It has been shown by a number of workers that hydrophilic peptides have a tendency to concentrate in the interior of a sample droplet while hydrophobic peptides tend to concentrate at the surface.[9,10] Thus, in mixture analysis, quantitative measurements of peptides with greatly differing hydrophobicities are problematic and some molecular species present may not be observed at all. This ion-suppression effect is greatly diminished for CF-FAB because of the mechanical mixing of the sample as it flows over the target surface in a thin film. This is shown in Table I for the analysis of the equimolar mixture of seven synthetic peptides.[11] For the analysis obtained for std-FAB, the $(M + H)^+$ ions for peptides I, III, IV, and V are significantly suppressed, presumably the result of the hydrophilic nature of these compounds which carry charges of either 2+ or 3+ at pH 1. The mass spectrum taken using CF-FAB shows that this effect is greatly decreased or eliminated for this same mixture of peptides.

The elimination or minimization of the ion-suppression effect is critical when the FAB map of a protein[12] is obtained or when the molecular

[9] W. D. Lehmann, M. Kessler, and W. A. Koenig, *Biomed. Mass Spectrom.* **11,** 217 (1984).

[10] S. Naylor, F. Findeis, B. W. Gibson, and D. H. Williams, *J. Am. Chem. Soc.* **108,** 6359 (1986).

[11] R. M. Caprioli, W. T. Moore, G. Petrie, and K. Wilson, *Int. J. Mass Spectrom. Ion Proc.* **86,** 187 (1988).

[12] H. R. Morris, M. Panico, and G. W. Taylor, *Biochem. Biophys. Res. Commun.* **117,** 299 (1983).

distributions of synthetic peptides are being measured. In the former technique, the masses of all the tryptic fragments of a protein are determined in order to deduce structural aspects of that protein, i.e., corroboration of the composition of the fragment by comparison to that inferred from the DNA sequence, identification of the presence and possible sites of postribosomal modifications, location of disulfide bridges, etc. In such applications, one would expect wide variations in the hydrophobic/hydrophilic nature of the fragments which would result in significant ion-suppression effects and, therefore, a compromised analysis.

Applications to Peptide and Protein Analysis

CF-FAB has been used quite effectively for the analysis of peptides derived from biochemical reactions, those isolated from biological tissues, and those prepared synthetically. Generally, there are two approaches to such an analysis: (1) direct injection of the sample with no prior purification or concentration, and (2) separation of the components prior to the analysis using liquid chromatography (LC), capillary electrophoresis (CE), or other suitable technique, either on- or off-line.

Direct analysis is fast and easy, and the method of choice, if buffers and salts do not interfere with the analysis and only selected peptides are of interest. This procedure would be suitable if one wanted to check the progress of a protease digestion, monitor a specific ion that was known to be recorded without difficulty under the conditions used, or quickly assess the nature of the products of a synthetic reaction.

Separation of compounds in a mixture prior to mass spectrometric analysis is recommended when it is important to record all of the ions in the sample, e.g., in the sequence analysis of a tryptic digest of a protein of unknown or uncertain sequence, or in the case where relatively high concentrations of salts, buffers, or other compounds would interfere with the analysis. Also, a separation process is advantageous where peptides of the same nominal mass occur in the sample. Although off-line techniques can be used effectively with individual samples, combining the separation process to the mass analysis process on-line is significantly more efficient and can be much more sensitive because it minimizes losses of sample incurred in manual processes. Specific applications are given in the following paragraphs which illustrate both the direct analysis and the combined separations approaches to mass spectrometric measurements of peptide mixtures.

Batch Sample Processing. One of the most common uses of the CF-FAB interface is the routine introduction of large numbers of aqueous samples. The procedure is essentially a flow-injection analysis whereby sample plugs are injected into a carrier stream which is flowing to the target. Of utmost concern in this batch sample processing is the amount of memory or carryover from sample to sample. We have found that at levels of about 100 pmol, samples can be injected every 2–3 min without incurring significant carryover or sample buildup.[13] This is illustrated in Fig. 4 for alternating injections of 100 pmol of the peptide substance P ($[M + H]^+ = 1348$) and tetradecapeptide renin substrate ($[M + H]^+ = 1761$). A portion of the total ion chromatogram is shown in Fig. 4A. After 1 hr of alternating injections, the molecular ion regions of each of the two peptides showed only insignificant traces of the other. This can be seen in Fig. 4B and C, which shows the spectrum for the last injection of each peptide.

A typical use of this technique is for checking the purity of a peptide made by either automated or manual synthetic methods. Literally, a 5-min analysis provides a comprehensive and specific analysis of the different molecules present after the peptide has been cleaved from the resin. Figure 5 shows the molecular distribution produced from the synthesis of the peptide Met-His-Arg-Gln-Glu-Ala-Val-Gly(NH$_2$), at $(M + H)^+ = 926.9$. The sample was the crude aqueous wash of the peptides cleaved from the resin by hydrogen fluoride treatment. Two major peptide contaminants can be identified. The des-Gly peptide homolog at m/z 869.7 and the methionyl sulfoxide analogs of the major peptides at m/z 885.8 and 942.8. A small amount of the des-Met product can be seen at m/z 795.7.

Another example of the use of the flow-injection method for batch sample processing is the time-course monitoring of enzymatic reactions. One of the major advantages of CF-FAB is the reproducibility of measurements achievable from sample to sample, allowing quantitative comparisons of separate samples without the use of internal standards. This can be done because, unlike std-FAB, the CF-FAB sample introduction probe is not disturbed and always presents the target in the same orientation for bombardment. For example, kinetic constants have been determined with CF-FAB in the flow-injection mode for the tryptic hydrolysis of the peptide Met-Arg-Phe-Ala.[5] The rate of hydrolysis of this peptide was determined at five different substrate concentrations (1.25, 1.0, 0.75, 0.5, and 0.25 mM) with the same amount of enzyme. Figure 6A shows the total ion chromatograms for the $(M + H)^+$ ion recorded from each of seven injections of the reaction mixture for all five substrate concentrations. Samples were injected at 3-min intervals. A double-reciprocal (Lineweaver–Burk) plot of the rates calculated from these measurements in which $1/S$ (where

[13] R. M. Caprioli, in "Biologically Active Molecules" (U. P. Schlunegger, ed.), p. 79. Springer-Verlag, Berlin, 1989.

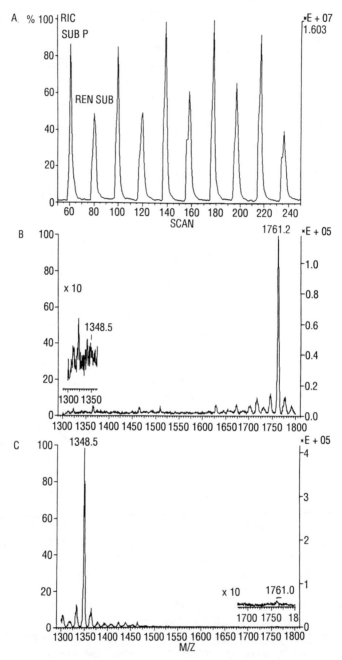

FIG. 4. Analysis of memory effects after 20 alternating injections, 3 min apart, of 100 pmol each of substance P ([M + H]$^+$ = 1348) and renin substrate tetradecapeptide ([M + H]$^+$ = 1761) using CF-FAB. (A) Total (reconstructed) ion chromatogram; (B) the mass spectrum from the last injection of substance P; (C) the mass spectrum from the last injection of renin substrate tetradecapeptide.

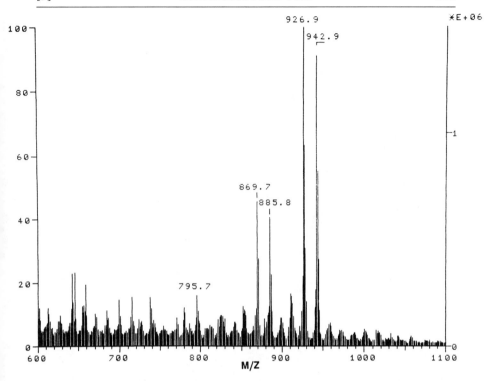

FIG. 5. The CF-FAB mass spectrum of the synthetic peptide Met-His-Arg-Gln-Glu-Ala-Val-Gly(NH₂). The sample was the aqueous extract after HF cleavage of the peptide from the resin used for the automated synthesis. The spectrum shows a significant amount of des-Gly product in addition to oxidation of the methionyl residue to the sulfoxide adduct.

S is the substrate concentration) is plotted against $1/V$ (where V is the velocity of the reaction) gives an X-intercept of $-1/K_m$ and a Y-intercept of $1/V_{max}$ (Fig. 6B). In this study, K_m was measured to be 1.85 mM, in close agreement to the value 1.90 mM previously reported. The major advantage of the technique is that an internal standard did not have to be used because absolute peak area measurements are sufficiently reproducible under stable operating conditions.

Liquid Chromatography–Mass Spectrometry. Although there are a number of different interfaces and ionization techniques that have been used to couple LC to MS, only a few involve FAB. A moving belt was first used as a transport mechanism onto which the column effluent[14,15]

[14] P. Dobberstein, E. Korte, G. Meyerhoff, and R. Pesch, *Int. J. Mass Spectrom. Ion Phys.* **46,** 185 (1983).

[15] J. G. Stroh, J. C. Cook, Jr., R. M. Milberg, L. Brayton, T. Kihara, Z. Huang, and K. L. Rinehart, Jr., *Anal. Chem.* **57,** 985 (1985).

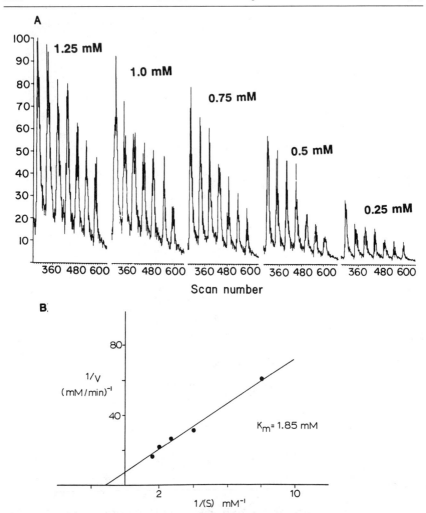

FIG. 6. Flow-injection analysis of the hydrolysis of the peptide Met-Arg-Phe-Ala by trypsin. (A) Selected-ion recording of the molecular species at m/z 524 at various substrate concentrations; (B) double-reciprocal plot of rate data obtained from measurements of areas under the peaks shown above. (Reproduced with permission from Ref. 5.)

was sprayed. After passing through a desolvation chamber, the belt then moved into the mass spectrometer ion source. Later, LC/CF-FAB was reported which utilized direct bombardment of the liquid column effluent inside the ion source.[16] A similar technique, frit-FAB (5), employs a fine

[16] R. M. Caprioli, B. DaGue, T. Fan, and W. T. Moore, *Biochem. Biophys. Res. Commun.* **146,** 291 (1987).

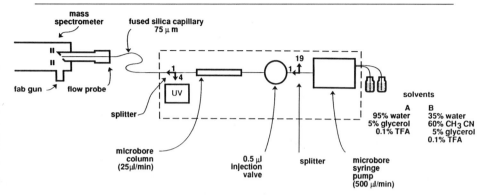

FIG. 7. Instrumental arrangement for microbore LC/CF-FAB mass spectrometric analyses. Details are given in the text. (Reproduced with permission from Ref. 18.)

mesh frit to disperse the column effluent at the target prior to atom bombardment.

The combination of liquid chromatography with CF-FAB has been demonstrated to be quite effective for peptide analysis. Since the flow rate for the CF-FAB interface is about 5–10 μl/min, it is desirable to use microbore (1 mm id) or capillary bore (0.3 mm id) columns which give good separations at low flow rates. The use of large- or regular-bore columns have been reported,[17] but the high ratio splitters (100 : 1) necessary to reduce the flow rate from 1 ml/min to 10 μl/min sacrifice a great deal of sensitivity. Microbore columns require only about a 3 : 1 split ratio, give excellent chromatographic performance, and are quite durable. The new packed capillary-bore fused-silica columns require no split since they operate at 5 μl/min normally and might prove to be quite effective, although they are more fragile and difficult to work with.

The typical instrumental arrangement used for microbore LC/MS with CF-FAB for the analysis of tryptic digests[18] is shown in Fig. 7. A splitter (19 : 1 ratio) is placed before the injector so that gradients may be formed quickly and dead volumes in the pump and associated plumbing cleared efficiently. Gradients are formed at about 500 μl/min, while the column flow is 25 μl/min. A second splitter (3 : 1) is placed after the column to allow about 8 μl/min to flow into the mass spectrometer. A 1 × 50 mm C_{18} reversed-phase microbore column is used for the separation with acetonitrile and aqueous solvent systems, except that 5% glycerol is added to each solvent. If, for some reason, it is not acceptable to add this glycerol

[17] D. E. Games, S. Pleasance, E. Ramsey, and M. A. McDowall, *Biomed. Environ. Mass Spectrom.* **15**, 179 (1988).
[18] R. M. Caprioli, B. B. DaGue, and K. Wilson, *J. Chromatogr. Sci.* **26**, 640 (1988).

to the solvent systems, it may be added postcolumn through the use of a coaxial interface.[19]

The effectiveness of this technique is illustrated for the analysis of the tryptic digest of 24 pmol of human apolipoprotein A-I (approximate MW 23,000).[20] Mass spectra were obtained on a Finnigan MAT 90 high-field mass spectrometer equipped with a CF-FAB interface. An Applied Biosystems model 130A microbore LC was employed using a 15-min gradient elution starting from 100% solvent A to 100% solvent B (where A consists of 95% water, 5% glycerol, and 0.1% TFA (v/v/v), and B, 60% acetonitrile, 35% water, 5% glycerol, and 0.1% TFA (v/v/v). Figure 8A shows the total ion chromatogram obtained from the analysis and Fig. 8B the mass spectrum of scan 80 from the chromatogram. A total of 86 molecular ion species of peptides were identified, although many had intensities of 10% or less of that of the major ions and presumably arose from impurities in the sample. The resolution of the microbore system is very good, as shown by the detailed mass analysis of a small region of the chromatogram in Fig. 9A. Figure 9B shows the specific ion chromatograms for the eight molecular species found in this region. This latter plot was made by superimposing all eight independently normalized ion chromatograms. All 27 peptides expected to be produced that fell within the m/z 400–2000 mass range scanned by the spectrometer were identified. These data are presented in Table II. It should be noted, however, that it is not always possible to identify all of the peptides in such a mixture, although typically >90% of these are found. Parameters such as the amount of sample, presence of impurities, and chemical properties of fragments, will influence the degree of success in such an analysis. For example, when the apolipoprotein A-I digest analysis was performed by injecting 2.4 pmol of digest onto the liquid chromatography column, only 21 peptides could be identified in the mixture by mass analysis.

For an LC/MS analysis such as this, one can produce a "summary spectrum" which is created by summing all the mass spectra taken in the analysis. For the apolipoprotein A-I digest, this summary spectrum is shown in Fig. 10 and is compared with that obtained from the direct analysis of the same amount of digest using std-FAB. The major ion species in the std-FAB spectrum are cluster ions from glycerol, with only some of the peptide ion observable among the peaks in the background. The summary spectrum from the LC/MS run is clear of the interfering

[19] J. S. M. deWit, L. J. Deterding, M. A. Moseley, K. B. Tomer, and J. W. Jorgensen, *Rapid Commun. Mass Spectrom.* **2,** 100 (1988).

[20] W. T. Moore, B. DaGue, M. Martin, and R. M. Caprioli, *Proc. 36th ASMS Conf. Mass Spectrom. Allied Topics, San Francisco, CA* p. 1081 (1988).

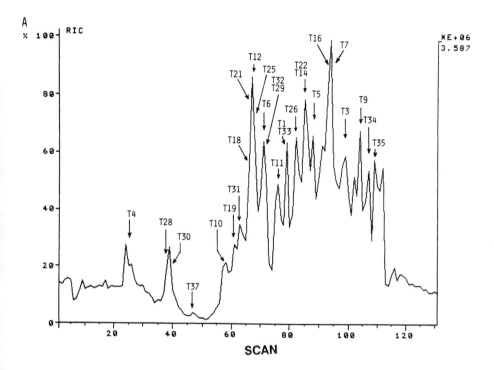

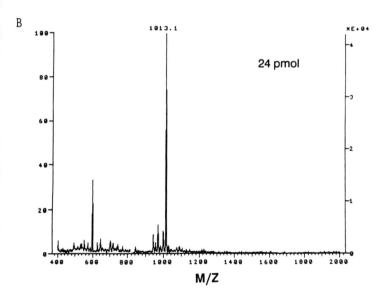

FIG. 8. Analysis of 24 pmol of the tryptic digest of human apolipoprotein A-I. (A) Total ion chromatogram, showing positions of identified tryptic fragments, and (B) mass spectrum of scan 80.

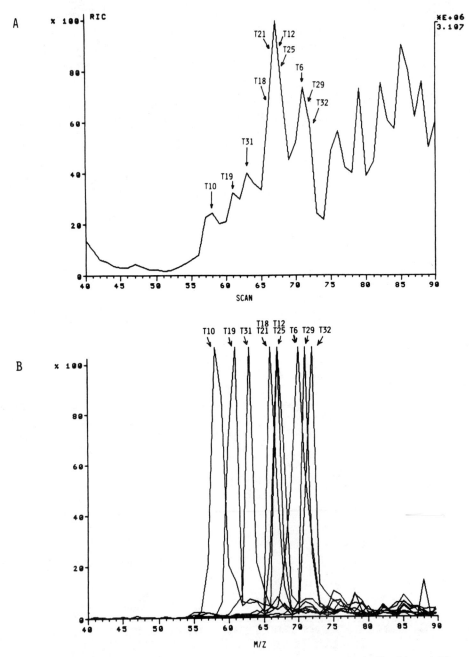

FIG. 9. (A) Expanded portion of the total ion chromatogram shown in Fig. 8A, and (B) selected-ion chromatograms of specific tryptic fragments located in scans 55–75. Each scan was independently normalized and plotted as an overlay with the others.

TABLE II

ANALYSIS OF TRYPTIC DIGEST OF APOLIPOPROTEIN A-I BY
LC/CF-FAB

Tryptic peptide	Calculated $(M + H)^+$	Measured $(M + H)^+$
T1 DEPPQSPWDR	1226.5	1226.9
T2 VK	246.2	—[b]
T3 DLATVYVDVLK	1235.7	1236.2
T4 DSGR	434.2	434.2
T5 DYVSQFEGSALGK	1400.7	1400.9
T6 QLNLK	615.4	615.4
T7 LLDNWDSVTSTFSK	1612.8	1613.8[a]
T8 LR	288.2	—[b]
T9 EQLGPVTQEFWDNLEK	1932.9	1933.9[a]
T10 ETEGLR	704.3	704.7
T11 QEMSK	622.3	622.6
T12 DLEEVK	732.4	732.7
T13 AK	218.1	—[b]
T14 VQPYLDDFQK	1252.6	1252.9
T15 K	147.1	—[b]
T16 WQEEMELYR	1283.6	1284.0
T17 QK	275.2	—[b]
T18 VEPLR	613.3	613.3
T19 AELQEGAR	873.4	873.9
T20 QK	275.2	—[b]
T21 LHELQEK	896.5	897.0
T22 LSPLGEEMR	1031.5	1031.5
T23 DR	290.1	—[b]
T24 AR	246.1	—[b]
T25 AHVDALR	781.4	781.8
T26 THLAPYSDELR	1301.6	1301.9
T27 QR	303.2	—[b]
T28 LAAR	430.3	430.1
T29 LEALK	573.3	573.6
T30 ENGGAR	603.3	603.5
T31 LAEYHAK	831.4	831.8
T32 ATEHLSTLSEK	1215.6	1215.8
T33 AKPALEDLR	1012.6	1013.1
T34 QGLLPVLESFK	1230.7	1231.1
T35 VSFLSALEEYTK	1386.9	1387.1
T36 K	147.1	—[b]
T37 LNTQ	475.2	475.2

[a] Mass measurement as centroid of isotopic envelope.
[b] Ion outside of mass range scanned.

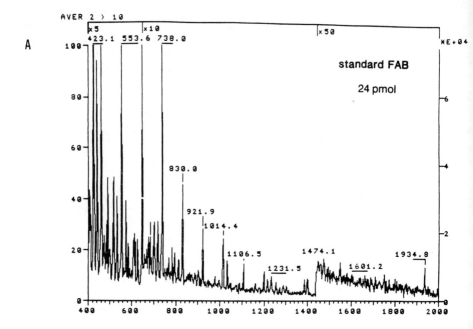

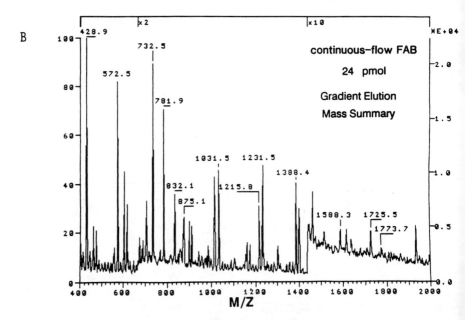

FIG. 10. (A) std-FAB mass spectrum of the apolipoprotein A-I tryptic digest in a high glycerol concentration matrix. (B) Reconstructed mass spectrum of the same amount of digest as in (A), obtained by summing all of the mass spectra taken throughout the LC/CF-FAB analysis.

peaks and shows the peptide ions to be the most intense. Comparison of these spectra clearly points to the LC/CF-FAB technique as being significantly more effective than std-FAB in identifying peptide fragments in protease mixtures.

Capillary Electrophoresis–Mass Spectrometry. Capillary electrophoresis (CE) has been shown to be an extremely effective separation device for compounds derived from biological origin because it provides high-resolution separations of charged molecules, with sample loads in the femtomole range using open tubular fused-silica capillaries. These capillaries are generally about 1 m in length, have inner diameters of 10–100 μm, and can achieve separations having 100,000 or more theoretical plates.[21]

The basic need for an interface between the end of the CE capillary and the mass spectrometer results from the incompatibility of the CF-FAB process and CE for liquid flow. The CF-FAB source requires a solvent which is maintained at a steady flow rate in the range of approximately 2–15 μl/min. Optimal performance of the CE process, in contrast, requires no mechanical flow at all, although a flow rate in the nanoliter/minute range results from electroosmotic effects. Several interface designs have been reported, the first simply being an open liquid junction in the cathodic reservoir,[22] and later, a coaxial design.[23] The interface shown in Fig. 11 is a hybrid of these two, allowing sufficient flow to stabilize the CF-FAB target as a result of the pressure differential between the cathodic reservoir and mass spectrometer source, while not causing mechanical flow in the CE capillary.[24] Thus, compounds eluting from the end of the CE capillary are pulled into the probe capillary and are subsequently analyzed. Comparison of an output trace of a UV detector placed prior to the interface and the total ion chromatogram trace from the mass spectrometer (positioned after the interface) shows that the band broadening as a result of the interface is small, essentially maintaining the high-resolution integrity of CE separation.

Achieving high-resolution separations by CE is an asset of the technique, although in this regard, there is some incompatibility with the mass spectrometer as a detector. Most mass spectrometers are scanning instruments that take generally 2–20 sec to scan a mass range, depending

[21] J. W. Jorgenson and K. D. Lukacs, *Science* **222,** 266 (1983).

[22] R. D. Minard, D. Chin-Fatt, P. Curry, Jr., and A. G. Ewing, *Proc. 36th ASMS Conf. Mass Spectrom. Allied Topics, San Francisco, CA* p. 950 (1988).

[23] K. Tomer, M. Moseley III, J. deWit, L. Deterding, and J. Jorgenson, *Adv. Mass Spectrom.* **11A,** 440 (1989).

[24] R. M. Caprioli, W. T. Moore, K. B. Wilson, and S. Moring, *J. Chromatogr.* **480,** 247 (1989).

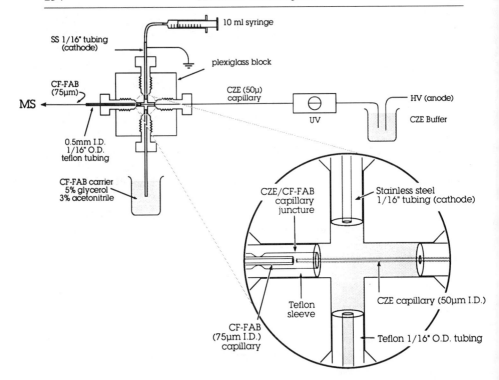

FIG. 11. Instrumental arrangement and enlargement of the liquid junction interface between the cathodic end of the electrophoresis capillary and the transfer capillary for the CF-FAB probe. The syringe is used to flush the reservoir of the interface block. (Reproduced with permission from Ref. 24.)

on the m/z values to be covered. Thus, a typical magnetic instrument with a fast-scanning laminated magnet would require approximately 10 sec or more to cover a mass range of m/z 300–3000 in order to achieve adequate signal-to-noise measurements. If a minimum of 3 to 4 scans across an eluting peak is desired, then the eluting peak width would need to be approximately 30 sec wide. The situation is not unlike that encountered in GC/MS applications where high-resolution capillary columns are employed. Of course, the narrower the scan range, the greater the number of scans that can be taken per unit time. Also, some new instruments have integrating array detectors that simultaneously collect masses in a window which may be 10–20% of the entire mass range. Such detectors will improve the situation markedly even if it must be stepped over several mass windows to cover the necessary mass range.

The sensitivity of the CE/CF-FAB method was measured for angiotensin II, a nonapeptide of molecular mass 1045 Da. The mass spectrometer was scanned over a five-mass window centered on m/z 1046, the $(M + H)^+$ ion. On injection of 368 fmol of the peptide, a signal-to-noise ratio of 12 : 1 was recorded; for 75 fmol a ratio of about 3 : 1 was recorded. However, this sensitivity is only relative and is influenced significantly by the specific operating conditions. For example, to a first approximation, the relative sensitivity is inversely proportional to the number of masses scanned in a given application. Thus, if one is analyzing tryptic peptides where it is desirable to scan a mass range of m/z 500–2500, then the amount of sample needed to give reasonable signal-to-noise measurements is more likely to be at the 1–10 pmol level. This is shown later for the analysis of the proteolytic digest of a protein.

Another potential incompatibility between CE and MS is high salt concentrations. Typically, 40 mM sodium chloride or sodium phosphate is employed in CE buffers. Unfortunately, the FAB process does not tolerate these salt concentrations well; sensitivity can be drastically decreased relative to a salt-free sample. In addition, even moderate amounts of NaCl in the sample, for example, produces mass spectra that are characterized by the presence of sodium adduct ions at M + 22, M + 44, M + 66, etc., mass units above the $(M + H)^+$ molecular species of the compound, corresponding to $(M + Na)^+$, $(M + 2Na)^+$, $(M + 3Na)^+$, etc., respectively. Furthermore, the limit of detection of a given compound is poorer because the molecular species are distributed into several molecular masses. To overcome this incompatibility, a discontinuous buffer system can be employed, i.e., moderate (20 mM) salt solutions can be used in the anode buffer reservoir and CE capillary, but only low concentrations of salt or other electrolyte are used in the interface. Since the CE flow is very small compared to the flow of buffer from the cathodic reservoir, the overall salt concentration going to the FAB target is quite low.

We have used CE/MS for checking the molecular distributions of synthetic peptides and for the analysis of mixtures generated by the hydrolysis of proteins with proteases. For example, the total ion chromatogram for the analysis of 10 pmol of the tryptic digest of β-lactoglobulin is shown in Fig. 12. The mass spectrometer was scanned from m/z 500–2000 at 8 sec/scan. Specific mass analysis of these data showed molecular species of the individual peptides as indicated in the figure. Two different peptide fragments of m/z 573 were identified, one at scan 155 and the other at scan 339. It is believed that the latter is

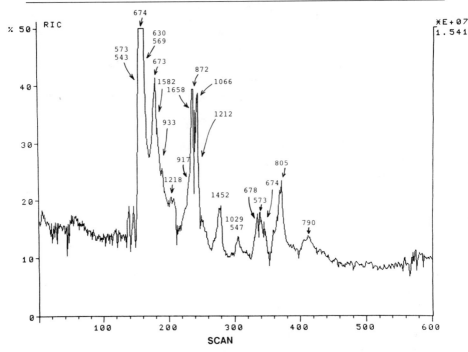

FIG. 12. The total ion chromatogram for the CE/CF-FAB analysis of 10 pmol of the tryptic digest of β-lactoglobulin. The (M + H)⁺ measurements for the peptides identified in the various chromatographic peaks are shown.

produced by the action of a contaminating protease on a tryptic fragment, and appears late because it is more negatively charged than the other fragment. It should be noted that the peak widths measured from specific ion chromatograms are about 20 sec wide, demonstrating the high resolution of the CE separation.

Conclusion

Within the past few years, several mass spectrometric techniques have been introduced which allow the direct analysis of aqueous systems. CF-FAB is one of these and has been shown to be of value in a wide variety of applications involving biological processes and systems. The analysis of peptides remains one of the most successful uses because of its high sensitivity for these compounds and ease of use. The technique is not without disadvantages, however, which arise mainly from the high pressures generated by the evaporation of the solvent in the ionization chamber of the mass spectrometer. Nevertheless, the technique provides the biochemist with a mass specific tool to directly measure molecular weights of compounds isolated in aqueous solutions and to follow the sometimes complicated chemical alterations of these compounds produced by enzymatic or other chemical reactions.

[10] Collision-Induced Dissociation

By ROGER N. HAYES and MICHAEL L. GROSS

Introduction

The determination of molecular structure by mass spectrometry requires that the molecule be ionized and undergo structurally informative fragmentations. Fragmentation may be induced by the ionization process itself or by some other means of excitation. Often the optimum methods of ion formation do not impart sufficient energy to cause appreciable fragmentation of the molecular ion. The internal energy of the ion in these circumstances may be increased by collisional activation (CA). In this chapter, a review of the methods of CA is presented.

Tandem mass spectrometry (MS/MS), in all its various configurations, has emerged as the most important technique for acquiring collisional-activated dissociation (CAD) [also termed collision-induced dissociation (CID)] mass spectra. CA has grown from its introduction[1,2] in mass spectrometry in 1968 to a widely used technique in the areas of chemical and biochemical analysis. Its acceptance can be easily gauged by the growing number of published studies[3-5] in which the technique was used, and by the commitment of instrument manufacturers to the design and marketing of tandem mass spectrometers.

The overall CAD process can be separated into two consecutive steps occurring on well-separated time scales. The first is a fast (ca. 10^{-15}–10^{-14} sec) CA step in which some portion of the initial translational energy of the accelerated ion is converted into internal energy of both ion and target (the target also acquires translational energy). The second step in this process is the dissociation of the now energized (and typically isolated)

---

[1] K. R. Jennings, *Int. J. Mass Spectrom. Ion Phys.* **1,** 227 (1968).

[2] W. F. Haddon and F. W. McLafferty, *J. Am. Chem. Soc.* **90,** 4745 (1968).

[3] K. B. Tomer, *Mass Spectrom. Rev.* **8,** 445 (1989).

[4] K. Biemann and S. Martin, *Mass Spectrom. Rev.* **6,** 75 (1987).

[5] N. J. Jensen and M. L. Gross, *Mass Spectrom. Rev.* **6,** 497 (1987).

ion. The yield of product ions after CA depends on the probability of unimolecular decomposition of the precursor ion after excitation. The quasi-equilibrium theory (QET) is usually invoked to explain the rates of such reactions.[6-10] QET postulates that unimolecular decomposition reactions depend upon the random distribution of the internal energy of the ion among all the vibrational modes of that ion. The rate of decomposition is, therefore, related to the probability of a given vibrational mode or modes acquiring enough energy to rupture bonds.[11] There are $3N - 6$ vibrational modes in a nonlinear ion with N atoms; thus, the number of vibrational modes grows in direct proportion to molecular mass for a given class of compounds. Because the random distribution of internal energy among the vibrational modes is required by QET, the average energy mode must decrease with increasing molecular mass. Because the decrease is related to the inverse of the mass of the ion, the fragment ion yield should decrease similarly beyond some threshold. A number of experiments[12-18] have been conducted to address this relationship.

Variations in fragment ion distributions occur because of differences in the amount of acquired internal energy, which in turn depends on the chosen method of activation.[7,19-22] Variations in ion distributions can also

[6] P. J. Todd and F. W. McLaffery, in "Tandem Mass Spectrometry" (F. W. McLafferty, ed.), p. 149. Wiley, New York, 1983.

[7] K. Levsen, "Fundamental Aspects of Organic Mass Spectrometry." Verlag Chemie, Weinheim, New York, 1978.

[8] A. G. Brenton, R. P. Morgan, and J. H. Beynon, Annu. Rev. Phys. Chem. 30, 51 (1979).

[9] H. M. Rosenstock, M. B. Wallenstein, A. L. Wahrhaftig, and H. Eyring, Proc. Natl. Acad. Sci. U.S.A. 38, 667 (1952).

[10] R. A. Marcus and O. K. Rice, J. Phys. Colloid Chem. 55, 894 (1951).

[11] T. Baer, "Mass Spectrometry," Vol. 6, p. 1. The Royal Society of Chemistry, Burlington House, London, 1981.

[12] W. A. Chupka, J. Chem. Phys. 30, 191 (1959).

[13] B. Andlauer and Ch. Ottinger, J. Chem. Phys. 55, 1471 (1971).

[14] B. Brehm and E. von Puttkamer, Z. Naturforsch. 22A, 8 (1967).

[15] T. Baer, in "Gas Phase Ion Chemistry" (M. T. Bowers, ed.), Vol. 1. Academic Press, New York, 1979.

[16] J. H. D. Eland, in "Mass Spectrometry, Specialist Periodical Report" (R. A. W. Johnstone, Sr. Reporter), Vol. 5, Chap. 3. The Chemical Society, London, 1979.

[17] F. W. McLafferty, T. Wachs, C. Lifshitz, G. Innorta, and P. Irving, J. Am. Chem. Soc. 92, 6867 (1970).

[18] G. M. Neumann, M. M. Sheil, and P. J. Derrick, Z. Naturforsch. 39A, 584 (1984).

[19] F. W. McLafferty (ed.), "Tandem Mass Spectrometry." Wiley, New York, 1983.

[20] F. W. McLafferty, P. F. Bente III, R. Kornfeld, S. C. Tsai, and I. Howe, J. Am. Chem. Soc. 95, 2120 (1973).

[21] R. B. Cody and B. S. Freiser, Anal. Chem. 51, 547 (1979).

[22] Md. A. Mabud, M. J. DeKrey, and R. G. Cooks, Int. J. Mass Spectrom. Ion Proc. 67, 285 (1985).

be attributed to such factors as dissimilar reaction times, the form of energy deposited, the amounts of ion scattering upon excitation, and the different angles over which the ions are collected.[7,19,23,24]

There are practical advantages to controlling the amount of internal energy uptake by an ion undergoing CA. Systematic variation of the amount of acquired energy by the fragmenting ions can be used, for example, in optimizing molecular ion abundances,[25–30] and differentiating isomers.[29,30–32] Unfortunately, few of the methods generally used to activate and dissociate gas-phase ions allow easy, well-defined control of ion internal energies over a wide energy range.[33] Nevertheless, with the advent of tandem mass spectrometry as a tool for mixture analysis[19,34,35] and for ion structural characterization,[19,36,37] the number of methods that can be used to activate ions has increased.

For the purposes of this chapter, we will focus on two methods of ion activation: collisions of accelerated ions with a "stationary" gas-phase target in the (1) high-energy (kiloelectron volt),[1,2,19,20,23,24] and (2) low-energy (electron volt)[19,38–41] ranges of laboratory ion kinetic energy. Three other methods of ion activation that will be briefly mentioned for compara-

[23] R. G. Cooks (ed.), "Collision Spectroscopy." Plenum, New York, 1978.

[24] K. Levsen and H. Schwarz, *Mass Spectrom. Rev.* **2**, 77 (1983).

[25] H. Budzikiewicz, C. Djerassi, and D. H. Williams, "Mass Spectrometry of Organic Compounds." Holden-Day, San Francisco, 1967.

[26] G. Spiteller-Friedmann, *Liebigs Ann. Chem.* p. 690 (1966).

[27] D. P. Stevenson and C. D. Wagner, *J. Am. Chem. Soc.* **72**, 5612 (1950).

[28] F. H. Field and S. H. Hasting, *Anal. Chem.* **28**, 1248 (1956).

[29] A. G. Harrison, "Chemical Ionization Mass Spectrometry." CRC Press, Boca Raton, FL, 1983.

[30] E. Lindholm, *in* "Ion Molecule Reactions" (J. L. Franklin, ed.), p. 457. Plenum, New York, 1972.

[31] J. Sunner and I. Szabo, *Int. J. Mass Spectrom. Ion Phys.* **26**, 241 (1977).

[32] A. N. H. Yeo and D. H. Williams, *J. Am. Chem. Soc.* **93**, 395 (1971).

[33] K. R. Jennings, *Gazz. Chim. Ital.* **114**, 313 (1984).

[34] R. G. Cooks and R. A. Roush, *Chim. Ind.* (*Milan*) **66**, 539 (1984).

[35] G. L. Glish, V. M. Shaddock, K. Harmon, and R. G. Cooks, *Anal. Chem.* **52**, 165 (1980).

[36] K. Levsen and H. Schwarz, *Angew. Chem., Int. Ed. Engl.* **15**, 509 (1976).

[37] S. Verma, J. D. Ciupek, and R. G. Cooks, *Int. J. Mass Spectrom. Ion Proc.* **62**, 219 (1984).

[38] F. Kaplan, *J. Am. Chem. Soc.* **90**, 4483 (1968).

[39] R. A. Yost and C. G. Enke, *Anal. Chem.* **51**, 1251A (1979).

[40] P. H. Dawson, J. B. French, J. A. Buckley, D. J. Douglas, and D. Simmons, *Org. Mass Spectrom.* **17**, 205, 212 (1982).

[41] Z. Herman, J. H. Futrell, and B. Friedrich, *Int. J. Mass Spectrom. Ion Proc.* **58**, 181 (1984).

tive purposes are surface-induced dissociation,[22,42] photodissociation,[43-46] and electron impact activation.

Depending upon the choice of experimental conditions, the individual methods allow some measure of control over the degree of excitation of a selected ion. The techniques of angle-resolved mass spectrometry (ARMS)[19,37,47-49] and energy-resolved mass spectrometry (ERMS) utilizing CA with gaseous targets[19,37,50,51] have proved to have some use in controlling the extent of activation.

High-Energy Collisional Activation

Instrumentation

CAD mass spectra may be acquired on any of four distinct classes of mass spectrometer (see [2] in this volume): magnet sector, quadrupole, time-of-flight (TOF), and Fourier transform (FT) mass spectrometers. There are also a number of hybrid configurations employing a combination of sector and quadrupole analyzers. High-energy collisional activation over a wide mass range, however, is currently accessible only on sector instruments and some hybrids. The most favorable configuration of a tandem sector mass spectrometer consists of two double-focusing mass spectrometers that combines high-resolution parent ion selection in MS-1 with at least unit mass resolution in MS-2 (see [6] in this volume). Alternative choices of instrument configuration include three-sector instruments of either *EBE* or *BEB* configuration, and double-focusing instruments of *EB* or *BE* configuration. In each of the latter alternative configu-

[42] M. J. DeKrey, Md. A. Mabud, R. G. Cooks, and J. E. P. Syka, *Int. J. Mass Spectrom. Ion Proc.* **67**, 295 (1985).
[43] R. C. Dunbar, in "Gas Phase Ion Chemistry" (M. T. Bowers, ed.), Vol. 2, Chap. 14. Academic Press, New York, 1979.
[44] D. S. Bomse, R. L. Woodin, and J. L. Beauchamp, *J. Am. Chem. Soc.* **101**, 5503 (1979).
[45] R. C. Dunbar, in "Gas Phase Ion Chemistry" (M. T. Bowers, ed.), Vol. 3, Chap. 20. Academic Press, New York, 1984.
[46] F. M. Harris and J. H. Beynon, in "Gas Phase Ion Chemistry" (M. T. Bowers, ed.), Vol. 3, Chap. 19. Academic Press, New York, 1984.
[47] J. A. Laramee, J. J. Carmody, and R. G. Cooks, *Int. J. Mass Spectrom. Ion Phys.* **31**, 333 (1979).
[48] D. M. Fedor and R. G. Cooks, *Anal. Chem.* **52**, 679 (1980).
[49] S. Verma, J. D. Ciupek, R. G. Cooks, A. E. Schoen, and P. Dobberstein, *Int. J. Mass Spectrom. Ion Phys.* **52**, 311 (1983).
[50] D. D. Fetterolf and R. A. Yost, *Int. J. Mass Spectrom. Ion Phys.* **44**, 37 (1982).
[51] S. A. McLuckey, G. L. Glish, and R. G. Cooks, *Int. J. Mass Spectrom. Ion Phys.* **39**, 219 (1981).

rations, the user is sacrificing either parent ion or product ion resolution, or both.

Regardless of the choice of instrument configuration, an obvious requirement is the presence of a collision cell that can be pressurized with a suitable target gas. The collision cell is usually positioned at or near a focal point (or plane) of the instrument.

Mechanism of High-Energy Collisional Activation

The earliest discussions of mechanisms of collisional activation are those of Durup[52,53] and Los and Govers[54] and were concerned mainly with small systems (2–5 atoms). Durup discussed four basic mechanisms of CAD: (1) vertical electronic excitation to a dissociative state, (2) adiabatic dissociation following momentum transfer, (3) complete inelastic collision with an orbiting complex formation, and (4) collision-induced predissociation. Russek[55] proposed an alternative mechanism for kiloelectron volt (keV) collisions that recognizes the adiabatic nature of electronic transitions leading to polarization-induced, vibrational–rotational excitation and dissociation.

Efficient conversion of ion translational energy into internal energy occurs when the collision interaction time (t_c) and the period of the internal mode that is being excited (τ) are comparable.[56,57] For an ion of m/z 1000 and with a translational energy of 8 keV, the value of t_c for an interaction path with a target molecule of several angstroms is ca. 10^{-14} sec. The Bohr period of an electron in a valence orbital of a polyatomic molecule is of a similar duration. Thus, collisions at kiloelectron volt energies are expected to result in excitation of internal electronic modes. Redistribution of the excitation energy to vibrational internal modes, a premise of QET, ultimately results in bond cleavage. The mechanism of coupling of electronic states has been described in terms of crossings of the various electronic states in the transient collision complex.[58–60] Excitation of rotational/vibra-

[52] H. Yamaoka, P. Dong, and J. Durup, *J. Chem. Phys.* **51**, 3465 (1969).
[53] J. Durup, *in* "Recent Developments in Mass Spectrometry" (K. Ogata and T. Hayakawa, eds.), p. 921. University of Tokyo Press, Tokyo, 1970.
[54] J. Los and T. R. Govers, *in* "Collision Spectroscopy" (R. G. Cooks, ed.), p. 289. Plenum, New York, 1978.
[55] A. Russek, *Physica* **48**, 165 (1970).
[56] S. Singh, F. M. Harris, R. K. Boyd, and J. H. Beynon, *Int. J. Mass Spectrom. Ion Proc.* **66**, 131 (1985).
[57] J. H. Beynon, R. K. Boyd, and A. G. Brenton, *Adv. Mass Spectrom.* **10**, 437 (1986).
[58] W. Weizel and O. Beeck, *Z. Phys.* **76**, 250 (1932).
[59] W. Lichten, *Phys. Rev.* **131**, 229 (1963).
[60] T. F. O'Malley, *in* "Advances in Atomic Molecular Physics" (D. R. Bates and I. Esterman, eds.), Vol. 7. Academic Press, New York, 1971.

tional modes is also possible but would be considerably less efficient; the interaction time is ca. 10 times shorter than the period of a molecular vibration.

Evidence that electronic excitation is indeed the dominant process for high-energy collisional activation is given in part by the close similarity between the CA and 50 eV electron-ionization mass spectra of a number of aromatic compounds.[1] Corroborative evidence is that electronic excitation in the limit (i.e., charge stripping or loss of an electron) occurs upon high-energy CA.

Relationship between Pressure and Number of Collisions

The pressure of collision gas in the collision cell is somewhat arbitrarily chosen; the actual pressure of gas within the collision cell is often not measured directly, but rather the pressure at some other nearby location is obtained. Instead, it is more convenient and precise to use that observed pressure that attenuates the main beam of ions by a given percentage. This is common practice with tandem magnet sector instruments. As the main ion beam becomes increasingly more attenuated, the probability of multiple collision encounters increases. Concomitantly, the yield of fragment ions resulting from high-dissociation energy pathways also increases. Figure 1 shows the total collision probability (P_n) for single and higher-order encounters for an ion with a nominal collision cross section of 5×10^{-16} cm^2, traversing a 1-cm collision region, as a function of collision gas pressure.[61,62] If the main ion beam is attenuated by 30%, 95, 5, and 0% of the encounters are single, double, and triple collisions, respectively; if the attenuation is 50%, the corresponding values are 70, 20, and 10%, respectively. Some workers use 70–90% attenuation, which corresponds to many multiple collisions.

Alternatively, an optimum pressure of collision gas for CA can be defined as that pressure that gives rise to the maximum abundances of fragment ions.[63] It is known that this optimum pressure is dependent on the incident ion mass because the optimum pressure seems to decrease as the mass is increased.[64]

Center-of-Mass Considerations

In a collision between a fast moving ion and a stationary target gas atom, the maximum amount of kinetic energy available (based on center-

[61] J. L. Holmes, *Org. Mass Spectrom.* **20**, 169 (1985).
[62] M. S. Kim, *Int. J. Mass Spectrom. Ion Phys.* **50**, 189 (1983).
[63] M. S. Kim and F. W. McLafferty, *J. Am. Chem. Soc.* **100**, 3279 (1978).
[64] G. M. Neumann and P. J. Derrick, *Org. Mass Spectrom.* **19**, 165 (1984).

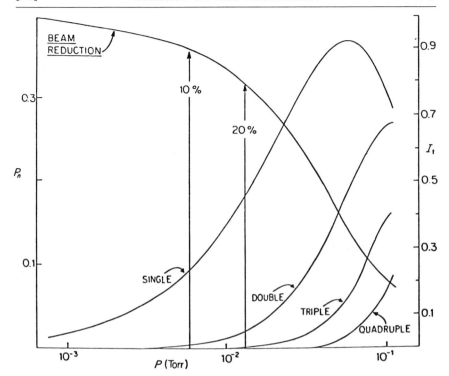

Fig. 1. Total collision probability, P_n, and the fractions of single- and multiple-collision processes as a function of collision gas pressure (for an ion of collision cross section 5×10^{-16} cm^2, collision path 1 cm). I_t is the transmission intensity of the ion beam. (Reprinted with permission from Ref. 61.)

of-mass considerations) for conversion into internal energy of the ion is [Eq. (1)]:

$$E_{com} = m_t E_{lab}/(m_t + m_i) \qquad (1)$$

where m_t and m_i are the masses of the target atom and ion, respectively, and E_{lab} is the kinetic energy of the ion in the laboratory frame (i.e., if the accelerating potential is 8 kV, E is 8 keV). Increasing the kinetic energy of the ion or increasing the molecular weight of the collision gas will increase the available energy, whereas increasing the mass of the ion will decrease the available energy approximately as a function of $1/m_i$.

In studies of internal energy deposition,[63,65,66] it was demonstrated that

[65] V. H. Wysocki, H. I. Kenttämaa, and R. G. Cooks, *Int. J. Mass Spectrom. Ion Proc.* **75**, 181 (1987).

[66] H. I. Kenttämaa and R. G. Cooks, *J. Am. Chem. Soc.* **107**, 1881 (1985).

an average energy of ca. 1–3 eV (1 eV = 23.06 kcal/mol) is deposited into an ion in a high-energy collision, but that the distribution of deposited energies exhibits a high-energy tail extending beyond 15 eV of internal energy.

The probability of a collision depends upon target gas pressure and the collision cross sections of both ion and target species. The latter factors, in turn, depend on the molecular masses of both ion and target species and on the kinetic energy of the ion. The efficiency of the collision process for converting kinetic energy into internal energy of the ion is also a function of the kinetic energy of the ion and the molecular weights of both the ion and target species.[6,18,64,67,68] Processes that compete with internal energy deposition include elastic collisions (for which there is no energy conversion), charge exchange, and charge stripping.

The collision cross section of an ion decreases with increasing kinetic energy, but for a given kinetic energy, the collision cross section increases with increasing molecular mass of both the ion and target species.[6,18] The latter effect is consistent with the observation that the pressure of collision gas required for a 50% attenuation of the ion beam decreases with increasing molecular mass of both ion and target gas.[69] Neumann and co-workers[18] determined the absolute collision cross section for a series of ions derived from peptides and found that the cross sections are approximately a linearly increasing function of mass.

Neumann and co-workers[18] also studied the energetics of collisions between a series of peptide ions, of 8 keV kinetic energy, and atomic gases, usually helium. They determined the kinetic energy loss, ΔE_i, of the ion and the internal energy, Q, taken up by the ion. Comparing these data with the available energy, E_{com}, as a function of ion mass shows that internal energy uptake, Q, increases with mass up to about 1500, and then decreases. It is thought that for peptide ions of ca. 1500 u, virtually all available energy is converted into internal energy.

On the other hand, the kinetic energy loss, ΔE_i, is a rapidly increasing function of mass. At lower masses (<600 u) ΔE_i and Q are nearly equal and much less than E_{com}, but at 1500 u, $Q \approx E_{com}$ and $\Delta E_i \approx 2Q$. On the basis of the few measurements reported, it is tentatively concluded that increasing the available energy, E_{com}, by either adding to the kinetic energy of the ion (increasing the accelerating voltage from 8 to 25 kV) or by increasing the mass of the target gas (from helium to argon), shifts the

[67] R. G. Gilbert, M. M. Sheil, and P. J. Derrick, *Org. Mass Spectrom.* **20**, 430 (1985).
[68] R. K. Boyd, *Int. J. Mass Spectrom. Ion Proc.* **75**, 243 (1987).
[69] M. L. Gross, K. B. Tomer, R. L. Cerny, and D. E. Giblin, *in* "Mass Spectrometry in the Analysis of Large Molecules" (C. J. McNeal, ed.). Wiley, New York, 1986.

anticipated mass where $Q \approx E_{com}$ to higher mass. Thus, more efficient collisional activation may be possible by using more massive collision gases, especially for precursor ions of $m/z > 2000$. Larger target gases should also lead to decreases in both the kinetic energy loss of the ion and the mass shift of the fragments.

The mass shift (sometimes called a Derrick shift[64]) is seen as a shift in the kinetic energy of the fragment ion when using an ESA as the fragment ion mass analyzer [i.e., the MIKES (mass-analyzed ion kinetic energy) technique[70]]. These energy shifts may be sufficiently large that mass assignments are in error by a few mass units. Mass shifts on tandem double-focusing mass spectrometers,[71] however, do not lead to misassignments of mass, because of the refocusing properties of the second analyzer. Some loss of fragment ion abundance, however, can arise due to energy-loss shifts away from the ideal linked-scan locus.[68]

Choice of Collision Gas

It is generally accepted that helium is the target gas of choice for high-energy collisions because its use minimizes the major processes competing with collisional-activated dissociation of the projectile ion (i.e., neutralization and scattering beyond the acceptance angle of the second mass spectrometer[70,72,73]).

The low efficiency of internal energy deposition in small angle collisions with helium, however, makes the use of heavier target gases such as argon or xenon of comparable value in high-energy experiments.[73] To address this issue, Gross and co-workers[69] studied the CA decompositions of a peptide and two polyethylene glycols employing both helium and xenon as collision gases. A small, but definite, effect of changing the collision gas from helium to xenon is seen for the lower molecular mass glycol and the peptide (see Tables I and II).

A more dramatic effect on ion yields is observed, however, for the higher molecular mass glycol. An explanation for this effect can be developed by considering the translational energy loss that is *not* accompanied by fragmentation. This loss leads to a low-energy "foot" for the precursor ion peak in the CAD spectra (Fig. 2). This "foot" is most significant for

[70] R. G. Cooks, J. H. Beynon, R. M. Caprioli, and G. R. Lester, "Metastable Ions." Elsevier, New York, 1973.

[71] K. Sato, T. Asada, M. Isihara, F. Kunihiro, Y. Kammai, E. Kubota, C. E. Costello, S. A. Martin, H. A. Scoble, and K. Biemann, *Anal. Chem.* **59**, 1652 (1987).

[72] P. H. Hemberger, J. A. Laramee, A. R. Hubik, and R. G. Cooks, *J. Phys. Chem.* **85**, 2335 (1981).

[73] J. A. Laramee, D. Cameron, and R. G. Cooks, *J. Am. Chem. Soc.* **103**, 12 (1981).

TABLE I
CAD Ion Yields as Function of Collision Gas

Compound	He (%)	Xe (%)
PEG 31 $[M + Na]^+ = 1405$	8.8	12.6
Substance P free acid $[M + H]^+ = 1349$	4.9	7.4
PEG 45 $[M + Na]^+ = 2021$	2.8	42.0

PEG 45, the data for which were obtained at 6 kV accelerating voltage. Nearly all of the reduction in main beam intensity stemming from interaction with the collision gas is accounted for by the increase in the number of ions that make up the "foot." When xenon is used as a collision gas, the "foot" becomes insignificant, and the yield of product ions increases. The fraction of product ions collected is ca. 25% of the molecular ions lost in the collision process.

The "foot" accompanying the precursor ion peak is likely due to ions that have converted insufficient collision energy into internal energy to cause decomposition during the time-of-flight from collision to detection. A plot of the internal energy, Q, vs. the translational energy lost by the ion, ΔE_i, shows that xenon is significantly more efficient at transforming ΔE_i into internal energy than is helium (Fig. 3). The use of xenon leads to a substantial decrease in the number of ions that have lost translational energy but have *not* decomposed. For a given parent ion mass, ΔE_i is reduced when xenon is used as collision gas relative to helium, which leads to reduced mass shifts. A comparison of the mass observed for a fragment ion arising from xenon and helium collision gas illustrates the decreased mass shift. Thus, the heavier collision gas decreases product ion mass shifts, ΔE_i, and increases the energy available, Q, for inducing fragmentations, at least for ion masses up to 2000.

TABLE II
CAD Ion Yields for Substance P Free Acid
as Function of Collision Gas
and Collision Energy

Acceleration voltage (keV)	He (%)	Xe (%)
8.0	4.9	7.4
6.15	3.5	8.2
4.0	3.1	7.0

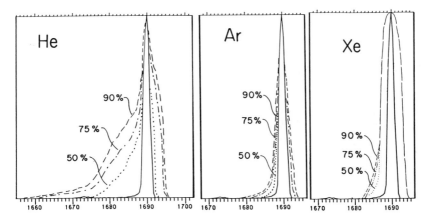

FIG. 2. Molecular ion region for polyethylene glycol (PEG) 45 with helium (HE), argon (Ar), and xenon (XE) collision gases.

Precision (Reproducibility) of Spectra

Precision or reproducibility of CAD spectra can be addressed on a day-to-day basis on a given instrument or on an instrument-to-instrument basis. High-energy CAD mass spectra are typically quite reproducible not only from day-to-day but also between the various instrumentation configurations that can be used to acquire CAD spectra. Some variation of the

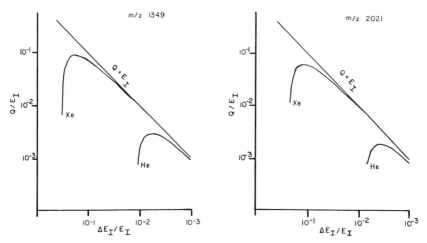

FIG. 3. Plot of available energy, Q, vs. translational energy lost, ΔE_i, at mass 2000 for helium and xenon collision gases. (Reprinted with permission from Ref. 69.)

TABLE III

COMPARISON OF PARTIAL HIGH-ENERGY CAD SPECTRA OF SUBSTANCE P
(Arg-Pro-Lys-Pro-Gln-Gln-Phe-Phe-Gly-Leu-Met-NH$_2$)[a]

m/z	Fragment[b]	Kratos TA[c]	ZAB-T[d]	JEOL HX110/HX110[e]	Kratos Concept IIHH[f]
1171.8	a_{10}	78	50	70	64
1129.7	d_{10}	76	50	71	53
1001.6	a_8	100	100	100	100
854.6	a_7	76	106	94	88
707.5	a_6	24	44	41	41
650.5	d_6	26	53	60	27
579.4	a_5	14	33	44	27
522.4	d_5	8	30	34	29
451.3	a_4	11	28	60	52
354.3	a_3	20	42	54	52
226.2	a_2	13	g	174	186

[a] Taken on various tandem sector mass spectrometers. The collision cell was *not* floated during the CAD experiment on the Kratos TA. For the ZAB-T and Concept IIHH instruments the collision cell was floated to +2 kV, and for the HX110/HX110 instrument the collision cell was floated to +3 kV. The accelerating voltage of MS-I was 8 kV for all instruments except the HX110/HX110 instrument for which the voltage was 10 kV. The relative abundances of low mass ions are typically attenuated in instances where the collision cell is not floated.

[b] See this volume, Appendix 5, for explanation of nomenclature.

[c] University of Nebraska triple analyzer mass spectrometer.

[d] University of Nebraska tandem double-focusing mass spectrometer at VG Analytical.

[e] MIT tandem double-focusing mass spectrometer.

[f] University of California–San Francisco tandem double-focusing mass spectrometer.

[g] Not determined.

relative abundances of fragment ions can be expected, but gross discrepancies will seldom be observed. This is illustrated by the CAD mass spectrum of substance P (Arg-Pro-Lys-Pro-Gln-Gln-Phe-Phe-Gly-Leu-Met-NH$_2$) recorded on four different instrument configurations (see Table III). The spectra are similar in both a qualitative and quantitative sense. Although there are variations in the relative abundances of product ions, no product ions are absent from the CAD spectra. Thus, one should be able to use the high-energy CAD spectra to confirm sample identity even if obtaining a CAD spectrum of authentic material is precluded because of unavailability of that material. This advantage should be important in regulation and forensics, for example.

Charge-Remote Fragmentations

In addition to enhancing the degree of fragmentation of an ionized molecule, CA can also open entirely new fragmentation channels not

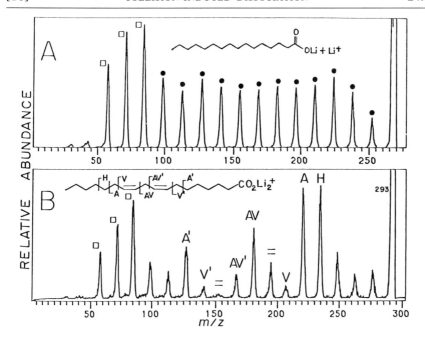

FIG. 4. CAD mass spectra of daughter ions produced by collisionally activating the $(M - H + 2Li)^+$ ions of (A) palmitic acid and (B) 8,11-octadecadienoic acid.

available with the energy deposited upon ionization. One particularly useful process is that involving fragmentation at a site remote from the charge.

This feature is most obvious for FAB-desorbed fatty acids, lipids, steroids,[74] and other materials that contain a long alkyl chain. A common characteristic of these materials is that the charge is rigidly located with little possibility of charge migration onto the alkyl chain. The absence of charge on the chain precludes obfuscating rearrangement reactions such as migration of double bonds or branch points. This leaves the alkyl chain amenable to unimolecular, thermal-like reactions such as 1,4-eliminations of H_2. For example, the product ion spectra of a saturated straight-chain fatty acid shows a smooth pattern of $C_nH_{2n} + H_2$ losses upon CA (Fig. 4).

If the hydrocarbon chain contains an unsaturation, the probability for charge-remote losses of H_2 and C_nH_{2n} are reduced in the vicinity of the double bond and change to a series of losses of $C_nH_{2n-2} + H_2$ once the double bond is part of the expelled group.[75] The perturbation in the spectral

[74] K. B. Tomer, N. J. Jensen, M. L. Gross, and J. Whitney, *Biomed. Environ. Mass Spectrom.* **13**, 265 (1986).

[75] K. B. Tomer, F. W. Crow, and M. L. Gross, *J. Am. Chem. Soc.* **105**, 5487 (1983).

pattern appears as a gap in the regular sequence. Similarly, various other functional groups cause unique perturbations in the pattern that can serve to locate the site of substitution.[76,77] Although certain charge-remote fragmentations occur at low energy or even as metastable ion processes,[78] many are most readily seen upon high-energy CA. The interested reader is directed to a recent review.[79]

The phenomenon is not limited to fatty acids and related materials. Peptides will also undergo charge-remote fragmentations.[80,81] This is particularly important for those peptides containing basic amino acid residues because, regardless of the position of charge, fragmentation will occur at (almost) any peptide bond. Charge-remote fragmentations have also been used to differentiate the isomeric amino acids leucine and isoleucine in those peptides that give rise to C-terminal ions.[81] This is virtually impossible to accomplish by either conventional mass spectrometric means or low-energy CA.

Angle-Resolved Mass Spectrometry

Angle-resolved mass spectrometry (ARMS) provides a means to vary in a systematic way the internal energy deposited upon CA. The technique relies on the fact that the amount of translational energy deposited internally into the ions as a result of collision with a neutral and relatively stationary target gas molecule is a function of the scattering angle associated with fragmentation.[49,56,82] Specifically, those collisions of a more violent nature that deposit sufficient energy to access higher energy processes result in a larger scattering angle of the projectile ion.[72,82,83] A detailed theoretical basis for the ARMS technique is beyond the scope of this chapter, and the reader is directed to a paper by Boyd and co-workers.[56,82]

[76] N. J. Jensen, K. B. Tomer, M. L. Gross, and P. A. Lyon, in "Desorption Mass Spectrometry. Are SIMS and FAB the Same?" (P. A. Lyon, ed.), Vol. 291, p. 194. ACS Symp. Ser., American Chemical Society, Washington, D.C., 1985.

[77] M. L. Gross, N. J. Jensen, D. L. Lippstreu-Fisher, and K. B. Tomer, in "Mass Spectrometry in the Health and Life Sciences" (A. L. Burlingame and N. Castagnoli, Jr., ed.), p. 209. Elsevier, Amsterdam, 1985.

[78] J. Adams and M. L. Gross, J. Am. Chem. Soc. 111, 435 (1989).

[79] J. Adams, Mass Spectrom. Rev. 9, 141 (1990).

[80] R. S. Johnson, S. A. Martin, and K. Biemann, Int. J. Mass Spectrom. Ion Proc. 86, 137 (1988).

[81] R. S. Johnson, S. A. Martin, K. Biemann, J. T. Stults, and J. T. Watson, Anal. Chem. 59, 2621 (1987).

[82] R. K. Boyd, E. E. Kingston, A. G. Brenton, and J. H. Beynon, Proc. R. Soc. London A392, 59, 89 (1984).

[83] S. A. McLuckey, S. Verma, R. G. Cooks, R. S. Mason, M. J. Farncombe, and K. R. Jennings, Int. J. Mass Spectrom. Ion Phys. 48, 423 (1983).

Unfortunately, the ARMS technique often requires specialized equipment for angular selection,[47,48,84,85] although it can be accomplished on most instruments that are capable of varying the potential on an exit deflector of a collision cell. The exit deflector can be used to redirect the angle-selected scattered ions to an appropriate trajectory for subsequent detection.[86-89] Alternatively, the scattering angle may be defined by precollision electrical deflection of the parent ion beam.[49] Angular selection is usually performed in the xy plane with a range of deflection of $0°$ to $6°$, although a range of $0°-1°$ can provide meaningful data.

Breakdown curves are constructed by plotting the normalized product ion abundances against scattering angle. These data can be used to probe the internal energy dependence of the fragmentation process under consideration.

One problem inherent to the ARMS technique is the poor signal-to-noise ratio often encountered in acquiring data. This is because most of the parent ion beam is rejected in order to maximize angular resolution. Moreover, for larger scattering angles, the cross section for fragmentation decreases dramatically. There have also been some questions raised recently concerning the interpretation of the results.[56,82,90]

Low-Energy Collisions

Instrumentation

Low-energy CA is most often carried out in quadrupole reaction chambers of triple quadrupole and hybrid sector-quadrupole mass spectrometers. The quadrupole reaction chamber typically uses a radio frequency (rf) field to contain and transmit ions (see [7] in this volume). In such containment devices, a number of parameters can be optimized including the collision energy and the number of activating collisions.[19,91] Reactant ions are selected by either another quadrupole (triple quadrupole) or a

[84] J. B. Hasted, "Physics of Atomic Collisions," 2nd Ed. Elsevier, Amsterdam, 1972.
[85] G. W. McClure and J. M. Peek, "Dissociation in Heavy Particle Collisions." Wiley, New York, 1972.
[86] T. Ast, T. W. Terwilliger, R. G. Cooks, and J. H. Beynon, *J. Phys. Chem.* **79,** 708 (1975).
[87] R. G. Cooks, T. Ast, and J. H. Beynon, *Int. J. Mass Spectrom. Ion Phys.* **11,** 490 (1973).
[88] M. Medved, R. G. Cooks, and J. H. Beynon, *Chem. Phys.* **15,** 295 (1976).
[89] D. E. Giblin, *Proc. 33rd ASMS Conf. Mass Spectrom. Allied Topics, San Diego, CA* (1985).
[90] P. J. Todd, R. J. Warmack, and E. H. McBay, *Int. J. Mass Spectrom. Ion Phys.* **50,** 299 (1983).
[91] K. L. Busch, G. L. Glish, and S. A. McLuckey, "Mass Spectrometry/Mass Spectrometry." VCH Publ. New York, 1988.

double-focusing sector analyzer (hybrid configuration), and product ions are typically analyzed by a quadrupole mass filter.

The transmission efficiency of the rf quadrupole is very high because of the strong focusing properties inherent in operating in the rf-only mode; ion trajectories are actually stabilized after collision and fragmentation. Fragmentation efficiencies can also be quite high because the ion traverses a considerable distance in the quadrupole (ca. 1.5 times the length of the quadrupole) and can undergo many collisions.

Mechanism of Activation

Although collisions at kiloelectron volt energies are expected to result in excitation of electronic internal modes (and to a lesser extent rotational–vibrational modes), collisions at low energies (<100 eV) will no longer result in efficient transfer of translational energy to electronic internal modes. A typical interaction time of a projectile ion of mass 200 and a translational energy of 30 eV with a target molecule over a few angstroms is on the order of 10^{-13} sec. This length of time is longer than the period of an electronic internal mode and thus, the probability of excitation of such a mode is reduced with respect to excitation by high-energy (keV) CA.[92]

The interaction time of ca. 10^{-13} sec is comparable to the reciprocal of most vibrational frequencies. Under such conditions, the collisions are nonadiabatic, and the interaction is said to have an impulsive character that can efficiently induce energy transfer.[92] The subsequent transfer of translational vibrational energy is believed[93] to occur by internuclear momentum transfer.

In the so-called binary or spectator model, the projectile ion and the target gas interact as essentially structureless elastic spheres. Momentum transfer occurs between the two impact partners, leading ultimately to rotational–vibrational excitation of the ion *and* the gas, together with a change in momentum of the center of mass of each. For larger precursor ions, the target atom interacts with only a portion of the projectile ion. The maximum amount of energy (center-of-mass kinetic energy, E_{com}) available for internal excitation is given by [Eq. (2)][82]

$$E_{com} = E_{lab} m_t / [m_p + m_t(m_p/m_{pi})] \qquad (2)$$

where m_p is the projectile mass, m_{pi} is the impact portion of the projectile, and m_t is the target mass. The elastic limit is thought to be reached when

[92] R. N. Schwartz, Z. I. Slawsky, and K. F. Herzfeld, *J. Chem. Phys.* **20**, 1591 (1952).
[93] D. J. Douglas, *J. Phys. Chem.* **86**, 185 (1982).

$m_p = m_{pi}$. This situation would correspond to CA of small organic ions, such as those that are typically formed by electron or by chemical ionization.

Significantly, as the precursor ion mass is increased to above ca. 600 u, the CA efficiency decreases markedly with increasing mass. One possible explanation is that this is a result of the simple disparity in physical sizes of the projectile ion and the target gas atom or molecule. A number of experiments were designed to address this relationship.[94–96]

Before applying the spectator mechanism for rotational–vibrational excitation to rationalize experimental results, one need also be aware of the consequence of any precollision internal energy of the projectile ion. If the polyatomic ions are appreciably energized, the vibrational periods are likely to be substantially reduced even to the extent of being shorter than the interaction time. In these situations, the conversion of translational to internal energy may well occur by some other mechanism rather than the spectator mechanism described above.

Collision Gas and Pressure Effects

In the low-energy regime, the nature of the target gas plays a more significant role than it does in high-energy activations. This is because a different excitation mechanism (vibrational excitation) is in operation. Furthermore, a larger fraction of the maximum available energy is converted into internal energy of the projectile ion. Bursey and co-workers[97,98] demonstrated that, in general, heavier targets are preferred over lighter targets because they provide a larger E_{com}. For example, collisions with helium were found to deposit only very small amounts of energy in comparison with collisions employing nitrogen, argon, or krypton.[96,97] Specifically, for every volt change in E_{lab} of the ion at a given pressure, there is an increase of 0.04, 0.25, and 0.32 eV, in maximum possible energy deposited when using helium, nitrogen, and argon, respectively.[96]

Increasing the pressure of target gas will also increase the average fraction of energy deposited per collision. In general, the amount of internal energy deposited in an ion can be described in terms of conventional rate equations governing an ion beam moving through a uniform

[94] A. J. Alexander and R. K. Boyd, *Int. J. Mass Spectrom. Ion Proc.* **90**, 211 (1989).

[95] K. L. Schey, K. I. Kenttämaa, V. H. Wysocki, and R. G. Cooks, *Int. J. Mass Spectrom. Ion Proc.* **90**, 71 (1989).

[96] A. J. Alexander and P. Thibault, *Rapid Commun. Mass Spectrom.* **2**, 224 (1988).

[97] J. A. Nystrom, M. M. Bursey, and J. R. Hass, *Int. J. Mass Spectrom. Ion Proc.* **55**, 263 (1983/1984).

[98] M. M. Bursey, J. A. Nystrom, and J. R. Hass, *Anal. Chim. Acta* **159**, 265, 275 (1984).

target [Eq. (3)][93]

$$I(S) = I^o e^{-\sigma S} \tag{3}$$

where $I(S)$ is the parent ion current after collision, S is the target thickness, and σ is the cross section of a fragmentation reaction.

At larger values of E_{com}, however, the increase in internal energy with increasing pressure may not be linear. This is particularly true for low mass ions because of the greater scattering efficiency of these ions by the target gas. Moreover, at high gas pressures (>10–20×10^{-7} torr) a large fraction of ions are likely to undergo multiple collisions.

Energy-Resolved Mass Spectrometry

The variation of ion energy in a controlled fashion is an added resource for determining fragmentation pathways and the structures of gas-phase ions. Such energy-resolved studies permit the construction of breakdown graphs expressing the variation of fragment ion yields with internal energy. From these data, the energy dependence of a particular fragmentation process can be determined. Moreover, the breakdown curves are characteristic of ion structure and are useful for distinguishing subtle structural features for systems that have similar CAD spectra at one energy.

Control of ion internal energy is accomplished in a number of ways including photodissociation, charge exchange, variation of ionizing electron energy, and collisional excitation or stabilization. The method of selecting ions with different internal excitation following high-energy collisions is based on the scattering angle of a particular ion [i.e., ARMS (see above)]. A simpler, and more accessible, method for controlling ion energies is to use variable low-energy collisional excitation.[37,50,51,99] Moreover, the degree of fragmentation is typically very sensitive to changes in the kinetic energy of the ion at low energies. This is because the low-energy CA process is very efficient,[100] and consequently a considerable amount of the available translational energy is converted into internal energy (see discussion on activation mechanism).

An important application of energy-resolved mass spectrometry is the differentiation of isomeric ion structures.[98] A notable example is the differentiation of the C_3H_6 radical cation isomers of propene and cyclopropane. The isomers were distinguished on the basis of their energy-dependent rates of hydrogen atom loss.[98]

[99] D. D. Fetterolf and R. A. Yost, *Int. J. Mass Spectrom. Ion Phys.* **44**, 37 (1982).
[100] R. A. Yost, C. G. Enke, D. C. McGilvery, D. Smith, and J. D. Morrison, *Int. J. Mass Spectrom. Ion Phys.* **30**, 127 (1979).

Precision (Reproducibility) of Spectra

The collision energy selected for low-energy activation is, to some degree, arbitrarily chosen. The amount of energy deposited is controlled by the collision energy, the pressure of collision gas within the collision chamber, and the choice of target gas. It has been demonstrated that the amount of internal energy deposited in an ion increases as a linear function of the kinetic energy of the molecular ion for a given pressure and gas.[96–98] The significance of this is that the CAD mass spectra depend on the parameters selected, and spectra are variable, especially from instrument to instrument. For example, pressure in the collision cell is seldom measured directly but rather by using an ion gauge located remote from the cell. Different instrument designs have different pumping conductances. As a result, the relation between reported and true pressure in the collision quadrupole is different from instrument to instrument and, therefore, difficult to reproduce between different types of instruments.

Control of the internal energies of the ions, both before and after activation, is critical if reproducible spectra are to be obtained. In 1983 in an international round robin study,[101] it was determined that the relative abundances of given pairs of ions observed in product ion scans varied by factors ranging into the hundreds even though nominal operating conditions were supposedly the same from instrument to instrument. Both triple sector quadrupole and hybrid instruments were used in the comparison. This comparison demonstrates the serious problem that can be encountered if precise CAD spectra are to be reported and used later in, for example, library searching.

Because there are over 400 tandem instruments with a qQ second-stage analyzer in use today,[102] it is clear that measures must be taken to ensure better precision. This is true even for uses such as peptide sequencing in which the principal information is product ion mass rather than relative abundance. There will be occasions when another sample of the peptide is isolated. Proof of structure is much less equivocal if the literature and unknown spectra agree. A number of approaches are under consideration to deal with the serious lack of precision that can be exhibited both on a day-to-day and instrument-to-instrument basis. Prospective users of triple quadrupole and hybrid instruments are urged to become familiar with these problems and the approaches to a solution that are currently under investigation.

The most encouraging approach has been developed at the National

[101] P. H. Dawson and W.-F. Sun, *Int. J. Mass Spectrom. Ion Proc.* **55,** 155 (1983/1984).

[102] R. I. Martinez, *J. Res. Natl. Inst. Stand. Tech.* **94,** 281 (1989); *ibid., Rapid Commun. Mass Spectrom.* **3,** 127 (1989).

Institute of Standards and Technology (NIST) in the United States. The NIST protocol was validated in a second international round-robin; the salient result is that at least half of the commercial QqQ instruments can provide an instrument-independent spectral representation if the NIST protocol is used.[102] The protocol is kinetics-based (i.e., it takes into consideration the kinetic processes of fragmentation) and involves careful selection of MS/MS parameter settings that govern number of collisions, the collision energy, the transmission through the collision and mass analysis quadrupole analyzers, and the uniformity of detector response over wide mass ranges. Internal energy of the precursor ion is controlled by using 70 eV electron ionization to prepare the ions. Effects of internal energy are removed by subtracting out the product ions produced in metastable ion decompositions. Once the instrument is "standardized," it is asserted that the instrument will produce reproducible spectra of ions produced by other methods (e.g., FAB). More study is needed to validate the protocol.

Another approach involves a triple-point calibration obtained by constructing breakdown curves of the product ions of m/z 69, 100, and 131 from the perfluorokerosene (PFK) precursor ion of m/z 231.[103] The relative abundances of the daughter ions are plotted on the ordinate and the corresponding collision energy is plotted on the abscissa. Ideally, the resulting triangle should be as small as possible. Alternatively, the intersect of fragment ions of m/z 131 and 181 can be used to select a collision energy. Various instruments are tuned until the "cross point" is reached and then CAD spectra are obtained under those conditions.

Another approach estimates the average internal energy from the distribution of a series of fragment ions of an appropriate model compound.[95] A particularly useful class of model compound is transition metal polycarbonyl compounds. The sequential loss of CO from the molecular ion is easily modeled and permits calculation of approximate internal energy distributions of the parent ion.

Comparison with High-Energy Activation

The most important observation in a comparison of low- and high-energy CA is the superiority of CAD at high energy over low energy for large organic ions.[104] For small organic precursor ions (<600 u), however, the CAD yields obtained by using rf-only quadrupole collision cells are

[103] C. J. Mirocha, A. E. Schoen, and K. L. Busch, personal communication (1990).
[104] A. J. Alexander, P. Thibault, R. K. Boyd, J. M. Curtis, and K. L. Rinehart, *Int. J. Mass Spectrom. Ion Proc.*, in press (1990).

actually higher than those from sector instruments by at least an order of magnitude.

The differences in the extent of fragmentation obtained by using low- and high-energy activation methods can be explained in terms of the amount of internal energy deposited by a collision.[65,105] High-energy CA deposits an average energy of 1 to 3 eV, but the distribution of deposited energies extends beyond 15 eV of inernal energy. Conversely, low-energy CA deposits low average energies with an energy distribution exhibiting no significant high-energy tail.[65,104]

Theory postulates that the rate of decomposition is related to the probability of a given vibrational mode or modes acquiring enough energy to rupture bonds.[11] Thus, large molecules, with many degrees of vibrational freedom over which to distribute energy, are less likely to undergo fragmentation upon low- than high-energy activation. Similarly, fragmentation processes with high activation energies (e.g., charge-remote fragmentation) are also less likely to occur upon low-energy collision.

The differences caused by using either low- or high-energy activation were demonstrated in a study of a series of polycyclic aromatic hydrocarbons.[106] The extent of fragmentation and even the types of dissociation products are very dependent on the activation method used.

The differentiation of isomeric structures is an important application of the CAD technique. Isomer differentiation typically requires that the amount of energy deposited in a single collision be maximized to avoid the possibility of isomerization occurring before dissociation.[59,63,107] If multiple low-energy collisions are required to induce fragmentation, there is a high probability of irreversible multistep isomerization of the ions by a collision before sufficient energy is deposited for dissociation.[59,108] In light of this, isomer differentiation is best tackled by employing high-energy CA.

On the other hand, at low collision energies it is experimentally easier to vary the internal energy of an ion by using ERMS than to use ARMS at high collision energies. Moreover, the effects of stepwise excitation on fragmentation can be determined if low-energy collisions are employed.[76,109]

[105] H. I. Kenttämaa and R. G. Cooks, *Int. J. Mass Spectrom. Ion Proc.* **64**, 79 (1985).
[106] S. J. Pachuta, H. I. Kenttämaa, T. M. Sack, R. L. Cerny, K. B. Tomer, M. L. Gross, R. R. Pachuta, and R. G. Cooks, *J. Am. Chem. Soc.* **110**, 657 (1988).
[107] T. Ast, *Adv. Mass Spectrosc.* **10A,** 471 (1986).
[108] H. I. Kenttämaa, *Org. Mass Spectrom.* **20**, 703 (1985).
[109] J. S. Brodbelt-Lustig and R. G. Cooks, *Int. J. Mass Spectrom. Ion Proc.* **86**, 253 (1988).

Collisional Activation in Hybrid, Fourier Transform,
and Ion Trap Mass Spectrometers

Hybrid Mass Spectrometers

The usual hybrid mass spectrometer configurations consist of a sector mass spectrometer (*EB* or *BE*) linked to two quadrupoles. The first quadrupole is operated in the rf-only mode and serves as the collision cell. The transmission of both precursor and fragment ions in a hybrid instrument is a particularly complex function of E_{lab} and the setting of the collision cell rf-amplitude (in this volume [7]).[99,110]

Briefly, parent ions are selected by the sector mass spectrometer and focused by a deceleration lens onto the entrance aperture of the rf-quadrupole (Q_1). In Q_1, the ions undergo low-energy CA at selected energies of up to approximately 100 eV. Fragment ions produced in Q_1 are passed into Q_2 where they are mass analyzed. A number of other scan modes can be accessed by differential scanning of one or more fields associated with the magnetic sector, the electric sector, and the quadrupole mass filter.[111]

Recently, Boyd and co-workers (Alexander *et al.*[112]) compared a sector-quadrupole hybrid with a tandem double-focusing mass spectrometer. The comparison highlighted the inability of the hybrid configuration to give mass spectra showing sequence-specific fragment ions for peptide ions above mass 800. This failing is attributed to the use of low-energy CAD to effect fragmentation. On the other hand, the structural information obtained on the tandem double-focusing instrument by using high-energy CAD was almost invariant regardless of the collision conditions employed.

Quadrupole Ion Trap Mass Spectrometers

The ion trap detector (ITD) is a quadrupole ion storage device.[113,114] A sequence of rf and dc voltages are used for CAD of mass-selected ions. The amplitude of the rf voltage applied to a ring electrode determines the mass range of the ions trapped (typically 10–600 u). Single-ion selection is carried out by raising the rf voltage level while applying an appropriate

[110] J. D. Ciupek, J. W. Amy, R. G. Cooks, and A. E. Schoen, *Int. J. Mass Spectrom. Ion Proc.* **65**, 141 (1985).
[111] J. N. Louris, L. G. Wright, R. G. Cooks, and A. E. Schoen, *Anal. Chem.* **57**, 2918 (1985).
[112] A. J. Alexander, P. Thibault, R. K. Boyd, J. M. Curtis, and K. L. Rinehart, submitted for publication.
[113] J. N. Louris, R. G. Cooks, J. E. Syka, P. E. Kelley, G. C. Stafford, and J. F. J. Todd, *Anal. Chem.* **59**, 1677 (1987).
[114] M. Weber-Grabau, P. Kelley, J. Syka, S. Bradshaw, and J. Brodbelt, *Proc. 35th ASMS Conf. Mass Spectrom. Allied Topics, Denver, CO* (1987).

dc voltage to the ring electrode (time interval, 3–5 msec). During the reaction time, a supplementary ac voltage ("tickle" voltage) is applied across the end caps at the fundamental frequency of the selected ions. This accelerates the ions and leads to CA with the helium buffer gas that is contained in the trap. After the reaction period, the fragmentation products are ejected from the trap by using mass-selective instability induced by an rf voltage ramp, and detected with an external electron multiplier.

It should be emphasized that the reaction times in conventional rf quadrupoles and the ITD are somewhat different, being less than 100 μsec for a triple quadrupole instrument and several milliseconds for the ITD device. Thus, considerably more collisions will occur in the ITD, and, in principle, higher energies can be deposited in an ion but in smaller increments.[76,115] This is often not the case because helium is used for collisions in the ion trap and the activating collisions are relatively inefficient.[94] Moreover, given the long reaction times, an ion that has accumulated sufficient internal energy to decompose by a low-energy process may not survive to undergo another activating collision. Thus, ITD mass spectra are typically dominated by fragment ions produced by low-energy processes.

Fourier Transform Mass Spectrometers

The Fourier transform ion cyclotron resonance mass spectrometers (FTMS) consist of a cubic cell that serves as an ion source, an ion trap, a mass analyzer, and detector.[116] This simple design, however, is deceptive. The exceptional capabilities of FTMS are only now being fully realized. Equipped with superconducting magnets, FTMS instruments can achieve high mass ranges (a CsI cluster of m/z 31,830 has been observed[117]), high mass resolution (up to 1×10^8 at m/z 18), and simultaneously recording of all the different mass ions in the spectrum. One disadvantage of FTMS is that the trap requires high vacuum ($<10^{-7}$ torr) to prevent collisions between ions and neutral gas molecules from degrading the resolving power of the device.

The capability of FTMS to perform CAD experiments is based on the double-resonance technique that ejects unwanted species from the cell.[118] Most CAD experiments in FTMS have been carried out at low collisional

[115] J. Brodbelt, H. I. Kenttämaa, and R. G. Cooks, *Org. Mass Spectrom.* **23**, 6 (1986).
[116] M. B. Comisarow, *Int. J. Mass. Spectrom. Ion Phys.* **37**, 251 (1981).
[117] C. B. Lebrilla, D. T.-S. Wang, R. L. Hunter, and R. T. McIver, Jr., *Anal. Chem.* **62**, 878 (1990).
[118] M. B. Comisarow, V. Grassi, and G. Parisod, *Chem. Phys. Lett.* **57**, 413 (1978).

energies (<100 eV) because of constraints of ion trapping.[119,120] The use of larger cells, higher magnetic fields, or other special designs, however, may allow kilovolt energies to be achieved. The problem is that different mass ions have different energies.

Limitations of CA and the Future

The continued development of methods for activating ions is assured given the success of applying the CAD technique to solving problems in both ion chemistry and analysis. The method is now routine for investigations of molecules with masses up to approximately 2000, although tandem sector instruments give a better performance at this mass. The CA fragmentation chemistry of biomolecules is relatively simple and serves to incise the molecular ion into component building blocks, a welcome property in structure studies.

There is, however, one principal shortcoming inherent in the CAD technique. This involves limitations on the amount of energy that can be deposited into a molecule, and hence the degree of fragmentation that can be induced. This limits the acquisition of MS/MS spectra in favorable cases to organic and organometallic ions of mass 3000. Increased instrument sensitivity may allow useful spectra to be obtained at even higher masses, but major improvements are not expected in the near future. A recent approach to improving instrument sensitivity is the utilization of a microchannel-plate detector that simultaneously detects product ions over a wide mass range.

Alternatives

Given the principal shortcoming of CAD, the development of other methods of ion activation is currently of high interest. Three approaches that may be promising are photoactivation, electron-impact activation, and collisions with surfaces. These methods may overcome the limitation of small energy transfer to projectile ions in collisional activation with small atomic gases. Another approach but less general that still involves target gas collisions is electron capture-induced dissociation. In this method, doubly charged ions are collided with suitable gases.

Surface-Induced Dissociation. Cooks and co-workers[22] have shown that it is possible to excite organic ions by allowing them to collide with a stainless steel surface. This method of ion excitation is known as surface-

[119] R. B. Cody, Jr., I. J. Amster, and F. W. McLafferty, *Proc. Natl. Acad. Sci. U.S.A.* **82**, 6367 (1985).

[120] R. B. Cody and B. S. Freiser, *Int. J. Mass Spectrom. Ion Phys.* **41**, 199 (1982).

induced dissociation (SID). Collision of an ion with a surface can lead to deposition of large amounts (in excess of 7 eV for 100 eV collisions) of internal energy into an ion.[21,43] Moreover, the internal energy distribution of the excited ion is considerably more narrow than that obtained by using gas-phase collisional activation.[121] Although the approach shows promise, general applicability for high mass species has not been demonstrated.

Electron Capture-Induced Decomposition. Electron capture-induced decomposition (ECID) is a novel technique recently developed for activating polyatomic ions.[21,122–125] The ECID process is as follows: A keV-beam of doubly charged ions $(M)^{2+}$ is collided on various targets (gas or surface), and among various processes, electron capture occurs, resulting in $(M)^{+}$ ions. If these ions are formed with sufficiently large internal energies, they will decompose to form fragment ions $[F_i]^{+}$. The internal energy distributions of the $(M)^{+}$ ions can be altered by selecting target gases with different ionization energies; collision gases of lower ionization energy result in higher excitation energies than collisions with gases of higher ionization energy.

In general, electron capture is accompanied by a significant degree of excitation that is subsequently reflected in the increased amount of fragmentation. An increase in average excitation energies of an ion often results in a higher probability of sampling a significant fraction of unisomerized ions. That being the case, ECID is well suited in its application to the study of isomeric systems.[122–125] A notable example is the differentiation of the $(C_6H_6)^{2+}$ isomeric ions of benzene, 2,4-hexadiene, and 1,5-hexadiyne.[125] Broad analytical applicability is limited because of the difficult problem of generating large numbers of doubly charged ions. Electrospray and other spray methods may be useful for overcoming this problem.

Photodissociation. Photodissociation of ions is an emerging technique that is particularly compatible with ion-trapping mass spectrometers, such as the FT instrument. These instruments are well-suited for photodissociation studies because ions have relatively long residence times within an ion trap, and sufficient photofragment ion yields are obtained using conventional arc lamp sources and monochromators. The ion containment

[121] M. J. DeKrey, H. I. Kenttämaa, V. H. Wysocki, and R. G. Cooks, *Org. Mass Spectrom.* **21**, 193 (1986).

[122] K. Vékey, A. G. Brenton, and J. H. Beynon, *Int. J. Mass Spectrom. Ion Proc.* **70**, 277 (1986).

[123] K. Vékey, A. G. Brenton, and J. H. Beynon, *J. Phys. Chem.* **90**, 3569 (1986).

[124] J. M. Curtis, K. Vékey, A. G. Brenton, and J. H. Beynon, *Org. Mass Spectrom.* **22**, 289 (1987).

[125] K. Vékey, A. G. Brenton, and J. H. Beynon, *Org. Mass Spectrom.* **23**, 31 (1988).

properties of an rf-only quadrupole have also been applied[126,127] to studies of photodissociation of small ions. Photodissociation has also been demonstrated in other types of mass spectrometers including sector instruments. Indeed, photodissociation of desorbed ions of biomolecules may become an important means for obtaining structural information on high mass molecules. Most of the work thus far has been either with small molecules or of a method development nature. Whether this method can be turned into a routine approach remains to be seen.

For maximum photofragment ion yields in a sector instrument, it is desirable to introduce the radiation in a manner that provides prolonged interaction between ion and laser beams. Axial coincidence is most easily achieved by merging the ion and laser beam over the entire length of a suitable field-free region. Typical photon fluxes required are on the order of 10^{18} sec^{-1}, corresponding to a laser power of 440 mW at a wavelength of 450 nm.[128]

Multiphoton absorption is certainly a possibility that takes advantage of a high-power pulsed laser. Activation caused by absorption of multiple infrared photons,[129] results predominantly in dissociation by the lowest accessible energy channel.

Electron Impact Activation. If photoactivation is a tenable means of ion activation, then it is expected that electron impact should have even greater opportunities. High flux, high-energy electron beams are more easily generated than are comparable flux and energy photon beams (e.g., lasers are not required). Moreover, electron impact excitation is probably more general and does not require a chromophore in the analyte ion. Although this method was demonstrated with FT mass spectrometers[130] and Wien filter mass spectrometers,[131] the generality and broad applicability of the method remains to be established.

Conclusion

Collisional activation is the simplest and most readily applied method of ion activation in mass spectrometry on tandem sector instruments, and can be used for activating ions up to m/z 2000–2500. The resulting CAD spectra are reproducible, certainly on a day-to-day or even from instru-

[126] M. L. Vestal and J. H. Futrell, *Chem. Phys. Lett.* **28,** 559 (1974).
[127] D. C. McGilvery and J. D. Morrison, *Int. J. Mass Spectrom. Ion Phys.* **25,** 81 (1978).
[128] D. H. Russell, *Mass Spectrom. Rev.* **9,** 405 (1990).
[129] L. R. Thorne and J. L. Beauchamp, *in* "Gas Phase Ion Chemistry," (M. T. Bowers, ed.), Vol. 3. Academic Press, New York, 1984.
[130] R. B. Cody and B. S. Freiser, *Anal. Chem.* **59,** 1056 (1987).
[131] W. Aberth and A. L. Burlingame, manuscript in preparation.

ment to instrument. On tandem quadrupole and hybrid instruments, the capabilities of CA do not appear to extend to as high masses, although impressive results from Hunt and co-workers[132] give some evidence to the contrary. The lack of precision of low-energy spectra is a problem that needs to be solved.

Collisional activation will always be limited by the inability to convert laboratory translational energy into internal energy to drive decompositions. This disadvantage is a major obstacle for activating high-mass ions. More massive targets (in the limit, a solid surface) may be used to overcome the disadvantage. Other approaches such as electron impact and photoactivation may be surrogates. Difficulty, expense of implementation, and lack of generality stand in the way of rapid and routine use of these methods. It appears that collisional activation will continue to play a major role in tandem mass spectrometry for the foreseeable future.

Acknowledgment

The support of the National Science Foundation (Grant CHE-8620177) during the preparation of this chapter is gratefully acknowledged.

[132] D. F. Hunt, J. R. Yates III, J. Shabonowitz, S. Winston, and C. R. Hauer, *Proc. Natl. Acad. Sci. U.S.A.* **83**, 6233 (1986).

[11] Principles of Californium-252 Plasma Desorption Mass Spectrometry Applied to Protein Analysis

By Ronald D. Macfarlane

Introduction

The complete characterization of a protein is a formidable challenge to the experimentalist, requiring the combination of several analytical methods that give fundamental composition, structure, and conformation. Mass spectrometry contributes to determining fundamental composition through measurements of molecular mass and amino acid sequence. It is an impressive fact that a single number, the molecular mass of a protein, can often tie together the results of more information-laden analyses by nuclear magnetic resonance (NMR) and X-ray crystallography. The more accurate the molecular mass determination is, the more restricted are the possible choices in the "solution" of a structure by NMR or X ray. The focus of this chapter is on one of the mass spectrometric methods, californium-252 plasma desorption mass spectrometry (^{252}Cf-PDMS), that has been used in the determination of molecular mass of proteins up to 45,000 u.[1] This technique was the first of the particle-induced desorption methods to demonstrate that gas-phase molecular ions of proteins could be produced from a solid matrix.[2] The other methods now in use for protein analysis [fast atom bombardment (FAB), laser desorption, and electrospray] are discussed elsewhere in this volume.

The ^{252}Cf-PDMS method has been a particularly attractive addition to the protein characterization laboratory because of its simplicity of operation, low cost, and high reliability. A unique feature of ^{252}Cf-PDMS for protein analysis is the ability to carry out chemical reactions with proteins directly on the surface of the ^{252}Cf-PDMS sample after a molecular mass determination has been made (see [23] in this volume). This capability perfectly matches the methods of enzymatic degradation and chemical modifications that protein chemists are accustomed to carrying out when a protein structure is being elucidated.

The introduction of the commercial Bio-Ion (Uppsala, Sweden) ^{252}Cf-PDMS in 1982 has greatly accelerated the adoption of this technology.[3] The care that has been given to automation and ease of operation of the commercial instrument combined with the inherent simplicity of the method has attracted scientists with no background in mass spectrometry to its use in a manner much the same as a UV/VIS or Fourier transform infrared (FT-IR) spectrophotometer is used for routine assay.

It is a simple procedure to obtain a ^{252}Cf-PD spectrum of a protein using the Bio-Ion instrument. A sample is prepared, placed into a small compartment, a few commands are entered at a keyboard, and in some minutes a mass spectrum appears. What is going on behind the electronic interface between the investigator and instrument, the principles underlying ^{252}Cf-PDMS, has been the subject of the work of a relatively small group of chemists and physicists from the nuclear and physical sciences with no previous experience in mass spectrometry. That we know so much about ^{252}Cf-PDMS has been an important factor in applying this technique to protein analysis because we understand what variables in the methodology are important to improving sensitivity and mass accuracy and we have clear directions on how the method can be improved beyond its present capabilities. This chapter focuses on some of the practical aspects of ^{252}Cf-PDMS, and the variables that influence performance, particularly those

[1] G. Jonsson, A. Hedin, P. Hakansson, B. U. M. Sundqvist, H. Bennich, and P. Roepstorff, *Rapid Commun. Mass Spectrom.* **3**, 190 (1989).

[2] R. D. Macfarlane and D. F. Torgerson, *Science* **191**, 920 (1976).

[3] Bio-Ion Nordic, AB, Box 15045, S-750 15 Uppsala, Sweden.

that have an effect on protein analysis and which can be influenced by the user. Knowledge of the basis of the operation might give the user a better feeling for the strengths and limitations of the method and how the user can influence (positively or negatively) the results that are obtained.

Another function of this chapter is to introduce and define some of the commonly used vocabulary associated with the ^{252}Cf-PDMS technology but which is not a part of the mass spectrometry literature. This vocabulary is a carryover from the early days of ^{252}Cf-PDMS when all the research and development were carried out by nuclear scientists. To assist the reader in this exercise, an *important term or phrase* that first appears in the text will be set in italics and defined and will subsequently be used throughout the remainder of the chapter in the context of its normal usage.

The Bio-Ion ^{252}Cf-PDMS system has now been installed in major protein research laboratories throughout the world and has been effectively utilized in a wide variety of applications. These researchers have provided valuable feedback on current performance levels, problems, and capabilities they would like to have incorporated in the future. Influence of sample purity, limitations on mass accuracy, possibilities for increasing the mass range and mass accuracy are among the topics most discussed; many of these subjects are addressed in this chapter.

Why Highly Purified Protein Fractions Are Necessary

Proteins, by virtue of their highly polar structure and strong intermolecular interactions, naturally exist as a solid or as a solute in an aqueous solution. Mass spectrometry requires gas-phase molecular ions of proteins in order to carry out the mass measurement. Twenty years ago, this was a problem because proteins decompose rather than volatilize when heated. The new particle-induced desorption methods have solved this problem for the general class of involatile molecules by exciting the matrix with a short intense burst of energy. The energizing process results in the formation of a high concentration of gas-phase species just above the surface of the sample that can undergo rapid chemical reactions. The desired gas-phase chemical reaction for protein mass spectrometry is the protonation of the molecule. The presence of impurities in the matrix introduces competing gas-phase reagents that can interfere with the protein protonation reaction sometimes leading to total quenching of protein molecular ion formation. Even when the protein is a major component of the sample, it may not be possible to obtain a mass spectrum because of this impurity-quenching effect. The problem with impurities is best handled by reducing their concentration to the lowest possible level. Information on sample preparation for ^{252}Cf-PDMS is given in [23] in this volume.

Preparing Targets for ^{252}Cf-PDMS

The most effective sample preparation method for the ^{252}Cf-PDMS of proteins is a variation of the reversed-phase high-performance liquid chromatography (HPLC) methodology where protein is adsorbed onto a lipophilic surface from an aqueous solution. In HPLC, the column packing provides the lipophilic surface, while for ^{252}Cf-PDMS, the surface is part of the sample that is to be analyzed. The sample that is inserted into the mass spectrometer is called a *target*. It consists of a triple layer comprised of a thin polymer support film (1.5 μm thick polyethylene terephthalate, Mylar), a vacuum-evaporated metal layer, and a 1-μm thick deposit of nitrocellulose.[4] The target is prepared by stretching the metallized Mylar over a ring and depositing the nitrocellulose uniformly over the Mylar surface by electrospraying. The metallic layer has two functions. It serves as a cathode surface when the nitrocellulose is electrosprayed onto the Mylar. When a high voltage is applied to the target after it has been inserted into the mass spectrometer, the metallic layer establishes the extremely flat contours of the electric field lines in the ion source region that are necessary to efficiently transmit ions through the mass spectrometer. For this reason, it is essential that the target surface be free from wrinkles. The protein is added to the target by adsorption onto the nitrocellulose from aqueous solution. Solution concentrations are on the order of 10^{-4} to 10^{-3} M and volumes as small as 0.5 μl are used. Because the adsorption is an equilibrium between protein adsorbed on the target surface and protein in solution this concentration range is necessary to fully occupy the adsorption sites on the nitrocellulose. The target containing the adsorbed protein is then mounted onto an assembly that can accommodate several targets at one time. The assembly is a wheel (Bio-Ion) or a rectangular block. The assembly, called a *target wheel* or *target stick,* is used to insert several samples into the spectrometer at one time, a feature that is useful when a series of samples is being analyzed. The target wheel also automatically gives a precise alignment for the targets and provides electrical contact when a high voltage is applied.

How a ^{252}Cf-PD Mass Spectrometer Functions

After the target wheel is mounted in the spectrometer, the target to be analyzed is positioned on the axis of a long cylindrical tube called the *flight tube*. The ions travel along the length of the flight tube when the mass

[4] G. P. Jonsson, A. B. Hedin, P. L. Hakansson, B. U. M. Sundqvist, B. G. Save, P. F. Nielsen, P. Roepstorff, K. E. Johansson, I. Kamensky, and M. Lindberg, *Anal. Chem.* **58,** 1084 (1986).

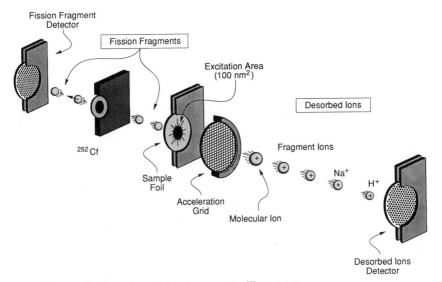

FIG. 1. Configuration of the elements of a ²⁵²Cf-PDMS mass spectrometer.

measurement is being made. When the target is in position for analysis, it is located between the *²⁵²Cf source* and an *acceleration grid*. The ²⁵²Cf source is radioactive and is predominantly an α emitter with a small, but essential, spontaneous fission decay component. The acceleration grid is a high-transmission nickel mesh maintained at ground potential providing a flat electrode that attracts ions emitted from the target. When the ions approach the acceleration grid, 90% of the ions pass through and enter the flight tube. The layout of the target/source region is shown in Fig. 1. Also shown are the two ion detectors used in the measurement; both are positioned on the axis of the spectrometer. The ion detector located close to the ²⁵²Cf source is called the *start detector*. Its function is to record the time of the fission event by detection of the charged particle radiation from the ²⁵²Cf source, and to resolve the *fission fragments* from the more abundant α particles. (Fission fragments are high-energy, highly charged heavy ions, e.g., 80 MeV cesium, $^{144}Cs^{+20}$.) The start detector has two components, a *conversion foil* and an *electron detector*. Separation of the fission fragments from the α particles is carried out using the conversion foil which is positioned in front of the electron detector. When a fission fragment or α particle passes through the conversion foil, electrons are ejected from the surface and are recorded by the electron detector. A fission fragment can be distinguished from an α particle because it ejects many more electrons. The second detector is located at the end of the flight tube and is referred to as the *stop detector*. Its function is to record the arrival of secondary ions that travel the length of the flight tube.

The general operation of the spectrometer is as follows. First, a high voltage is applied to the target wheel (± 15 kV, the polarity determined by the polarity of the ions to be analyzed). When a fission event takes place in the ^{252}Cf source, two fission fragments are formed which travel in opposite directions. If one of the fission fragments travels in the direction of the target, its partner (the *complementary* fission fragment) is recorded by the start detector. When the fission fragment passes through the target, several secondary ions are emitted and are accelerated to high energy in the electric field between the target and the acceleration grid. The ions pass through the acceleration grid, travel the length of the flight tube, and their arrival recorded by the stop detector. The time an ion takes to traverse the flight tube is dependent on its mass/charge ratio, the basic principle of the time-of-flight (TOF) mass spectrometry method. (A description of the TOF mass analyzer is presented in the chapter by Jennings [2].) The precise time interval that is recorded for an individual secondary ion is the time between the detection of the complementary fission fragment by the start detector and the arrival time at the end of the flight tube. A layout of the electronics is shown in Fig. 2. Both the start and stop detectors produce a transient electronic pulse when an ion is detected. The individual pulses have a wide range of shapes and amplitudes and consequently cannot be used directly for precise timing measurements. This problem is solved by transmitting the pulses to a module called a *constant fraction discriminator* (*CFD*) for processing before the time interval measurement is made. The circuitry of the CFD is designed to convert these randomly shaped pulses to a set of smooth identically shaped pulses (*fast logic pulses*) without disturbing the time relationship between the pulses from the start and stop detectors. The fast logic pulses from the CFD are transmitted to the *time-to-digital converter* (*TDC*) for the time interval measurement. Its function is to measure the time separation between the start and stop pulses. For protein molecular ions, the time interval (TOF) is on the order of tens of microseconds. A TOF value is recorded for all of the secondary ions resulting from a single fission fragment. Approximately 5–15 ions are emitted per incident fission fragment. Since the incident fission fragment intensity is on the order of 2000/sec, this sequence of events is repeated every 500 μsec. On the average, a final TOF spectrum (which is converted to a mass spectrum in the data analysis) represents the accumulative contribution of a million fission fragments passing through the sample generating approximately 10 million secondary ions. Only about 10,000 of these will be molecular ions of the protein, but it is more than sufficient to produce an interpretable mass spectrum.

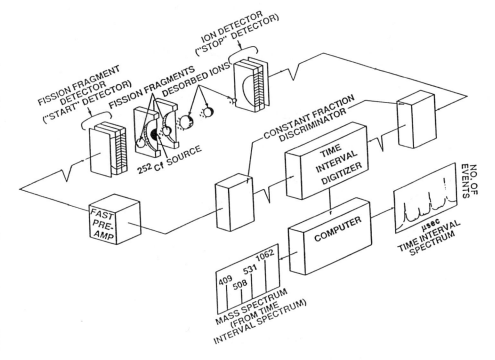

FIG. 2. Configuration of the electronics modules used in ^{252}Cf-PDMS.

Operational Features Relevant to Protein Analysis

Molecular Mass Determination

The most important application of ^{252}Cf-PDMS for protein analysis has been the measurement of the molecular mass of the native protein, a derivative, or a degradation product. In principle, it is feasible to measure molecular mass with an accuracy of ± 1 u in the mass range up to 50,000 u using ^{252}Cf-PDMS because of the high accuracy of the modern TDC used in making the TOF measurements. However, published molecular masses for proteins in this range are at least an order of magnitude less accurate than what is possible. In order to understand the factors that influence the accuracy of the molecular mass determination, it is necessary to examine how the molecular mass measurement is made. The basic measurement is the TOF of the protein molecular ion. This value is converted to m/z using the expression:

$$\text{TOF} = c_1(m/z)^{1/2} + c_2 \tag{1}$$

The constant, c_1, contains the instrumental parameters such as acceleration voltage, distance between the target and acceleration grid, and length of the flight tube while the constant, c_2, is a *"time offset,"* an instrumental electronics correction factor that accounts for the inherent difference in the timing of the pulses from the start and stop detector circuitry. Both of these constants can be calculated from first principles but a more reliable determination is by the accurate measurement of the TOF for ions with known m/z values that are present in the TOF spectrum.[5] The values of these two constants are determined for each target during the ^{252}Cf-PDMS analysis as part of the standard protocol so that each analysis generates its own calibration curve at the same time that the TOF spectrum is recorded. This procedure elmininates errors due to small perturbations of the instrumental parameters that influence the TOF.

The TOF of an ion type in the TOF spectrum does not have a single value but, rather, is a distribution of values that closely resembles a Gaussian distribution. The reason for this is that the influence of the *initial kinetic energy distribution* of the ion is superimposed on the TOF peak. The ion receives an initial kinetic energy during desorption from the target surface that is dependent on the desorption mechanism as well as collisional interactions that occur between other desorbed species. This effect is significant. For an insulin molecular ion with a typical average TOF (denoted as $\langle TOF \rangle$) value of 20 μsec, the width of the distribution, in the absence of initial kinetic energy effects, would be expected to be 0.5 nsec but it is actually 100 times broader. In addition, the $\langle TOF \rangle$ value is shifted by an amount that is dependent on the average value of the initial kinetic energy distribution. This effect is one of the major contributors to decrease of mass accuracy in molecular mass determination. In order for Eq. (1) to be precisely valid, the initial kinetic energies of the calibration ions and the protein molecular ions must be equal. The calibration ions that are commonly used are intense low-molecular peaks in the spectrum (m/z 1–100) with initial average kinetic energy values ranging from 1 to 10 eV, while protein molecular ions have a narrower and more uniform range of values on the order of 2 to 3 eV. This means that there is a small but systematic error in the mass determination when Eq. (1) is used due to a mismatch in the initial kinetic energies of the calibration ions and the protein molecular ions. However, there is a way to solve this problem. Instead of selecting light ions to evaluate the constants in the calibration equation, a more accurate calibration curve can be obtained using as calibration ions, species that have initial kinetic energies that match those of proteins. The best choice is another protein, one that is well-character-

[5] R. D. Macfarlane, *Anal. Chem.* **55**, 1247A (1983).

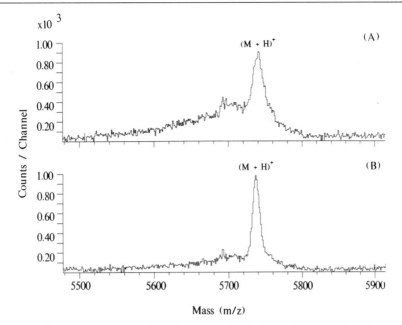

FIG. 3. ^{252}Cf-PD mass spectra of insulin in the molecular ion region. (A) Using a nitrocellulose substrate; (B) using an anthroic acid substrate.

ized, such as, insulin. These "mass-marker proteins" can be introduced into the adsorbed layer along with the protein that is being studied.

The evaluation of ⟨TOF⟩ from the TOF distribution for the protein molecular ion can also be complicated by the presence of multiple components in the molecular ion peak due to metastable decay. Metastable decay of the molecular ion is the result of the transfer of too much internal energy to the protein during the desorption process which is influenced by choice of substrate. One of the advantages of the use of nitrocellulose as a substrate is that metastable ions are formed in much lower abundance. However, even for nitrocellulose, there still can be a significant metastable component in the molecular ion peak that introduces errors in the evaluation of ⟨TOF⟩. An example of this problem is shown in the spectrum of the insulin molecular ion (Fig. 3A) which has a shoulder on the low mass side. The presence of this component influences the value of ⟨TOF⟩ for this peak. We have recently found a new substrate, anthroic acid, that reduces even further the intensity of the metastable component of the insulin molecular ion and also gives a more symmetric, higher resolution peak (Fig. 3B). Consequently, a more accurate determination of the ⟨TOF⟩ of this peak can be made.

Sensitivity

In the context of a ^{252}Cf-PDMS analysis, the sensitivity for the detection of a particular protein is the minimum number of molecular ions required to resolve the peak in the mass spectrum from background. Sensitivities at the picomole level have been reported in the literature[6] using ^{252}Cf-PDMS, but this is clearly not a limiting value. The ^{252}Cf-PDMS method has some unique features regarding the question of sensitivity because some variables are available that other mass spectrometric methods do not possess. The two most important features are the fact that individual ions are detected and the analysis can be continued for several hours if the molecular ion yield is very low. However, the large background in the high mass region places severe limitations on the ability to resolve a weak peak from the background. An example of the magnitude of the problem is shown in Fig. 4A which is the spectrum in the molecular ion region of ribonuclease. The background is more intense than the molecular ion peak. In the data analysis, this background is subtracted using a computer-based algorithm before the $\langle TOF \rangle$ measurement is made as shown in Fig. 4B. The background is due to secondary ions and neutral particles that were detected by the stop detector but which were not correlated in time with any of the pulses from the start detector. We now have developed a very effective method that eliminates 98% of this uncorrelated background and the neutral component using a novel electrostatic ion switching deflector,[7] which will be used in future studies. The use of the new anthroic acid matrix has also improved sensitivity because the molecular ion peaks are considerably sharper and more clearly discernible from the background. An example is shown in Fig. 4C which is a spectrum of the molecular ion region of ribonuclease adsorbed on an anthroic acid substrate. The peak is considerably sharper than what was obtained using the nitrocellulose substrate (Fig. 4B).

Any effect that influences the number of detected protein molecular ions for a selected number of incident fission fragments passing through the target is a potential source for improvement of sensitivity. Listed below are the most important effects.

Sample Purity. As already discussed, the quenching effect in molecular ion formation caused by sample impurities is a major cause of loss of sensitivity. This effect was recently borne out in a study involving a protein with a molecular mass of 13,000 u that was purified as a single fraction by HPLC just prior to ^{252}Cf-PDMS analysis. An aliquot of the sample was

[6] B. Sundqvist, A. Hedin, P. Hakansson, M. Salehpour, and G. Sawe, *Int. J. Mass Spectrom. Ion Proc.* **65**, 69 (1985).
[7] B. Wolf, P. Mudgett, and R. D. Macfarlane, *J. Am. Soc. Mass Spectrom.* **1**, 28 (1990).

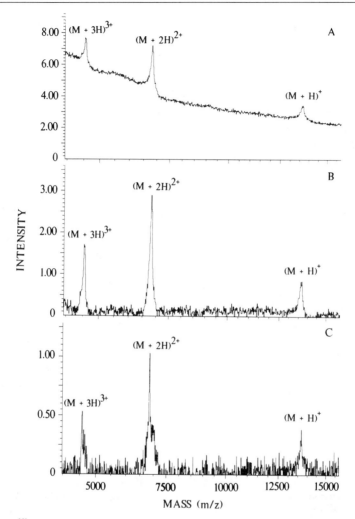

FIG. 4. ^{252}Cf-PD mass spectra of ribonuclease. (A) Gross spectrum using a nitrocellulose substrate; (B) after background subtraction; (C) using an anthroic acid substrate and after background subtraction.

used in the first analysis. A reasonably intense molecular ion peak was observed. The unused part of the sample was then recycled by HPLC using a different solvent system and a slice of the main fraction was again analyzed by ^{252}Cf-PDMS. The molecular ion intensity increased by a factor of 2.

Many proteins are only stable in salt or buffer solutions and the presence of these salts in the target could present a problem if they are not removed. However, by employing the adsorption method to prepare targets, proteins are first immobilized on the target and then rinsed with water to remove salt impurities.

Increasing Fission Fragment Flux. The fission fragment flux is the number of fission fragments that pass through the target per second and per unit area. The molecular ion intensity is proportional to the product of this flux and the number of protein molecules exposed to the fission fragments. The effective flux is determined by the strength and area of the ^{252}Cf source and the distance between the source and the target. By being able to reduce the area of the target, a smaller number of protein molecules is required to obtain an equivalent mass spectrum. Optimizing the source/ target geometry has made it possible to achieve sensitivity at the subpicomole level. With the ^{252}Cf source closer to the target, there is an additional enhancement in molecular ion intensity because the fission fragments that pass through the target at oblique angles desorb more molecular ions.[8]

Sensitivity of Stop Detector for Ion Detection. The stop detector operates on the electron multiplication principle in which the impact of the incident ion results in the ejection of one or more electrons. These electrons trigger the formation of an avalanche of electrons that are collected and form an electronic pulse that is transmitted to the stop CFD. The key step is the initial interaction of the ion with the detector surface because if an electron is not ejected the ion is not detected. This process is dependent on ion velocity and when the ion velocity is low (e.g., <21,000 m/sec), as it is for the heavier protein molecular ions, the probability for electron emission can be extremely small. (Light ions have velocities greater than 100,000 m/sec when acceleration voltages of 10 to 15 kV are used.) The problem is partially solved by increasing the ion velocity. Larger acceleration potentials are employed and an additional "postacceleration" electric field is used to enhance the ion velocity at the point of impact with the stop detector. But even after increasing ion velocity by the use of stronger electric fields, the problem is still present for proteins with a molecular mass greater than 5000 u and it is a major contributor to loss of overall sensitivity. We recently found that the intensity of the $(M + H)^+$ ions of insulin is enhanced by a factor of three by increasing the acceleration voltage from 10 to 15 kV. However, even at 15 kV, only 33% of the singly protonated molecular ions are detected while in the same spectrum, the doubly protonated $(M + 2H)^{2+}$ molecular ion with an energy of 30 keV is detected with 100% efficiency. A method for calculating

---

[8] P. Hakansson, I. Kamensky, and B. Sundqvist, *Nucl. Instrum. Meth.* **198,** 43 (1982).

detection efficiencies for protein molecular ions has been published based on this work.[9]

²⁵²Cf Source

Chemical Properties and Occurrence

In ²⁵²Cf-PDMS, the ²⁵²Cf source is the heart of the method, providing a compact, portable pulsed energy source in the form of the nuclear fission fragments that are emitted. The nuclide, ²⁵²Cf, is a radioactive isotope of the element, californium, which is a member of the actinide series. The chemistry of the lanthanides and the actinides are closely related because both series involve the filling of an inner f-elecrtronic shell. The element dysprosium (Dy) is the analog of Cf in the lanthanide series and is used as a model in developing chemical purification procedures for Cf. The most stable oxidation state for both elements is $+3$. None of the Cf isotopes exist in nature because they are unstable toward nuclear decay. The isotope, ²⁵²Cf, is manufactured from ²³⁹Pu (plutonium) in a high neutron-flux nuclear reactor by successive capture of 13 neutrons which increases the mass number from 239 to 252 and by 4 β decay steps that increase the atomic number from 94 to 98. The high-flux reactor that is located at Oak Ridge National Laboratory (TN) produces 500 mg of ²⁵²Cf per year. (Approximately 0.01 μg of ²⁵²Cf are in the source used for ²⁵²Cf-PDMS.) For ²⁵²Cf-PDMS applications, the ²⁵²Cf must be free of inert materials because their presence in the source could attenuate the energy of the fission fragments that are emitted. Final purification of the Cf from the other elements is carried out by complexation with α-aminohydroxybu-tyric acid followed by cation-exchange chromatography. The ²⁵²Cf sources are prepared by vacuum evaporation or electrodeposition of the purified material onto a thin metal backing.

Nuclear Decay Properties

The nuclide, ²⁵²Cf, has a half-life of 2.6 years and has two nuclear decay modes. The major decay mode (96.9%) is α decay to ²⁴⁸Cm (curium); 3.1% is by spontaneous nuclear fission. This is a decay mode where the ²⁵²Cf nucleus spontaneously divides into two smaller nuclei. All of the isotopes of the elements above lead (Pb) are unstable toward this mode of decay because it is a highly exothermic process. However, there is a substantial energy of activation (the *fission barrier*) that strongly attenuates the rate

[9] P. W. Geno and R. D. Macfarlane, *Int. J. Mass Spectrom. Ion Proc.* **92**, 195 (1989).

of decay. The rate of the fission process is enhanced by nuclear excitation as in the case of the thermal neutron capture of ^{235}U (uranium) which excites the ^{236}U nucleus to 6.4 MeV. This is an example of a thermal neutron-induced fission which is the principal nuclear reaction that is carried out in a nuclear reactor. As the atomic number increases, the fission barrier decreases due to increased Coulomb repulsion in the nucleus. From U to the heaviest of the new elements, spontaneous fission, the division of a nucleus while in its ground state, becomes a common decay mode. From Cf to hahnium (Ha) $(Z = 105)$, 19 spontaneously fissioning isotopes are known.

Spontaneous fission occurs in two modes. One of the modes is called *symmetric fission* where the nucleus divides into two equal masses. The nuclide, mendelevium, (^{260}Md), is an example. The ^{252}Cf nucleus undergoes *asymmetric fission* which means that the nucleus splits into two nuclei with unequal masses. Some nuclei such as ^{257}Fm (fermium) exhibit both modes of fission. In asymmetric fission, the mass division does not occur the same way each time. For the spontaneous fission of ^{252}Cf, approximately 130 different fission fragment pairs have been detected. The peaks of the mass distributions center around ^{108}Tc (technetium) for the lighter mass fragment and at ^{144}Cs for the heavy fragment. The average total energy released is 212 MeV. Most of this energy (190 MeV) is transformed into the kinetic energy of the fission fragments. The remainder is in the form of nuclear excitation of the fission fragments. The excitation of some of the fission fragments is high enough that neutrons are emitted from the fragment. The number of neutrons emitted per fission event is called the *neutron multiplicity;* for ^{252}Cf, the neutron multiplicity is 3.86. Although these neutrons do not participate in ^{252}Cf-PDMS, they are an important component of the nuclear radiation surrounding the ^{252}Cf source and their presence constitutes a potential health hazard. This effect will be discussed later in this chapter.

Properties of Nuclear Fission Fragments

During the fission process, there is also a considerable amount of electronic excitation of the fission fragment ions resulting in loss of electrons. The average atomic charge (z) of the fission fragments is on the order of $+20$. The combination of high velocity (v) and high charge state are the principal properties of the fission fragments that determine how much energy is deposited when a fission fragment passes through the target. The energy deposition density is given in terms of a linear energy transfer (LET) or *"dE/dX"* which varies as z^2 and inversely with v^2. The *dE/dX* of the fission fragment is expressed as energy loss per unit length in the fission track. The *range* is the average distance that a fission fragment can travel in a medium before it loses all of its energy. For targets composed of organic molecules, the average range is 14 μm and the average initial dE/dX is 10,000 eV/nm. The molecular ion desorption yield is sensitively dependent on the dE/dX of the fission fragment as it emerges from the surface of the target so that any factors that influence dE/dX also have an effect on sensitivity. When the fission fragment passes through thin films, it is continually losing energy and capturing electrons which reduce the atomic charge of the ion. Consequently, the dE/dX of the ion as it emerges from the surface is significantly attenuated. For example, the dE/dX of a fission fragment that enters the target at an angle of 45° is attenuated to 80% of its original value and its average energy is reduced to 29 from 95 MeV. By utilizing ultrathin target materials, the dE/dX at the target surface can be restored to 98% of its initial value, with a reduction of only 12 MeV in kinetic energy.

Radiological Aspects of ^{252}Cf

Kinds of Radiation. The radiation from a ^{252}Cf source is composed of α particles, fission fragments, neutrons, β particles, and high-energy photons (X and γ rays). When the source is outside of the ^{252}Cf-PDMS system, all of these forms of radiation contribute to an overall radiation field. This radiation field is markedly reduced once the source is mounted inside the mass spectrometer because only the neutrons and high-energy photons can penetrate through the walls of the vacuum chamber.

Radiation Levels. A typical source strength of a ^{252}Cf source used in ^{252}Cf-PDMS is 5 μCi or 1.85 $\times$ 10^5 Bq (becquerels). This corresponds to an α disintegration rate of 1.8 $\times$ 10^5/sec, a fission rate of 5700/sec, and a neutron emission rate of 22,000/sec. Only 35% of the fission events are utilized in a typical ^{252}Cf-PDMS analysis because many of the fission fragments from the source are emitted in a direction that does not intercept the target.

Radiation Exposure. The greatest potential danger from the ^{252}Cf source occurs when it is outside the spectrometer and when the ^{252}Cf becomes airborne. This can occur if the ^{252}Cf is an *open source* which means that the ^{252}Cf deposit is on the surface of the source. Under these conditions, ^{252}Cf can become airborne by a process called *self-transfer* where the fission fragments eject ^{252}Cf atoms from the source into the atmosphere probably by the same mechanism that is responsible for the ejection of protein molecules from the target. The danger of airborne ^{252}Cf is that it can be inhaled. Once inside the body, ^{252}Cf concentrates in the bone and liver resulting in continuous radiation of these organs by the heavily ionizing α particles and fission fragments. The annual limit on intake by inhalation established by the Nuclear Regulatory Commission is only 700 Bq, a reflection of the potential seriousness of this type of exposure. Fortunately, self-transfer can be completely eliminated by the use of a *cover foil*. This is a thin metal foil that is placed over the ^{252}Cf source immediately after the source is prepared forming a *sealed source*. Under these conditions, all of the ^{252}Cf atoms that would be self-transferred into the atmosphere are trapped by the cover foil. As long as the cover foil remains intact and pinhole free (i.e., the source remains sealed), self-transfer cannot occur. Nickel foils are generally used because of their mechanical stability, low chemical reactivity, and availability as ultrathin foils (0.25–1 μm). Sources sealed with nickel cover foils are always stored in a vacuum or inert atmosphere to maintain their integrity. When exposed to air for an extended period, the reaction of the fission fragments with the nitrogen, oxygen, and water in the air produces free radicals that are transformed into corrosive molecules that attack the foil and create pinholes. When this happens, the seal is broken and self-transfer of ^{252}Cf into the atmosphere can take place. Our laboratory is monitored on a weekly basis for the presence of airborne ^{252}Cf to ensure that the integrity of the ^{252}Cf source in the instrument is maintained. In the 17 years that we have been involved with ^{252}Cf-PDMS studies, we have never had an incident where the seal on the ^{252}Cf source was broken. It has happened, however, at one ^{252}Cf-PDMS facility where the ^{252}Cf source was stored improperly in air, resulting in low-level contamination of the laboratory. Several commercial companies throughout the world produce properly sealed ^{252}Cf sources that can be directly used in a ^{252}Cf-PD mass spectrometer. In the United States, sources can be purchased from Isotope Products Laboratory, Burbank, CA. A government-controlled license is required to procur a ^{252}Cf source. In order to obtain a license, it is necessary to prove that adequate facilities and experienced personnel are available for handling and monitoring of the source.

Dose Levels from a ^{252}Cf Source. When the ^{252}Cf source is fixed in the mass spectrometer, only the fast neutrons and γ rays contribute to the ionizing radiation field outside the spectrometer. The amount of radiation that an individual receives is measured quantitatively in terms of what is precisely defined as the *radiation absorbed dose* (rad). The fast neutrons trigger the more adverse effects on biological tissue. After accounting for differences in ionizing power for fast neutrons and γ rays, the radiation absorbed dose is finally converted to another precisely defined term, the *dose equivalent,* which is the fundamental quantity used to define maximum exposure limits for individuals. The Standard International unit for dose equivalent is the Sievert (Sv). The maximum exposure limit for an individual, established by the Nuclear Regulatory Commission, is 0.0017 Sv (1.7 mSv) per year. Since our environment (cosmic rays, soil, air, food) contributes 1.3 mSv/year, it takes very little additional exposure to reach the maximum exposure limit. (Incidently, smoking at the rate of 1.5 packs per day subjects the individual to a dose equivalent of 80 mSv/year, 47 times the maximum exposure limit!).

The *dose equivalent rate* is a measure of how many Sv/s an individual receives when in the vicinity of a radioactive source such as ^{252}Cf and is used to calculate the dose equivalent for a particular exposure time. It depends on source strength as well as how close the individual is to the source. For example, if the source strength is 5 μCi and the distance is 1 m, the dose equivalent rate from a ^{252}Cf source is calculated to be approximately 1 nSv/s. An individual would reach the annual limit (including contributions from the unavoidable natural sources of radiation) after a total time of 100 hr exposure at a distance of 1 m. The amount of exposure received is reduced exponentially with distance so that by locating the spectrometer in the laboratory in a relatively remote site the average exposure can be easily reduced to well below the natural background level.

Concluding Remarks

The simplicity and reliability of ^{252}Cf-PDMS for protein molecular weight determination is the major attraction for the methodology in the protein laboratory. The design of the commercial Bio-Ion ^{252}Cf-PDMS system enhances the ease of operation by providing computer control of its function. It is important for the user to know the basic principles of the method and how much the quality of the protein sample influences the results. A general knowledge of the inner workings and associated terminology gives the user an insight into some of its strengths and weaknesses and the basic langauge needed to communicate with other scientists who are also involved with ^{252}Cf-PDMS. Finally, the development of the capabilities of ^{252}Cf-PDMS for protein analysis is on-going, involving improvements in instrumentation as well as target design and utilizing some of the basic principles that have been developed from fundamental studies of fission fragment/solid interactions.

Acknowledgments

I wish to acknowledge the students and colleagues in my laboratory who have, over the years, made substantial contributions to the development of the ^{252}Cf-PDMS methodology. I would like to especially cite D. F. Torgerson, C. J. McNeal, and R. Martin. I would also like to thank B. Wolf for providing some results in advance of publication. The financial support of the U.S. Atomic Energy Commission in the first stages of development, the National Institutes of Health (GM 26096), the National Science Foundation (CHE-8604609), and the Welch Foundation (A-258) are gratefully acknowledged.

[12] Mass Spectrometry of Peptides and Proteins by Matrix-Assisted Ultraviolet Laser Desorption/Ionization

By FRANZ HILLENKAMP and MICHAEL KARAS

Introduction

Over the last two decades the analysis of organic compounds has developed into one of the major applications of mass spectrometry, even though until rather recently the accessible mass range was limited to the order of 1000 Da, this limit set primarily by the ability to generate large mass ions without extensive and often unspecific fragmentation. The various desorption techniques [field desorption (FD-MS), secondary ion mass spectrometry (SIMS), fast atom bombardment (FAB-MS), plasma desorption (PDMS) and, most recently, laser desorption/ionization (LDI-MS)] have extended the accessible mass range into the range of several hundred thousand daltons. Electrospray has also been shown to be capable of generating (multiply charged) ions of intact large bioorganic molecules. If there is any common denominator to all these techniques, it is the fact that the separation of the molecules from their surroundings in the condensed state and their ionization are both achieved in one single step. This chapter discusses the status of laser desorption/ionization of large biomolecules; the other techniques are discussed in several other chapters of this volume (see, in particular, [1] and [22]).

Three features make mass spectrometry an attractive analytical technique for biorganic molecules: very high sensitivity, potential for very highly accurate mass determination, and ability to elucidate structural information of the molecules under investigation. It will be shown in this chapter that the goal of extremely high sensitivity has already been achieved in LDI-MS. It also yields important structural information, namely, on the level of the quaternary and, possibly, tertiary structures more than on that of the primary and secondary ones. The attainable accuracy in molecular mass determination still leaves much to be desired, but substantial improvements can be expected.

Principles of Laser Desorption/Ionization

Most lasers throughout the wavelength range from the far-ultraviolet to the far-infrared and with pulse durations from continuous wave down to the femtosecond (10^{-15} sec) range have at some point been used to irradiate solid or liquid material in vacuum or under ambient pressure with the intent to couple energy from the laser beam into the sample via linear or nonlinear absorption and to transfer some of the sample material into the gas phase, a prerequisite for a mass spectrometric analysis. A variety of lasers have also been combined with mass spectrometers (for a review see Ref. 1). Two ranges of laser wavelengths have evolved as being most useful for laser desorption of organic materials: the far-infrared, where the CO_2 laser emits at a wavelength of 10.6 μm, at which radiation can efficiently couple into rovibrational modes of the molecules and the far-ultraviolet, so far preferentially at wavelengths below ca. 300 nm, where at least aromatic molecules can become electronically excited. The most commonly used wavelength in the latter regime is 266 nm of the frequency-quadrupled neodymium/yttrium–aluminum–garnet (Nd–YAG) laser. In both wavelength ranges, pulses of 100 nsec or less duration are used, because longer exposure times will lead to classical thermal heating resulting in the well-known pyrolytic decomposition of organic molecules. Time-of-flight (TOF) mass spectrometers and Fourier transform ion cyclotron resonance (FT-ICR) mass spectrometers have so far been most widely used. Laser beams can be focused down to spots of submicrometer diameter by suitable microscopic objectives. This good spatial resolution, combined with the high detection sensitivity of mass spectrometers, has led to the development of various laser microprobes for trace element analysis, and also detection of organic molecules (for a review see Ref. 2).

The fact that the desorption techniques under suitable conditions succeed to transfer essentially nonvolatile, fragile molecules into the gas phase with no or only limited fragmentation has led to use of the term of "soft desorption." This is somewhat misleading, as all the methods actually deposit a large amount of energy per unit volume into the sample, leading to a strong perturbation of the system. Such a strong perturbation

[1] F. Hillenkamp, in "Ion Formation from Organic Solids II" (A. Benninghoven, ed.), Springer Series in Chem. Phys. No. 25, p. 190ff. Springer Verlag, Heidelberg, 1983.

[2] A. H. Verbueken, F. J. Bryunseels, and R. E. van Grieken, *Biomed Mass Spectrom.* **12**, 438 (1985).

seems, in fact, to be the prerequisite for the establishment of thermodynamic nonequilibrium reaction channels of the system, which lead to the intact desorption of molecules rather than a thermal decomposition. Nonetheless it appears that the amount of energy deposited into the samples is a rather critical parameter. In LDI its value is determined by the laser irradiance (W cm^{-2}), often called the intensity, the pulse width, and, particularly, the sample absorption at the laser wavelength, if any. For nonabsorbing samples, nonlinear absorption can be induced, if the laser irradiance is high enough (typically ca. 10^8 W cm^{-2}). This does not, however, appear to be a suitable mode of operation for LDI of large molecules, because the deposited energy per unit volume cannot be controlled under these circumstances and will in all likelihood exceed the desirable value thus leading to excessive fragmentation.

Direct laser desorption of intact bioorganic molecules without a matrix seems to be limited to molecular masses of about 1000 Da or a little more. Phospholipids, isolated from bacterial membranes, have been successfully analyzed with both a CO_2 laser[3] and a Nd-YAG laser at 266 nm wavelength.[4] This mass range limitation gave rise to the development of the matrix-assisted LDI. Two somewhat contradictory observations led the way. First, the process is more controllable if the energy is transferred to the sample via resonant absorption; second, the fragmentation depends critically on the amount of energy deposited into the molecule under investigation: the less energy deposited, the less the degree of fragmentation. It was observed by the authors several years ago that small, nonabsorbing molecules could be successfully desorbed intact if immersed in a suitably absorbing surrounding.[5,6] This eliminates the (technically difficult) need for tuning the laser wavelength to the absorption band of the molecule under analysis, and enables the otherwise impossible analysis of mixtures of molecules with different absorption bands. Moreover, and more importantly, it opens up the possibility for a selective deposition of the energy into a matrix in which the sample is dissolved. This leaves the molecule to be analyzed essentially without alteration, provided that a suitable combination of laser wavelength, analyte absorption, and matrix absorption is chosen. A more systematic investigation revealed that the matrix can serve several functions to enhance the desorption of large molecules:

[3] N. Qureshi, J. P. Honovich, H. Hara, R. J. Cotter, and K. Takayama, *J. Biol. Chem.* **263,** 5502 (1988).

[4] U. Seydel, B. Lindner, U. Zaehringer, E. T. Rietschel, S. Kusumoto, and T. Shiba, *Eur. J. Biochem.* **145,** 505 (1984).

[5] M. Karas, D. Bachmann, and F. Hillenkamp, *Anal. Chem.* **57,** 2935 (1985).

[6] M. Karas, D. Bachmann, U. Bahr, and F. Hillenkamp, *Int. J. Mass Spectrom. Ion Proc.* **78,** 53 (1987).

(1) present in large excess relative to the analyte, it separates single analyte molecules from each other, thereby limiting aggregation, which would otherwise prevent molecular ion formation (matrix isolation); (2) absorb the laser energy from the beam and transfer it into excitation energy of the solid system, e.g., via electron–phonon coupling, to lead to the desired strong perturbation and disintegration of at least some top molecular layers of the sample; (3) aid ionization of analyte molecules via suitable excited state photochemical reactions or two-photon photoionization in the condensed state. This function has not yet been fully established for the matrix–protein system, but can be inferred from systematic investigations on smaller molecules.[7]

Among a number of different organic matrices tested, the authors have found nicotinic acid to give the best results at the wavelength of the quadrupled Nd-YAG laser at 266 nm. Beavis and Chait have also tested a number of different matrices and have found 2-pyracinecarboxylic acid, vanillic acid, thymine, and thiourea to work as good or even better at this wavelength under their experimental conditions.[8] They also speculate that a relatively low temperature of evaporation or sublimations is desirable, as long as the vacuum is not too strongly perturbed. Tanaka et al.[9] have taken another approach to matrix desorption of proteins. They use a slurry of metal particles of ca. 10 nm size in glycerol as absorbing matrix and an N_2 laser, emitting radiation at 337 nm, for desorption. Use of an identical matrix for a large variety of different analyte molecules does, of course, simplify the analysis, but it should be stressed again that a careful control of the laser irradiance on the sample is very critical to meet the fine line between desorption of the large molecular ions and their excessive fragmentation.

Materials and Methods

Instrumentation

All examples discussed here were obtained with a LAMMA 1000 laser microprobe, described in detail elsewhere,[10] and shown diagrammatically in Fig. 1. The beam of a Q-switched ($\tau = 10$ nsec) frequency-quadrupled ($\lambda = 266$ nm) Nd-YAG laser is focused by the Ultrafluar 10 UV-transmitting microscopic objective **(1)** of 0.2 NA at 45° onto the sample. Collinear

[7] M. Karas and F. Hillenkamp, in "Ion Formation from Organic Solids IV" (A. Benninghoven, ed.), p. 103. Wiley, New York.

[8] R. C. Beavis and B. T. Chait, Rapid Commun. Mass Spectrom. 3, 233 (1989).

[9] K. Tanaka, H. Waki, Y. Ido, S. Akita, Y. Yoshida, and T. Yoshida, Rapid Commun. Mass Spectrom. 8, 151 (1988).

[10] P. Feigl, B. Schueler, and F. Hillenkamp, Int. J. Mass Spectrom. Ion Phys. 47, 15 (1983).

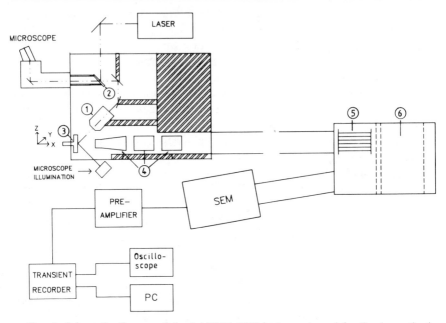

FIG. 1. Schematic diagram of the LAMMA 1000 instrument used for the desorption/ionization mass spectrometry of proteins. (1) 10/0.2 microscopic objective; (2) dichroic mirror; (3) sample stage; (4) three-element Einzel lens; (5) ion deflector; (6) ion reflector.

microscopic sample observation is possible via a dichroic mirror (2) in the beam path. The smallest nominal focus is about 3 μm in diameter; the affected sample area under this condition has been shown to be about 1 μm in diameter.[11] For LDI of large biomolecules out of a reasonably homogeneous sample, the laser beam is usually defocused to a diameter of typically 10–30 μm in diameter. Best results are obtained at an irradiance of about 10^7 W cm^{-2}. The sample is mounted in vacuum on a x,y,z movable microscopic stage (3) at ground potential. The ions are collimated and accelerated to an energy of 3 keV by a three-element immersion Einzel lens (4) to be analyzed by a time-of-flight (TOF) mass spectrometer, equipped with a beam deflector (5) and a second-order ion reflector (6). The equivalent length of the total flight path of the ions is 2.5 m. For more detailed information about TOF mass spectrometers, the reader is referred to [2] in this volume. The ion detector is a standard EMI 9643 secondary electron multiplier (SEM) (Thorn EMI Electron Tube Ltd. Ruislip, Middlesex, UK). To enhance ion detection, a separate Venetian-blind conversion

[11] F. Hillenkamp, in "Microbeam Analysis 1989" (P. E. Russell, ed.), p. 277. San Francisco Press, San Francisco, 1989.

electrode, held at a potential of typically 20 kV, is mounted in line in front of the SEM. Even at this potential, large protein ions have velocities below 10^4, the largest ones even below 3.5×10^3 msec^{-1}. At such low-impact velocities the ion to electron conversion efficiency of such electrodes should be exceedingly low, according to the accepted theory. Quite in contrast to this, strong multiplier signals are observed. Apparently different conversion processes prevail for the large molecular ions. In LDI a relatively large number of ions is generated for each laser pulse of 10 nsec in duration. These ions must be detected in an analog fashion rather than by pulse counting as is typically done in other TOF mass spectrometers. The SEM signal at a 50-ohm load resistor is therefore preamplified by a factor of typically 10 and then digitized at 10- to 500-nsec time intervals by a LeCroy 9400 (Chesnut, NY) transient oscilloscope. The data is then transferred to an PC-AT for spectrum averaging, mass calibration, and storage. All spectra shown in this publication are original data without any data manipulation such as background subtraction or filtering. Some of the features of the LAMMA 1000 microprobe as described above and used by the authors are not a necessary prerequisite for LDI of large proteins as has been demonstrated by Beavis and Chait.[8] They irradiate an area of ca. 100×300 μm with a laser of otherwise essentially equal parameters. No provisions for *in situ* sample observation is included in their instrument. The beam strikes the sample at an angle of ca. 80° to the surface normal; best results are claimed to be obtained for this geometry with irradiances between 5×10^5 and 10^6 W cm^{-2}. The sample is held at a potential of 20 kV relative to the straight tube TOF-MS at ground potential. Chevron dual multichannel plates without substantial further postacceleration are used for ion detection. Digitization and electronic data processing are essentially the same as described above.

Sample Preparation

Proper sample preparation is essential for good LDI results. Even though matrix and analyte will normally form a homogeneous (typically aqueous) solution, they may well separate upon solvent evaporation before or upon transfer into the vacuum. Sample homogeneity at least on the microscopic level, if not on the molecular one, is quite critical and will mostly depend on the properties of the matrix and the analyte. Small amounts of additives such as trifluoroacetic acid or ethanol have been found to be helpful in some instances. The protein to be analyzed (0.5–1 μl of a typically 0.1 g/liter aqueous solution) is mixed with an equal amount of a 5×10^{-2} mol/liter solution of nicotinic acid, dripped onto an inert metal substrate (e.g., silver or stainless steel), and dried in a gentle stream

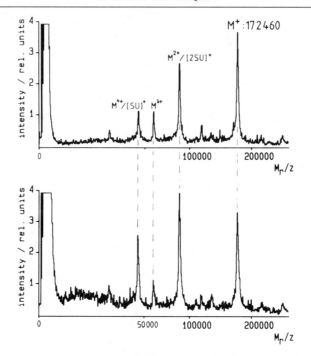

FIG. 2. LDI spectra of the enzyme glucose isomerase; 10^{-6} mol/liter protein, 5×10^{-2} mol/liter aqueous nicotinic acid solutions. Top spectrum: purified enzyme; bottom spectrum: 10^{-3} mol fraction of NaCl added to protein. The peak at $M_r/Z = 172,460$ represents the tetramer parent molecular ion of the functional enzyme, consisting of four equal, noncovalently bound subunits. The peak at $M_r/Z = 86,200$ comprises the contribution from the doubly charged parent molecular ion as well as from the dimer of two subunits. All other smaller peaks can be assigned in a similar fashion to combinations of a given number of subunits and charges.

of warm air before being introduced into the vacuum for analysis. For samples with molecular masses up to typically 20,000–30,000 Da a signal of typically 100–300 mV is registered by the transient oscilloscope for each single laser shot. Many successive spectra can usually be obtained from the same spot and/or from different areas of the sample. For larger molecules, or when the signal is lower for other reasons, up to 100 spectra are summed and averaged. The whole process of sample preparation and analysis takes typically 10 to 15 min. Samples need to be carefully purified, and, particularly, salt-free. Figure 2 shows two spectra of the enzyme glucose isomerase. The top spectrum was taken from a salt-free sample, the bottom one from a sample with an (intentional) NaCl contamination of 10^{-3} mol fraction relative to the protein. Dissociation into subunits is somewhat increased and the signal-to-noise (S/N) ratio is slightly worse

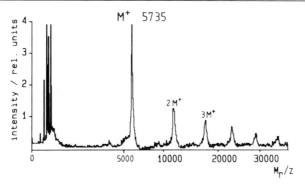

FIG. 3. Single-exposure LDI spectrum of insulin, 10^{-5} mol liter^{-1} protein, 5×10^{-2} mol/liter aqueous nicotinic acid solutions. For general rules of peak assignments, see legend to Fig. 2.

in the latter case. Further salt contamination quickly leads to a total loss of the protein signal. The exact cause for this observation and the question as to whether the system is equally sensitive to all types of salts and/or other contaminants awaits further investigation.

Characteristics of Laser Desorption/Ionization Mass Spectra

Ions of either polarity have been obtained by LDI; in agreement with former observations for direct desorption of smaller molecules they seem to have basically about the same abundance, somewhat dependent, of course, on the features of the molecule under investigation. To what extent the protein spectra of the two polarities contain supplementary information still needs to be explored.

The positive single-exposure spectrum of insulin, shown in Fig. 3, exhibits some of the basic features of LDI spectra of proteins. The insulin monomer forms the base peak, but a multitude of oligomers is seen as well. The dominant monomer signal is somewhat surprising in view of the fact that insulin is known to predominantly exist as a dimer under physiological conditions. It can be speculated that the rather nonphysiological pH of 3.8 of the aqueous solution of nicotinic acid causes a predominantly monomeric distribution already in solution. The distribution of oligomers certainly also depends on the specific molecular structure; as a rule, more oligomers of higher order are usually seen in spectra of small molecules like insulin, but this may well just reflect the decreasing probability of ion formation and detection with increasing mass. This distribution also seem to depend on the purity of the sample (see, for example, Fig. 2), the physicochemical state of the prepared sample, and the laser irradiance. These questions need further investigation to obtain a better control over the influence of the various parameters, but the potential for obtaining information about the quaternary structure of proteins, consisting of noncovalently bound subunits is clearly there.

The authors rarely observed ion signals attributable to chains of proteins, linked by low-energy covalent bonds like sulfur bonds. This, however, again seems to depend somewhat on the ion polarity and the experimental conditions; Beavis and Chait have reported a negative ion spectrum of insulin, showing signals for the A and B chains.[8] A spectrum, obtained by the authors from urease seems to imply that under favorable conditions one can see ion signals of functional groups[12]; this observation, however, needs further clarification, as such signals in the lower mass range could also originate from minor contaminations, too small in concentration to be detected in chromatographic or electrophoretic separations. In their exploratory work, the authors have found such contaminations to be a very common problem for many presumably highly pure samples, commercially available or supplied by collaborating laboratories. Unspecific fragmentation is virtually absent in LDI spectra. Under proper laser irradiance all signals in the mass range below 500 Da can be assigned to matrix ions. Under the experimental conditions, as used by the authors, these signals typically are of a magnitude comparable to that of the proteins. Beavis and Chait seem to encounter substantially higher matrix signals. It is not clear whether this reflects differences in the ion formation, or differing ion transmission of the respective spectrometers as a function of the ion mass and, particularly, the radial initial velocity distribution for the two classes of ions. Most of the surprisingly small noise background, seen across the whole spectrum, is not electronic in nature; it is caused by the LDI process itself, most probably by fragments of ions which result from collisions with surfaces in the spectrometer.

Multiply charged ions are frequently observed, though at relatively low abundances. For the rather common proteins with a number of equal, noncovalently bound subunits, most of them coincide in apparent mass with subunit peaks, thereby complicating the interpretation of the spectrum.

From Figs. 2 and 3 the mass resolution of the ion peaks can be inferred to be about $m/\delta m = 50$. This is about a factor of ten below that of the spectrometer for low mass ions; the peaks of the matrix ion have, in fact, a much higher resolution. Several factors are believed to contribute to this

---

[12] F. Hillenkamp and M. Karas, *Proc. 37th ASMS Conf. Mass Spectrom. Allied Topics, Miami Beach, FL* p. 1168 (1989) (extended abstr.).

poor resolution. The very process of ion formation with its intrinsic spread of initial ion energies and possibly a time error due to cluster decay as well as the geometry of the postacceleration detector are believed to contribute to this effect. Beavis and Chait have demonstrated a mass resolution of about 600 for insulin in their spectrometer; yet even under their conditions the mass resolution degraded to much lower values for the higher mass ions. Because the reason for the excessive width of the ion peaks is as yet only very incompletely understood, it is not clear to what extent mass resolution can still be improved by an intelligent variation of the experimental parameters. It is to be expected, and has in fact been demonstrated by Beavis and Chait for insulin, that part of the peak width results from nonresolved adduct ion peaks on the high mass side and a peak due to cleavage of one COOH group on the low mass side.[8]

Absolute mass determination, most probably the primary application of the technique, even though directly influenced by the mass resolution, can be considerably improved through the use suitable centroiding programs and by taking advantage of oligomeric and multiply charged ions, having integer multiple or fractional masses. Depending mainly on the attained signal-to-noise ratio, the precision of mass determination for proteins is typically in the range of 5×10^{-3} to 10^{-3}, i.e., uncertainty is 10–50 mass units at 10,000 Da and 100 to 500 mass units for a protein of 100,000 Da mass. Absolute accuracy is yet another problem. It is quite conceivable that multiple adducts could be formed of the proteins with matrix molecules or their photoproducts. Unless this would be balanced by a cleavage of multiple small side groups from the protein, peak centroiding would be expected to result in measured masses systematically too high. Multiple cleavage of side groups must, however, be considered to be rather rare, considering the virtually nonexisting cleavage of molecules into larger fragments. Though in some cases measured masses seem to have been too high, it does, so far, not appear to be generally the case. Systematic investigations in this direction have been severely hampered by the non-availability of an alternative technique for control and consequently by a lack of accuracy to which the actual mass of a given protein is really known. Even for proteins which have been sequenced, it is not always clear whether or not the protein really exists as the pure amino acid sequence in the samples provided and to what extent this might be influenced by the conditions of sample preparation. Again, this question needs substantial further investigation. Provided they are not saturated and a good enough time resolution is used to sample the low mass matrix ions, mass calibration based on these ions seems to be as good as those based on low mass proteins such as insulin and myoglobin digests. This can also be taken to suggest that as yet there is not undiscovered new effect in the formation process of the large protein ions which would then principally limit accuracy of mass determination. In view of all this it is expected that mass resolution and accuracy of mass determination will improve by about one order of magnitude in the foreseeable future.

Sensitivity of the technique is another important issue. As was already pointed out for the microprobe applications, laser desorption in conjunction with TOF mass spectrometry is an intrinsically very sensitive technique. Consequently, the minimum amount of sample material needed for a protein analysis is entirely limited by the amount that can still reasonably be handled. In a first test of sensitivity, spectra of several proteins have been obtained from total amounts of used material of 50 fmol (100 nl of a 5×10^{-7} mol/liter aqueous solution put on the substrate).[13] The actual amount of material consumed for the spectra is much less, as can be inferred from the many spectra (typically 100–1000, depending on the sample thickness at the analyzed spot) which can be obtained from the same or adjacent spots on the sample. It has been estimated that the actual amount per spectrum is in the range of 10^{-18} to 10^{-16} mol. From the observation that many successive identical spectra can be obtained from the same spot, it can also be concluded that the nonconsumed material does not suffer measurable radiation damage and can therefore be regained from the sample, or used, for example, for on-site enzymatic digestion as has been done in PDMS. It would therefore appear that LDI will evolve as the most sensitive analytical technique in protein analysis.

Selected Applications

Over about one hundred different peptides and proteins have so far been successfully analyzed by LDI with catalase at mass 236,000 Da being the largest functional unit and the trimeric subunit of urease at mass 275,000 Da the largest ion detected. These results can be found in Refs. 8, 12–17. Except for the expected decrease of detection sensitivity with increasing mass there seems to be very little discrimination among the greatly varying groups of proteins. The real upper mass limit has not yet been explored because it is difficult to obtain proteins with masses much

[13] M. Karas, A. Ingendoh, U. Bahr, and F. Hillenkamp, *Biomed. Environ. Mass Spectrom.* **18,** 841 (1989).

[14] M. Karas and F. Hillenkamp, *Anal. Chem.* **60,** 2299 (1988).

[15] M. Karas, U. Bahr, A. Ingendoh, and F. Hillenkamp, *Angew. Chem., Int. Ed. Engl.* **28,** 760 (1989).

[16] M. Karas, U. Bahr, and F. Hillenkamp, *Int. J. Mass Spectrom. Ion Proc.* **92,** 231 (1989).

[17] M. Salehpour, I. Perera, J. Kjellberg, A. Hedin, M. A. Islamian, P. Hakanson, and B. U. R. Sundquist, *Rapid Commun. Mass Spectrom.* **3,** 259 (1989).

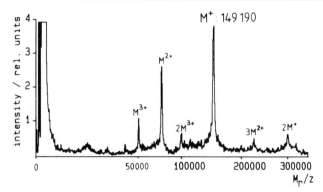

FIG. 4. LDI spectrum of monoclonal antibody (IgG of mouse against a specific human lymphokine). Average of 25 spectra. For general rules of peak assignments, see legend to Fig. 2.

above that of catalase, which do not dissociate into subunits of lower mass already in the solution at pH 3.8. In fact, the trimeric subunit of urease, seen in the spectra, is reported in the literature to be relatively stable.

LDI is not only suited for the analysis of very large proteins, but its sensitivity, speed, and ease of application make it also very well suited for the analysis of peptides in the range of several hundred to several thousand daltons. In this lower mass range it offers some unique advantages; in particular, it seems to be more independent of the properties of a given protein than other desorption methods such as FAB-MS and PDMS. The successful analysis of a Zn-binding protein of mass 11,373 Da (isolated from rat liver by a group at the University of Goettingen) serves as a good example. Among the 100 amino acids in the protein are 37 glutamic acids and 11 aspartic acids. All attempts to analyze this molecule with FAB-MS or PDMS failed.

Figure 4 shows the spectrum of a monoclonal antibody, an example of a very carefully purified large mass protein. It contains the typical signals for oligomers and multiply charged ions; no signals for the small and large chains are detected. The signal-to-noise ratio is excellent.

Two spectra of a violet phosphatase, extracted from beef spleen, are shown in Fig. 5. The top spectrum from a single exposure of the untreated molecule demonstrated the ability of LDI to desorb glycoproteins without major problems. The lower spectrum, averaged over 20 single shots (hence the better signal-to-noise ratio) is taken after cleavage of the carbohydrate part with endoglycosidase H. The two peaks below 10,000 Da are believed to represent impurities in the cleaving enzyme, rather than fragments of the deglycosylated violet phosphatase. Figure 6 is another example of a

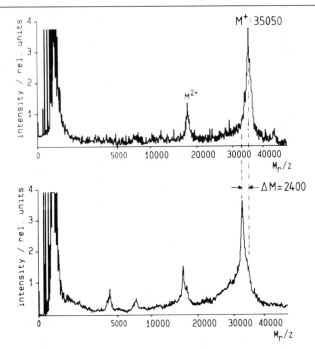

FIG. 5. LDI spectra of violet phosphatase from beef spleen. Top: whole enzyme, single shot; bottom; enzyme after cleavage of carbohydrate fraction with endoglycosidase H; average of 25 spectra.

glycoprotein with a carbohydrate fraction of several weight percent. It demonstrates the possibility of determining the molecular weights of two components of very similar mass in a mixture after electrophoretic separation with an accuracy of better than 30 Da. The two lipases OF1 and OF2 were isolated from the yeast *Candida cylindracea*.

Three different heme proteins have been analyzed to elucidate the influence of the absorption of prosthetic groups and their state of binding to the protein on the spectra.[17] In agreement with the effect of resonant excitation of molecules or parts thereof on the likelihood of fragmentation or cleavage as explained above, it was observed that the heme group was cleaved mostly from myoglobin and hemoglobin, to which it is not covalently bound. In contrast, the covalent bond of the heme to cyto-chrome *c* resisted cleavage on laser desorption despite the strong heme absorption at 266 nm, exceeding that of the matrix by almost one order of magnitude. Care was taken to ascertain by chromatography and centrifuga-tion following precipitation that in myoglobin and hemoglobin the heme

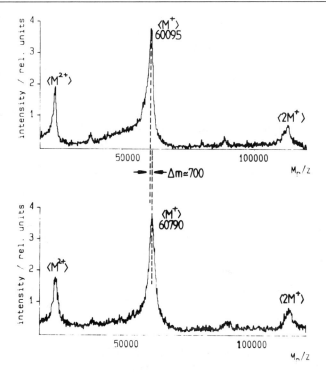

FIG. 6. LDI spectra of two components of a lipase obtained from *Candida cylindracea* (top: OF1; bottom: OF2); average of 25 spectra. For general rules of peak assignments, see legend to Fig. 2.

had not already been cleaved off the protein in the nicotinic acid solution (hemoglobin did, however, dissociate into two α/β dimers, both retaining two hemes). Very preliminary results for the analysis of polynucleotides support these findings. Though signals were obtained for a polynucleotide with a mass as high as 39,000 Da, mass resolution and the signal-to-noise (S/N) ratio were very poor. This is to be expected because the strong absorption of every single base in the chain contradicts the basic requirements for matrix absorption. The analyte should only have weak absorption at the wavelength used, compared to that of the matrix.

Future Developments

The method for LDI analysis of proteins, as described, was discovered only recently and other biomolecules are being studied. The whole field is, therefore, still in a state of very rapid development. Neither its potential nor its limitations have been fully explored. Important developments are, therefore, expected to take place within the next few years; interested readers are strongly advised to consult the most recent literature on the subject.

Some of the most urgent needs for improvement are obvious from the limitations discussed above. The method of sample preparation has a very strong influence on the quality of the spectra. More standardized preparation techniques, leading to a more uniform sample quality are therefore highly desirable; this is particularly true as one seeks new matrices having special properties, which may cause more severe problems with respect to sample homogeneity. A large variety of usable matrices is of interest for numerous reasons. Different pH values in aqueous solution may open possibilities to systematically study the quarternary structure of molecules, as explained above. Matrices having their absorption maximum at wavelengths other than 266 nm are of high interest for at least two reasons. Use of longer wavelengths would avoid the problem of strong resonant absorption of the analyte for many classes of biomolecules; the use of shorter wavelengths may induce more systematic fragmentation and thus give access to more structural information. Lysozyme has already been successfully desorbed at a wavelength of 337 nm with an alternative organic matrix[18] (as Tanaka et al. have done with their metal matrix[9]). The true sample consumption should be determined more carefully and possibilities for on-site sample processing as well as sample recapture should be developed. The first very encouraging results obtained for carbohydrates and polynucleotides certainly suggest that examination of other classes of biomolecules will be worthwhile. Above all, mass resolution needs to be improved. This may turn out to be a multifaceted, difficult problem and it may take some time for a major breakthrough; nonetheless a resolution in the range of about $m/\delta m = 1000$ should be achievable. As is often true in a new and rapidly developing field, the most exciting breakthroughs may, in the end, occur in areas not even anticipated at the time of this writing.

Acknowledgments

The authors would like to thank the Deutsche Forschungsgemeinschaft and the Bundesministerium fuer Forschung und Technologie, represented by the VDI-TZ, for their financial support as well as Finnigan MAT, Bremen, for their technical help. The protein samples for the spectra, reported in this chapter were supplied by Prof. H. Witzel, Department of

[18] F. Hillenkamp, M. Karas, A. Ingendoh, and B. Stahl, in "Biological Mass Spectrometry" (A. L. Burlingame and J. A. McCloskey, eds.), p. 49. Elsevier, Amsterdam.

Biochemistry, University of Münster, the monoclonal antibody by Prof. A. Sorg, Department of Experimental Dermatology, University of Münster. The analysis of polynucleotides was a joint project with Prof. J. A. McCloskey and Dr. P. Crain, Department of Medicinal Chemistry, University of Utah.

[13] Utility of Exact Mass Measurements

By Klaus Biemann

Introduction

High-resolution mass spectrometry and exact mass measurement have been synonymous with the determination of elemental composition of an organic compound or a fragment ion ever since Beynon, in the early 1960s, demonstrated the practical utility of these techniques.[1] By precisely measuring (to about 1–10 ppm) the relative ratio of the accelerating voltages required to focus, at fixed magnetic field, an ion of unknown mass and a reference ion onto the collector slit of a double-focusing mass spectrometer, it is possible to measure the m/z value of the unknown ion with an accuracy that permits the calculation of its elemental composition.

Such measurements serve a variety of purposes: the elucidation of fragmentation pathways of compounds of known structures; the determination of the elemental composition of synthetic compounds in lieu of combustion analysis; the determination of the structure of natural products; etc. The demand for accuracy of the measurement increases in that order. Fragments are obviously of lower mass than those of the molecules from which they are derived. The lower the mass, the smaller the number of atoms and elements present in the ion; therefore, fewer combinations need to be considered at any given nominal mass. At the same time, it is easier to measure the mass to ± 0.001 u for an ion of m/z 100 (which requires 10 ppm accuracy) than at m/z 500 (which requires 2 ppm). Also, at low mass the number of carbon atoms must be relatively small and the abundance of ions containing one ^{13}C is therefore very low. Thus, one need not consider this isotope when computing elemental compositions from accurate mass measurements of abundant ions.

When determining the exact mass of a synthetic compound, further constraints come into play: one has a predicted value, assuming that the proper compound has been synthesized, and if the measured value falls

---

[1] J. H. Beynon, in "Advances in Mass Spectrometry" (R. M. Elliott, ed.), Vol. 2. Pergamon, London, 1962.

within the accepted range (generally within ± 0.003 u in the range below m/z 500), one can be rather certain that the expected compound has been synthesized, keeping in mind, however, that the method cannot differentiate isomers and may be insensitive to the presence of impurities.

The third area, the determination of the previously unknown structure of a compound, is the most demanding: the type of elements present is not known (although the origin and the chemical properties of the substance may provide some hints) and there is no guarantee that the ion being measured is indeed a molecular ion. Thus, combinations of atoms not representing structures with fully paired electrons must also be considered, and the composition of fragment ions must be determined so that they can be used to derive the final structure of the molecule.

Measurements at Low Mass

For 30 years, the vast majority of mass measurements carried out involved compounds of molecular mass below 500 Da, which were ionized using electron ionization (EI). Such measurements are (1) more easily carried out by this ionization method, because it is easy to use an internal mass standard, generally a perfluorinated alkane (perfluorokerosene, PFK), triperfluoroalkyltriazines or phosphazines (see this volume, Appendix 2); (2) as the mass range to be considered increases (± 0.003 u, or ± 6 ppm, at m/z 500 corresponds to 0.006 u; ± 0.005 u, or ± 5 ppm, at m/z 1000 corresponds to 0.01 u), the number of combinations of atoms (and of elements) that fit within this range dramatically increases with mass; and (3) it becomes more and more difficult to ionize larger and larger molecules by EI-MS.

An example illustrating these various problems is the determination of the structure of a highly modified nucleoside isolated from rat liver tRNA.[2] In order to obtain a good signal, the compound had to be trimethylsilylated (TMS); the mass of the M^+ ion was found to be 818.3184. Even though the origin of the compound limited the kind and number of elements to within a certain range (at least 4 oxygens from the furanose, at least 2 nitrogens from the base, and at least 3 silicons from the TMS derivative of the three hydroxyls of the furanose), it was impossible to select a single (or a few) elemental composition from the very large number that would fit within a reasonable error limit, e.g., ± 0.005 at that mass. However, the perdeutero-TMS derivative had a molecular ion at m/z 863 (from a low-resolution spectrum), which indicated the incorporation of 45 deuterium atoms and thus the presence of five TMS groups. Subtracting the calcu-

---

[2] Z. Yamaizumi, S. Nishimura, K. Limburg, M. Raba, H. J. Gross, P. F. Crain, and J. A. McCloskey, *J. Am. Chem. Soc.* **101,** 2224 (1979).

lated mass of these from 818.3184 and adding 5×1.007825 (to restore the five hydrogen atoms of the native compound) gave a value of 458.1203 for the free nucleoside. During permethylation of the nucleoside with CH_3I and CD_3I, a degradation product was formed that exhibited molecular ions of m/z 383.1611 and 398.2558, respectively. This data indicated that this part, which apparently had lost a substituent (side chain?), contains five methyl groups. Again, subtracting the mass of $5 \times CH_2$ from 383.1611 resulted in 313.0831 for the exact mass of this central moiety. Within the constraints for a nucleoside in the number of C, H, N, and O atoms and double bonds or rings, all compositions were outside the acceptable range of error and too high in fractional mass. Only when sulfur was also included in the calculations, a good composition, $C_{16}H_{10}D_{15}N_5O_4S$, was obtained for the perdeuteromethyl product and, again indirectly, $C_{11}H_{15}N_5O_4S$ for the central nucleotide system. The presence of five nitrogen atoms indicates that it must be an adenosine derivative containing a sulfur atom and an additional CH_2 group. Subtracting the exact mass of this composition (less one hydrogen) from 458.1203 (see above) gives 146.0464 for the side chain that replaces a hydrogen on the purine system of the modified nucleoside. Although this indirect value is subject to the combined experimental error of two measurements (the molecular ions of the trimethylsilylated nucleoside and the permethylated degradation product), the low mass greatly restricts the number of elemental compositions for this substituent which must have one free valence. In fact, only $C_5H_8NO_4$ makes sense which, in consideration of the methylation and trimethylsilylation data, best fits a CO-threonyl group. Taking all other mass spectrometric information into account, the structure of this novel nucleoside was determined to be represented by **1**.

1

This example, which combines the consideration of the exact masses of the molecular ions of a number of derivatives with some of their fragment ions, not only illustrates the power of such measurements but also indicates the limitations when dealing with unknown compounds approaching or exceeding molecular mass 500 Da.

Limitations at Higher Mass

Beyond the range of molecular mass of 500 Da, it becomes practically impossible to narrow down the number of elemental compositions falling within an acceptable error to relatively few possibilities, without extensive knowledge of the nature of the compound under consideration. Such is the case when one merely needs this information to differentiate among a few possibilities. An early example is the determination of the elemental composition of the carcinostatic indole alkaloid vinblastine, for which a molecular weight of 810.4219 was established by high-resolution mass spectrometry, using perfluorokerosene as an internal standard and photographic recording with a Mattauch–Herzog-type spectrometer (CEC 21-110B).[3] The compound was known to consist of two covalently linked indole alkaloids. Therefore, it had to contain four nitrogen atoms, in addition to C, H, and O. Since the mass difference between O and CH_4 is 0.0364 mass units, the mass accuracy of ±0.005 was more than sufficient to narrow down the various $C_nH_mN_yO_x$ possibilities to a single one, $C_{46}H_{58}N_4O_9$. This finding was crucial for establishing the structure of this complex alkaloid.[4]

Both the information content of an exact mass measurement and the experimental difficulties in obtaining the data, change dramatically as the mass of the molecule or the fragment increases. First, because of the exponentially increasing number of combinations of atoms and elements that will fit the mass measurement within the also increasing error limits, the unique positive information content vanishes. In other words, under these circumstances an exact mass measurement can only be used to rule out a possible structure (because it does not agree within experimental error) but not to unambiguously prove the correctness of a structure (or, more exactly, of an elemental composition). For this reason, such a measurement is mainly used to support (but not, by itself, prove) a proposed structure or to differentiate between two or more possibilities, if they differ in mass by more than the error range expected under the conditions of the measurement.

For example, in the identification of the posttranslationally modified N terminus of Met-tRNA synthetase (methionine–tRNA ligase) from yeast

---

[3] P. Bommer, W. J. McMurray, and K. Biemann, *J. Am. Chem. Soc.* **86**, 1439 (1964).
[4] N. Neuss, M. Gorman, W. Hargrove, N. J. Cone, K. Biemann, G. Buchi, and R. E. Manning, *J. Am. Chem. Soc.* **86**, 1440 (1964).

a tryptic peptide was found, the fast atom bombardment (FAB) mass spectrum of which indicated a molecular weight $[(M + H)^+ = m/z$ 998] but did not match any of those predicted from the amino acid sequence derived by translation of the open reading frame of the cDNA sequence. However, it did correspond to the N-terminal tryptic peptide, assuming that the initiating Met had been removed and the Ser that follows was acetylated. This plausible assumption was supported by the exact mass of the $(M + H)^+$ ion of this peptide (998.5210), which agreed well with 998.5199 calculated for the $(M + H)^+$ ion of CH_3CO-Ser-Phe-Leu-Ile-Ser-Phe-Asp-Lys and corresponding to positions 2–9 according to the cDNA sequence.[5]

Such mass measurements are even more important in the determination of the structure of peptides and related compounds if they contain unusual amino acids or other building blocks. Exact mass measurements [by EI-MS, as well as field desorption (FD-MS)] revealed the presence of hydroxyisovalerylpropionic acid as a component of didemnin A, a member of a class of antiviral and cytotoxic cyclic depsipeptides.[6] This was concluded from the mass difference (156.0787, $C_8H_{12}O_3$) between the sum of the amino acids known to be present and the exact mass measured for the molecular ion (m/z 942.5678).

One of the largest molecules for which the molecular weight has been measured and for which the highest accuracy was obtained is a tryptic peptide derived from hemoglobin.[7] The value obtained for the monoisotopic $(M + H)^+$ ion was found to be 1734.8005 ($\pm$0.0038). By coincidence, the isotopic cluster of this ion was bracketed by the cluster of an accompanying peptide for which the monoisotopic $(M + H)^+$ ion is of the composition $C_{77}H_{137}N_{24}O_{21}$ (m/z calculated: 1734.0385). This circumstance greatly simplified precise mass measurement, except that it required a resolution of 1 : 10,000 to provide exact peak positions. The measurement was carried out by slowly scanning the magnetic field repeatedly to cover the range from m/z 1500 to 2000, and using the peak positions of the monoisotopic ion and the $^{13}C_1$ and $^{13}C_2$ isotope $(M + H)^+$ peak of the reference peptide to determine the mass of the monoisotopic and $^{13}C_1$ ion of the "unknown" peptide $(M + H)^+$ ion by interpolation.

To put this measurement (1733.7932, after subtraction of one proton) into perspective, it should be noted that there were a total of 28 elemental compositions that fit within the experimental error range, even if one

[5] B. W. Gibson and K. Biemann, *Proc. Natl. Acad. Sci. U.S.A.* **81,** 1956 (1984).

[6] K. L. Rinehart, Jr., J. B. Gloer, and J. C. Cook, *J. Am. Chem. Soc.* **103,** 1857 (1981).

[7] T. Sakurai, T. Matsuo, M. Morris, Y. Fujita, H. Matsuda, and I. Katakuse, *in* "Advances in Mass Spectrometry" (P. Longevialle, ed.), Vol. 11A, p. 220. Heyden & Son, London, 1989.

restricted them to values for C (62–89), H (93–144), N (15–31), O (15–34), S (0–2), and 19–44 double bonds or rings, parameters logical for a peptide in this mass range. The known composition of this "unknown," $C_{74}H_{115}N_{19}O_{27}S$ (calculated mass: 1733.7931), was, of course, one of these 28 but it would have required an accuracy of ±0.0001 u to make it a unique solution. This would be impossible even under these rather ideal conditions, namely, a set of reference ions differing only by a few tenths of a mass unit from the isotope cluster of the unknown, and a concentration ratio of the two peptides to give about equal ion abundances.

From the foregoing, it is quite clear that exact mass measurements with an accuracy of a few millimass units are very useful to establish the elemental composition of an ion at $m/z < 500$ and that this can be extended to about m/z 1000 if part of the molecule is known and thus can be subtracted to give an indirect measurement of the mass of the much smaller remainder. Beyond m/z 1000, exact mass measurements are of little practical use, considering the effort and amount of material required. The most likely application would be the differentiation among two or a few possible structures if they differ sufficiently in mass. A hypothetical case would be a peptide for which the sequence has been determined mass spectrometrically, but where Lys could not be distinguished from Gln. Since these two amino acids differ by 0.0364 u, a mass measurement accurate to ±0.010 u would be adequate to determine the number of lysines and glutamines present. The practitioner will, however, almost always prefer to acetylate the peptide instead and deduce the number of lysines from the increase in molecular weight by 42 u for each Lys (in addition to the acetylation of a free N terminus). Similarly, if one needs to confirm the identity of a peptide (as in the identification of the N-terminal peptide of Met-tRNA synthetase discussed earlier), a collision-induced dissociation (CID) spectrum, even if only a linked scan with a two-sector instrument, would give sufficient sequence information to support or reject the assumed structure.

However, above m/z 1000 there arises an entirely different reason for determining the mass of an ion with reasonable accuracy, namely, within a few tenths of a mass unit. At low mass (<500) it is common practice to use the "nominal" mass (corresponding to the sum of integer atomic weights), i.e., the nearest integer of the observed value. Once the ion contains more than about 65 hydrogen atoms (or more if a mass deficient element is present), this is no longer true and the nominal mass becomes difficult to determine. This is clearly shown by one of the examples discussed above. The two $(M + H)^+$ ions of m/z 1734.0385 and 1734.8005 would appear to both correspond to nominal mass 1734 if the integer were used (by dropping the fractional mass), or 1734 and 1735, respectively, if rounded to the nearest integral mass. Neither is correct because the sums of the integer weights of the atoms present in the $(M + H)^+$ ions are 1733 and 1734, respectively, a discrepancy which is chiefly due to the large number of hydrogens present in these ions.

Above mass 1000, the nominal mass value, therefore, is meaningless and the determination of the exact mass rapidly becomes very difficult to carry out; the usefulness of the result also decreases rapidly. There is, however, an important feature of a double-focusing mass spectrometer: medium resolving power at relatively high ion transmission permits the determination of the mass of an ion to within ±0.1 u, if checked with a calibration standard (usually CsI clusters) before and/or after the unknown. Even without that check, an initially well-calibrated instrument routinely gives mass assignments within ±0.3 u at least up to m/z 5000 if there is sufficient signal. This is important because it provides unambiguous unit mass assignment and thus allows differentiation of ions differing by one mass unit, such as a peptide in which one asparagine has been deamidated to aspartic acid. It should be noted that at m/z 3000, 0.3 u represents an accuracy of 100 ppm while 0.1 u corresponds to 33 ppm.

Other examples of the advantage of knowing the mass of the fragment ion with a reliability of 0.1 to 0.3 u are molecules that consist of structural elements differing widely in elemental composition, such as glycolipids. Because of the great difference in the number of hydrogens in the lipid portion compared with the carbohydrate moiety, a high fractional mass is indicative of the former and a low fractional mass of the latter.

Monoisotopic vs. Average Mass

The other extreme of exact mass measurement should be mentioned briefly, namely, the determination of the "average mass" of an isotopic cluster recorded at such low resolution (1 : 500) that the cluster is completely unresolved. Since the value obtained is the average of the isotopic masses of each element present, weighted by the relative abundance of the isotopes, it corresponds to the sum of the "chemical" atomic weights. This parameter is generally measured for compounds of high molecular weight, where it is difficult to obtain the high ion current needed to produce a well-shaped peak with the narrow slits required to achieve unit resolution.

While it would seem that mass measurement using unit resolution is always preferable, another practical problem arises. This is the need to correctly identify the $(M + H)^+$ ion in the isotopic cluster, which is very complex at high mass because of the presence of a large number of carbon atoms and the resulting high probability that an ion contains a number of ^{13}C atoms (the presence of other polyisotopic elements further aggravates

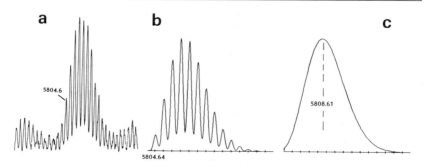

FIG. 1. (a) Recording of the $(M+H)^+$ ion region of the FAB mass spectrum of human insulin at near unit resolution; (b) theoretical distribution of the isotope cluster corresponding to the elemental composition of protonated human insulin $(C_{257}H_{384}N_{65}O_{77}S_6)$ at resolution $1:6000$ (10% valley definition), monoisotopic mass $= 5804.6455$ (theoretical); (c) unresolved cluster, average mass $= 5808.61$ (theoretical); dashed line indicates centroid.

the situation). This problem is illustrated in Fig. 1. It is not easy to identify the ^{12}C species from the molecular ion cluster of human insulin $(M + H = C_{257}H_{384}N_{65}O_{77}S_6)$ produced by FAB ionization in a large double-focusing mass spectrometer (Fig. 1a), even when comparing it with the theoretical cluster (Fig. 1b) which is, of course, possible only if one knows the elemental composition, at least approximately. This disadvantage disappears if one degrades the resolution to obtain a single, smooth envelope (Fig. 1c). It has to be kept in mind, however, that the average mass obtained with that measurement is sensitive to the presence of a significant contribution of M^+ ions (which sometimes are quite abundant when m-nitrobenzyl alcohol is used as the FAB matrix) or partial reduction of —S—S— bonds, possibly caused by the use of SH-containing matrices [thioglycerol or dithiothreitol (DTT)] and prolonged bombardment with the primary atom or ion beam. The former case leads to an erroneously low average molecular weight, while the latter situation increases it.

The difference between monoisotopic and average mass increases with increasing mass. The difference is greater for molecules containing a larger number of atoms with a large mass excess (such as 1.007825 for hydrogen) or a relatively abundant heavy isotope. Therefore, the magnitude of the difference at a given molecular mass is characteristic for the compound type. It is largest for peptides and proteins because of their high C + H content, less for carbohydrates because of the high oxygen content, and least for oligonucleotides because of their low C + H content and the large number of P atoms, an element which is monoisotopic as well as mass deficient.[8]

[8] S. C. Pomerantz and J. A. McCloskey, *Org. Mass Spectrom.* **22,** 251 (1987).

Experimental Procedures

Exact mass measurements are most simply carried out by subjecting a mixture of the sample and an appropriate reference compound (see this volume, [14])[9] to EI-MS. The most precise method suitable for determining the mass of one or a few ions in the spectrum of the unknown is the determination of the ratio of the acceleration voltages required to alternately focus a reference ion beam and the ion to be measured on the collector slit. The beam profiles are displayed on an oscilloscope screen and the voltage ratio varied until they are superimposed (hence the term "peak matching"). This ratio, which equals the ratios of the two masses, is set with a voltage divider accurate to $1 : 10^6$. The reference ion should be within 10%, preferably 2%, of the mass of the ion to be measured. Long-chain perfluorinated compounds (see [14] and Appendix 2 in this volume) fragment well by EI-MS and provide reference ions over a wide mass range.

The alternative to peak matching is to record the entire spectrum of the mixture and use a computer to recognize the ions from the reference compound; their known masses (Appendix 2) are used to compute the m/z values of the ions generated by the unknown. The most reliable and accurate method used photographic plates and focal plane mass spectrometers (Mattauch–Herzog geometry) for the recording of the spectra and a microdensitometer for the precise meaurement of line positions.[10] The data for the structure determination of the modified nucleoside (1) discussed earlier in this chapter was recorded in this manner. Such mass spectrometers are no longer commercially available and now the spectra are recorded by scanning the magnet of the spectrometer and digitizing the signal for further processing by a computer. Over short mass ranges, scanning the accelerating voltage at fixed magnetic field usually gives more accurate mass values.

These methods work well and are generally applicable if the sample is sufficiently volatile to use EI-MS. For larger and more polar molecules, field desorption (FD)-MS or fast atom bombardment (FAB)-MS has to be used. With these ionization methods it is more difficult to produce simultaneous reference and sample ion beams, which are relatively close in mass.

For FD-MS this is further complicated by the fact that both compounds would have to ionize at the same emitter current (anode temperature) to allow either peak matching or simultaneous recording. One alternative is the use of a reference compound (such as PFK, or the fluoroalkyltriazines

[9] R. M. Milberg, this volume [14].
[10] K. Biemann, *Pure Appl. Chem.* **9,** 95 (1964).

and phosphazines) that is sufficiently volatile to be vaporized into the FD ion source and thus field-ionizes independent of anode temperature.

In the case of FAB ionization, different problems have to be overcome. The major one is the lack of extensive fragmentation which requires the use of molecular ions or cluster ions as mass standards. The simplest are the clusters of the matrix itself and these are quite useful if sufficiently close in mass and the signals are intense. Sometimes one may choose a particular matrix based on this consideration.

More widely applicable are salt clusters, either CsI or a mixture of alkali halides (see this volume, Appendix 2). The difficulties with these are that salts very efficiently suppress the ionization of organic molecules by FAB and thus cannot be used as a mixture of reference and unknown. This incompatibility can be overcome by recording the spectra of the reference and the sample one after the other, while avoiding any changes of the focusing conditions, electrical potentials, and magnetic field between the two scans. Exact mass values are obtained either by merging the two data files and processing the resulting "simultaneous" spectrum as described above for the EI mode, or calculating the mass values for each spectrum based on precalibration of the mass spectrometer and correcting the values obtained for the unknown by the deviations found for the spectrum of the reference material.[11] Alternatively, the reference ions can be brought close to the sample ions (but not overlapping) using the peak-matching circuit and taking the accelerating voltage ratios into account when calculating the mass of the unknown.[12]

To avoid removing the sample probe and reinserting it, various dual probes have been developed that keep the two samples physically separated but allow alternate positioning into the primary beam, either by rotation (180°)[13] or lateral movement of a suitably configured probe tip consisting of two parallel, but separate, segments. Care has to be taken that reference and sample do not mix by "creeping" or sputtering.

Internal standards more suitable for mixing with the sample are polyethylene or polypropylene glycols (this volume, Appendix 2), which consist of mixtures of oligomers covering relatively wide mass ranges, or their alkylphenyl-substituted analogs, which extend to higher mass ($\sim$2100).[14] They can be added to the matrix and, in general, rarely interfere with the ionization of the compound to be analyzed. The main difficulty lies in achieving a strong enough signal from the reference compound, because

[11] C. Fenselau, J. Yergey, and D. Heller, *Int. J. Mass Spectrom. Ion Phys.* **53**, 5 (1983).
[12] M. Barber, R. S. Bordoli, G. J. Elliott, A. N. Tyler, J. C. Bill, and B. N. Green, *Biomed. Mass Spectrom.* **11**, 182 (1984).
[13] J. M. Gilliam, P. W. Landis, and J. L. Occolowitz, *Anal. Chem.* **55**, 1531 (1983).
[14] M. M. Siegel, R. Tsao, S. Oppenheimer, and T. T. Chang, *Anal. Chem.* **62**, 322 (1990).

it is a complex mixture and each component represents a very small amount. For this reason, it is advantageous to prepare simplified cuts by vacuum distillation or high-performance liquid chromatography (HPLC) fractionation of commercially available mixtures of oligomers. Because these nonbasic substances preferably form cationized rather than protonated molecules, a small amount of alkali halide (i.e., KCl) is added to the matrix.

The most practical and probably the most accurate method is to mix the unknown with a compound of the same chemical characteristics (i.e., a hydrophobic peptide, if the unknown is a peptide that is likely to be hydrophobic) and which is of closely related mass (within <10%). In this case, peak matching is the most convenient and accurate mode of measurement, although scanning (preferably electrically) can also be used. This approach requires the availability of a relatively large collection of reference compounds. A minimal set of peptides is listed in this volume, Appendix 2.

Any mass measurement involving a sample dissolved in a matrix, as is necessary for FAB-MS, must be carried out with a resolution setting that resolves the sample ion to be measured from matrix clusters at the same nominal mass. If this precaution is not taken, the result will be the average of the two contributions, weighted by the relative abundances of the two ion beams.

[14] Selection and Use of Mass Reference Standards

By Richard M. Milberg

Reference compounds are required for data system and instrument calibration, high-resolution mass measurements, instrument tuning, and sensitivity/resolution performance checks. A reference compound should be readily available, should provide a series of regularly spaced peaks of good intensity over the desired mass range in the selected ionization mode, and should be volatile enough to be pumped away or otherwise easily removed from the ion source. Exact mass values of reference ions are given in Appendix 2 at the end of this volume.

Electron Ionization

Perfluorokerosene (PFK) is a mixture of perfluoronated alkanes which is widely employed in electron ionization (EI) mass spectrometry and is

METHODS IN ENZYMOLOGY, VOL. 193

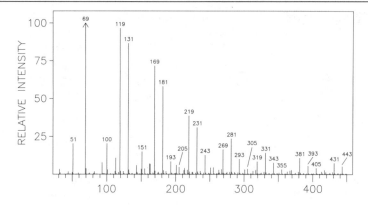

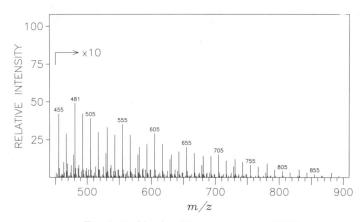

FIG. 1. Positive ion EI mass spectrum of PFK.

available in low (PFK-L) and high (PFK-H) boiling mixtures.[1] PFK-L has ions useful to about 600 mass units and PFK-H (Fig. 1) is useful to about 900 mass units. PFK can be introduced through a heated inlet system or from a direct insertion probe. Absorbing the PFK on a small amount of charcoal in the direct-probe crucible gives better control over the rate of volatilization.

Perfluorotributylamine[1] (molecular weight 671, also known as PFTBA, FC-43, or heptacosa) is frequently used for EI (Fig. 2). PFTBA is volatile enough to be introduced without a heated inlet system and has relatively more intense peaks at the high mass end than PFK. The ratio

[1] Available from PCR Incorporated, P.O. Box 1466, Gainesville, FL 32602.

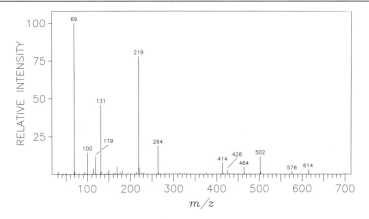

FIG. 2. Positive ion EI mass spectrum of perfluorotributylamine.

of m/z 69 to m/z 502 is often used as a measure of tuning in quadrupole instruments.

Negative Ion Electron Ionization

PFK is useful for negative ion EI calibration.[2] It may be necessary to pretune the instrument on an abundant Cl^- m/z 35 peak from some compound such as hexachlorobutadiene introduced through the reference inlet because negative ion tuning can be very different from positive ion tuning. A separate negative ion PFK reference table is needed for data system calibration since the negative ion PFK spectrum is very different from the positive one.

High-Mass Electron Ionization

High-mass compounds providing reference peaks to over 3000 mass units, such as the triazines[3] (Fig. 3) and the Ultramark[4] compounds, are readily available.[1] The Ultramark F series are Fomblin oil mixtures (perfluoro polyethers) and Ultramark 1621 is a fluorinated phosphazine compound. These can be difficult to pump away and are best introduced by direct probe where the temperature can be controlled and the unused compound removed.

Ultramark 2500F is routinely employed in this laboratory to calibrate the VG 11250J data system/70-SE4F mass spectrometer from m/z 30 to

[2] R. S. Gohlke and C. H. Thompson, *Anal. Chem.* **40**, 1004 (1968).
[3] T. Aczel, *Anal. Chem.* **40**, 1917 (1968).
[4] W. V. Ligon, Jr., *Anal. Chem.* **50**, 1228 (1978).

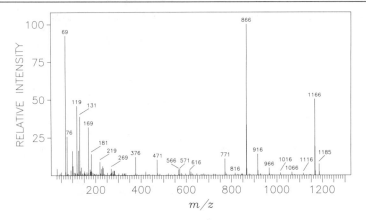

FIG. 3. Positive ion EI mass spectrum of tris(perfluoroheptyl)-s-triazine.

m/z 2000 for probe and gas chromatography/mass spectrometry (GC/MS) work. The 2500F is introduced by direct probe and pumps away quickly from the ion source.

The difference in intensities between the high- and low-mass peaks in the Ultramark and triazine reference compounds can make data system calibration difficult. A manual calibration, where several reference peaks are identified manually, may have to be performed and the data system calibration parameters altered. The reference table relative intensity values may have to be adjusted to match a particular batch of compound due to variations in the components which constitute the mixture.

PFK is sometimes used in conjunction with a second reference compound, such as a triazine, to calibrate over a wide mass range when the peaks from other reference compounds are too far apart at lower mass values. This is necessary for scanning high-resolution GC/MS measurements where it is important that the reference peaks be closely spaced and on-scale. Triazines cannot be mixed with the PFK because of differing volatility; the PFK is introduced through the reference inlet and the triazine by direct sample probe. The temperature of the direct probe is adjusted to balance the intensities of the PFK peaks and high-mass peaks of the other compound. A table must be created which combines the peaks of the two reference compounds.

An alternative method of high-mass calibration is to install a fast atom bombardment (FAB) ion source and calibrate with CsI over the desired mass range. The EI source is reinstalled (except in the case of combined EI-FAB sources), the sample run, and the data mass converted employing the FAB calibration file with a background lock-mass peak in the EI spectra. This works quite well on the VG mass spectrometers in this laboratory.

Chemical Ionization

Fluorine-containing compounds work well for chemical ionization (CI) calibration. PFK can be used in positive ion CI employing methane as the reagent gas. The methane CI spectrum of PFK is similar to that in EI except that the peaks at higher mass are relatively more intense in CI than EI. Depending on the data system calibration procedure, the same PFK table may be used for EI and CI. Other compounds such as FC-70 and FC-71[5] can also be employed for positive ion CI. The triazines are not very useful in CI since many of the peaks in EI are not present in the CI mode. The Ultramark F series give similar EI and CI spectra except that the peaks at higher mass are relatively more intense in CI than EI. The same reference table can be employed in CI as in EI over most of the mass range.

Negative Ion Chemical Ionization

PFK as well as PFTBA, FC-70, and FC-71 works quite well in negative ion CI using methane reagent gas. Negative ion CI reference tables must be used because the negative ion spectra are quite different from the positive ones. Perfluoronated compounds have few useful ions below m/z 150 in negative ion CI but this is usually not a problem. Ion source cleanliness is very important in negative ion CI because processes that complete for electrons can reduce the intensity of, or eliminate, the reference peaks. It may be necessary to pretune on an abundant Cl^- ion (as previously mentioned in the section on negative ion EI) or on the F^- m/z 19 peak from the reference compound.

Field Ionization

Field ionization (FI) produces mainly molecular ions with very little fragmentation. PFK tends to irreversibly degrade the sensitivity of the FI emitter wire unless it is maintained at an elevated heating current. Usually one calibrates in EI using PFK or other reference compounds and employs this for FI. Instruments with FI/FD (field desorption)/FAB combination ion sources can be calibrated in FAB using CsI deposited on a FAB target

---

[5] Available from Scientific Instrument Services, R.D.#3, Box 593, Ringoes, NJ 08551.

used in place of the FI emitter or on a special direct-probe tip and this calibration then employed for FI. The m/z 58 peak of acetone, introduced through the reference inlet and used for tuning purposes in FI/FD, can be employed as a lock mass for mass conversion.

High-Performance Liquid Chromatography–Mass Spectrometry (HPLC/MS)

Polyethylene glycol (PEG) mixtures are employed for calibration of thermospray and other similar interfaces. Usually a 1% solution of a PEG is employed through the interface by postcolumn addition.

Desorption Ionization Methods

Many compounds are sufficiently polar or have high molecular weights and cannot be introduced into the gas phase directly for EI, CI, or FI ionization but work by various desorption techniques.

Field Desorption

One generally does not calibrate directly in the FD mode since most compounds produce mainly M^+ or $(M + H)^+$ ions with little fragmentation. Polystyrene and polypropylene glycol mixtures have been employed and produce FD spectra with a series of ions 104 and 44 u apart, respectively.[6] In those instruments with combination EI/FI/FD ion sources the calibration is performed in the EI mode and this calibration used for FD. Many newer mass spectrometers have combination FI/FD/FAB ion sources. Calibration is performed in the FAB mode using CsI introduced on a FAB target in place of the FD emitter or from the direct-sample probe with the salt deposited onto the end of a tab. Calibration in the FAB mode using CsI is quite adequate for use in FI or FD out to at least 15,000 amu.

High-Resolution Field Desorption

It is difficult to find suitable FD reference compounds that will desorb at the same emitter current as a given sample for high-resolution measurements. High-resolution FD measurements are usually performed with the reference compound ionized in FI employing signal averaging of electrical scans over a narrow mass range or by peak matching. We have employed

[6] T. Matsuo, H. Matsuda, and I. Katakuse, *Anal. Chem.* **51,** 1329 (1979).

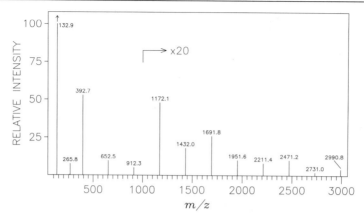

FIG. 4. Positive ion FAB mass spectrum of CsI.

readily available volatile compounds that give intense molecular ions in FI as well as the triazines and phosphazines.[7] Occasionally a suitable FI reference peak is not available, in which case the reference material can be PFK or some other compound ionized in the EI mode in instruments in combination with EI/FI/FD ion sources. The FD probe must be withdrawn slightly so that the emitter is not damaged by the ionizing electrons in the EI mode. The reference signal is recorded by signal averaging in the EI mode and the sample peak is run in the FD mode and the two spectra combined.

*Fast Atom Bombardment and Liquid Secondary
   Ion Mass Spectrometry*

Saturated aqueous solutions (CsI in 2 ml of water in a vial with an excess of the salt on the bottom) of CsI provide positive and negative ion fast atom bombardment (FAB) mass spectra consisting of $(Cs_nI_{n-1})^+$ (Fig. 4) or $(Cs_{n-1}I_n)^-$ ions. Clusters can be observed beyond m/z 40,000. High-purity (99.9% or better) CsI should be used otherwise cloudy solutions will result and the CsI spectrum will have impurity peaks in it which can confuse the data system. A minimum amount of CsI should always be used on the FAB target and it is important to use a fresh coating of CsI on the target each time. The glycerol matrix peaks can also be used for calibration for lower mass work (<1200 mass units).

[7] K. L. Olson, K. L. Rinehart, Jr., and J. C. Cook, Jr., *Biomed. Mass. Spectrom.* **4**, 284 (1977).

Mixtures of cesium, rubidium, and sodium halide salts (see Appendix 2, this volume) have been employed for calibration with mass spectrometer data systems to fill in the gaps of the CsI peaks at lower masses.

Negative Ion Fast Atom Bombardment. CsI works quite well for negative ion calibration in FAB. The tuning for negative ion FAB will be different than that for positive ion in most mass spectrometers. A 2:1 mixture of CsI in water (260 mg/ml) and glycerol has also been used for negative ion FAB calibration.[8]

High-Resolution Fast Atom Bombardment (HRFAB). CsI is not generally used for high-resolution accurate mass measurements because the peaks are too far apart. Glycerol matrix peaks, if close enough to the unknown peak, are frequently used as peak matching standards. In this laboratory, a set of standard reference compounds is maintained for use in peak matching HRFAB. These were samples originally submitted for FAB analysis that gave strong molecular ion peaks. These compounds are kept in solution in the freezer at a concentration of 10 μg/μl and 1 μl is added to the target containing the matrix and sample. Since the supply of many of these compounds has been exhausted, we employ (see Appendix 2, this volume) commercially available peptides for use as standards.

Polyethylene glycols (PEG 300, 600, 1000, and 1450) and PEG 750 monomethyl ether (available from Aldrich Chemical Co., Milwaukee, WI, and Sigma Chemical Co., St. Louis, MO) are also employed for HRFAB employing electrical or magnetic scanning over a limited mass range with signal averaging.[9] The different PEG mixtures cover the mass range from m/z 63 to m/z 2036 with peaks 44 mass units apart. The desired PEG at a concentration of 1 to 2% is dissolved in the liquid FAB matrix. One complication with PEG is that, depending on impurities in the sample and which matrix is employed, the PEG peaks can appear either as PEG + H^+, PEG + NH_4^+ or PEG + Na^+ species (PEG in thioglycerol tends to give mainly PEG + NH_4^+ peaks). Separate reference tables must be maintained and the spectra examined manually to determine which species of PEG peaks are present.

Tandem Mass Spectrometry

Calibration of the second analyzer in tandem mass spectrometry (MS/MS) employing two magnetic sector instruments generates a linked scan of the magnetic field (B) and electric section (E) voltage such that the

---

[8] K. Sato, T. Asada, M. Ishihara, F. Kumihiro, Y. Kammei, E. Kubota, C. E. Costello, S. A. Martin, H. A. Scoble, and K. Biemann, *Anal. Chem.* **59,** 1652 (1987).

[9] B. N. Green, Application Note No. 23, "Accurate Mass Determination by FAB/LSIMS," and accompanying table of PEG masses, VG Analytical Ltd., Floates Road, Wythenshawe, Manchester M23 9LE, U.K.

TABLE I
COMMONLY USED REFERENCE COMPOUNDS

Compound	Upper mass limit	Ionization mode
Perfluorotributylamine (FC-43)	600	EI/CI
Perfluorotripentylamine (FC-70)	750	EI/CI
Perfluorotrihexylamine (FC-71)	920	EI/CI
PFK-L (perfluoronated alkane mixtures)	600	EI/CI
PFK-H (perfluoronated alkane mixtures)	900	EI/CI
Ultramark F series (Fomblin oil fractions, perfluoro polyethers)		
1600F	1600	EI/CI/FAB
1960F	1960	EI/CI/FAB
2500F	2500	EI/CI/FAB
3200F	3200	EI/CI/FAB
Ultramark 1621 (fluorinated phosphazine)	1600	EI/CI
Tris(perfluoroheptyl)-s-triazine	1185	EI/FI
Tris(perfluorononyl)-s-triazine	1485	EI/FI
CsI	>25,000	FAB
CsI/NaI/RbI	>3,000	FAB
Glycerol/CsI	>3,000	FAB negative ion
CsI/NaI/LiI/KI/RbI/LiF	>3,000	FAB-MS/MS
CsF/LiF/NaI	>3,000	FAB-MS/MS negative ion
Polyethylene glycol (300, 600, 1000, and 1450) and polyethylene glycol 750 monomethyl ether	>2,000	FAB/thermospray/ electrospray
Polystyrene mixtures	>10,000	FD
Polypropylene glycol mixtures	>2,000	FD

ratio B/E remains constant. If the gas collision cell in our VG 70-SE4F spectrometer is operated at ground potential with a precursor ion at m/z 1000, we then have to scan to m/z 6 in order to produce a linked scan for daughter ions calibrated down to m/z 75.[10] This requires that a mixture of salts be used, including Li (with isotopes at ^{6}Li and ^{7}Li). Other ions must be added to fill the gaps between the CsI peaks at lower masses to provide a correct B/E-linked scan and accurate daughter ion mass assignments. We employ a mixture of CsI (0.45 g), NaI (0.7 g), LiI (0.2 g), KI (0.02 g), RbI (0.05 g), and LiF (0.2 g) dissolved in enough water (1–2 ml) to make a saturated solution for calibration in positive ion MS/MS. This mixture is used in the calibration-only second FAB source between MS-1 and MS-2. We use a saturated solution mixture of CsF, LiF, and NaI for negative ion MS/MS calibration. Glycerol/CsI has also been employed for negative ion MS/MS calibration purposes.[8]

[10] R. K. Boyd, *Int. J. Mass Spectrom. Ion. Proc.* **75**, 243 (1987).

We have found that these salt mixtures age and their FAB spectra will change over time. It is, therefore, necessary to adjust the salt concentrations and the reference table when a new calibration mixture is made.

Conclusions

There are wide varieties of reference compounds available for use in mass spectrometry and it is impossible in this brief space to give all the details of their use. Table I summarizes the commonly used reference compounds and Appendix 2 at the end of the volume includes tables with accurate masses of reference compound peaks.

[15] Chemical Derivatization for Mass Spectrometry

By DANIEL R. KNAPP

Introduction

Chemical derivatization has played an imporant role in organic mass spectrometry from the earliest practice of the technique.[1] Although the major impetus for its early use was to confer the necessary volatility to sample compounds, chemical derivatization was also used to increase the information yield from mass spectral data. With decades of progress in mass spectrometry and the development of a wide range of new techniques, chemical derivatization still plays an important role. This chapter will give an overview of the use of chemical derivatization in mass spectrometry with particular emphasis on analytical strategy aspects. General considerations with respect to the practical aspects of chemical derivatizations and some general methods will also be described. More specific applications are given in later chapters in this volume, as well as in a recent review[2] and monograph.[3]

[1] K. Biemann, "Mass Spectrometry, Organic Chemical Applications." McGraw-Hill, New York, 1962.
[2] R. J. Anderegg, *Mass Spectrom. Rev.* **7,** 395 (1988).
[3] D. R. Knapp, "Handbook of Analytical Derivatization Reactions." Wiley, New York, 1979.

METHODS IN ENZYMOLOGY, VOL. 193

Reasons for Use of Chemical Derivatization in Mass Spectrometry

The reasons for utilizing chemical derivatization in relation to mass spectrometry can be summarized as follows: (1) enhancement of volatility; (2) degradation of the sample molecule to smaller subunits; (3) enhancement of detectability; (4) enhancement of separability; (5) modification of fragmentation: (a) enhancement of molecular weight related ions and (b) enhancement of structurally informative ions; (6) determination of functional groups.

Volatility enhancement was a requirement for many samples in the early days of mass spectrometry when the available inlet systems required significant vapor pressure. In current practice, derivatization for volatility enhancement is used primarily in relation to interfaced gas chromatography–mass spectrometry (GC/MS) where such derivatization serves to enable or improve the gas chromatography. Vapor pressure of a compound is influenced by intermolecular attractions due to dispersion (van der Waals) forces, ionic interactions, and hydrogen bonds. The total dispersion forces increase with molecular size and little can be done with respect to these forces to increase volatility other than reduce the size of the molecule (see below). Chemical derivatization for volatility enhancement is aimed at reducing ionic and hydrogen bond interactions by conversion of ionizable groups to nonionizable derivatives (e.g., carboxyl groups to esters); replacing hydrogens bound to heteroatoms (N–H, O–H, S–H) with alkyl, acyl, silyl, or other groups; and reducing the polarity of hydrogen bond accepting groups (e.g., conversion of carbonyls to methoximes). An example of a multistep derivatization to increase volatility is the conversion of peptides (**I**) to the volatile polyamino alcohol trimethylsilyl ethers (**II**).[4] Not only are these derivatives sufficiently volatile for GC/ MS, but they also have the added advantage of yielding stable, structurally informative ammonium-type ions in the mass spectrum. Thus, these derivatives (as is often the objective) serve a dual purpose.

$$\underset{\substack{\text{peptide} \\ \textbf{(I)}}}{H_2N-(CH-C)_n-OH} \xrightarrow[\text{3. trimethylsilylation}]{\substack{\text{1. acylation (R'CO-)} \\ \text{2. reduction}}} \underset{\substack{\text{trimethylsilyl (TMS)} \\ \text{polyamino alcohol} \\ \textbf{(II)}}}{R'CH_2NH-(CH-CH_2)_n-OTMS}$$

Degradation of sample molecules by chemical means was originally employed for volatility consideration, i.e., large molecules were degraded to smaller constituent molecules with sufficient vapor pressure to be ana-

---

[4] K. Biemann, this volume [18].

lyzed. In current practice, degradations are still used to identify constituent subunits, but molecular size reductions are driven more by instrument mass range limits than volatility. In the case of tandem mass spectrometry, the need for degradation to smaller size fragments may be motivated by molecular size limits of current ion dissociation technology rather than the analyzer mass range. An example of a contemporary use of degradation is the acetolysis of glycoproteins to cleave off the oligosaccharide components as peracetylated carbohydrates[5]:

$$\text{Glycoprotein} \xrightarrow{\text{acetolysis}} \text{Peracetylated oligosaccharides}$$

Derivatization for *detectability enhancement* involves chemical conversions to promote the formation of stable charged species with either little fragmentation or characteristic fragmentation behavior to yield intense well-defined ions. An example is the attachment of an electrophoric moiety to promote the formation of negative ions in "negative ion chemical ionization" (NICI) mass spectrometry. This approach has been used, in particular, for eicosanoids for high sensitivity quantitation.[6] For example, prostaglandin E_2 **(III)** is converted in three steps to the methoxime (MO), trimethylsilyl (TMS), pentafluorobenzyl (PFB) derivative **(IV)** which exhibits good gas chromatographic behavior. Under NICI conditions the derivative captures an electron to yield a negatively charged molecular radical anion which readily loses a pentafluorobenzyl radical to form a relatively stable carboxylate anion. The resulting mass spectrum exhibits an intense $(M - 181)^-$ peak which can be monitored by GC/MS selected-ion monitoring for very sensitive, and selective, detection and quantitation of eicosanoids in biological samples.

Prostaglandin E_2

(III)

MO-PFB-TMS Derivative

(IV)

1. methoximation
2. pentafluorobenzyl esterification
3. silylation

The advent of fast atom bombardment (FAB) ionization[7] has led to the use of chemical derivatization for detectability enhancement by the

[5] A. Dell, this volume [35].

[6] I. Blair, this series, Vol. 187, p. 13.

[7] In this chapter, the term, FAB, is used generically to include ion bombardment of samples in liquid matrices.

intentional introduction of a charged group into the molecule, which is just the opposite of the traditional masking of polar groups. For example, the ketosteroid androsterone **(V)**, which is undetectable under normal FAB conditions (glycerol, 4 μg/μl) in the native state, gives an intense molecular ion as the quaternary ammonium derivative **(VII)** resulting from derivatization with Girard's reagent T **(VI)**.[8]

androsterone   quaternary ammonium derivative

(V)   (VII)

Derivatization can be used in conjunction with tandem mass spectrometry (MS/MS) to detect specific compound classes in complex mixtures. Collision-induced dissociation (CID) of molecular ions from phenols (e.g., **VIII**) is relatively inefficient, but the carbamate derivatives (e.g., **IX**) prepared with methyl isocyanate undergo ready dissociation with the characteristic neutral loss of methyl isocyanate (57 mass units) from the protonated molecular **(X)** ion formed under chemical ionization (CIMS) conditions. Thus an MS/MS neutral loss scan can be used to detect phenols in complex mixtures.[9]

(VIII)   (IX)   (X)   (XI)

When mass spectrometry is interfaced to a separation technique (e.g., gas chromatography, high-performance liquid chromatography, capillary zone electrophoresis) derivatization may be used for *enhancement of separability* of the compounds of interest or to separate compounds of interest from other compounds present that interfere with the analysis. For example, if methyl ester derivatives are inadequately separated in a GC/MS analysis, other higher alkyl esters can be examined. Trimethylsilyl

[8] G. C. DiDonato and K. L. Busch, *Biomed. Mass Spectrom.* **12,** 364 (1985).
[9] D. F. Hunt, J. Shabanowitz, T. M. Harvey, and M. L. Coates, *J. Chromatogr.* **271,** 93 (1983).

derivatives have been replaced with other alkylsilyl derivatives (e.g., iso-propyldimethylsilyl) to facilitate the chromatographic separation in the simultaneous analysis of multiple eicosanoids by GC/MS.[10]

Derivatization for *modification of the fragmentation* behavior of mole-cules in the mass spectrometer is a powerful tool for structure elucidation. Such derivatization can be used to obtain molecular weight information or structural information. Compounds which fragment extensively and give low abundance molecular ions can be converted to derivatives which give more stable molecular ions by virtue of introduction of a charge-stabilizing center. Alternatively, derivatives can be formed which fragment readily in a well-defined manner to give an abundant fragment ion charac-teristic of the molecular weight. An example of the latter is the *tert*-butyldimethylsilyl (TBDMS) derivative (e.g., **XIII**) which readily loses a *tert*-butyl radical from the molecular ion to give a very abundant $(M - 57)^+$ ion **(XIV)**. This fragment ion retains the entire sample molecule structure and thus is indicative of the molecular weight.

$$
\underset{\text{(XII)}}{\text{R-OH}} \xrightarrow{\text{derivatization}} \underset{\substack{\text{TBDMS derivative}\\ \text{(XIII)}}}{\overset{\substack{\text{CH}_3\ \ \text{CH}_3\\|\ \ \ \ |}}{\underset{\substack{|\ \ \ \ |\\ \text{CH}_3\ \ \text{CH}_3}}{\text{R-O-Si-C-CH}_3}}} \xrightarrow{\text{MS}} \underset{\substack{\text{(M-57)}^+\\ \text{(XIV)}}}{\overset{\substack{\text{CH}_3\\+\ |}}{\underset{\substack{|\\ \text{CH}_3}}{\text{R-O=Si}}}}
$$

Fragmentation modifying derivatives are also used extensively to *en-hance* the formation of *structurally informative ions*. A classic problem is the identification of double-bond positions and branching points in long-chain hydrocarbon groups. These chains fragment extensively with no position significantly favored for charge retention. A large number of derivatives have been examined for identification of double-bond posi-tions, with two general approaches. One approach has been to carry out a chemical conversion on the double bond itself to introduce charge-stabilizing, fragmentation directing groups. The second approach which is also useful for localization of branching and other substituents has been to introduce a charge-stabilizing group at the end of the chain. The pyridine ring appears to be the best group identified so far for this purpose. For fatty alcohols **(XV)** it is introduced as the nicotinic ester derivative **(XVI)**[11] and for fatty acids **(XVII)** as the picolinyl ester **(XVIII)**.[12] These deriva-tives give a series of ions due to cleavage at each carbon position allowing

[10] H. Miyazaki, M. Ishibashi, K. Yamashita, Y. Nishikawa, and M. Katori, *Biomed. Mass Spectrom.* **8,** 521 (1981).

[11] W. Vetter and W. Meiser, *Org. Mass Spectrom.* **16,** 118 (1981).

identification of positions of unsaturation and branching as well as other features, such as, hydroxyl groups, additional carboxyl groups, and cyclopropane rings.

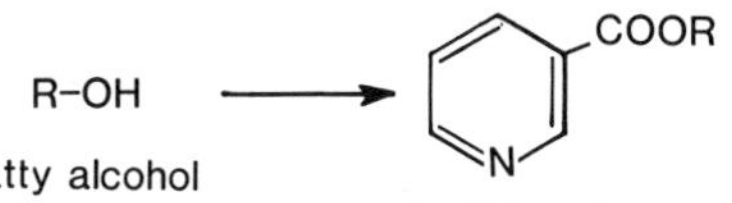

R–OH
fatty alcohol
(XV)

nicotinic ester
(XVI)

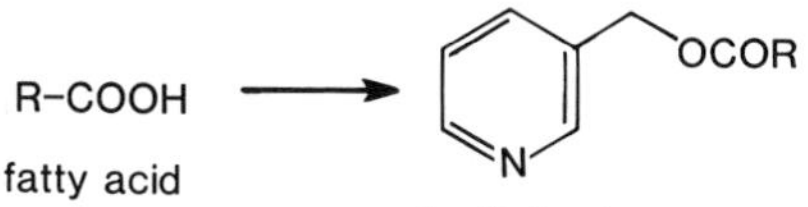

R–COOH
fatty acid
(XVII)

picolinyl ester
(XVIII)

A similar approach has been used to enhance the formation of sequence-specific ions in peptide sequencing using FAB ionization where a charge-stabilizing group (e.g., prolyl)[13] or a preformed positive charge (via quaternization of the terminal amino group with methyl iodide)[14] is introduced at the N terminus. These groups stabilize the positive charge on the N terminus of the peptide promoting the formation of a series of N-terminal sequence ions.

Chemical derivatizations can also be used to determine the number and types of *functional groups* in a molecule. Examples from the peptide field include determining the number of carboxyl groups (and therefrom the number of acidic amino acid residues) by measuring the mass shift upon methyl ester formation, and determining the number of amino groups (and therefrom the number of lysine residues) by the mass shift following N-acetylation. Fragment ions containing the derivatized functional groups can be identified using the isotope doublet technique. For example, if acetylation is carried out using an equimolar mixture of acetic anhydride and hexadeuterated acetic anhydride, any ion containing an acetyl group will appear as a visually conspicuous equal-intensity doublet of peaks separated by three mass units. A similar approach can be used in conjunction with accurate mass measurements to yield an isotopic peak to corrobo-

[12] D. J. Harvey, *Biomed. Mass Spectrom.* **9**, 33 (1982).
[13] D. F. Hunt, J. Shabanowitz, J. R. Yates, P. R. Griffin, and N. Z. Zhu, in "Mass Spectrometry of Biological Materials" (C. N. McEwen and B. S. Larsen, eds.), p. 169. Dekker, New York, 1990.
[14] K. Biemann, this volume [25].

rate an elemental composition determination.[15] For example, a monoacetylated ion would have one peak whose elemental composition has only ^{1}H and a corresponding peak 3 mass units higher with three ^{2}H and three less ^{1}H.

General Considerations in Chemical Derivatization

A useful chemical derivatization reaction should be specific in that it modifies only the intended part of the molecule and should yield preferably a single stable product. It should be reasonably fast and simple to carry out and amenable to very small sample sizes. Ideally the derivatization reagents should also be reasonably stable and safe to handle. Few derivatization procedures meet all of these criteria.

Chemical derivatization is a microscale form of synthetic organic chemistry where the objective is to achieve a 100% yield of a single product. While this goal is rarely achieved in macroscale synthesis, the use of high-purity, highly reactive reagents, often in large excess, makes it possible to approach this goal in many analytical derivatizations. The qualities of high purity and high reactivity of reagents present a difficult combination since highly reactive reagents are prone to decomposition and reaction with extraneous materials (e.g., atmospheric moisture). Great care is required to meet these criteria. Specially purified reagents packaged in small aliquots are preferred. Although reagents are less expensive in larger packages, such purchase is false economy if the reagents deteriorate before use. Use of large excesses of reagents to drive reactions to completion results in the need to remove the excess reagent. A common approach to this removal is evaporation resulting in concentration of any less volatile contaminants. Solvent evaporation also results in concentration of impurities. Therefore, solvents should be of the highest quality and volumes used kept to a minimum. The problems of concentration of low-volatility impurities in solvents and reagents are avoided with the use of vapor-phase derivatization (see below).

Practical Aspects of Chemical Derivatization

Most derivatization reactions are carried out in solution phase. The most widely used containers for such reactions are the tapered interior reaction vials.[16] They are available in sizes ranging from 0.1 to 10.0 ml in

---

[15] K. Biemann, this volume [13].

[16] Available from a variety of sources, e.g., React-Vials from Pierce Chemical Co., Rockford, IL; V-Vials from Wheaton, Millville, NJ.

both clear and amber glass. The caps accommodate a variety of different types of liners, but the Teflon-faced rubber disks are most amenable to use with derivatization reactions. The caps are perforated to allow passage of a syringe needle through the liner without opening the container and exposing the contents to the atmosphere. Such perforation, however, destroys the integrity of the Teflon face and allows contact of the vial contents with the less chemically resistant bulk material of the cap liner. An alternative closure method is the use of all-Teflon valve closures.[17] In the closed position, only Teflon is exposed to the vial interior. When the valve is opened, a rubber backup seal, which can be pierced with a syringe needle, seals out atmospheric moisture. To avoid the use of rubber, the backup seal (a cylindrical piece of silicone rubber) can be removed from its hole and a flow of inert gas purged through, which is used to protect the contents from the atmosphere.

A variation of the conical reaction vial, which also facilitates the use of solvent extraction, is the Keele microreactor vial.[18] The interior of this container contains a constriction which forms an "hourglass." Phase separation of microvolumes of liquid is facilitated by drawing the sample into the narrow constricted region using a syringe and needle. Separation of phases in microscale extractions can also be carried out in volumetric pipettor tips. Both phases are drawn into the pipettor allowing them to separate. The phases are then expelled sequentially into separate containers.

The tapered reaction vials have smooth cylindrical sides and ground flat bottoms which allow close fit into bored aluminum heating blocks. Reaction mixtures can be stirred magnetically using miniature stirring paddles; heating block units are available commercially with built-in magnetic stirrers.[16] The units can also be fitted with multiport evaporator units to evaporate samples under a stream of inert gas.

An alternative method to evaporate samples to dryness employs the vacuum centrifuge device.[19] This device spins tubes in a centrifuge under vacuum to evaporate samples while continuously forcing the remaining solution to the bottom of the tube. This method has the advantage of concentrating the sample to a small area of the bottom of the tube rather than being dried over a larger surface area as occurs with evaporation under an inert gas stream. It also eliminates "bumping" during evaporation.

[17] Mininert valves, Pierce Chemical Co., Rockford, IL.
[18] A. B. Attygalle and E. D. Morgan, *Anal. Chem.* **58**, 3054 (1986); available commercially from Wheaton, Millville, NJ.
[19] Speedvac, Savant Instruments, Farmingdale, NY.

Tapered reaction vials have the disadvantage of being relatively expensive. For many applications, especially where large numbers of containers are used, autosampler vials with disposable glass inerts can be used. These vials are available in a wider range of sizes and styles (including smaller sizes) than can currently be found for tapered reaction vials.

Another type of container, which is used for some of the less rigorous reactions in peptide analysis, is the polypropylene microcentrifuge tube. These are available either with snap tops or screw caps.[20] The latter type are available both with and without O-ring seals. For carrying out derivatizations, the screw caps which seal without O rings are preferable since the rubber O rings can yield contaminants.

Polypropylene tubes are preferable to glass for those compounds which are prone to loss by adsorption on glass surfaces (e.g., basic peptides). Alternatively, the glass surfaces can be silanized to mask the adsorptive sites with methylated silyl groups. The most convenient way to silanize glassware is to use the vacuum oven technique where the glassware is heated in the presence of hexamethyldisilazane vapor.[21] The solution silanization procedure can be used for small quantities of glassware which do not warrant the dedication of a vacuum oven apparatus (some workers feel that the solution method is better regardless). In the solution method, the thoroughly cleaned and oven-dried glassware is treated for 30 min with a 5% solution of dimethyldichlorosilane in toluene. It is then washed sequentially with toluene and anhydrous methanol. After air-drying under a fume hood, the glassware is oven-dried before use.

Some containers, such as autosampler vials and polypropylene tubes, are disposable. More expensive containers such as the tapered reaction vials must be cleaned and reused. For high-sensitivity analysis, conventional cleaning is often followed by baking to pyrolyze any remaining organic material. Laboratories processing large numbers of containers have found it convenient to use a standard domestic self-cleaning oven for this purpose. Pyrolytic cleaning destroys surface silanization thus necessitating resilanization prior to use.

Reagent measurement and delivery of less reactive reagents and those not sensitive to atmospheric exposure can be performed with any of the many types of micropipettes. More reactive reagents are easily transferred using microliter syringes. The "gas-tight" type with Teflon-tipped plungers are easy to use and do not suffer from reagent soaking into the space around the metal plunger. If all exposure to metal must be avoided, the

[20] Sarstedt, Newton, NC; Bio-Rad, Richmond, CA.
[21] D. C. Fenimore, C. M. Davis, J. H. Whitford, and C. A. Harrington, *Anal. Chem.* **48**, 2289 (1976).

removable needle type can be used with a disposable glass micropipette substituted for the needle. An alternative measuring device where the reagent contacts only glass is a modified Eppendorf-type pipettor where the disposable polypropylene tip has been cut off and a melting point capillary whose tip has been drawn out in a flame[18] is forced into the opening. When this device is used with corrosive reagents, the metal piston inside the pipettor must be cleaned after each use to prevent its deterioration.

Derivatization with Vapor-Phase Reagents

An alternative to derivatization in solution phase is to use vapor-phase reagents acting upon a sample film dried on a surface. This approach, used to acylate steroid samples on platinum gauze 20 years ago,[22] has been reintroduced by Biemann's laboratory.[23] The use of vapor-phase reagents acting upon a dried sample allows multistep reactions to be carried out without evaporation of reagents and solvents (and attendant accumulation of impurities) and without sample transfers (and attendant losses). In the recent embodiment of the technique,[24,25] a few microliters of the sample solution are placed in a melting point capillary which contains a small indentation to hold the liquid in place. Evaporation of the solvent under vacuum deposits the sample on a small area of the inner surface of the capillary. The sample-containing capillaries are then placed in a reaction vessel[25] made from a commercially available vacuum hydrolysis tube.[26] The vacuum hydrolysis tube is used in a horizontal position with indentations added to support the capillaries lying horizontally (so that their ends do not contact the tube surface) and with the addition of a reagent well. The reagent (ca. 200 μl) is added to the well and the sample-containing tubes inserted into the vessel. After cooling the reagent, the vessel is evacuated, sealed, and then placed in an oven at the appropriate temperature. After the reaction period, the reagent in the well is cooled again and the excess vapor removed by evaporation. Multiple derivatizations can be carried out by placing the sample capillaries in successive vessels containing different reagents. Following derivatization the sample tubes can be directly transferred to a Caplan and Cronin[27]-type gas chromatograph injector for GC/MS analysis or dissolved in a small volume of solvent for probe analysis. Transfers of very small volumes of solution from these

[22] E. C. Horning and B. F. Maume, *J. Chromatatogr. Sci.* **7,** 411 (1969).
[23] J. E. Vath, M. Zollinger, and K. Biemann, *Fresenius Z. Anal. Chem.* **331,** 248 (1988).
[24] K. Biemann, this volume [18].
[25] C. E. Costello and J. E. Vath, this volume [40].
[26] Pierce Chemical Co., Rockford, IL.
[27] P. J. Caplan and D. A. Cronin, *J. Chromatogr.* **267,** 19 (1983).

tubes can be carried out with a microliter syringe with a fused-silica capillary "needle," such as is used for on-column injections in capillary gas chromatogrpahy.

Derivatization reactions which have been carried out by this method include trifluoroacetylation, acetylation, esterification (of a peptide via an intermediate azalactone formed with acetic anhydride), quaternization of amino groups, borane reduction of peptides and gangliosides, and trimethylsilylation.[23] A variety of other reactions should also be amenable to this approach.

On-Probe Derivatization in FAB Analysis

For mass spectrometric analysis by FAB techniques, derivatization reactions may also be carried out directly on the sample in the liquid matrix on the probe. Examples of such reactions include acetylation, esterification, disulfide reduction, and oxidations where the reagents are added directly to the sample matrix. In some cases, the matrix itself can serve as a derivatization reagent, i.e., the reduction of disulfide bonds in a thioglycerol matrix. Using on-probe derivatization, it is possible to obtain spectra of a compound and one or more derivatives from the same sample loading on the probe. Desirable properties in a derivatization reagent for this purpose are rapid reaction at ambient temperature, relative insensitivity to atmospheric oxygen and moisture, and volatility for removal of excess reagent. Excess reagents can be removed by evaporation in the vacuum lock during probe insertion or, to protect the mass spectrometer from excessive exposure to reagents, in a separately pumped vacuum system.

Derivatization Methods

The following is a sampling of general-purpose derivatization reaction procedures. These methods have been generalized from a collection of methods optimized for specific sample types,[3] from recent literature, and from collected "recipes" whose original sources are difficult to determine. In most cases, exact quantities are not stated. The reagents are generally used in large excess but on an analytical scale (e.g., nano- to picomole) even microliter quantities of reagents usually represent a large excess. The user should, however, check that the reagent amount used is in excess of the amount of sample being derivatized. Different compound types will require specifically optimized conditions for optimum derivative yields, especially for high-sensitivity quantitation; for these specific conditions the reader is referred to later chapters in this volume and to the previously referenced sources.[2,3] The methods given here should be useful for qualitative work and as a starting point to develop optimized methods.

Methyl Ester Formation

A solution of 2 N methanolic HCl is prepared by adding dropwise 150 μl of acetyl chloride to 1 ml of methanol. The acetyl chloride is conveniently measured using a disposable glass micropipette and added to the methanol in a glass test tube with mixing on a Vortex mixer. The solution is allowed to stand at room temperature for 5–10 min and then added to the dried sample. After standing 1–2 hr at room temperature (a much shorter time may be adequate), the excess reagent is removed by evaporation under an inert gas stream or under vacuum. The esterification reagent should be freshly prepared; old solutions may contain chloromethane which could cause extraneous reactions.

Methyl esters can also be prepared using diazomethane. This method avoids the risk of methanolysis of susceptible functional groups (e.g., amide groups in peptides). Diazomethane is highly toxic and explosive but can be handled safely in small quantities in a hood. Small quantities (less than 1 mmol) for derivatization can be prepared using a generator as described by Fales et al.[28] (now commercially available)[26,29,30] which avoids the inconvenience and hazards of distillation. The diazomethane is produced by adding alkali via a syringe port in the generator to an aqueous solution of N-methyl-N-nitroso-N'-nitroguanidine (MNNG) contained in the inner of two concentric reservoirs (the more common Diazald precursor should not be used because it requires heating). The generated diazomethane dissolves in diethyl ether contained in the outer reservoir. Diazoethane can be prepared in the same manner using the homologous precursor. Deuterodiazomethane (which yields [^{2}H$_2$]methyl derivatives) can be prepared from MNNG by substituting ^{2}H-labeled (-O^{2}H) 2-(2-ethoxyethoxy)ethanol for the water and sodium deuteroxide as alkali.

The derivatization is carried out by dissolving the sample in the ether solution of diazomethane. If necessary, methanol can be used to dissolve the sample. The reaction with carboxyl groups is very rapid and the persistence of the yellow color of diazomethane indicates excess reagent. For larger sample sizes, the by-product nitrogen can be observed as bubbles. The excess reagent and solvent are removed by evaporation (in a

[28] H. M. Fales, T. M. Jaouni, and J. F. Babashak, *Anal. Chem.* **45**, 2302 (1973).
[29] Wheaton, Millville, NJ.
[30] Aldrich Chemical Co., Milwaukee, WI.

hood). Diazomethane can also react with carbonyl groups and double bonds to give extraneous products.

Acetylation

Acetylations of OH (both alcohol and phenol) and NH (amine) groups can be carried out using acetic anhydride with either base or acid catalysis. Common reagents include 1 : 1 (v/v) acetic anhydride/pyridine or acetic anhydride/acetic acid. Dimethylaminopyridine, a much stronger catalyst than pyridine, can be used in much smaller amounts (e.g., 2% in acetic anhydride). Reactions are usually complete in 30–60 min (often less) at room temperature or with only mild heating. Peptide amino groups can be acetylated in a few minutes at room temperature with an equal volume mixture of acetic anhydride and methanol. Acetylated peaks can be identified as M,M + 3 doublets by use of 1 : 1 acetic anhydride/[^{2}H$_6$]acetic anhydride.

Trifluoroacetylation

Trifluoroacetyl derivatives are commonly prepared from OH and NH groups; the amide derivatives are, in general, more stable than the ester derivatives. Trifluoracetic anhydride (TFAA), trifluoroacetylimidazole (TFAI), or N-methylbis(trifluoroacetamide) (MBTFA) can be used as reagents. TFAA is used with either acidic (typically TFA) or basic (typically trimethylamine) catalysis. Both TFAA and TFAI methods yield nonvolatile by-products which normally require isolation of the derivative by solvent extraction. The most convenient general-purpose reagent is MBTFA. The sample is dissolved in the reagent or in a mixture (1 : 1, v/v) of the reagent and pyridine and heated at 60° for 1 hr or more. For GC/MS analysis, the mixture can be injected directly since the by-product is the innocuous and volatile N-methyltrifluoroacetamide. Alternatively, the excess reagent and by-product can be evaporated under vacuum to isolate the derivative (as long as the derivative is less volatile than the by-product).

Methyl trifluoroacetate can be used to trifluoroacetylate amino groups (e.g., in peptide methyl esters) under mild conditions.[31] The sample is dissolved in an equal volume mixture of methanol and methyl trifluoroacetate. Triethylamine is added to adjust the pH to above 8 (checked with moistened pH paper) and the solution allowed to stand at room temperature overnight. The excess reagents can be removed by evaporation.

[31] S. A. Carr, W. C. Herlihy, and K. Biemann, *Biomed. Mass Spectrom.* **8,** 51 (1981).

Trimethylsilylation

A variety of reagents are available for trimethylsilylation. A good general-purpose reagent which will silylate alcohols, phenols, amines, carboxylic acids, and thiols is *N,O*-bis(trimethylsilyl)trifluoroacetamide (BSTFA). The by-products trimethylsilyltrifluoroacetamide and trifluoroacetamide are quite volatile, as is the reagent itself. The sample to be silylated is dissolved in the reagent or, if necessary, in silylation-grade acetonitrile, dimethylformamide, or pyridine followed by addition of excess reagent. The reaction often proceeds well at room temperature but may require heating. A more powerful reagent contains 1% trimethylchlorosilane (TMCS) in BSTFA. For GC/MS analysis, the derivatization mixture is normally injected directly.

tert-Butyldimethylsilylation

The most convenient reagent for forming *tert*-butyldimethylsilyl (TBDMS) derivatives is *N*-methyl-*N*-(*tert*-butyldimethylsilyl)trifluoroacetamide (MTBSTFA). Its use is similar to BSTFA but often requires more vigorous reaction conditions to transfer the bulkier silyl group. MTBSTFA containing 1% of the corresponding chlorosilane (available commercially)[32] is probably the best first choice for most compounds. The TBDMS group is much more stable toward hydrolysis than the TMS group; unlike TMS derivatives, the TBDMS derivatives can be isolated, if desired, for direct probe analysis.

Methoxime Formation

Methoxime derivatives are used to reduce the polarity and hydrogen bond acceptor potential of aldehyde and ketone carbonyl groups. A disadvantage of the derivative is the possible formation of chromatographically separable *syn* and *anti* products due to restricted rotation about the methoxime double bond.

The derivative is prepared by adding a 2% solution of methoxyamine hydrochloride in pyridine to the sample. Reaction conditions vary for different compounds but are typically several hours to overnight at room temperature or one to a few hours at 60°. The excess reagent is not volatile, therefore the derivatized sample is usually recovered by adding water (or aqueous NaCl) and extracting with ethyl acetate. In a multistep derivatization involving water-sensitive derivatives, the methoximation must be performed first.

---

[32] Regis Chemical Company, Morton Grove, IL.

On-Probe Derivatization Methods

On-probe derivatization methods are commonly used in FAB analysis, but can be carried out on sample in the probe capillary in EI and CI probe analysis. Thus, reaction conditions are less stringent than for conventional derivatizations. The following listed below are commonly used on-probe procedures.

Methyl Ester Formation. One microliter of 2*N* methanolic HCl is added to the sample (or in the case of FAB, the sample/matrix mixture) on the probe tip and allowed to stand for a few minutes at room temperature. The sample is then reinserted into the vacuum lock and the excess reagents pumped away. If the ester peaks are not observed, the derivatization can be repeated for a longer period.

Acetylation of Amino Groups in Peptides. A 0.5-μl aliquot of 1 : 1 (v/v) methanol/acetic anhydride and 0.5 μl of pyridine are added to the sample on the probe tip and allowed to stand at room temperature for a few minutes. The excess reagents are then removed in the vacuum lock prior to analysis.

Reduction of Disulfide Bonds in Peptides. Raising the pH of a sample in thioglycerol or dithiothreitol (DTT)/dithioerythritol (DTE) matrix by addition of 0.5 μl 28% aqueous ammonia will promote reduction by the thiol matrix. Alternatively, 1 μl of either 1 : 1 (v/v) 0.2 *M* *N*-ethylmorpholine buffer (pH 8.5)/1 *M* DTT or 1 : 1 (v/v) aqueous ammonia (28%)/ thioglycerol is added and allowed to stand for a few minutes at room temperature.[33] The sample is then acidified with HCl or trifluoroacetic acid, and FAB matrix added if necessary, prior to removal of excess reagent in the vacuum lock and analysis.

Performic Acid Oxidation. Performic acid is prepared by adding 5% (by volume) of 30% hydrogen peroxide to 99% formic acid and allowing the solution to stand at room temperature for 2 hr. One microliter of the solution is added to the sample in *glycerol* matrix and allowed to stand for several minutes prior to analysis.

Summary and Future Prospects

Chemical derivatization in relation to mass spectrometry has evolved from primarily volatility considerations to a primary emphasis upon manipulation of the ion chemistry in the mass spectrometer to increase the information yield. The advent of FAB methods at first appeared to obviate the need for derivatization, but now it is recognized that FAB presents

---

[33] H. Rodriquez, B. Nevins, and J. Chakel, *in* "Techniques in Protein Chemistry" (T. E. Hugli, ed.), p. 186. Academic Press, San Diego, 1989.

new opportunities to exploit derivatization. Likewise, the rapidly developing field of tandem mass spectrometry creates additional opportunities for the creative use of chemistry in conjunction with mass spectrometry to increase the ability to solve structural problems. In the area of protein structure, for example, mass spectrometry has already established itself as the "third leg of the stool" of protein primary structure methods (along with Edman and DNA sequencing). To date, however, mass spectrometry has been little exploited for other than determination of primary covalent structure. In conjuction with creative uses of chemical modifications (derivatizations), mass spectrometry offers the potential to yield three-dimensional information as well. Chemical linkage experiments can yield distance geometry measurements which, in conjunction with computer molecular graphics, can be used to refine three-dimensional structures. As a further extension, chemical trapping of intermediate states offers the potential to gain information on molecular dynamics. Thus, there appears to be a wide range of future applications for chemical derivatization in the broadest sense of the term. Just as with protein chemistry, whose demise was prematurely announced some years ago,[34] chemical derivatization in relation to mass spectrometry is far from dead.

[34] A. D. B. Malcolm, *Nature (London)* **275**, 90 (1978).

[16] Introduction of Deuterium by Exchange for Measurement by Mass Spectrometry

By James A. McCloskey

The chemical incorporation of deuterium followed by measurement of the intact product by mass spectrometry has for a number of years played a major role in three, often overlapping areas: the structural characterization of molecules of unknown structure; to gain information on the mechanisms of chemical or biological reactions; and as an aid in the interpretation of mass spectra.[1] In each of these areas mass spectrometry offers several general advantages: (1) relatively routine measurements can be made in the sample quantity range 10^{-11}–10^{-6} g, sometimes in mixtures or without extensive sample purification; (2) the sites of labeling can often be deter-

[1] H. Budzikiewicz, C. Djerassi, and D. H. Williams, "Structure Elucidation of Natural Products by Mass Spectrometry," Vol. 1, Alkaloids, Chap. 2. Holden-Day, San Francisco, 1964.

mined from the resulting mass spectrum; (3) in addition to the overall extent of labeling, the molar distribution of labeled species can be measured, which is not possible using methods in which the sample is combusted or similarly degraded prior to measurement. On the other hand, measurement of incorporation levels at several mole percent and below may be difficult (depending on molecular weight and method of sample introduction into the mass spectrometer) due to the necessity to correct for background ions and for naturally occurring heavy isotopes.

The present chapter describes methods for the direct introduction of deuterium by simple exchange of heteroatom-bound protium, or "active" hydrogen [Eq. (1)].

$$ ROH \xrightarrow[\text{active D}]{\text{excess}} ROD $$

$$ RCO_2H \xrightarrow[\text{active D}]{\text{excess}} RCO_2D $$

$$ \backslash\,N-H \xrightarrow[\text{active D}]{\text{excess}} \backslash\,N-D \tag{1} $$

$$ RSH \xrightarrow[\text{active D}]{\text{excess}} RSD $$

These reactions are generally quantitative in terms of incorporation level due to use of an excess of labeled solvent such as D_2O, and are essentially instantaneous without use of acid, base, or catalyst, thus generally excluding exchange of carbon-bound hydrogen.

Although microscale deuterium exchange procedures are experimentally simple, they suffer from the principal disadvantage of ease of reexchange, which can occur readily due to adsorbed protium (water) on glassware or mass spectrometer surfaces, or in solvents or air. Compared with incorporation and measurement of nonactive deuterium or of other isotopes, sample handling and introduction into the mass spectrometer is usually a critical experimental issue, as described in later sections.

Primarily due to reexchange, mass spectrometric methods employing thermal vaporization are generally limited to compounds having 4–6 exchangeable hydrogens, while with ion desorption methods which use deuterated matrix, the useful range may be extended to compounds having over 20 active hydrogen atoms.

Some related methods not covered in detail in the present chapter include on-column exchange of active hydrogen with GC/MS[2] and ther-

[2] J. A. McCloskey, this series, Vol. 14, p. 438.

mospray LC/MS,[3] direct liquid introduction of deuterated solutions by thermospray,[4] gas-phase exchange of active hydrogen by chemical ionization reagents[5,6] such as D_2O,[7] and the exchange of active hydrogen[8] and the hydrogen alpha to carbonyl groups[9] by preparative gas chromatography. Fetizon and Gramain,[10] and, in particular Thomas,[11] have cataloged numerous methods and reactions for the incorporation of deuterium into organic compounds.

Reagents

D_2O. 98–99 atom%. D_2O is widely available commercially with D concentrations above 98 atom % at reasonable cost due to its extensive use as an NMR solvent. Suppliers include: Cambridge Isotope Laboratories, Woburn, MA; Isotec, Inc., Miamisburg, OH; Merck Sharp & Dohme Isotopes, St. Louis, MO; and Icon Services, Summit, NJ. Although very highly labeled water (e.g., >99.998 atom %D) is available at higher cost, there is rarely any justification for use of solvents over 98 atom %D for exchange reactions to be monitored by mass spectrometry. A number of other deuterated solvents and reagents, such as CD_3OD (or CH_3OD at higher cost) and ND_3, are also available from the above suppliers.

Glycerol-$(OD)_3$. 98.8 atom %D, MSD Isotopes, St. Louis, MO.

3-Nitrobenzyl Alcohol-OD.[12] 3-Nitrobenzyl alcohol (NBA) is widely used as a matrix for desorption mass spectrometry. NBA (1 ml) is vigorously shaken by hand in a 1 : 1 (v/v) mixture with D_2O (1 ml 99 atom %D) for 1 min in a 3-ml Reacti-Vial (Pierce Chemical Co., Rockford, IL). The resulting two-phase system is allowed to separate for 5 min and the top D_2O layer removed by aspiration. The procedure is repeated twice, for a total of three times. The resulting level of deuterium incorporation is reported as approximately 84%.[12] (No special precautions were taken

[3] C. G. Edmonds, S. C. Pomerantz, F. F. Hsu, and J. A. McCloskey, Anal. Chem. **60**, 2314 (1988).

[4] M. M. Siegel, Anal. Chem. **60**, 2090 (1988).

[5] D. F. Hunt, C. N. McEwen, and R. A. Upham, Tetrahedron Lett. p. 4539 (1971).

[6] W. Blum, E. Schlumpf, J. G. Liehr, and W. J. Richter, Tetrahedron Lett. p. 565 (1976).

[7] D. F. Hunt, C. N. McEwen, and R. A. Upham, Anal. Chem. **44**, 1292 (1972).

[8] G. J. Kallos and L. B. Westover, Tetrahedron Lett. p. 1223 (1967).

[9] M. Senn, W. J. Richter, and A. L. Burlingame, J. Am. Chem. Soc. **87**, 680 (1965).

[10] M. Fetizon and J.-C. Gramain, Bull. Soc. Chim. Fr. p. 651 (1969).

[11] A. F. Thomas, "Deuterium Labeling in Organic Chemistry." Appleton Century Crofts, New York, 1971.

[12] A. M. Reddy, V. V. Mykytyn, and K. H. Schram, Biomed. Environ. Mass Spectrom. **18**, 1087 (1989).

to exclude traces of H_2O, so that higher incorporation levels should be attainable.)

Storage of Reagents

Active deuterated reagents are subject to significant back-exchange in a period of minutes or less when reagents are in use or storage vessels are uncapped, and may undergo slow reexchange when stored for long periods unless precautions are taken. Ampules or containers should be very tightly sealed and stored in a desiccator. In areas of high humidity, subdivision into smaller storage containers, or other manipulations, should be carried out in a dry glove-box if very high levels of enrichment are to be maintained.

After preparation by exchange of active deuterated compounds, storage or transfer of the labeled product in organic solvents should be made with caution. Traces of H_2O or other protic compounds in solvents can provide sufficient protium on a molar ratio basis to effect significant back-exchange. Thomas has discussed methods for drying of glassware and maintenance of high levels of active deuterium.[13]

Procedures

Direct Exchange for Measurement by Electron Ionization (EI) or Chemical Ionization (CI) Mass Spectrometry

Sample (<200 ng) is transferred as a solid or in an easily evaporated solution to the mass spectrometer direct inlet sample cup or holder (and if necessary, dried under vacuum after insertion by direct probe into the vacuum lock of the mass spectrometer). D_2O (1 μl) or other labeled solvent such as CD_3OD is added by syringe and the mixture is rapidly stirred (if necessary to effect solution) using the tip of the syringe and is inserted without delay into the vacuum lock. Light vacuum is immediately applied to remove moisture-containing air from the vicinity of the sample. D_2O has a tendency to form bubbles during vacuum evaporation in the vacuum lock. If the sample holder is narrow (e.g., the shape of a melting point capillary) and larger solutions are used (>2 μl), these bubbles may displace the solution from the capillary, therefore the pumping should proceed uniformly and slowly over several minutes in order to reduce bubble formation. Because this occasional problem is dependent in part on the sample holder and pumping geometry of the inlet system, it is advisable

[13] A. F. Thomas, "Deuterium Labeling in Organic Chemistry," Chap. 1. Appleton Century Crofts, New York, 1971.

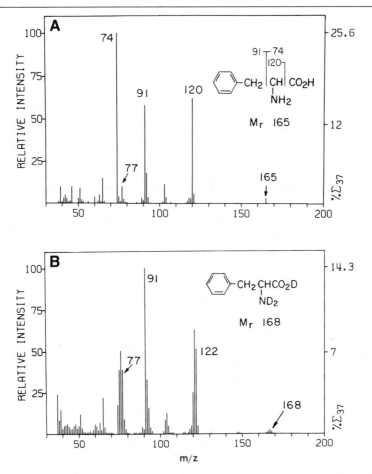

FIG. 1. Electron ionization mass spectra of phenylalanine (M_r 165). (A) Unlabeled; (B) following exchange with 2 μl D_2O in the mass spectrometer direct-probe sample holder.

to run a model compound to test the overall exchange method. Alternatively, the sample holder can be dried under vacuum in a glass desiccator, where D_2O evaporation can be visually observed, and the dried sample holder transferred immediately for introduction into the mass spectrometer.

Figure 1 shows a typical result of a simple one-pass exchange, in which three atoms of protium in phenylalanine were exchanged in an EI probe sample holder. Incorporation of up to three deuterium atoms is clearly shown by shifts of the m/z 120 and 165 ions, but limitations due to back-exchange are also evident. Mass shifts of fragment ions (m/z 74, 91, 92,

120) are in accord with the assignments shown, plus McLafferty rearrangement of amino hydrogen to form m/z 92.

Reduction in Back-Exchange

A significant portion of observed re-exchange results from collisions between vaporized sample molecules and ion source surfaces which contain layers of ubiquitous adsorbed water and organic material. This effect can be reduced by admitting a volatile reagent such as CD_3OD from a batch inlet, during vaporization of the labeled sample. Following the experiment the mass spectrum of an unlabeled test substance should be determined, to establish whether "re-exchange" of the ion source surfaces (i.e., removal of deuterium) is necessary. Although generally less effective than coadmission of a volatile-labeled reagent in achieving higher exchange levels, multiple exchanges of the sample in the direct probe holder can also be carried out, and the sample holder can be first cleaned by flaming and then washed with D_2O before sample loading.

Analysis of Labeled Compounds by CI. Caution must be used if exchanged compounds containing active deuterium are exposed to CI reagent gas such as NH_3 or H_2O which contain active hydrogen, since quantitative loss of deuterium will occur. This process is the reverse of a method whereby the initial exchange reaction can be effectively carried out in the gas phase by the use of D_2O or similar compounds as reagent gas.[7] However, as a result of collision between sample ions and reagent gas neutrals, deuterium may be incorporated at nonactive carbon atoms, particularly in aromatic compounds.[14,15] Similar effects have been observed in thermospray ionization,[3,4] in which CI reactions presumably operate to some extent. The reader is referred to a review of the problem of deuterium exchange in CI by Harrison.[16]

Matrix Exchange for Measurement by Fast Atom Bombardment (FAB) or Liquid Secondary Ion Mass Spectrometry (LSIMS)

Procedures.[17] For experiments that require high deuterium incorporation levels ($>\sim95\%$), the sample probe tip is rinsed with ethanol and then cleaned by fast atom (or ion) bombardment of the dry surface for 2 to 3

---

[14] B. S. Freiser, R. L. Woodin, and J. L. Beauchamp, *J. Am. Chem. Soc.* **97**, 6895 (1975).

[15] D. F. Hunt and S. K. Sethi, *J. Am. Chem. Soc.* **102**, 6953 (1980).

[16] A. G. Harrison, "Chemical Ionization Mass Spectrometry," p. 129. CRC Press, Boca Raton, FL, 1983.

[17] S. Verma, S. C. Pomerantz, S. K. Sethi, and J. A. McCloskey, *Anal. Chem.* **58**, 2898 (1986).

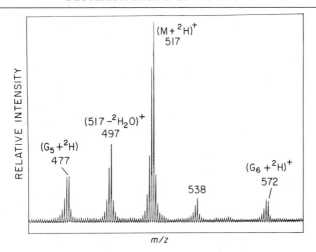

FIG. 2. Fast atom bombardment mass spectrum of melezitose following deuterium exchange. M is defined as the molecule in which all 11 hydroxyl hydrogens have been replaced by deuterium, and similarly for the glycerol (G) cluster ions. (Adapted with permission from Sethi *et al.*[18])

min, then rinsed with D_2O immediately before loading the sample. Samples are prepared off-line in D_2O solutions, at concentrations dependent on the ionization efficiency of the sample, but generally 0.5–3 $\mu g/\mu l$). One microliter of the D_2O solution is mixed on the probe tip with a 25% solution of glycerol-$(OD)_3$ in D_2O and then transferred to the probe vacuum lock in less than 10 sec. Syringes, glass capillaries, or other devices used for sample or solvent transfer should be prerinsed with D_2O immediately before use.

If lower deuterium incorporation levels are satisfactory, for example, when the sample molecule contains few ($<\sim 6$) exchangeable hydrogens or a solvent of lower deuterium content is used, some of the above precautions can be relaxed, in particular, the procedure for cleaning the probe by ion or atom bombardment.

The FAB mass spectrum of melezitose following deuterium exchange in glycerol-$(OD)_3$ and D_2O is shown in Fig. 2, and is representative of cases in which the number of hydrogens exchanged can be ascertained by visual inspection. The isotopic pattern in the cluster m/z 514–518 is asymmetric because m/z 517 represents the maximum exchange possible (D_{11} in melezitose, plus the ionizing deuteron), with m/z 518 corresponding very closely to the required ^{13}C isotope peak for 12 carbon atoms. The decrease in abundances of m/z 516–514 corresponds to statistical decrease in deuterium content dictated by the H/D ratio in the sample at the time of ionization.

Caution in interpretation of the data should be exercised when the resulting isotope pattern is sufficiently symmetrical that the mass of the fully exchanged ion cannot be determined by visual inspection. This circumstance results when the number of exchangd hydrogens becomes large ($> \sim 18$), or the extent of deuteration falls below ~ 85–90%. In those cases, the correct number of deuterium atoms introduced can often be determined by comparison of the experimentally observed isotope pattern with theoretical patterns, calculated using test values of n, the number of deuterium atoms present. This method[17] requires knowledge of the approximate elemental composition of the unlabeled molecule, and the level of deuterium exchange. This latter value can be determined simply and accurately from matrix cluster ions in the spectrum,[17] for example, the m/z 477 region in Fig. 2.[18]

Calculation of Deuterium Content

Both the molar distribution of deuterium and the overall percentage of deuterium incorporated can be readily calculated from the resulting mass spectrum.[19] The method is illustrated below using the data from phenylalanine in Fig. 1b. The same principles can be extended to other stable isotopes, such as ^{18}O,[20] and to more complex patterns as might result from presence of multiisotopic elements (Br, Cl). For extended treatment of isotopic abundance calculations, the reader is referred to earlier literature.[17,21]

The tabulation below, with data taken from Fig. 1b, illustrates how deuterium content can be calculated.

Ion		Relative intensity (%)
m/z 165	M	0.70
m/z 166	M + 1	1.92
m/z 167	M + 2	2.60
m/z 168	M + 3	1.63
m/z 169	M + 4	0.16

1. Estimate the abundance of naturally occurring heavy isotopes in the molecule (see also Appendixes 1 and 4 at the end of this volume).

[18] S. K. Sethi, D. L. Smith, and J. A. McCloskey, *Biochem. Biophys. Res. Commun.* **112**, 126 (1983).

[19] K. Biemann, "Mass Spectrometry: Organic Chemical Applications," p. 233. McGraw-Hill, New York, 1962.

[20] R. C. Murphy and K. L. Clay, this volume [17].

[21] H. Yamamoto and J. A. McCloskey, *Anal. Chem.* **49**, 281 (1977).

Depending on the accuracy desired, the second isotope peak (due to $^{13}C_2$, etc.) can usually be ignored for molecules under $M_r \sim 400$ if only C, H, N, and O are present.

Phenylalanine = $C_9H_{11}N$

First isotope peak = $(9 \times 1.1) + (1 \times 0.4) = 10.3\%$

2. Correct peak M + 1 for natural heavy isotopes.

$0.70 \times 0.103 = 0.072$

$1.92 - 0.07 = 1.85\% = (M + 1)_{corr}$

3. Using the value of $(M + 1)_{corr}$, calculate the natural heavy isotope contribution to M + 2, and correct the abundance of M + 2.

$1.85 \times 0.103 = 0.19$

$2.60 = 0.19 = 2.41\% = (M + 2)_{corr}$

4. Make similar corrections to M + 3 and M + 4.

$2.41 \times 0.103 = 0.25$

$1.63 - 0.25 = 1.38\% = (M + 3)_{corr}$

$1.38 \times 0.103 = 0.14$

$0.16 - 0.14 = 0.02\% = (M + 4)_{corr}$

5. The sum of corrected abundance values from steps 2–4 is 6.36. Each ion can be expressed as a percentage of all molecular species.

M	0.70%	11% D_0
M + 1	1.85%	29% D_1
M + 2	2.41%	38% D_2
M + 3	1.38%	22% D_3
M + 4	0.02%	0% D_4
		100%

The calculated distribution, therefore, confirms that phenylalanine has three exchangeable hydrogen atoms, even though the extent of labeling is incomplete due to re-exchange. A similar calculation can be applied to the data in Fig. 2, showing that a maximum of 12 deuterium atoms have been incorporated, or 11 in the neutral molecule.

The deuterium content can be calculated as a percentage of the maximum possible, as follows:

6. Calculate the percentage deuterium in each molecular species, relative to the maximum value (three in phenylalanine).

M	0%
M + 1	33%
M + 2	66%
M + 3	100%

7. The percentage of deuterium is determined by multiplying the above values by the appropriate mole fraction (step 5 above).

D_0	$0 \times 0.11 = 0.0$
D_1	$33 \times 0.29 = 9.6$
D_2	$66 \times 0.38 = 25.0$
D_3	$100 \times 0.22 = 22.0$

The sum of these values, 57%, represents the total extent of deuteration. When applied to the data for MD^+ from melezitose in Fig. 2, a similar calculation shows 93% deuteration, indicating a much lower level of re-exchange, reflecting presence of an excess of active deuterium in intimate contact with solute molecules at the time of ionization, and on ion source surfaces.

[17] Preparation of Labeled Molecules by Exchange with Oxygen-18 Water

By ROBERT C. MURPHY and KEITH L. CLAY

The combined use of stable isotope-labeled molecules and mass spectrometry has been a powerful means to address numerous problems in chemistry and biochemistry which had been difficult or impossible to probe by other techniques. Molecules labeled with ^{18}O have been used in numerous mass spectrometric studies from the classic determination of the fate of the carboxyl oxygen atoms in ester hydrolysis,[1] mechanistic studies of gas-phase ion decompositions, internal standards for quantitative mass spectrometric analysis, and detailed studies of chemical and biochemical reaction mechanisms.

This chapter will emphasize the synthesis of ^{18}O-labeled molecules to equilibrium exchange reactions with $H_2^{18}O$. Several functional groups containing oxygen have the property of undergoing exchange of their

[1] M. L. Bender, *J. Am. Chem. Soc.* **73**, 1626 (1951).

oxygen atoms with water and these reactions can be catalyzed by both chemical and enzymatic means. Examples of such moieties are carboxylic acids, ketones, phenols, and phosphate esters, and it should be noted that the rates of ^{18}O exchange differ for each of the functional groups. In general, the rank for rate of exchange is as follows: $C=O > COOH > Ar-OH > PO_4$ (where Ar means aromatic). However, it is important to realize that several enzymatic systems have been found to greatly accelerate some of these exchange–equilibrium reactions out of this above rank order.

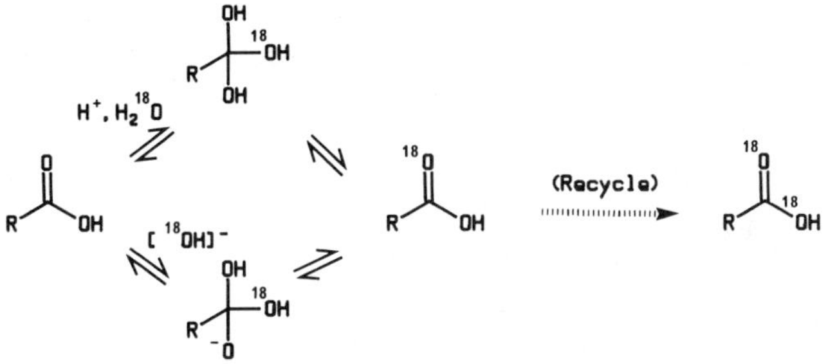

(1)

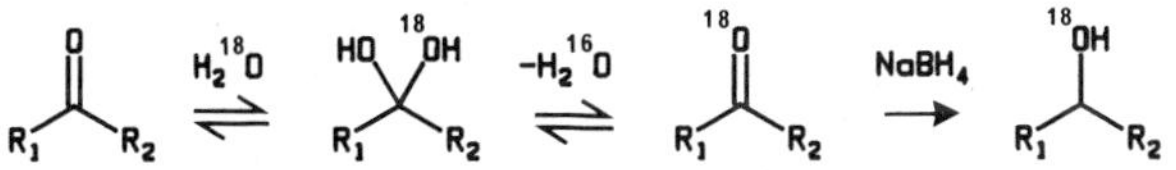

(2)

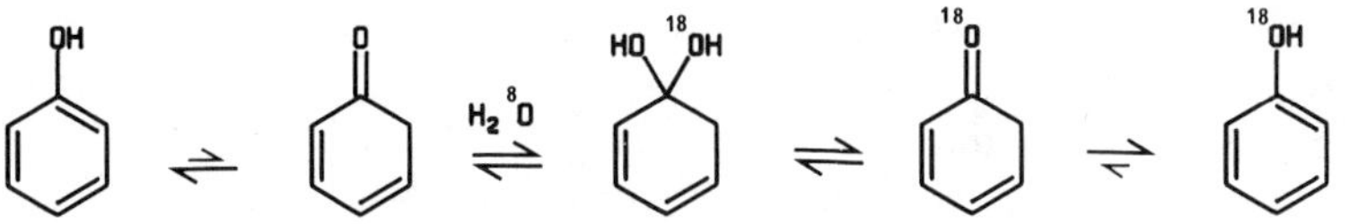

(3)

As seen in reaction (1), carboxylic acids can undergo either acid- or base-catalyzed exchange with $H_2{}^{18}O$ through the formation of an ortho acid intermediate. With a large excess of $H_2{}^{18}O$ one can realize almost complete incorporation of two atoms of ^{18}O into the carboxylate moiety. For some applications the advance in mass by 4 daltons (Da) is quite useful, e.g., internal standard for quantitative mass spectrometry. It is often necessary to employ high temperatures and strong acidic or basic conditions in order to achieve full equilibrium in a reasonable length of time (from 1 to 5 days) and for those carboxylic acids which are stable to such conditions this is a straightforward and inexpensive source of highly enriched material. Unfortunately, many carboxylic acids are not sufficiently stable to survive such rigorous treatment. Two strategies have been developed to overcome this limitation. First, one could employ an esterification–deesterification cyclic process using diazomethane and ^{18}O-labeled hydroxide.[2] Second, several enzymatic systems catalyze carboxyl

---

[2] R. C. Murphy and K. L. Clay, this series, Vol. 86, p. 547.

oxygen exchange process. The simple carbonyl can undergo oxygen exchange through the formation of an intermediate *gem*-diol as shown in reaction (2). This reaction is typically quite rapid and, even under mild conditions, excellent oxygen incorporation can be achieved within several hours at room temperature. The rapidity of this reaction is also an attractive route in the synthesis of ^{18}O-labeled alcohols through exchange of the carbonyl followed by metal hydride reduction.[3]

Exchange of the oxygen atom of alcohols with water is not a practical reaction for most alcohols. However, phenolic compounds are a special case. These molecules can undergo keto–enol tautomerization with some small extent of keto character at the phenoic C—O bond as can be seen in reaction (3). This keto form, even though highly unfavored, can undergo formation of the *gem*-diol in the presence of water which leads ultimately to incorporation of ^{18}O in the phenolic oxygen position. Simple phenols do not readily exchange, however, this has been observed with catechols.[4] As is the case for simple carbonyls and carboxylic acids, other structural features in a complex molecule can facilitate or inhibit exchange reactions. Pyrimidines exist predominately in the keto form and exchange their carbonyl oxygen much more rapidly than a simple phenol.[5]

Although the specific uses of ^{18}O-labeled molecules for mass spectrometric investigations will not be discussed in this chapter, there are two important issues to consider when using ^{18}O-labeled molecules. First, many complex molecules are too unstable for direct chemical exchange and one must find mild conditions, typically through the use of purified enzymes, to catalyze equilibrium with the excess $H_2^{18}O$. Second, one must consider back-exchange or loss of label, if the ^{18}O-labeled compound will be exposed to an aqueous environment during use. Back-exchange is particularly important in biochemical studies when the labeled molecules are added to a physiological fluid. Many physiological fluids contain esterases which can catalyze the back-exchange process. A general rule to follow would be: the easier ^{18}O can be incorporated into a molecule through an exchange reaction, the easier it can be lost by back-exchange.

Reagents

$H_2^{18}O$. 97–99 atom %. Several commercial firms now offer $H_2^{18}O$ at reasonable prices. These include Isotec, Inc., Miamisburg, Ohio; Yeda Research and Development Company, Rehovot, Israel; Norsk-Hydro, Oslo, Norway; and Cambridge Isotope Labora-

---

[3] J. Goto, H. Miura, M. Ihada, and T. Nambara, *J. Chromatogr.* **452**, 119 (1988).
[4] K. C. Clay and R. C. Murphy, *Biomed. Mass Spectrom.* **7**, 345 (1980).
[5] G. Puzo, K. H. Schram, and J. A. McCloskey, *Nucleic Acids Res.* **4**, 2075 (1977).

tories, Woburn, MA. It is important to ascertain whether $H_2{}^{18}O$ contains any deuterium; some older methods for purification of $H_2{}^{18}O$ led to enrichment of the labeled water with deuterium which required the need to "normalize" the water back to its natural protium content.

12 N HCl in $H_2{}^{18}O$. Anhydrous HCl (Matheson Gas, East Rutherford, NJ) is bubbled slowly through a thick-walled capillary Teflon tube into 1 ml $H_2{}^{18}O$ held in a 13 mm × 120 mm glass tube kept at 4° in an ice bath. At first no bubbles emerge to the surface, but toward saturation vigorous bubbling occurs. This solution is approximately 12 N HCl and little dilution of the ^{18}O enrichment occurs if exposure to ambient atmosphere is avoided during the process. The precise normality can be determined from a small aliquot (10 μl) by titration.

Procedures

Method I. Acid-Catalyzed Exchange of Amino Acids.[4] Most of the 20 naturally occurring amino acids can be labeled with two ^{18}O atoms in the carboxyl moiety through simple exchange with $H_2{}^{18}O$ containing 1 N HCl. The amino acid for exchange (4 mg) is placed in a 0.3-ml vial to which is added 100 μl 99 atom % $H_2{}^{18}O$. Anhydrous HCl is bubbled slowly into the suspension to just effect solution of the amino acid. The vial is then sealed with a Teflon-lined capped. The concentration of the hydrochloric acid in the solution prepared by this method is not critical and has been estimated to be 1 N.[6] Alternatively, one can add 100 μl 1 N HCl in $H_2{}^{18}O$ (obtained by dilution of a saturated solution of HCl in $H_2{}^{18}O$) to 1–4 mg of the amino acid. This solution is then placed in a heating block at 60° for 2–3 days. Aliquots are removed at various time intervals to follow the extent of the oxygen exchange reaction. Figure 1 is a mass spectrum of [^{18}O]tris(trimethylsilyl)tyrosine synthesized by exchange in 1 N HCl.[4] The abundant ion at m/z 222 indicates the incorporation of two ^{18}O atoms in the carboxylic acid moiety. Experiments with tyrosine have indicated that there is no loss of optical activity during this exchange reaction. Table I lists 16 amino acids which have been exchanged with ^{18}O by these conditions.[6]

Using ^{18}O-labeled amino acids a rather curious phenomenon was discovered suggesting that red blood cells catalyze the rapid back-exchange of ^{18}O-labeled amino acids most likely through the process of transport across the red cell plasma membrane.[7] The mechanism for this reaction is

[6] R. C. Murphy and K. L. Clay, *Biomed. Mass Spectrom.* **6**, 309 (1979).

[7] K. L. Clay and R. C. Murphy, *Biochem. Biophys. Res. Commun.* **95**, 1205 (1980).

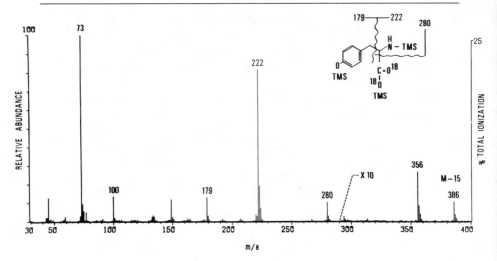

FIG. 1. Mass spectrum of [^{18}O]tris(trimethylsilyl)tyrosine synthesized by exchange with 95 atom % H$_2$^{18}O in dilute acid. The relative abundance of all ions above m/z 290 has been multiplied by a factor of 10 (with permission[4]).

not entirely understood; however, suggestions for this oxygen exchange process are shown in Fig. 2.

Some carboxylic acids undergo exchange under relatively mild conditions, e.g., 3-[1,1-^{18}O$_2$]hydroxyhexadecanoic acid was obtained by dissolving 1 mg in 0.2 ml of dioxane and adding 0.2 ml of H$_2$^{18}O (95 atom % ^{18}O). This solution was shaken for 5 days at room temperature resulting in greater than 90 atom % 3-[^{18}O$_2$]hydroxyhexadecanoic acid.[8]

Method II. Enzyme-Catalyzed Exchange of Carboxylic Acids. Highly enriched [1,1-^{18}O$_2$]leukotriene B$_4$ can be synthesized from LTB$_4$ using commercially available butyrylcholinesterase.[9] Just prior to use, this enzyme (pseudocholinesterase, Type IX) is suspended in 0.2 ml H$_2$^{18}O (98 atom %) for a final concentration of 100 enzyme activity units/ml. After placing the enzyme in a water bath at 37°, LTB$_4$ (20 μg) is added in 20 μl of methanol. Figure 3 shows the incorporation of ^{18}O into the carboxylic oxygen atoms during the course of a 26-hr incubation period. After 19 hr, 89 atom % [^{18}O$_2$]LTB$_4$ is obtained with less than 2 atom % [^{16}O$_2$]LTB$_4$.

Other sensitive biomolecules have been synthesized by use of esterases and H$_2$^{18}O, including various peptides,[10] eicosanoids,[9] and phosphate es-

[8] J. A. Zirrolli and R. C. Murphy, *Org. Mass Spectrom.* **11**, 1114 (1976).
[9] J. Westcott, K. L. Clay, and R. C. Murphy, *Biomed. Mass Spectrom.* **12**, 714 (1985).
[10] D. M. Desiderio and M. Kai, *Biomed. Mass Spectrom.* **10**, 471 (1983).

TABLE I

OXYGEN-18-LABELED AMINO ACIDS, SYNTHESIZED BY A SINGLE ACID-CATALYZED
EXCHANGE[a,b] WITH $H_2^{18}O$ WITH MASS SPECTRAL ANALYSIS
OF THE FINAL ISOTOPIC CONTENT[c,d]

Amino acid	$^{16}O_2$	$^{16}O^{18}O$	$^{18}O_2$	Temperature (°)	Time (days)
Alanine	1.0	27.6	71.5	69	3
Aspartic acid	2.6	17.7	79.7	60	2
Cysteine	0.4	3.4	96.2	69	3
Dopa	1.0	3.9	95.1	60	2
Glutamic acid	3	18	80	60	2
Glycine	—	5	95	60	2
Histidine	0.4	3.4	96.2	69	3
Isoleucine	1.0	9.7	89.3	69	3
Leucine	0.8	6.2	93.0	60	2
Lysine	0	3.3	96.7	60	2
Methionine	0	2.8	97.2	69	3
Phenylalanine	0.2	9.1	90.7	69	3
Proline	5.2	6.5	88.3	69	3
Serine	0.3	4.5	95.2	60	2
Threonine	—	6	94	60	2
Tryptophan	—	6	94	60	2
Tyrosine	1.7	5.0	93.2	60	2
Valine	1.0	27.6	71.4	69	3

[a] With hydrochloric acid [N].
[b] $H_2^{18}O$ used was 99 atom % ^{18}O.
[c] ^{18}O content evaluated from the mass spectrum of the trimethylsilyl derivative of amino acid molecular ion $[M]^+$, molecular ion less methyl ($[M-15]^+$), or the amine fragments m/z 218.
[d] Taken from Ref. 6 with permission.

ters.[11] Table II (see Refs. 2, 8–10, 12, 13, 18, 20) summarizes reports of the synthesis of biomolecules containing a highly enriched ^{18}O-carboxyl group through the use of acid- and enzyme-catalyzed reactions in $H_2^{18}O$.

Method III. Acid-Catalyzed Exchange of Keto Groups. The time-dependent exchange of ^{18}O atoms in phosphoenolpyruvate can be studied by adding 100 μl of 0.1 M phosphoenolpyruvate solution at pH 8 to 2 ml of $H_2^{18}O$ which is either 0.01 N, 0.08 N, 1 N, or 5 N, in HCl.[13] The incorporation of ^{18}O proceeds smoothly in these acid-catalyzed processes with up to five ^{18}O atoms incorporated in the 5 N HCl in less than 5 min. This rapid exchange of the phosphoryl as well as carboxyl oxygens in phosphoenol-

[11] T. F. Walseth, R. M. Graeff, and M. D. Goldberg, this series, Vol. 159, p. 60.
[12] C. Fenselau, P. C. C. Feng, T. Chen, and L. P. Johnson, *Drug Metab. Dispos.* **10,** 316 (1982).
[13] C. C. O'Neal, G. S. Bild, and L. T. Smith, *Biochemistry* **22,** 611 (1983).

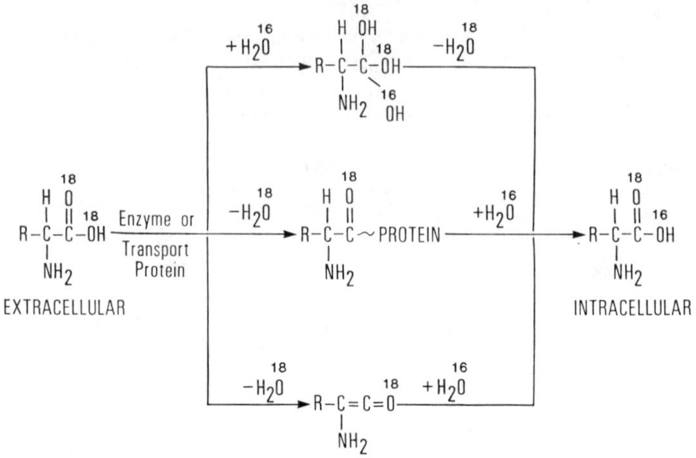

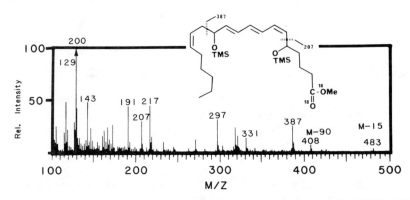

FIG. 2. Possible intermediates in the enzymatically catalyzed back-exchange of carboxyl
^{18}O-labeled amino acids by the red blood cell (with permission of Ref. 6).

FIG. 3. Electron ionization mass spectrum of the methyl ester, trimethylsilyl derivative
of $[^{18}O_2]LTB_4$ synthesized by pseudocholinesterase IX exchange with 98 atom % $H_2^{18}O$.
LTB_4 (20 μg) was incubated with 0.2 ml $H_2^{18}O$ at 37° for 19 hr (with permission of Ref. 9).

pyruvic acid is explained by the intermediate formation of a cyclic phosphate ester and its reversible cleavage to the original phosphoenolpyruvate
or a transient acyl phosphate. In the opening of the cyclic phosphate
ester, ^{18}O from water is incorporated into the molecule. This incorporation
occurs orders of magnitude faster than that observed for incorporation of
oxygen atoms into phosphoric acid or other carboxylic acid.[13] α-Keto
acids also undergo rapid exchange of their keto group.[14]

[14] T. S. Viswanathan, C. E. Hignite, and H. F. Fisher, *Anal. Biochem.* **123,** 295 (1982).

TABLE II
ACID- AND ENZYME-CATALYZED EXCHANGE OF CARBOXYLIC ACID
OXYGEN ATOMS WITH $H_2^{18}O$

Compound	Catalyst	Atom % $^{18}O_2$	Ref.
Acid-Catalyzed			
Arachidonic acid	HCl (1 N)	97.1%	2
UDPglucuronic acid	HCl (5 N)	31% $^{18}O_2$,	12
		41% $^{18}O_3$,	
		12% $^{18}O_4$	
Quinolinic acid	3 N HCl		18
3-Methoxyhexadecanoic acid	(self)	>95%	8
Phosphoenolpyruvic acid	0.1–5 N HCl	40% $^{18}O_4$,	13
		44% $^{18}O_5$	
Enzyme-Catalyzed			
Methionine enkephalin	Porcine liver esterase II	80 ± 4%	10
Leucine enkephalin	Porcine liver esterase II	84 ± 4%	10
12-HETE	Pseudocholinesterase IV	96%	9
15-HETE	Pseudocholinesterase IV	86%	9
5-HETE	Pseudocholinesterase IV	80%	9
LTB$_4$	Pseudocholinesterase IX	89%	9
PGD$_2$	Porcine liver esterase	85%	2
PGF$_{2\alpha}$	Porcine liver esterase	90%	20
6-Keto-PGF$_{1\alpha}$	Porcine liver esterase	71%	2
TxB$_2$	Porcine liver esterase	62%	2
PGB$_2$	Porcine liver esterase	44%	2
PGE$_2$	Porcine liver esterase	70.5%	2

The synthesis of ^{18}O-labeled oxosteroids and bile acids has been carried out in more than 50 different species.[3,15-17] Typically, the steroid (10 mg) is dissolved in 200 μl of 2-propanol followed by the addition of 100 μl of $H_2^{18}O$ (98 atom %) and concentrated HCl to yield a final concentration of 10^{-6} M HCl.[16] The reaction is carried out for 48 hr under nitrogen at 24°. Following purification, the steroid typically contains 95 atom % ^{18}O. In this procedure, the acid catalyst is prepared from aqueous concentrated hydrochloric acid which contains $H_2^{16}O$. Since only micromolar amounts of HCl are needed, a very low dilution of the $H_2^{18}O$ occurs. 2-Propanol is chosen as solvent because of the general solubility of common steroids

[15] A. M. Lawson, F. A. J. M. Leemans, and J. A. McCloskey, *Steroids* 14, 603 (1969).
[16] C. G. Eriksson and P. Ensroth, *J. Steroid Biochem.* 28, 549 (1987).
[17] P. Vouros and P. J. Harvey, *Biomed. Mass Spectrom.* 7, 217 (1980).
[18] M. P. Heyes and S. P. Markey, *Biomed. Environ. Mass Spectrom.* 15, 291 (1988).

TABLE III
^{18}O INCORPORATION INTO NUCLEOSIDES AND BASES BY
EXCHANGE WITH $H_2^{18}O$

Compound	Incorporation (atom % ^{18}O)	Ref.
Uridine	90%	5
5-Methyluridine	90%	5
5-Fluorouridine	77%	5
5-Hydroxyuridine	70%	5
Pseudouridine	38%	5
5-Fluorouracil	75.4%	19

and also the bulky nature of the alcohol reduces competition of the alcohol in the formation of the *gem*-diol [reaction (3)]. Lawson *et al.*[15] studied several steroid ketones and found a wide variance in the relative rate of ^{18}O incorporation. These relative rates illustrate the sensitivity of the ^{18}O exchange reaction to other structural features in the molecule.

The incorporation of ^{18}O into nucleosides and their corresponding bases have been found to be rather facile.[5,19] A pyrimidine base (uracil, 1 mg) is dissolved in 5 μl of $H_2^{18}O$ and placed in a melting point capillary tube. After the addition of 0.5 μl of concentrated (12 N) HCl, the tube is sealed and heated for 18 hr at 100°. The solvent is then removed from the capillary tube under reduced pressure in the presence of P_2O_5 in a vacuum desiccator. Exchange of the oxygen atom with $H_2^{18}O$ follows a pseudo-first-order process and results in 94 atom % excess ^{18}O). Table III lists incorporation of ^{18}O into several nucleosides using this procedure. It should be noted that these experiments employ aqueous concentrated HCl as the source of acid which diluted the $H_2^{18}O$ content as described above. In order to maximize the atom % ^{18}O one could dilute 12 N HCl made in $H_2^{18}O$ as described.

Method IV. Acid-Catalyzed Exchange of Oxygen Atoms of Phenols. One example of phenolic oxygen exchange is that found for 3,4-dihydroxyphenylalanine (Dopa).[4] Dopa (10 mg) is dissolved in 300 μl of 3 N HCl in 99 atom % $H_2^{18}O$. (This solution is prepared by diluting 12 N HCl in $H_2^{18}O$ with the appropriate volume of $H_2^{18}O$.) The sample is thoroughly flushed with argon and sealed with a Teflon cap and placed in a heating block maintained at 135°–150° for up to 20 days. At intervals, samples are removed, lyophilized to dryness, and then derivatized with bis(trimethylsilyl)trifluoroacetamide and acetonitrile. Figure 4 shows the mass spectra of the TMS derivative of Dopa after 1 and 20 days.[4]

[19] M. C. Cosyns-Duyck, A. A. M. Cruyl, A. P. DeLeenheer, A. De Schryuer, J. V. Huys, and F. M. Belpaire, *Biomed. Mass Spectrom.* **7,** 61 (1980).

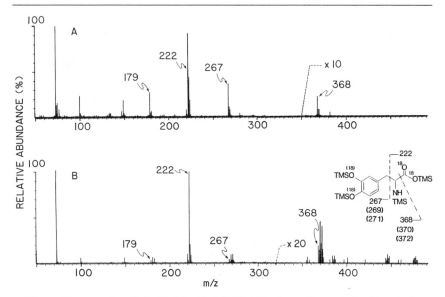

FIG. 4. Mass spectra of tetrakis(trimethylsilyl)3,4-dihydroxyphenylalanine (Dopa). Dopa was heated at approximately 150° in $H_2^{18}O$, 3 N HCl for (A) 1 day, and (B) 20 days as indicated in the structure. The carboxyl oxygen atoms are fully exchanged with ^{18}O while the catechol oxygens are partially exchanged indicated by (18) over that oxygen atom. The ions which include the partially exchanged oxygen atoms are listed in parentheses (with permission of Ref. 4).

It is widely appreciated that phenols can undergo the keto–enol tautomerization [reaction (3)] as evidenced by the ease of formation of $[^2H_2]$tyrosine when heated in ^2HCl solutions. However, tyrosine itself does not exchange its phenolic oxygen atom under the conditions described above for Dopa. It is postulated that the electron-donating property of the orthohydroxy oxygen in Dopa might facilitate formation of the *gem*-diol intermediate necessary for oxygen exchange. It may be necessary to have a neighboring group participating in lowering the activation energy necessary for the formation of this intermediate in phenols in order to observe exchange of oxygen atoms. Anthraquinone undergoes rapid exchange with $H_2^{18}O$.[20]

Method V. Enzymatic Hydrolysis of Peptide Esters. Several neuropeptides have been labeled with ^{18}O at the free carboxyl terminus through a combination of acid-catalyzed exchange and enzymatic hydrolysis of the corresponding methyl ester derivative.[10] The pentapeptide methionine enkephalin (450 μg) is esterified with 3 N methanolic HCl at 37° in $H_2^{18}O$

[20] C. J. Proctor, B. Kralj, E. A. Lorka, C. J. Porter, A. Maquestiorii, and J. H. Beynon, *Org. Mass Spectrom.* **16**, 312 (1981).

(99 atom %). Approximately 32% of the starting peptide is recovered unesterified after high-performance liquid chromatography (HPLC) purification and it is 87 atom % doubly labeled with ^{18}O. The methyl ester derivative (35%) is added to 50 μl $H_2^{18}O$ buffer solution containing 2 mg porcine liver esterase (type II, Sigma, St. Louis, MO). The buffer solution is made by first lyophilizing 50 μl (50 mM, pH 8) potassium phosphate containing the dissolved esterase followed by reconstituting with 50 μl $H_2^{18}O$. After 2 hr incubation at 37° methionine enkephalin is recovered with 77 atom % $^{18}O_2$ and 22 atom % $^{18}O^{16}O$ species.

There have been numerous enzymes reported to facilitate $H_2^{18}O$ equilibration with substrates. Some enzymes have broad specificities, such as, porcine liver esterase and pseudocholinesterase used for peptides[10] and prostaglandins.[21] Others are quite specific, such as acid phosphatase (for phosphate esters)[22] and carbonate dehydratase (for bicarbonate).[23] Our current state of knowledge concerning the use of enzymatic systems is limited since it is still empirical, but there is great promise in the area to improve our ability to rapidly and completely exchange functional oxygen groups with the oxygen atoms in $H_2^{18}O$.

Acknowledgments

This work was supported, in part, by grants from the National Institutes of Health (HL25785 and RR01152).

[21] W. C. Pickett and R. C. Murphy, *Anal. Biochem.* **111,** 115 (1981).
[22] R. L. Van Etten and J. M. Risby, *Proc. Natl. Acad. Sci. U.S.A.* **75,** 4784 (1978).
[23] N. Bitterman, L. Lin, and R. E. Forster, *J. Appl. Physiol.* **65,** 1902 (1988).

Section II

Peptides and Proteins

[18] Peptides and Proteins: Overview and Strategy

By Klaus Biemann

Introduction

During the decade of the 1980s, mass spectrometry experienced a quantum leap that propelled it into the midst of protein chemistry. Prior to that period, the methodology was limited to small molecules, generally of molecular weights below 500. Although in the 1970s, field desorption (FD) ionization, developed by Beckey,[1] made it possible to ionize nonvolative compounds up to a molecular weight of 1500 to 2000, it required considerable experimental skill to obtain reliable results, particularly with large, polar peptides. Another novel method introduced during this time, plasma desorption mass spectrometry (PDMS) pioneered by Macfarlane,[2] permitted the mass spectrometric determination of molecular masses ranging to many kilodaltons. Because that method used the fission products of californium-252 to ionize the molecules of interest, commercial instruments were not available for many years and this fact hampered the widespread use of this very useful technique. However, for the first time it demonstrated the fact that very large and polar molecules can be ionized and the resulting heavy ions detected in a mass spectrometer.

The situation changed dramatically, when M. Barber et al.[3] developed a very simple approach to the ionization of large and polar molecules and used peptides as examples.[1] Irradiation of a solution of the compound of interest in a liquid matrix of low vapor pressure, such as glycerol, with fast [a few kiloelectron volt (keV) kinetic energy] particles (argon atoms) produced protonated molecules, $(M + H)^+$, that were ejected into the mass analyzing system of a mass spectrometer and were recorded after mass separation. The process, termed fast atom bombardment (FAB) ionization, was very easy to implement on existing mass spectrometers, particularly those that were equipped for FD, and over which it had the advantage of much greater ease of operation. This fact, combined with the demonstration that it was applicable to peptides of molecular masses in the range of a few thousand daltons (insulin soon became the yardstick of performance), immediately opened up the field of peptide identification.

[1] A. G. Harrison and R. J. Cotter, this volume [1].
[2] R. D. Macfarlane, this volume [11].
[3] M. Barber, R. S. Bordoli, R. D. Sedgwick, and A. N. Tyler, J. Chem. Soc. Chem. Commun. p. 325 (1981).

It should be noted that it was not the neutral particles but the liquid matrix that led to success, where similar attempts involving direct bombardment with keV ions (i.e., Ar^+) of even much smaller molecules deposited on a metal surface (secondary ion mass spectrometry, SIMS) had failed. It was later found that bombardment of a glycerol solution of a peptide with Cs^+ ions works equally well; the methodology is often also referred to as liquid secondary ion mass spectrometry (LSIMS).[4]

At the time FAB ionization was discovered, the mass range of the larger, commercially available magnetic mass spectrometers barely reached 2000 u at full accelerating voltage. This provided the impetus to design instruments of higher mass range without sacrificing ion transmission and thus sensitivity, which resulted in the commercial availability of mass spectrometers having a mass range of up to 15,000 u at 8 or 10 kV. This approaches the practical limit of keV Cs^+ ions to generate a sufficient number of $(M + H)^+$ ions to be detectable. Beyond that size, other ionization methods (see below) are much more efficient.

While FAB-MS (or LSIMS) quite efficiently generates protonated molecules, these are chemically very stable and have little excess energy. As a result, not much fragmentation takes place and the signal (i.e., the "mass spectrum") consists mainly of the isotope cluster of $(M + H)^+$. Fragment ion abundances are generally ten times lower and the spectrum is, therefore, dominated by the signal for the protonated molecule. Thus, such spectra contain little structural information beyond the molecular weight. In spite of this fact, the literature abounds with FAB mass spectra obtained with large samples (high concentrations in glycerol of commercially or otherwise available peptides), which are then "interpreted" in terms of their known sequence. Scattered among them are, indeed, a few peptides of partly unknown structure but, in these cases, the problem was most often one of distinguishing between a choice of candidate structures. The problem with normal FAB-MS spectra obtained with realistic sample sizes is not only the low abundance of the fragment ions but the fact that the matrix itself, and in combination with the sample, creates a spectrum that contains a peak at every mass. Thus, it is tempting to select those peaks that fit a known or preconceived structure while ignoring those that do not.

However, the lack of fragmentation has great advantages as it permits the determination of the molecular weights of all the components present in a mixture, as long as they differ sufficiently in mass to be resolved. Alternatively, a FAB mass spectrum of a compound assumed to be homogeneous may reveal that it is a mixture. As a consequence of this lack of

[4] W. Aberth, K. M. Straub, and A. L. Burlingame, *Anal. Chem.* **54,** 2029 (1982).

fragmentation and the accompanying lack of structural information, other methods were developed to induce cleavage of bonds. Foremost among these is tandem mass spectrometry,[5-7] where the species, generally $(M + H)^+$, in the first mass spectrometer (MS-1) is fragmented upon collision with an inert atom (collision-induced dissociation, CID, or collisionally activated dissociation, CAD[8]) or some other energy-transfer process, and the fragment ions are recorded in the second mass spectrometer (MS-2).

This principle had been used previously, mainly for studies of ion structure, elucidation of fragmentation mechanisms, or the qualitative and quantitative analysis of relatively small molecules for which mass range and resolution were not a problem.[9] For this study of the CID spectra of peptides of the size produced by proteolysis of proteins, larger instruments were required, either a pair of double-focusing mass spectrometers,[5] a triple quadrupole mass spectrometer, or a combination of magnetic and quadrupole analyzers.[6]

As has been mentioned earlier, the determination of molecular weights beyond 10,000–15,000 becomes increasingly difficult by FAB ionization in combination with a magnetic deflection mass spectrometer, and PDMS may be limited to molecular weights of ≤45,000.[2] Two novel methods for the mass spectrometric determination of molecular weights, apparently particularly well-suited for proteins, were developed very recently.

Hillenkamp has shown that a protein molecule embedded in a UV-absorbing matrix, such as nicotinic acid, can be ionized by irradiation with a neodymium/yttrium–aluminum–garnet (Nd-YAG) laser pulse at 266 nm.[10] Mass measurements are carried out with a time-of-flight (TOF) mass spectrometer.[11] To date, protein molecular weights exceeding 300,000 have been measured with a ±0.1–0.5% accuracy of mass assignment. Another novel ionization method, electrospray (ES), is based on the ability to generate multiply protonated, and therefore multiply charged, molecules by generating a spray of very small droplets of a solution of the protein emerging from the tip of a needle held a few keV above the entrance aperture of a quadrupole mass spectrometer.[12] The success of this method

[5] M. L. Gross, this volume [6].
[6] R. A. Yost and R. K. Boyd, this volume [7].
[7] K. Biemann, this volume [25].
[8] R. N. Hayes and M. L. Gross, this volume [10].
[9] For a review, see "Tandem Mass Spectrometry" (F. W. McLafferty, ed.). Wiley, New York, 1983.
[10] F. Hillenkamp and M. Karas, this volume [12].
[11] K. R. Jennings and G. G. Dolnikowski, this volume [2].
[12] C. G. Edmonds and R. D. Smith, this volume [22].

is due to the presence of well distributed basic sites (amino acids) along the protein chain, as first demonstrated by Fenn *et al.*[13] The general distribution of basic amino acids in most proteins is such that a multiplicity of peaks, one for each $(M + H_n)^{n+}$ ion, centers around $m/z \sim 1000$. Therefore, a quadrupole mass spectrometer with a mass range of $m/z \sim 1500$ suffices to record the data and the molecular weight can be redundantly determined from each pair of adjacent peaks.[12-14] Averaging the results leads to a mass accuracy of ± 0.05–0.01%, even if the m/z value of each peak is only determined to within one mass unit. To date, the ES method has been applied to proteins of molecular weights as high as 130,000, but above 75,000 the multiply charged peaks come so close together that their reliable mass measurement is increasingly difficult. A more detailed discussion of these considerations is presented elsewhere in this volume.[14]

At the present time, it appears that ES is the method of choice for molecular weights up to 75,000, where matrix-assisted laser desorption (LSD) takes over. All these measurements, and generally also those above m/z 5000 from magnetic deflection mass spectrometers, provide the average molecular weight,[15] because both ES and LSD lack the resolution required to separate the isotopic clusters of the protonated molecules. A double-focusing magnetic mass spectrometer is capable of unit resolution up to at least 10,000 molecular weight but the effort required to achieve this is rarely worth it, because at that mass range there is little difference in the information content of monoisotopic and polyisotopic mass measurements.

Utility of Molecular Weight Determinations

There are many problems in peptide and protein structure that can be simply solved by determining molecular weights to better than one mass unit. One of the first applications of FAB-MS to protein structure was the verification and, if necessary, correction of the DNA sequence of a gene coding for the corresponding protein.[16] It merely requires translating the open reading frame of the gene sequence into the corresponding protein sequence and calculating the molecular weights of the tryptic peptides expected from the locations of lysine and arginine. Matching these values with the actual molecular weights, determined by FAB-MS, of the compo-

[13] J. B. Fenn, M. Mann, C. K. Meng, S. F. Wong, and C. M. Whitehouse, *Science* **246**, 64 (1989).

[14] I. Jardine, this volume [24].

[15] K. Biemann, this volume [13].

[16] K. Biemann, *Int. J. Mass Spectrom. Ion Phys.* **45**, 183 (1982).

nents of the tryptic digest of the protein reveals whether or not the DNA is correct. Those values that do not match indicate either an error in the DNA sequence or a posttranslational modification of the gene product. Sequence errors are usually misidentifications of a base that leads to a codon representing a different amino acid, or a deletion or insertion of a base causing a frame shift. In the former case, there will be an unmatched peptide, the molecular weight of which differs from the predicted one by the mass difference of two amino acids. If a frame shift has occurred, there will be one or usually more peptides that do not match the open reading frame but fit those expected for one of the other two. It is then easy to determine where a base has been missed, and where another one inadvertently has been inserted elsewhere, to return to the open reading frame.

Such a verification process is even more efficient if the mass spectrometric information is acquired in parallel with or ahead of the DNA sequencing. Each preliminary set of DNA data can then be checked immediately after it has been obtained. If correct, no further effort needs to be devoted to this section of the DNA; if not, the region where the error occurred is pinpointed and easily corrected by reinspection of the sequencing gel. This parallel approach was used to advantage during the determination of the sequence of the gene coding for glycyl-tRNA synthetase (glycine–tRNA ligase) from *Escherichia coli*.[17]

In contrast to these relatively early days of DNA sequencing, the methodology is now so fast and redundant that an error in the final result is rare. However, mass spectrometry is often still required to settle some remaining questions concerning the gene or its transcription. An example that illustrates a number of points made in the introduction involves the structure of the gene coding for the cytoplasmic valyl-tRNA synthetase (valine–tRNA ligase) (Val-RS) from yeast. After the DNA sequence had been determined, the question of the site of initiation of transcription remained because there were two ATG codons for Met near the putative N terminus, separated by only 45 codons. The first stop codon was 1103 triplets downstream of the first ATG and the protein could therefore be either 1058 or 1104 amino acids long, a difference indistinguishable by molecular weight determinations based on chromatography or electrophoresis. The obvious solution, a number of Edman steps, failed, indicating that the N-terminal amino acid is acylated.

A tryptic digest of Val-RS was partially separated by high-performance liquid chromatography (HPLC) (shown in Fig. 1). No particular effort was made to completely separate all components and 27 fractions were

[17] T. A. Webster, B. W. Gibson, T. Keng, K. Biemann, and P. Schimmel, *J. Biol. Chem.* **258**, 10637 (1983).

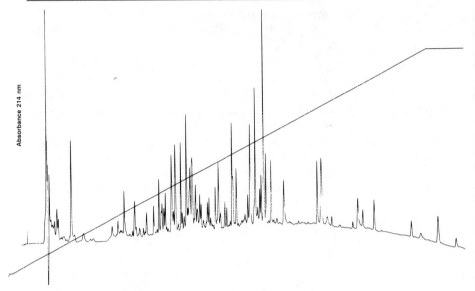

Fig. 1. HPLC trace of the tryptic digest of 2 to 3 nmol of cytoplasmic valyl-tRNA synthetase (valine–tRNA ligase) from yeast. The diagonal line indicates the solvent gradient from water to acetonitrile (both containing 0.1% trifluoroacetic acid) at 1%/min. [From K. Biemann, *in* "Protein Sequencing: A Practical Approach" (J. B. C. Findlay and M. J. Geisow, eds.), p. 99. IRL Press, Oxford, 1989.]

collected. A FAB mass spectrum was recorded for each of the fractions and the molecular weights of a total of 80 peptides were determined from the resulting data. Of these, 79 corresponded to most of those predicted for the tryptic peptides expected for the protein sequence beginning at the second Met-47, but none fit the region between Met-1 and Met-47. This indicated that (1) the DNA sequence is essentially correct because the peptides identified by molecular weight covered about 80% of the protein sequence, and (2) the translation started at the second ATG codon. However, none of the molecular weights corresponded to the peptide Met-47–Lys-59. However, among the 11 abundant peaks in HPLC fraction 10 (Fig. 2) there was one at *m/z* 1351.7 that did not fit any of the predicted tryptic peptides; it was in the range but less than that expected for the protonated Met-47–Lys-59 segment. Both its mass and CID spectrum[7] indicated that transcription began at Met-47 that had been posttranslationally removed, followed by N-acetylation of the new N-terminal amino acid serine.[18,19] It is of interest to note that the mitochondrial Val-RS in yeast was shown to be derived from the same gene except that translation began

[18] B. Chatton, P. Walter, J. P. Ebel, F. Lacroute, and F. Fasiolo, *J. Biol. Chem.* **263**, 52 (1988).
[19] K. Biemann and H. A. Scoble, *Science* **237**, 992 (1987).

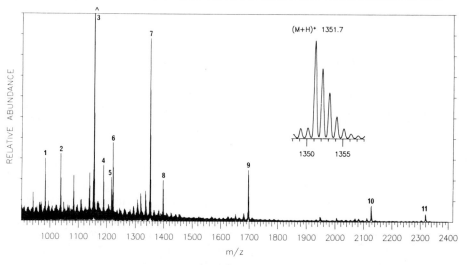

Fig. 2. FAB mass spectrum of fraction 10 collected during the HPLC separation shown in Fig. 1. Raw peak profiles representing the region from m/z 1348 to 1358 are shown expanded in the upper right. The m/z values of peaks 1–11 are: 982.5, 1037.6, 1154.6, 1188.7, 1216.6, 1221.6, 1351.7, 1397.7, 1696.9, 2126.9, and 2319.1, respectively. (From Ref. 19.)

at Met-1, representing an example of the use of the same genetic material for different purposes.[18]

The need for the verification of the amino acid sequence of a protein deduced from the correct DNA sequence of the corresponding gene may gain added significance by the recent findings of the incorporation of an amino acid other than that coded for by the DNA sequence.[20]

The Val-RS example illustrates the strategies followed when searching for or confirming a change in an otherwise known amino acid sequence of a protein. Such cases may involve natural mutations,[21] site-specific modifications introduced by cDNA technology,[22] determination of glycosylation sites by removing all but the N-acetylglucosamine moiety with endo-β-N-acetylglucosaminidase H (Endo-H) or generation of Asp from Asn linked to the carbohydrates by treatment of the glycoproteins with peptide N^4-(N-acetyl-β-glucosaminyl)asparagine amidase (PNGase F),[23] or other modifications such as phosphorylation and sulfation.[24] Molecular

[20] W. Leinfelder, K. Forchhammer, B. Veprek, E. Zehelein, and A. Böck, *Proc. Natl. Acad. Sci. U.S.A.* **87,** 543 (1990).

[21] Y. Wada, T. Matsuo, and T. Sakurai, *Mass Spectrom. Rev.* **8,** 379 (1989).

[22] P. L. Ahl, L. J. Stern, D. During, T. Mogi, H. G. Khorana, and K. J. Rothschild, *J. Biol. Chem.* **263,** 13594 (1988).

[23] S. A. Carr, J. R. Barr, G. D. Roberts, K. R. Anumula, and P. B. Taylor, this volume [27].

weight determinations are also crucial for the assignment of disulfide bridges.[25]

Posttranslational modifications of the amino acid sequence specified by the base sequence of the corresponding gene are widespread. In addition to the specific modifications mentioned above, there are the actions of exopeptidases upon the primary product. The aforementioned removal of a methionine and acylation of the next amino acid is quite common. Processing at the C terminus by carboxypeptidases also occurs frequently. This is more difficult to recognize by the conventional Edman degradation, because it requires either careful monitoring of the yields of the PTH-amino acid of each cycle, or the isolation of all C-terminal proteolytic peptides followed by Edman degradation of each and making certain that the reaction sequences are successful and complete to the last step in each case as was done in the discovery of such a "ragged end" of human γ-interferon.[26] The same result is accomplished much more easily and efficiently by mass spectrometry, which would reveal that there are, in the proteolytic (or CNBr) digest, a number of peptides whose molecular weights correspond to the expected C-terminal peptide and those lacking one, two, three, etc., amino acids. For this purpose, it is not even necessary to separate these peptides from each other or from other peptides in the digest.

Any modification of an amino acid that leads to a change in mass will, if incorporated in a peptide, cause a corresponding change of the molecular weight of the peptide and can thus be easily detected. If it is necessary to identify the amino acid and its position, a CID spectrum[7] will reveal it. Such a scenario is encountered, for example, when probing the active site of an enzyme by photoaffinity labeling. The attachment of the moiety activated by photolysis is easily recognized because the mass increment is predictable. In this manner, the incorporation of a benzyladenine group on the histidine nearest the C terminus of a cytokinin-binding protein was demonstrated by an MS/MS experiment on the appropriate tryptic peptide.[27]

Whereas all these considerations pertained to naturally occurring, biosynthetic or chemical modifications of proteins, it is important to also recognize the utility of mass spectrometry for the confirmation of the correctness of the structures of peptides prepared by chemical synthesis.

[24] B. W. Gibson and P. Cohen, this volume [26].

[25] D. L. Smith and Z. Zhou, this volume [20].

[26] Y.-C. E. Pan, A. S. Stern, P. C. Familetti, F. R. Kahn, and R. Chizzonite, *Eur. J. Biochem.* **166,** 145 (1987).

[27] A. C. Brinegar, G. Cooper, A. Stevens, C. R. Hauer, J. Shabanowitz, D. F. Hunt, and J. E. Fox, *Proc. Natl. Acad. Sci. U.S.A.* **85,** 5927 (1988).

These methods are now automated and make it possible to produce quite large molecules. There are a great number of pitfalls that can be encountered, such as failure to incorporate an amino acid, to incorporate it twice, or to generate γ- instead of α-Asp bonds. Particularly when synthesizing larger peptides, the product often consists of more than one compound, some of which may be due to the complications enumerated above, while others may simply be the desired peptide from which not all of the protecting groups have been completely removed. Either simple molecular weight determinations or a tandem mass spectrum quickly reveal the nature of the problem.[19,28]

Problems Requiring Sequence Information

In the preceding section, it was pointed out that the determination of the molecular weights of peptides derived from a protein often suffices to answer a specific question. Frequently, however, sequence information or the identification of the specific site of a modification is required. As described in more detail elsewhere in this volume,[7] CID of the appropriate protonated peptide molecule $[(M + H)^+]$ provides this information. In addition to the interpretation of such CID spectra with the emphasis on protein sequencing, there are many other areas in which this methodology is useful.

The determination of the primary structure of a protein requires the reliable identification of each amino acid along the protein chain and nothing less. On the other hand, there are situations where even partial sequence information is useful. Foremost among them is the search for a short sequence that preferably consists of amino acids coded for by only one base triplet (Met, Trp) or two (Phe, Tyr, His, Gln, Glu, Asn, Asp, Cys), so that one can construct a short oligonucleotide useful for the isolation of the gene that codes for the protein of interest.

This information is relatively easily achieved by tandem mass spectrometry. From the discussion elsewhere in this volume,[7] it is quite obvious that partial sequences can be obtained from a single digest of a smaller protein sample because abundant fragment ions will still be recognized with smaller samples. Furthermore, a second or third digest is not needed because neither complete sequence information nor overlap information is required. Knowing that ions at low mass in a CID spectrum indicate the presence of certain amino acids, one should pay particular attention to spectra that exhibit peaks at m/z 104 (for Met) or 130 and 159 (for Trp). Large peptides ($M_r > 3000$) often result in CID spectra that do not reveal

[28] I. A. Papayannopoulos and K. Biemann, *Biomed. Environ. Mass Spectrom.,* in press.

the entire sequence but still show a portion (usually the middle),[19] and thus may contain sufficient information to construct a useful oligonucleotide probe.

Most of these discussions were related to problems of protein chemistry in the "precloning" stage. This field is usually plagued by a scarcity of starting material that has to be tediously isolated from its natural source, although a certain level of impurities can be tolerated. The situation is almost completely reversed in the "postcloning" stage, where the protein produced is generally abundant but the questions asked are also very different. The two most important ones are: How homogeneous is it? Were all posttranslational modifications carried out properly? Some of the chapters that follow touch on these questions to varying degrees and describe how at least parts of these complex questions can be answered. The last chapter[29] of this section is specifically devoted to this subject, using as an example a protein produced on a large scale by recombinant technology for clinical trials. Quality control and product integrity are of extreme importance for pharmaceuticals, particularly for those that are used over extended periods of time. For these reasons, it can be expected that mass spectrometry will attain an increasingly important role in this area and may become a methodology required by the agencies formulating the prerequisites of approval and the regulations of such biologicals for human use.

Acknowledgment

The portion of the methodologies and results described that were developed and carried out in the author's laboratory were supported by grants from the National Institutes of Health (RR00317 and GM05472). The author is also indebted to all his present and former associates, particularly B. W. Gibson, S. A. Martin, I. A. Papayannopoulos, and H. A. Scoble.

---

[29] H. A. Scoble and S. A. Martin, this volume [28].

[19] Enzymatic and Chemical Digestion of Proteins for Mass Spectrometry

By TERRY D. LEE and JOHN E. SHIVELY

Introduction

The standard approach for obtaining partial or complete structures of proteins without resorting to cloning cDNAs is to digest the purified protein with one or two selected endoproteases or by chemical cleavage methods, then to fractionate the resulting peptides by reversed-phase high-performance liquid chromatography (HPLC), and to sequence the resulting peptides. In the past, peptides have been primarily sequenced by automated Edman chemistry, which currently has a sensitivity in the mid to low picomole range. If possible, structures were confirmed by amino acid compositional analysis which revealed the amount of sample present and the relative molar ratios of each amino acid. With the advent of mass spectrometric techniques such as fast atom bombardment mass spectrometry (FAB-MS), it has been possible to obtain molecular weights for the majority of peptides analyzed, even at the low picomole range. For laboratories with access to MS/MS approaches (see [24] and [25] in this volume), it is often possible to obtain complete structures by mass spectrometric methods alone. Historically, nanomole amounts have been required for sequencing by mass spectrometry. However, as instrumentation and methods improve, more and more problems can be solved with sample amounts in the 100 pmol range.

With these objectives in mind, we present here methods for obtaining peptide maps at the 100 pmol level on widely available standard proteins (horse cytochrome c and myoglobin). In order to simplify the protocol, standards were chosen that do not require reduction and S-alkylation prior to digestion. However, it should be noted that many proteins are protease-resistant unless this step is performed first. For each method described, a peptide map is shown (see Fig. 1), from which selected peaks were analyzed by FAB-MS to demonstrate that the method is compatible with standard FAB-MS procedures (representative spectra are given later in Fig. 2). For convenience, the sequences for horse heart cytochrome c and horse heart myoglobin are given in Fig. 3 and Fig. 4 (see later), respectively; those peptides that were characterized are indicated. It should be noted that this is not an exhaustive treatment of all available methods, but is a compilation of the more commonly encountered approaches. While it

METHODS IN ENZYMOLOGY, VOL. 193

is undoubtedly easier to obtain results at >1 nmol level, it was thought that the majority of researchers already obtain good results at these higher levels, but often encounter difficulties when working at the 100 pmol level or below. Furthermore, it is good practice to validate a method on a readily available standard protein before attempting the experiment on a precious unknown. Thus, these protocols will serve as a starting point to confirm that the method is working, and the hardware involved (HPLC, mass spectrometer, sequencer, amino acid analyzer, etc.) is functioning properly. No attempt is made to optimize or discuss the HPLC peptide mapping procedures (this topic is also covered in [21] in this volume).

Materials and Sources

Horse heart cytochrome *c*, Sigma (St. Louis, MO)
Horse muscle (apo)myoglobin, Sigma
Bovine trypsin (TPCK-treated), Sigma
Chymotrypsin, Sigma
Staphylococcus aureus protease V8 (endoproteinase Glu-C), Boehringer Mannheim (Indianapolis, IN)
Endoproteinase Lys-C (*Lysobacter enzymogenes*), Boehringer Mannheim
Endoproteinase Lys-C (*Achromobacter lyticus*), Wako Pure Chemicals (Dallas, TX; Osaka, Japan)
Endoproteinase Arg-C (mouse submaxillary gland), Boehringer Mannheim
Endoproteinase Asp-N (*Pseudomonas fragi*), Boehringer Mannheim
Pepsin, Sigma
Thermolysin, Sigma
N-Chlorosuccinimide, Sigma
Cyanogen bromide, Sigma
Ammonium bicarbonate, Mallinckrodt (St. Louis, MO)
Trifluoroacetic acid (sequencer-grade), Pierce (Rockford, IL)
Acetonitrile (HPLC-grade), Baker (Phillipsburg, NJ)
Dimethyl sulfoxide (DMSO), Baker
Dithiothreitol (DTT)/dithioerythritol (DTE), Calbiochem (San Diego, CA)
Disodium phosphate, Sigma

Alternative enzyme sources include Calbiochem, Worthington, and Pierce. Except as noted, the enzymes were used without further purification.

Solutions and Preparation

Ammonium bicarbonate, 1.0 *M* (pH 8.5): 3.95 g NH_4HCO_3 in 500 ml water

Ammonium bicarbonate, 0.1 M (pH 8.5): dilute above 1/10 in water
Trifluoroacetic acid, 10% (TFA): add 1 ml of TFA to 9 ml of water
Tris-HCl, 0.1 M (pH 9.0): add 12.1 g of Tris base to 900 ml of water,
    titrate to pH 9.0 with 12 M HCl, and dilute to 1 liter
Sodium phosphate, 0.10 M, EDTA, 4 mM (pH 7.4): 1.42 g of disodium
    phosphate and 0.15 g sodium EDTA in 90 ml water, adjust pH to 7.4
    with 10% phosphoric acid, and dilute to 100 ml
HPLC, solvent A: 1 ml TFA in 999 ml water
HPLC, solvent B: 1 ml TFA, 99 ml water, 900 ml acetonitrile
    Water was obtained from a MilliQ system. All solutions were stored
at 4° unless otherwise noted.

Mass Spectrometry

All mass spectrometric analyses are performed on a JEOL HX100HF
double-focusing magnetic sector mass spectrometer having modified
Nier–Johnson geometry. The instrument is operated at 5 kV accelerating
voltage and a nominal resolution setting of 3000. Sample ionization is
accomplished using a xenon atom beam having 6 keV translational energy.
Unless otherwise noted, dried, HPLC-purified samples in polypropylene
tubes are taken up in 2 μl of DMSO and applied to a 1.5 mm × 6 mm
stainless steel sample stage which had previously been covered with ca.
2 μl of the DTT : DTE (5 : 1) matrix. Spectra are recorded over the mass
range of 300 of 3500 m/z and mass assignments made using a JEOL DA5000
data system. Unless otherwise noted, mass values reported are for the
monoisotopic protonated molecular ion.

Endoproteinase Lys-C (Achromobacter lyticus). The enzyme (3.3 μg)
is dissolved in 1 ml water, divided into 100-μl aliquots, and stored at $-20°$.
The enzyme is diluted (5 μl in 1 ml) with 0.1 M Tris buffer just before use.

The sample, cytochrome c (1.2 μg, 100 pmol) in 20 μl of 0.1 M Tris
buffer (pH 9.0), is mixed with 100 μl of diluted enzyme (1 : 200 mol/mol;
0.00165 μg), and digested for 6 hr at 30°. The sample is neutralized with
10 μl of 10% TFA and chromatographed on a Vydac C_{18} column (2.1 ×
150 mm) on a Beckman System Gold Chromatograph using a linear gradient from 98% A to 70% B over 60 min with a flow rate of 0.2 ml/min
(214 nm detection, 0.08 AUFS) (Fig. 1). Peaks are manually collected in
1.5-ml polypropylene tubes and stored at $-20°$ prior to mass spectrometric
analysis (Table I) (Fig. 2).

This enzyme is more aggressive than the enzyme from *Lysobacter*. It
can be used at an enzyme to substrate ratio of 1/200 to 1/400, and tolerates
moderate amounts of urea (1–4 M) and SDS (0.1%).[1]

[1] K. Morihara, T. Oka, H. Tsuzuki, Y. Tochino, and T. Kanaya, *Biochem. Biophys. Res. Commun.* **92**, 396 (1980).

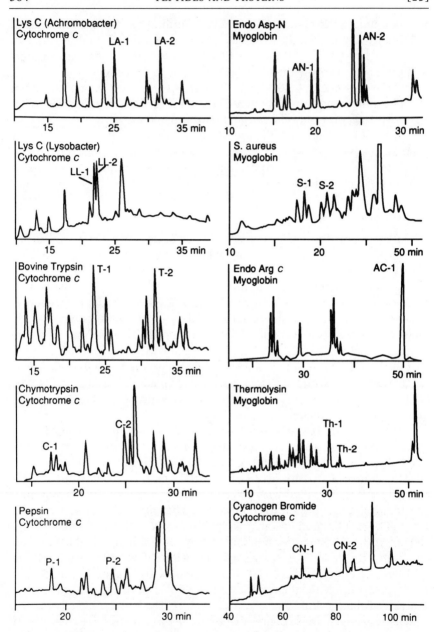

FIG. 1. HPLC separations of enzyme and chemical digests of cytochrome c and myoglobin. Labeled peaks were analyzed by FAB mass spectrometry (see Tables I and II). The sequence of the peptides is shown in Figs. 3 and 4.

TABLE I
EXPERIMENTAL AND CALCULATED MASS VALUES FOR PROTONATED
MOLECULAR IONS IN MASS SPECTRA OF SELECTED DEGRADATION PRODUCTS
FOR HORSE HEART CYTOCHROME c

Enzyme	HPLC fraction	MH+ Observed m/z	MH+ Calculated m/z
Endoproteinase Lys-C	LA-1	604.3	604.4
(Achromobacter)	LA-2a	1350.8	1350.7
	LA-2b	1478.8	1478.8
Endoproteinase Lys-C	LL-1	779.5	779.5
(Lysobacter)	LL-2	1296.7	1296.8
Trypsin	T-1	1470.7	1470.7
	T-2a	1350.8	1350.7
	T-2b	1478.8	1478.8
Chymotrypsin	C-1	803.4	803.5
	C-2	1162.5	1162.6
Pepsin	P-1	966.7	966.6
	P-2	1162.6	1162.6
Cyanogen bromide	CN-1	1763.2	1762.9
	CN-2	2780.0	2779.6

Endoproteinase Lys-C (Lysobacter enzymogenes). The enzyme (3 U, 100 μg) is mixed with 200 μl of water. The enzyme is active for at least 1 month when stored at 4° and is diluted 1 : 100 with water just before use.

The sample, cytochrome c (1.4 μg, 120 pmol) in 50 μl of water, is mixed with 2–3 μl of 1.0 M ammonium bicarbonate buffer (pH 8.5) and 5 μl of diluted enzyme solution (0.05 μg; 1 : 30, w/w) in a 1.5-ml polypropylene tube, and digested for 24 hr at 37°. The sample is directly chromatographed on a Vydac C_4 column (2.1 × 250 mm) on a Brownlee Microgradient System using a linear gradient from 98% buffer A to 100% buffer B over 60 min with a flow rate of 0.15 ml/min (214 nm detection, 0.16 AUFS) (Fig. 1). Peaks are manually collected in 1.5-ml polypropylene tubes stored at 4° prior to analysis by mass spectrometry (Table I).

Standard-grade enzyme is used rather than the more expensive sequencing-grade enzyme. This enzyme cleaves on the C-terminal side of lysine residues.

Trypsin. Trypsin is further purified by reversed-phase HPLC according to Titani et al.[2] Trypsin (250 μl; 1 μg/μl) is purified on a Brownlee Aquapore C_4 column (4.6 × 30 mm) on a Beckman System Gold HPLC using a linear gradient from 100% A to 100% B over 60 min with a flow rate of

[2] K. Titani, T. Saagawa, K. Resing, and K. A. Walsh, *Anal. Biochem.* **123,** 408 (1982).

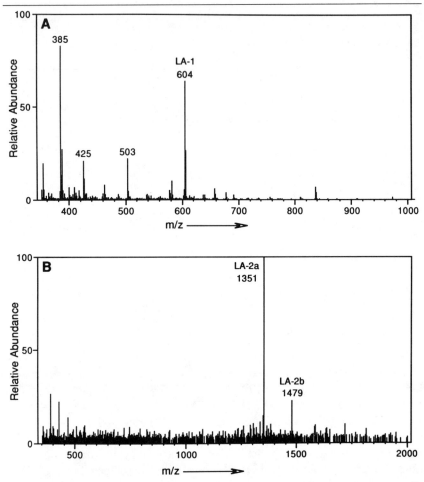

FIG. 2. Positive ion FAB mass spectra of (A) HPLC fraction LA-1 and (B) HPLC fraction LA-2. Spectra are representative of those obtained from other samples.

0.5 ml/min (214 or 280 nm detection). The major peak eluting at 43% B is manually collected into a 1.5-ml polypropylene vial, dried in a vacuum centrifuge, redissolved in 1 ml of water, and stored frozen in 20-μl aliquots. The amount collected is verified by amino acid analysis (20 μl gave 2.5 μg). The enzyme is diluted (20 μl in 1 ml) of 0.1 M ammonium bicarbonate just before use.

The sample, cytochrome c (1.2 μg, 100 pmol) in 20 μl of 0.1 M ammonium bicarbonate buffer (pH 8.5) in a 1.5-ml polypropylene tube, is mixed with 100 μl of HPLC-purified, diluted trypsin (1 : 50 w/w; 0.025 μg), and digested for 18 hr at 37°. The sample is neutralized with 10 μl of 10% TFA, and chromatographed on a Vydac C_{18} column (2.1 × 150 mm) on a Beckman System Gold using a linear gradient from 98% A to 70% B over 60 min with a flow rate of 0.2 ml/min (214 nm detection, 0.08 AUFS) (Fig. 1). Peaks are manually collected in 1.5-ml polypropylene tubes and stored at −20° prior to FAB-MS analysis (Table I).

Trypsin cleaves on the C-terminal side of lysine and arginine residues, but not at Lys-Pro, and occasionally not at Arg-Pro. The trypsin must be purified to avoid spurious cleavages due to contamination with chymotrypsin. Trypsin can tolerate small amounts of urea (1–2 M), but is inhibited by sodium dodecyl sulfate (SDS) and guanidine-HCl. If the sample is treated with urea, the urea must be freshly deionized. Trypsin may also tolerate small amounts of propanol or acetonitrile (10–20%, see Ref. 3).

Chymotrypsin. Chymotrypsin is HPLC-purified as described above for trypsin. The final concentration is 0.05 μg/μl. Although Titani and co-workers[2] included calcium chloride in the HPLC buffer to maintain the activity of chymotrypsin, we have found it unnecessary.

The sample, cytochrome c (1.2 μg, 100 pmol) in 25 μl of 0.1 M ammonium bicarbonate buffer (pH 8.5) in a 1-ml polypropylene tube, is mixed with 3 μl (0.0075 μg) of HPLC-purified chymotrypsin (1 : 160, w/w), and digested for 18 hr at 37°. The sample is neutralized with 3 μl of 50% acetic acid, diluted to 100 μl with 0.1% TFA, and chromatographed on a Vydac C_{18} column (2.1 × 250 mm) on a Brownlee Microgradient System using a linear gradient from 98% A to 100% B over 60 min with a flow rate of 0.15 ml/min (214 nm detection, 0.08 AUFS) (Fig. 1). Peaks are manually collected in 0.5-ml polypropylene tubes stored at −20° prior to analysis by mass spectrometry (Table I).

Chymotrypsin generally cleaves on the C-terminal side of hydrophobic and aromatic residues, such as, leucine, phenylalanine, tyrosine, tryptophan, and, occasionally, at histidine. HPLC purification of chymotrypsin removes contaminating trypsin.

The sequence of horse heart cytochrome c is shown in Fig. 3.

Endoproteinase Asp-N. The enzyme (2.0 μg) is dissolved in 50 μl of water and stored at −20°. The enzyme remains active when tested over a period of 1 to 2 months.

Horse muscle myoglobin is prepared at a concentration of 0.7 μg/μl in 5% acetic acid and stored at 4°. The sample, myoglobin (1.7 μg, 110 pmol) in 2.5 μl of 5% acetic acid and 40 μl of 0.05 M sodium phosphate buffer (pH 8.0) in a 1-ml polypropylene tube, is mixed with 1.0 μl of enzyme (1 : 50 w/w; 0.04 μg), and digested for 18 hr at 20°. The sample is neutralized

---

[3] K. Gjesing Welinder, *Anal. Biochem.* **174,** 54 (1988).

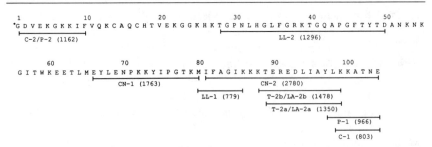

FIG. 3. Sequence of horse heart cytochrome c and location of the fragments analyzed by FAB mass spectrometry. The N-terminal glycine (*G) is acetylated.

with 10 μl of 10% TFA and chromatographed on a Vydac C_{18} column (2.1 × 100 mm) on a Beckman System Gold Chromatograph using a linear gradient from 95% A to 65% B over 65 min with a flow rate of 0.2 ml/min (214 nm detection, 0.16 AUFS) (Fig. 1). Peaks are manually collected in 1.5-ml polypropylene tubes and stored at −20° prior to analysis by mass spectrometry (Table II).

This enzyme cleaves on the N-terminal side of aspartic acid, and sometimes glutamic acid residues. In performic acid oxidized samples, it will also cleave at cysteic acid residues. The enzyme can tolerate small amounts of SDS (0.1%).

Endoproteinase Glu-C (Staphylococcus aureus). The enzyme (50 μg) is dissolved in 50 μl of water and stored at 4°.

The sample, myoglobin (1.7 μg, 100 pmol) in 2.5 μl of 5% acetic acid in a 0.5-ml polypropylene tube, is mixed with 12.5 μl of 0.1 M sodium phosphate, 4 mM EDTA buffer (pH 7.4), 4 μl of water, and 6 μl of the enzyme solution (0.06 μg) diluted 1 : 100 with water (1 : 30 w/w), and digested for 16 hr at 37°. The sample is acidified with 0.5 μl of 50% acetic acid and diluted to 100 μl with 0.1% aqueous TFA prior to fractionation over a Vydac C_{18} column (2.1 × 250 mm) on a Brownlee Microgradient System. Digestion products are eluted using a linear gradient from 98% buffer A to 100% buffer B over 60 min with a flow rate of 0.15 ml/min (214 nm detection) (Fig. 1). Peaks are manually collected into 0.5-ml polypropylene tubes and stored at −20° prior to mass spectrometric analysis (Table II).

In the conditions described above, the enzyme cleaves on the C-terminal side of glutamic and aspartic acid residues. In 0.5 M ammonium acetate (pH 4.0), the enzyme is reported to cleave exclusively after glutamic acid.[4]

---

[4] J. Houmard and G. R. Drapeau, *Proc. Natl. Acad. Sci. U.S.A.* **69**, 3506 (1972).

TABLE II
EXPERIMENTAL AND CALCULATED MASS VALUES FOR PROTONATED
MOLECULAR IONS IN MASS SPECTRA OF SELECTED DEGRADATION PRODUCTS
FOR HORSE HEART MYOGLOBIN

		MH$^+$	
Enzyme	HPLC fraction	Observed m/z	Calculated m/z
Endoproteinase Asp-N	AN-1	1439.6	1439.7
	AN-2	1858.0	1857.9
Staphylococcus aureus	S-1	822.6	822.5
protease	S-2	664.4	664.3
Endoproteinase Arg-C	AC-1	3405.6[a]	3405.7[a]
Thermolysin	Th-1	1345.4	1345.6
	Th-2	1420.4	1420.8

[a] Average mass. Other values are monoisotopic masses.

With myoglobin, however, no reaction occurs under these conditions. The enzyme can tolerate small amounts of SDS (0.05–0.1%).

Endoproteinase Arg-C (Mouse Submaxillary Gland). The enzyme (100 units, 265 μg) is mixed with 100 μl of water. The enzyme is active for 2 to 3 months when stored at $-20°$. The enzyme solution is further diluted 1 : 250 just before use.

The sample, apomyoglobin (1.7 μg, 100 pmol) in 2 μl of water in a 0.5-ml polypropylene tube, is mixed with 100 μl of 0.1 M ammonium bicarbonate buffer (pH 8.0) and 3 μl of the diluted enzyme solution (0.032 μg; 1 : 50, w/w), and digested for 24 hr at 37°. The sample is directly chromatographed on a Vydac C$_{18}$ column (2.1 $\times$ 250 mm) on a Beckman System Gold. Digestion products are eluted using a linear gradient from 95% buffer A to 60% buffer B over 60 min with a flow rate of 0.25 ml/min (214 nm detection, 0.16 AUFS) (Fig. 1). Peaks are manually collected into 1.5-ml polypropylene tubes and stored at $-20°$ prior to analysis by mass spectrometry (Table II).

The enzyme cleaves at the C-terminal side of arginine residues.[5] In this example, some cleavage at lysine is also observed.

Thermolysin. Thermolysin is prepared at a concentration of 1.0 μg/μl in water and stored at $-20°$. The enzyme remains active for 1 to 2 months. The enzyme is diluted 1 : 20 in water just prior to use.

The sample, myoglobin (1.7 μg, 110 pmol) in 2.5 μl of 5% acetic acid and 40 μl of 0.1 M ammonium bicarbonate buffer (pH 8.5) in a 1-ml

[5] M. Levy, L. Fishman, and I. Schenkein, this series, Vol. 19, p. 672.

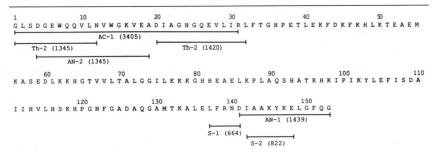

FIG. 4. Sequence of horse muscle (apo)myoglobin and location of the fragments analyzed by FAB mass spectrometry.

polypropylene tube, is mixed with 2.0 μl of 1 : 20 diluted enzyme (1 : 20, w/w; 0.08 mg), and digested for 20 min at 40°. The sample is neutralized with 10 μl of 10% TFA and chromatographed on a Vydac C_{18} column (2.1 × 100 mm) on a Beckman System Gold using a linear gradient from 95% A to 65% B over 65 min with a flow rate of 0.2 ml/min (214 nm detection, 0.16 AUFS) (Fig. 1). Peaks are manually collected into 1.5-ml polypropylene tubes and stored at −20° prior to mass spectrometric analysis (Table II).

Thermolysin cleaves at the N-terminal side of leucine, isoleucine, valine, phenylalanine, methionine, and alanine, often resulting in the formation of very small peptides. Thus, digestion conditions must be adjusted accordingly. The enzyme can be used on proteins resistant to proteases or for cleavage between disulfide bonds. It can be seen that digestion is incomplete for this example; however, longer digestion conditions may result in degradation of the peptides to di- to tetrapeptides. Appropriate digestion times may be ascertained by removing aliquots at hourly intervals, followed by analytical HPLC analysis.

The sequence of horse muscle myoglobin is given in Fig. 4.

Pepsin. Pepsin (636 μg) is dissolved in 20 ml of 1 mM ammonium acetate buffer (pH 7). The enzyme is made fresh each time before use.

The sample, cytochrome c (1.2 μg, 100 pmol) in 95 μl of 0.01 M HCl (pH 2.0), is mixed with 5 μl of the enzyme solution (1 : 7.5, w/w; 0.16 μg by amino acid analysis), and digested for 4 hr at 37°. The sample is directly chromatographed on a Vydac C_{18} column (2.1 × 150 mm) on a Beckman System Gold using a linear gradient from 98% A to 70% B over 60 min with a flow rate of 0.2 ml/min (detection at 214 nm, 0.16 AUFS) (Fig. 1). Peaks are manually collected in 1.5-ml polypropylene tubes and stored at −20° prior to analysis by mass spectrometry (Table I). Samples are dissolved in 2 μl of 1% TFA for mass spectrometric analysis and scanned over the mass range of 700 to 3000.

Pepsin cleaves at the C-terminal side of phenylalanine, methionine, leucine, and tryptophan residues adjacent to hydrophobic residues. The digestion of cytochrome c is incomplete, likely due to an impurity as reported by Yuppy and Paleus.[6] Pepsin has maximal activity at pH 2, but may autodigest when stored for extended periods at pH <4.

Cyanogen Bromide. Cyanogen bromide (CNBr) (2 mg) is dissolved in 100 μl of 70% TFA and freshly made before each use. CAUTION: cyanogen bromide is extremely toxic and should be handled with gloves in a well-ventilated fumehood. Since extremely small amounts are used here, the samples may be later dried in a vacuum centrifuge equipped with a dry ice trap. Carefully clean the trap following this procedure. Do not follow this procedure on scaled-up runs. See Vol. XI in this series for further details.[7]

The sample, cytochrome c (1.2 μg, 100 pmol) in 5 μl of 70% TFA, is mixed with 100 μl of cyanogen bromide (2 mg), and digested for 2 hr at 25°. The sample is dried on a vacuum centrifuge equipped with a dry ice trap, redissolved in 100 μl of 70% TFA, and chromatographed on a Vydac C_{18} column (2.1 × 250 mm) on a Beckman System Gold Chromatograph using a linear gradient from 98% A to 60% B over 120 min with a flow rate of 0.25 ml/min (214 nm detection, 0.16 AUFS). The chromatograph is shown in Fig. 1. Peaks are manually collected in 1.5-ml polypropylene tubes and stored at −20° prior to mass spectral analysis (Table I).

CNBr cleaves at the C-terminal side of methionine residues, converting the methionines to a mixture of homoserine and homoserine lactone. Acidic conditions favor the lactone, while basic conditions favor the free acid. If desired, the interconversion can be affected prior to, or on the stage, during FAB-MS analysis. Some degradation of tyrosine and tryptophan residues may be observed. If the reaction is performed in formic acid, one may also observe reduction of disulfide bonds.[8] If the reaction is performed with heptafluorobutyric acid, some cleavage at tryptophan may occur.[9] CNBr fragments are often large and hydrophobic, and, therefore, tend to aggregate, making them difficult to separate by reversed-phase HPLC. It is suggested that they be dissolved in a strongly denaturing solvent such as hexafluoroacetone trihydrate (HFA), 70% TFA, or 6 M

[6] H. Yuppy and S. Paleus, *Acta Chim. Scand.* **9**, 353 (1955).
[7] E. Gross, this series, Vol. 11, p. 238.
[8] S. Villa, G. De Fazio, and U. Canosi, *Anal. Biochem.* **177**, 161 (1989).
[9] J. Ozols and C. Gerhard, *J. Biol. Chem.* **252**, 5986 (1977).

guanidine-HCl before injection onto the reversed-phase column. Even under these circumstances, the peaks may be broad and obtained in low yields. In this example, better results are obtained with 70% TFA compared to 90% formic acid.

N-Chlorosuccinimide. N-Chlorosuccinimide (NCS) (13.3 mg) is dissolved in 10 ml of 0.3 M dimethylformamide (DMF). The solution is freshly made before each use.

The sample, cytochrome *c* (1.2 μg, 100 pmol) in 5 μl of water, is mixed with 10 μl of 50% acetic acid, 5 μl of 0.01 M NCS in DMF, and digested for 2 hr at 25°. The reaction is terminated with 4 μl of 0.08 M N-acetyl-L-methionine, and directly chromatographed on a Vydac C_4 column (2.1 × 250 mm) on a Beckman System Gold using a linear gradient from 90% A to 100% B over 60 min with a flow rate of 0.25 ml/min (214 nm detection, 0.04 AUFS). Peaks are collected manually in 1.5-ml polypropylene tubes and stored at −20° prior to microsequence analysis. The chromatogram is not shown. Peaks are concentrated in a vacuum centrifuge to an approximate volume of 2–5 μl, and subjected to microsequence analysis. The sequence obtained is KEETLMEYLEN-PKKYIPGTKMIFAG. . . . The fragment corresponds to residues 60–104 of cytochrome *c*. The N-terminal yield is 10 pmol, corresponding to an overall yield of 10% (losses may occur during the cleavage reaction, HPLC, and sequence analysis).

NCS cleaves at the C-terminal side of tryptophan residues.[10] The reagent may also oxidize methionine to methionine sulfone, and may destroy some tyrosine residues. In this example, neither methionine nor tyrosine is affected. Other reagents which have been used to cause oxidative cleavage at tryptophan are CNBr/formic acid/HFBA[9] and iodosobenzoic acid.[11] Due to the rare occurrence of tryptophans in proteins, the expected fragments are very large, usually beyond the mass range of most instruments. The same arguments about solubility and low yields which apply to CNBr fragments also apply here. Due to the large expected fragment sizes in this example, no attempt is made to perform FAB-MS analysis.

Additional Comments

Each of these examples included an enzyme/buffer or reagent/solvent control (not shown). The control allows the subtraction of spurious peaks (should they occur), or an estimation of autodigestion in the case of

[10] Y. Shechter, A. Patchornik, and Y. Burstein, *Biochemistry* **15,** 5071 (1976).
[11] W. C. Mahoney, P. K. Smith, and M. A. Hermodson, *Biochemistry* **20,** 443 (1981).

proteolysis. The conditions outlined here should only be used as a starting point. If volumes and concentrations are varied, one should first verify the appropriateness of the conditions on standards. Many proteins are protease-resistant, and will require prior reduction and S-alkylation (or at least denaturation). Unfortunately, this step usually requires desalting or repurification of the protein before proceeding to enzymatic digestion. The sample may be S-alkylated with iodoacetic acid[12] or 4-vinylpyridine.[13] A general protocol for the latter is given below. A reversed-phase HPLC method for desalting and repurifying the S-alkylated protein is given by Pan et al.[12] Another approach is to denature the protein in formic acid and break the disulfide bonds by performic acid oxidation.[14] This approach will also oxidize methionine and tryptophan residues, thus preventing cleavage by CNBr, but may be useful for Asp-N cleavage at cysteic acid residues. A representative protocol for performic acid oxidation is given below. Finally, in order to locate disulfide bonds, it will be necessary to maintain them intact through the digestion process. A separate chapter in this volume deals with this problem.

Reduction/S-alkylation. Dissolve the sample in 100 μl of 6 M guanidine-HCl, 0.1 M Tris-HCl (pH 8.0), and 10 mM EDTA. Flush the tube (serum-capped Reacti-Vial) with argon, add 1 μl of 100 mg/ml DTT, and reduce for 60 min at 30° to 50°. Add 1 μl of 4-vinylpyridine and alkylate for 30 to 60 min at 30° to 50°. Desalt the sample by reversed-phase HPLC.[12]

Performic Acid Oxidation. Dissolve the sample in 100 μl of 90 to 98% formic acid, chill to 0°, and add 1–10 μl of freshly prepared performic acid (add 250 μl of 30% hydrogen peroxide to 4.75 ml of 90 to 98% formic acid for 2 hr at 25° and chill to 0°). Allow the reaction to proceed for 5 to 30 min and dry on a vacuum centrifuge. Redissolve the sample in an appropriate solvent such as 70% TFA or 100% HFA.

This chapter presents methods which have been documented in our laboratory. Obviously, it is not comprehensive. Excellent protocols are available for other proteases, and for specific cleavage at aspartyl-proline peptide bonds, cysteine peptide bonds, etc. It is hoped that this will provide a starting point for newcomers to the field, or for researchers who have previous experience with digesting nanomole amounts of protein, but need a protocol for starting with picomole amounts of protein. Most importantly, the protocols are designed with further analysis by mass spectrometry in mind.

[12] Y.-C. Pan, J. Wideman, R. Blacher, M. Chang, and S. Stein, *J. Chromatogr.* **297,** 13 (1984).
[13] A. S. Inglis, this series, Vol. 91, p. 26.
[14] C. H. W. Hirs, this series, Vol. 11, p. 197.

Acknowledgments

The authors express their sincere thanks for the many hours of work and helpful discussions contributed by Stan Hefta, Gottfried Feistner, Kay Rutherfurd, Mitsuru Haniu, Jimmy Calaycay, Michael Ronk, Michael Davis, and Shane Rutherfurd. This work was supported in part by NIH grants CA37808, GM40673, DK33155, HD14900, and CA33572.

[20] Strategies for Locating Disulfide Bonds in Proteins

By DAVID L. SMITH and ZHONGRUI ZHOU

Disulfide bonding in proteins is one of the most frequently encountered posttranslational modifications of proteins. Because of the important role played by disulfide bonds in establishing and maintaining the three-dimensional character of proteins, it is important to know whether a protein or peptide contains disulfide bonds, and, for proteins that contain more than two half-cystinyl residues, it is important to know which residues are joined by disulfide bonds. Although there is a good method for quantifying the number of disulfide bonds in protein,[1] the unambiguous determination of the locations of disulfide bonds continues to challenge protein chemists.

Disulfide cross-linkages have often been located by cleaving a protein between the half-cystinyl residues (i.e., with disulfide bonds intact) and identifying the disulfide-containing peptides by their amino acid compositions or sequences.[2-4] An alternative method, which is based on the observation that disulfide bonds may be reduced sequentially under specific conditions, has also been used successfully.[5-7] Although most of our present understanding of disulfide bonding in proteins can be traced to the successful application of these methods, they are often inadequate for the most challenging problems. For example, disulfide-containing peptides can be identified by their amino acid composition or sequence only if they

[1] T. W. Thannhauser, Y. Konishi, and H. A. Scheraga, this series, Vol. 143, p. 115.
[2] P. E. Staswick, M. A. Hermodson, and N. C. Nielsen, J. Biol. Chem. 259, 13431 (1984).
[3] T. W. Thannhauser, C. A. McWherter, and H. A. Scheraga, Anal. Biochem. 149, 322 (1985).
[4] L. Haeffner-Gormley, L. Parente, and D. B. Wetlaufer, Int. J. Pept. Protein Res. 26, 83 (1985).
[5] J. R. Reeve, Jr., and J. G. Pierce, Int. J. Pept. Protein Res. 18, 79 (1981).
[6] W. R. Gray, F. A. Luque, R. Galyean, E. Atherton, R. C. Sheppard, B. L. Stone, A. Reyes, J. Alford, M. McIntosh, B. M. Olivera, L. J. Cruz, and J. Rivier, Biochemistry 23, 2796 (1984).
[7] T. E. Creighton, this series, Vol. 107, p. 305.

are purified to homogeneity. This requirement may not be achieved in the case of large proteins, or when the quantity of protein is very small. It is likewise important to note that the conditions used to sequentially reduce disulfide bonds are also the conditions that frequently lead to disulfide exchange. As a result, the locations of disulfide bonds in the partially reduced protein may not be the same as the locations of disulfide bonds in the native material. The problem of disulfide exchange plagues all attempts to locate disulfide bonds in proteins, irrespective of the general method!

The advent of fast atom bombardment mass spectrometry (FAB-MS) has made the task of locating disulfide bonds in proteins more tractable. Morris and Pucci[8] were the first to recognize the potential of FAB-MS for identifying disulfide-containing peptides from which the locations of disulfide bonds in proteins may be determined. Since the amino acid sequence of a protein is usually known, at least in the regions of the half-cystinyl residues, before an attempt is made to locate the disulfide bonds, FAB-MS can be used to identify disulfide-containing peptides, even if they have not been purified to homogeneity. The ability to identify peptides in mixtures constitutes a significant advantage of FAB-MS. This feature is especially important when only small quantities of protein are available.

FAB-MS has now been used to locate disulfide bonds in a variety of proteins.[9-13] As a result, this application will likely become another standard tool for protein chemists. The goal of this chapter is to describe a general strategy based on this technique for locating disulfide bonds in proteins.

Interconversion of Thiols and Disulfides

Disulfide bonds are easily reduced to give the corresponding thiols under some conditions [Eq. (1)]. This reaction may be highly undesirable because it is the first step in the formation of new disulfide bonds [Eq. (2)].

[8] H. R. Morris and P. Pucci, *Biochem. Biophys. Res. Commun.* **126,** 1122 (1985).

[9] T. Takao, M. Yoshida, Y.-M. Hong, S. Aimoto, and Y. Shimonishi, *Biomed. Mass Spectrom.* **11,** 549 (1984).

[10] R. Yazdanparast, P. Andrews, D. L. Smith, and J. E. Dixon, *Anal. Biochem.* **153,** 348 (1986).

[11] R. Yazdanparast, P. C. Andrews, D. L. Smith, and J. E. Dixon, *J. Biol. Chem.* **262,** 2507 (1987).

[12] F. Raschdorf, R. Dahinden, W. Maerki, W. J. Richter, and J. P. Merryweather, *Biomed. Environ. Mass Spectrom.* **16,** 3 (1988).

[13] S. Akashi, K. Hirayama, T. Seino, S.-i. Ozawa, K.-i. Fukuhara, N. Oouchi, A. Murai, M. Arai, S. Murao, K. Tanaka, and I. Nojima, *Biomed. Environ. Mass Spectrom.* **15,** 541 (1988).

For proteins that have more than one disulfide bond, these reactions may lead to disulfide exchange. When controlled, this reaction is useful for identifying disulfide-containing peptides. As a result, it is important to understand the interconversion of thiols and disulfides.

$$
\left(\begin{array}{c} S \\ | \\ S \end{array}\right. \xrightleftharpoons{RSH} \left(\begin{array}{c} S-SR \\ \\ SH \end{array}\right. \xrightleftharpoons{RSH} \left(\begin{array}{c} SH \\ \\ SH \end{array}\right. + RS-SR \tag{1}
$$

$$
\left(\begin{array}{c} S \\ | \\ S \end{array}\right. + \left.\begin{array}{c} HS \\ \\ HS \end{array}\right) \rightleftharpoons \left(\begin{array}{cc} S-S \\ \\ SH & HS \end{array}\right) \tag{2}
$$

As indicated in Eq. (1), disulfide bonds are converted into thiols by addition of a reducing agent, such as dithiothreitol or mercaptoethanol. Dithiothreitol is usually preferred because it is a stronger reducing agent, and because it is also an excellent matrix for FAB-MS. However, mercaptoethanol is more volatile and can be removed from a sample placed under vacuum. As a practical note, disulfide bonds can generally be reduced after incubation at pH 8 for 4 hr with an excess of dithiothreitol.[14] Alkaline conditions are required because the reactive intermediate in the reduction reaction is the thiolate anion. Most thiols have pK_a values in the range 7–9.[15]

Disulfide exchange is a two-step process [Eqs. (1) and (2)], both of which involve the thiolate anion. As a result, acidic conditions are always preferred for investigations of disulfide cross-linkages. Since the resistance of a disulfide bond to reduction is strongly dependent on the conformation of the protein in which it is located, it is generally not possible to predict the upper limit of pH at which particular disulfide bonds are stable. For this reason, it is important to be cognizant of the possibility of disulfide interchange, and to choose the analytical procedures accordingly. It is also noted that concentrated acid (e.g., 8 M sulfuric acid) can cause disulfide exchange.[16] More detailed discussions of disulfide/thiol/disulfide interconversion have been published.[7,17]

The facile reduction of disulfide bonds may be used advantageously in assays for their detection and identification. Yazdanparast et al.[10] have shown that disulfide bonds in peptides are reduced in situ during analysis

[14] W. Konigsberg, this series, Vol. 25, p. 185.
[15] M. Friedman, "The Chemistry and Biochemistry of the Sulfhydryl Group in Amino Acids, Peptides and Proteins." Pergamon, New York, 1973.
[16] A. P. Ryle, F. Sanger, L. F. Smith, and R. Kitai, Biochem. J. 60, 541 (1955).
[17] D. B. Wetlaufer, this series, Vol. 107, p. 301.

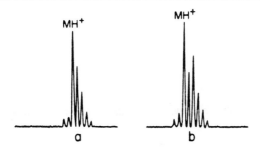

FIG. 1. Mass spectrum of the molecular ion region of [Arg⁸]vasopressin: (a) zero time; (b) after 6 min of continuous xenon atom beam bombardment. (From Ref. 10; reprinted with permission.)

by FAB-MS. This phenomenon is illustrated in Fig. 1, which shows the molecular ion region of the FAB mass spectrum of [Arg⁸]vasopressin. This peptide has a molecular weight of 1083 and one disulfide bond. The relative intensities of the molecular ion peaks recorded shortly after bombardment with xenon atoms are as expected for the natural distribution of isotopes (Fig. 1a). However, after continuous bombardment of the sample for 6 min, the intensity of the "second isotope" peak is elevated. Based on experiments performed with a variety of matrices, with and without continuous bombardment with xenon, it was demonstrated that the increased intensity of the "second isotope" peak was due to the molecular ion (MH⁺) of the reduced peptide. It follows that the mass spectrum in Fig. 1b is a superposition of the isotope peaks of the MH⁺ ions of the nonreduced (disulfide-containing) and reduced (thiol-containing) forms of [Arg⁸]vasopressin. The peculiar "isotope pattern" illustrated in Fig. 1b is diagnostic for peptides which have intramolecular disulfide bonds.

Peptides with intermolecular disulfide bonds also undergo reduction during FAB-MS analysis. In this case, the FAB mass spectrum may have molecular ions for both of the constituent peptides of the disulfide-containing peptide. This is illustrated in Fig. 2, which is the FAB mass spectrum of the 1–17/120–129 disulfide-bonded peptide (MH⁺ 3086.6) isolated from the partial acid hydrolysate of hen egg-white lysozyme. The peaks at m/z 1202.6 and 1887.0 are the molecular ions of the constituent peptides 1–17 and 120–129.

Although disulfide-containing peptides may be reduced at different rates, the FAB mass spectra of most peptides give evidence for reduction of the disulfide bonds within 5 min of continuous bombardment with xenon. This evidence may be an increase in the intensity of a peak two mass units higher than the molecular ion of peptides with an intramolecular

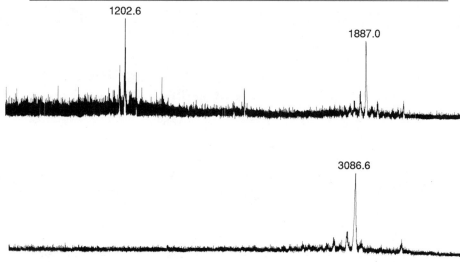

FIG. 2. FAB mass spectrum of a peptide (MH$^+$ 3086.6) which contains an intermolecular disulfide bond. Peaks at m/z 1202.6 and m/z 1887.0 are the molecular ions of constituent peptides which are formed during FAB-MS analysis. This peptide, which comprises the 1–17/120–129 segments of hen egg-white lysozyme, was isolated from a partial acid hydrolysate of the protein.

disulfide bond, or the appearance of the molecular ions of the constituent peptides for peptides that contain intermolecular disulfide bonds. The molecular ion of one of the constituent peptides may not be apparent, because its m/z is too low to be distinguished from the matrix peaks. In addition, hydrophilic peptides often give weak FAB-MS signals when mixed with hydrophobic peptides.[18,19]

Specific Fragmentation of Proteins

Disulfide cross-linkages may be located by cleaving a protein between half-cystinyl residues to give peptides that contain only one disulfide bond. The molecular weights of these peptides are determined by FAB-MS and related to specific segments of the parent protein. This latter step, identification of the disulfide-containing peptide, is facilitated greatly if a cleavage reagent which cleaves the protein at specific sites is used. For

[18] S. Naylor, A. F. Findeis, B. W. Gibson, and D. H. Williams, *J. Am. Chem. Soc.* **108,** 6359 (1986).
[19] D. H. Williams, A. F. Findeis, S. Naylor, and B. W. Gibson, *J. Am. Chem. Soc.* **109,** 1980 (1987).

example, tryptic peptides are easily identified because the C-terminal residue of most of the peptides is either Arg or Lys.

Proteins may be cleaved selectively with a variety of enzymes, such as trypsin, chymotrypsin, *Staphylococcus aureus* V8, or pepsin.[20,21] Trypsin has been used most frequently because it is highly specific for Arg and Lys residues. Unfortunately, trypsin has maximum activity at pH 8.3, and is not active in acid. As a result, disulfide interchange may occur during enzymolysis. Chymotrypsin is less specific than trypsin, and is likewise active only under alkaline conditions. *Staphyloccocus aureus* V8 is useful because it has maximum activity at pH 4.0 and 7.8, cleaving on the C-terminal side of Glu or Glu and Asp, respectively. Pepsin has maximum activity around 3.0, and is, therefore, ideal for preserving disulfide cross-linkages. Unfortunately, the sites at which pepsin cleavage is likely are not highly predictable.

Cyanogen bromide is also used frequently to cleave proteins on the C-terminal side of Met.[20,21] This reaction is attractive because it is highly specific for Met, it is generally free of side reactions, and because the reagents (cyanogen bromide and formic acid) are volatile. In addition, disulfide bonds are stable during treatment with cyanogen bromide. Since proteins in their native state are often resistant to enzymatic attack, cleavage with cyanogen bromide is used to open up or unfold the protein, rendering it susceptible to enzymolysis. Because Met is not a particularly abundant constituent of proteins, cleavage with cyanogen bromide usually gives large peptides, frequently containing more than two half-cystinyl residues or too large for analysis by FAB-MS.

In practice, multiple chemical and enzymatic cleavage reagents will likely be required to produce disulfide-containing peptides from which all of the disulfide cross-linkages in large proteins may be located. If the sample must be exposed to alkaline conditions, the results should be examined carefully for evidence of disulfide interchange. When less specific cleavage reagents, such as chymotrypsin or pepsin are used, methods for systematically relating the molecular weights of peptides to specific segments of the protein must be used.

The general approach we currently use to locate disulfide bonds in proteins may be illustrated with ribonuclease A, which is a small protein with 124 amino acids and four disulfide bonds. The structure of ribonuclease A, is given in Fig. 3. The protein was cleaved initially on the C-terminal side of Met with cyanogen bromide. Although there are only two

[20] T. D. Lee and J. E. Shively, this volume [19].
[21] G. A. Allen, *in* "Laboratory Techniques in Biochemistry and Molecular Biology," 2nd Ed. Elsevier, Amsterdam, 1989.

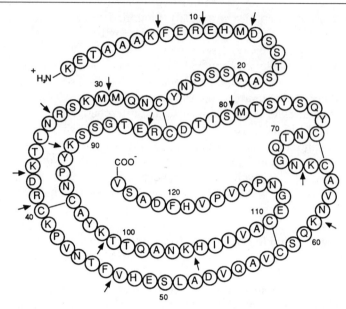

FIG. 3. Structure of ribonuclease A illustrating sites at which it is cleaved by trypsin and CNBr. (Adapted from Ref. 11 with permission.)

Met residues in ribonuclease A, their cleavage is important because it makes the protein susceptible to attack by proteases such as trypsin or *Staphylococcus aureus* V8. The m/z values of molecular ions of peptides found in the cyanogen bromide/tryptic digest of ribonuclease A are given in Table I. Coordinates relating these peptides to specific segments of the protein are included. Although these assignments were made originally by assuming that cleavage had occurred after Met, Lys, and Arg, it has now become apparent that a systematic search of all possible assignments via computer-assisted analysis is desirable, even when highly specific proteases are used to fragment the protein. For example, trypsin usually does not cleave on the carboxyl side of Arg or Lys when it is followed by Pro. As a result, peptides that would result from cleavage of the peptide bond joining Lys-41 to Pro-42 in ribonuclease A were not formed (see Fig. 3). A molecular ion corresponding to the 40–61 segment, which includes Lys-41 and Pro-42, was found at m/z 2403 (Table I).

Correlation of the peptides listed in Table I with the structure of ribonuclease A (Fig. 3) indicates that peptides which were derived from nearly all of the parent protein were found in the digest. Unfortunately, none of these peptides changed molecular weight when the digest was treated with a reducing agent, suggesting that none of the peptides found by FAB-MS

TABLE I

ASSIGNMENTS AND MASS-TO-CHARGE RATIO OF
THE MOLECULAR IONS FOUND BY FAB-MS
ANALYSIS OF TRYPTIC DIGEST OF CNBr-
TREATED RIBONUCLEASE A[a]

Residues	m/z
1–7	718
8–10	451
11–13	367
34–37	475
40–61	2403
62–66	534
80–85	694
86–91	608
92–98	858
99–104	662
105–124	2167

[a] Adapted from Ref. 11.

analysis of the digest contained disulfide bonds. It is interesting to note that a peptide (MH[+] 2403, residues 40–61) which has two cysteinyl residues (Cys-40 and Cys-65) was found. Had the mass of this peptide been determined (incorrectly) to be two mass units lower, we would have concluded (incorrectly) that Cys-40 and Cys-65 are joined by a disulfide bond.

The relative sensitivity of FAB-MS for peptides that are analyzed in a mixture, such as an enzymatic digest, depends substantially on chemical and physical properties of the peptides.[18,19] Further analysis of this digest after fractionation by reversed-phase high-performance liquid chromatography (HPLC) demonstrated that disulfide-containing peptides were indeed present in the digest. The chromatogram of this digest is given in Fig. 4. Chromatographic peaks F1–5 were collected and analyzed by FAB-MS. Molecular weights and assignments of disulfide-containing peptides found in these fractions are given in Table II. All of these fractions gave FAB mass spectra with peaks that disappeared when the fraction was treated with a reducing agent, indicating that these peaks are due to disulfide-containing peptides. Fraction 3 may be used to illustrate a problem frequently encountered when proteases are used to fragment the protein. The peak at m/z 1644 was assigned to the 40–46/92–98 peptide, which was formed by cleavage after Phe-46. Despite the high specificity of trypsin for cleaving after Arg and Lys, it occasionally displays chymotryptic behavior, cleaving on the C-terminal side of Phe, Trp, Tyr, Leu, or

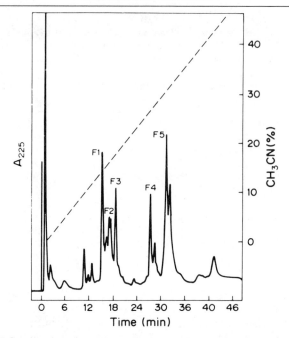

FIG. 4. HPLC separation of the tryptic peptides of CNBr-treated bovine ribonuclease A. F1–F5 indicate the elution of peptides that have disulfide bonds. (From Ref. 11; reprinted with permission.)

Met.[21,22] Chymotryptic-like cleavage reactions may occur efficiently even though the trypsin has been chemically treated to inhibit any chymotrypsin that may be present as a contaminant. Because such unexpected cleavage is frequently encountered, it is important to systematically search for all possible assignments of each molecular ion. In the case of fraction 3, a computer-assisted search of the amino acid sequence of ribonuclease A indicates that there is only one contiguous segment (i.e., 40–46) whose molecular ion would be at m/z 808. A similar systematic search for segments which could give a peak at m/z 858 indicate that there are three potential assignments. However, only one of the segments would be formed by cleavages consistent with the specificity of trypsin.

Fractions 4 and 5 illustrate other types of problems that may be encountered when analyzing proteolytic digests for disulfide-containing peptides. Although constituent peptides of the disulfide-containing peptide 47–61/105–124 (m/z 1614 and 2167) were present, a peak corresponding to the intact parent peptide (m/z 3778) was not detected. The results for fraction 5 were similar in that constituent peptides were found, but the molecular

[22] S. Maroux, M. Rovery, and P. Nesnuelle, *J. Biol. Chem.* **255,** 4786 (1966).

TABLE II
SEGMENTS AND MASS-TO-CHARGE RATIO OF MOLECULAR IONS OF
PEPTIDES FOUND BEFORE AND AFTER CHEMICAL REDUCTION OF
SELECTED HPLC FRACTIONS OF TRYPTIC DIGEST OF CNBr-TREATED
RIBONUCLEASE A[a]

		Mass of MH[+]	
		---	---
Fraction	Residues	Nonreduced	Reduced
1	62–66/67–79	1979	534
			1449
2	14–29/80–85	2296	1604
			694
3	40–46/92–98	1664	808
			858
4	47–61/105–124	—	1614
			2167
5	92–98/40–61/105–124	—	858
			2403
			2167

[a] Adapted from Ref. 11.

ion of the intact disulfide-containing peptide (m/z 5423) was not detected, probably because it was beyond the effective mass range of the mass spectrometer. This peptide has two cystinyl residues, and could not, therefore, be used to locate disulfide bonds uniquely.

From the results presented in Table II, it is apparent that two of the disulfide bonds (Cys-65/Cys-72 and Cys-26/Cys-84) can be located uniquely by examination of the appropriate chromatographic fractions of the tryptic digest of CNBr-treated ribonuclease A. The same procedure was used to locate the remaining disulfide bonds from peptides found in a *Staphylococcus aureus* V8 digest. Locating disulfide bonds from the corresponding disulfide-containing peptides is not difficult if the protein is fragmented between the half-cystinyl residues with highly specific enzymes. Although some peptides may be formed by incomplete cleavages, or by unexpected cleavage reactions, the high specificity of the enzyme is an important aid for identifying the peptides. In addition, multiple cleavage reactions may be required to definitively locate all disulfide linkages.

Nonspecific Fragmentation of Proteins

Peptides formed by highly specific cleavage of proteins are relatively easy to identify because the N- or C-terminal residues are known. However, proteins are often resistant to specific cleavage reagents, either because the half-cystinyl residues are not separated by residues that are required by the reagent for cleavage, or because the peptide bonds are not accessible to the cleavage reagent due to protein folding. The latter problem is frequently encountered with proteins that are tightly coiled, as is often the case when the disulfide bonds are intact. As a result, cleavage reactions that occur readily for proteins in which the disulfide bonds have been reduced and carboxymethylated may not proceed appreciably for the native protein. Proteins may also be fragmented with reagents that are not highly sequence-specific. For proteins which are highly folded, peptide bonds exposed to the solution are cleaved most easily. After a few cuts have been made on the surface of the protein, partial denaturation exposes more sites for cleavage. This process may continue until a complex mixture of peptides or amino acids has been produced.

Partial acid hydrolysis is an example of a nonspecific cleavage reaction. Since the hydrolysis conditions may be varied so that the primary products are large peptides or amino acids, it should be possible to fragment any protein between the half-cystinyl residues to give disulfide-containing peptides from which the locations of disulfide bonds can be determined. Partial acid hydrolysis, which is also attractive because disulfide bonds are stable in acid,[7,17] has been used to investigate disulfide bonding.[11,16,23,24] Unfortunately, the low specificity of partial acid hydrolysis of even a small protein results in the formation of a highly complex mixture of peptides from which extraction and identification of disulfide-containing peptides may be difficult. However, new analytical methods, which include HPLC, FAB-MS, and computer-assisted analysis of mass spectrometric information have made it possible to reliably identify peptides diagnostic of disulfide cross-linking. This approach, which is given in Scheme 1, will be illustrated with hen egg-white lysozyme (HEL).

The first step is to determine hydrolysis conditions that provide peptides suitable for locating disulfide bonds. Dilute acids are used for partial acid hydrolysis because concentrated solutions of strong acids give mostly small peptides. Previous reports of the variables of partial acid hydrolysis[25-27] should be consulted for more information. The sensitivity of lysozyme to hydrolysis conditions was investigated by varying the temperature and time of hydrolysis for hydrochloric ($0.03\ M$), acetic ($0.25\ M$), and oxalic ($0.25\ M$) acids. The extent of hydrolysis was monitored by reversed-phase HPLC. This survey indicated that hydrolysis with oxalic acid at

[23] Z. Zhou and D. L. Smith, *J. Protein Chem.* **9**, in press (1990).
[24] J. G. Pierce and T. F. Parsons, *Annu. Rev. Biochem.* **50**, 465 (1981).
[25] A. Light, this series, Vol. 11, p. 417.
[26] A. S. Inglis, this series, Vol. 91, p. 324.
[27] J. Schultz, this series, Vol. 11, p. 255.

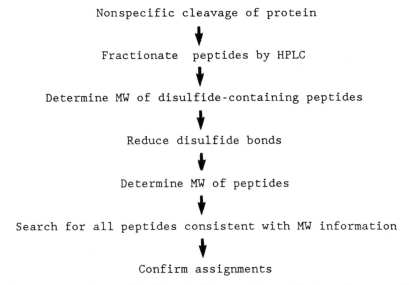

SCHEME 1. Procedure for identifying disulfide-containing peptides formed by nonspecific cleavage of a protein.

100° for 8 hr gives many disulfide-containing peptides with molecular weights in a range accessible to FAB-MS. Hen egg-white lysozyme (1 mM) in 0.25 M oxalic acid was sealed in an evacuated glass tube (5 mm id) and heated. Aliquots of the hydrolyzate were analyzed directly by HPLC.

Partial acid hydrolysis of a relatively small protein, such as HEL, gives several hundred peptides under these conditions. Although the hydrolysate is fractionated by reversed-phase HPLC to group the peptides by hydrophobicity, no attempt is made to isolate the peptides to homogeneity. Mass spectra are recorded for each fraction before and after chemical reduction of the disulfide bonds. Peaks that are not present in the mass spectra of reduced aliquots are assumed to be due to disulfide-containing peptides. Molecular weight information obtained by FAB-MS is used with the known amino acid sequence of the parent protein to identify disulfide-containing peptides. Computer-assisted analysis is used to match the molecular weights of these peptides with all segments of the protein that could possibly be joined by a disulfide bond. The computer program, which allows for hypothetical cleavage at any peptide bond, restricts the output to peptides that have two cysteinyl (i.e., half-cystinyl) residues. As a result, peptides which contain one disulfide bond as well as an additional

TABLE III
MASS-TO-CHARGE RATIO OF MOLECULAR IONS
FOUND BY FABMS ANALYSIS BEFORE AND
AFTER CHEMICAL REDUCTION OF ONE HPLC
FRACTION OF PARTIAL ACID HYDROLYSATE OF
HEN EGG-WHITE LYSOZYME

m/z before reduction	m/z after reduction
2038	—[a]
1901	—[a]
1844	—[a]
1562	1562
1456	1456
1203	1203
1047	1047
904	904
789	789

[a] The peak was not present after reduction.

cysteinyl residue are rejected. In addition, peptides with molecular weights exceeding the mass range of the mass spectrometer are excluded.

The m/z values of all of the prominent molecular ion peaks found in the FAB mass spectrum of one fraction of a partial acid hydrolysate of HEL are listed in Table III. Peaks below m/z 700 are not included. Peaks due to disulfide-containing peptides were distinguished from others by chemically reducing an aliquot of the sample and noting which peaks were no longer present in the FAB mass spectrum. This experiment showed that the three highest mass ions were due to disulfide-containing peptides. A systematic search of the amino acid sequence of HEL for pairs of segments that could be joined by a disulfide bond to give a peptide with a molecular ion at m/z 2038 ± 1 shows that there are at least 579 possibilities! This calculation assumes that cleavage of any peptide bond is possible, that any pair of half-cystinyl residues may be joined by a disulfide bond, and that each segment contains only one half-cystinyl residue. Because these assumptions are predicated only on knowing the sequence of the protein, this approach can be used for any protein and any cleavage reagent. Unfortunately, the approach is so unrestrictive that there are always a very large number of possible assignments. However, the specificity of the search is increased greatly if it is assumed that at least one of the other peaks in the FAB mass spectrum is due to one of the constituent peptides of the disulfide-containing peptides. Results obtained when the search is limited in this way are presented in Table IV for the three disulfide-containing peptides listed in Table III. For the disulfide-contain-

TABLE IV
COMPUTER-ASSISTED ANALYSIS OF MOLECULAR WEIGHT
INFORMATION GIVEN IN TABLE III[a]

Segment	Disulfide linkage	m/z of S–S peptide[b]	m/z of constituent peptide[b]	m/z of S–S peptide after Edman[b]
Peak at m/z 2038				
1–7	6/127	2038.1	838.4	1810.9
120–129			1202.7	
111–119	115/127	2037.0	1136.5	1664.9
123–129			903.5	
Peak at m/z 1901				
5–17	6/76	1900.0	1455.8	1640.8
76–79			447.2	
5–17	6/80	1900.0	1455.8	1627.8
77–80			447.2	
5–17	6/94	1902.0	1455.8	1642.9
94–97			449.2	
5–17	6/127	1901.0	1455.8	1631.8
124–127			448.2	
5–17	6/127	1901.0	1455.8	1687.9
126–129			448.2	
22–30	30/127	1900.0	999.5	1656.8
123–129			903.5	
Peak at m/z 1844				
5–17	6/30	1845.0	1455.8	1584.8
30–33			392.2	
5–17	6/127	1844.0	1455.8	1691.8
127–129			391.2	
23–30	30/127	1842.9	942.4	1493.8
123–129			903.5	

[a] Tentative assignments of peaks due to disulfide-containing peptides and their constituent peptides are given. In addition, the mass-to-charge ratios of the molecular ions for disulfide-containing peptides after one step of Edman degradation are included.

[b] See Table III for a list of mass-to-charge ratios actually found before and after reduction.

ing peptide with a molecular ion at m/z 2038, this restriction reduces the number of tentative assignments from 579 to 2. The correct assignment could be established by determining the m/z of the molecular ion more accurately (e.g., to the nearest whole mass unit) or by removing the N-terminal residues of both segments by a one-step manual Edman degradation and determining the molecular weight of the peptide. After one step of Edman degradation, there was a new peak at m/z 1811, clearly showing that the peak at m/z 2038 was due to the 1–7/120–129 disulfide-containing peptide (Table IV). Since each of the segments (1–7 and 120–129) contains only a single half-cystinyl residue, it is concluded that these residues are joined by a disulfide bond.

A similar systematic search was made for segments of HEL which might be assigned to the other two disulfide-containing peptides with molecular ions at m/z 1901 and 1844. For the peptide at m/z 1901, there are six tentative assignments within a mass range of ± 1 unit, and only two assignments (5–17/124–127 and 5–17/126–129) with a molecular ion with a nominal mass of 1901. Since both assignments include Cys-6 and Cys-127, the disulfide bond is located by determining the m/z of the molecular ion to the nearest nominal mass. When the fraction was subjected to one step of manual Edman degradation, a new peak appeared at m/z 1688. As indicated in Table IV, this observation is consistent only with the 5–17/126–129 peptide. The remaining disulfide-containing peptide present in this chromatographic fraction was assigned to segments 5–17/127–129 by the unique m/z of the molecular ion (m/z 1844). It is noted that a peak corresponding to the removal of both N-terminal residues via Edman degradation (m/z 1535.8) was not found. We have generally had difficulty performing Edman degradation on peptides in which half-cystine is an N-terminal residue.

To locate all of the disulfide bonds in a protein, this strategy may be applied to each chromatographic fraction. Some fractions will likely contain no peptides with disulfide bonds. Other fractions will likely contain peptides with multiple disulfide bonds, or cysteinyl residues. By systematically searching for disulfide-containing peptides which could possibly be consistent with the peaks in the FAB mass spectrum, attention is focused on a small number of tentative assignments from which the locations of disulfide bonds may be uniquely deduced. When this approach was extended to all fractions of a partial acid hydrolysate of HEL, the molecular ions of 72 peptides disappeared upon chemical reduction.[23] Forty-six of these disulfide-containing peptides were definitively assigned to specific segments of HEL. These assignments were verified by changes in the molecular weights of the peptides after Edman degradation, or after treating the fractions with a residue-specific protease, such as trypsin or *Staphylococcus aureus* V8. The peptides that contain disulfide bonds but could not be definitively assigned to specific segments of HEL are likely due to impurities in the starting material, or to modifications of the protein which occur to a limited extent during partial acid hydrolysis. For example, Met may be oxidized during isolation or hydrolysis of a protein. Because the conditions used to hydrolyze HEL lead to cleavage of the protein at many sites, a very large number of peptides are produced. As a result, there are many peptides which may be related to the same disulfide linkage. Although this redundancy of information increases the confidence with which the locations of disulfide bonds may be located, it does lower the sensitivity of the method. Partial acid hydrolysis will be most useful for locating disulfide bonds in proteins for which highly specific reagents fail to give the appropriate disulfide-containing peptides.

Acknowledgments

This work was supported by grants GM 40384 and EY 07609 from the National Institutes of Health. The authors wish to acknowledge the assistance of Ted Smith and Jean Tien Pham for writing computer programs used for analysis of mass spectrometric data.

[21] Reversed-Phase High-Performance Liquid Chromatography for Fractionation of Enzymatic Digests and Chemical Cleavage Products of Proteins

By Kathryn L. Stone, James I. Elliott, Glenn Peterson, Walter McMurray, and Kenneth R. Williams

The extremely high resolving power and speed of reversed-phase high-performance liquid chromatography (HPLC) make it the current method of choice for fractionating complex mixtures of peptides derived from the enzymatic and chemical cleavage of proteins. With its high peak capacity, reversed-phase HPLC can readily bring about a 100- to 125-fold purification of a typical tryptic peptide with a gradient time of only about 90 min.[1,2] The excellent reproducibility of reversed-phase HPLC has fostered the use of comparative HPLC tryptic peptide mapping for identifying sites of posttranslational[3] and *in vitro* chemical modifications,[4] establishing precursor/product relationships,[5] confirming the identity of proteins,[6] and readily

[1] K. L. Stone, M. B. LoPresti, and K. R. Williams, in "Laboratory Methodology in Biochemistry" (C. Fini, A. Floridi, and V. Finelli, eds.), in press. CRC Press, Boca Raton, FL, 1989.

[2] K. L. Stone, M. B. LoPresti, J. M. Crawford, R. DeAngelis, and K. R. Williams, in "HPLC of Peptides and Proteins: Separation, Analysis and Conformation" (R. S. Hodges, ed.), in press. CRC Press, Boca Raton, FL, 1989.

[3] T. C. Terwilliger and D. E. Koshland, J. Biol. Chem. 259, 7719 (1984).

[4] K. R. Williams, K. L. Stone, M. K. Fritz, B. M. Merrill, W. H. Konigsberg, M. Pandolfo, O. Valentini, S. Riva, S. Reddigari, G. L. Patel, and J. W. Chase, in "Proteins: Structure and Function" (J. J. L'Italien, ed.), p. 45. Plenum, New York, 1987.

[5] A. Kumar, K. R. Williams, and W. Szer, J. Biol. Chem. 261, 11266 (1986).

[6] K. R. Williams, S. Reddigari, and G. L. Patel, Proc. Natl. Acad. Sci. U.S.A. 82, 5260 (1985).

detecting single amino acid substitutions.[7] In one recent instance,[8] comparative HPLC tryptic peptide maps succeeded in identifying a single asparagine to serine change in a particular isolate of the bacteriophage T4 DNA polymerase. This 104,000 Da DNA polymerase contains a total of 898 amino acids.

Although reversed-phase HPLC provides an invaluable tool for the protein chemist, its utility is enhanced even further by coupling it with mass spectrometry. A reversed-phase HPLC tryptic peptide/FAB mass spectrometric approach provides an elegant and rapid means to accurately determine the molecular weights of the resulting peptides. This information can, in turn, be used to rapidly verify the primary structures of proteins that have been deduced from their DNA sequences.[9,10] With the use of a tandem mass spectrometer, this reversed-phase HPLC/mass spectrometric approach can be extended further so that selected tryptic peptides can be sequenced and sites of posttranslational modifications precisely identified.[11] Recent advances in the sensitivity of tandem MS/MS clearly indicate its potential for sequencing peptides at the low picomole level.[12,13] As predicted by Gibson et al.,[13] "as further improvements are made in the methodology for mass spectrometric protein and peptide sequencing, it is almost certain that mass spectrometry will become an essential part of the 'classical' strategy now being used by most protein chemists."

Despite the relative simplicity of reversed-phase HPLC, there are several decisions that have to be made regarding the appropriate choice of an HPLC system; mobile phase, column manufacturer and dimensions, gradient parameters, and loading conditions can all affect the resulting chromatogram. Each of these questions will be dealt with in turn in this chapter with the goal of providing practical suggestions that will work well with nanomole and subnanomole amounts of peptide mixtures. While most of the studies that will be described were carried out on 50-pmol aliquots derived from a large-scale tryptic digest of transferrin, identical conditions can be used for separating tryptic and other cleavage peptides derived from 25 pmol or larger quantities of virtually any other protein.

[7] K. R. Williams, J. B. Murphy, and J. W. Chase, *J. Biol. Chem.* **259,** 11804 (1984).

[8] J. Rush and W. H. Konigsberg, unpublished data (1989).

[9] B. W. Gibson and K. Bieman, *Proc. Natl. Acad. Sci. U.S.A.* **81,** 1956 (1984).

[10] H. A. Scoble, *in* "A Practical Guide to Protein and Peptide Purification for Microsequencing" (P. Matsudaira, ed.), p. 91. Academic Press, San Diego, CA, 1989.

[11] K. Biemann and H. A. Scoble, *Science* **237,** 992 (1987).

[12] P. R. Griffin, J. Shabanowitz, J. R. Yates, N. Z. Zhu, and D. F. Hunt, *in* "Techniques in Protein Chemistry" (T. E. Hugli, ed.), p. 160. Academic Press, San Diego, CA, 1989.

[13] B. W. Gibson, Z. Yu, B. Gilece-Castro, W. Aberth, F. C. Walls, and A. L. Burlingame, *in* "Techniques in Protein Chemistry" (T. E. Hugli, ed.), p. 135. Academic Press, San Diego, CA, 1989.

Performance Criteria for Reversed-Phase High-Performance
Liquid Chromatography

Resolution, sensitivity of detection, and reproducibility are three important HPLC parameters that can be quantitated most accurately by chromatographing an aliquot of an enzymatic digest of a reasonably large protein. A tryptic digest of carboxamidomethylated transferrin, which has a molecular weight of approximately 75,000, serves as a convenient standard for assessing reversed-phase HPLC performance. In preparation for carrying out the digest, a large-scale carboxamidomethylation reaction is started by dissolving 300 mg of human transferrin in 2 ml of 6 M guanidine-HCl, 0.5 M Tris-HCl, pH 8.1, and 2 mM EDTA. After incubating at 50° for 30 min, the protein is reduced by adding 174 mg dithiothreitol and then incubating at 50° for an additional 4 hr. The transferrin is then alkylated by adding a twofold molar excess (416 mg) of iodoacetamide over dithiothreitol and incubating in the dark for 20 min at room temperature. After extensive dialysis against 1 mM NH$_4$HCO$_3$, the precipitated protein is dried *in vacuo,* redissolved in 2 ml 8 M urea, 0.4 M NH$_4$HCO$_3$, and an aliquot subjected to hydrolysis and ion-exchange amino acid analysis to accurately determine the protein concentration. After diluting to 2 M urea, 0.1 M NH$_4$HCO$_3$ with water, a 1:25 (w/w) ratio of trypsin:substrate is added and the digest continued for 24 hr at 37°. After acidifying, the digest is stable for at least 6 months at −20° and suitable dilutions are made in 0.05% trifluoroacetic acid (TFA). It should be noted that this large-scale transferrin digestion procedure is not suitable for use with subnanomole or low nanomole amounts of proteins where the carboxamidomethylation reaction and subsequent trypsin digestion are best carried out in the same tube without any intervening buffer exchange. With small amounts of denatured proteins, the latter step is likely to lead to partial or complete loss of the sample. Chapter [19] in this volume, as well as previous publications, provides detailed methodologies for enzymatically digesting subnanomole amounts of proteins[1,10,14,15] and for carrying out chemical cleavages.[10,16]

In order to test a particular HPLC parameter, at least five consecutive injections of a dilution of the digest were made in a volume of 20 μl each. In the case of an analytical-size column (3.9–4.6 mm id), a 250-pmol aliquot

[14] K. L. Stone, M. B. LoPresti, J. M. Crawford, R. DeAngelis, and K. R. Williams, *in* "Techniques in Protein Chemistry" (T. E. Hugli, ed.), p. 377. Academic Press, San Diego, CA, 1989.

[15] K. L. Stone, M. B. LoPresti, J. M. Crawford, R. DeAngelis, and K. R. Williams, *in* "Protein Sequence from Microquantities of Proteins and Peptides" (P. Matsudaira, ed.), p. 31. Academic Press, San Diego, CA, 1989.

[16] A. Fontana and E. Gross, *in* "Practical Protein Chemistry—A Handbook" (A. Darbre, ed.), p. 67. Wiley, New York, 1986.

was normally used and the column was eluted at a flow rate of 1.0 ml/min. These parameters were reduced to 50 pmol and a flow rate of 0.15 or 0.20 ml/min in the case of a narrow-bore (2.0–2.1 mm id) column. In all instances, the column was equilibrated with 98% buffer A (0.05% TFA) and 2% buffer B (0.05% TFA in 80% CH_3CN) and except where noted the following gradient was used: 0–63 min, 2–37% B; 63–95 min, 37–75% B; 95–105 min, 75–98% B. Resolution was quantitated in terms of the number of absorbance peaks at 210 nm detected by a Nelson Analytical Model 4416X Multi-Instrument Data System using an area threshold of 50 μV-sec, a noise threshold of 1 μV, and a sampling rate of 0.434 points/sec. In those cases where gradient times were varied, the sampling rate was also proportionately changed in order to keep the number of data points per run constant. Reproducibility was quantitated in terms of the average variation in retention time of 10 absorbance peaks that were chosen on the basis of their ease of identification in repetitive runs and the fact that they were well distributed throughout the chromatogram.

High-Performance Liquid Chromatography

Certainly the first and one of the most important decisions that must be made prior to carrying out a reversed-phase HPLC peptide separation is selecting an appropriate HPLC system. In the following section four commercial HPLC systems—the Hewlett-Packard (HP) 1090 (Palo Alto, CA), the Applied Biosystems (ABI) Model 130 (Foster City, CA), a Waters Associates Analytical System (Milford, MA) (see Fig. 4 below), and the Perkin-Elmer Peptide Mapping System (Emeryville, CA) consisting of a Model 250 binary pump, a Model 7125 injector, and the LC-135 diode array detector—were used as test instruments to illustrate how the relative sensitivity, resolution, and reproducibility of different HPLC systems can be objectively compared. The ABI Model 130 is equipped with a manual sample injector while the remaining three systems that were tested had automatic injectors. Both the HP 1090 and Perkin-Elmer systems had diode array detectors while the remaining two systems had variable wavelength detectors. Rather than recommending any one of these or any other commercial HPLC system, our intent is to demonstrate relatively straightforward approaches that can be used for systematically comparing different HPLC systems with the goal of determining their suitability for a given application. In this regard it should be emphasized that in the following studies no attempt has been made to determine the extent of variability between different serial number HPLC systems and columns obtained from the same manufacturer. Hence, it is quite possible that if a HPLC system or column (other than the particular one used in this study) with different serial numbers were tested, the results might be substantially different. It is also important to note that the HPLC system requirements that are needed to enable a few relatively simple separations to be carried out at 1.0 ml/min on nanomole quantities of material are far less stringent than are required for comparative HPLC tryptic peptide mapping at 0.15 ml/min of 50-pmol quantities of enzymatic digests of high-molecular weight proteins.

The detector flow cell path length is one system variable that can directly and predictably affect sensitivity. The ABI Model 130 and the Kratos detector used in the Waters Associates HPLC system that was tested had 8-mm path length flow cells while the Perkin Elmer detector had a 10-mm path length flow cell. Since the HP 1090 detector was equipped with a 6-mm path length flow cell, the 10-mm Perkin-Elmer flow cell should have provided a 66% improvement in sensitivity; in practice, however, this increased sensitivity was not required for any of the studies that will be described. Even though the "noise level" of the detector and system can potentially limit the sensitivity that can be achieved, particularly below a level of about 10 pmol of digest, the noise level of the detectors tested in this study was sufficiently low that it also did not impose a limitation in the studies that follow. A reasonable test of the noise level would be to carry out a blank run at 0.15 ml/min and verify that there is no significant detector or system noise after plotting out the run using a full-scale absorbance at 210 nm of 0.015 (see Fig. 6 below). This latter value would be appropriate for plotting out a 50-pmol tryptic digest of a protein that had been chromatographed on a system such as the HP 1090 that is equipped with a 6-mm path length flow cell.

Figure 1 shows a comparison of the resolution that was obtained on three of the HPLC systems tested. Although the "dead volume" between the point of gradient mixing and the inlet of the column is considerably greater in the Waters Associates HPLC (largely as a result of the WISP Model 710 automatic sample injector) as compared to the ABI or HP systems tested, the overall resolution obtained on the Waters Associates HPLC was nonetheless still comparable to that on the ABI 130 system (Fig. 1). In contrast to the Waters HPLC system, the ABI Model 130 uses a manual injector which imposes a severe limitation. In general, prior to fractionating a digest on a "new" protein, it is best to carry out one control chromatogram on a comparable size aliquot of the transferrin tryptic digest (to verify that the HPLC system is operating well), one blank analysis (to detect absorbance peaks that may have carried over from one or more previous runs), and, whenever possible, one preliminary HPLC run on 10% of the digest so that the sample can potentially be redigested with the same or a different enzyme in the event the digest did not proceed well.

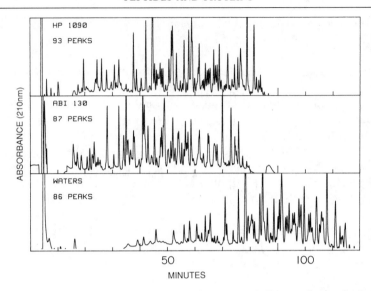

FIG. 1. HPLC separation of tryptic peptides from 50-pmol aliquots of a transferrin digest that were chromatographed on the three different HPLC systems indicated above. In all cases the same 2.1 mm × 25 cm Vydac C_{18} column was used that was eluted at a flow rate of 0.15 ml/min, as described in the text. The indicated number of absorbance peaks were detected by a Nelson Analytical Model 4416X Multi-Instrument Data System and the 210-nm full-scale absorbance settings used were 0.036 in the case of the Applied Biosystems and Waters Associates HPLC Systems and 0.022 in the case of the Hewlett-Packard System. (Taken from Stone et al.[14] with permission.)

With an automated HPLC system these three runs can be conveniently carried out the night before the actual "collection run" is done. It would appear that the particular HP 1090 HPLC used in Fig. 1 provided slightly better resolution and a slightly longer gradient delay than was obtained on the ABI 130 unit that was tested.

Excellent run-to-run reproducibility, as evidenced by a low variability in peak retention time, is essential for comparative HPLC peptide mapping. Since there is often more variability in peak retention times at low as compared to higher flow rates, it is essential that when evaluating this parameter, that it be checked over the entire intended flow rate range. In the case of the Waters Associates and HP 1090 systems evaluated in Fig. 1, the average peak retention time variability on both these units increased only slightly from a value of about ±0.22% at 1.0 ml/min to a value of ±0.26% at 0.15 ml/min. As shown in Fig. 2, this latter value is still sufficiently small to afford good reproducibility. In contrast, the comparable parameter for the particular Perkin-Elmer system tested in Fig. 3 increased from a value of about ±0.26% at 1.0 ml/min to a value of ±1.0%

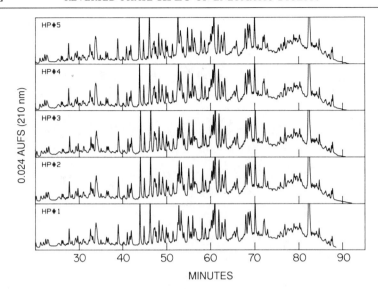

FIG. 2. HPLC separation of tryptic peptides from five consecutive 50-pmol injections of a transferrin digest that were chromatographed on a 2.1 mm × 25 cm Vydac C_{18} column at 0.15 ml/min using an HP 1090 HPLC.

at 0.15 ml/min. As shown in Fig. 3, this latter extent of variability is too large to permit comparative HPLC peptide mapping.

In order to maintain the excellent resolving power of reversed-phase HPLC, it is essential that the "dead volume" in between the detector flow cell and the fraction collector be reduced to an absolute minimum and that fractions be collected by peak rather than by time. In the Waters Associates 510-based system depicted in Fig. 4, the 55-cm length of 0.005 inch id stainless steel tubing connecting the outlet of the flow cell to the fraction collector not only introduces minimal dead volume but also provides sufficient back-pressure to prevent solvent degassing and consequent bubble formation in the detector flow cell. When operated at 0.15 ml/min, the "delay" in between when a peak is first detected in the flow cell and when it actually reaches the outlet at the fraction collector is nearly 28 sec. For comparison, the peak delay on the HP 1090 system used in Fig. 2 was approximately 40 sec at 0.15 ml/min when the factory-welded stainless steel tubing emerging from the outlet of the diode array detector flow cell was directly connected to the ISCO "FOXY" fraction collector (Lincoln, NE) via a union and the same length 0.005 inch id stainless steel tubing that was used in Fig. 4.

Although the peak delay can be easily programmed into the ISCO peak detector used in Fig. 4, it is essential that it be accurately determined. One

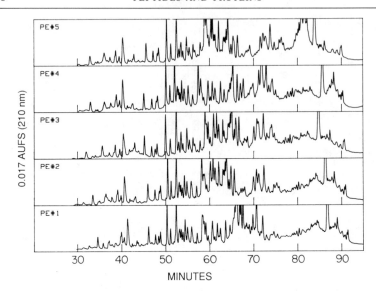

Fig. 3. HPLC separation of tryptic peptides from five consecutive 50-pmol injections of a transferrin digest that were chromatographed on a 2.1 mm × 25 cm Vydac C_{18} column at 0.15 ml/min using the Perkin-Elmer Peptide Mapping System. This HPLC system consisted of a Model 250 binary pump, an LC-135 dual-channel diode array detector that was connected to a Nelson Analytical Model 4416X Multi-Instrument Data System and a Model 7125 injector.

simple approach that can be used to evaluate this parameter is to remove the column from the HPLC system and then inject a sample of ATP or some other chromophore and collect the resulting fractions by time using 0.1 min/fraction. The fractions are then diluted with water, their absorbance read on a spectrophotometer, and the resulting absorbance plot then compared with the chart recorder tracing in order to calculate the delay. In terms of automated peak collection, the system delay actually imposes a lower limit on the flow rate. That is, those peak detectors that we are aware of are only capable of counting the delay down for one peak at a time. As a result, if during the first 28-sec delay on the Waters system depicted in Fig. 4, another absorbance peak enters the detector flow cell, it will reactivate the peak detector which will, in turn, simply re-start counting down the delay. The result, of course, is that the two peaks will be pooled together. On the Waters 510 based system (Fig. 4), this problem seldom occurs at 0.15 ml/min assuming a 105-min gradient run time and a 28-sec delay, but it often occurs if the delay is increased to the 42 sec that is required in order to run this HPLC system with the same 105-min gradient run time at 0.1 ml/min. Unless the system "peak delay time" is

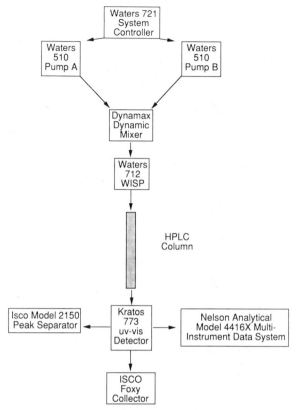

FIG. 4. Configuration of the Waters Associates HPLC system used for automated peak collection. All connections prior to the column, including the inlet line on the "WISP," utilized minimal lengths of 0.010 inch id stainless steel tubing while the column was connected to the Kratos detector with 4 cm of 0.005 inch id stainless steel tubing. The detector was, in turn, connected to the fraction collector via a 55-cm length of 0.005 inch tubing followed by a 2-cm length of 0.010 inch Teflon tubing that was connected to the ISCO "FOXY" fraction collector. The delay between the detector flow cell and the fraction collector was determined by injecting ATP to be 28 sec at 0.15 ml/min.

electronically compensated for prior to the peak detector or a peak detector is designed that can simultaneously "track" more than one peak, the practical upper limit for a peak delay time is about 25 to 30 sec. Hence, the Waters 510 based system in Fig. 4 cannot be used in an automatic peak collection mode below a flow rate of about 0.15 ml/min and the HP 1090 system used in Fig. 2 is limited in this mode to flow rates above about 0.2 ml/min. In our experience, the "peak delay volume" for a given HPLC system cannot be accurately calculated but rather must be experimentally

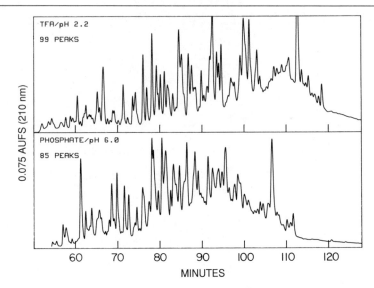

FIG. 5. HPLC separation of tryptic peptides from 50-pmol aliquots of a transferrin digest that were chromatographed at a flow rate of 0.15 ml/min on a Waters Associates HPLC system equipped with a 2.0 mm × 15 cm Delta Pak C_{18} column. Peptides were eluted in the upper chromatogram with the 0.05% TFA buffer system described in the text. In the lower chromatogram, a phosphate-buffered mobile phase was used in which buffer A was 5 mm potassium phosphate, pH 6.0, and buffer B consisted of 20% buffer A and 80% acetonitrile.

determined and represents another important parameter that must be considered in selecting and configuring an HPLC system.

Mobile-Phase Considerations

While the low ultraviolet absorbance, high-resolution, excellent solubilizing properties, and high volatility of the TFA/acetonitrile system have made it an almost universal mobile phase for reversed-phase HPLC peptide separations, there are occasions when a different mobile phase might prove advantageous. Hence, although Fig. 5 confirms the predicted higher retention and consequent higher peak capacity of the pH 2.2 TFA mobile phase as compared to the pH 6.0 phosphate mobile phase, it is also apparent that the two systems differ substantially in selectivity. Much of this difference results from the greater effect that an increase in mobile-phase pH has on a peptide containing ionizable glutamic and/or aspartic acid side chains. One application of the pH 6.0 mobile phase then is for resolving peptide mixtures that cannot be completely separated with the pH 2.2 system. With either the pH 6.0 or pH 2.2 mobile phases it may be

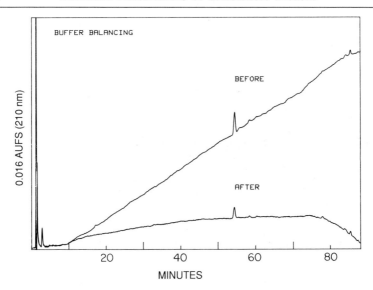

FIG. 6. Reversed-phase HPLC "blank" chromatogram obtained on a Waters Associates HPLC system equipped with a 2.0 mm × 15 cm Delta Pak C_{18} column before and after balancing the apparent absorbance at 210 nm of the two mobile-phase buffers. In the upper chromatogram both buffers contained 0.05% TFA while in the lower chromatogram an additional 0.011% TFA was added to buffer B which also contained 80% acetonitrile. The amount of TFA to add to either buffer to exactly balance the baseline is initially best determined by running a blank chromatogram, such as that shown above, and then rerunning another blank injection cycle after adding an additional 0.01% TFA to either buffer as appropriate to increase its absorbance. After the effect of this incremental increase in TFA concentration on the baseline slope is determined, one can, then, at any time use this reference curve to calculate the exact amount of TFA that needs to be added to either the A or B buffer to bring them into balance. After determining for a particular HPLC system the appropriate TFA concentrations to include in the two buffers to give a balanced baseline, it is usually not necessary to make any further adjustments provided that reasonable care is taken in making the mobile-phase buffers.

necessary to "balance the baseline," especially in the case of very high sensitivity applications such as below the 100 pmol range. As shown in Fig. 6 this task can easily be accomplished with the low pH system by adding a small excess of TFA to either buffer as needed to increase its apparent absorbance at 210 nm.

Choice of Appropriate Reversed-Phase HPLC Column

Most commercially available reversed-phase HPLC supports for peptide mapping have a particle size near 5 μm and a pore size of approximately 300 Å which appears to be nearly optimal for carrying out peptide

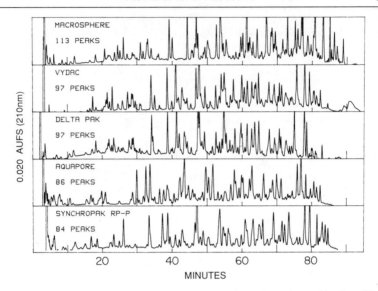

Fig. 7. Comparison of reversed-phase HPLC separations of tryptic peptides from 50 pmol of a transferrin digest that was chromatographed at 0.2 ml/min on an HP 1090 HPLC system. The 2.1 mm × 25 cm Alltech Macrosphere and Vydac C_{18} columns were packed with a 5-μm support as was the 2.0 mm × 15 cm Waters Associates Delta Pak C_{18} column. The 2.1 mm × 22 cm Brownlee Aquapore cartridge column was packed with a 7-μm C_8 support while the 2.1 mm × 25 cm Synchropak RP-P column had a 6.5 μm C_{18} support. In all cases the pore size was 300 Å and the number of absorbance peaks detected by the Model 4416 Multi-Instrument Data System are as indicated above. (Taken from Stone *et al.*[2] with permission.)

separations. The bonded phase is typically either C_4 or C_{18} with the latter being the most common. A direct comparison of these two bonded phases from one manufacturer indicated little difference in selectivity between the two. The only difference was that the elution profile was shifted earlier in the case of the C_4 column and hydrophilic peptides, that in general eluted in the first one-third of the gradient, were better resolved on the C_{18} support.[17] Although there may be instances where a particularly hydrophobic peptide might chromatograph better on a C_4 as opposed to a C_{18} support, in general, we routinely use the latter support. Figure 7 indicates that even when the particle size, bonded phase, and pore size are kept reasonably constant, there still appears to be considerable variation in the ability of commercially available narrow-bore HPLC columns to resolve complex peptide mixtures. All of the 2.0–2.1 mm id columns used in Fig. 7 have a C_{18} bonded phase except for the Aquapore column which has a

[17] K. L. Stone and K. R. Williams, *J. Chromatogr.* **359**, 203 (1986).

C_8 bonded phase. Although a 2.1 mm × 22 cm Aquapore C_8 cartridge column was used in Fig. 7, we have obtained somewhat better resolution (at similar flow rates, data not shown) on a 1.0 mm × 25 cm Aquapore C_8 column and, therefore, use this latter column for peptide rechromatography (see below).

A visual comparison of the chromatograms in Fig. 7 suggests that of the columns tested, the Vydac and Delta Pak C_{18} columns are most similar to each other in terms of their resolving power and selectivity. The similar resolution on these two columns is surprising in view of the fact that the Delta Pak C_{18} column has a length of only 15 cm as compared to the 25-cm Vydac C_{18} column. When these two columns were compared using the same gradient times as in Fig. 7 but at a flow rate of 0.15 ml/min, as opposed to the 0.20 ml/min that was used in Fig. 7, the results were somewhat different in that now better resolution was achieved on the Vydac C_{18} column. That is, at 0.15 ml/min approximately 124 peaks were detected on the Vydac C_{18} column as opposed to 102 peaks on the Delta Pak C_{18} column (data not shown). Since, in both cases, columns with different serial numbers were used for the 0.15 and 0.20 ml/min experiments, the better resolution obtained on the Vydac C_{18} column at 0.15 ml/ml may be dependent on either flow rate or column serial number. That Vydac and Delta Pak C_{18} columns have similar selectivities is apparent from the relative ease with which most corresponding peaks can be identified on these two chromatograms. In contrast, it is more difficult to identify corresponding peaks in the Aquapore C_8 and Vydac C_{18} chromatograms. By "screening" small aliquots or actually repurifying peptides that were originally isolated on a Vydac C_{18} column on an Aquapore C_8 column, it is possible to utilize the different selectivity of these two columns to detect or separate peptides that coelute on a Vydac C_{18} column.[18] Figure 8 shows an example of an absorbance peak (labeled No. 1 on the upper chromatogram) eluted from a Vydac C_{18} column that appeared to represent a homogeneous peptide but that on further chromatography on an Aquapore C_8 column proved to contain at least five major components as shown in the lower chromatogram in Fig. 8.

The final decision that must be made in selecting an HPLC column relates to column dimensions. Since Fig. 9 shows a direct relationship between resolution and column length it is best to use a 25-cm column if that length is available. In the case of the Delta Pak C_{18} column, however, the longest length available is 15 cm. The choice between a conventional 3.9–4.6 mm id and a 2.0–2.1 mm narrow-bore column is dictated primarily

---

[18] K. L. Stone and K. R. Williams, *in* "Macromolecular Sequencing and Synthesis" (D. H. Schlessinger, ed.), p. 7. Liss, New York, 1988.

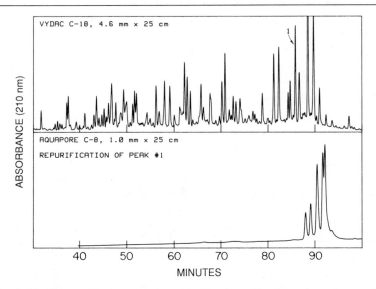

FIG. 8. Peptide repurification on an Aquapore C_8 column. The upper chromatogram shows the separation obtained at 0.5 ml/min on a 4.6 mm × 25 cm Vydac C_{18} column for a tryptic digest carried out on 2 nmol of a 66,000-Da protein. In the lower chromatogram the peak labeled No. 1 (in the upper chromatogram) was mixed with an equal volume of 8 M urea and then reinjected onto the same Waters Associates HPLC system that was now eluted at 0.15 ml/min with a 1.0 mm × 25 cm Aquapore C_8 column in place. The full-scale absorbance was 0.4 for the Vydac C_{18} and 0.15 for the Aquapore C_8 chromatogram.

by the amount of digest that is being chromatographed. Since a significant loss in resolution begins to occur when more than about 500 pmol of a complex enzymatic digest is fractionated on a narrow-bore column (data not shown) amounts greater than about 500 pmol are best chromatographed on conventional 3.9–4.6 mm id columns. By using reduced flow rates (see below), it is possible to easily extend the "working range" of a conventional 3.9–4.6 mm id column down to at least 250 pmol.[2,14]

Optimum Gradient Times and Flow Rates

Within reasonable limits, gradient time appears to be a more important determinant of resolution than gradient volume. As shown in Fig. 10, resolution increases as the gradient time is increased from 21 to at least 105 min, while the flow rate is kept constant at 1.0 ml/min. That this increase in resolution results more from the increased gradient time as opposed to the increased gradient volume is suggested by the observation that when the gradient time is held constant at 105 min, no significant

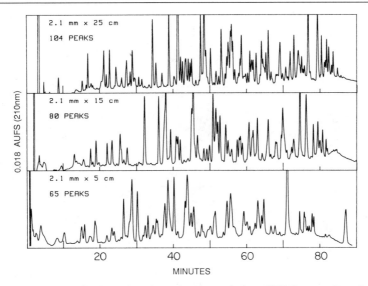

FIG. 9. Influence of column length on the reversed-phase HPLC separation of tryptic peptides from 50 pmol carboxamidomethylated transferrin. Three different 2.1 mm id Vydac C_{18} columns that were packed with the same 300 Å pore size, 5 μm particle size resin and that measured 25, 15, and 5 cm in length, respectively, as indicated above, were eluted at 0.2 ml/min on an HP 1090 HPLC. The number of absorbance peaks detected in each chromatogram is indicated above. (Taken from Stone et al.[1] with permission.)

increase in resolution is seen when the flow rate is increased from 0.4 to 1.0 ml/min.[2] The relative unimportance of flow rate and gradient volume is demonstrated by comparing the top two panels in Fig. 11, where even a fivefold reduction in flow rate (and gradient volume) resulted in only a 14% decrease in resolution as measured by the number of peaks detected. Hence 116 peaks were detected in the top panel at 1.0 ml/min compared to the 100 peaks that were detected in the middle panel at 0.2 ml/min. A comparison of the bottom two chromatograms in Fig. 11 does, however, indicate that flow rate is a major determinant of the sensitivity of detection. Hence, when the gradient time is kept constant and the flow rate is decreased from 1.0 ml/min in the bottom chromatogram to 0.2 ml/min in the middle chromatogram, there is an approximately fourfold increase in peak height. That the sensitivity of detection is more a function of flow rate than column diameter is shown in Fig. 12 where approximately equal sensitivity was achieved when 50 pmol of a tryptic digest of transferrin was eluted at 0.2 ml/min from either a 2.0 mm id or a 3.9 mm id reversed-phase column. The suboptimal linear flow velocity on the 3.9 mm id column probably accounts for the slightly lower resolution on this column,

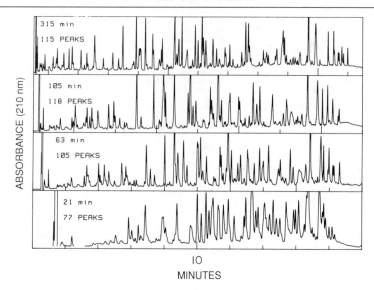

FIG. 10. Influence of gradient length on the reversed-phase HPLC separation of tryptic peptides from 250 pmol transferrin on an HP 1090 HPLC eluted at 1.0 ml/min from a 3.9 mm × 25 cm Delta Pak C_{18} column. The 105-min gradient was previously described in the text while in the case of the 315, 63, and 21 min gradients, all gradient step times listed in the text were proportionately increased or decreased to give the appropriate overall gradient time. The number of absorbance peaks detected at 210 nm is as indicated above.

that is, 93 peaks were detected on the 3.9 mm id column as compared to the 106 peaks detected on the 2.0 mm id narrow-bore column when both were eluted at 0.2 ml/min (Fig. 12). Based on comparisons carried out with both Vydac and Delta Pak C_{18} columns, better resolution is achieved (when the linear flow velocity and sample load/column volume ratio is kept constant) on a 3.9–4.6 mm id conventional as compared to a 2.0–2.1 mm id narrow-bore column. The slightly lower resolution obtainable on narrow-bore columns probably relates to the increased difficulty of packing these columns. In terms of maximum resolution, then, if the amount of digest available is above about 250 pmol it is probably best to use a conventional-size column eluted at 0.4–0.5 ml/min whereas, below this amount, it is best to use a narrow-bore column that is eluted at 0.15–0.20 ml/min.[2]

Choice of Sample Volume and Buffer

Samples for reversed-phase HPLC can be loaded in a wide range of salts and buffers. While enzymatic digests are routinely loaded in 2 M urea, samples can also be loaded in 8 M urea, 6 M guanidine-HCl, or

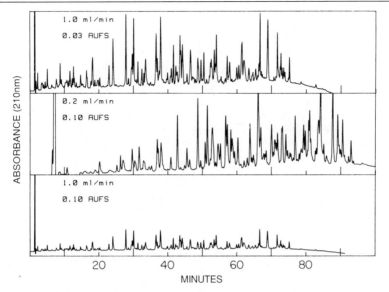

FIG. 11. Effect of decreased flow rate on the reversed-phase HPLC separation of tryptic peptides from 250 pmol transferrin on a 3.9 mm × 15 cm Delta Pak C$_{18}$ column that was eluted with an HP 1090 HPLC at the indicated flow rates using the 105-min gradient described in the text. The top and bottom chromatograms are identical except for the use of different full-scale absorbance settings as shown. (Taken from Stone et al.[14] with permission.)

aqueous TFA, acetic or formic acid mixtures with little loss in resolution (data not shown). When repurifying peptides, the eluted peak can be mixed with an equal volume of 8 M urea (to reduce the acetonitrile concentration) and then reinjected in a volume that may exceed 1.0 ml onto a 1.0 mm × 25 cm Aquapore C$_8$ column that has a total column volume of only about 0.2 ml. In contrast to most nonvolatile salts, detergents such as SDS (sodium dodecyl sulfate) can pose problems both in terms of carrying out an enzymatic digest and with respect to the subsequent reversed-phase HPLC. As shown in Fig. 13, amounts of SDS greater than about 5 μg result in a substantial loss in resolution on a reversed-phase support. Either a simple acetone precipitation[1] or blotting[19] and then eluting[14,15] a protein from polyvinylidene difluoride (PVDF) membranes offers two viable approaches for decreasing the SDS concentration in SDS–polyacrylamide gel purified samples to a level that no longer interferes with either enzymatic digestion or reversed-phase HPLC.

[19] P. Matsudaira, J. Biol. Chem. 262, 10035 (1987).

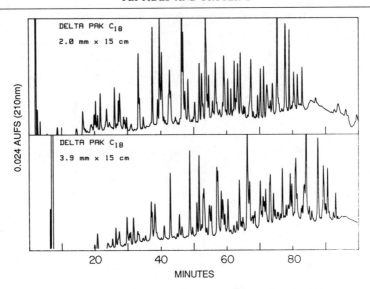

FIG. 12. Reversed-phase HPLC separation of tryptic peptides from 50 pmol transferrin on an HP 1090 HPLC using either a 2.0 mm × 15 cm (top) or a 3.9 mm × 15 cm (bottom) Delta Pak C_{18} column eluted at 0.2 ml/min. (Taken from Stone *et al.*[14] with permission.)

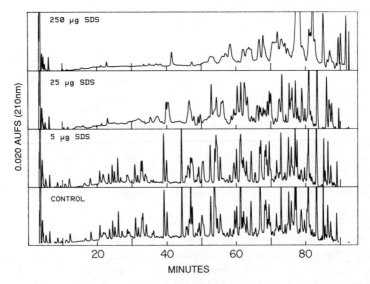

FIG. 13. Effect of SDS on reversed-phase HPLC. The indicated amounts of SDS were added to 50-pmol aliquots of a large-scale trypsin digest of carboxamidomethylated transferrin that was then injected onto an Alltech Macrosphere 2.1 mm × 25 cm column that was eluted at a flow rate of 0.2 ml/min with an HP 1090 HPLC.

Practical Guidelines for Performing Reversed-Phase HPLC
  Peptide Separations

The 0.05% trifluoroacetic acid/acetonitrile mobile phase and the 105-min gradient given previously provide near optimum resolution with a total run time, including column reequilibration, of less than 3 hr. In general, amounts of peptides less than about 250 pmol are chromatographed on 2.0–2.1 mm id narrow-bore columns eluted at 0.15 to 0.20 ml/min while larger amounts are chromatographed on 3.9–4.6 mm id columns that are best eluted at about 0.5 ml/min. Several 300 Å pore size, 5-μm particle size, C_{18} columns are commercially available, including the Macrosphere, Vydac, and Delta Pak columns that can resolve 100 or more different peptides in a single run. When available, the 25-cm length version of these columns should be used. Enzymatic digests of proteins can be conveniently injected in 2 M urea while cyanogen bromide and other peptides resulting from chemical cleavage of proteins are best loaded in 6–8 M guanidine-HCl. Using a commercial peak detector, it is possible to automatically collect (in 1.5-ml capless Eppendorf tubes positioned on top of 13 × 100 mm test tubes) peptides from as little as 50 pmol of a tryptic digest of a protein. Provided that the Eppendorf tubes are capped immediately following the HPLC run (to prevent the acetonitrile from evaporating) these peptides may be stored for periods of up to a year prior to sequence analysis.

Protein Chemistry/Mass Spectrometric Approach to Primary Structure
  Determination: Structure of STP-2 Symposium Test Peptide

Approximately 3 nmol of the STP-2 synthetic peptide was provided as an "unknown" structure to the participants at the first symposium of the Protein Society. The STP-2 synthetic peptide was designed to not only test the capabilities of state-of-the-art analytical instruments, but also the ingenuity of the protein chemists that operate these systems.[20] Upon receipt of the sample, we identified residues 1–37 in STP-2 by direct gas-phase sequencing of 440 pmol of this peptide. Several tryptic peptides were then isolated and sequenced from a digest carried out on 330 pmol of STP-2 that confirmed the NH_2-terminal sequence data and that extended it to include Gly-38. Although the amino acid sequencing data we obtained on STP-2 seemed qualitatively conclusive and the sequence of the proposed 38-residue peptide agreed well with its experimentally determined amino acid composition, there were two quantitative problems with the

[20] J. Rivier, in "Techniques in Protein Chemistry" (T. E. Hugli, ed.), p. 549. Academic Press, San Diego, CA, 1989.

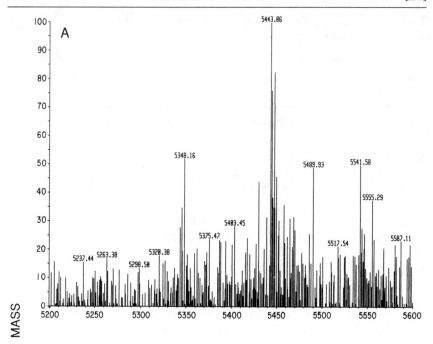

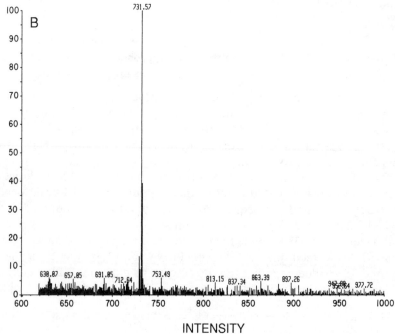

STP-2 NH$_2$-terminal sequencing data that could not be readily explained. That is, there was an unexplained twofold drop in sequencing yield at cycle 20 and the coupling yield, based on the yield of the isoleucine at position 2 and assuming that STP-2 was a 38-residue peptide with a molecular mass of about 4200 Da, was 120% instead of our normal 50–75%. Finally, instead of the expected peak near m/z 4200, FAB mass spectrometry (Fig. 14A) of the STP-2 peptide indicated an average protonated molecular ion of m/z 5445.8. This finding was best explained by assuming that STP-2 actually consisted of a 38-residue peptide that is connected via an ε-NH$_2$ linkage at Lys-19 to an 18-residue peptide that has the identical sequence as the first 18 amino acids in the 38-residue parent peptide. Taken together, these data suggested the following structure beginning with hydroxyproline (HPr): (HPr-I-I-S-T-A-G-R-G-G-S-E-V-P-V-S-H-K)$_2$-K-H-S-V-P-V-E-S-G-G-R-A-G-T-S-I-I-G-G-G. The predicted chemical molecular weight of this peptide is 5447.12 for the singly protonated species which is just slightly larger than the value of 5445.8 that we determined (Fig. 14A).

In what was anticipated to be a final proof of the proposed structure, a synthetic peptide was made, using a bis-tBOC-lysine derivative at position 19 to introduce the branch point that corresponded to the above 56-residue peptide. As shown in Fig. 15, comparative HPLC demonstrated that this synthetic 56-mer did not have exactly the correct structure because it consistently eluted from a reversed-phase support just slightly later than STP-2. As shown in Fig. 16, comparative HPLC tryptic peptide mapping quickly demonstrated that STP-2 and the 56-mer differ at some position within the tryptic peptide spanning residues 30–38. This tryptic peptide elutes at about 33 min, in the case of the 56-mer, as compared to about 31.5 min in the case of STP-2 (Fig. 16). FAB mass spectrometry of the residue 30–38 tryptic peptide from STP-2 indicated a protonated molecular ion that was 0.7 u *less* than expected, that is, two separate determinations gave values of 731.57 and 731.63 compared to the predicted value of 732.28 (Fig. 14B). In contrast, mass spectra on the NH$_2$-terminal tryptic peptide from STP-2, residues 1–8, gave a mass that was only 0.3 units *larger* than expected. If the COOH terminus of STP-2 was amidated, then the corresponding mass expected for the residue 30–38 peptide would

---

FIG. 14. (A) Positive ion FAB mass spectrometry of approximately 175 pmol of the STP-2 peptide. Based on the above results, the average protonated molecular ion of STP-2 was determined to have an m/z value of 5445.8. (B) FAB mass spectrometry of the STP-2 tryptic peptide corresponding to residues 30–38. The observed m/z of 731.57 for the protonated molecular ion was sufficiently close to the predicted value of 731.3 to demonstrate that STP-2 ends with an amidated carboxy terminus.

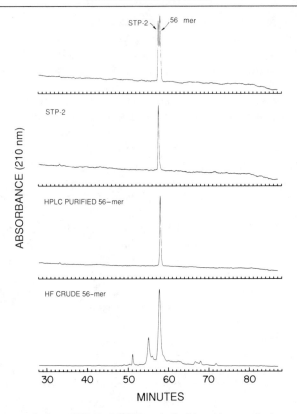

Fig. 15. Reversed-phase HPLC of STP-2 and of a 56-residue synthetic peptide, denoted as the 56-mer above, whose structure was identical to that for STP-2 except that the 56-mer did not end with an amidated carboxy terminus. The above data were obtained using a 4.6 mm × 25 cm column that was eluted at a flow rate of 1.0 ml/min on a Waters Associates HPLC. The full-scale absorbance settings used for the above chromatograms were 0.225, 0.060, 0.045, and 0.048, respectively, going from the bottom to the top. In the top chromatogram approximately an equimolar mixture of the 56-mer and STP-2 was made in order to confirm that the two peptides did not coelute.

be 731.3 which, as in the case of the NH_2-terminal tryptic peptide, would again be 0.3 u less than the observed average value of 731.6. As final proof of the STP-2 structure, the COOH-terminal tryptic peptide, A-G-T-S-I-I-G-G-G, was synthesized with an amidated COOH terminus and, as expected, it now coeluted on reversed-phase HPLC with the corresponding tryptic peptide from STP-2 (data not shown). In addition to the studies described here, two other correct solutions for the STP-2 structure were also submitted.[20] All three groups that solved the STP-2 structure relied on a combined mass spectrometric/protein chemistry approach.

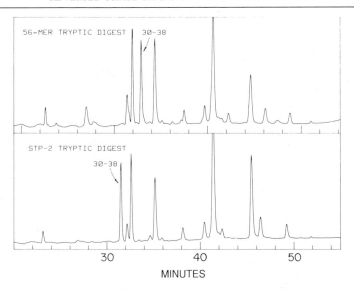

FIG. 16. Comparative HPLC tryptic peptide mapping of 1.0 nmol STP-2 (top) versus that for a 56-residue synthetic peptide (bottom) which had the initially presumed STP-2 structure. The HPLC was carried out on a Waters Associates HPLC System equipped with a 4.6 mm × 25 cm Vydac C_{18} column that was eluted at 1.0 ml/min. The lack of cochromatography of the peptide spanning residues 30–38 is due to the carboxy-terminal amidation that is present in the STP-2 peptide but that is not present in the 56-mer.

Comments

The standard peptide workshops sponsored by the Protein Society provide a preview of the potential contribution that mass spectrometry can make in elucidating the primary structures of proteins. While mass spectrometry proved invaluable for determining the actual size of the STP-2 peptide, it appeared from the 1988 studies with the STP-3 peptide that mass spectrometry was in "strong contention for being the most sensitive and most economical (in terms of time) technique for the determination of peptide sequences."[20] In terms of solving the complete primary structures of proteins, it would seem that a combined mass spectrometric/DNA sequencing approach might provide the best possible choice. In view of the fact that most eukaryotic proteins have blocked NH_2 termini and cannot be directly sequenced[21] one ideal approach then would be to carry out a tryptic digest, resolve the resulting peptides by reversed-phase HPLC, and then use mass spectrometry to determine each of their molecu-

[21] J. Brown and W. K. Roberts, *J. Biol. Chem.* **251**, 1009 (1976).

lar weights and to sequence a few of the moderate-sized peptides. These short stretches of peptide sequence could then serve as the basis for synthesizing oligonucleotide probes that, in turn, could be used to isolate the corresponding cDNA clone. The complete amino acid sequence of the protein could then be predicted from the resulting cDNA sequence and the predicted and actual molecular weights for all expected tryptic peptides could be compared. In this way peptides could be selected for mass spectrometric sequencing that either correspond to cDNA regions of sequence that contain DNA sequencing errors, such as single base "deletions," or that contain posttranslational modifications that, in many cases, play an essential role in protein function. It is clear that by rapidly bringing about either a complete or partial resolution of complex enzymatic and chemical digests of proteins, reversed-phase HPLC can make an essential contribution toward elucidating the primary structures of proteins.

Acknowledgments

The authors wish to thank the Sep/a/ra/tions (Hesperia, CA) and the Nest Group (Southborough, MA), as well as the Waters Chromatography Division, Millipore Corporation (Milford, MA), for providing some of the HPLC columns used in this study. We also thank the Perkin-Elmer Corporation (Emeryville, CA) for the temporary loan of an HPLC system.

[22] Electrospray Ionization Mass Spectrometry

By CHARLES G. EDMONDS and RICHARD D. SMITH

Electrospray ionization (ESI) occurs during the electrostatic nebulization of a solution of charged analyte ions by a large electrostatic field gradient (approximately 3 kV/cm). Highly charged droplets are formed in a dry bath gas, at near atmospheric pressure. These charged droplets shrink as neutral solvent evaporates until the charge repulsion overcomes the cohesive droplet forces leading to a "Coulombic explosion." In the popular model by Iribarne and Thomson[1,2] the smaller droplets continue to evaporate and the process repeats until the droplet surface curvature is sufficiently high to permit the field-assisted evaporation of charged solutes.

[1] J. V. Iribarne and B. A. Thomson, *J. Chem. Phys.* **64**(6), 2287 (1976).
[2] B. A. Thomson and J. V. Iribarne, *J. Chem. Phys.* **71**, 4451 (1979).

Details of this mechanism await full experimental verification. However, it is clear that molecular ions (with an unknown initial extent of solvation) are produced from liquid solution under mild conditions by ESI. These ions generally arise by attachment of proton, alkali cation, or ammonium ions for positive ion formation or, with reversal of the nebulizing field, by proton or other cation abstraction for negative ion formation.

ESI was originally described by Dole et al.[3] in studies of the intact ions from synthetic and natural polymers of molecular mass in excess of 100,000 Da based on gaseous ion mobility measurements. In the present context it is impressive to note that these workers, employing the so-called "plasma chromatograph," with detection by Faraday-cage current, tentatively (and correctly) interpreted experiments on lysozyme as demonstrating a multiple charging phenomenon.[4] After a 10-year hiatus, these experiments were extended by Fenn et al.[5] at Yale University (New Haven, CT) employing atmospheric pressure sampling of ions and analysis with a quadrupole mass spectrometer, and essentially simultaneously by researchers in the Soviet Union[6] using a magnetic sector instrument. This work outlined the fundamental aspects of ESI, demonstrating its utility for the analysis of biomolecules of modest molecular weight and as an interface for the combination of a liquid chromatograph with a mass spectrometer.[7] Access to higher molecular weights through the production of molecular ions bearing multiple charges was demonstrated by the American workers for polyethylene glycol oligomers of nominal molecular mass of 17,500 Da bearing a net charge of up to $+23$.[8] In an important extension of these experiments, the multiple protonation of basic residues in the ESI mass spectra of oligopeptides and proteins showed the extension in the mass range for the analysis up to 40,000 Da analyzable using a quadrupole

[3] M. Dole, L. L. Mack, R. L. Hines, R. C. Mobley, L. D. Ferguson, and M. B. Alice, *J. Chem. Phys.* **49**(5), 2240 (1968); L. L. Mack, P. Kralik, A. Rheude, and M. Dole, *J. Chem. Phys.* **52**(10), 4977 (1970); G. A. Clegg and M. Dole, *Biopolymers* **10**, 821 (1971).

[4] M. Dole, H. L. Cox, Jr., and J. Gieniec, *in* "Advances in Chemistry Series, No. 125" (E. A. Mason, ed.), p. 73. American Chemical Society, Washington, D.C., 1973; J. Gieniec, L. L. Mack, K. Nakamae, C. Gupta, V. Kumar, and M. Dole, *Biomed. Mass Spectrom.* **11**(6), 259 (1984).

[5] M. Yamashita and J. B. Fenn, *J. Phys. Chem.* **88**, 4451 (1984); *ibid.*, p. 4671; J. B. Fenn, M. Mann, C. K. Meng, S. F. Wong, and C. M. Whitehouse, *Science* **246**, 64 (1989); *ibid.*, *Mass Spectrom. Rev.* **9**, 37 (1990).

[6] M. L. Alexandrov, L. N. Gall, N. V. Krasnov, V. I. Nikolaev, V. A. Pavlenko, and V. A. Shkurov, *Dokl. Akad. Nauk S.S.S.R.* **277**, 379 (1984); *ibid.*, *J. Anal. Chem. U.S.S.R.* **40**(9/1), 1227 (1985-transl. 1986) and references therein.

[7] C. M. Whitehouse, R. N. Dreyer, M. Yamashita, and J. B. Fenn, *Anal. Chem.* **57**, 675 (1985).

[8] S. F. Wong, C. K. Meng, and J. B. Fenn, *J. Phys. Chem.* **92**, 546 (1988).

mass spectrometer of mass/charge limit 1600.[9] These results were rapidly confirmed by work in other laboratories.[10,11] The utility of this multiple-charging phenomenon has been demonstrated to extend to more than 130,000 Da,[12] employing quadrupole mass analysis of conventional mass/charge (m/z) range, and to permit the measurement of relative mass in the range of 5 to 40 kDa with precision of better than 0.01%.[11-13]

Practice of Electrospray Ionization

Atmospheric Pressure Electrospray Ionization Source

The ESI "source" may be simply a metal capillary (e.g., stainless steel hypodermic needle). Such an arrangement with a cylindrical counterelectrode is the basis of source arrangements described by Whitehouse et al.[7] A 1–20 μl/min flow of solution, typically water–methanol mixtures, containing the analyte and often other additives, such as acetic acid, is delivered to the capillary terminus from infusion syringes or liquid chromatographic columns. Alternatively, the ESI process can be accompanied by pneumatic nebulization accomplished by a high-velocity annular flow of gas at the liquid exit of the injection capillary and has been termed ionspray.[14] This method has the advantage of accommodating flow rates up to approximately 100 ml/min, making the method particularly attractive for liquid chromatography (eluent stream split) with gradient elution. The principal disadvantage of such an arrangement is the reduction in sensitivity which accompanies the reduction in the charge/mass for the resulting droplets.

A modified ESI source developed at our laboratory for combined capillary electrophoresis-MS[15] is shown in Fig. 1. An organic liquid sheath (typically pure methanol or acetonitrile which may be augmented by small proportions of acetic acid or water) flows in the annular space between

[9] C. K. Meng, M. Mann, and J. B. Fenn, Z. Phys. D—Atoms, Molecules Clusters 10, 361 (1988); ibid., Proc. 36th ASMS Conf. Mass Spectrom. Allied Topics, San Francisco, CA p. 771 (1988).

[10] J. A. Loo, H. R. Udseth, and R. D. Smith, Biomed. Environ. Mass Spectrom. 17, 411 (1988).

[11] T. R. Covey, R. F. Bonner, B. I. Shushan, and J. Henion, Rapid Commun. Mass Spectrom. 2(11), 249 (1988).

[12] J. A. Loo, H. R. Udseth, and R. D. Smith, Anal. Biochem. 179, 404 (1989).

[13] M. Mann, C. K. Meng, and J. B. Fenn, Anal. Chem. 61, 1702 (1989).

[14] A. P. Bruins, T. R. Covey, and J. D. Henion, Anal. Chem. 59, 2642 (1987).

[15] R. D. Smith, J. A. Olivares, N. T. Nguyen, and H. R. Udseth, Anal. Chem. 59, 436 (1988); R. D. Smith, C. J. Barinaga, and H. R. Udseth, Anal. Chem. 60, 1948 (1988).

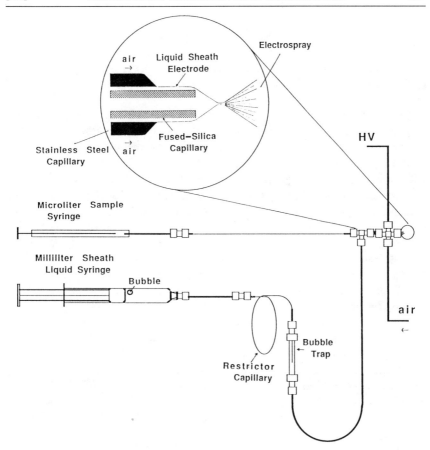

FIG. 1. Schematic of the sheath flow interface for ESI mass spectrometry showing the conventional arrangement for direct infusion of analyte solutions.

100 μm id fused-silica capillary and a cylindrical stainless steel electrode (500 μm id) which delivers analyte solution to the ion source. An electrospray voltage of 4–6 kV is applied to this surrounding electrode from which the fused-silica tube protrudes approximately 0.2 mm. For experiments in which sample solution is directly infused to the ESI source, syringe pumps deliver controlled flows of analyte and sheath liquids at rates of 0.5–1 and 2–4 μl/min, respectively. The best ESI performance requires a matching of solution conductivity and flow rate as discussed below. Stability of the ESI source depends critically on the stability of these flows. For analyte flows in our laboratory, microliter syringes (50–250 μl gas-tight, Hamilton Co., Reno, NV) installed in a syringe infusion pump (Pump 22, Harvard

Apparatus, Cambridge, MA) delivered through 30–70 cm lengths of fused-silica capillary are used. Electrical isolation through this length of fused silica is adequate for electrospray voltages of ±6 kV. Sheath flows from 1.5- to 5-ml disposable (Becton Dickinson and Co., Rutherford, NJ) or glass gas-tight syringes (Hamilton Co., Reno, NV) originate in a syringe pump (Model 341B, Sage Instruments, Cambridge, MA). In the case of this flow, further stabilization is accomplished with a fused-silica flow restrictor (20 cm × 50 μm) in combination with the compliance of a deliberately introduced bubble in the delivery syringe of approximately 200 μl. Problems which arise from the formation of bubbles in the sheath solvent in the connecting lines are minimized by the inclusion of a trapping volume. Additional stability may be obtained with degassing (by brief sonication) of the organic solvents used in the sheath and avoiding unnecessarily high ambient temperatures. A cooling of the entire assembly has been found advantageous for the combination of capillary electrophoresis with ESI mass spectrometry. Within this general scheme, many satisfactory arrangements for capillary tubing and ESI components are possible.

Positively or negatively charged droplets and ions may be produced depending upon the capillary bias. In the negative ion mode, an electron scavenger, such as oxygen, is required to inhibit electrical discharge at the capillary terminus. This is accomplished using an air flow (250 ml/min) through a third 1/8 inch id annular passage in the electrospray ion source. This flow is also used in positive ion operation. Its velocity is much lower than required for nebulization and is observed to minimize the effects of air circulation around the ion source. It should be noted that power supplies for electrospray application should be selected to minimize the hazard to operators from electrical shock. These should be of the radio frequency oscillator design where stored energy is minimized (generally less than 1 joule). Sample solutions of proteins and peptides are typically prepared in 5% (v/v) acetic acid in good-quality (>17 MΩ/cm) deionized water. Useful analysis may be obtained on oligopeptide and small protein sample concentrations of 1 to 100 μM with better performance obtainable in favorable cases.

The use of electric fields in the droplet nebulization (i.e., without a nebulizing gas flow) results in the optimum deposition of charge on the droplet affording the most efficient production of analyte ions. However, this leads to some practical restrictions on the range of solution conductivity and dielectric constant properties for stable operation at useful flow rates. Solution conductivities of $\leq 10^{-5}$ mΩ, corresponding to an aqueous electrolyte solution of $\leq 10^{-4}$ N utilize sample most efficiently. Fluids with higher surface tensions require a higher threshold voltage for electrospray production (e.g., ~8 kV for water vs. ~4 kV for methanol given a 1-cm capillary counterelectrode distance), while higher dielectric liquids produce higher total ESI currents. For direct infusion experiments with typical 5% (v/v) aqueous acetic acid and a methanol sheath, total electrospray currents of 0.1 to 0.5 μA are typical. Solution electrolyte concentrations above about 10^{-3} N result in disabling instability in electrospray operation.

We have found that varying the solvent composition of the electrospray experiment (sheath liquid and/or sample solvent) can sometimes change the charge distribution for a given molecule. For example, oligopeptides and protein ESI mass spectra using a 2-propanol sheath reduces the relative intensity of ions in higher charge states and results in at least an order of magnitude drop in sensitivity relative to a methanol or acetonitrile sheath.[16] In addition, "tailing" occurs on the high m/z side of peaks of ESI mass spectra. This is most evident with a 2-propanol sheath and due to incomplete desolvation, yielding solvent–molecule complexes. However, since the mass spectrum is obtained only after extensive desolvation, both at atmospheric pressure and during introduction to the mass spectrometer, the degree of solvent association for any single step in the ESI process is not readily ascertained.

Atmosphere–Vacuum Interface

The critical feature of an ion source operated at atmospheric pressure is the efficient sampling and transport of ions into the mass spectrometer at operating vacuum (e.g., $>5 \times 10^{-5}$ torr for a quadrupole mass filter). The oldest and simplest method (available as a commercial instrument from Sciex, Thornhill, Ontario, Canada) utilizes a single laser-drilled pinhole orifice (100 to 130 μm) entering directly into the high-vacuum region of the mass spectrometer and cryopumping capable of very high pumping speeds. In this arrangement (relevant voltages as indicated in parentheses), positive ions produced at atmospheric pressure in an electrospray ion source (-4 kV) drift toward the plate on which the pinhole sampling aperture (35 V) is mounted. This is against a countercurrent flow of dry gas directed through an axial plenum (650 V). This curtain of dry nitrogen serves to exclude large droplets and particles and to dry the droplets and decluster the ions. As the ions pass through the orifice into the vacuum, further declustering is accomplished by acceleration of the ions into the first radio frequency (rf)-only focusing quadrupole (30 V) of a tandem quadrupole mass spectrometer.

Alternatively, instruments may be based on differentially pumped vacuum technology. Such a system which employs a capillary inlet-skimmer

[16] J. A. Loo, H. R. Udseth, and R. D. Smith, *Rapid Commun. Mass Spectrom.* **2**(10), 207 (1988).

inlet system has been described by Fenn and co-workers.[5] In this arrangement the electrospray needle, maintained at a few kilovolts with respect
to the surrounding cylindrical electrode, affords the electrospray plume
which encounters a countercurrent flow of bath gas (typically nitrogen at
slightly above atmospheric pressure, 45°–75°, 100 ml/sec) which sweeps
away uncharged material and solvent vapor from the mass spectrometer
inlet. As droplets drift toward the end wall, ions are formed and some are
entrained in the flow of gas entering the glass capillary at the end of this
chamber, emerging as a free jet expansion in the first stage of differential
pumping. Ions are then transmitted through a skimmer into the inlet optics
of a (quadrupole) mass analyzer. The electrically insulating nature of the
connecting capillary is an advantage in providing a flexible choice of
operating voltages in the system. For example, in contrast to the conventional experiment employing fused-silica capillary tubing to deliver sample
to the electrospray ion source at high potential, this approach allows the
outlet of a liquid chromatographic apparatus in an LC/ESI-MS experiment
to be at ground potential. In that case, the electrode and cylinder end plate
may be floated at high voltage and hydrodynamic flow through the glass
capillary serves to efficiently deliver ions to the skimmer region. Ion
mobilities at atmospheric pressure are sufficiently low that the viscous
drag may be sufficient to deliver ions from (or back to) ground potential
across gradients of as much as 15 kV, allowing voltages appropriate for
injection of ions into a magnetic sector instrument.[17] Typical voltages for
use with a quadrupole instrument are indicated in parentheses: electrospray needle (ground), cylindrical electrode (13.5 kV), capillary entrance
(-4.5 kV), capillary exit (40 V), skimmer (-20 V), and quadrupole entrance (ground).

The two ESI-MS instruments developed at our laboratory utilize a
stage of differential (mechanical) pumping, as well as cryopumping, to
increase the ion current sampled from atmospheric pressure. One of these
is a modified TAGA 6000E (Sciex, Thornhill, Ontario, Canada). As with
other electrospray interface arrangements, a countercurrent flow of nitrogen gas (20°–70°) at a rate of 4–6 liter/min, passing through a plenum
between the ESI source and the nozzle, aids the desolvation process of
the electrospray-produced charged droplets, prevents solvent introduction
to the vacuum system, and minimizes clustering during expansion into the
mass spectrometer. The exit of this plenum (5 mm id) held at 650 V (with
ESI source at 4 kV for positive ion production) defines the first focusing

[17] C. M. Whitehouse, M. Yamashita, C. K. Meng, and J. B. Fenn, *in* "Proceedings of the
14th International Symposium on Rarefied Gas Dynamics" (H. Oguchi, ed.), p. 857. Univ.
of Tokyo Press, Tokyo, 1985.

aperture of the ion optical system. The ESI source is mounted 1–2 cm from the entrance of the quadrupole mass spectrometer. Ions are sampled from atmospheric pressure through a 1-mm nozzle orifice to a 2-mm diameter beam skimmer (Beam Dynamics, Inc., Minneapolis, MN) in front of a rf-only quadrupole. Nozzle voltages (V_n) range between 100 and 400 V for positive ion studies, while the skimmer is held at 70 V and the offset of the rf-only entrance quadrupole is maintained at 60 V. The nozzle–skimmer region is pumped using a 50 liter/sec single-stage Roots blower maintaining this region at 1 to 10 torr. The cryopumped tandem quadrupole chamber typically reaches 5×10^{-5} torr in normal operation. The ability to "collisionally heat" ions by the application of electric fields in the differentially pumped regions is useful for both desolvation and, at higher voltages, collisional activation and collision-induced dissociation (CID) of analyte ions, as discussed in the final section of this chapter.

The operation of the ESI source in combination with the atmosphere–vacuum interface (of whatever design) requires the empirical adjustment of a number of parameters according to the details of the particular apparatus, the species being analyzed, and the object of the experiment. For example, if the counter flow of desolvating gas is too high, analyte ions may be swept away from the entrance of the interface with an unacceptable decrease in sensitivity. Conversely, if this flow is too low analyte solvation may be excessive, with broadened peaks and decreased signal. Similarly, the selection of voltages on the nozzle and skimmer elements vary the collision energy in this high-pressure region and affect the extent of analyte solvation observed and also ion focusing. A certain amount of optimization by trial and error is advisable for these and other important parameters.

Electrospray Ionization—Mass Spectrometry

Operation in Positive Ion Mode

For peptides such as bradykinin and glucagon shown in Fig. 2, only singly or multiply protonated, and in a few cases multiply sodiated ions (depending on solution buffer) are observed in ESI, with a general absence of fragment ions.[12,18,19] The number of possible protonation sites is the principal factor affecting the multiple charging observed in ESI mass spectra. As the analyte oligopeptide increases in relative molecular mass (M_r) the typical bell-shaped distribution of multiply charged molecular

[18] J. A. Loo, H. R. Udseth, and R. D. Smith, *Rapid Commun. Mass Spectrom.* **2**, 207 (1988).
[19] C. J. Barinaga, C. G. Edmonds, H. R. Udseth, and R. D. Smith, *Rapid Commun. Mass Spectrom.* **3**, 160 (1989).

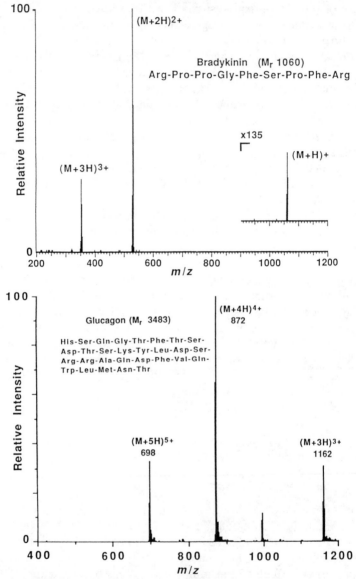

FIG. 2. ESI mass spectra of bradykinin (upper) and glucagon (lower). The additional ion at m/z 996 is attributed to a contaminating oligopeptide.

ions appears with the multiple charging having the effect of extending the mass range of the mass spectrometer by a factor equal to the number of charges. For most of the compounds so far examined (prepared in aqueous solution, pH <4), an approximate linear correlation is observed between the maximum number of charges and the number of basic amino acid residues (e.g., arginine, lysine, and histidine).

Electrospray Ionization—Mass Spectrometry of Proteins and M_r Determination

Large oligopeptides and small proteins so far successfully examined by ESI-MS show a distribution of multiply charged molecular ions arising (in positive ion mode) by either proton or alkali ion attachment and no evidence of fragment ions due to fragmentation unless dissociation is induced during transport into mass spectrometer vacuum by collisions at higher energy. The largest protein so far examined is the native covalent dimer of bovine serum albumin with a M_r of more than 130 kDa.[12] It should be noted that in the case of noncovalently bonded species, such as protein subunits, the mass spectra show only the contributions of the individual subunits. These mass spectra show a distribution of charge states which usually spans about one-half the number of the most abundant charge state. Figure 3 shows a spectrum for equine myoglobin which illustrates the features of such a spectrum. The distinctive bell-shaped distribution of charge states arises due to the quantum nature of the charge, i.e., adjacent peaks differ by one charge. For the example of myoglobin, the highest m/z ion observed is at 1305 Da, corresponding to a net charge of $13+$. The most abundant multiply protonated ion is observed at m/z 893 with $19+$ and the highest charge state observed is $28+$ (a small peak at m/z 606 not labeled in Fig. 3). A striking feature of ESI mass spectra of proteins is that the average charge state increases in an approximately linear fashion with molecular weight, extending the mass range of the experiment (by the addition of compensating charge) with the increase in mass of the analyte.

For the positive ion ESI spectra of proteins, the determination of M_r is straightforward given two assumptions: (1) that adjacent peaks of a series differ by only one charge and (2) that charging is due to cation attachment (a proton, or rarely for lower M_r oligopeptides, mixed H^+ and Na^+) to the molecular ion. With these it follows that the observed m/z ratios for each member of the distribution of multiply protonated molecular ions (in the case of oligopeptides and proteins) are related by a series of simple linear simultaneous equations where M_r and charge are unknown. Any two are sufficient to uniquely determine M_r and charge as is indicated

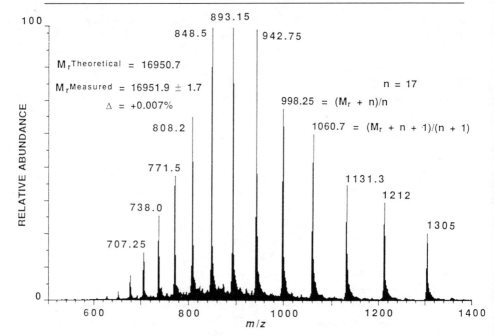

FIG. 3. ESI mass spectrum of equine myoglobin. Peaks m/z 707.25 through m/z 1305 are centroid values for these multiply protonated (24+ through 13+) ions. The simultaneous relation of one pair, m/z 998.25 and m/z 1060.7, is illustrated where the integer value of n is 17. Calculation of M_r as discussed in the text.

by the relations shown for the adjacent peaks at m/z 998.25 and m/z 1060.7 in Fig. 3. These m/z values are obtained from the centroid of the peak, consistent with fact that isotopic contributions for these large multiply charged ions are not resolved and the average mass defined by M_r. Workers have presented more or less elaborate methods based on this fundamental principle for extracting M_r from these data with high precision and accuracy.[11-13] Reduced to its simplest form, given the assumptions mentioned above, Eq. (1) describes the relationship between a multiply charged ion at m/z p_1 with charge z_1 and the M_r

$$p_1 z_1 = M_r + M_a z_1 = M_r + 1.0079 z_1 \qquad (1)$$

where we assume that the charge-carrying species (M_a) is a proton. The molecular weight of a second multiply protonated ion at m/z p_2 (where

$p_2 > p_1$) that is j peaks away from p_1 (e.g., $j = 1$ for two adjacent peaks) is given by

$$p_2(z_1 - j) = M_r + 1.0079(z_1 - j) \qquad (2)$$

Equations (1) and (2) can be solved for the charge of p_1:

$$z_1 = j(p_2 - 1.0079)/(p_2 - p_1) \qquad (3)$$

The molecular weight is obtained by taking z_1 as the nearest integer value, and thus the charge of each peak in the multiple charge distribution.

This approach can easily be generalized to determine the mass of the charged adduct species M_a (e.g., H^+, Na^+, K^+).

$$p_1 = (M_r/z_1) + M_a \qquad (4)$$

Calculation of M_r for each of the observed m/z values provides enhanced precision. Thus for the data shown in Fig. 3, a standard deviation of ± 1.7 Da is observed for 12 m/z measurements from one spectrum. In this example, the error in the determination, i.e., $[(M_r^{theoretical} - M_r^{measured})/ M_r^{theoretical}] \times 100$, is 0.007%. Molecular weights have now been demonstrated to be measurable to better than 0.005% with quadrupole mass spectrometers. Typical results are summarized in Table I. Such measurements are generally two to three orders of magnitude more accurate than obtainable using electrophoretic methods, and perhaps an order of magnitude better than time-of-flight (TOF)-based laser desorption or plasma desorption methods.

Operation in Negative Ion Mode

Preliminary experiments have demonstrated the successful application of ESI mass spectrometry to oligonucleotides.[11,20] The negative ion ESI mass spectra of small oligodeoxyribonucleotides are characterized by multiply charged molecular anions of the form $(M - nH)^{n-}$. Small oligonucleotides ($n = 3–8$) afford molecular ions at near the maximum possible charge state (ionization of the internal phosphodiester and/or terminal phosphate moieties). As the M_r of larger oligomers increases, the extent of such multiple charging compensates, which maintains the m/z of members of the distribution of molecular anions in the range of our quadrupole mass analyzer. In addition, with the increase in level of polymerization and with decreasing charge state for a given molecular anion, we observe an increased contribution of species arising from the substitution of sodium

[20] C. G. Edmonds, C. J. Barinaga, J. A. Loo, H. R. Udseth, and R. D. Smith, *Proc. 37th ASMS Conf. Mass Spectrom. Allied Topics, Miami Beach, FL* p. 844 (1989).

TABLE I

CALCULATED AND MEASURED RELATIVE MOLECULAR MASSES BY ELECTROSPRAY
IONIZATION—QUADRUPOLE MASS SPECTROMETRY FOR A REPRESENTATIVE GROUP OF
OLIGOPEPTIDE AND SMALL PROTEIN STANDARDS

| Sample | Relative molecular mass (M_r) | | Mass measurement error (%) |
	Calculated	Measured	
LHRH (luteinizing hormone releasing hormone	1182.3	1182.0	−0.02
Bovine insulin	5733.6	5733.9	+0.01
Porcine insulin	5777.6	5776.5	−0.02
Porcine lactobinoylinsulin	6232.8	6234.5	+0.03
Porcine dilactobinoylinsulin	6688.0	6688.4	+0.01
Equine heart cytochrome c	12,360.1	12,358.7	−0.01
Bovine ribonuclease A	13,682.2	13,681.3	−0.01
Bovine α-lactalbumin	14,175.0	14,173.3	−0.01
Carboxymethylated bovine α-lactalbumin	14,647.4	14,645.9	−0.01
Chicken egg white lysozyme	14,306.2	14,304.6	−0.01
Human lysozyme	14,692.8	14,695.2	+0.02
Soybean trypsin inhibitor	20,090.6	20,097.6	+0.03
Bovine trypsin	23,290.2	23,291.9	+0.01
Bovine chymotrypsin	25,233.5	25,226.6	−0.03
Bovine carbonate dehydratase	29,024.6	29,021.6	−0.01

for the proton on the un-ionized phosphates. This substitution of Na for H among the increasing number of possible phosphate moieties (which accompanies increasing polymerization level and/or decreasing charge state) reduces the measurable current for any given ion and thus the sensitivity of the analysis. Where resolution permits, the separation of the members of these suites of molecular anions appear as shown in Fig. 4 for a synthetic deoxyribonucleotide 21-mer. In these circumstances the M_r of the parent molecule may be determined with the same precision and accuracy as observed for proteins under positive ionization conditions. For larger synthetic deoxyribonucleotides, we observe multiply charged molecular anions with broadened peaks apparently due to this alkali attachment. Measurement of M_r of oligonucleotides is significantly affected by alkali metal association when resolution is insufficient for distinction among the sodium-associated peaks. In such cases, determinations based on the maximum or centroid of a broadened, sodium-associated peak will tend to overestimate the M_r. With our present experimental regime for synthetic oligodeoxyribonucleotides $n = 20–60$ we have observed an

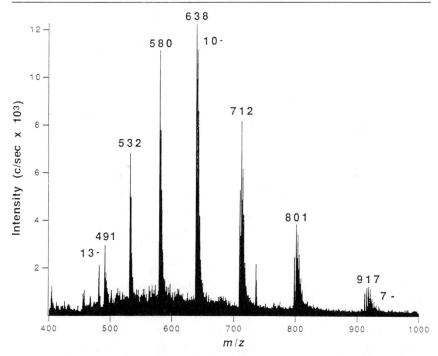

FIG. 4. Negative ion ESI mass spectrum of the synthetic deoxyribonucleotide 21-mer, d(AAATTGTGCACATCCTGCAGC) ($M_r^{\text{Calculated}}$ 6390.3). Peak multiplicity due to sodium association is particularly noticeable in the lower charge states of the molecular anion, e.g., ions grouped around at m/z 712 (M − 10H + Na)$^{9-}$, and m/z 801 (M − 11H + Na)$^{10-}$. M_r^{Measured} 6390.6, Δ = +0.005%.

error of the order of 1% for such measurements. Presently, the largest nucleotides so far successfully ionized are natural small oligoribonucleotide 76-mer tRNAs.

Sample solutions of oligonucleotides are prepared in deionized water at concentrations of 1 to 2 μg/μl. Sample delivery to the electrospray ion source optimizes at approximately 0.5 μl/min and the mass spectrometer is scanned with best integration and signal/noise to cover the full mass range in 2.2 min. Thus, a typical experiment might consume 1 μg (i.e., 3 nmol or 0.3 OD_{260} for a deoxyoligonucleotide 10-mer), although much smaller sample sizes are accommodated. The spectrum shown in Fig. 4 is the average of five such consecutively acquired scans. For this direct infusion of microliter volumes, the typical minimum sample must be increased by approximately tenfold to accommodate the volume holdup of connecting fused-silica tubing and the microliter syringe. This excess is recoverable at the end of the experiment as the uncontaminated analyte in aqueous solution. Alternatively, a reversible syringe pump may be used to minimize sample consumption by drawing a much smaller volume from a sample container and delivering this to the ion source. Electro-osmotic sampling methods offer prospects for very efficient sample manipulation with minimum sample consumption and loss. Presently, for samples available in the minimum amount of 10 μg (i.e., 30 nmol or 3 OD_{260} for a deoxyoligonucleotide 10-mer), this arrangement permits practical analysis of oligonucleotide samples with sensitivities of a factor 3–5 superior to those typically required for FAB and PD experiments. However, with these techniques the unconsumed portion of the analyte may also be recovered (with some difficulty) from FAB matrix or the PD Mylar foil sample stage.

Tandem Mass Spectrometry of Doubly Charged Tryptic Fragments

The major characteristic of ESI mass spectra for oligopeptides is the predominance of multiply charged molecular ions and the lack of fragmentation. While this allows accurate molecular weight determinations, little information regarding molecular structure is obtained. This characteristic is shared by the other desorption ionization techniques and extensive MS/MS studies of peptides have been reported; their scope and utility are highlighted elsewhere in this volume. However, to date tandem mass spectrometry of singly charged ions generated by these methods has not been fruitful beyond about m/z 3000 due, in part, to the decreased primary ion signal as molecular weight increases and the decreasing efficiency for CID with increasing M_r.[21,22] Large multiply charged oligopeptide and protein ions can be efficiently dissociated by collision with a neutral gas,[16,19,23,24] but interpretation presents special difficulties due to the fact that mass spectrometry does not directly provide information on charge state. This complication is largely avoided for doubly charged ions. In the case of the majority of tryptic peptides, the two charges are expected to reside on opposite ends of the molecule, at the N terminus and on the arginine or lysine residues at the C terminus. The obvious exception is the fragment containing the C terminus of the original molecule which is protonated only at the N terminus. Additional variations occur when a

[21] M. L. Gross, K. B. Tomer, R. L. Cerny, and D. E. Giblin, in "Mass Spectrometry in the Analysis of Large Molecules" (C. J. McNeal, ed.), p. 171. Wiley, New York, 1986.

[22] G. M. Neumann and P. J. Derrick, Org. Mass Spectrom. 19, 165 (1984).

[23] R. D. Smith, C. J. Barinaga, and H. R. Udseth, J. Phys. Chem. 93, 5019 (1989).

[24] R. D. Smith, J. A. Loo, C. J. Barinaga, C. G. Edmonds, and H. R. Udseth, J. Am. Soc. Mass Spectrom. 1, 53 (1990).

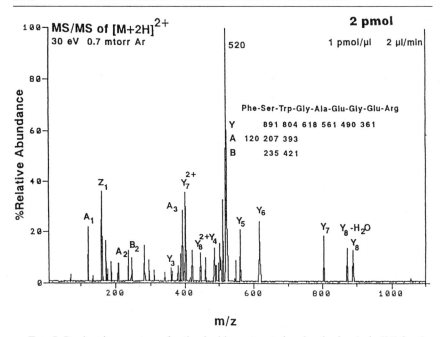

FIG. 5. Product ion spectrum for the doubly protonated molecular ion (m/z 520) for the small peptide Phe-Ser-Trp-Gly-Ala-Glu-Gly-Glu-Arg (M_r 1037.5). (Reproduced from Ref. 25 with permission of the author and publisher.)

histidine residue is present which has the effect of increasing the charge state of the molecule by one. In the CID of such doubly charged molecules, primarily singly charged product ions result. This typical result is illustrated in Fig. 5 for the small peptide Phe-Ser-Trp-Gly-Ala-Glu-Gly-Glu-Arg (M_r 1037.5).[25] In related studies, nearly complete sequence information was obtained from the MS/MS analysis of a doubly charged tryptic peptide (M_r 2315) of the protein, α-casein, generated by FAB[26] with a triple quadrupole mass spectrometer and ionspray LC/MS/MS experiments have been performed from doubly charged tryptic fragments of large proteins.[27,28]

[25] I. Jardine, in "Current Research in Protein Chemistry" (J. Villafranca, ed.). Academic Press, San Diego, CA, in press.

[26] D. F. Hunt, N.-Z. Zhu, and J. Shabanowitz, *Rapid Commun. Mass Spectrom.* **3**, 122 (1989).

[27] E. D. Lee, J. D. Henion, and T. R. Covey, *J. Microcolumn Sep.* **1**, 14 (1989).

[28] A. Treston, P. Kasprzyk, F. Cuttitta, T. Covey, and J. Mulshine, *Proc. 37th ASMS Conf. Mass Spectrom. Allied Topics, Miami Beach, FL* p. 889 (1989).

Developing Techniques

High-Precision M_r Determination at High Mass

Up to the present, ESI has been widely applied only with quadrupole mass analysis, although workers have demonstrated the method with magnetic sector[6,29] and Fourier transform ion cyclotron resonance instruments (FT-ICR).[30] It is reasonable to project improvements in the precision of measurement of M_r using ESI: quadrupole mass spectrometers under favorable conditions are capable of mass precision measurements of $\pm 0.0006\%$,[31] and conventional double-focusing mass spectrometers or FT mass spectrometers routinely achieve 1 ppm (0.0001%) precision.

Qualitative Investigations of Larger Oligopeptides and Proteins by the Collision-Induced Dissociation (CID) of Multiply Charged Molecular Ions

It is our original observation that the voltage applied between the nozzle and skimmer (nozzle–skimmer voltage bias, V_n) in the atmosphere–vacuum interface to the mass spectrometer greatly influences the maximum number of charges observed for a particular molecule. For large molecules, the more highly charged ions are more susceptible to dissociation. This observation, together with its converse, i.e., that the more highly charged species (at lower m/z) are favored at lower V_n (less energetic collisions), is consistent with a CID process. It is generally observed that as parent ion m/z values increase, the efficiency of CID processes decreases. Since collision energy is proportional to the number of charges at a given m/z (i.e., $E_{tr} = qV$), it is reasonable to assume that the greater translational energies for ions with a greater number of charges is the primary reason for this increased CID efficiency. In addition, it is clear that significant collisional heating of the molecular ion may occur in the nozzle–skimmer region of the atmosphere–vacuum interface. We have demonstrated the production of sequence-specific fragment ions from the multiply charged molecular ions of oligopeptides larger than might be efficiently dissociated as singly charged species[19] and that unique multiply charged product ions are observed in such experiments on proteins.[23,24]

Experiments by CID on the $(M + 3H)^{3+}$ to $(M + 6H)^{6+}$ molecular ions of melittin (M_r 2845) from ESI yields multiply charged daughter

[29] C. K. Meng, B. S. Larsen, C. N. McEwen, C. M. Whitehouse, and J. B. Fenn, *Proc. 2nd Int. Symp. Mass Spectrom. Health Life Sci., San Francisco, August* (1989).

[30] K. D. Henry, E. R. Williams, B. H. Wang, F. W. McLafferty, J. Shabanowitz, and D. F. Hunt, *Proc. Natl. Acad. Sci. U.S.A.* **86,** 9075 (1989).

[31] J. Roboz, J. F. Holland, M. A. McDowall, and M. J. Hillmer, *J. Rapid Commun. Mass Spectrom.* **2,** 64 (1988).

ions (to 4 +) that can be readily ascribed to the known sequence of the polypeptide.[19] Dramatic differences among the spectra of the various charge states and the large number of fragment ions in various charge were observed. Although CID of multiply charged ions from larger peptides yield product ions that can be readily correlated to the sequence, *complete* sequence assignments are generally lacking. Studies to date indicate that a progressively smaller portion of the molecule is susceptible to CID with increasing analyte size. For example, human parathyroid hormone (1–44, M_r 5064) yields multiply charged molecular ions from 4 + to 9 +. Collisional analysis of these parent ions, producing primarily singly charged b and y sequence ions (see end of this volume, Appendix 5) from both C and N termini, affords sequence information for approximately one-third of the molecule. For still larger protein molecules [e.g., bovine pancreatic ribonuclease A (M_r 13,682), composed of 124 amino acids and 4 disulfide bridges], only small portions of the molecule are directly probed by CID using the present quadrupole instrumentation. Nonetheless, this CID can be efficient for such multiply charged species and product ion spectra may be interpreted on the basis of known sequence and modes of fragmentation.[32] In addition, differences in the product ion spectra of the same charge state for the native and reduced forms of cystinyl-bridged proteins suggest that higher-order (secondary and tertiary) structure of these ions in the gas phase may influence the activation and dissociation processes.[32] The use of such differences to probe higher-order protein structure is an important possibility.

In proteins where fragmentation is particularly diverse and abundant, but not necessarily fully interpretable, CID spectra may provide a qualitatively unique "fingerprinting." The CID mass spectra for the $(M + 15H)^{15+}$ molecular ion of cytochrome c proteins from nine different species (cow, chicken, dog, horse, pigeon, rabbit, rat, tuna, yeast) with a moderate M_r range have been obtained and compared.[24] Strong differences in the spectra are observed, even between species differing by as little as 4 amino acid residues (of over 100). Although presently requiring 10–100 pmol of material, such results suggest the possibility of applying CID mass spectral patterns for rapid characterization of unknown proteins or for comparison among related molecular species (e.g., mutations and post-translational modifications). While conventional CID is normally performed in the collision quadrupole chamber (rf-only quadrupole, Q_2), collisional processes in the differentially pumped atmosphere–vacuum interface can also occur prior to mass analysis. Collisional product ions thus become available for MS/MS analysis (i.e., CID/MS/CID/MS) permitting the confirmation of their structure or the probing of regions not

[32] J. A. Loo, C. G. Edmonds, and R. D. Smith, *Science* **248**, 201 (1990).

susceptible to fragmentation in initial CID. Confirmation of the product assignments can be obtained by further dissociation of the fragment ions formed using the ESI interface.[24]

The possibilities of MS/MS of multiply charged ions have only begun to be exploited. New instrumentation and methods addressing, among other challenges, the assignment of charge state or mass of multiply charged product ions will be essential. Higher-energy CID or alternative activation methods, such as photo- and surface-induced dissociation or sequential "tandem" MS methods (MS[n], with $n \geq 3$) should generate more sequence information from larger species.

Interface with Capillary Separation Schemes

Liquid chromatography is a mainstay of biochemical analysis. As already mentioned, the now commonplace method of "microbore" (1–2 mm id columns with flow rates of 10 to 100 μl/min) liquid chromatography is particularly well adapted to ESI practiced with pneumatic-assisted nebulization, i.e., ionspray. The examination of tryptic digests of proteins is a straightforward application of ESI methods with facile and immediate prospects. The practice of liquid chromatography in very narrow-bore columns and capillary tubes (id ≤ 250 μm and flow rates of 0.1 to 1 μl/min) is a more demanding variation which a few laboratories have pursued.

Capillary electrophoresis (CE) in its various manifestations (free solution, isotachophoresis, isoelectric focusing, polyacrylamide gel, micellar electrokinetic "chromatography") is attacting attention as a method for rapid high-resolution separations of very small sample volumes of complex mixtures. Combination with the inherent sensitivity and selectivity of mass spectrometry, makes CE-MS a potentially powerful bioanalytical technique. The correspondence between capillary zone electrophoresis (CZE) and ESI flow rates and the fact that both are primarily used for ionic species in solution provide the basis for facile combination. Small peptides are easily amenable to CZE/ESI-MS analysis with good reproducibility. For example, high-efficiency separations of biologically active peptides, [e.g., dynorphin and enkephalin peptides with over 250,000 theoretical plates,[33] and of tryptic digests (as discussed above)] have been demonstrated. High-efficiency CZE with detection by ESI-MS at the picomole level has been demonstrated.[34] Capillary isotachophoresis (CITP) based on ESI-MS has been de-

[33] E. D. Lee, W. Mück, J. D. Henion, and T. R. Covey, *J. Chromatogr.* **458**, 313 (1988).
[34] J. A. Loo, H. K. Jones, H. R. Udseth, and R. D. Smith, *J. Microcolumn Sep.* **1**, 223 (1989).

scribed.[35,36] CITP is an attractive complement to CZE, and is ideally suited for combination with mass spectrometry. CITP is well suited to low-concentration samples where the amount of solution is relatively large, whereas CZE is ideal for the analysis of minute quantities of solution. Sample sizes which can be addressed by CITP are much greater (>100-fold) than CZE. CITP results in concentration of analyte bands, which is in contrast to the inherent dilution with CZE. Electromigration injection allows effective sample volumes to be much larger still due to enrichment during migration into the capillary from low ionic strength samples.[36] Detection limits of approximately 10^{-11} M have been demonstrated for quaternary phosphonium salts and substantial improvements appear feasible.[36] Analytes elute in CITP as bands where the length of the analyte band provides information regarding concentration. Most important, however, is that CITP provides a relative pure analyte band to the ESI source, without the large concentration of a supporting electrolyte demanded by CZE. This latter characteristic circumvents the principal research challenge in the development of CZE/ESI-MS, the disadvantageous effect of constituents of the supporting electrolyte on ESI. Thus, CITP/MS has the potential of allowing much greater sensitivities (and analyte ion currents) than feasible with CZE/MS due to more efficient analyte ionization. The relatively wide and concentrated separated bands in CITP facilitate MS/MS experiments (which often require more concentrated samples than provided by CZE). These characteristics make CITP/MS/MS potentially well suited for characterization of enzymatic digests of proteins.[37,38] Significant research effort in all areas of the separation of biopolymers and their constituents by capillary electrophoresis in all its several modes is required to capitalize on the potential which their combination with ESI mass spectrometry embodies.

Acknowledgment

The authors are pleased to acknowledge the important contributions of our colleagues Drs. J. A. Loo, C. J. Barinaga, and H. R. Udseth, and the kind donation of oligonucleotide samples by Dr. T. Keough, Proctor and Gamble, Inc., Cincinnati, OH. This work was supported by the U.S. Department of Energy (Contract DE-AC06-76RLO 1830), the National Institute of General Medical Sciences (GM 42940), and the National Science Foundation (DIR 8908096). Pacific Northwest Laboratory is operated by Battelle Memorial Institute.

[35] R. D. Smith, J. A. Loo, C. J. Barinaga, C. G. Edmonds, and H. R. Udseth, *J. Chromatogr.* **480**, 211 (1989).

[36] H. R. Udseth, J. A. Loo, and R. D. Smith, *Anal. Chem.* **61**, 228 (1989).

[37] C. G. Edmonds, J. A. Loo, S. M. Fields, C. J. Barinaga, H. R. Udseth, and R. D. Smith, *in* "Biological Mass Spectrometry" (A. L. Burlingame and J. A. McCloskey, eds.). Elsevier, Amsterdam, in press.

[38] R. D. Smith, S. M. Fields, J. A. Loo, C. J. Barinaga, and H. R. Udseth, *Electrophoresis,* in press.

[23] Sample Preparation for Plasma Desorption Mass Spectrometry

By PETER ROEPSTORFF

Sample amounts are often very limited in practical biochemical studies, and the sample quantity available for mass spectrometric analysis may represent tedious and costly preparation procedures, or may even be the ultimate sample amount. It is, therefore, of the utmost importance that the mass spectrometric procedures employed give optimal chances for success at the first trial, and extract maximum information from a given sample amount.

Plasma desorption mass spectrometry (PD-MS)[1,2] has recently been very successful for the analysis of peptides and proteins up to a molecular weight of approximately 35,000,[3] and it seems to be a promising routine method in the protein chemistry laboratory, because the instrumentation is relatively inexpensive and very simple to operate.[4] Because of this simplicity the only way to improve the results is by improving the quality of the sample. The method by which the sample is prepared is, therefore, of major importance. As in all desorption ionization methods, the presence of low-molecular weight contaminants, and among these especially alkali metal salts, may reduce the spectrum quality considerably, and often leads to complete suppression of the signals from high-molecular weight samples. Unfortunately, most methods in protein chemistry are based on the use of salt-containing aqueous buffers, necessitating the development of sample application methods and alternative solvent systems, acceptable for use in mass spectrometry. The PD mass spectra of peptides and proteins most frequently contain only molecular ions and no structurally meaningful fragment ions. It is, however, possible to obtain structural information by combining chemical or enzymatic cleavage or modification procedures known from classical protein chemistry with the mass spectrometric analysis.

This chapter will describe the methods currently used in protein studies

---

[1] D. F. Thorgerson, R. P. Skowronski, and R. D. Macfarlane, *Biochem. Biophys. Res. Commun.* **60**, 616 (1974).

[2] R. D. Macfarlane, this volume [11].

[3] For a recent review see R. J. Cotter, *Anal. Chem.* **60**, 781A (1988).

[4] Commercial instruments are available from BioIon Nordic AB, Box 15045, S-750 15 Uppsala, Sweden.

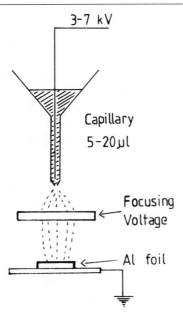

FIG. 1. Principle of the electrospray method for sample application.

by PD-MS in the author's laboratory for sample application and for obtaining structural information on the sample.

Sample Application by Electrospray

The first method used for sample application in PD-MS was the electrospray method,[5] the principle of which is shown in Fig. 1.

Procedure

Sample solution (3–10 μl) containing 0.1–10 μg/μl of peptide is placed in the capillary and the high voltage gradually increased until the spray begins (visual observation in a parallel light beam). The focusing voltage is then regulated to obtain a well-defined sample deposition on the aluminum foil with a spot-size of the area exposed to fission fragments (approximately one-half the diameter of the aluminum foil). The spray stops when all of the sample solution is consumed.

[5] C. J. McNeal, R. D. Macfarlane, and E. L. Thurston, *Anal. Chem.* **51,** 2036 (1979).

Comments

Electrospray is only possible with reasonably volatile organic solvents, e.g., acetic or trifluoroacetic acid, chloroform, acetone, methanol, or ethanol. A water content up to 30% may be acceptable. This limits its applicability in protein chemistry because most proteins are best dissolved in aqueous solvents. The electrospray technique, furthermore, requires samples of a very high purity with respect to low-molecular weight impurities and considerably higher sample amounts than the nitrocellulose-based techniques described below.

Improved results can be obtained by cospraying peptide and protein samples with an excess of reduced glutathione (GSH).[6] This results in improved molecular ion yields, increased abundance of multiply charged molecular ion species, and often also in a reduction of the influence of alkali metal ions. GSH must be present in a 25–100 times molar excess relative to the sample. In practice, this is achieved by making the spray solvent 50 mM with respect to GSH. Typically, the sample is dissolved in a solution of 50 mM GSH in acetic acid/trifluoroacetic acid, 10:1 (v/v), to a concentration of 1 μg/μl; 3–7 μl is electrosprayed on the aluminum target.

Sample Application by Adsorption on Nitrocellulose

A considerable improvement in sample handling procedures has been obtained by the introduction of nitrocellulose as a matrix.[7] The principle (see Fig. 2) is based on adsorption of the protein or peptide to a thin nitrocellulose layer placed on the aluminum foil, followed by removal of salt contaminants by washing with ultrapure water or dilute acid. The use of nitrocellulose compared to electrospray results in improved molecular ion yields (Fig. 3), increased abundance of multiply charged molecular ions, and sharper peaks. The effect is qualitatively similar, but quantitatively better, than that obtained by adding GSH to the electrospray solution.

Three different methods have been used for application of the sample onto the nitrocellulose layer. The original method was based on adsorption of the protein from a few microliters of aqueous solution followed by washing of the surface with 1–2 ml of pure water or dilute acid (Fig. 2, left-hand side). The sample solution is sandwiched between the nitrocellu-

[6] M. Alai, P. Demirev, C. Fenselau, and R. J. Cotter, *Anal. Chem.* **58,** 1903 (1986).

[7] G. P. Jonsson, A. B. Hedin, P. L. Håkansson, B. U. R. Sundqvist, G. S. Säve, P. F. Nielsen, P. Roepstorff, K. E. Johansson, I. Kamensky, and M. S. L. Lindberg, *Anal. Chem.* **58,** 1084 (1986).

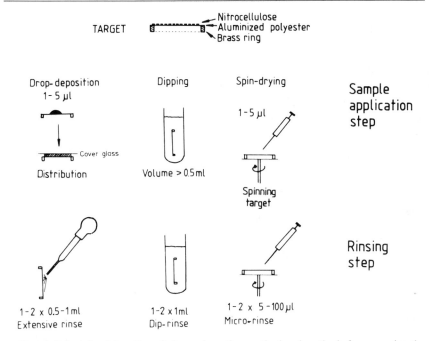

FIG. 2. Principle of the nitrocellulose adsorption method and methods for preparing the sample. *Left*, adsorption followed by extensive wash; *center*, the dipping technique; *right*, the spin technique.

lose layer and a microscope cover glass in order to spread it on the surface. This application technique requires sample concentrations of the order of 0.5 to 1 μg/μl. An alternative method, which allows application from very dilute solutions, is to dip the nitrocellulose-covered target in the protein solution for 0.5 to 1 min, followed by drying and washing of the surface (Fig. 2, center). This method has, for example, been successfully used to extract lutenizing hormone releasing hormone from a $5 \times 10^{-7} M$ solution.[7] The third method, now the standard in the author's laboratory, applies the sample on a spinning nitrocellulose target[8] (Fig. 2, right-hand side). This method was first introduced for application of small peptides, which do not bind strongly to nitrocellulose, and therefore might be removed in the washing procedure. The spin-drying procedure combines the adsorption and washing in a single step, because a thin sample layer is deposited in the central area (the area exposed to fission fragments), whereas the excess

[8] P. F. Nielsen, K. Klarskov, P. Højrup, and P. Roepstorff, *Biomed. Environ. Mass Spectrom.* **17,** 355 (1988).

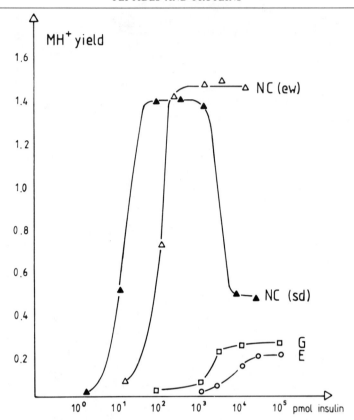

FIG. 3. Molecular ion yield of insulin (expressed as number of molecular ions recorded per 100 fission events) as a function of the sample amount and sample application technique. ▲, Spin drying (sd) on nitrocellulose (NC); △, adsorption on nitrocellulose followed by extensive wash (ew); □, electrospray with glutathione (G); and ○, electrospray (E). (Adapted from Ref. 12 with permission from the publisher.)

sample and the more soluble salts migrate with the solvent to the periphery of the target. The spin-drying method has been found to give better sensitivity, not only for small peptides and for other types of compounds, which do not bind to nitrocellulose, but also for large peptides and proteins (Fig. 3). The following sections describe the procedures currently used as standard in the author's laboratory.

Preparation of Nitrocellulose Targets

A stock solution of nitrocellulose in acetone (2 μg/μl) is prepared by dissolving a small piece of nitrocellulose blotting membrane (Bio-Rad Laboratories, Richmond, CA) in an appropriate volume of acetone (analytical-grade). This solution (25–50 μl) is electrosprayed onto the aluminized Mylar target. The focusing voltage is adjusted to give a spot size of approximately 7 mm in diameter. Each target is visually inspected for homogeneity. A thick homogeneous nitrocellulose layer is increasingly important with increasing molecular weight of the sample, or for the analysis of very small sample amounts. The best targets are, therefore, used for such samples, whereas lower quality targets may be used for less critical samples.

Application of Sample onto Nitrocellulose

The peptide or protein may, in principle, be dissolved in any solvent which does not cause damage to the nitrocellulose film, e.g., water, dilute acid or base, salt-containing buffer solutions, 2-propanol and ethanol–water or acetonitrile–water mixtures. The most frequently used solvent in the author's laboratory is ultrahigh-quality (UHQ) water (see Comments) containing 0.1% trifluoroacetic acid (TFA) and 15% ethanol or acetonitrile. If necessary, the pH may be adjusted by addition of ammonia. The nitrocellulose-covered target is placed in a holder mounted horizontally on the shaft of a variable-speed motor. The sample solution (2–5 μl) containing 0.01–1 μg/μl of sample is placed in the center of the target, and the motor speed is gradually increased to distribute the solvent on the entire surface, followed by drying at full speed (2500 rpm). The drying, which propagates from the center, is easily observed. Sometimes the sample solution is simply allowed to dry without spinning, and distribution on the surface omitted or effected in the following washing step.

From Fig. 3 it is seen that the application of washing procedures may lead to reduced sensitivity, probably due to removal of sample in the washing step. On the other hand, application of too much sample by the spin technique leads to reduced molecular ion yields. Fortunately, PD-MS is essentially a nondestructive method, which means that it is possible to record a spectrum of the sample after application, for example, by the spin technique. If indicated upon examination of the spectrum, a washing procedure may then be applied and the sample reanalyzed. A poor sample ion yield indicates the need for washing. Most frequently, poor yields are due to a too high content of alkali metal ions, which can be ascertained by observation of the ions for Na$^+$ and K$^+$ at m/z 23 and 40, respectively. If the summed abundance of these ions is more than one-half of the abundance of the H$^+$ ion, washing is indicated. Another reason for low sample ion yield may be that too much or too little sample has been applied. If the former is suspected, washing may also improve the result, whereas improved results in the latter case can only be obtained by prolonged recording of the spectrum in order to obtain better ion statistics.

Washing

Washing of the nitrocellulose-bound sample may be performed by either of the methods shown in Fig. 2. Microwashing, performed by applying 2 × 5–10 μl of a 0.1% TFA solution to a slowly spinning target, followed by drying at full speed, is usually preferred. If indicated by a high content of alkali metal ions, or if a salt-containing sample buffer has been used, washing with 2 × 100 μl of washing solution on a spinning target may be preferable, or an extensive wash may be applied. The latter is effected by literally sluicing the nitrocellulose surface with 2 × 0.5 to 1 ml of solvent with a Pasteur pipette as illustrated in Fig. 2.

Comments

The quality of the water used for the sample solution, the washing solvents, and also for the HPLC solvents, if HPLC is the last purification step prior to mass spectrometric analysis, is of the utmost importance. In the author's laboratory, 15–18 MΩ/cm resistivity water (UHQ-water) prepared with an Elgastat UHQ apparatus (Elga Ltd., High Wycombe Bucks, UK) is always used.

In Situ Reactions to Obtain Further Information

Plasma desorption mass spectra of peptides and proteins are dominated by molecular ion species, whereas structurally meaningful fragment ions are sparse or absent. However, as mentioned above, most of the sample is undamaged after recording a spectrum. The remaining nitrocellulose-bound sample can, therefore, after removal of the target from the mass spectrometer, be subjected to chemical or enzymatic reactions *in situ*,[9,10] and structural information thus be obtained. The most commonly used reactions are reduction of disulfide bonds with dithiothreitol (DTT),[9,11,12] cleavage with endopeptidases,[10,12,13] resulting in so-called PD maps, and cleavage with exopeptidases, especially carboxypeptidases,[10,14] which can

[9] B. T. Chait and F. H. Field, *Biochem. Biophys. Res. Commun.* **134,** 420 (1986).
[10] B. T. Chait, T. Chaudhary, and F. H. Field, *in* "Methods in Protein Sequence Analysis 1986" (K. A. Walsh, ed.), p. 483. Humana, Clifton, NJ, 1987.
[11] P. F. Nielsen and P. Roepstorff, *Biomed. Environ. Mass Spectrom.* **18,** 138 (1989).
[12] P. F. Nielsen, P. Roepstorff, I. G. Clausen, A. B. Jensen, I. Jonassen, A. Svendsen, P. Balschmidt, and F. B. Hansen, *Protein Eng.* **2,** 449 (1989).
[13] P. Roepstorff, P. F. Nielsen, K. Klarskov, and P. Højrup, *in* "The Analysis of Peptides and Proteins by Mass Spectrometry" (C. J. McNeal, ed.), p. 55. Wiley, New York, 1988.
[14] K. Klarskov, K. Breddam, and P. Roepstorff, *Anal. Biochem.* **180,** 28 (1989).

give C-terminal sequence information. An example of an *in situ* reduction is shown in Fig. 4.

Procedures

Reduction of disulfide bonds is effected by placing 2 μl of a solution of 0.08 M DTT in 0.1 M ammonium bicarbonate, pH 7.8, on the sample surface. A microscope cover glass is placed on top to prevent evaporation, or the target is placed in a small, closed plastic box containing a moist piece of filter paper. After reaction at room temperature for 10 min, the target is spin-dried and reinserted into the mass spectrometer.

Enzymatic cleavages are performed in a similar way. In order to obtain acceptable cleavage yields in a reasonably short time, it is necessary to use a high enzyme-to-substrate ratio, e.g., between 1 : 2 and 1 : 1. Typical standard reaction conditions are with trypsin, *Staphylococcus aureus* protease, or carboxypeptidases: 15–30 min at 37° with 2 μl of a solution containing 1 μg/μl of enzyme in 0.1 M ammonium bicarbonate, adjusted to an appropriate pH with acetic acid.

Monitoring of Reactions in Solution

In situ reactions are often incomplete, most likely because some sample molecules embedded in the matrix are inaccessible to the reagents, or because the enzymes are inactivated by contact with the nitrocellulose. Therefore, if sufficient sample amounts are present, it may be advantageous to monitor reactions carried out in solution. Likewise, mass spectrometric monitoring of a preparative reduction and alkylation reaction or an enzymatic cleavage might often be useful.[13,15]

Procedures

The standard procedures used in protein chemistry can normally be applied. It is, however, preferable to exchange alkali metal salts in buffers with ammonium salts, because the latter do not have a negative effect on the quality of the plasma desorption spectra.[16] If this is possible, aliquots of 2 to 10 μl are withdrawn from the reaction solution and applied onto a nitrocellulose target by the spin-drying procedure, followed by mass

[15] P. Roepstorff, P. F. Nielsen, K. Klarskov, and P. Højrup, *Biomed. Environ. Mass Spectrom.* **16,** 9 (1988).

[16] M. Mann, H. Rahbek-Nielsen, and R. Roepstorff, in IFOS V, "Proceedings of the Fifth Symposium on Ion Formation from Organic Solids" (A. Hedin, B. U. R. Sundquist, and A. Benninghoven, eds.), p. 47. Wiley, New York, 1990.

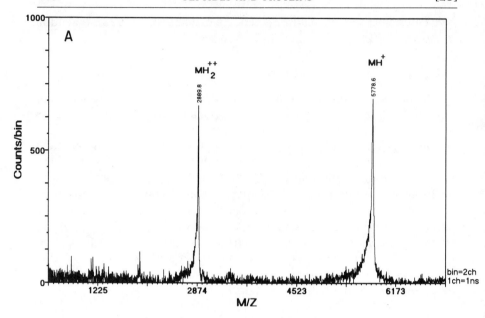

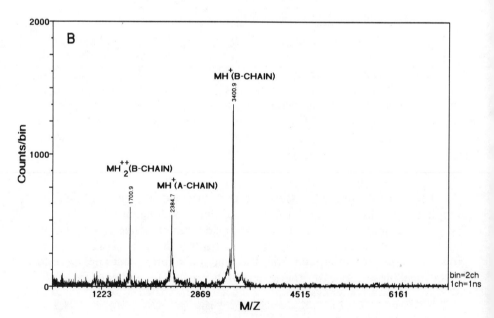

Fɪɢ. 4. Positive ion PD-mass spectra of human insulin showing the intact molecule (A) and the same sample after *in situ* reduction showing the A and B chains (B).

spectrometric analysis. UHQ-water is, of course, used for all solvents. If metal salts cannot be eliminated, or if the solution contains high concentrations of urea or guanidine-HCl, extensive washing is required, a procedure that may lead to loss of small peptides.

[24] Molecular Weight Analysis of Proteins

By Ian Jardine

Introduction

The molecular weight of a protein has always been recognized as an important analytical parameter in biochemistry. Sodium dodecyl sulfate–polyacrylamide gel electrophoresis (SDS–PAGE) is universally used to purify proteins; molecular weights are routinely determined after separations by comparison of the migration of the protein of interest to that of a set of standard proteins. A typical protein molecular weight experiment would be to analyze the molecular weight before and after removal of carbohydrate to estimate the carbohydrate content of the glycoprotein and the mass of the deglycosylated protein. Estimates of the accuracy of SDS–PAGE protein molecular weight determination range from a few percent for well-behaved proteins in this system up to 20 or 30% for heavily glycosylated proteins. Clearly, high accuracy (e.g., to <1.0%) in protein molecular weight measurement has never been achieved with SDS–PAGE, so even some of the molecular weight standards used are not well characterized.[1]

After many years of effort to ionize, vaporize, and, subsequently, mass analyze proteins at high sensitivity via ionization techniques such as plasma desorption (PD) and fast atom bombardment (FAB), the protein molecular weight determination scenario has now changed dramatically with the recent discovery of two new highly sensitive mass spectrometric ionization techniques, electrospray (ESI)[2,3] and matrix-assisted laser desorption (LD).[4,5] These methods have both been demonstrated to allow

[1] H. D. Kratzin, J. Wiltfang, M. Karas, V. Neuhoff, and N. Hilschmann, *Anal. Biochem.* **183,** 1 (1989).

[2] C. M. Whitehouse, R. M. Dryer, M. Yamashita, and J. B. Fenn, *Anal. Chem.* **57,** 675 (1985).

[3] C. G. Edmonds and R. D. Smith, this volume [22].

[4] M. Karas, U. Bahr, A. Ingendoh, and F. Hillenkamp, *Angew. Chem., Int. Ed. Engl.* **28,** 760 (1989).

[5] F. Hillenkamp and M. Karas, this volume [12].

the ionization and subsequent precise mass analysis of proteins to over 100,000 Da at low picomole levels and below and with accuracies of better than 0.1%.

To date, electrospray ionization has been most readily coupled to quadrupole mass analyzers (ESI-quad), although coupling to a magnetic sector instrument, an Fourier transform ion cyclotron resonance system (FT-ICR), and a quadrupole ion trap, have all been demonstrated in at least preliminary fashion. Matrix-assisted laser desorption ionization has been demonstrated using time-of-flight mass analysis (LD-TOF), but numerous other mass analyzers, such as magnetic sector systems with array detectors, and ion traps, are, in principle, also options as mass analyzers for this ionization method. In other words, these ionization techniques might soon be wed to either low- or high-resolution and performance mass analyzers, although the problems associated with coupling these ionization techniques with other types of mass analyzers are not trivial. A liquid introduction atmospheric pressure ionization technique such as electrospray, for example, creates serious electrical discharge problems when coupled to high-voltage accelerating systems operating at high vacuum, such as magnetic sector mass spectrometers.

The following discussion addresses some of the issues which must now be considered in the determination of protein molecular weights by these new techniques in mass spectrometry. The ionization techniques of electrospray and matrix-assisted laser desorption are discussed elsewhere in this volume.[3,5] The general issue of the utility of exact mass measurement is also covered.[6]

Experimental Procedure

Electrospray mass spectra of proteins are obtained on a Finnigan MAT TSQ 700 triple quadrupole mass spectrometer. The proteins are dissolved in 50:50 (v/v) methanol–water solutions, sometimes containing up to 5% acetic acid, to a concentration of 10 pmol/μl and infused into the hollow electrospray needle at a flow rate of 5 μl/min. The mass spectrometer is repetitively scanned at a rate of 3 sec/scan and profile data is accumulated over a number of scans (usually about 10 or 20 scans is sufficient). The resulting m/z versus %relative abundance profile data is directly converted into mass versus %relative abundance profile data by the data system. This conversion directly identifies the components of mixtures, at least to the resolution of the mass spectrometer (usually set at $M/\Delta M$ of 1000) and assigns a mass value to the components. This deconvolution procedure

---

[6] K. Biemann, this volume [13].

follows the algorithms of Mann *et al.*[7] with modifications implemented in order to reveal components of mixtures. Identification of the mixture components allows reexamination of the original data in terms of which multiply charged envelope belongs to each component, if desired. Recalculation of the mass of each component can then be carried out using all the ions from each component. Use of this multiple point data set allows a calculation of the relative standard deviation ($\pm$RSD%) or precision of the mass measurement of each component.

The proteins used for the data presented here were obtained from various sources and are used without further purification. The Volga hemoglobin variant and the prolactin were obtained from Dr. T. Lee of the City of Hope Medical Center (Los Angeles, CA); the insulin-like growth factor and its oxidized methionine analog were from Dr. P. Griffin of Genentech (So. San Francisco, CA); plasminogen-activating inhibitor (PAI) and interleukin 6 (IL-6) were from Dr. L. Poulter of I.C.I. Great Britain (Manchester); bovine red blood cell ubiquitin and recombinant *Escherichia coli* thioredoxin were from Calbiochem (La Jolla, CA); chicken egg-white lysozyme and recombinant sperm whale myoglobin were from Sigma Chemical Co. (St. Louis, MO).

Results and Discussion

Isotopic Distributions and Average Mass

It is possible to calculate the molecular weight of small organic compounds by using the nominal masses of the elements (carbon, C = 12; hydrogen, H = 1; nitrogen, N = 14; oxygen, O = 16; etc.) but, since mass spectrometers can measure mass more accurately than nominal mass, the monoisotopic masses of the elements, comprising the most abundant isotope of each atom (C = 12.0000; H = 1.0078; N = 14.0030; O = 15.9949; etc.), are commonly used. Of course, the mass spectrometer detects all isotopes of every element in a molecular weight-associated ion, and so an isotope distribution is obtained if the mass resolution of the instrument is sufficient to separate the isotopes at the molecular mass of the molecule. At high enough resolution, the individual isotopes are distinguished, but at lower mass spectrometer resolution (e.g., as with ESI-quad and LD-TOF) the ion profiles coalesce to form a single asymmetric peak. Determining the mean value of this peak by centroiding provides the average mass of the molecule (after subtracting the mass of a hydrogen atom, if the ion corresponds to the MH$^+$ species). The average mass of a molecule is

[7] M. M. Mann, C. K. Meng, and J. B. Fenn, *Anal. Chem.* **61**, 1702 (1989).

calculated by using the average chemical or atomic weight of the individual elements (i.e., for the elements commonly encountered in protein molecular weight analysis: C = 12.011; H = 1.008; N = 14.007; O = 15.999; S = 32.060; P = 30.974; and so on).

As discussed by Yergey *et al.*[8] and Biemann,[6] there is no choice but to use atomic weights of the elements and the average chemical weight for proteins. It is observed that the isotope pattern of proteins becomes more symmetrical and extends over many masses (at the 1% relative abundance level, approximately 15 u at 10,000 Da; 40–50 u at 100,000 Da), the nominal mass value quickly becomes irrelevant, and even the monoisotopic mass diminishes rapidly to obscurity. In the molecular weight analysis of *pure* proteins, therefore, it is not necessary to mass analyze with very high mass resolution both because the monoisotopic ion becomes undetectable for all practical purposes, and since even the most abundant peak (even if it were possible to pick out the most abundant peak in a real resolved spectrum from the many peaks close to the same intensity) contains overwhelming combinations of isotopes which would require extremely high mass spectrometer resolution ($>10^6$) for accurate analysis. This conclusion is fortunate since, for example, quadrupole (for electrospray) and TOF (for laser desorption) mass analyzers are currently of relatively low resolution ($M/\Delta M$ of approximately 200–2000). In addition, many mass spectrometers, especially high-resolution magnetic systems, commonly trade sensitivity for resolution. That is, higher resolutions are only obtained at the expense of considerably decreased sensitivities. Sometimes such sensitivity decreases can result in poorer ion statistics and less precision in the molecular weight measurement and so operating at low resolution will often provide better mass accuracy and precision for average molecular weight measurements than at high resolution using the same mass spectrometer.

A high-performance quadrupole instrument would be expected to obtain a mass accuracy of 100 ppm or 0.01%, for an ion beam of sufficient intensity. This means that for pure proteins of masses 10,000 and 100,000 Da, mass accuracies of approximately ±1 and ±10 Da, respectively, would be about the best that could be expected. This is borne out in Table I where some calculated and experimentally determined molecular weights of what were considered to be, at least by ESI-quad analysis, reasonably pure proteins, are compared.

The quadrupole was calibrated in the electrospray mode using clusters of cesium iodide [$(CsI)_nCs^+$, $n = 1$ to approximately 20] generated in the electrospray by infusing a concentrated CsI solution. This method of

[8] J. Yergey, D. Heller, G. Hansen, R. J. Cotter, and C. Fenselau, *Anal. Chem.* **55**, 353 (1983).

TABLE I

COMPARISON OF CALCULATED MOLECULAR WEIGHTS OF
PROTEINS VERSUS DETERMINED BY ESI-MS

| Protein | Molecular weight | |
	Calculated	From ESI-MS
Ubiquitin	8564.9	8563.0 ± 1.0
Thioredoxin	11,673.4	11,671.2 ± 0.6
Lysozyme	14,306.2	14,307.4 ± 0.9
Myoglobin	17,199.1	17,196.8 ± 2.1
Interleukin 6 (IL-6)	20,907.3	20,901.5 ± 4.5
Prolactin	22,660.0	22,662.5 ± 1.4
Plasminogen-activating inhibitor (PAI)	42,769.9	42,777.5 ± 14.1

calibration is preferred at the present time over employing a "known" protein as a calibrant, since it is likely that the protein may not be pure, or may not stay pure over time and repeated use (see discussion on protein mixtures below). In addition, the cesium iodide clusters are monoisotopic and provide symmetrical ion abundance profiles which do not change in width over the mass range of interest for electrospray (m/z 100–2500) in contrast to multiply charged ions from proteins. An example of the calibrated multiply charged profile data from a typical protein analysis is shown in Fig. 1. This recording is of high quality and, indeed, when automatically converted into molecular weight data,[7] two relatively symmetrical molecular weight peaks are generated (Figs. 2 and 3). The experimentally determined molecular weights for these proteins agreed to within 0.02% of the expected values. Some additional peaks are apparent in this analysis. The species with an asterisk are approximately 23 Da above their companion species at lower mass and are attributed to sodium adducts.

The accuracy and precision of molecular weight measurement by ESI-quad are very good compared to SDS–PAGE, and reflects the use of the multiple measured data points of the original multiply charged protonated ion envelope. Similarly, it is anticipated that the mass accuracy obtained by LD-TOF may approach that of electrospray on a quadrupole instrument. This will result from the excellent signal-to-noise obtained in LD-TOF analysis of proteins, the ease of calibration of such systems, and the availability of some multiply charged as well as dimeric and trimeric species to increase precision (see discussion by Hillenkamp and Karas[5]). The use of internal standards has not yet been invoked to any significant extent for either ESI-quad or LD-TOF molecular weight analysis of proteins, probably because of the difficulty of certifying proteins as being well

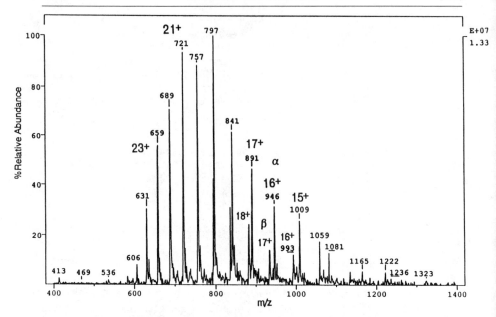

FIG. 1. ESI mass spectrum of a hemoglobin variant (hemoglobin Volga). The calculated charge states of the α and β chains of this hemoglobin variant are indicated.

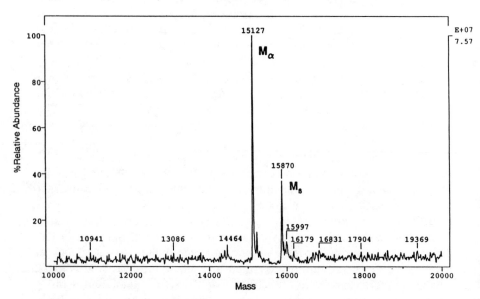

FIG. 2. Deconvoluted ESI-MS of the profile data from Fig. 1 of the Volga hemoglobin variant. The x axis is now a mass scale instead of a m/z scale.

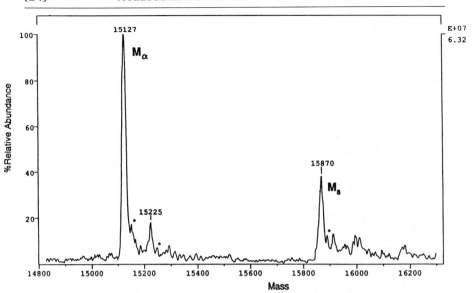

FIG. 3. Expansion of the mass spectrum of Fig. 2 at the mass range around the two major protein species (α and β chains of Volga hemoglobin variant) revealing the relatively pure nature of these species as well as the presence of other minor components. The peaks marked with an asterisk (∗) are approximately 23 Da above their immediate lower mass partners and are thus probably M + Na species.

enough characterized and of sufficient purity for the task. It is doubtful, in any event, whether such an approach would enhance the molecular weight accuracy in such low-resolution analyses.

The ability of ESI-quad and LD-TOF to allow molecular weight determination of *pure* proteins which do not contain any heterogeneity (e.g., of carbohydrate) at high mass needs to be considered separately. That is, LD-TOF will retain its ability to measure such proteins to masses well in excess of 100,000 u while for ESI-quad the situation is not so obvious. That is, to determine the mass of a protein by ESI-quad the adjacent multiply protonated ions must be resolved from each other, otherwise the mass of these distinct species cannot be determined, the charge state cannot be determined, and the protein mass, therefore, cannot be calculated. Consider a theoretical situation where no matter what the mass of the protein is, the number of protons which it contains always produces a *m/z* ion envelope centered around 1000 u. This seems a reasonable extrapolation given the data obtained thus far on proteins successfully analyzed by electrospray.[3] The following discussion pertains. If the protein weighs 10,000 and has at least one abundant species with ten protons attached $[M + 10H]^{10+}$, this species will occur at m/z 1001. The $[M + 11H]^{11+}$ species will occur at m/z 910.1. Assuming the isotope envelope at mass 10,000 Da spreads over about 15 Da, the peak width at base (1%) at m/z 1000.1 will be 15 u/10 or 1.5 u wide. The resolution required to separate such peaks (1.5 u wide peaks at m/z 1001 and 910.1) is very low ($\ll$100). Continuing this analysis for a protein of 100,000 Da: 100,000 Da plus 100 protons = 100,100 Da; 100,100 Da divided by 100 charges $[M + 100H]^{100+} = m/z$ at 1001; the $[M + 101H]^{101+}$ species will fall at 991.1 u. Assuming the isotope envelope at mass 100,000 spreads over about 50 Da, the peak width at m/z 1001 will be 50 u/100 or 0.5 u wide. The resolution required to separate such peaks (0.5 u wide peaks at m/z 1001 and 991.1) is still low (<100). For a protein weighing 1,000,000 Da the peaks to be measured would fall at m/z 1001 and 1000, respectively, but they will be considerably less than 0.5 u wide, and the resolution of a quadrupole ($M/\Delta M$ of approximately 1000) will still allow clear separation and measurement of these ions.

This analysis suggests that as the protein molecular weight increases even to 1,000,000 Da, the resolution required to separate adjacent charge states begins at a very low level and increases, but not rapidly, and to well within the resolution capabilities of a standard quadrupole system, i.e., <1000. Of course, this analysis holds only for the *extremely unlikely event* that the protein would be completely pure and homogeneous and no adducts of any significance such as $M + Na^+$ are formed.

Analysis of Mixture Spectra of Proteins

In the normal course of research, proteins are seldom completely pure or homogeneous. An investigator would perhaps like to use the mass spectrometer to monitor protein purification and to analyze proteins which contain normal biological levels of heterogeneity. An example of such an analysis is provided in Fig. 4. The deconvolution of the prolactin ESI spectrum (Fig. 5) provides an interpretable molecular ion region, where a major component at 22,663 Da and a significant component at 22,763 Da are clearly distinguished. From the protein sequence, the mass of the prolactin is calculated as 22,659.97 Da, compared to that determined from the ESI/MS data of 22,662.5 ± 1.4 Da, and so the minor impurities do not seem to have greatly affected the mass measurement in this case. The difference in mass of the two significant components (approximately 100 Da) suggests possibly a prolactin species with an additional amino acid (perhaps valine or threonine which have residue masses of 99.13 and 101.1 Da, respectively), but the mass accuracy is not good enough to come to a definitive conclusion. Of course, the minor component may not relate

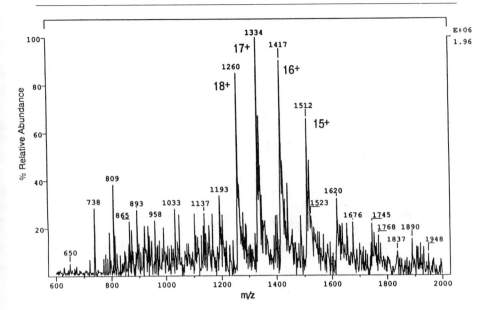

FIG. 4. ESI mass spectrum of prolactin. The signal-to-noise indicates that the material may not be pure.

to prolactin at all. It is also clear that some other impurities are present in this sample. The mass spectrometric analysis has, therefore, been useful; it confirms that the major anticipated prolactin species is present but that the sample contains at least one significant impurity and other minor ones.

Considering examples at substantially higher mass, the spectrum shown in Fig. 6 of a highly purified recombinant plasminogen-activating inhibitor (PAI) protein converts to the molecular weight profile shown in the inset of Fig. 6. The profile clearly tails extensively to higher mass. The experimentally determined mass of 42,777.5 ± 14.1 (0.033%) Da seems to be somewhat high relative to the calculated mass of 42,769.9 Da and may reflect the contribution of higher mass impurities or $M + Na^+$ species to the mass determination.

In comparison to the PAI protein, consider the glycoprotein ovalbumin which weighs approximately 45,000. The electrospray profile data of Fig. 7 consists of relatively broad peaks (compare to Fig. 6). This material is clearly not homogeneous and, indeed, the molecular mass profile is not only difficult to deconvolute to a smooth spectrum (inset in Fig. 7) but the fact that there is more than one component is now obvious. The deconvoluted profile clearly indicates at least two major components at

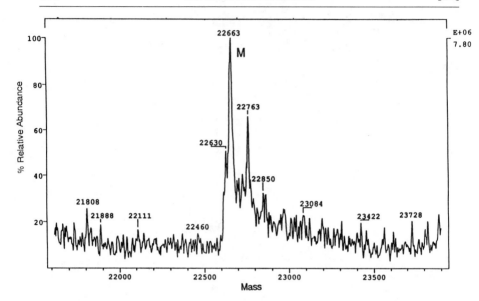

FIG. 5. Deconvoluted ESI-MS of prolactin. The major prolactin species is clearly visible at 22,663 Da (cf. calculated M_r of 22,660 Da). A major impurity is also evident at 22,763 Da, as are other minor impurities around the major species.

44,308 and 44,559 Da (which may be due to carbohydrate heterogeneity), as well as many others. For a reasonable opportunity of obtaining accurate molecular weight information on the major species, the glycoprotein will clearly have to be purified further, which may be very difficult for such closely related heterogeneous glycoproteins. Alternatively, it may in the future be possible to collect the ESI data with much higher resolution mass analysis to resolve the components of the mixture. Data from a higher resolution ESI mass spectrum will be analyzed by the same deconvolution and molecular weight analysis programs.

Mass Differences Resolved in Protein Molecular Weight Analysis

There are, of course, many types of heterogeneity which can occur in proteins. Examples of the differences in mass which must be considered for some of these are listed in Table II. The absolute mass difference, however, is not the only criterion. Other factors which must be considered are the relative amounts of components in the mixture as well as the mass at which the measurement is being made. To resolve a component at 10% concentration relative to the major material, a resolution requirement based on a 10% valley definition can be used. This resolution, however,

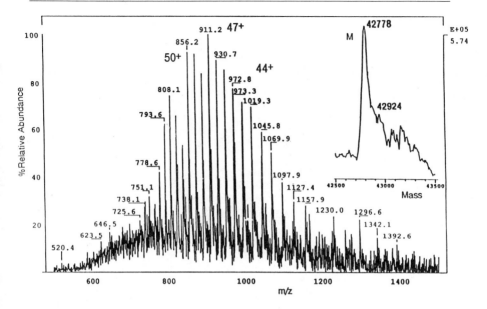

FIG. 6. ESI mass spectrum of plasminogen-activating inhibitor (PAI) protein; *inset*, the molecular weight region of the deconvoluted data.

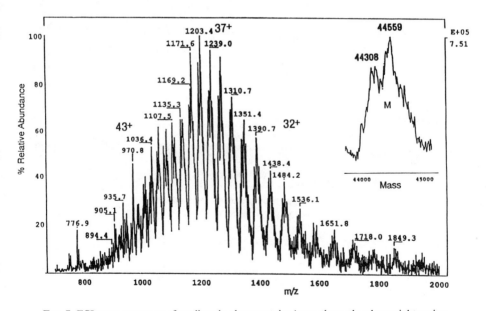

FIG. 7. ESI mass spectrum of ovalbumin glycoprotein; *inset*, the molecular weight region of the deconvoluted data.

TABLE II
MASS DIFFERENCES CONSIDERED IN MOLECULAR WEIGHT ANALYSIS OF PROTEINS

Reaction	Protein	Daltons
Amino acid substitution	Lys for Gln	0.04
	Asn for Leu	0.94
	Glu for Lys	0.95
	Asp for Asn	0.99
	Glu for Gln	0.99
	Thr for Val	1.97
	Val for Pro	2.01
	Cys for Thr	2.04
	Met for Glu	2.07
	His for Met	5.95
Amino acid additions	Gly	57.05
	Ala	71.08
	Ser	87.08
	Pro	97.12
	Val	99.13
	Thr	101.10
	Cys	103.14
	Leu, Ile	113.16
	Asn	114.10
	Asp	115.09
	Gln	128.13
	Lys	128.17
	Glu	129.12
	Met	131.19
	His	137.14
	Phe	147.18
	Arg	156.19
	Tyr	163.18
	Trp	186.21
Posttranslational and chemical modifications	SO_3 versus HPO_3	0.08
	—S—S— to —SH HS—	2.02
	CH_2 (e.g., methyl ester)	14.03
	O (e.g., methionine sulfoxide; hydroxyproline; oxidized tyrosine)	15.99
	H_2O (e.g., homoserine lactone/homoserine after CNBr cleavage; pyroglutamic acid)	18.02
	CO (e.g., N-terminal formylation)	28.01
	CH_2CO (e.g., acetylation)	42.04
	CO_2 (e.g., γ-carboxyglutamic acid)	44.01
	HPO_3	79.98
	SO_3	80.06
	Hexoses	160.14
	N-Acetylaminohexoses	201.19
	Sialic acid	291.26

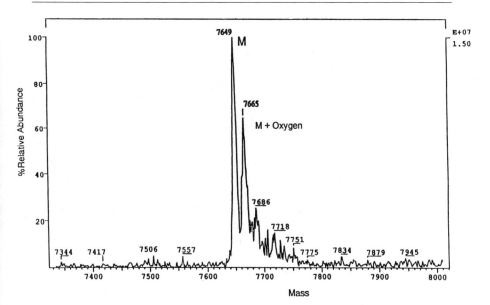

FIG. 8. Molecular weight region of the deconvoluted ESI mass spectrum of insulin-like growth factor (M_r 7649) which had been mixed with some of the methionine sulfoxide analog (M_r 7665) of the same protein.

will be insufficient for the same component at the 1% level. In addition, as mass measurement is made at higher mass, the width of the isotope envelope will place further constraints on the resolution required.

An oxidation (plus one oxygen or 16 Da) requires a resolution of <1000 at mass 10,000 Da, which is well within the capability of a quadrupole system. The molecular species and methionine sulfoxide species of insulin-like growth factor are clearly resolved after deconvolution (Fig. 8), but a resolution approaching 10,000 will be necessary at mass 100,000 Da, which would require a higher resolution analyzer, such as a magnetic system. Each situation has to be considered with these constraints in mind. There will be many situations where low- to medium-resolution quadrupole systems will be sufficient to satisfactorily analyze protein mixtures. Such solvable problems can be estimated from Table II. At mass 10,000, for example, any mass difference >10 Da, such as all the incremental amino acid and carbohydrate differences, addition of H_2O, CO, CH_3CO, CO_2, HPO_3, SO_3, and so on, should be measurable, but at best down to about the 10% level. As the mass difference to be resolved becomes smaller and as the mass at which the measurement is being made increases, the constraints on such analyses will increase.

Mass spectrometry obviously cannot distinguish the Leu/Ile pair by molecular weight measurement alone. Distinguishing Lys from Gln would require a resolution of at least 250,000 at mass 10,000; distinguishing SO_3 from HPO_3 requires at least one-half of that resolution at that mass; distinguishing among the 1 Da difference pairs Asn/Leu, Glu/Lys, Asp/Asn, and Glu/Gln requires a resolution of *at least* 10,000 at m/z 10,000; the 2 Da pairs require one-half that resolution, as does distinguishing —S—S— from —SH HS—. Clearly, and particularly at high mass, even very high-resolution mass analysis will be insufficient to solve many mixture problems. In such cases, the combination of mass spectrometry with other techniques will be essential. To estimate the amount of deamidation (particularly Asn to Asp) present in a purified protein, for example, it would be prudent to try to separate the components of the mixture by electrophoresis or, failing that, to enzymatically digest the protein and analyze the resulting peptides by capillary electrophoresis, which may separate the deamidated peptides on the basis of charge.

So far only a relatively small number of proteins have been analyzed by ESI/MS. In general, these have been exploratory experiments where the protein is relatively pure, quite soluble in convenient ESI solvent systems such as water/methanol/acetic acid, and where salts have been by and large avoided since they may well interfere with the analysis. Much additional work needs to be carried out, therefore, on proteins with difficult solubility characteristics to match perhaps with other ESI solvent systems. Furthermore, it is *highly recommended* that for the best opportunity for successful molecular weight analyses of proteins by ESI/MS, the interface between the biochemist and the mass spectroscopist should be a close knit one so that the characteristics of the particular protein to be studied (mass, heterogeneity, sample size, solubility, salt content, and so on) can be carefully evaluated and controlled (e.g., a desalting step may well be essential) before analysis.

Conclusions

In summary, the exciting new mass spectrometric techniques of electrospray ionization and matrix-assisted laser desorption ionization, which are now available for protein molecular weight analyses on relatively cost-effective mass spectrometers (quadrupoles and TOF), promise a revolution in the accuracy and precision to which such measurements can be made. Nevertheless, some proteins may prove to be difficult to analyze without a clear understanding of the solubility and salt content of the sample. The methods will allow rapid assessment to be made of the purity of analyzed proteins, at least regarding the presence of other protein materials. The resolution capability of the mass analyzer, however, will determine the ultimate level to which the analyses of mixtures can be carried out. Even the currently highest available mass spectrometer resolution (perhaps 10^5 or so) will be insufficient for some applications, and the mass spectrometer will be just one of the many techniques available to the protein biochemist for molecular weight and purity analysis.

[25] Sequencing of Peptides by Tandem Mass Spectrometry and High-Energy Collision-Induced Dissociation

By Klaus Biemann

Introduction

To ionize large polar molecules, such as peptides of a size encountered in the study of various aspects of protein structure, methods are used that impart little excess energy on the molecule. Furthermore, most of these "soft" ionization techniques[1] produce the charged particle by addition or removal of a proton to form the very stable, even-electron ions $(M + H)^+$ or $(M - H)^-$, which have little tendency to fragment. Of the various methods, bombardment with fast (kiloelectron volt, keV) atoms (Xe°) or ions (Cs^+) is most useful for the production of well-defined protonated or deprotonated peptide ion beams. The first demonstration of the generation of large protonated peptides involved fast atom bombardment (FAB),[2] but the data and methodology discussed in this chapter utilize either Xe° or Cs^+ as primary particles.

As has been pointed out in the overview,[3] this lack of fragmentation is very useful for the unambiguous determination of molecular weights of peptides, even in quite complex mixtures, but the resulting absence of structural information is a definite drawback. This can be overcome by transferring additional energy sufficient to force fragmentation of the stable molecular ion formed in the initial ionization process. Collision with a neutral gas is presently most widely used leading to collision-induced dissociation (CID), sometimes also referred to as collisionally activated

[1] A. G. Harrison and R. J. Cotter, this volume [1].
[2] M. Barber, R. S. Bordoli, R. D. Sedgwick, and A. N. Tyler, *J. Chem. Soc. Chem. Commun.* p. 325 (1981).
[3] K. Biemann, this volume [18].

dissociation (CAD).[4] The resulting fragment ions (product ions) are then mass analyzed in a second mass spectrometer, hence the term tandem mass spectrometry or MS/MS.[5,6] As discussed earlier,[4] the necessary energy can be acquired by the precursor ion, either by multiple collisions at low energy (eV) or single collisions at high energy (keV). The latter is more reliable but requires acceleration of the precursor ion by kilovolt potentials necessitating the use of a magnetic deflection mass spectrometer as the second of the two analyzers. For peptide sequencing, and structure determination in general, both mass spectrometers must have unit mass resolution and good ion transmission over the mass range of interest. Above molecular weight 1000, this requirement presently is best met by MS/MS systems consisting of two double-focusing mass spectrometers of high mass range.[5]

Such an arrangement permits the selection of the ^{12}C-only component of the isotopic multiplet of the precursor ion, which (after CID) produces a monoisotopic product ion spectrum that is devoid of the complexity of isotope clusters, particularly at high mass (large number of carbon atoms). Assuming that fragmentation of the protonated peptide molecule follows a rational pattern, the amino acid sequence and other structural features can be deduced from the m/z values of the product ion spectrum.

As will be discussed in more detail, the major effects directing fragmentation are the sites for proton attachment. For this reason, positive ion MS/MS is the method of choice, which has the additional advantage of higher sensitivity for peptides, because they seem to form $(M + H)^+$ ions more easily than $(M - H)^-$ ions. An exception are those that contain strongly acidic groups, such as phosphorylated or sulfated peptides. In these cases, negative ion CID spectra are often more informative.[7]

Fragmentation of Protonated Peptide Molecules and keV Collisions

Peptides derived from proteins are linear molecules (at least after reduction of any disulfide bridges originally present) that consist of repeating units having an identical backbone, —NH—CH(R)—CO—, but differing in the nature of the side chain, R (see Appendix 6 at end of volume). There are only 20 different amino acids that occur in proteins (and are defined by the genetic code), in addition to the occasional occurrence of others formed by posttranslational processes. An enormous number of

[4] R. N. Hayes and M. L. Gross, this volume [10].
[5] M. L. Gross, this volume [6].
[6] R. A. Yost and R. K. Boyd, this volume [7].
[7] B. W. Gibson and P. Cohen, this volume [26].

structurally different peptides are, therefore, possible and any notion of developing a ''library'' of all of them for use in any matching-type identification must be immediately discarded.

However, the structural features mentioned above are the unique reason why tandem mass spectrometry is so successful for peptide sequencing and so generally applicable: cleavage of the same bond, for example, the CO—NH bond, of consecutive peptide linkages and retention of the positive charge at the CO— group would generate a series of ions of increasing mass. The mass difference between consecutive pairs would reveal the identities of the consecutive amino acids (with the exception of the differentiation of Leu and Ile, which are isomeric, and Gln and Lys, which are isobaric).

Some CID mass spectra are almost as simple as implied above but they are the exception. An example is shown in Fig. 1, which is the CID mass spectrum of protonated Val^1-gramicidin A. The peaks labeled b_n are indeed due to the process anticipated above (see Appendix 5 at end of volume for structures of ions formed on high-energy CID of protonated peptides). It should be noted that this peptide has some unique features: the N terminus is acylated, the C terminus is amidated, and there is no basic or acidic amino acid present. The most basic centers in this molecule are, therefore, the amide groups, and it is there that protonation takes place. In any one $(M + H)^+$ ion colliding with a He atom, there is one protonated amide nitrogen (they are all of about the same basicity) which can then undergo fragmentation in either one of two ways:

$$\cdots\text{NH}-\overset{\overset{\text{R}}{|}}{\text{CH}}-\text{CO}^+ \quad \overset{\alpha}{\longleftarrow} \quad \cdots\text{N}-\overset{\overset{\text{R}}{|}}{\text{CH}}-\text{CO}-\overset{+}{\text{NH}_2}-\overset{\overset{\text{R}}{|}}{\text{CH}}-\text{CO}\cdots \quad \overset{\beta}{\longrightarrow}$$

$$b_n$$

$$\overset{+}{\text{H}_3}\text{N}-\overset{\overset{\text{R}}{|}}{\text{CH}}-\text{CO}\cdots$$

$$y_n$$

The spectrum clearly shows an uninterrupted series of peaks, starting with b_2 (cleavage, denoted α above, of the Gly-2—Ala-3 bond) to b_{15} (cleavage at the Trp-15 carbonyl group). There is, however, another set of peaks, labeled y_4–y_{14}. These are due to the other fragmentation pathway (β) depicted above, which involves the transfer of a hydrogen atom from the N-terminal side leading to charge (proton) retention at the C-terminal portion.[8] The y_{12}–y_{14} ions provide the necessary information to establish

[8] D. R. Mueller, M. Eckersley, and W. J. Richter, *Biomed. Environ. Mass Spectrom.* **23**, 217 (1988).

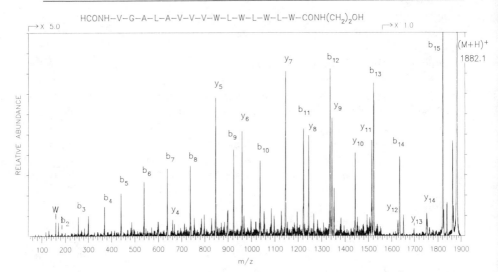

FIG. 1. The CID mass spectrum of the $(M + H)^+$ ion (m/z 1882.1) of Val^l-gramicidin A taken at 7 keV collision energy. The mass differences of the consecutive b_n ions and their correspondence to the amino acid sequence are shown. (From Ref. 11.)

the N-terminal sequence Ac-Val-Gly-Ala, in spite of the absence of a b_1 ion and the low abundance of b_2.

Unfortunately, most of the peptides encountered in proteolytic digests of proteins are not that simple. Although the presence of —COOH, —OH, —SH, or —S— groups does not alter the pattern of protonation of the amide groups, the much more basic amino or guanidino groups of lysine and arginine protonate predominately and thus have a profound effect on the CID spectra.[9]

A set of very closely related peptides, generated by tryptic cleavage of three different thioredoxins, is shown in Fig. 2. The first pair (Fig. 2a,b) both have a C-terminal lysine, which effectively competes with the amide groups for the proton in the $(M + H)^+$ ion. Charge retention on the C-terminal fragments is thus favored and the y_n ions are more pronounced than in Fig. 1 and complete (only y_6 is of very low abundance). In addition, there is another set of ions present, w_{3-5}, w_7, and w_{8a}. These are also C-terminal ions and are due to loss of part of the side chain (see the end of this volume, Appendix 5).

The mechanism of formation of these ions involves cleavage of the

[9] K. Biemann, *Biomed. Environ. Mass Spectrom.* **16**, 99 (1988).

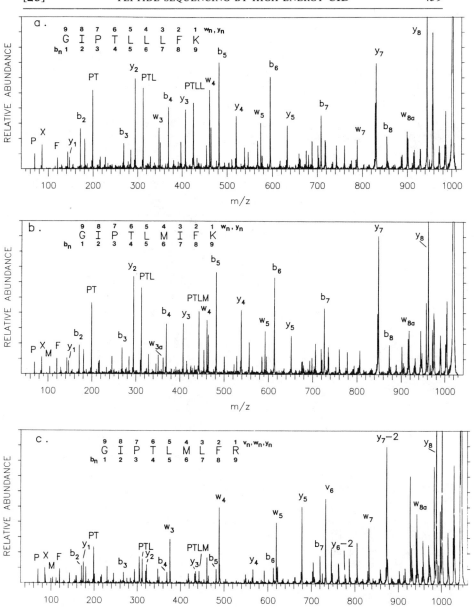

FIG. 2. Comparison of the CID mass spectra of analogous tryptic peptides from three different thioredoxins: (a) *E. coli*, (b) *Rhodospirillum rubrum*, and (c) *Chromatium vinosum* taken at 10 keV collison energy (cell potential at ground). (From Ref. 9.)

β,γ-bond of the side chain (R_n) of the corresponding $z_n + 1$ ion.[10] This process requires the presence of such a β,γ-bond that can be easily cleaved and, therefore, does not occur with glycine, the aromatic amino acids, and alanine (loss of H is apparently not favored). For these reasons, w_n ions rarely represent a contiguous series, in contrast to the backbone cleavages that are relatively independent of the nature of the side chain. An exception to the latter statement is the fragmentation of the peptide bonds at glycine, which generally exhibits sequence ions of much lower abundance than its neighbors (see b_2 and y_{13} in Fig. 1).

While this additional fragmentation process complicates the spectrum and decreases the overall signal intensity (by competing for the fragmenting MH$^+$ ions), it is of great utility by providing a means for the differentiation of the isomeric amino acids leucine and isoleucine: the w_n ion of Leu is always 14 u lower than that of Ile (the notations w_{na} and w_{nb} refer to the retention of the smaller and larger substituent, respectively, of an amino acid with two different substituents at the β-carbon, such as Ile and Thr). Since position 3 (from the C terminus) is Leu in Fig. 2a but Ile in 2b, the w_3 ion is at m/z 348 in the former but at m/z 362 in the latter, while y_3 is at the same mass (m/z 407) in both. The replacement of Leu-6 in Fig. 2a by Met-6 in Fig. 2b causes the expected shift by $+18$ u for all ions that retain this amino acid.

The spectrum is further complicated by the appearance of so-called internal sequence ions (see end of this volume, Appendix 5), PT, PTL, and PTLL (PTLM in Fig. 2b), which require fragmentation at two peptide bonds corresponding to a y_n cleavage at proline and a b_n cleavage further toward the C terminus. These peaks confirm that part of the sequence but may be more of a nuisance than an advantage. However, for the determination of a previously unknown sequence, it is important to be able to assign as many significant peaks in a spectrum as possible, to be confident that the correct answer has been found.

Finally, there are the peaks at low mass, labeled P, X (for L or I), and F in Fig. 2a and P, X, M, and F in Fig. 2b. These are due to the immonium ions (see Appendix 5 at the end of this volume) of the amino acids represented by their single letter code. They point to the type of some (but not all) of the amino acids present in the peptide and thus also help in the interpretation. The peak at m/z 104 in Fig. 2b indicates the presence of Met, which is clearly absent in the peptide represented by Fig. 2a.

Noteworthy is the great similarity of these two spectra with respect to fragmentation and relative intensities, taking into account the mass shifts

[10] R. S. Johnson, S. A. Martin, K. Biemann, J. T. Stults, and J. T. Watson, *Anal. Chem.* **59,** 2621 (1987).

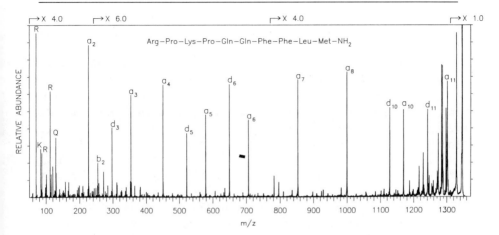

FIG. 3. CID mass spectrum of the decapeptide substance P. Collision energy 7 keV. From this spectrum the numerical data in the last column of Table III, chapter [10] in this volume, were derived.

discussed above. It demonstrates that the fragmentation behavior is not changed by replacing Leu with Ile or Met. However, a drastic change results when the C-terminal Lys is replaced by the much more basic Arg. In Fig. 2c, the vast majority of the fragments are due to C-terminal ions (v_n, w_n, y_n), because in this peptide the proton is most strongly attached to the C-terminal Arg. The y_7-2 ion (an imine ion, see Appendix 5, end of this volume) is always particularly dominant for proline; it can also be seen in Fig. 2a and b. The ions denoted v_n (v_6 in Fig. 2c) are also C-terminal ions that have further fragmented, in this case by loss of the entire side chain (cleavage of the α,β-bond) and an additional hydrogen from a y_n ion.[11]

The effect of the presence and position of a basic amino acid is also apparent in the CID spectrum of substance P, which carries arginine at the N terminus and lysine in position 3 (Fig. 3). As can be surmised from the foregoing discussion, in this case the spectrum consists almost exclusively of peaks resulting from fragments that retain the proton at the N terminus, namely, ions of type a_n and d_n (see end of volume, Appendix 5). The simplicity of the spectrum is most likely due to the N-terminal arginine, and not to the less basic lysine. With the charge retained at the protonated guanidino group, the a_n ions are produced by charge-remote

[11] R. S. Johnson, S. A. Martin, and K. Biemann, *Int. J. Mass Spectrom. Ion Proc.* **86,** 137 (1988).

fragmentation[4,12] via the corresponding $a_n + 1$ ion, which also leads to the d_n ion by cleavage of the β,γ-bond of the side chain,[11] analogous to the formation of the w_n ions from $z_n + 1$ ions.

Not only is the formation of d_n ions similar to that of w_n ions, but so is their information content. They permit the differentiation of the isomers leucine and isoleucine and do not form a continuous series because, again, there must be present an easily cleavable β,γ-bond in the amino acid that represents the C terminus of the fragment. At this point, it should be noted that one sometimes encounters d_n or w_n ions that have undergone cleavage of the β,γ-bond of the amino acid one (or even two) position away from the backbone cleavage at a point where there is an amino acid that does not have an easily cleaved β,γ-bond. An example is shown in this volume [28] in Fig. 3,[13] which depicts the CID spectrum of LYTGEACRTGD. It exhibits an ion labeled w_{8E}, denoting a C-terminal fragment formed by cleavage of the NH—CH$_2$ bond of Gly at position 8 (from the C terminus) and cleavage of the β,γ-bond of the Glu side chain (position 7). Such "secondary" w_n and d_n ions are observed when the primary cleavage is at an amino acid not capable of side-chain fragmentation (such as Gly) but an amino acid is nearby that favors this cleavage (like Glu).[10,11] This is not confusing as long as one is aware of the process. The d_n and w_n ions are most useful for the differentiation of Leu and Ile, for the confirmation of the interpretations based on backbone fragmentation (a_n, b_n, c_n, x_n, y_n, and z_n ions), and provide an indication of the location (N-terminal or C-terminal) of basic amino acids. The higher energy requirements of charge-remote fragmentation processes are the reasons for the absence of d_n, v_n, and w_n ions in low-energy ($<$200 eV) collision spectra of peptides[4,6] which, therefore, do not permit the differentiation of leucine from isoleucine.

The peaks at low mass are again indicative of the presence of certain amino acids, while those near the precursor ion are due to loss of the various side chains and also help in determining the nature of the amino acids in the peptide.

In this context, the reader should be reminded of the fact that the amino acid composition is usually not known for the peptide being analyzed, particularly if it is produced by chemical or enzymatic cleavage of a protein and thus occurs in a mixture to begin with. It is one of the major advantages of the mass spectrometric protein-sequencing strategy that individual peptides do not need to be isolated and purified, a step that would be necessary to obtain an interpretable amino acid analysis. The only compositional data available are that of the intact protein, which is useful because they

[12] K. B. Tomer, F. W. Crow, and M. L. Gross, *J. Am. Chem. Soc.* **105**, 5487 (1983).
[13] H. A. Scoble and S. A. Martin, this volume [28].

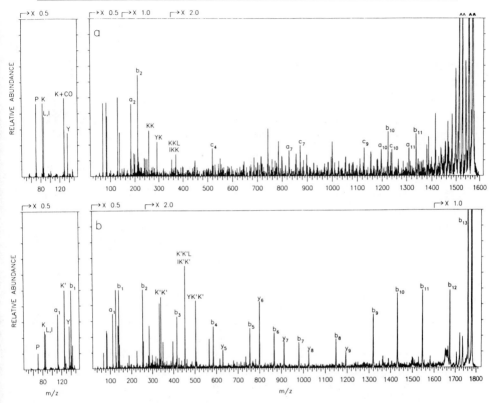

FIG. 4. CID mass spectra of a synthetic peptide, Pro-Leu-Tyr-Lys-Lys-Ile-Ile-Lys-Lys-Leu-Leu-Gln-Ser, (a) before and (b) after acetylation of all five basic amino groups.

obviously eliminate the need for consideration in the interpretation of the CID spectrum of any peptide of an amino acid known to be absent in the protein.

Not all CID spectra of protonated peptides are readily interpretable. Aside from the general quality of the spectrum (good signal-to-noise ratio), the most important prerequisite is the presence of one or more complete, or at least overlapping, ion series due to the backbone fragmentation. As was pointed out in the foregoing, the presence and position of basic amino acids have a significant effect on fragmentation. If at the C terminus, as is the case in tryptic peptides (except the original C terminus of the protein), there is no problem (see Fig. 2); also peptides with a basic amino acid at the N terminus (Fig. 3) or none at all (Fig. 1) give very simple spectra. However, a problem may arise with internal basic amino acids because this could lead to the formation of only C-terminal ions, due to backbone

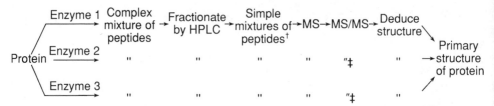

SCHEME 1. Strategy for sequencing of proteins. †, If necessary, derivatize certain fractions; ‡, may not be necessary for all components.

cleavage between the N terminus and the basic amino acid and vice versa. The spectrum would probably still be interpretable, and the structure could be confirmed by further digestion of the peptide (or the fraction of the mixture in which it occurs) with trypsin. This generates two smaller peptides, one with a C-terminal basic amino acid and the other without any.

A more complex, but also somewhat contrived, example is shown in Fig. 4a, the CID spectrum of a protonated peptide containing a total of four lysines in two rather symmetrically distributed pairs. This is a synthetic peptide of $(M + H)^+ = m/z$ 1573.0. As Fig. 4 implies, it is not interpretable because of the absence of significant specific fragmentation. Surmising that this behavior is due to the many lysines, the peptide was acetylated to eliminate the effect of the basic ε-amino groups of this amino acid. Although the derivatization causes an increase in mass by 210 u (five hydrogen atoms replaced by five acetyl groups at the N terminus and the four lysines), the resulting CID spectrum is very simple and typical for that expected for a peptide lacking any basic functionality. Similar to Fig. 1, it displays a complete b_n series and a partial y_n series. At the same time, the change in mass reveals there are four lysines present and eliminates the need to differentiate this amino acid from glutamine, which is no longer isobaric with acetyllysine.

Strategy for Protein Sequencing by Tandem Mass Spectrometry

The speed with which CID spectra of protonated peptides can be recorded and the reliability with which they can now be interpreted make tandem mass spectrometry an attractive method for protein sequencing. The first step of the strategy, outlined in Scheme 1, is the digestion, after reduction and alkylation of disulfide bonds, with an appropriate enzyme.[14] Because of the well understood effect of a C-terminal basic amino acid and the potential difficulties created by internal basic amino acids, trypsin

[14] T. D. Lee and J. E. Shively, this volume [19].

is the enzyme to use first. An additional advantage is its high specificity for lysine and arginine, which are moderately abundant in most proteins. The majority of tryptic peptides are therefore usually 5–15 amino acids long, a size that is amenable to efficient and complete fragmentation by high-energy CID (presently limited to peptides up to 20–25 amino acids in length).

The digest is then partially separated by reversed-phase high-performance liquid chromatography (HPLC), for a number of reasons: (1) to remove buffer salts and remaining enzyme; (2) to simplify the mixture; and (3) to avoid the simultaneous presence of hydrophobic and hydrophilic amino acids in a fraction to be analyzed by FAB-MS/MS. The first is to eliminate salts that interfere with FAB ionization. The second is to avoid wasting too much material that is continuously ionized while one $(M + H)^+$ ion after the other is selected by the first mass spectrometer $(MS-1)$ and analyzed in the second $(MS-2)$; three peptides per HPLC fraction is a convenient maximum and this is easily achieved by a standard solvent program without the need for time-consuming optimization of the HPLC conditions. The third is to eliminate the problem of the suppression of ionization of hydrophilic by hydrophobic peptides,[15] which may lead to low precursor ion currents for the former or result in missing some of them altogether.

Many of the resulting CID spectra will be completely interpretable in terms of sequence. Others will leave some ambiguities: (1) Sometimes the peptide bonds between the two N-terminal and/or two C-terminal amino acids do not fragment and their order can thus not be determined, or there may even be more than one pair that corresponds to the mass of these two amino acids (the sum of their mass is, of course, evident from the spectrum). (2) If the appropriate d_n or w_n ions are absent, leucine and isoleucine cannot be distinguished. (3) Even in tryptic peptides, the specificity of the enzyme[14] cannot always be relied upon to differentiate lysine from glutamine.

There is also the possibility that a small peptide is formed and escapes detection because it elutes in the void volume of the HPLC column, or its $(M + H)^+$ ion signal is swamped by the abundant peaks at low mass due to the matrix. An example of such a case is the dipeptide Leu—Lys in the tryptic digest of erythropoietin discussed elsewhere in this volume.[13] Since trypsin has little or no exopeptidase activity, it is unlikely that a single lysine or arginine is ever missed.

For all these reasons, one or more additional digests are needed. An

[15] S. A. Naylor, A. F. Findeis, B. W. Gibson, and D. H. Williams, *J. Am. Chem. Soc.* **108,** 6359 (1986).

enzyme of high specificity for one or a few amino acids other than lysine and arginine is chosen to generate a new mixture of peptides, which is entirely different from the tryptic digest. It is highly probable that information not available from the CID spectra of the tryptic peptides will be contained in the new set. If endoproteinase Glu-C is used at low pH,[14] the C-terminal amino acid will be glutamic acid (except for the original C terminus), and none of the deficiencies of the CID spectra of the tryptic peptides enumerated above will cause exactly the same ambiguities. For example, what were the two N-terminal and two C-terminal amino acids for a tryptic peptide cannot possibly be the same in a peptide generated from the same protein by Endo Glu-C. Similarly, a small tryptic peptide that may have escaped detection will be part of a larger Endo Glu-C peptide and vice versa. Finally, at least the molecular weights of the peptides in a second digest are needed to properly align all the tryptic peptides in a unique manner.

There are other enzymes that are very useful in special cases. Lys-C does not cleave at arginine, but does cleave Lys—Pro bonds, and AspN cleaves at the N terminus of aspartic acid. Often the choice of the enzyme is dictated by the amino acid composition of the protein. For example, for an arginine-poor, lysine-rich protein, Lys-C is preferable to trypsin because the former is more efficient and more specific.

An example of this strategy is the determination of the primary structure of the glutaredoxin isolated from rabbit bone marrow.[16] A total of three digests (trypsin, α-chymotrypsin, and Endo Glu-C) and one subdigest with thermolysin of selected fractions of the tryptic digest were required to deduce the entire structure, solely by tandem mass spectrometry. The results are summarized in Fig. 5, where the heavy underlining represents the portion of each proteolytic peptide for which the amino acid sequence could be unambiguously deduced from the corresponding CID spectrum. Thin underlining indicates portions for which no specific cleavage took place (or large peptides that were not subjected to MS/MS). It will be noted that these are chiefly the two N-terminal amino acids but the missing information is always provided by one or more peptides from another digest. Figure 5 reveals the high degree of redundancy of the sequence information, which leads to a correspondingly high degree of confidence in the correctness of the primary structure derived from this 106 amino acid long protein. Admittedly, there is no overlap for Glu-55 and Ile-56 but this is a region of high homology with the two known mammalian glutaredoxins of presently known sequence.[16] If needed, one would further search the digests for an indication of an appropriate overlapping peptide

[16] S. Hopper, R. S. Johnson, J. E. Vath, and K. Biemann, *J. Biol. Chem.* **264**, 20438 (1989).

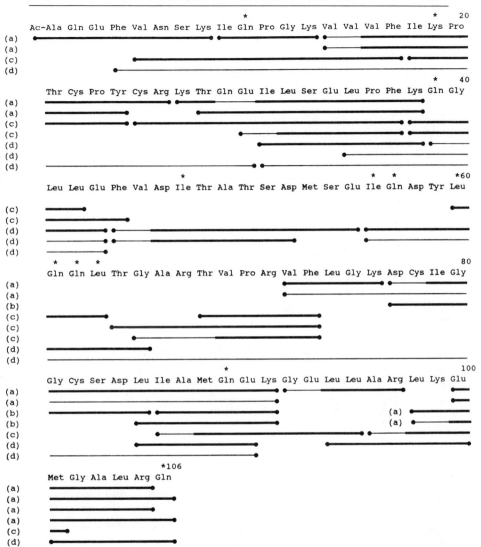

FIG. 5. The amino acid sequence of rabbit bone marrow glutaredoxin determined by tandem mass spectrometry. (a) Tryptic peptides; (b) peptides resulting from digestion of tryptic HPLC fractions with thermolysin; (c) α-chymotryptic peptides; (d) endoproteinase Glu-C peptides. Heavy underlining indicates sequence derived from CID mass spectra; asterisks indicate ambiguities solved by derivatization. (From Ref. 16.)

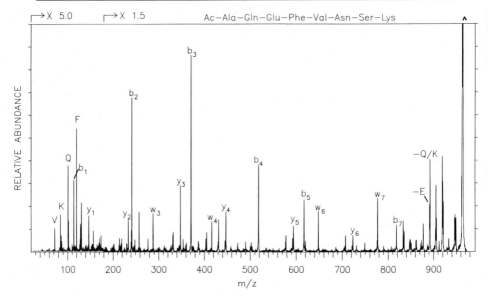

Fig. 6. CID mass spectrum of the N-terminal peptide of the glutaredoxin isolated from rabbit bone marrow. Collision energy 10 keV. (From Ref. 16.)

or carry out another digest, in this case, with endoproteinase Asp-N to generate the 51–57 peptide that should be easy to find.

A few of the pertinent CID spectra shall be discussed because they involved a particular point or question that arose in the course of this work. The one shown in Fig. 6, derived from the $(M + H)^+$ ion of m/z 964.6 in fraction 6 of the HPLC of the tryptic digest was, in principle, compatible with a sequence Leu(or Ile)-Gln-Glu-Phe-Val-Asn-Ser-Lys. It consisted of only b_n, y_n, and w_n ions, consistent with the presence of a basic C-terminal amino acid. However, there were no a_n ions at all, implying that there is no basic N-terminal group (as in Figs. 1 and 4), and there was an abundant b_1 ion which we had observed only in peptides with an N-terminal acyl group (Fig. 4 and Ref. 17). Since N-acetylalanine is of the same nominal mass as the leucines, it is more likely that the correct structure is as shown in Fig. 6. This peptide must represent the N-terminal sequence of the protein, which was known to be resistant to the Edman degradation.

Differentiation of Lysine/Glutamine and Leucine/Isoleucine

Lysine and glutamine differ in mass by only 0.0364 u, which is too little to be distinguishable in a CID spectrum unless it were in a fragment at

[17] K. Biemann and H. A. Scoble, *Science* **237**, 992 (1987).

relatively low mass. The simplest but indirect method of differentiation makes use of the specificity of trypsin. As was shown in Fig. 5, tryptic peptides due to cleavage at lysine take place very clearly, which can be taken as a reliable identification of this amino acid at these positions. However, this does not mean that all other residues of $\Delta m = 128$ must be glutamines, because there may be a structural factor that prevents tryptic cleavage at lysine. For example, enzyme specificity cannot be used to argue that the amino acids at positions 2 and 106 must be Gln rather than Lys, because trypsin is not an exopeptidase and may not cleave at or very near the N or C terminus of the protein. The fact that trypsin does not cleave the 19–20 bond does not require that the amino acid at position 19 must be Gln, because it is known that trypsin does not cleave a Lys—Pro bond.

It is therefore prudent to confirm the Lys/Gln assignments at positions that are not readily cleaved by trypsin. This is accomplished by acetylation of the peptide in question (or the HPLC fraction in which it occurs). As discussed previously in connection with Fig. 4a and b, each lysine present in the peptide adds 42 mass units upon acetylation (in addition to the acetylation of the N-terminal amino group), while glutamine is unaffected. For this test, it suffices to simply determine the shift in molecular weight. A CID spectrum is required only if there is more than one residue of $\Delta m = 128$ present and the mass increase indicates that some of them are lysines, while others are glutamines. The Lys and Gln residues marked by an asterisk in Fig. 5 were finally identified, as shown in the sequence, by such acetylation experiments on the HPLC fractions that contained peptides representing these regions.[16] For this reason, it is advisable not to use up all the material for the first MS/MS experiment.

Another problem is the differentiation of leucine and isoleucine. The specificity of chymotrypsin, which cleaves only at the former, can be used for this purpose and, as has been stated earlier, these two amino acids can also be recognized from certain fragment ions, the N-terminal d_n-type and the C-terminal w_n-type. Using this feature, 15 of the 19 leucines or isoleucines in this glutaredoxin could be assigned on the basis of at least one CID spectrum. The remaining four (positions 47, 56, 60, and 63), again indicated by asterisks in Fig. 5, occurred in Endo Glu-C peptides not containing any basic amino acids and, therefore, were devoid of d_n or w_n ions.

From the earlier discussion of the effect of charge localization on the basic terminal amino acids and the particular simplicity of CID spectra of $(M + H)^+$ ions of peptides with arginine at the N terminus, which also triggers the formation of d_n ions, it is clear that changing the charge distribution toward this type would favor the fragmentation processes permitting the differentiation of leucine and isoleucine. Rather than adding an arginine to the N terminus, we went one step further and incorporated a quaternary ammonium ion group by reaction of the peptide with $ClCH_2COCl$, followed by treatment with triethylamine. This reaction sequence incorporates a $(CH_3)_3N^+CH_2$—CO— group at the N terminus. FAB ionization ejects this cation directly into the ion source of MS-1, and CID causes it to fragment for mass analysis in MS-2. The accompanying dramatic change, with the desired effect, is shown in Fig. 7a and b. The former reveals the general sequence (as it turned out, positions 56–65) from the abundant b_n and y_n ions, but the absence of d_n or w_n ions does not allow the differentiation of the isomeric leucines. Fixing the positive charge at the N terminus dramatically changes the fragmentation (Fig. 7b) and the spectrum now consists almost exclusively of a_n and d_n ions, reminiscent of Fig. 2. The m/z values of d_1, d_5, and d_8 indicate that the amino acid at position 56 is Ile, while 60 and 63 must be Leu. In a similar experiment, position 47 was found to be occupied by Ile. To be of practical utility for protein sequencing, it is important to be able to carry out this reaction at the subnanomole level. This is accomplished by treating a thin film of the peptide with reagent vapor (see Procedures section below).

It should be noted that hydroxyproline is of the same nominal mass as the leucines, all of which give rise to mass differences of 113 u (see Appendix 6 at end of this volume). They differ by 0.0364 u, but as has already been pointed out in the discussion of the lysine/glutamine pair, exact mass measurement[18] of fragments produced by CID is difficult. The 3- and 4-hydroxyprolines are produced by posttranslational hydroxylation of proline and their occurrence is therefore limited to certain proteins, such as the collagens and some other structural proteins. The amino acid composition of the intact protein will reveal its presence and only then does one need to consider the differentiation of the hydroxyprolines from the leucines. One major difference is the fact that the former cannot produce d_n ions because, as with proline, the 5-membered ring connects the β-carbon with the nitrogen atom.[11] The prolines do, however, form w_n ions, which also permit the differentiation of the 3- and 4-hydroxy isomers.[19] The elimination of H_2O from fragment ions, indicating the presence of an amino acid of residue mass 113, further identifies it as a hydroxyproline.

The rules governing the characteristic relationship of the degree of charge localization and the various fragmentation processes are summarized in Scheme 2.[11] These generalizations always have to be kept in mind

[18] K. Biemann, this volume [13].
[19] D. B. Kassel and K. Biemann, *Anal. Chem.* **62**, 1691 (1990).

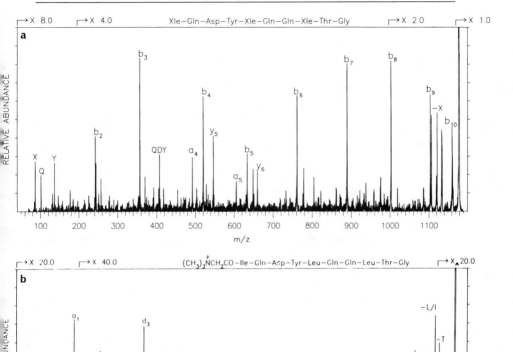

FIG. 7. (a) CID mass spectrum of the 56–65 peptide generated from rabbit bone marrow glutaredoxin by proteolysis with endoproteinase Glu-C. Xle, X = Leu or Ile. Collision energy 10 keV. (b) CID mass spectrum of the same peptide after conversion to a derivative carrying a quaternary ammonium moiety at the N-terminal amino acid which allows the differentiation of leucine and isoleucine.

when interpreting a CID mass spectrum of a peptide in terms of an amino acid sequence. The latter must always be in agreement with these conditions, which are based on the experience accumulated in our laboratory during the interpretation and correlations of many hundreds of CID spectra of peptides derived from various sources.

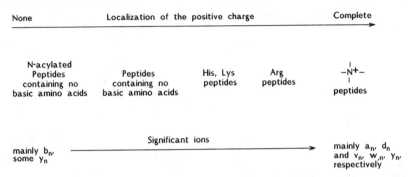

SCHEME 2. Summary of rules relating degree of charge localization and fragmentation process. (From Ref. 11.)

Computer-Aided Interpretation of High-Energy CID Spectra of Peptides

The earlier discussion of the information content of CID spectra of peptides shows that the mode of interpretation is based on the mass difference between certain fragment ions formed by analogous processes. From the raw spectrum, however, it is by no means obvious which peaks are related in this manner. In the course of manual interpretation, one often quickly recognizes that a few abundant ions show consecutive mass differences by one of the 18 "residue masses" (NH—CHR—CO), characteristic of the 20 common protein amino acids (see this volume, Appendix 6). This partial sequence is then extended by searching for further, perhaps not so obvious, peaks that also differ by one of these mass values. Finally, one has to recognize whether the particular ion series is an N- or a C-terminal one. This becomes clear when one reaches the first or last amino acid. There are also other indications: for example, a consistent difference of 28 u points to a_n and b_n series. An alternative is to search the region below mass 300 for peaks of an m/z value that would correspond to $a_{1,2}$, $b_{1,2}$, or $y_{1,2}$ ions, of any one or a combination of any two of the 18 residue masses and then search for the third amino acid, etc.

All these approaches involve the evaluation of the mass differences of many peaks in the spectrum. For a simple one, such as shown in Figs. 1 or 3, this is quite straightforward, but the more complex spectra (i.e., Fig. 2) cause more difficulties and their interpretation is therefore more time consuming.

These numerical correlations are, of course, ideally suited for a computer. Two quite different approaches have been taken. One is an interactive display program, PLTSEQ,[20] that connects all peaks in the spectrum

[20] H. A. Scoble, J. E. Biller, and K. Biemann, *Fresenius Z. Anal. Chem.* **327,** 239 (1987).

that differ by one of the 18 (or more, if modified amino acids are present) residue masses and labels the interconnecting line accordingly (i.e., D for a mass difference of 115, indicating aspartic acid). Such a display is helpful, as it speeds up the manual interpretation and may reveal some connections that are missed by visual inspection of the spectrum.

The second type of interpretive algorithm involves a systematic and exhaustive check for all the ions that can possibly be formed, based on our now rather complete understanding of the fragmentation of protonated peptide molecules on high-energy collision. The algorithm, SEQPEP,[21] interprets the sequence starting at the C-terminal amino acid, which is identified by searching for y_1, a_{m-1}, and b_{m-1} ions [where m is the number (at this point unknown) of amino acids in the peptide] that would be predicted for each of the 20 amino acids (18 mass values). All matches are retained, then extended by each of the 18 residue masses, and the other predicted ions (c_n, d_n, v_n, w_n, x_n, y_n, and z_n) searched for. A scoring procedure keeps track of the peaks found and the process is continued until the mass of the precursor ion, $(M + H)^+$, is reached. In each step there may be a large number of possible partial sequences, of which the 200 highest scores are continuously retained.

There are always, of course, many sequences that would potentially fit the data. These are then ranked with respect to the fraction of the spectral information (masses and abundances of the ions) accounted for by those peaks that are found to match the ones predicted for the sequences, because the one that accounts for all of the peaks in the spectrum is more likely to be correct than another sequence that leaves some of them unassigned.

However, there may still be a number of closely related sequences that have a similar high score and one should not expect that the absolute highest one corresponds to the correct sequence. For example, as was stated earlier, cleavage of peptide bonds at glycine is not favored and leads to fragment ions of very low abundance, thus not affecting the final score very much. The sequence Gly-Gly is of the same residue mass as Asn, and Gly-Ala or Ala-Gly similarly correspond to Gln. The algorithm will therefore often find both with about equal probability, which requires closely scrutinizing the spectrum to select the correct sequence or to devise another test to distinguish them.

Sometimes it is obvious from the characteristics of the spectrum which one of the closely high scoring sequences can be ruled out. Since the computer prints the assignments of all peaks in the spectrum, a sequence that includes an arginine at the N terminus but apparently produces mainly

[21] R. S. Johnson and K. Biemann, *Biomed. Environ. Mass Spectrom.* **18,** 945 (1989).

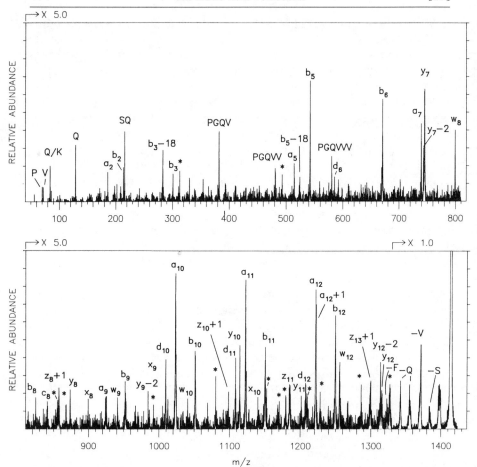

FIG. 8. CID mass spectrum of the chymotryptic peptide 5–17, Val-Asn-Ser-Lys-Ile-Gln-Pro-Gly-Lys-Val-Val-Val-Phe, interpreted by the computer program SEQPEP. Peaks exceeding the intensity threshold for the input data for SEQPEP but not identified by the algorithm are labeled with an asterisk. (From Ref. 21.)

v_n, w_n, and y_n ions cannot possibly be the correct one. As always, experience and common sense have to be used to evaluate solutions presented by a computer. For specific examples, results and evaluation of the output of SEQPEP, the reader is referred to the original paper.[21] It should be noted, however, that the program is sufficiently fast (0.5–4 min per spectrum on a VAX 750, much less on higher level VAXs) to be compatible with the speed by which the mass spectrometric data are acquired.

The result of one such computer interpretation of the spectra recorded in the course of the determination of the primary structure of the rabbit bone marrow glutaredoxin is presented here. The peak assignments for the highest scoring sequence (VNSQIQPGQVVVF) in the SEQPEP output are transferred to the spectrum and is shown in Fig. 8. The input consisted of the 71 peaks that were at least 5% as abundant as the most abundant fragment ion and of these, all but those labeled with an asterisk were in agreement with the computer-assigned sequence. It should be noted that this spectrum is of somewhat low quality (high signal-to-noise ratio) and could not be interpreted by even an experienced experimentor. It took only 4 min on a VAX 750 computer to arrive at what turned out to be the correct sequence, which was finally assigned position 5–17 of the glutaredoxin (see Fig. 5).[21]

Finally, it should be recognized that one should not consider the CID spectrum of each peptide just by itself, but remember that all of them are derived from one original protein sequence. As a matter of practicality, one should not try to squeeze the maximum of information out of a single spectrum (which may be quite time consuming, if not frustrating) but consider all the data together. As alluded to earlier, an ambiguity in one spectrum may be clearly resolved in the spectrum of another peptide (usually from another digest). These correlations are either made manually or by a computer program that makes candidate sequences of peptides from within one digest or compares data from different digests, a process that is also required to properly align all the peptides to a final structure as depicted in Fig. 5.

Procedures

Enzyme Digests Suitable for Mass Spectrometry

Enzymatic digestion, reduction of disulfide bonds, alkylation of cysteine, cleanup, and partial fractionation of the digest are described by Lee and Shively.[14] For the work used as examples in this chapter, 2–4 nmol of protein was used for each digest because this work was carried out before we had constructed our array detector.[22–24] With the 50- to 100-fold increase in sensitivity, very good CID spectra can be obtained when starting with much smaller amounts of protein. The procedures and precautions outlined elsewhere in this volume[14] then become important because

[22] J. A. Hill, J. E. Biller, S. A. Martin, K. Biemann, K. Yoshidome, and K. Sato, *Int. J. Mass Spectrom. Ion Proc.* **92,** 211 (1989).

[23] J. A. Hill, J. E. Biller, S. A. Martin, K. Biemann, and M. Ishihara, *Proc. 37th ASMS Conf. Mass Spectrom. Allied Topics, Miami Beach, FL* p. 1077 (1989).

[24] S. Evans, this volume [3].

they minimize loss of part or all of the digestion products and avoid the introduction of contaminants at a level interfering with the collection and interpretation of the mass spectrometric data. For the alkylation of cysteine, we now prefer addition of 4-vinylpyridine over reaction with iodoacetic acid or amide. The chromophore of the resulting ethylpyridine derivative aids in the detection of cysteine-containing fragments during fractionation by HPLC, and this derivative is also more hydrophobic which aids ionization by FAB. The ethylpyridine group produces an abundant ion at m/z 106, which is therefore indicative of the presence of cysteine. Finally, the alkylation can be carried out with volatile reagents, triethylphosphine (as the reducing agent) and 4-vinylpyridine, and thus requires only minimal clean-up.

As mentioned elsewhere in this volume,[14] many proteolytic enzymes undergo autodigestion. This is more pronounced when a protein that is more resistant to digestion is treated with an enzyme : substrate ratio higher than the usual 1 : 100 and for prolonged periods of time. These are quite easily recognized simply on the basis of their molecular weights. If in doubt, a CID spectrum is measured to confirm the assignment.

Operation of MS-1 for Molecular Weight Determinations

The sample (0.01–1.0 nmol) dissolved in glycerol with addition of some aqueous acid is placed on the sample holder of the JEOL HX110/HX110 tandem mass spectrometer (for a description, see Ref. 5). This sample solution is generally prepared by adding 1–2 μl of glycerol to the preconcentrated HPLC fraction, followed by removal of the remaining HPLC solvent in a vacuum centrifuge. A fraction of the residual glycerol mixture is used for the determination of the molecular weights of the peptides present, by placing it on the sample holder of the FAB-MS ion source. If necessary, some thioglycerol or m-nitrobenzyl alcohol (NBA) is added. Residual water and volatile organic solvent are removed in a separate small vacuum chamber to avoid excessive contamination of the vacuum lock of the spectrometer. In the author's experience, NBA is found to be more useful, particularly for peptides of higher molecular weights (>3000), than thioglycerol, which evaporates faster and reduces disulfide bonds.[25] The properties of these and other matrices are discussed elsewhere.[26]

The sample is then irradiated by either an 8 keV Xe° beam or a 15 keV Cs$^+$ ion beam. Accelerating voltage is 10 kV and the magnet of MS-1 is scanned at a rate that covers m/z 300–5000 or m/z 1000–10,000 in 1.5 min. Resolution in the scanning mode is set either for unit resolution for the

[25] D. L. Smith and Z. Zhou, this volume [20].
[26] E. De Pauw, this volume [8].

expected molecular weight range or to 1 : 1000 to obtain average molecular weights.[18] Spectra are recorded by the JEOL DA5000 data system and displayed or plotted as raw data profiles. It should be noted that at higher mass ($\geq$750 u), conversion to the traditional "bar graph" format is neither necessary nor meaningful, mainly because of the high fractional mass of these heavier ions.[18] Mass calibration is carried out with CsI or other salt mixtures.[27,28]

Operation in MS/MS Mode

A larger fraction of the sample is introduced into the FAB ion source of MS-1 and its magnet set, using the mass display, to an m/z value corresponding to the ^{12}C-only component $(M + H)^+$ isotope cluster to be subjected to CID. The cluster is displayed on the oscilloscope, the ^{12}C species centered, and the source and collector slits proportionally opened to pass this beam with the highest efficiency. Because static resolution permits slit settings twice as wide as required when scanning across a resolving slit, appreciable increase in precursor ion current is gained. The exact and proper setting of the magnetic field is checked by observing the peak profile while scanning the resulting beam across the detector of MS-2 to check that it is a single peak. The collision gas (He, or Xe if the collision energy is less than 1 keV) is then admitted to the collision cell at a pressure sufficient to reduce the precursor ion current to 10–30% of its original value. It should be noted that while the ^{13}C$_1$ species of $(M + H)^+$ is appreciably more abundant than the ^{12}C analog for peptides above m/z 2000, nothing is gained by using it as the precursor because, in that case, most of the fragment ion peaks would be split into doublets, thus reducing their ion relative currents.

The collision cell is operated at potentials ranging from ground to near the accelerating potential of MS-1. To obtain the spectra characteristic of high-energy collisions discussed in this chapter, the cell potential should be no more than the accelerating potential minus 0.5 kV. Generally, we operate the cell at 3 kV, except when using the array detector when the cell potential is set for 5 kV,[22] 8 kV,[23] or even higher.

The resolution of MS-2 is set anywhere between 500 and 2500, depending on the intensity of precursor ion currents. Since the spectrum is monoisotopic and—at least in the CID spectra of peptides—the occurrence of structurally significant doublets differing by only one mass unit is relatively rare (it does not matter whether the presence of an $a_n + 1$ ion next to a

[27] K. Sato, T. Asada, M. Ishihara, F. Kunihiro, Y. Kammei, E. Kubota, C. E. Costello, S. A. Martin, H. A. Scoble, and K. Biemann, *Anal. Chem.* **59,** 1652 (1987).
[28] R. M. Milberg, this volume [14].

generally much more abundant a_n ion is still recognized), unit resolution is not so important, as long as one displays and evaluates the resulting CID spectra in the form of raw data profiles. Although already mentioned earlier, it cannot be overemphasized that conversion to bar graphs, a practice that has persisted from the time when mass spectrometry was limited to relatively low mass, is useless and should be avoided because it degrades, if not falsifies, the data.

MS-2 is scanned from m/z 50 to the precursor ion in 0.5 to 2 min, unless an array detector is used that generally requires 0.2–2.0 sec for each step, depending on the intensity of the product ion current. Scanning or stepping of E_2 and B_2 is controlled by the data system, based on prior calibration in the normal scanning mode (scan B_2 at constant E_2) with a salt mixture in the ion source of MS-2.[27] The data system then calculates the appropriate values for E_2 and B_2 to be scanned or stepped, based on the m/z of the precursor ion, the accelerating voltage of MS-1, and the potential of the collision cell. In the extreme cases of the cell potential being either zero or equalling the accelerating potential, the ratio of B_2/E_2 either remains constant or only B_2 is scanned (E_2 constant), respectively. At intermediate cell potential, the relationship between the values for B_2 and E_2 varies appropriately.

N-Acetylation of Peptide Fragments

Portions of the digest fractions containing Gln and Lys ambiguities are dried in vacuo and dissolved in 50 μl of glacial acetic acid. Acetic anhydride (5 μl) is added and the solution reacted for 6 hr at room temperature. The sample is then dried in vacuo and dissolved in 50 μl of 1 M NH$_4$OH for 30 min at room temperature to hydrolyze the O-acetyl groups. After removal of the ammonia in a vacuum centrifuge, glycerol (1 μl) is added and the solution frozen in nitrogen, concentrated in vacuo, and analyzed by FAB-MS.

Formation of 2-Trimethylammonium Acetyl N-Terminal Derivatives

The appropriate fractions of the *Staphylococcus aureus* digest, which contain peptides 44–55 and 56–65, are concentrated to a volume of about 30 to 50 μl. These solutions are dried in the bottom of a melting point capillary in vacuo in 10-μl aliquots. The melting point capillaries are placed into an apparatus which is designed to react small quantities of sample with reagent vapor.[29] In this case, the dried digest fractions are reacted

[29] J. E. Vath, M. Zollinger, and K. Biemann, *Fresenius Z. Anal. Chem.* **327**, 239 (1988).

with chloroacetyl chloride vapor for 1 hr at 45° and subsequently with vapor from a 25% solution of trimethylamine in water for 2 hr at 45°. To dissolve the product, 10 μl of a 10% aqueous glycerol solution are placed into the capillary and the derivatized samples removed from the capillary using a syringe with a fused-silica needle (J & W Scientific, Folsom, CA). About 0.5 μl of the solutions is placed onto the FAB probe and subjected to tandem mass spectrometry.

Conclusions

In the foregoing, it has been demonstrated that peptide sequencing by high-energy collision tandem mass spectrometry is now a practical method that leads to complete protein sequences with no ambiguities. In speed and efficiency, it far exceeds that of the automated, gas-phase Edman degradation and, when using array detectors,[22-24] it also comes close in sensitivity, if not exceeding it. Tandem mass spectrometry even has some advantages because it is not one-directional and is independent of the need of a basic primary or secondary N-terminal amino group and of modifications of amino acids. On the other hand, the Edman method can provide N-terminal sequence information on an intact protein (if not blocked), can be carried out by less highly trained operators, and is less expensive per unit, although the latter feature is outweighed by the much longer time required per residue (1 day vs. a few minutes that it takes to generate the data for a 15- to 20-amino acid peptide). Another advantage of tandem mass spectrometry is the ability to sequence peptides present in mixtures without the need of prior complete separation. Thus one can expect that tandem mass spectrometry will become a widely used component of peptide and protein structure methodology, complementary to the, up until now, almost exclusively used Edman degradation.

Acknowledgment

The portion of the methodologies and results described that were developed and carried out in the author's laboratory were supported by grants from the National Institutes of Health (RR00317 and GM05472). The author is also indebted to all his present and former associates, particularly R. S. Johnson, S. A. Martin, I. A. Papayannopoulos, H. A. Scoble, and J. E. Vath.

[26] Liquid Secondary Ion Mass Spectrometry of Phosphorylated and Sulfated Peptides and Proteins

By Bradford W. Gibson and Philip Cohen

Introduction

The phosphorylation and sulfation of amino acids are two of the most common posttranslational modifications known to occur in proteins and peptides, and are frequently critical to the function of these molecules by increasing or decreasing their activity. In the case of protein phosphorylation, enzymes called protein kinases are known to transfer the terminal γ-phosphate moiety in ATP to the nucleophilic hydroxyl group of a target serine, threonine, or tyrosine residue. These modifications are reversed by the action of protein phosphatases, a group of enzymes that catalyze the hydrolysis of these ester bonds *in vivo*. Hormones and other stimuli control the phosphorylation states, and hence the activities, of regulatory proteins by activating or inhibiting protein kinases and phosphatases. A number of proteins are phosphorylated at multiple sites by several protein kinases. For example, glycogen synthase, the rate-limiting enzyme in the synthesis of glycogen, is phosphorylated at up to nine serine residues by at least five distinct protein kinases.[1] In the analogous case of protein sulfation, sulfate groups have been found attached to tyrosine residues only through the action of tyrosyl sulfotransferases.[2] Sulfation of tyrosine residues occurs at fairly well-defined recognition sequences and is thought to be irreversible *in vivo*.[3] In many cases, such as in the gastrin and cholecystokinin peptide family, sulfation is an important maturation step which is essential for full functional activity of these peptide hormones.[4]

Despite the importance and widespread occurrence of phosphorylation and sulfation of hyroxyl amino acids in proteins and peptides, the methodology for determining their presence and sequence positions has been less than adequate. For example, a frequent method of determining phosphorylation stoichiometries *in vivo* is to label cells or tissues to steady state by incubation with radiolabeled inorganic [^{32}P]phosphate.[5] However, due to

---

[1] L. Poulter, S.-G. Ang, B. W. Gibson, D. H. Williams, C. F. B. Holmes, F. B. Caudwell, J. Pitcher, and P. Cohen, *Eur. J. Biochem.* **175,** 497 (1988).

[2] R. H. W. Lee and W. B. Huttner, *Proc. Natl. Acad. Sci. U.S.A.* **82,** 6143 (1980).

[3] W. B. Huttner, *Annu. Rev. Physiol.* **50,** 363 (1988).

[4] V. Mutt, *in* "Gastrointestinal Hormones" (G. B. J. Glass, ed.), p. 169. Raven, New York, 1980.

[5] J. C. Garrison, this series, Vol. 99, p. 20.

multiple pools of phosphate within cells (inorganic phosphate, phospholipids, RNA, DNA, etc.), a true steady state is rarely, if ever, achieved. In addition, different phosphorylation sites on the same protein will turn over at different rates, and it is virtually impossible to assess when and whether each site has reached isotopic equilibrium. Furthermore, estimation of the stoichiometry of phosphorylation at a particular site assumes isotopic equilibrium has been reached between intracellular ATP and that site, and that the specific radioactivity of the γ-phosphate of the pool of intracellular ATP which labels that site can be measured. It is extremely difficult to test the validity of either of these assumptions, and where they have been tested, they have been shown to be invalid.[6]

Serious problems are also encountered in the sequence determination of proteins or peptides that contain phosphorylated and sulfated amino acids. The phospho ester bonds of serine and threonine are not stable to the chemical conditions required in sequential Edman degradation, and the covalently attached phosphate group is rapidly hydrolyzed or eliminated to inorganic phosphate.[7,8] Under special circumstances, phosphotyrosine residues can be identified as phenylthiohydantoin-Tyr(P), but low recoveries and high carryover into neighboring cycles greatly limit the general utility of the Edman sequencing approach.[9] Even more problematic are the extremely labile tyrosyl sulfate ester bonds which are not amenable to automated Edman degradation since they are quickly hydrolyzed to unsulfated tyrosine during the first cycle of acidic cleavage.[10]

Nonetheless, several strategies have been developed to overcome some of these inherent analytical limitations. For example, if the peptide is ^{32}P-labeled, the cycle of Edman degradation at which phosphoserine and phosphothreonine are converted to inorganic phosphate can be identified.[11] Alternatively, phosphoserine can be converted specifically to S-ethylcysteine whose phenylthiohydantoin derivative is stable to Edman degradation and easily identified by HPLC techniques.[12,13] However, the latter method is not applicable to phosphothreonine, phosphotyrosine, or sulfated tyrosine.

[6] S. E. Mayer and E. G. Krebs, J. Biol. Chem. 245, 3153 (1970).
[7] D. B. Rylatt and P. Cohen, FEBS Lett. 98, 71 (1979).
[8] C. G. Proud, D. B. Rylatt, S. J. Yeaman, and P. Cohen, FEBS Lett. 80, 435 (1977).
[9] T. Patschinsky, T. Hunter, F. S. Esch, J. A. Cooper, and B. M. Sefton, Proc. Natl. Acad. Sci. U.S.A. 79, 973 (1982).
[10] G. Horton, H. Sims, and A. M. Strauss, J. Biol. Chem. 261, 1786 (1986).
[11] Y. Wang, A. W. Bell, M. A. Hermodson, and P. J. Roach, in "Methods in Protein Sequence Analysis 1986" (K. A. Walsh, ed.), p. 479. Humana, Clifton, NJ, 1987.
[12] H. E. Meyer, E. Hoffmann-Pororske, H. Korte, and L. M. G. Heilmeyer, Jr., FEBS Lett. 204, 61 (1986).
[13] C. F. B. Holmes, FEBS Lett. 215, 21 (1987).

More recently, liquid secondary ion mass spectrometry (LSIMS) has been applied to these same analytical problems and has been shown to be a highly sensitive and direct technique for sequencing peptides containing phosphorylated and sulfated amino acids.[1,14-18] LSIMS methodologies are especially powerful when several phosphorylation or sulfation sites occur within a single peptide, because the precise number of phosphorylated or sulfated residues can be readily determined. In addition, the sequence positions and identity of phosphorylated and sulfated amino acids can be determined by LSIMS, tandem mass spectrometry (MS/MS), or a combination of manual Edman degradation and LSIMS. LSIMS methods can also be used to assess phosphorylation stoichiometries, often in conjunction with reversed-phase high-performance liquid chromatography (HPLC) and amino acid analysis,[16] and to a more limited extent, sulfation stoichiometries.

Determination of Presence of Sulfate or Phosphate

In principle, the molecular weight of an intact protein can be determined directly by mass spectrometry using LSIMS or some other soft ionization technique. This information can then be used to determine if any posttranslational modifications exist that can be accounted for by phosphorylation or sulfation without the need for radiolabeling the protein. This, however, requires that the amino acid sequence of the protein is known and that deviations from the predicted and observed mass can be determined with enough accuracy to be accounted for by shifts of one or more 80 Da, the mass difference corresponding to the presence of phosphorylated or sulfated amino acids. Although no published attempts have yet been detailed for such intact phospho- or sulfoprotein analysis, several recombinant hirudin peptides ($M_r \approx 6700$) have been analyzed with such a purpose in mind.[19]

Given recent advances in high mass analysis of biopolymers, the direct analysis of proteins is likely to become a relatively routine technique as

[14] B. W. Gibson, A. M. Falick, L. Poulter, D. H. Williams, and P. Cohen, in "Methods in Protein Sequence Analysis 1986" (K. A. Walsh, ed.), p. 463. Humana, Clifton, NJ, 1987.
[15] B. W. Gibson, A. M. Falick, A. L. Burlingame, L. Nadasdi, A. C. Nguyen, and G. L. Kenyon, J. Am. Chem. Soc. 109, 5343 (1987).
[16] C. F. B. Holmes, N. K. Tonks, H. Major, and P. Cohen, Biochim. Biophys. Acta 929, 208 (1987).
[17] K. Biemann and H. A. Scoble, Science 237, 992 (1987).
[18] E. Arlandini, B. Gioia, G. Perseo, and A. Vigevani, Int. J. Pept. Res. 24, 386 (1984).
[19] A. Van Dorsselaer, P. Lepage, F. Bitsch, O. Whitechurch, N. Riehl-Bellon, D. Fraisse, B. Green, and C. Roitch, Biochemistry 28, 2949 (1989).

appropriate instrumentation such as laser desorption, time-of-flight (TOF), and electrospray mass spectrometers become commercially available. Recent reports have suggested that these newer techniques will be capable of carrying out sensitive and accurate high mass measurements (10–100 kDa) with as little as a few picomoles of protein.[20,21] Presently, a more accessible approach for determining the presence or absence of phosphorylated and sulfated amino acids is to analyze smaller tryptic or other proteolytic fragments (<10 kDa) generated from these larger proteins. In principle, amino acid composition data of an unknown peptide can be compared with the precise molecular weight determined by LSIMS to deduce the presence or absence of phosphate and sulfate. Such procedures have been used successfully to identify phosphorylated amino acids, as well as other posttranslational modifications such as C-terminal amides and blocked N termini.[17,22–24] However, the inaccuracies of amino acid analysis often limits this approach, and this type of comparison has been more useful when the sequence surrounding the suspected sites is already known. Several examples of such analyses have been published[1,14,16,25] and a few pertinent examples are described below. It should be noted that in the case of peptides suspected of containing tyrosine sulfate, it is essential that the mass spectra be first acquired in the negative ion mode. This is due to the propensity of the MH$^+$ ions to lose SO_3 and erroneously suggest that the peptide is not sulfated.[15,18,26]

Tryptic Peptides from Glycogen Synthase

In an extensive study to determine the *in vivo* sites of phosphorylation in rabbit muscle glycogen synthase and how its phosphorylation state altered in response to adrenaline, numerous peptides were isolated from tryptic digests of glycogen synthase and analyzed by LSIMS. In almost all cases, the amino acid sequence around the sites of suspected phosphorylation sites was known from previous sequencing studies.[27] HPLC frac-

[20] M. Karas and F. Hillenkamp, *Anal. Chem.* **60**, 2299 (1988).
[21] J. B. Fenn, M. Mann, C. K. Meng, S. F. Wong, and C. M. Whitehouse, *Science* **246**, 64 (1989).
[22] B. W. Gibson, L. Poulter, and D. H. Williams, *Peptides* **6** (Suppl. 3), 23 (1985).
[23] B. W. Gibson, D. J. Daley, and D. H. Williams, *Anal. Biochem.* **169**, 217 (1988).
[24] K. Biemann and S. A. Carr, this series, Vol. 106, p. 29.
[25] V. Vaughn, R. Wang, C. Fenselau, and H. B. White III, *Biochem. Biophys. Res. Commun.* **147**, 115 (1987).
[26] L. J. Miller, I. Jardine, E. Weissman, V. L. W. Go, and D. Speicher, *J. Neurochem.* **43**, 835 (1984).
[27] P. Cohen, in "The Enzymes" (P. D. Boyers and E. G. Krebs, eds.), 3rd Ed., Vol. 17, p. 361. Academic Press, Orlando, FL, 1986.

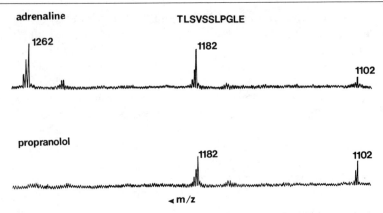

FIG. 1. Partial LSIMS spectrum of the N5–N15 peptides from glycogen synthase isolated from rabbits injected with adrenaline (upper panel) or propranolol (lower panel) showing the dephosphorylated (MH⁺ 1102), monophosphorylated (MH⁺ 1182), and diphosphorylated (MH⁺ 1262) forms. The matrix used was glycerol in 1% HCl. (Reprinted from Ref. 1 with permission from the *Eur. J. Biochem.*)

tions encompassing the elution volume of both the phosphorylated and dephosphorylated forms of a given peptide were pooled, and aliquots of the samples subjected to LSIMS analysis. If molecular ion signals were observed that were 80 Da (or multiples of 80 Da) higher than those of the nonphosphorylated peptides, it could be clearly determined that the peptides had been phosphorylated and the number of phosphorylation sites was clearly evident by the increase in mass.

In one experiment, a peptide thought to contain the dephosphorylated and monophosphorylated forms of the N-terminal region of glycogen synthase (residues N5–N15) was isolated by reversed-phase chromatography on a Vydac C_{18} column after successive treatment with trypsin and *Staphylococcus aureus* V8 proteinase.[1] The dephosphorylated form of this peptide (Thr-Leu-Ser-Val-Ser-Ser-Leu-Pro-Gly-Leu-Glu), was predicted to have a MH⁺ = 1102 (M_r 1101). The LSIMS spectrum of an aliquot of the total mixture was run in glycerol with 1% HCl and scanned at 300 sec/decade. The resulting spectrum of the peptide isolated from animals injected with the adrenaline antagonist propranolol (Fig. 1), clearly showed the molecular ion for the dephosphorylated species at m/z 1102, as well as an additional MH⁺ species at m/z 1182 which corresponds to the monophosphorylated form. After injection of adrenaline, an unexpected additional MH⁺ ion was observed at m/z 1262, which corresponds to the diphosphorylated derivative. The latter result gave the first clear evidence for the presence of two separate sites of phosphorylation within this deca-

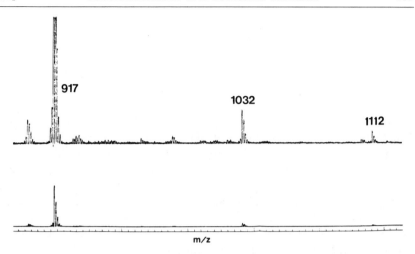

FIG. 2. Molecular ion region of the LSIMS spectrum of dephosphorylated (MH⁺ 1032) and phosphorylated (MH⁺ 1112) angiotensin II obtained after *in vitro* phosphorylation by pp60^{src}.

peptide region. LSIMS analysis also led to the discovery of a second phosphorylation site in a further peptide from glycogen synthase,[1] as well as in a peptide isolated from inhibitor 2,[16] a regulator of protein phosphatase 1.

Phosphorylation of Tyrosine in Angiotensin by pp60^{src}

In a continuing effort to evaluate peptide substrates of the oncogenic tyrosine kinase, pp60^{src}, [Val⁵]-angiotensin II (Asp-Arg-Val-Tyr-Val-His-Pro-Phe) was incubated with pp60^{src} in the presence of [γ-³²P]ATP.[14] The resulting peptide mixture was then separated by reversed-phase chromatography using a Vydac C_{18} column and fractions were assayed for the [γ-³²P]radiolabel by liquid scintillation counting. A small increase in background counts was found on the leading edge of the nonphosphorylated angiotensin peak, which was subsequently analyzed by positive ion LSIMS on a Kratos MS50 mass spectrometer fitted with a Cs⁺ SIMS source as previously described.[28] The spectrum of this fraction (Fig. 2) shows the presence of three MH⁺ ions at m/z 917, 1032, and 1112. The major peak with MH⁺ 917 was found to correspond to desAsp[Val⁵]angiotensin II, and was subsequently determined to be a trace contaminant in the original angiotensin that was being inadvertently enriched in the

[28] A. M. Falick, G. H. Wang, and F. C. Walls, *Anal. Chem.* **58,** 1308 (1986).

radiolabeled fraction. The two smaller peaks, MH^+ 1032 and 1112, correspond to [Val^5]angiotensin II and its monophosphorylated tyrosine form in a ratio of approximately 3 : 1, respectively.

Sulfated Tyrosine-Containing Peptides

Sulfated tyrosine residues are extremely acid labile,[29] and this lability is also apparent during LSIMS analysis where they readily undergo desulfation.[15,26] This effect is most prominent in the positive ion mode where the $(MH - SO_3)^+$ ion is much more abundant than the protonated molecular ion (MH^+), which is of similar abundance to the $(MH - SO_4)^+$ peak. In the negative ion mode, the situation is more like that expected for a less labile linkage, and the $(M - H)^-$ ion is by far the most abundant peak. While the loss of sulfate can cause problems in identifying sulfated peptides and in quantitating the relative proportion of sulfated and nonsulfated peptides, it can also aid in the identification of peptides that contain this modification. For example, in the negative ion mode the related $(M - HSO_3)^-$ peak is still abundant in most spectra (ca. 20% relative to the deprotonated molecular ion), and can thus serve as a diagnostic marker for the presence of sulfated tyrosine. While the presence of this signal could also be interpreted as the loss of HPO_3 from tyrosine, serine, or threonine, the magnitude of this loss from phosphorylated residues is much less (1–5%). A second mass spectrum can also be acquired in the positive ion mode which would shift the MH^+ ion to the nonsulfated form. In addition, limited acid treatment of sulfated peptides (1 N HCl, 100° for 5 min or 6 N HCl, 37°, 5 min) can also quantitatively remove the suflate group(s) from tyrosine,[29] but should have little or no effect on the more acid-stable phosphorylation sites. Alternatively, treatment with *Helix pomatia* arylsulfatase for 6 hr at pH 5.2 in 100 mM ammonium sulfate will also remove the sulfate from sulfopeptides.[30] Thus, reanalysis by LSIMS of a peptide suspected to contain sulfated tyrosine after acidic hydrolysis of the labile sulfate moieties can be used to support the presence of sulfation as opposed to phosphorylation. Figure 3 shows both the negative and positive ion spectra in the molecular ion region for caerulein run in a mixture of thioglycerol and glycerol. As discussed in more detail later, peptides screened by MS/MS in the negative ion mode will yield an abundant immonium-type ion for Tyr(S) at m/z 214, the same mass as the Tyr(P) immonium ion in the negative mode.

[29] W. B. Huttner, this series, Vol. 107, p. 200.
[30] G. Hortin, K. F. Fok, P. C. Toren, and A. W. Strauss, *J. Biol. Chem.* **262,** 3082 (1987).

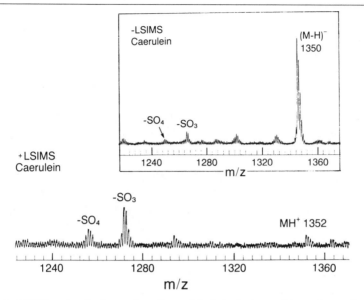

FIG. 3. LSIMS molecular ion region of caerulein in both the positive and negative ion modes. (Reprinted from Ref. 15 with permission from the *J. Am. Chem. Soc.*)

Quantitation of Phosphate or Sulfate Content

Chemical methods are available for determining the amount of phosphate bound covalently to proteins,[31,32] but these methods give no information concerning the location of these phosphates. In contrast, if a protein is digested with a specific proteinase and fractionated to yield mixtures containing both the modified and unmodified peptides, the extent of phosphorylation (or sulfation) and its approximate location can, in principle, be determined directly by LSIMS analysis. However, several problems are encountered in such a scheme. First, LSIMS is a surface sampling technique and the molecular ion abundance for peptides in mixtures reflect, to a large extent, their relative surface activities.[33] Second, LSIMS is also a dynamic technique, and molecular ion abundances for peptides that have a surface excess or deficit relative to the interior (or bulk) concentration reach their maximum molecular ion abundances at different times. As a result, there is the danger that a signal from a phosphorylated peptide will be suppressed by the presence of the more hydrophobic

[31] P. S. Guy, P. Cohen, and D. G. Hardie, *Eur. J. Biochem.* **115,** 399 (1981).

[32] J. E. Buss and J. T. Stull, this series, Vol. 99, p. 7.

[33] S. A. Naylor, A. F. Findeis, B. W. Gibson, and D. H. Williams, *J. Am. Chem. Soc.* **108,** 6359 (1986).

dephosphorylated derivative, leading to underestimation of the phosphorylation stoichiometry. Third, the relatively large loss of SO_3 from the $(M - H)^-$ ions of sulfated peptides make any measurement of a mixture of the two species highly misleading. Nonetheless, there are conditions under which the relative proportions of phosphorylated and dephosphorylated peptides in a mixture can be determined with relative accuracy. The peptides must have a considerable surface excess or deficit, with an average hydrophobicity/hydrophilicity index ($\Delta F/n$) of greater than $+0.2$ kcal (large surface deficit) or less than -0.2 kcal (large surface excess), as estimated by the method of Bull and Breese[33,34] where $\Delta f = +0.65$ kcal is used for the phosphate moiety. Peptides should also be relatively large (≥ 15 residues) since they tend to suppress each other to a lesser extent, due to the smaller contribution of their phosphorylation sites to difference in their surface activity. Simply put, this means that both the phosphorylated and dephosphorylated peptides must either run well (surface excess) or run poorly (surface deficit) to ensure that the presence of a phosphate group will not have a great effect either way on the relative surface activity. These criteria can often be met by choosing a proteinase of appropriate specificity. Alternatively, peptides can be peracetylated with acetic anhydride/pyridine ($1 : 1$, v/v) or esterified with various alcohols in the presence of HCl or 0.2 M acetyl chloride[33,35] to yield more hydrophobic derivatives that will reduce or eliminate any differences in their relative ionization efficiencies. Molecular ion abundances should also be integrated over an extended period of time (5–10 min) to minimize any remaining difference in their relative ion yields as a function of time. In addition, certain of the more hydrophobic hydroxyl-containing matrices, such as 1,2,6-hexanetriol, have been shown[36] to yield molecular ion abundances for hydrophilic peptides with $+\Delta F$ values (surface deficit) that more accurately reflect the relative concentrations of phosphorylated peptides. A similar effect has been found when certain strong acids (e.g., $HClO_4$) are added to the matrix, but these approaches should be tested with the peptides in question to confirm the validity of the method. More recently, continuous-flow interfaces for LSIMS have been developed and shown to reduce the differences in relative molecular ion abundances for peptides with different surface activities,[37] but to date this technique has not been applied to the determination of phosphorylation stoichiometries.

[34] H. B. Bull and K. Breese, *Arch. Biochem. Biophys.* **161**, 665 (1974).
[35] A. M. Falick and D. Maltby, *Anal. Biochem.* **182**, 165 (1989).
[36] L. Poulter, S.-G. Ang, D. H. Williams, and P. Cohen, *Biochim. Biophys. Acta* **929**, 296 (1987).
[37] R. M. Caprioli, W. T. Moore, G. Petri, and K. Wilson, *Int. J. Mass Spectrom. Ion Proc.* **86**, 187 (1987).

TABLE I
RATIOS OF MH$^+$ ABUNDANCES FOR PHOSPHORYLATED AND DEPHOSPHORYLATED
PEPTIDES (P/D) IN DIFFERENT MATRICES

Peptide	ΔF cal/mol (P/D)	Ratios of molecular ion abundances (P/D) in different matrices with 1% HCl[a]			
		G	T/G	DT/DE	HT
SPQPSRRGSESSEE	+506/+460	1.0	1.0	1.0	1.0
LRRASLG	+81/−11	0.6	0.5	0.4	0.5
LRRASLG-OEt	−111/−203	0.8	0.6	0.5	1.0
PLSRTLSVAA-NH$_2$	−53/−118	0.5	0.4	0.2	0.8
Ac-PLSRTLSVAA-NH$_2$	−98/−163	0.7	0.4	0.3	0.9
PLSRTLSVAA-OEt	−188/−253	1.0	1.0	1.0	—

[a] G, Glycerol; T, thiogylcerol; DT, dithioglycerol; DE, dithioerythritol; HT, hexanetriol.

Estimation of Phosphorylated/Dephosphorylated Ratio by LSIMS of Peptides from Glycogen Synthase

As shown previously in Fig. 1, a mixture of phosphorylated and dephosphorylated peptides from the N-terminal region of glycogen synthase were analyzed directly by LSIMS in a single spectrum. In that example, the relative molecular ion abundances between the dephosphorylated (MH$^+$ 1102), monophosphorylated (MH$^+$ 1182), and diphosphorylated (MH$^+$ 1262) peptides were seen to exist in a 19:35:46% ratio for the adrenaline-treated animals and a 41:59:0% ratio for the animals treated with propranolol. These values were then used to estimate the phosphorylation stoichiometries, but reflected relative molecular ion abundances after integration of many scans during the sample lifetime (ca. 10 min) and not just the single scan shown in Fig. 1. As stated previously, it is essential to integrate the molecular ion abundance during the entire sample lifetime due to differences in surface activities between the phosphorylated and nonphosphorylated peptide forms.

Estimation of Phosphorylated/Dephosphorylated Ratio by LSIMS: Studies with Model Peptides

A series of model peptides with a single phosphorylation site were analyzed by LSIMS to determine the accuracy of equating molecular ion abundances to relative concentrations of phosphorylated and dephosphorylated forms.[36] In Table I, equimolar amounts of several phosphorylated/dephosphorylated peptide mixtures were analyzed by LSIMS in several different matrices. In the case of the peptide SPQPSRRGSESSEE, which has a small predicted surface deficit for both forms (ΔF +460/+506 cal/mol, respectively), a relative ion abundance ratio of 1.0 was observed in all four matrices tested. For the heptapeptide LRRASLG (ΔF +0.081/−0.011 kcal/mol), which shifts between a predicted slight surface deficit for the phosphorylated form to a small surface excess for the dephosphorylated form, the relative molecular ion abundance ratios tend to favor the hydrophobic dephosphorylated peptide. However, this could be largely overcome if the peptides were first converted to the more hydrophobic ethyl ester derivatives (ΔF values of −0.11/−0.20 kcal/mol, respectively), or run in a matrix such as hexanetriol.

Determination of Phosphorylated/Dephosphorylated Ratio by LSIMS in Conjunction with Reversed-Phase HPLC and Amino Acid Analysis

It is frequently possible to separate the phosphorylated form of a peptide from its dephosphorylated counterpart by reversed-phase HPLC, the phosphopeptide eluting at a slightly lower concentration of acetonitrile. Separation is enhanced if chromatography is carried out in ammonium acetate, pH 6.5, instead of the more commonly employed 0.1% trifluoroacetic acid. The phosphorylated and dephosphorylated peptides are identified by LSIMS, and their proportions can then be determined accurately by either quantitative amino acid analysis or by integration of the UV-absorbing peaks on the chromatogram. This procedure has been applied successfully in the analysis of the *in vivo* phosphorylation states of inhibitor-2[6] and the glycogen-binding subunit of protein phosphatase 1.[38]

Sequence Determination

Sequence Determination by LSIMS

The sequence determination of phosphopeptides or sulfopeptides by LSIMS can be carried out under similar conditions to other peptides, but at least 1 or 2 nmol of peptide is generally required to obtain enough fragment ions to determine the sequence of modified and unmodified amino acids by analysis of the mass differences between ions. In the case of serine, threonine, or tyrosine phosphorylation, a mass difference of 167, 181, or 243 Da, will be observed in place of the unmodified mass differences of 87, 101, and 163 Da, respectively. For example, in the partial spectra of the C-98–C-123 ($MH^+ = 2713$) tryptic peptide from glycogen synthase

---

[38] C. MacKintosh, D. G. Campbell, A. Hiraga, and P. Cohen, *FEBS Lett.* **234**, 189 (1988).

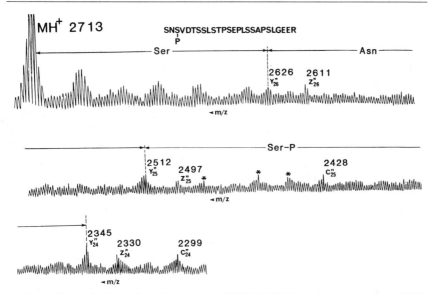

FIG. 4. Molecular ion and partial sequence of C-98–C-123 from glycogen synthase (MH$^+$ 2713) showing the N-terminal sequence Ser-Asn-Ser(P)-. (Reprinted from Ref. 1 with permission from the *Eur. J. Biochem.*)

(Fig. 4), a Y_n'' (or y_n) and Z_n'' (or z_n) ion series are observed with a mass difference of 167 (Ser-P) for the y_{25} to y_{24} ions, yielding an N-terminal sequence of Ser-Asn-Ser(P)-. An analogous shift is also observed for sulfated tyrosine (243 Da) relative to the corresponding nonsulfated derivative (163 Da), as long as the spectrum is acquired in the negative ion mode.

Sequence Determination by Combined LSIMS/Manual Edman Degradation

Due to the low extent of fragmentation that is often observed in LSIMS spectra, one does not always identify the site of modification and some other technique becomes necessary. Although direct identification of phenylthiohydantoin amino acids for phosphorylated residues is generally not possible during automated Edman sequence analysis, manual Edman degradation followed by molecular weight determination of the N-terminally truncated peptide(s) by LSIMS can be used to sequence these phosphopeptides. Since the released amino acid derivative is not analyzed as in traditional Edman sequencing, a slight modification of the manual Edman procedure published by Tarr[39] that is better suited to subsequent LSIMS analysis is required, and is described below.

Approximately 0.1–2 nmol of phosphopeptide is dried in a 0.5 ml Reacti-Vial fitted with a modified Mininert valve (Pierce Chem., Rockford, IL) and nitrogen purge line.[39] A solution of 90% ethanol/triethylamine (TEA)/phenyl isothiocyanate (PITC) (7 : 2 : 1, v/v/v) is aged for 5 min and the 10-μl aliquot is added to the dried phosphopeptide and incubated at 50° for 5 min after purging with nitrogen. The sample is then dried *in vacuo*, followed by an optional washing step. This consists of one or more washes of the coupled peptide (dissolved in 3–4 μl of 0.25% aqueous trimethylamine) with 100 μl of heptane/ethyl acetate mixtures starting at a ratio of 15 : 1 and followed by a 7 : 1 ratio. The wash step is recommended if more than four cycles of Edman degradation are planned, and should be omitted altogether if the peptide is hydrophobic. Alternatively, a single wash step (100 μl of heptane/ethyl acetate, 15 : 1) can be used successfully. The sample is then dried *in vacuo* and cleaved with 5 μl of concentrated HCl for 5 min at 37° after first purging the vial with nitrogen. At this point the sample can be dissolved in an appropriate solvent, such as H_2O or 1–5% trifluoroacetic acid, and an aliquot saved for LSIMS analysis. Additional cycles of Edman degradation can then be performed beginning with the addition of the ethanol/TEA/PITC solution. A typical example of such a procedure is shown in Fig. 5 where LSIMS spectra of the molecular ion regions are shown before and after the Edman cycles where phosphoserine residues are located.

Identification of Phosphoserine by Combination of LSIMS and Automated Sequencing

If a peptide has been shown to contain a phosphorylated residue by LSIMS analysis, another way to identify the position of phosphoserine is to convert these residues to *S*-ethylcysteine by β-elimination in 0.1 *N* NaOH followed by addition of ethanethiol.[12,16] The phenylthiohydantoin derivative of *S*-ethylcysteine can be easily detected by conventional automated Edman sequencing. This method is simple and extremely sensitive, but is not applicable to peptides containing phosphothreonine, phosphotyrosine, or sulfotyrosine, or when phosphoserine is the N-terminal or C-terminal residue of a peptide.[40,41] It has been used to identify the phosphoserine residues in several proteins.[16,38,41]

[39] G. E. Tarr, *in* "Methods in Protein Sequence Analysis" (M. Elzinga, ed.), p. 223. Humana, Clifton, NJ, 1984.

[40] H. E. Meyer, K. Swiderek, E. Hoffmann-Posorske, H. Korte, and L. M. G. Heilmeyer, Jr., *J. Chromatogr.* **397**, 113 (1987).

[41] A. J. Garton, D. G. Campbell, P. Cohen, and S. J. Yeaman, *FEBS Lett.* **229**, 68 (1988).

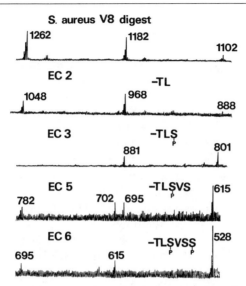

Fig. 5. Positive LSIMS spectra of N-5–N-15 after two, three, five, and six cycles of manual Edman degradation. A loss of 167 Da (ΔSer-P) rather than 87 Da (ΔSer) after cycles three and six indicates the presence of phosphoserine at those positions. (Reprinted from Ref. 1 with permission from the *Eur. J. Biochem.*)

Sequence Determination by Tandem Mass Spectrometry (MS/MS)

One of the newer and most powerful techniques for the direct sequencing of both phosphorylated and sulfated peptides is tandem mass spectrometry (MS/MS). As described in [25] in this volume, this procedure involves the selection of the monoisotopic ^{12}C peptide molecular ion from the molecular ion cluster in the first mass spectrometer, either from a single purified peptide or from a mixture of peptides, and collisionally activated by helium in an intermediate collision cell. The resulting daughter ions are then separated in a constant B/E scan in the second mass spectrometer and acquired either on a conventional postacceleration detector or on an optically coupled diode array detector.[42] In either case, the MS/MS spectrum is interpreted to yield the amino acid sequence following the generalized rules described elsewhere in this volume [24], except that the mass values for the phosphorylated amino acids [Ser(P) = 167 Da, Thr(P) = 181 Da, and Tyr(P) = 243 Da] and sulfated tyrosine [Tyr(S) = 243 Da] must also be used. However, there are several important features that should be considered as important and unique features of the MS/MS spectra of phosphorylated and sulfated peptides, and these are described below for some representative cases.

[42] J. Cottrell and S. Evans, *Anal. Chem.* **59**, 1990 (1987).

MS/MS Spectra of Leu-Arg-Arg-Ala-Ser(P)-Leu-Gly. In the MS/MS spectra of peptides containing phosphoserine, it has been noted that several definitive fragment ions are formed that can be used to both identify and sequence these types of peptides. Biemann and Scoble[17] have pointed out that abundant losses of H_3PO_4 (98 Da) are observed from the protonated parent molecular ion in the positive ion mode and is consistent with the accompanying formation of dehydroalanine. In addition, a low mass fragment ion corresponding to $H_2PO_3^-$ was postulated to exist in the corresponding negative ion mode. Indeed, some of these important features can be seen in the positive and negative ion spectrum of phosphorylated kemptide shown in Fig. 6. In the positive ion mode, both HPO_3 and H_3PO_4 losses are observed from the MH^+ ion, as well as the low mass phosphoserine immonium ion at m/z 140. In the negative ion MS/MS spectrum[43] of phosphorylated kemptide, a loss of HPO_3 and H_3PO_4 can be seen from the deprotonated molecular ion at m/z 770 and 752. However, the postulated $H_2PO_3^-$ ion (m/z 81) is not observed, but instead two abundant low mass ions at m/z 79 and 97 are found which correspond to PO_3^- and $H_2PO_4^-$, respectively. Unlike the positive ion spectrum, ion fragmentation is largely directed by the phosphoserine in the negative ion mode, and most fragments contain this residue as might be expected from negative charge stabilization on the phosphate moiety (see Fig. 6b). In both the positive and negative ion MS/MS spectra, the position of the phosphoserine can be readily deduced from the expected mass difference (167 Da) in the relevant daughter ions as can be seen between the a_4 (m/z 469) and a_5 (m/z 636), and d_5 (m/z 540), and d_6 (m/z 707) ions in the positive ion spectrum shown in Fig. 6a.

MS/MS Spectra of [Tyr(P)]⁵-Bradykinin and Arg-Tyr(P)-Val-Phe. Tyr⁵-Bradykinin was purchased from Peninsula Laboratories (Belmont, CA) and was converted to the phosphotyrosine form using a novel synthetic procedure that involves the preparation of an adenylylated intermediate.[44] The tetrapeptide Arg-Tyr(P)-Val-Phe was synthesized by the Merrifield method using *N*-(*tert*-butoxycarbonyl)-*O*-(dibenzylphosphono)-L-tyrosine

---

[43] In the negative ion mode, fragment ions are observed that are two mass units lower than the corresponding fragments generated from the molecular ion in the positive ion mode. To simplify the nomenclature, the lower case letters corresponding to fragments with charge retention at the N terminus (a_n, b_n, c_n, and d_n) and C terminus (x_n, y_n, z_n, v_n, and w_n) will also be used for negative ion MS and MS/MS spectra, but will represent ions two mass units lower than those defined for the positive ion fragments shown in Appendix 5 at the end of this volume.

[44] B. W. Gibson, W. Hines, Z. Yu, G. L. Kenyon, L. Sager, and J. J. Villafranca, *in* "Current Research in Protein Chemistry" (J. J. Villafranca, ed.), p. 151. Academic Press, San Diego, 1990.

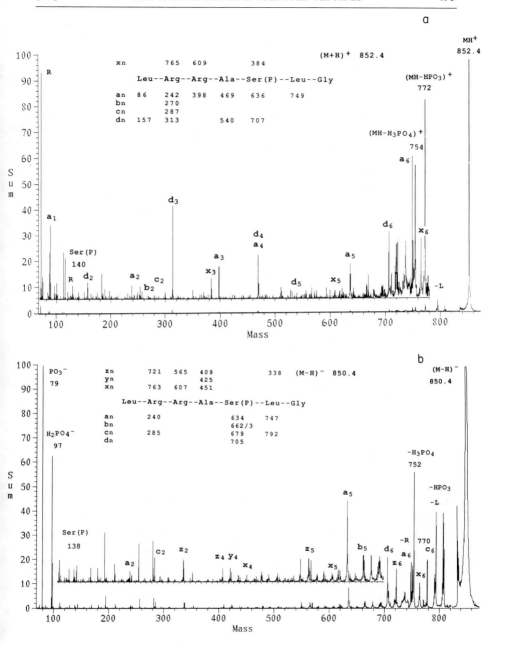

FIG. 6. MS/MS spectrum of Leu-Arg-Arg-Ala-Ser(P)-Leu-Gly (M_r 851) in the (a) positive and (b) negative ion mode.

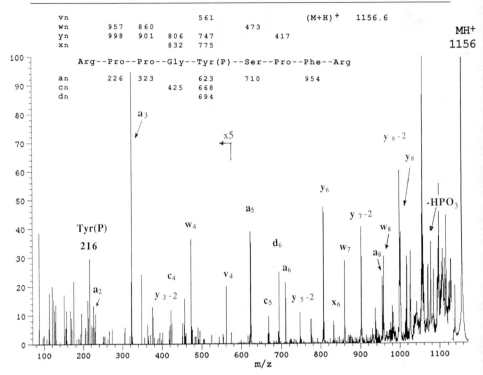

FIG. 7. Positive ion MS/MS spectrum of phosphorylated $[Tyr(P)]^5$-bradykinin.

for phosphotyrosine.[15] Approximately 1 nmol of the resulting phosphopeptides were dissolved in 1% trifluoroacetic acid and applied to a stainless steel probe containing a mixture of thioglycerol and glycerol (1 : 2). A MS/MS spectrum was then acquired for each peptide on a Kratos Concept tandem MS/MS instrument with a diode array detector on MS-2. Collision-induced dissociation (CID) of the parent molecular ion was obtained using helium in the collision cell whose pressure was adjusted to transmit a 65% attenuation of the ^{12}C monoisotopic molecular ion. As can be seen from the MS/MS spectrum of $[Tyr(P)]^5$-bradykinin (Fig. 7), fragment ions are clearly defined that give a prominent N-terminal a_n series and a C-terminal x_n and w_n series that clearly pinpoint the site of phosphorylation as $Tyr(P)^5$. In addition, a prominent immonium ion is seen at m/z 216 which always appears when tyrosine is phosphorylated. In the high mass region, reasonably abundant ions are seen at m/z 1076/5 and to a lesser extent at 1059/58 that correspond to loss of $HPO_3/H_2PO_3\cdot$ and $HPO_4/H_2PO_4\cdot$, respectively. The abundant loss of phosphate might explain the absence of a complete side-chain loss for phosphotyrosine, i.e., $(MH-CH_2C_6H_4-PO_4H_2)^+$ at m/z 969, in this same high mass region.

A similar MS/MS spectrum can also be seen in the positive ion mode for Arg-Tyr(P)-Val-Phe, where abundant daughter ions are seen at m/z 584 ($-HPO_3$) and 583 ($-H_2PO_3$), as well as the diagnostic immonium ion for Tyr(P) at m/z 216 (see Fig. 8a). In the corresponding negative ion spectrum (Fig. 8b), abundant ions are observed for the loss of HPO_3 (m/z 582) and H_2PO_3 (m/z 581) as well as an analogous deprotonated immonium-type ion for phosphorylated tyrosine at m/z 214 ($NH=CHCH_2-C_6H_4OPO_3H^-$). In addition to the deprotonated immonium ion for phosphotyrosine, two other related ions were present at m/z 199 and 186 whose structures are consistent with $CH_2=CHC_6H_4OPO_3H^-$ and the radical anion side-chain cleavage product, $\cdot CH_2C_6H_4OPO_3H^-$, respectively. However, the most dominant ion fragments in the negative ion mode appear at m/z 97 ($H_2PO_4^-$) and m/z 79 (PO_3^-) as is the case for phosphoserine-containing peptides. Indeed, fragmentation in the negative ion mode is largely directed by the presence of phosphotyrosine in a manner similar to that observed in phosphoserine-containing peptides. In the case of Arg-Tyr(P)-Val-Phe, this results in a strong N-terminal series with the a_2 ion [m/z 370, Arg-Tyr(P)$^-$], one of the most abundant daughter ion fragments.

It is important at this point to consider the difference between the MS/MS negative ion spectra of phosphotyrosine and phosphoserine-containing peptides. As previously mentioned, the predominant losses observed in phosphoserine peptides in the positive ion mode are (MH $-$ $H_3PO_4)^+$, which is not observed for phosphotyrosine, and the abundant $(M - H - H_3PO_4)^-$ and $(M - H - HPO_3)^-$ ions in the negative ion spectra, only the latter of which is also seen in the spectra of phosphotyrosine-containing peptides. However, the low mass fragments $H_2PO_4^-$ (m/z 97) and PO_3^- (m/z 79) in negative ion spectra are present in the negative ion MS/MS spectra of both phosphotyrosine- and phosphoserine-containing peptides. In contrast, the deprotonated immonium ion and side-chain fragments for phosphotyrosine (m/z 214, 199, and 186) are clearly absent in the MS/MS spectra of phosphoserine peptides. Thus, there are clearly unique daughter ions in both the positive and negative ion MS/MS spectra for phosphoserine and phosphotyrosine peptides that can be used to unambiguously distinguish between these phosphorylation types. It should also be pointed out that the negative ion MS/MS spectra of both phosphotyrosine and phosphoserine peptides yield the most abundant fragment ions at low mass, and these important ions can be used to easily identify the presence of these phosphorylated residues with high

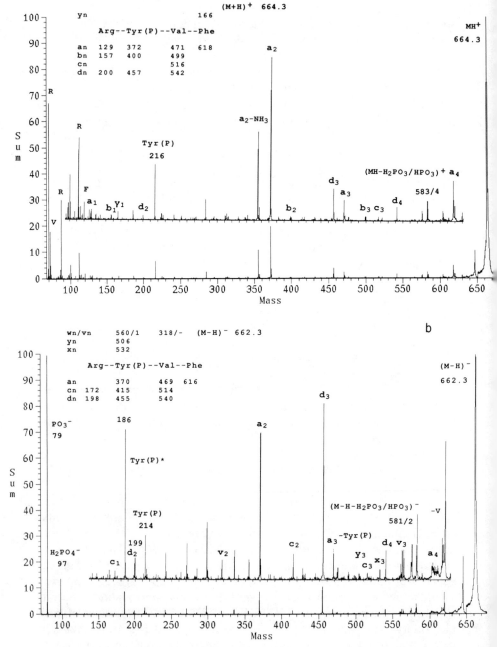

FIG. 8. MS/MS spectrum of Arg-Tyr(P)-Val-Phe (M_r 663) in the (a) positive and (b) negative ion mode.

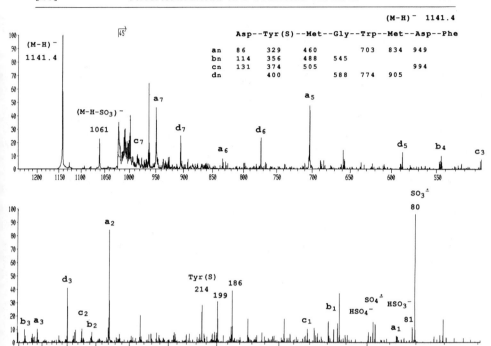

FIG. 9. Negative ion MS/MS spectrum of CCK-8 showing the dominance of the (M − H − SO$_3$)$^-$ and SO$_3$$^{\cdot-}$ daughter ions.

sensitivity. Conversely, the positive ion MS/MS spectra of these phosphorylated peptides contain more sequence-specific fragment ions throughout the entire mass range, and, therefore, may turn out to yield better overall sequence data.

MS/MS Spectrum of CCK-8, Asp-Tyr(S)-Met-Gly-Trp-Met-Asp-Phe-NH$_2$. One might expect some similarities to exist between the MS/MS spectra of peptides containing phosphotyrosine and sulfated tyrosine. Since abundant molecular ions for sulfated tyrosine peptides are obtained only in the negative ion mode, a meaningful comparison can only be made in this ion mode. Indeed, if one considers the tandem negative ion spectrum of CCK-8 (Fig. 9), analogous losses of sulfate as SO$_3$ and to a much smaller extent SO$_4$ (or SO$_2$ and O$_2$) are readily apparent from the (M − H)$^-$ at *m/z* 1141, although the extent of SO$_3$ loss is at least an order of magnitude larger compared to the analogous loss of HPO$_3$ from phosphotyrosine moieties. One also observes a deprotonated immonium-type ion at *m/z* 214, NH=CHCH$_2$C$_6$H$_4$OSO$_3$$^-$, and two equally abundant ions at *m/z* 199 and 186 that are consistent with a hydrogen-rearranged product CH$_2$=CHC$_6$H$_4$OSO$_3$$^-$ and the radical anion side-chain cleavage product, $\cdot CH_2C_6H_4OSO_3{}^-$. These latter three ions are nominally the same mass as the phosphotyrosine immonium ion (m/z 214) and its related side-chain cleavage products (m/z 199 and 186). However, a major difference is seen at lower masses where a very abundant ion is observed for the sulfur trioxide radical anion ($SO_3{}^-$). In addition, at much lower abundances one can also discern several related sulfur oxide species at m/z 81 ($HSO_3{}^-$), m/z 96 ($SO_4{}^-$), and m/z 97 ($HSO_4{}^-$). All of these ions appear to be general features of negative ion MS/MS spectra of sulfated tyrosine peptides, and clearly allow one to distinguish the presence of this modification relative to peptides containing phosphotyrosine (or phosphoserine). It is also apparent from the spectrum of CCK-8 and other peptides containing sulfotyrosine that fragmentation is greatly influenced by the presence of the sulfate group. Indeed, in the case of CCK-8 (Fig. 9) and sulfated Leu-enkephalin,[45] the presence of sulfotyrosine at or near the N terminus results in a complete domination of N-terminal daughter ions, i.e., a_n, b_n, c_n, and/or d_n.

Concluding Remarks

Over the last few years, mass spectrometry has made substantial contributions to protein chemistry, particularly in the identification and structure determination of posttranslational modifications. This is largely due to the failure or inadequacy of Edman chemistry and some of the "classical" techniques for the identification of most posttranslational modifications, and the inherent versatility of mass spectrometric methods to accommodate essentially all types of modifications. Since protein sulfation and phosphorylation are two of the most common and important modifications that are encountered, it is appropriate that mass spectrometric methods have been developed and continue to be improved for the analysis of these modifications. One of the most promising recent advances has been in the application of MS/MS to the analysis of phosphorylated and sulfated peptides, which, in principle, can lead to the determination of entire sequences for peptides containing these residues with masses up to m/z 2500 in a single experiment. Indeed, it is worth pointing out that MS/MS techniques have been used quite recently to identify the labile phosphorylation site in the chemotactic protein CheY at aspartic acid after conversion to the stable homoserine analog.[46] Another exciting development is in the

[45] B. W. Gibson, in "Biological Mass Spectrometry" (A. L. Burlingame and J. A. McCloskey, eds.), p. 315. Elsevier, Amsterdam, 1990.

[46] D. A. Sanders, B. L. Gillece-Castro, A. L. Burlingame, and D. E. Koshland, Jr., *J. Biol. Chem.*, **264**, 21170 (1989).

analysis of proteins by electrospray mass spectrometry and laser desorption time-of-flight (TOF) mass spectrometry (see [12] and [22] in this volume). Both of these methods are currently capable of measuring the molecular mass of a protein up to 60 kDa or larger with a degree of accuracy high enough to potentially distinguish between phosphorylated and nonphosphorylated forms. Thus, it should be possible to obtain information on the phosphorylation (or sulfation) states of intact proteins by these methods in the near future, although peptide fragments will still have to be generated and analyzed by more traditional LSIMS techniques to localize the actual sites of modification.

Acknowledgments

We would like to thank Drs. Dudley H. Williams and Linda Poulter for their help in the analysis of several phosphorylated peptides and in many discussions regarding the quantitation of phosphate in peptides by LSIMS. We would also like to acknowledge the financial support of the National Institutes of Health (B.G.), National Science Foundation (B.G.), National Cancer Institute (B.G.), Wellcome Trust (P.C.), Medical Research Council (P.C.), and the Royal Society (P.C.).

[27] Identification of Attachment Sites and Structural Classes of Asparagine-Linked Carbohydrates in Glycoproteins

By Steven A. Carr, John R. Barr, Gerald D. Roberts,
Kalyan R. Anumula, and Paul B. Taylor

Glycosylation of specific asparagine (Asn) residues is one of the most common posttranslational modifications of proteins. However, its importance is just now becoming understood and widely appreciated. Current theories concerning the roles of Asn-linked oligosaccharides in modulating the biological activity of glycoproteins have been recently reviewed[1-5] (also see [29] in this volume). The attachment-site Asn residues are with few exceptions present in an Asn-X-Ser/Thr consensus sequence in which

[1] T. W. Rademacher, R. B. Parekh, and R. A. Dwek, *Annu. Rev. Biochem.* **57,** 785 (1988).
[2] J. U. Baenziger, *Am. J. Pathol.* **121,** 382 (1985).
[3] M. Fukuda, *Biochim. Biophys. Acta* **780,** 119 (1985).
[4] T. Feizi and R. A. Childs, *Trends Biol. Sci.* **10,** 24 (1985).
[5] N. Sharon and H. Lis, "Lectins." Chapman and Hall, London, 1989.

METHODS IN ENZYMOLOGY, VOL. 193

"X" may be any amino acid except proline.[6-8] However, only selected Asn residues present in the required consensus sequence are targets for glycosylation by the oligosaccharyltransferase; presumably this is due to as yet undefined recognition properties of this glycosyltransferase.[9] Thus, techniques are required for pinpointing which of the many potential glycosylation sites are utilized.

In contrast to the peptide backbone of the protein, structural heterogeneity of the Asn-linked oligosaccharides is the rule rather than the exception. Glycoproteins commonly have several different structural classes of oligosaccharide attached (i.e., oligomannose, hybrid, or complex[8]). Furthermore, each glycosylation site often has a heterogenous population of oligosaccharides of the same structural class, but differing in the number and type of sugar residues attached to the ends of the chains or antennae.[8] The compositions and structures of the carbohydrate chains attached to a glycoprotein are determined by the glycosylation machinery available in the specific cellular expression system used, and changes in the cellular expression system can significantly alter the structures of the pendant oligosaccharides. The term glycoform has recently been coined to describe this structural diversity which is a result of tissue- and cell-specific biosynthesis and processing of the Asn-linked oligosaccharide chains of glycoproteins.[1] To date, most information about the structures of N-linked carbohydrates in glycoproteins has been an average or composite structural view obtained on the pool of carbohydrates released by a variety of endoglycosidases or hydrazinolysis.[1,10]

Here we describe two interrelated mass spectrometry-based strategies for: (1) identifying the attachment sites of Asn-linked sugars in glycoproteins and extent of glycosylation at each site, and (2) defining the compositions and molecular heterogeneity of carbohydrates at each specific attachment site in glycoproteins.[11-13] The methods used for identification of attachment sites are known as carbohydrate mapping, while those employed for site-specific carbohydrate structure analysis are collectively referred to as carbohydrate fingerprinting. These techniques can be used to assess the number, location, structural class (including branching), and site heterogeneity of the carbohydrates present in glycoproteins.

[6] D. K. Struck and W. J. Lennarz, in "The Biochemistry of Glycoproteins and Proteoglycans" (W. J. Lennarz, ed.), p. 35. Plenum, New York, 1980.
[7] E. Bause and H. Hettkamp, FEBS Lett. 108, 341 (1979).
[8] R. Kornfeld and S. Kornfeld, Annu. Rev. Biochem. 54, 631 (1985).
[9] M. Geetha-Habib, R. Noiva, H. A. Kaplan, and W. J. Lennarz, Cell 54, 1053 (1988).
[10] J. A. Welply, Trends Biotechnol. 7, 5 (1989).
[11] S. A. Carr and G. D. Roberts, Anal. Biochem. 157, 396 (1986).
[12] S. A. Carr, G. D. Roberts, A. Jurewicz, and B. Frederick, Biochemie 70, 1445 (1988).
[13] J. R. Barr, S. A. Carr, K. Anumula, M. B. Vettese, and P. Taylor, Anal. Biochem., submitted (1990).

Identification of Attachment Sites of Asn-Linked Carbohydrates
in Glycoproteins

Principle

Carbohydrate mapping by fast atom bombardment mass spectrometry
(FAB-MS) has been used to identify the N-linked glycosylation sites in
tissue plasminogen activator expressed in chinese hamster ovary cells[11,12]
and a *Drosophila* insect cell line,[14] in the pre-S2 region of the hepatitis B
surface antigen,[15] in spermine binding protein from rat prostate,[16] and in
a recombinant, soluble form of the CD4 receptor glycoprotein.[17] A schematic outline of this part of the methodology is shown in Fig. 1. The initial
part of the strategy for generation of peptides and glycopeptides from
glycoproteins (left-hand side of Fig. 1) parallels that used for peptide
mapping by FAB-MS which is described in more detail elsewhere in this
volume (see in this volume [18]). The protein is usually reduced and
alkylated (e.g., carboxymethylation or pyridylethylation) prior to proteolysis in order to increase its susceptibility to digestion. Chemical cleavage
methods that denature the protein and produce large proteolytic fragments
must be used with extreme caution since the *N*-glycosidic or *O*-glycosidic
linkages may be labile under the cleavage conditions normally used. For
example, sialic acid and fucose are acid labile and may be cleaved during
the acidic reaction conditions employed for cyanogen bromide digestion.

A number of proteolytic enzymes are available that will cleave predictably and with reasonable specificity. The ones we most commonly employ
are trypsin (TPCK-treated), endoproteinase Lys-C (a Lys-specific enzyme), *Staphylococcus aureus* V8 protease (Staph. V8 or endoproteinase
Glu-C), and endoproteinase Asp-N.[14] Endoproteinase Asp-N cleaves selectively on the amino-terminal side of aspartic acid and cysteic acid, but
will also cleave on the amino-terminal side of glutamic acid with prolonged
incubation (S. Carr, G. Roberts, unpublished observation).

The molecular weights of the peptides released from the glycoprotein
are determined by FAB-MS and fitted to the known or deduced sequence
of the protein with the aid of computer programs (see [18] in this volume).
Glycopeptides are seldom observed by FAB-MS in the analyses of com-

[14] S. A. Carr, G. D. Roberts, and M. E. Hemling, *in* "Mass Spectrometry of Biological
Materials" (C. N. McEwen and B. Larsen, eds.), p. 87. Dekker, New York, 1990.
[15] S. A. Carr, R. J. Anderegg, and M. E. Hemling, *in* "The Analysis of Peptides and Proteins
by Mass Spectrometry" (C. J. McNeal, ed.), p. 95. Wiley, New York, 1988.
[16] R. J. Anderegg, S. A. Carr, I. Y. Huang, R. A. Hiipakka, C. Chang, and S. Liao,
Biochemistry **27,** 4214 (1988).
[17] S. A. Carr. M. E. Hemling, G. Folena-Wasserman, R. W. Sweet, K. Anumula, J. R. Barr,
M. J. Huddleston, and P. Taylor, *J. Biol. Chem.* **264,** 21286 (1989).

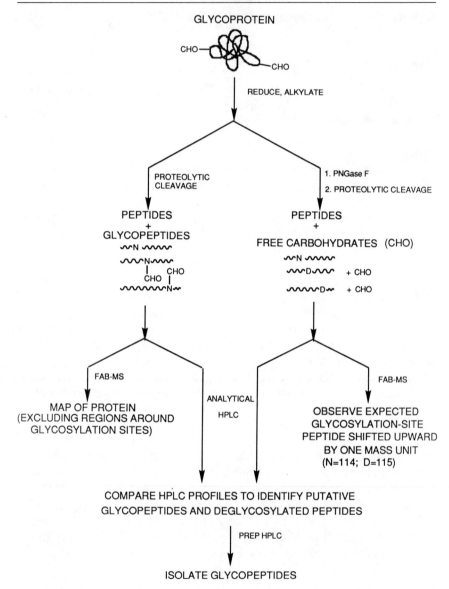

FIG. 1. Strategy for identifying attachment sites of Asn-linked oligosacharides in glycoproteins.

plex proteolytic digests of glycoproteins. The poor FAB-MS response of glycopeptides relative to the corresponding peptides is due to a number of factors including their greater mass, heterogeneity of the carbohydrate (which distributes the molecular ion signals among a number of species), and the hydrophilic character of the carbohydrate moiety which lowers the surface activity of the glycopeptide in the liquid matrix, and consequently, its detectability. Glycopeptides are usually only observed in highly simplified mixtures [such as those obtained following reversed-phase high-performance liquid chromatography (HPLC) separation, see below] and, even then, the glycopeptides may be difficult to detect unless the peptide is hydrophobic and the carbohydrate portion of the molecule is of relatively low molecular weight.

In order to detect peptides containing Asn-linked carbohydrate, the FAB-MS-derived peptide maps obtained before and after treatment of the glycoprotein with an endoglycosidase are compared (Fig. 1, right-hand side).[11–13] The most generally useful enzyme for this purpose is peptide N-glycosidase F (PNGase F; also known as N-glycanase and peptide-N^4-[N-acetyl-β-glucosaminyl]asparagine amidase) because it can be used to hydrolyze all of the commonly encountered N-linked oligosaccharides.[18–20] In contrast to endo-β-N-acetylglucosaminidase H and F (Endo H or Endo F),[20] PNGase F cleaves the β-aspartylglycosylamine linkage of all known types of Asn-linked sugars and converts the attachment site Asn to Asp.[19] An example of the appearance of the mass spectra is found in Fig. 2 for trypsin vs. PNGase/trypsin digests of reduced and alkylated soluble CD4 receptor glycoprotein.[17] New peaks appearing in the mass spectra after treatment with the glycosidase correspond to formerly glycosylated peptides. Because Asp weighs 1 Da more than Asn, the masses of these peptides are shifted upward by 1 Da compared to the mass calculated for the Asn-containing peptide. The technique will thus detect and locate Asn residues to which carbohydrate is attached independent of whether the Asn is present in the expected consensus sequence or not. In general, sequence coverage of the protein is also increased. Occasionally the signal for a former glycosylation site peptide may not be observed in the complex peptide mixture. These sites are detected by FAB-MS of simplified mixtures obtained by preparative fractionation by reversed-phase HPLC (Fig. 1), or by analysis of putative glycopeptide-containing fractions after deglycosylation (see below).

[18] T. H. Plummer, Jr., J. H. Elder, S. Alexander, A. W. Phelan, and A. L. Tarentino, *J. Biol. Chem.* **259,** 10700 (1984).

[19] A. L. Tarentino, C. M. Gomez, and T. H. Plummer, Jr., *Biochemistry* **24,** 4665 (1985).

[20] F. Maley, R. B. Trimble, A. L. Tarentino, and T. H. Plummer, Jr., *Anal. Biochem.* **180,** 195 (1989).

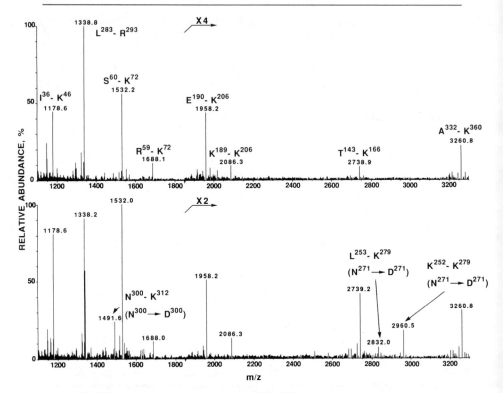

FIG. 2. FAB mass spectrum (mass range 1100–3300 displayed) of reduced and carboxymethylated soluble CD4 digested with (*top*) trypsin and (*bottom*) PNGase F, and then trypsin. Former glycosylation-site peptides are noted with arrows in the bottom panel.

In cases where signals for both the Asn- and Asp-containing forms of the peptides are observed in the FAB-MS data following PNGase F digestion, a semiquantitative estimate of the extent of glycosylation at that specific site is indicated by the ratio of the $(M + H)^+$ peaks. This analysis assumes that the relative ionization efficiencies of the Asp- and Asn-containing peptides are nearly equal. In cases where both peaks are present, serial dilutions of the sample should be analyzed by FAB-MS to ensure that one of the two forms of the peptide is not saturated in the matrix which could distort the relative peak intensities, and therefore the apparent concentrations.

Carbohydrate may be released from the glycoprotein at a number of points in the mapping strategy: (a) prior to reduction and alkylation, (b) after reduction and alkylation, but prior to proteolysis, and (c) after proteolysis. More complete release of carbohydrate using smaller amounts of glycosidase can be achieved if the glycoprotein is first denatured by reduction and alkylation of disulfides.[11,19,21] Proteolytic degradation to the glycopeptides also potentiates glycosidase activity provided that the carbohydrate is not attached to an Asn located at the N or C terminus of the peptide. These molecules are poor substrates for the enzyme.[19,20] However, when necessary, release can be affected by extended incubation using high concentrations of the enzyme,[17] or by using Endo H or F which do not have this limitation.

We have noted that commercially available PNGase F is contaminated with low levels of protease(s) with trypsin-like activity. This is usually not a problem since trypsin is often the enzyme of choice for carbohydrate or peptide mapping by FAB-MS. However, in cases where cleavage at Lys and Arg is undesirable, addition of protease inhibitors such as phenylmethylsulfonyl fluoride (PMSF), aprotinin, or leupeptin is recommended.[20] Chelating agents such as EDTA or 1,10-phenanthroline are also added to inhibit metalloproteases that have been found in preparations of PNGase F at trace levels.[20] Recombinant PNGase is now available which should be protease-free (Genzyme, Boston, MA).

The FAB-MS carbohydrate mapping strategy may be adapted for use with other glycosidases or chemical methods. The key requirements are that the enzyme or reagent be capable of efficient release of all types of carbohydrate of a given linkage and that the peptide be affected (if at all) in a predictable and reproducible manner. PNGase F appears to be ideally suited for use in the carbohydrate mapping strategy because it has very broad substrate specificity and it forms a mass spectrometrically unique derivative of the former attachment-site peptide. In contrast, Endo H and Endo F are not as broadly reactive. For example, neither are capable of releasing triantennary complex oligosaccharides.[20] For those glycoproteins, which are susceptible to digestion with Endo H or Endo F, a mass spectrometrically unique derivative of the attachment site is also produced.[11,22] The reaction leaves one pendent GlcNAc attached to the glycopeptide increasing its mass by 203 Da. Similarly, deglycosylation with trifluoromethanesulfonic acid[23] leaves one or two pendent N-acetylglucosamine residues on the attachment-site Asn which are readily detected by the FAB-MS mapping approach.[24]

[21] S. Hirani, R. J. Bernasconi, and J. R. Rasmussen, *Anal. Biochem.* **162**, 485 (1987).

[22] V. A. Reddy, R. S. Johnson, K. Biemann, R. S. Williams, F. D. Ziegler, R. B. Trimble, and F. Maley, *J. Biol. Chem.* **263**, 6978 (1988).

[23] A. S. B. Edge, C. R. Faltynek, L. Hof, L. E. Reichert, Jr., and P. Weber, *Anal. Biochem.* **118**, 131 (1981).

[24] R. J. Paxton, G. Mooser, H. Pande, T. D. Lee, and J. E. Shively, *Proc. Natl. Acad. Sci. U.S.A.* **84**, 920 (1987).

Procedure

Unless otherwise noted, only the highest purity reagents, solvents (generally glass-distilled, HPLC-grade), and enzymes ("sequencing-grade" when available) are employed. HPLC mobile phases as well as any solvent or buffer with which the sample has contact prior to analysis must be carefully chosen for the best compatibility with FAB-MS. In general, only volatile buffer systems should be used that may include acetic, formic, or trifluoroacetic acids, ammonium acetate, ammonium bicarbonate, borates, or pyridinium acetate. Nonvolatile buffers and additives such as sodium and potassium salts, phosphates, sulfates, citrates, and heptafluorobutyric acid should be avoided or removed by dialysis or chromatography. However, low millimolar concentrations of Tris-HCl and EDTA can be tolerated in FAB-MS.

The method describes the procedure as applied to soluble CD4 receptor, a 45-kDa glycoprotein containing two sites of glycosylation.[17]

Reduction and S-Carboxymethylation. Protein (1–5 mg) is reduced in 0.5 M Tris-HCl, pH 8.3, containing 2 mM EDTA and 6 M guanidine-HCl with dithiothreitol in a 50-fold molar excess over cysteinyl residues. The mixture is purged with argon (99.998%, Liquid Carbonic, Chicago, IL) which has been passed through an oxygen trap, and reacted at 50° for 4 hr under argon. After reduction, the protein solution is cooled to room temperature, and an aqueous solution of iodoacetic acid (argon-purged and pH adjusted to 8.2 with 5% NH$_4$OH) is added in a twofold molar excess over dithiothreitol. The pH of the mixture is checked and adjusted to 8.2 to 8.3, if necessary, and the reaction allowed to proceed in the dark at 37° for 1 hr. The sample is then dialyzed against 3.5 liter of 50 mM NH$_4$HCO$_3$ for 15 hr at 4°. When small amounts (≤1 mg) of low-molecular weight proteins (≤35 K) are reduced and alkylated, it is recommended that isolation of the protein be done by reversed-phase HPLC on a C$_4$ column (Vydac, 2.1 or 4.6 mm × 100 mm, 300 Å pore, 5 μm packing) using an acetonitrile/water [0.1% (v/v) trifluoroacetic acid] gradient. After loading, the gradient is held at 10 to 15% acetonitrile for 15 min to desalt. The gradient is then ramped to 80% acetonitrile in 10 min to elute the protein.

Selective Fragmentation of Proteins and Glycoproteins. For tryptic digests, reduced and alkylated protein is dissolved in 50 mM NH$_4$HCO$_3$ buffer at pH 8.5. An aliquot of trypsin (TPCK-treated; Worthington, Freehold, NJ) dissolved in the same buffer is added to the protein solution to give an initial enzyme : substrate ratio of 1 : 100. After 3 hr, a second aliquot of enzyme is added bringing the enzyme : substrate ratio to 1 : 50. Trypsin digests are incubated at 37° for 6 to 12 hr. Digests with *S. aureus* V8 protease (endoproteinase Glu-C, sequencing-grade, Boehringer-Mannheim, Indianapolis, IN) and endoproteinase Asp-N (sequencing-grade,

Boehringer-Mannheim) are performed in a similar manner at pH 7.8, 37°, for 12 to 24 hr. Digestions are stopped by lowering the pH to between 5 and 6 with 5% acetic acid. Aliquots of the digests are examined directly by FAB-MS or after concentration on a vacuum centrifuge (Savant, Farmingdale, NY).

Digestion with Glycosidases. Reduced and alkylated glycoprotein is dissolved with the aid of ultrasonication in 50–250 μl of 100 mM NH$_4$-HCO$_3$, 0.005% EDTA (free acid), and the pH adjusted to 8.5 with 5% NH$_4$OH. Approximately one unit (4 μl of a 0.25 units/μl solution of enzyme in glycerol) of PNGase F (Genzyme, Boston, MA) in 50 mM ammonium bicarbonate buffer at pH 8.5 is added, and the vial argon-flushed and shaken at 37° for 14 to 24 hr. If an enzymatic digest of the glycoprotein is being deglycosylated, protease inhibitors are added. Typically we add 1 μl of a freshly prepared solution (0.01 μg/μl) of aprotinin, a trypsin inhibitor from bovine lung (Boehringer-Mannheim or Serva Biochemicals, Westbury, NY) or a cocktail of leupeptin and PMSF at a final concentration of 1 μM and 200 μM, respectively. After ca. 6 hr, additional aliquots of PNGase F (1–2 μl) and protease inhibitor are added. The digests are stopped by addition of 10 μl of 5% acetic acid. Putative glycopeptide-containing HPLC fractions (containing 0.1–10 nmol of peptides and glycopeptides) are digested with PNGase F in the same manner, but only 0.25–0.5 units of enzyme is used and protease inhibitors are not added. Addition of detergents such as sodium dodecyl sulfate (SDS) should be avoided. SDS is not essential for deglycosylation of reduced and alkylated or proteolyzed glycoproteins with PNGase F, and it can be detrimental to the mass spectrometry experiments at even very low levels.

Fast Atom Bombardment (FAB) Mass Spectrometry. FAB mass spectra are obtained using either (a) the first double-focusing portion (MS-1) of a VG ZAB SE-4F tandem magnetic deflection mass spectrometer (accelerating voltage 10 kV, mass range 12,500) equipped with a standard-flow FAB ion source and high-voltage Cs ion gun or (b) a VG ZAB-HF magnetic deflection mass spectrometer (accelerating voltage 8 kV, mass range 3000) equipped with a standard flow FAB ion source and fast atom gun. For highest sensitivity at masses above 3000 Da, the ZAB SE equipped with the Cs ion gun is used. The high-voltage Cs gun is operated at 35 kV with an emission of 2 to 4 μA, while the FAB gun is typically operated at 8 kV and a discharge current of 1 mA. A VG 11-250J data system is used to acquire and process all data. Approximately 1–3 μl of an enzymatic digest containing ca. 0.1–1 nmol of digested protein is dispersed on the stainless steel target in a matrix of monothioglycerol (3-mercapto-1,2-propanediol, Sigma, St. Louis, MO) except as noted. Samples are initially examined using a low-resolution ($R \sim 800$), wide mass range "survey scan." For masses above ca. m/z 1200, in which the molecular ion cluster is unresolved, a mass value obtained in this manner corresponds to the chemical average rather than the monoisotopic mass. Each sample is analyzed in separate experiments over two or more overlapping mass ranges, for example (1) m/z 140–1800 at 40 sec/decade ($R \sim 2000$), (2) m/z 850–4100 at 40 sec/decade ($R \sim 800$ or 3000), and (3) m/z 2800–5650 at 80 sec/decade ($R \sim 800$).

Identification and Preparative Fractionation of Glycopeptides

Principle

Potential glycopeptides are identified by comparing the reversed-phase HPLC profiles of the digests obtained prior to and after digestion with the glycosidase.[25,26] An example of this is shown in Fig. 3 for the digests of soluble CD4 described above and illustrated in Fig. 2. Peaks which disappear or are greatly attenuated in the chromatogram of the sample after digestion with the glycosidase are likely to be glycopeptides with carbohydrate linked to Asn. New peaks appearing after deglycosylation correspond to former glycosylation-site peptides. Putative glycopeptide- and deglycosylated peptide-containing fractions (which also will contain coeluting peptides) are preparatively isolated from the digests for analysis by FAB-MS (see below).

Procedure

Reversed-Phase HPLC. Peptide mixtures derived from digests of 1 to 30 nmol of protein are analyzed and fractionated by reversed-phase HPLC on a Beckman System Gold equipped with a model 126 programmable solvent module and model 166 variable wavelength detector (Beckman Instruments, Inc., San Ramon, CA). An IBM-PC AT computer is used to control the system and to store and manipulate data via the System Gold software. Reversed-phase HPLC fractionation is done on fully end-capped Vydac or Bakerbond wide-pore C_{18} or C_8 columns (25 cm × 4.6 mm id, 5 μm, 300 Å pore size) with a Brownlee RP 300 guard column (30 × 4.6 mm, 7 μm, 300 Å pore size) using gradient elution at 1 ml/min. Smaller amounts of protein digests (<1 nmol) are chromatographed on 2.1 mm id columns (Vydac or Brownlee C_{18}) using a Brownlee MicroGradient HPLC system equipped with syringe pumps and a 200-μl internal volume dynamic mixer (Applied Biosystems, Foster City, CA). Solvent A is 0.1% (v/v)

[25] S. A. Carr, and G. D. Roberts, in "Methods in Protein Sequence Analysis 1986" (K. A. Walsh, ed.), p. 423. Humana, Clifton, NJ, 1987.
[26] J. J. L'Italien, *J. Chromatogr.* **359**, 213 (1986).

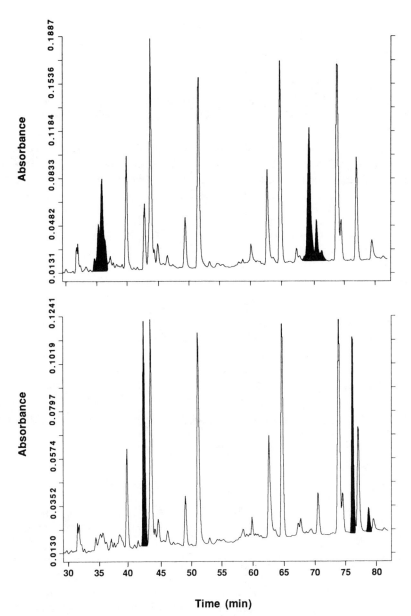

FIG. 3. HPLC trace of 2 nmol of trypsin-digested, reduced, and carboxymethylated (RCM)-soluble CD4 (*top*) for comparison with the HPLC trace of the trypsin digest of PNGase F-treated RCM-soluble CD4 (*bottom*). Only the regions of the chromatograms that exhibited peak shifts are shown. Mobile phases were as described in the text. Gradient: 5 min hold at 5% B; 5% B to 18% B in 18 min; 18% B to 45% B in 90 min; 45% B to 90% B in 32 min. Peaks containing glycopeptides (*top*) and former glycopeptides (*bottom*) are filled in.

aqueous trifluoroacetic acid; solvent B is 90% acetonitrile in water (v/v) containing 0.1% trifluoroacetic acid (v/v). A typical gradient is as follows: 0% B to 60% B in 90 to 150 min, then 60% B to 95% B in ca. 30 min. Detection is accomplished by UV absorbance at 215 nm. Chromatograms obtained before and after digestion with a glycosidase are compared on a light box to identify glycopeptides. One-minute fractions are collected in preparative analyses using a Gilson model 203 fraction collector (Gilson, Middleton, WI). One or two fractions to either side of the glycopeptide-containing fraction are pooled with the principal component to ensure that minor glycoforms from that site are not discarded.

Structural Classification of Oligosaccharides from Specific Attachment Sites

Principle

This part of the methodology is shown schematically in Fig. 4. HPLC fractions containing glycopeptides are analyzed by FAB-MS over a mass range of ca. 1800 to 6000 Da. As noted above, glycopeptides may be observed in favorable cases. Heterogeneity in the carbohydrate moiety provides a characteristic signature of peaks separated by 146, 162, 203, or 291 Da indicating the addition or subtraction of units of deoxyhexose (dHex), hexose (Hex), N-acetylhexosamine (HexNAc), and N-acetylneuraminic acid (NeuAc), respectively (Table I). The carbohydrate composition in terms of these sugar units for each of the glycopeptide molecular ions is defined by the in-chain mass of the carbohydrate (calculated as described below).

The identity of the former glycosylation-site peptide in the particular HPLC fraction is determined by FAB-MS following release of the carbohydrate by PNGase F as described above. A portion of the mixture of peptides and oligosaccharides is subsequently permethylated,[27] and aliquots are analyzed by methylation analysis[28,29] and FAB-MS (Fig. 4, bottom, left-hand side).[12,13] Free oligosaccharides may be permethylated in the presence of peptides using the method of Ciucanu and Kerek.[27] Oligosaccharides that contain sialic acid often fail to completely form the methyl ester of the acid moiety. Treatment with diazomethane usually completes this transformation and yields the desired permethylated species. Permethylated oligosaccharides are easily separated from peptides and salts via a

[27] I. Ciucanu and F. Kerek, *Carbohydr. Res.* **131,** 209 (1984).
[28] R. Geyer, H. Geyer, S. Kuhnhardt, W. Mink, and S. Stirm, *Anal. Biochem.* **133,** 197 (1983).
[29] S. B. Levery and S. Hakomori, this series, Vol. 138, p. 13.

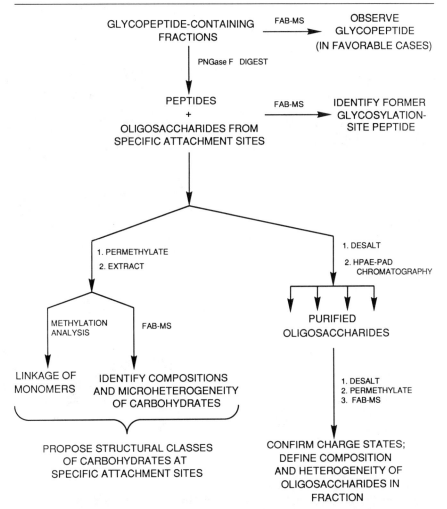

Fig. 4. Strategy for structural classification of oligosaccharides at specific attachment sites.

water–chloroform partition. Methylation analysis provides the linkages of the sugar units, while FAB-MS of the permethylated oligosaccharides provides compositions, molecular heterogeneity, and limited sequence information (as described in [35], this volume) for the carbohydrates from the specific attachment site. From this information, the structural classes (oligomannose, hybrid, or complex) and branching (biantennary, etc.) at the given site may be proposed. The relative intensities of the parent ion

TABLE I

MASSES FOR CALCULATION OF IN-CHAIN AND MOLECULAR MASSES OF COMMON
Asn-LINKED CARBOHYDRATES[a]

Sugars	In-chain mass			
	Underivatized		Permethylated	
	Monoisotopic	Chemical average[b]	Monoisotopic	Chemical average[b]
Deoxyhexose (dHex)	146.0579	146.144	174.0892	174.198
Hexose (Hex)	162.0528	162.143	204.0998	204.224
Hexosamine (HexN)	161.0688	161.158		
N-Acetylhexosamine (HexNAc)	203.0794	203.196	245.1263	245.277
N-Acetylneuraminic acid (NeuAc)	291.0954	291.259	361.1737[c]	361.394[c]
Hex-HexNAc	365.1322	365.339	449.2261	449.501
NeuAc-Hex-HexNAc	656.2276	656.598	810.3998[c]	810.896[c]
(Hex)$_3$-HexNAc-HexNAc	892.3172	892.821	1102.5520	1103.224
(Hex)$_3$-HexNAc-(dHex)HexNAc	1038.3751	1038.965	1276.6412	1277.422

End groups[d]	Underivatized		Permethylated	
	Monoisotopic	Chemical average[b]	Monoisotopic	Chemical average[b]
H + OH	18.0106	18.015		
CH$_3$ + OCH$_3$			46.0419	46.069

[a] Chemical average and monoisotopic masses for underivatized and permethylated forms are shown.
[b] Calculated using H = 1.0080, C = 12.0110, N = 14.0067, O = 15.9994.
[c] Carboxylic acid as methyl ester.
[d] End group mass + in-chain mass = molecular mass.

signals of the permethylated oligosaccharides provide a semiquantitative estimate of the relative amounts of each glycoform.

The underivatized carbohydrates from each given attachment site are also preparatively fractionated by high-performance anion-exchange chromatography (HPAE) using pellicular quaternary amine-bonded resins at high pH, and detected using a pulsed amperometric detector (Fig. 4, bottom, right-hand side and Fig. 5B).[30] This separation allows the microheterogeneity at each attachment site to be better defined. Under the chromatographic conditions used, carbohydrates are separated according to charge state (which is governed primarily by the number of NeuAc

[30] M. R. Hardy and R. R. Townsend, *Proc. Natl. Acad. Sci. U.S.A.* **85,** 3289 (1988).

residues), size, and other sources of heterogeneity including branching and linkage.[30] General rules governing the elution order of complex carbohydrates under the above conditions have been defined.[31]

Purified oligosaccharide fractions are desalted, permethylated, and analyzed by FAB-MS.[13,17] These data confirm the charge state and define the composition and heterogeneity of the oligosaccharides in the fraction as described above (for example, see Fig. 5c). In addition, the FAB-MS spectra of the purified fractions often exhibit more structurally informative fragmentation than the spectra of the complex mixture of carbohydrates. Methylation analysis of each fraction is useful for corroborating linkages assigned previously by analysis of the mixture of oligomers derived from a specific attachment site.

Procedure

FAB-MS of Reversed-Phase High-Performance Liquid Chromatography Fractions. Aliquots of the reversed-phase HPLC fractions before and after digestion with PNGase F are analyzed by FAB-MS as described above. Fractions obtained by reversed-phase HPLC are either sampled directly for FAB-MS or concentrated first to ca. 50–100 μl using a Savant vacuum centrifuge (no heating) following addition of ca. 100 μl of *n*-propanol to prevent peptides from precipitating during concentration. Samples that were taken to dryness are generally redissolved in 20–150 μl of 20% acetonitrile in water (v/v) containing 0.1% trifluoroacetic acid (v/v).

Calculation of Masses of Carbohydrates. The in-chain mass of the carbohydrate associated with a specific signal is calculated as follows: $(M + H)^+$ of glycopeptide $-$ $(M + H)^+$ of former glycosylation site peptide (Asn converted to Asp) = in-chain mass of the carbohydrate. The molecular weights of the permethylated carbohydrates are calculated by summing the in-chain masses of the permethyl monosaccharides and the masses of the methyl end groups (Table I). For ease of calculation, the monosaccharides and their in-chain masses (Table I) are entered in the table of elements in the elemental composition program on the VG data system.

Permethylation of Oligosaccharides. An aliquot (100 pmol–10 nmol) of PNGase F-treated glycopeptide is placed in a 1 ml thick-walled glass reaction vial (which has been silylated with dichlorodimethylsilane) and

[31] J. D. Olechno, S. R. Carter, W. T. Edwards, D. G. Gillen, R. R. Rownsend, Y. C. Lee, and M. R. Hardy, *in* "Techniques in Protein Chemistry" (T. E. Hugli, ed.), p. 364. Academic Press, San Diego, 1989.

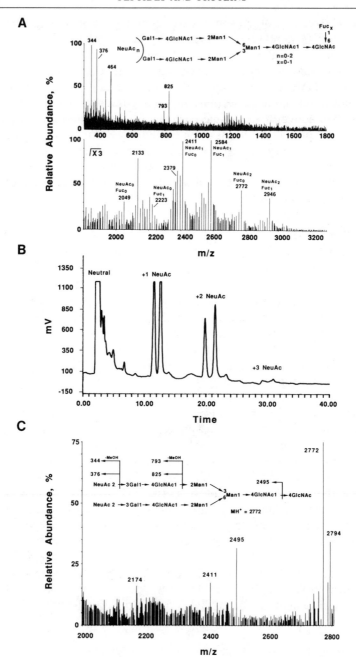

concentrated to dryness under reduced pressure. The resulting residue is then redissolved in 100 μl of DMSO. Argon is bubbled through the solution for ~5 min and 8.0 mg (200 μmol) of sodium hydroxide and a magnetic stir bar are added to the system. Methyl iodide (25 μl, 400 μmol) is then added to the reaction mixure and the resulting solution is stirred under argon for 20 to 30 min. The reaction is then quenched by the addition of 200 μl of 30% acetic acid and extracted with three portions of 200 μl of chloroform. The organic extracts are combined, washed with two portions of 200 μl of 30% acetic acid, and concentrated to dryness under reduced pressure on a vacuum centrifuge.

If sugars containing carboxylic acid moieties are known or suspected to be present, the methylated oligosaccharide is redissolved in 200 μl of dichloromethane and treated with diazomethane (ca. 6 mM in 2 : 1 ether/ethanol) dropwise until a yellow color persists. The resulting solution is then stirred at room temperature for 15 min. A second aliquot (2–3 drops) of the diazomethane solution is then added and the reaction mixture allowed to stir for an additional 10–15 min. This solution is then concentrated to dryness under reduced pressure.

High-Performance Anion-Exchange Chromatography. Glycopeptide-containing reversed-phase HPLC fractions (1–50 nmol) are treated with PNGase F (see above) to release the carbohydrate portion and the reaction mixture is concentrated to dryness under reduced pressure. The sample is desalted by dissolving the residue in water and applying it to a 2 ml column of AG 50W-2X (Bio-Rad, Richmond, CA; H$^+$ form, 200–400 mesh, 0.6 mEq of salt/ml of wet resin). The oligosaccharides are eluted with 5 column volumes of water and lyophilized. Fractionation of the mixture of carbohydrates is accomplished by HPAE on a Dionex Bio-LC (Dionex, Sunnyvale, CA). This system consists of a gradient pump and a gold working electrode for pulsed amperometric detection. A computer inter-

---

Fig. 5. (A) FAB mass spectrum of ca. 200 pmol of the permethylated carbohydrates released from Asn[271] attachment site by PGNase F digestion of glycopeptide fractions 69–72 from the tryptic digest of reduced and carboxymethylated soluble CD4 [Fig. 3 (*top*)]. The structures shown (biantennary oligosaccharides with varying numbers of NeuAc, some with fucose attached to the reducing-end GlcNAc[17]) are derived from the combination of the FAB-MS data and methylation analysis data (see Fig. 4 and text). (B) Preparative fractionation of ca. 10 nmol of PNGase F-released oligosaccharides from soluble CD4 Asn[271] site using HPAE with pulsed amperometric detection (see text for conditions). Within each charge group the fucosylated oligosaccharide elutes earlier than the nonfucosylated analog. (C) FAB mass spectrum of the peak eluting at 21.5 min in the HPAE chromatogram shown in (B), above. The fraction was desalted and permethylated prior to analysis by FAB-MS. The fucosylated analog (eluting at 20.0 min) also exhibited the fragment at m/z 2495, indicating that fucose is attached to the terminal GlcNAc.

face with A1450 software is employed on a IBM-compatible PC for data collection and handling. A polymeric pellicular anion-exchange column (HIPC-AS6 [Dionex], 4.6 × 250 mm) is used in all experiments. Samples are injected using a Spectra Physics SP8780 autosampler (Spectra Physics, San Jose, CA) equipped with a Tefzel rotor seal in a Rheodyne injection valve (Rheodyne, Cotati, CA). Chromatographic separations of the oligosaccharides are performed at 1.0 ml/min using the following gradient: isocratic at 175 mM NaOH and 44 mM NaOAc for 2 min followed by a linear gradient to 200 mM NaOAc and 175 mM NaOH at 55 min. The pH of the NaOAc solution is 5.5. Detection is achieved with a pulsed amperometric electrochemical detector with a gold working electrode at the following potentials; $E_1 = 0.01$ V ($t_1 = 0.3$ sec), $E_2 = 0.7$ V ($t_2 = 0.12$ sec), $E_3 = -0.3$ V ($t_3 = 0.3$ sec). A response time of 3 sec is used.[32]

The carbohydrate-containing fractions are then generally desalted on AG 50W-2X (Bio-Rad) using 10 ml of resin per milliliter of collected fraction. After drying under reduced pressure, the fractions are permethylated as described above. Samples desalted prior to permethylation are easier to handle because smaller amounts of insoluble residue are present during the extraction step. However, fractions from the HPAE system can be concentrated to dryness under reduced pressure and directly permethylated employing the same system as described above. The reaction mixture is then acidifid with 30% acetic acid and the permethylated oligosaccharides are separated from salts with a water–chloroform partition.

FAB-MS of Permethylated Carbohydrates. The FAB-MS spectra of the purified permethylated oligosaccharides are recorded in a similar fashion as described above with the following exceptions. A matrix of 1:1 thioglycerol/m-nitrobenzyl alcohol plus 1% trifluoroacetic acid is employed. This matrix appears to afford the greatest sensitivity for the permethylated oligosaccharides. It is important to note that sample-related ions are often not detected in the first 1–3 min of bombardment using this matrix. In general, ~100 pmol of permethylated oligosaccharide is sufficient to obtain analytically useful mass spectra; in favorable cases as little as 1 pmol has afforded useful data. Oligosaccharides lacking sialic acid residues generally can be analyzed using ~tenfold less sample.

Acknowledgment

This work was supported in part by a grant from the National Institutes of Health, GM-39526 to S. A. Carr.

[32] K. R. Anumula and P. B. Taylor, *Glyconjugate J.* **6**, 414 (1989).

[28] Characterization of Recombinant Proteins

By HUBERT A. SCOBLE and STEPHEN A. MARTIN

Introduction

The therapeutic utility of a recombinant protein is profoundly influenced by its macromolecular structure. The structural features which influence biological response may vary widely depending on the cell type used as the host expression system and the conditions under which the protein is expressed. Variation in biological response (*in vitro* or *in vivo*) may be attributed to the primary structure of the protein or to posttranslational modification or processing by the host expression system. Indeed, even within a specific cell line, a number of physiological and nonphysiological conditions influence the extent and type of posttranslational processing, and, hence, the therapeutic utility of the expressed protein.

Similarly, the fermentation or cell culture conditions and the purification process can influence the structure of the final product. Depending on the specific conditions of fermentation or cell culture and the method of cell harvesting, the protein may be exposed to a number of endogenous endo- or exoproteases or glycosidases. In addition, if the purification process includes affinity steps or other procedures of high selectivity, protein molecules with certain structural features may be excluded.

In order to better understand the functional implications of structural variation, it is important to assess in detail the structural features of recombinant proteins. These structural features include both host-specific modifications (e.g., glycosylation, acylation, and phosphorylation) and those that may be induced during expression or purification (e.g., oxidation, deamidation, and N- and C-terminal heterogeneity). The following strategy is designed to assess many of the structural questions that arise in the expression and purification of a recombinant protein.

General Strategy

The general strategy which has been adopted for the structural characterization of recombinant proteins is primarily a mass spectrometric-based approach with additional information from ancillary techniques. This strategy is shown in Scheme 1. The first steps in this approach involve specific proteolytic digestion of the protein followed by reversed-phase (RP) high-performance liquid chromatography (HPLC) of the resulting digest. Depending on the complexity of the digest, each RP-HPLC fraction may

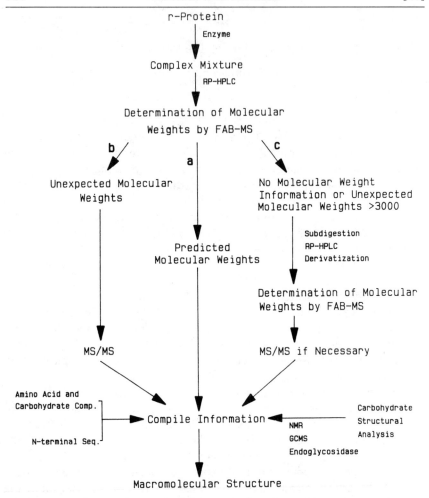

SCHEME 1. Strategy for structural characterization of recombinant glycoproteins.

contain many components, even if the chromatographic peak appears symmetrical. The collected fractions are next analyzed by fast atom bombardment mass spectrometry (FAB-MS).[1] This technique is advantageous in that mixtures of peptides can be analyzed in a single mass spectrometric experiment provided that they have different molecular masses, which is usually the case. The experimentally determined molecular masses are compared to those expected based on the specificity of the digest and the

[1] M. Barber, R. S. Bordoli, and A. N. Tyler, *J. Chem. Soc. Chem. Commun.* **7,** 325 (1981).

cDNA sequence. Following mass spectrometric analysis of the chromato-graphic fractions, a number of situations can result as outlined in the scheme. In the first situation (a), all experimentally determined molecular masses agree with those predicted. This is rarely the case due to posttrans-lational modification of the protein and to the occurrence of secondary cleavages from even the most specific proteases. The second situation (b) involves ions that were not expected and have molecular masses less than 3000 Da. In these situations the protonated molecules are subjected to MS/MS.[2–4] Depending upon the extent of sequence information derived from the collision-induced dissociation (CID) spectrum, the sample may be treated as outlined in the final pathway (c). Samples analyzed via this route are those having unexpected molecular masses in excess of 3000 Da or those samples exhibiting good absorbance at 214 nm, yet yielding poor or no molecular mass information. In such cases, samples are subdigested with a protease of differing specificity. This involves some knowledge of other characteristics of the peptide, such as the retention time of the peptide under reversed-phase conditions, the absorbance of the peptide at different wavelengths, N-terminal sequence, amino acid composition data, or from inference based on information one has gathered on other peptides in the digest. Following subdigestion and chromatographic sepa-ration, the samples are reexamined mass spectrometrically. Molecular masses determined at this step are correlated with peptides predicted from the cDNA sequence based on the specificity of the two proteases used for digestion. Peptides which remain unidentified at this step are analyzed by MS/MS with or without selective chemical derivatization.

The molecular mass and primary sequence information obtained in the various mass spectrometric experiments is then combined with the structural information generated using other techniques (amino acid and carbohydrate composition, N-terminal sequence, etc.). Any sites which are identified as locations of glycosylation are subjected to carbohydrate structural analysis. Finally, all protein and carbohydrate structural infor-mation is combined and interpreted to assemble the macromolecular struc-ture of the recombinant protein.

Proteolytic Digestion

As in the case of structural characterization using conventional protein chemistry, the mass spectrometric approach centers on the use of specific endoproteases for the degradation of the intact protein. Selection of an

[2] K. Biemann and H. A. Scoble, *Science* **237**, 992 (1987).
[3] K. Biemann, *Anal. Chem.* **58**, 1288A (1986).
[4] K. Biemann, this volume [25].

appropriate protease is largely dependent on the expected molecular mass of the resulting peptides and the specificity of the enzyme. In two- and four-sector mass spectrometry, the ion yield associated with peptides tends to decrease as a function of increasing molecular mass. Therefore, in order to achieve maximum sensitivity in the two-sector mode and primary structure information in the four-sector mode, peptides with molecular masses less than 3500 Da are desirable. Due to the natural frequency of lysine in most proteins and the availability of specific proteases, cleavage at this site results in peptides with molecular masses that, with few exceptions, fall below this value. Furthermore, by limiting the molecular mass of the fragments, this approach, in principle, can be used with most types of mass spectrometers.

Depending upon the protein under investigation and the information desired, proteolytic digestion is usually preceded by reduction and alkylation of cysteine sulfhydryl groups. A number of thiol alkylating reagents are available, the most common of which include: iodoacetic acid, iodoacetamide, and 4-vinylpyridine (see [19] in this volume). The use of 4-vinylpyridine has the advantage of introducing an additional chromophore into peptides containing cysteine. Monitoring the reversed-phase chromatogram at 254 nm may help in identifying those peptides containing this amino acid.

Reversed-Phase High-Performance Liquid Chromatography (RP-HPLC)

Under most conditions, peptides or proteolytic digests are chromatographed under reversed-phase conditions using eluents consisting of water/trifluoroacetic acid and acetonitrile/trifluoroacetic acid. This solvent system minimizes the contribution of sodium and potassium to the protonated molecule signal observed in FAB-MS. In some cases, it may be necessary to use other chromatographic modes (reversed-phase with phosphate or hexafluorobutyric acid-containing mobile phases; affinity, ion-exchange, or size-exclusion chromatography, etc.) to isolate certain peptides; however, isolated peptides are usually rechromatographed using acetonitrile/trifluoroacetic acid mobile eluents.

Typically, conventional analytical columns (4.6 mm id) are used for the separation of peptides in proteolytic digests, although when sample size is limited or a minimal volume is required, narrow-bore columns (2 mm id) can be used with little modification to the chromatographic hardware.[5]

---

[5] K. L. Stone, J. I. Elliott, G. Peterson, W. McMurray, and K. R. Williams, this volume [21].

Following collection directly into plastic microvials, samples are vacuum concentrated to near dryness. At this point, an aqueous solution of glycerol [typically 10% (v/v) in water] is added to the microvial, vortexed vigorously, and vacuum centrifuged to remove the remainder of the solvent. The glycerol solution is not added during the initial vacuum concentration since glycerol/peptide esters are easily formed in the acidic solution. This complicates the preliminary interpretation of the mass spectra of the mixtures and reduces the secondary ion yields of the protonated molecule. The final sample/matrix volume is a few microliters which is sufficient for two to three sample loadings for mass spectrometry. Care must be taken to ensure that the peptide dissolves in the glycerol matrix, either at this stage or upon the addition of matrix modifiers. Typical modifiers for FAB-MS of peptides include acetic acid, thioglycerol, dithiothreitol/dithioerythritol (DTT/DTE), 3-nitrobenzyl alcohol, etc.[6]

Mass Spectrometry

As described in Scheme I, mass spectrometry plays a central role in the structural characterization of recombinant proteins. Although, in principle, these analyses can be conducted on a variety of mass spectrometers, the work discussed in this chapter was carried out on a JEOL HX110/HX110 tandem double-focusing mass spectrometer of *EBEB* geometry that operates at an accelerating voltage of 10 kV.[7] The first objective of mass spectrometric characterization of a recombinant protein is to identify as many peptides as possible from the specific proteolytic digestion. In an effort to search for the occurrence of minor peptides due to proteolysis or processing events, mass spectrometric scans are conducted over the mass range extending from 100 to 6000 Da with a resolution of 1 : 1000. Further enhancement in sensitivity can be accomplished by using a Cs^+ ion gun operated between 25 and 35 kV. Ion signals that are weak are signal-averaged prior to determination of the average molecular weight, whereas those species with abundant secondary ion signals may also be scanned at unit resolution, yielding both an average and monoisotopic molecular weight. Since it is well documented that there is preferential detection of certain peptides from mixtures,[8,9] it is routine to add another matrix modifier, such as 3-nitrobenzyl alcohol, to the probe tip and to acquire another

[6] E. DePauw, this volume [8].
[7] K. Sato, T. Asada, M. Ishihara, F. Kunihiro, Y. Kammei, E. Kubota, C. E. Costello, S. A. Martin, H. A. Scoble, and K. Biemann, *Anal. Chem.* **59**, 1652 (1987).
[8] S. Naylor, A. F. Findeis, B. W. Gibson, and D. H. Williams, *J. Am. Chem. Soc.* **108**, 6359 (1986).
[9] S. Naylor, G. Moneti, and S. Guyan, *Biomed. Environ. Mass Spectrom.* **17**, 393 (1988).

mass spectrum. This frequently alters the relative ratio of the molecular ions enhancing those which were weak or absent from the mass spectrum without the modifier.

In those cases in which primary structural information is required (pathways b, c in Scheme I), the molecule of interest is transmitted with unit resolution into the collision region between the two mass spectrometers. The molecular ion is subjected to high-energy collisions[10] with helium producing fragment ions characteristic of the primary sequence of the peptide.[2-4] These data are combined with structural information generated from other techniques to arrive at a structure from the recombinant protein.

The following examples demonstrate the utility of this strategy for the characterization of recombinant proteins.

Characterization of Recombinant Human Erythropoietin

Characterization of Posttranslational Modifications and Processing Events

Erythropoietin (rEPO) is a glycoprotein hormone which stimulates proliferation and differentiation of erythroid precursor cells to mature erythrocytes.[11] The structure of human EPO produced by transfection of Chinese hamster ovary (CHO) cells with a cDNA has been investigated in an attempt to understand erythropoiesis and to investigate its use as a therapeutic agent. Based upon preliminary N-terminal sequence data, the protein was known to be N-terminally processed via the removal of a hydrophobic 27-amino acid signal sequence. In addition, the molecule has three consensus sites (Asn-X-Ser/Thr) for N-linked glycosylation.

Digestion of the reduced and alkylated rEPO with a lysine-specific protease should generate nine peptides ranging in length from 2 to 45 amino acids. These expected peptides, labeled K1 through K9, are shown in Fig. 1. As outlined in the generalized strategy, the choice of enzyme is dictated primarily by the mass of peptides that are generated and by the specificity and effectiveness of a particular enzyme in degrading the protein of interest. Figure 2 shows the reversed-phase chromatogram that results when CHO-derived rEPO is degraded with the lysine-specific protease endoproteinase lysine-C (Endo Lys-C). The chromatogram is characterized by eight major peaks and several minor peaks. Although there are

[10] S. A. Martin, R. S. Johnson, C. E. Costello, and K. Biemann, in "Analysis of Peptides and Proteins by Mass Spectrometry" (C. J. McNeal, ed.), p. 135. Wiley, New York, 1988.
[11] M. A. Recny, H. A. Scoble, and Y. Kim, J. Biol. Chem. 262, 17156 (1987).

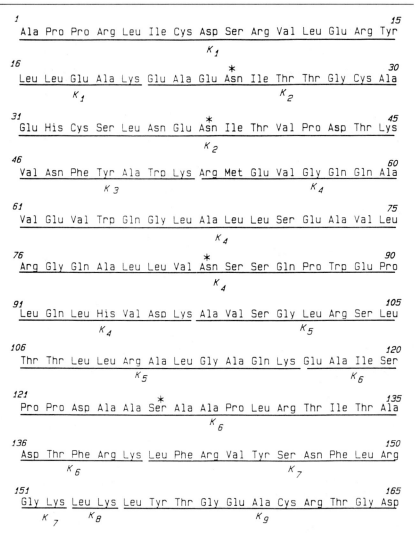

Fig. 1. Amino acid sequence of mature rEPO. Underlined peptides are those expected based on digestion of the glycoprotein with the lysine-specific protease, endoprotease lysine-C. Sites of N- and O-linked carbohydrate are marked with an asterisk. The cDNA sequence coded for an additional arginine at the C terminus of the protein (see text).

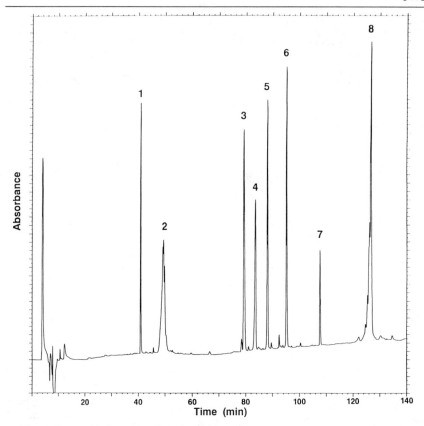

FIG. 2. Reversed-phase HPLC analyses of peptides resulting from endoprotease lysine-C digestion of reduced and pyridylethylated rEPO. Following digestion, peptides were separated on a Bio-Rad C_{18} column (HiPore RP318, 4.6 × 250 mm). Proteolytic digests were eluted in a series of linear gradients from solvent A (0.1% trifluoroacetic acid/water) to solvent B (0.1% trifluoroacetic acid/95% acetonitrile:5% water). Chromatographic profile is shown at 214 nm. Peaks are identified as follows: peak 1, K9; peak 2, K2; peak 3, K3; peak 4, K6; peak 5, K7; peak 6, K1; peak 7, K5; peak 8, K4.

eight peaks whereas nine were expected, the map is not very different from that expected, since one of the peptides that should result from the digest, K8, is a small dipeptide which probably elutes in the void volume.

Although the protein was known to be N-terminally processed, little was known concerning possible C-terminal processing. Preliminary amino acid composition data on the intact protein, indicated a lower recovery of arginine than was predicted based upon the cDNA sequence, which coded for an arginine at the C terminus of the glycoprotein. Analysis of peak 1 from the chromatogram yielded a $(M + H)^+$ ion at m/z 1290.6, which was

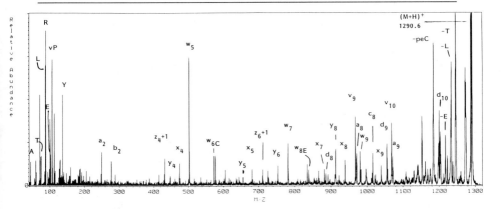

FIG. 3. FAB-MS/MS spectrum of the C-terminally processed peptide from an endopro-tease lysine-C digest of the mature rEPO (peak 1, Fig. 2). The molecular ion, $(M + H)^+$ at m/z 1290.6, and product ions confirm that the C terminus of the protein is LYTGEACRTGD. The peaks labeled w_{6C} and w_{8E} refer to w ions formed via the loss of the side chain of the amino acid denoted in the subscript.[4] vP, Vinylpyridine; peC, pyridylethylcysteine.

not expected based on the specificity of the digest. This peptide was further analyzed in an MS/MS experiment using a four-sector mass spectrometer. The MS/MS spectrum of this ion (Fig. 3) reveals a complex series of fragment ion types from both the N and C termini which is characteristic of a peptide without a basic amino acid at the C terminus to direct fragmentation. Careful examination of the spectrum reveals a series of C-terminal ions (v, w, x, y, z) extending from positions 4 (Arg) to 10 (Tyr) numbering from the C terminus and N-terminal ions (a and d) extending from positions 8 (Arg) to 10 (Gly) numbering from the N terminus (see Appendix 5 at end of volume). Together, these ion series confirm that the peptide is K9 which has been C-terminally processed by removal of arginine at position 166. A detailed discussion of the interpretation of four-sector collision spectra of peptides is presented in [25] in this volume.

Although amino acid analysis revealed a lower arginine recovery than expected, this result is difficult to strictly interpret. An arginine which is posttranslationally modified could escape detection in conventional amino acid analysis if it were to survive the conditions of hydrolysis. Therefore, in order to further assess possible C-terminal heterogeneity, the K9 peptide with and without C-terminal arginine was synthesized by solid-phase synthesis. Under the chromatographic conditions of our analysis the peptides coeluted, however, both peptides were easily detected mass spectrometrically. Since no peptide containing C-terminal arginine was detected mass spectrometrically, C-terminal processing of mature rEPO, originally

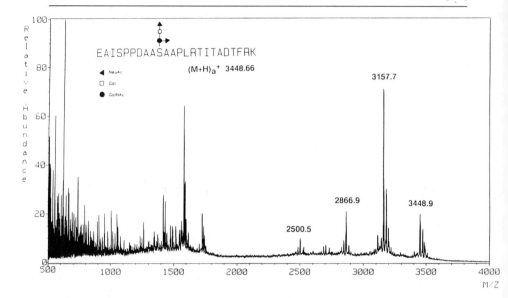

FIG. 4. Two-sector FAB-MS spectrum of peak 4 (Fig. 2) from an endoprotease lysine-C digest of rEPO. The molecular ion region exhibits a series of peaks characteristic of carbohydrate substituents. Based on the peaks between m/z 2500 and m/z 3500, this fraction is assigned to be O-linked glycopeptide, K6 (Fig. 1; see also text). The structure and calculated average molecular weight $[(M + H)_a^+]$ of the fully glycosylated species is shown in the upper corner of the mass spectrum.

thought to be 166 amino acids in length, involves specific and complete removal of this single residue. Thus, the rEPO precursor macromolecule undergoes both N-terminal and C-terminal processing to form mature erythropoietin ending in Asp-165. The K9 peptide labeled in Fig. 1 is that expected for the mature protein.

Mass spectrometric analysis of peak 6 corresponded to the endoprotease lysine-C peptide from mature N terminus. No additional processing of the N terminus was detected mass spectrometrically, supporting the earlier results obtained by N-terminal sequence analysis of the intact protein. Of the remaining major peaks, three (3, 5, 7) were identified as EPO peptides which were expected from lysine-specific digestion of the molecule. Peak 4 exhibited the mass spectrum shown in Fig. 4. The mass spectrum is dominated by a series of ion clusters at m/z 3448.9, 3157.7, and 2866.9 with a consecutive mass difference of 291 Da, highly suggestive of the presence of N-acetylneuraminic acid. In addition, the weak ion at m/z 2500.5 corresponds to the subsequent loss of a hexosamine plus a hexose from m/z 2866.9. This is the predicted $(M + H)^+$ ion for the K6 peptide of rEPO. It is clear from these data that the K6 peptide contains O-linked carbohydrate, however, there are several potential serine and threonine residues for O-linked attachment. Since N-terminal sequence analysis fails to give a PTH-Ser signal at Ser-126, this site was identified as the site containing O-linked carbohydrate. It is difficult to assess from these data alone whether the apparent heterogeneity observed at this site is due to the sample or whether the sample has undergone fragmentation during the FAB ionization process. Carbohydrate heterogeneity is typically observed in glycoproteins, although at a given glycosylation site, a well-defined set of structures is usually present. The structure shown in Fig. 4 is that commonly found at O-linked sites in CHO-expressed proteins and is consistent with our mass spectrometric data. N-Linked carbohydrates are considerably more diverse, consisting of a larger number and wider variety of substituents. These substituents result in the addition of several thousand daltons to the mass of the peptide.

Peaks 2 and 8, the remaining unidentified peaks in the chromatogram, are broad and reflect possible heterogeneity. As shown in in Fig. 1, the three potential N-linked sites would be found in two Endo Lys-C peptides, K2, containing two possible N-linked sites and, K4, containing one N-linked site. As has been reported previously, mass spectrometric analysis of intact glycopeptides is difficult without prior derivatization.[12] The mass spectra of these fractions revealed little meaningful information. These peaks are probably those containing the N-linked sites. Peptide K4 can be further subdigested with trypsin (Arg) to give an N-terminal 24-amino acid peptide and a C-terminal 21-amino acid glycopeptide. Peptide K2 can be subdigested with *Staphylococcus aureus* V8 protease (Glu-C) in a similar manner. Chromatographically, intact K2 and K4 can be identified based upon their differential absorbance at 280 nm, since peptide K4 contains two tryptophan residues and K2 contains no aromatic amino acids. Examination of the chromatogram at this wavelength revealed that peak 2 had little absorbance at this wavelength, while peak 8 had appreciable absorbance. Thus, on the assumption that peak 8 was the 45-amino acid N-linked K4 peptide, the fraction was subdigested with trypsin. The resulting chromatogram (Fig. 5) showed two prominent peaks, an earlier eluting peptide showing possible heterogeneity (fraction 8a) and a later eluting symmetrical peak (fraction 8b). The latter eluting peak exhibited an average $(M + H)^+$ ion at m/z 2684.5 which corresponded to the N-terminal portion of the K4 peptide (K4T1T2). The earlier eluting peak produced a mass spectrum of which the molecular ion region is shown in Fig. 6. As

[12] S. A. Carr, R. J. Anderegg, and M. E. Hemling, *in* "Analysis of Peptides and Proteins by Mass Spectrometry" (C. J. McNeal, ed.), p. 95. Wiley, New York, 1988.

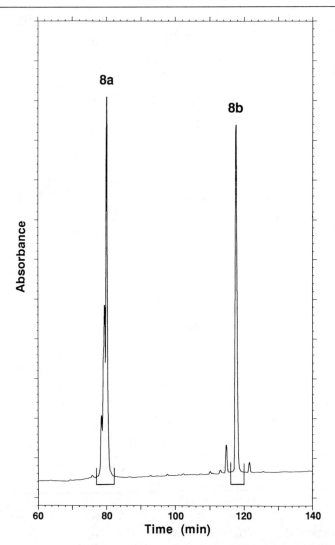

FIG. 5. Reversed-phase HPLC chromatogram resulting from the subdigestion of peak 8 (Fig. 2) from the endoprotease lysine-C digest of rEPO with trypsin. The subdigestion produced two main fractions, 8a and 8b, which were characterized mass spectrometrically (Fig. 6). The brackets under the fraction indicate those portions of the chromatogram which collected for mass spectrometric analysis.

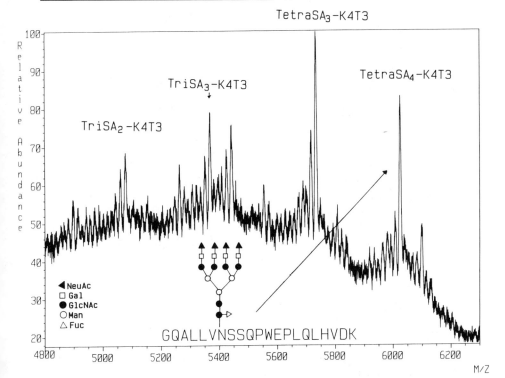

FIG. 6. Two-sector FAB mass spectrum of the molecular ion region of the K4 tryptic peptide (peak 8a, Fig. 5). The distribution of molecular ions observed correspond to the various glycosylated forms of the peptide. The ion with highest molecular weight, $(M + H)^+$ ion at m/z 6026, corresponds to the tetraantennary tetrasialylated glycopeptide (TetraSA4-K4T3) whose structure is shown. In addition to this structure there is a distribution of various other species, such as triantennary trisialylated glycopeptide (TriSA3-K4T3).

in the case of the O-linked K6 peptide, the ion clusters in this mass spectrum are separated by 291 Da (N-acetylneuraminic acid) and 366 Da (hexosamine and hexose) which are suggestive of a carbohydrate-containing peptide. The molecular mass of 6026 Da is consistent with that expected for a tetraantennary, tetrasialylated glycan attached to the C-terminal K4 tryptic peptide (TetraSA4-K4T3). Based on this molecular mass information, the mass of the K4 peptide is in excess of 8700 Da. Treatment of the intact K4 with the glycosidase PNGase F, which cleaves the carbohydrate from the protein and converts the N-linked asparagine to aspartic acid,[13] yielded an observed $(M + H)^+$ ion at m/z 5026.2 Da.

[13] S. A. Carr, J. R. Barr, G. D. Roberts, K. R. Anumula, and P. B. Taylor, this volume [27].

This is the average molecular mass expected for the deglycosylated peptide. Similar treatment of K2 (peak 2) yields the expected average molecular mass of 2903.4 Da.

Identification of C Termini and Peptides Resulting from Nonspecific Cleavage

Mass spectrometric analysis of the reversed-phase map (Fig. 2) revealed a number of peptides which could be attributed to nonspecific cleavage by the lysine-specific protease. Alternatively, these minor species may represent C-terminal truncations of the purified protein. Minor C-terminal proteolysis is difficult to assess, especially when the N terminus of the protein remains intact. However, if the proteolytic digest is performed in buffers prepared with ^{18}O-labeled water (H$_2$[18]O), then secondary proteolytic cleavages, which result during the generation of a set of peptides for primary structure analysis, can be distinguished from those occurring during cell culture or purification.[14] Peptides resulting from cleavage during digestion, whether specific or nonspecific, incorporate the label in the new C-terminal carboxylic acid moiety and shift in mass while cleavages occurring prior to digestion would not incorporate [18]O. Since the C terminus of the protein does not incorporate the label, all C termini of the protein can be identified.

Using this strategy, many of the smaller peaks in the map, as well as those coeluting with major peaks, could be attributed to nonspecific cleavage during digestion, even though a lysine-specific protease was used. Furthermore, all these peptides were generated via cleavage on the C-terminal side of arginine. This may represent a secondary activity of the enzyme or may result from contamination with another protease. As expected, the C-terminal K9 peptide did not shift in mass.

Figure 7 shows the molecular ion region from the K3 peptide enzymatically generated by a lysine-specific protease in the absence of H$_2$[18]O (A) and in the presence of H$_2$[18]O (B). The isotope distribution in the bottom panel shows the incorporation of one atom of [18]O (m/z 929.8) and two atoms of [18]O (m/z 931.8). This result can be viewed as enzyme-catalyzed hydrolysis (one [18]O atom, +2 Da) and enzyme-catalyzed exchange (one [18]O atom, +2 Da). Thus, the combination of these two processes can result in an increase in molecular mass of 4 Da. Figure 8 compares a selected region of the MS/MS spectrum from the (M + H)$^+$ ion at m/z 927.6 (A) with that of (M + H)$^+$ ion at m/z 931.6 (B). As expected, all ions originating from the N terminus of these peptides have the same mass

[14] K. Rose, M. G. Simona, R. E. Offord, C. P. Prior, B. Otto, and D. R. Thatcher, *Biochem. J.* **215,** 273 (1983).

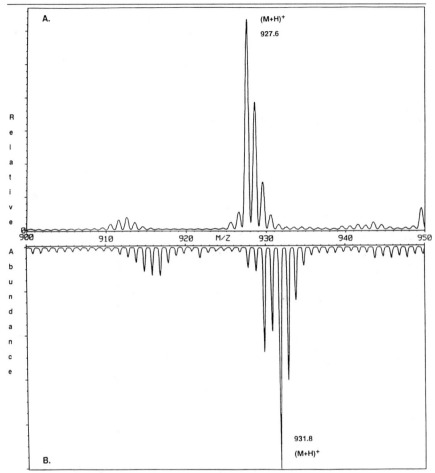

FIG. 7. Comparison of the FAB-MS molecular ion regions of the K3 peptide of rEPO, VNFYAWK (peak 3, Fig. 2) from a lysine-specific digest carried out in H_2O (A) and $H_2{}^{18}O$ (B), respectively. The protonated molecule, $(M + H)^+$ at m/z 927.6, in (A) is shifted by 4 Da to m/z 931.8 (B) as a result of the enzyme-catalyzed incorporation of two atoms of ^{18}O in the C-terminal carboxyl group of lysine.

(these ions are circled), while those originating from the C terminus are shifted in mass by 4 Da. This technique has the added benefit of aiding in the assignment of fragment ions in the MS/MS spectrum to the N or C terminus of the peptide.

The incorporation of the oxygen label into the side-chain carboxylic acid moieties of aspartic and glutamic acid has never been observed in applications of this strategy. The rate of incorporation for this process is

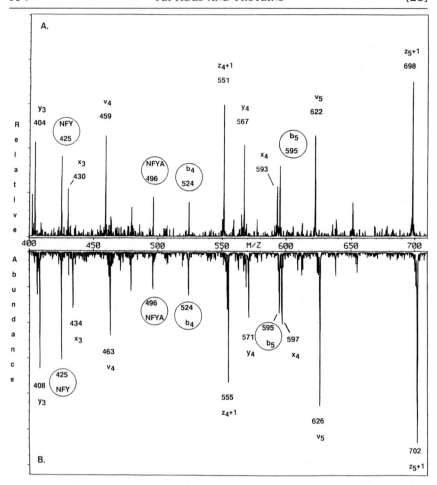

FIG. 8. Comparison of regions of the FAB-MS/MS spectra of the ^{12}C protonated molecules from the K3 peptides shown in Fig. 7. The enzyme-catalyzed incorporation of two atoms of ^{18}O occurs at the C-terminal carboxyl group, therefore, all product ions associated with the C terminus of the molecule will be shifted up in mass by 4 Da. Fragment ions related to the N terminus of the peptide will not shift in mass. The ions circled in (A) and (B) do not shift in mass, whereas the remainder of the ions in (B) shift up in mass by 4 Da, indicating that they are related to C-terminal fragment ion series.

very slow compared to that resulting from enzyme-catalyzed hydrolysis and is also pH dependent. In the presence of a lysine-specific enzyme, incorporation of the label into the C-terminal carboxylic acid group of lysine has been observed. However, in these cases the rate of incorporation is much slower and the isotope distribution pattern is very different

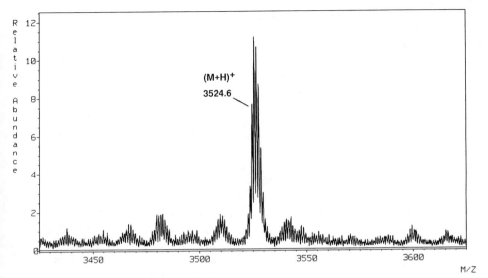

FIG. 9. Two-sector FAB mass spectrum of the molecular ion region of a peptide isolated from a nonreduced digest of rEPO. The molecular ion, $(M + H)^+$ at m/z 3524.6, corresponds to the mass of a disulfide-linked species consisting of rEPO peptides K1 and K9 (Fig. 1). This was confirmed by the detection of molecular ions corresponding to intact K1 and K9 and disappearance of the ion signal at m/z 3524.6 after reduction of the sample on the probe tip with triethylphosphine.

from that resulting from enzyme-catalyzed hydrolysis. Nonetheless, if the enzyme has an affinity for the C-terminal amino acid, incorporation of the label could occur.

Disulfide Bond Analysis

The proper formation of disulfide bonds is often critical for biological activity. This is especially true in the case of bacterially expressed proteins, where it may be necessary to refold the protein through a series of redox reactions in order to obtain the proper conformation. Refolding is not required for secreted proteins, however, it is important to ensure that once formed, disulfide linkages remain intact and in the proper orientation throughout the purification process. Figure 9 shows the two-sector FAB-MS molecular ion region of a peptide isolated from a nonreduced lysine-specific digest of rEPO. The retention time of this peptide was different from that of other peptides in the digest of the reduced and alkylated protein. The mass spectrum revealed an $(M + H)^+$ ion at m/z 3524.6 which corresponds in mass to the K1 and K9 peptide linked by a disulfide bond. On addition of 1 μl of 5% triethylphosphine in methanol to the probe tip, two new peaks appeared at m/z 2342.2 and m/z 1185.5, which corresponded to the reduced K1 and K9 peptides, respectively. In a similar manner, reduction on the sample probe tip can be carried out using NH_4OH and thioglycerol as has recently been reported.[15]

Other Covalent Modifications

In addition, to the characterization of natural posttranslational modifications, a number of other events which could occur during cell culture or purification may compromise the structural integrity of the protein. These include deamidation of asparagine and glutamine to their respective acids, or oxidation of amino acids such as methionine or tryptophan. If the protein has been exposed to high concentrations of urea, the possibility of carbamylation of lysine residues must be considered. In addition, if the N-terminal amino acid in a protein or peptide is glutamine, cyclization can occur to form pyroglutamic acid which results in N-terminal blockage. Since all of these events lead to covalent modification of the protein, mass spectrometric-based approaches can, in principle, detect these changes.

Conclusion

Detailed structural analysis of proteins is crucial for the in-depth understanding of biological function. This is especially true in the case of recombinant proteins since many factors, such as the host expression system, cell culture conditions, and purification, influence the final structure of the purified protein. Strategies based on mass spectrometry offer unique advantages over conventional approaches for the structural characterization of glycoproteins. Most notable among these is the ability to identify peptides which have been naturally posttranslationally modified or processed, or which have been covalently modified after expression. The most powerful structural approaches, however, are those that combine structural information that can be generated using a number of techniques, including those discussed in this chapter.

Acknowledgments

The authors thank Godfrey Amphlett, Michael Brenner, and James E. Vath for helpful discussions throughout the course of this work.

[15] D. F. Hunt, J. Shabanowitz, J. R. Yates, and P. R. Griffen, in "Mass Spectrometry of Biological Materials" (C. N. McEwen and B. S. Larsen, eds.), p. 169. Dekker, New York, 1990.

Section III

Glycoconjugates

[29] Glycoconjugates: Overview and Strategy

By ROGER A. LAINE

The chapters in this section reflect the latest and most useful methods in application of mass spectrometry to carbohydrates from biological samples. Tools for interpreting the data include example and standard spectra, and details of fragmentation mechanisms and pathways. This section should provide a good working knowledge in the use of mass spectrometry in structure analysis of carbohydrates, although, it is recommended that the primary references be consulted. In the interest of quoting the earliest relevant reference: "Teach him what has been said in the past; then he will set a good example . . . and judgement and all exactitude shall enter into him. Speak to him, for there is none born wise." [1]

In this introduction there will be no attempt to review the background associated with the methods in all of the chapters. Instead I will point to an earlier historical perspective and address some practical aspects of carbohydrate analysis.

The practical application of mass spectrometric techniques to the difficult task of complete structural elucidation of oligosaccharides, polysaccharides, and glycoconjugates has been influenced strongly by instrument development, namely, gas–liquid chromatography/mass spectrometry, soft ionization, and high-field magnets. First, methylation linkage analysis developed principally because of the first working interface between a gas–liquid chromatograph and a mass spectrometer (GC/MS) which was assembled by Ragnar Ryhage at the Karolinska Institute in Stockholm[2] on what would become the LKB-9000 instrument. The two-stage molecular separator, which constituted this interface, was ingenious and enabled on-line mass spectrometric analysis of gas–liquid chromatography effluents. The synergy of an on-line chromatographic and mass spectrometric system was astounding and still constitutes one of the most powerful analytical tools for volatiles.

At about the same time, around 1964, Hakomori adapted Cory's methylsulfinyl carbanion reagent to efficiently and completely methylate sug-

---

[1] Ptahhotpe, "The Maxims of Ptahhotpe," Introduction. (2350 B.C.)
[2] R. Ryhage, *Anal. Chem.* **36,** 759 (1964).

ars.[3] Stimulated by these two developments and faced with difficulties of elucidating bacterial polysaccharides, Bengt Lindberg and colleagues at the University of Stockholm developed methods for separation of partially methylated alditol acetates by gas–liquid chromatography and identification of the linkage positions by electron ionization (EI)-MS[4] (reviewed in [30], this volume). This technology enabled rapid linkage position analysis on samples of 0.1 to 1 μmol, and with a few improvements, is still the method most used for determination of linkage. Thus, we included the methodology for this crucial technique.[4] Although paper, thin-layer, and gas chromatographic analysis of methylated sugars predated this technology, larger quantities of sample were required for those techniques, and characterized standards were necessary. Much later, improvements in the methods included fused quartz capillary gas chromatography, which cut by one-third the time necessary for gas–liquid chromatography separations.[4] Chemical ionization mass spectrometry added tenfold to the sensitivity,[5] and a few improvements in the methods for methylation have been published.[6] However, the technique has remained fairly standard since 1970, practiced in only a handful of laboratories throughout the world.

Fragmentation schemes for derivatized oligosaccharides were intensely studied in the years prior to 1975, using trimethylsilyl, methyl, and acetyl derivatives for direct evaporation and GC/MS applications. Principal researchers in this field were Biemann, DeJongh, and Schnoes,[7] and Chizhov and Kochetkov of the USSR, the latter having written early reviews on this subject.[8,9] The Baltic seemed to connect researchers in the area of oligosaccharide and glycoconjugate mass spectrometry, including Karlsson[10] of Sweden, and Karkkainen from Finland. The latter showed that these compounds could be separated on a gas–liquid chromatograph. All of this work was extensively and thoroughly reviewed in 1974 by Lonngren and Svensson.[11] This review provides a valuable text for those who intend to acquire a background in mass spectrometry related to carbohydrate chemistry and provides a surprisingly rich lesson for those who believe that carbohydrate mass spectrometry began with fast atom bombardment ionization (FAB)-MS methods. Identification of nearly all

[3] S.-I. Hakomori, *J. Biochem.* **55,** 205 (1964).

[4] C. G. Hellerqvist, this volume [30].

[5] R. A. Laine, *Anal. Biochem.* **116,** 383 (1981).

[6] I. Ciucanu and F. Kerek, *Carbohydr. Res.* **131,** 209 (1984).

[7] K. Biemann, D. C. DeJongh, and H. K. Schnoes, *J. Am. Chem. Soc.* **85,** 1763 (1963).

[8] O. S. Chizhov and N. K. Kochetkov, *Adv. Carbohydr. Chem.* **21,** 29 (1966).

[9] O. S. Chizhov and N. K. Kochetkov, *Methods Carbohydr. Chem.* **6,** 540 (1972).

[10] B. E. Samuelsson, W. Pimlott, and K.-A. Karlsson, this volume [34].

[11] J. Lonngren and S. Svennson, *Adv. Carbohydr. Chem. Biochem.* **29,** 42 (1974).

of the principal types of fragmentation in monosaccharides and oligosac-charides was established before 1974 as reviewed and described by Lonn-gren and Svennson.[11] Surprisingly, few modern authors give proper credit to this large body of previous work.

Another significant advance mentioned above was the success in mass spectrometry of derivatized oligosaccharides and glycoconjugates. Many of the results from which FAB-MS of carbohydrates takes root are based on the early work of K.-A. Karlsson[10] and colleagues. EI spectra of very large permethylated,[12] intact, and amide-reduced glycosphingolipids were first reported as early as 1973, and are nearly as informative as any FAB-MS spectra published today.[10] Another early method with enduring quali-ties is the combination of high-temperature gas–liquid chromatography with mass spectrometry. Soft ionization applications of this technique for oligosaccharide measurements are described elsewhere in this volume.[13]

Since 100 nmol were often used for analysis by the EI methods, the third major impact on carbohydrate analysis has been the improvement in sensitivity using the new soft ionization methods related to FAB-MS. FAB-MS produces stable spectra with an intense molecular ion and an increase in sensitivity of 10- to 100-fold.[14,15] Array detectors increase the molecular ion detectability by another astonishing two orders of magni-tude, yielding picomole sensitivity.[16] A review of FAB-MS development, which accelerated progress regarding all polar biological compounds, is included in several of the chapters in this volume and need not be repeated here.

The pioneering work by Kotchetkov, Chizov, and Karlsson as men-tioned above has been extended to larger compounds by Dell,[14] Peter-Katalinić and Egge,[15] Costello,[17] and Burlingame and colleagues[16,18] and many others too numerous to mention here. Novel lipoidal derivatives at the reducing end of oligosaccharides have made them more amenable to FAB ionization.[18] FAB is also used to map the location of carbohydrates on the protein chain.[19]

Collision-induced dissociation (CID), when combined with FAB-MS to generate the molecular ion, gives a significant advance in assigning some of the many details of glycoconjugate structure. Early work in this

[12] K.-A. Karlsson, FEBS Lett. 32, 317 (1973).
[13] G. C. Hansson and H. Karlsson, this volume [39].
[14] A. Dell, this volume [35].
[15] J. Peter-Katalinić and H. Egge, this volume [38].
[16] B. L. Gillece-Castro and A. L. Burlingame, this volume [37].
[17] C. E. Costello and J. E. Vath, this volume [40].
[18] A. L. Burlingame and L. Poulter, this volume [36].
[19] S. A. Carr, J. R. Barr, G. D. Roberts, K. R. Anumula, and P. B. Taylor, this volume [27].

area has been reviewed in several recent publications and the preambles of relevant chapters in this volume[16,17,20,21,22] and need not be repeated here. High-energy collision (magnetic sector instruments) and low-energy collision (triple quadrupole instruments) may generate somewhat different sets of daughter ions. Costello's systematic nomenclature of the CID daughter ions[17] is supplanting the older assignments reviewed by Lonngren and Svensson,[11] and is also used by Gillece-Castro and Burlingame in this volume.[16] Statistical treatment of glycosidic cleavages based on likelihood of cleavage of more hindered bonds (which may not easily dissipate collision energy) has been suggested by Laine et al. and supported by molecular modeling.[20,22] A few other investigations have suggested that linkage information may be directly obtained from FAB-MS.[23–27] A particularly complete study of the use of FAB-MS and linked scanning for determination of linkage position of terminal saccharides in a set of 19 oligosaccharides was recently published by Garozzo et al.[28]

A detailed interpretation of a number of CID glycoconjugate spectra based on traditional ion fragment-cleavage mechanisms is described elsewhere in this volume.[16,17] Richter shows that epimer forms of sugars can be discerned in reducing terminal groups by CID methods.[21,29]

Californium-252 fission-fragment plasma desorption time-of-flight (TOF) mass spectrometry (PDMS) has also been developed for complex carbohydrates, principally by Jardine in collaboration with Brennan.[30] Figure 1, taken from Jardine et al.,[30] gives an example of the spectral appearance and type of information available from PDMS.

As currently practiced, PDMS only has mass resolution of about 0.1% and therefore at mass 1000, the accuracy is only within about 3 mass units. In practice, centroids of the isotope peak envelope can usually be estimated within one mass unit at mass 1000 and within 2 u at mass 2000.

[20] R. A. Laine, this series, Vol. 179, p. 157.
[21] W. J. Richter, D. R. Müller, and B. Domon, this volume [33].
[22] R. A. Laine, K. M. Pamidimukkala, A. D. French, R. W. Hall, S. A. Abbas, R. K. Jain, and K. L. Matta, J. Am. Chem. Soc. 110, 6931 (1988).
[23] Y. Chen, N. Chen, M. Li, F. Zhao, and N. Chen, Biomed. Mass Spectrom. 14, 9 (1987).
[24] J. P. Kamerling, W. Heerma, F. G. Vliegenthart, N. B. Green, I. A. S. Lewis, G. Strecker, and G. Spik, Biomed. Mass Spectrom. 10, 420 (1983).
[25] Z. Lam, M. B. Comisarow, G. G. S. Dutton, D. A. Weil, and A. Biarnson, Rapid Commun. Mass Spectrom. 1, 83 (1987).
[26] J. C. Prome, M. Aurelle, D. Prome, and D. Savagnac, Org. Mass Spectrom. 22, 6 (1987).
[27] Z. Lam, M. B. Comisarow, and G. G. S. Dutton, Anal. Chem. 60, 2304 (1988).
[28] D. Garozzo, K. M. Giuffrida, G. Impallomeni, A. Ballistreri, and G. Montaudo, Anal. Chem. 62, 279 (1990).
[29] D. R. Muller, B. Doman, and W. J. Richter, Adv. Mass Spectrom. 11B (1989).
[30] I. Jardine, G. Scanlan, M. NcNeil, and P. J. Brennan, Anal. Chem. 61, 416 (1989).

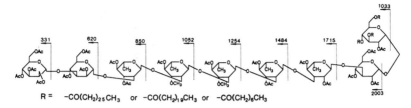

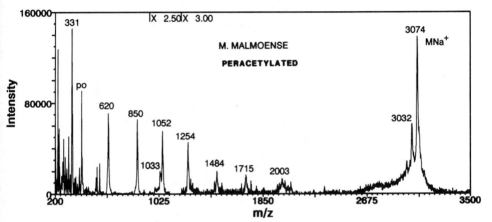

FIG. 1. Positive ion [252]Cf fission fragment ionization time-of-flight (TOF) mass spectrum (plasma desorption mass spectrometry, PDMS) of peracetylated *Mycobacterium malmoense* lipooligosaccharides (LOS). Recorded on a BIN-10K PDMS (Bio-Ion Nordic, Uppsala, Sweden) using accelerating voltage of 20 kV. LOS (3 μg) was coated on nitrocellulose-layered aluminized Mylar foils. Spectra were acquired for 1 hr. (Reproduced from Ref. 30 with permission of the authors.)

This approach has been extensively used in peptide analysis.[31] Although mass spectra of peptides of 34,000 have been recorded with this technique, successful high-molecular weight analyses are rare and often dependent on double or triply charged ions. For peptides, in practice, masses below 6000 Da are often attainable and masses between 1000 and 3000 Da routine. McNeal *et al.*[32] have used PDMS with ion-pairing reagents to examine sulfated oligosaccharides. We have also examined the usefulness of PDMS for saccharides of different types and derivatives, and have found a limit of DP (degree of polymerization) 13 for an underivatized dextran series.[33] Upon acetylation, the upper limit under established conditions appears to

[31] P. Roepstorff, this volume [23].

[32] C. J. McNeal, R. D. Macfarlane, and I. Jardine, *Biochem. Biophys. Res. Commun.* **139,** 18 (1986).

[33] C. M. David and R. A. Laine, unpublished (1990).

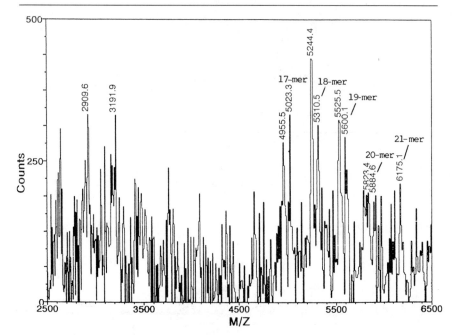

FIG. 2. Positive ion ^{252}Cf fission fragment ionization time-of-flight (TOF) mass spectrum (plasma desorption mass spectrometry, PDMS) of peracetylated Pharmacia T-10 dextran oligomers fractionated by gel permeation chromatography. Spectra were recorded on a Bio-Ion Nordic BIN-20 PDMS using accelerating voltage of 18 kV. Sample (1 μg) was coated on nitrocellulose-layered aluminized Mylar foils. Spectra were acquired for 1 hr.

be in the range of m/z 6200 for MH$^+$ and (MH-60)$^+$ ions at a DP of 21 as shown in Fig. 2.[33] Chitooligomers gave good results up to DP 12, where solubility became a problem. Acetylation did not improve the performance for chitooligomers.[33] The technique was obviously useful for the mycobacterial glycoconjugates[30] and heparin oligosaccharides,[32] and may be a valid alternative for FAB-MS in molecular weight determination.

N-Linked glycopeptides in the mass range 7000 to 12,000 Da, are found on glycoproteins such as human erythrocyte band 3[34] and human placental glycoproteins.[35] No one has yet defined an ionization technique that can show molecular weights or polymer distributions for carbohydrates of this mass. Matrix-assisted laser desorption TOF mass spectrometry may be one possibility to explore. Also, electrospray ionization may be possible

[34] J. Jarnefelt, J. Rush, Y.-T. Li, and R. A. Laine, *J. Biol. Chem.* **253**, 8006 (1978).
[35] B. C.-R. Zhu, S. F. Fisher, H. Pande, J. Calaycay, J. E. Shively, and R. A. Laine, *J. Biol. Chem.* **259**, 3962 (1984).

if the *N*-acyl groups are removed from the carbohydrate, creating multiply charged sites.

The reader will note there is a bit of the skills of the advertiser applied in each of the articles. Although there are a few pertinent caveats sprinkled in the chapters, each author is at least slightly optimistic about the extent of usefulness of his particular method.[36] *"Navita de ventis, de tauris nattat arator, enumerat miles vulnera, pastor oves."*[37] Thus, nothing has changed in four millenia with regard to the value[1] and the prejudice[37] of experts transmitting and extolling their experiences. In Section III of this volume you will find a selection of the world's specialists in carbohydrate analysis using mass spectrometry, each with a favorite approach. None, however, can be taken alone. Combinations of more than one method described here are necessary and synergistic.

The complete elucidation of the structure of a complex carbohydrate requires analytical input from a number of methods, including wet chemical, enzymatic, antibody or lectin affinity, radiochromatography, thin-layer, paper, gas–liquid, or high-performance liquid chromatography. Sugar composition, ring size, anomeric configuration, position of linkage, internal sequence, branching, and noncarbohydrate substitution cannot all be gained from using one technique, although high-field two-dimensional (2D) NMR comes the closest. For pure samples in the range of 5 μmol containing a reasonable number (fewer than 10) of monosaccharide units, NMR can provide complete structural data using a combination of two or more two-dimensional techniques including relay COSY, NOESY, HOHAHA, or TOCSY and other pulse sequence methods. Since NMR is nondestructive, and, as a lower limit (at least currently), if a quantity more than 50 nmol is available, NMR spectra should be recorded on the sample before any destructive analytical procedure is used.

Most problems in carbohydrate analysis arise not from synthesis nor from large-scale preparations, but from investigations of biologically active binding-recognition systems. Specificity usually dictates the existence of a rare sequence and, usually only picomole to nanomole quantities are readily available.

Mass spectrometry finds its niche in high sensitivity studies. In Section III the reader will find the work of many investigators who apply mass spectrometry to nanomole to picomole amounts of complex carbohydrates. The hard fact is, however, that mass spectrometry alone cannot

---

[36] F. Bollum, "I take a craftsman's view of the practice of science and of the scientist." Personal communication (ca. 1977).

[37] Propertius, "Elegies." Book ii (i), p. 43. (51 B.C.) Roughly, "The sailor recants his experiences of great storms, the farmer carries forth about oxen, the soldier talks about wounds and the shepherd discusses sheep."

yet give a complete structure. It is part of the goal of this introduction and overview to point out the current practical limitations to the analytical methods. Another goal is to suggest strategies and approaches for nanomolar amounts of oligosaccharides that would employ mass spectrometry to lead to complete elucidation of a structure. Many investigators not quoted in this chapter have contributed significantly in this field.

It is unrealistic to expect total characterization of intact structures from mass spectra. In the literature, as Biemann notes elsewhere in this volume,[38] examples are often taken from known structures which are then "interpreted" according to the ions which happen to occur in the spectra and can be readily explained, often ignoring the unassigned ions. Mass spectra of complete unknowns may not be so simple to interpret. From the empirical descriptions of fragmentation accompanying mass spectra of oligosaccharides, it is readily seen that this is a science in its embryonic stages of development. Despite the extensive early literature, there still seems to be the element of surprise in the origin and mechanism of formation of some fragments. Predictability of fragment ions in spectra according to detailed structure of oligosaccharides is becoming more exact from collision-induced dissociation and tandem mass spectrometry,[16,17,20–29] but it is far from perfect.

Example of an Approach to a Biological Carbohydrate

The author's particular prejudice lies in the following approach to a total carbohydrate structure (oligosaccharide): "The road to resolution lies by doubt . . . ."[39]

1. Gain strong evidence for purity. It is virtually impossible to begin with biological extracts and proceed easily to an absolutely pure sample. This derives from the many similar or related structures that are usually present in a biological system.[40] One must be assiduous about this aspect of the analysis. (a) Utilize at least three chromatographic systems, preferably at least one with a derivatized sample, permethylated or peracetylated. (See this series, Volumes 28, 83, 138, 179.) (b) Pay strict attention to composition analysis (methods are described in earlier volumes of this series, including Volumes 28, 83, 138, 179) for estimation of size, complexity, and purity.

2. If 50 nmol or more are available, at least obtain a one-dimensional ¹H NMR spectrum.

---

[38] K. Biemann, this volume [18].

[39] F. Quarles, "Emblems." Book iv (No. 2), Epig. 2 (1635).

[40] C. L. M. Stults, C. C. Sweeley, and B. A. Macher, this series, volume 179, p. 167. Note origins and subtle structural differences among 250 glycolipids.

3. Application of mass spectroscopy: (a) Use 1–20 nmol of the intact compound depending on sensitivity of the instrument, for FAB-MS. (b) Methylate 10–25 nmol (more is desirable). Use 1 or 2 nmol of the methylated derivative for FAB-MS.[14] Process the remainder for partially methylated alditol acetates,[4] or alternatively use reductive cleavage[41] and process samples by GC/MS, preferably chemical ionization (CI).[5] (c) Utilize Nilsson's technique[42] of periodate oxidation of the intact oligosaccharide, reduction, methylation (or a modification of this method using acetylation,[43] and FAB-MS or direct chemical ionization (DCI). CID may dramatically help this technique, but no report of its application is yet published. One may need 20–50 nmol due to the necessity of carrying the sample through several chemical steps.

4. Determine the remainder of your strategy. For example, you will not know the anomeric configurations or the relative location of linkage positions and anomeric configurations within the oligosaccharide.

After following steps 1–3 above, the reader will have knowledge concerning: (a) A reasonable estimation of degree of purity, but not proof. (b) Which sugar(s) belong to nonreducing termini. If the reader has a free oligosaccharide, it can be reduced with borohydride and the sugar located at the reducing end can be determined. If one methylates this sample, one will also obtain the linkage of the reducing end sugar. Otherwise one must do it the hard way. (c) Sugar composition and stoichiometry, which will give clues regarding purity. (d) Linkage positions and quantity of each sugar type in the molecule (very valuable in estimating purity), but not their relative order in the chain. (e) Whether a branched chain exists in the sample. (f) Whether any sugars have nonsaccharidic substitutions (FAB-MS of the intact saccharide plus FAB-MS of the methylated sample). (g) Size (molecular weight).

The investigator may have some probable knowledge regarding: (a) Location of branch, if any (by FAB-MS and CID cleavages). (b) Anomeric set if NMR has been performed, but not anomeric sequence order. (c) Reducing terminal if oligosaccharide has been reduced. (d) Some information from FAB-MS on the internal order of the sugars, i.e., if it is a heterooligosaccharide, e.g., methylpentose–hexose-N-acetylhexosamine–hexose. . . .

The application of mass spectrometry will not provide the investigator with information concerning the following: (a) Internal sugar order (man-

[41] G. R. Gray, this volume [31].
[42] A.-S. Angel and B. Nilsson, this volume [32].
[43] C. G. Hellerqvist, personal communication (1990).

nose–galactose–glucose) for oligomers where adjacent constituents have the same molecular weight, other than nonreducing terminal(s). (b) If identical sugars with different internal linkages exist, the internal order of the linkages will not be clear. However Nilsson's technique (this volume [32][43] may reveal some information regarding this aspect. (c) Anomeric configurations and their internal order among the constituent sugars.

The following steps are necessary to establish the structure:

1. Determine the ring size: As pointed out by Gray in this volume,[41] the 4-substitution of a hexopyranose will give the same partially methylated alditol acetate as the 5-linked furanose. Notwithstanding that 5-linked furanoses are either extremely rare or nonexistent in most biological systems, we must prove the structure. Therefore, a few methods exist which can assist with this determination. The one most often used is sequential release of the terminal sugars by glycosylhydrolases (see other volumes in this series: Vols. 28, 83, 138, 179) specific for pyranoses after the terminal residues are identified as to their ring size by methylation linkage analysis.

Testing a terminal sialic acid for anomeric linkage is relatively simple. Since biosynthesis has mainly used the α bond in sialic acids, testing α-sialidases from one or more sources will usually result in a released sialic acid, revealing the penultimate sugar.

When the enzymatically shortened oligosaccharide product is methylated, the ring form of the new nonreducing terminus sugar will be revealed by the Lindberg method.[4] Enzymes specific for this newly revealed terminal sugar can then be used to determine its anomeric specificity and ring size. Another glycosidase will reveal the next sugar, etc.

Gray offers an alternative method[41]: reductive cleavage of the methylated oligosaccharide resulting in preservation of the ring form which can be identified by gas–liquid chromatography/EI-MS.

2. Anomeric set in the absence of NMR data: The most common microscale anomeric determinations are made by α- and β-glycosylhydrolases when available. Once the sugar composition is known, the enzymes can be tested on the sample followed by chromatography to reveal mobility differences caused by loss of a sugar. As each successive saccharide is removed, a new nonreducing end is available for another exoglycosidase. Occasionally, endoglycosidases are useful.[34] Methylation linkage analysis after each successive enzyme release will establish ring size and absolute sequence. Branching or tandem repeated identical residues complicate the problem. Two or more sugars may be removed at once. This raises the question: Were they connected in tandem or on different branches? Obviously, this must be examined by repetitive continuous analysis during enzyme treatment by FAB-MS, preferably with CID.[16,17]

Chemically, chromium trioxide oxidation of the peracetylated oligosaccharide in glacial acetic acid has allowed discrimination between α-D and β-D bonds.[44] The β-D-glycosidic linkages has two oxygens in close proximity, the ring and glycosidic oxygen which renders it more susceptible to oxidation by CrO_3 than the α bonds.[44] Thus, sugar composition before and after oxidation will assign the survivors as α. Partial survivors suggest mixtures of α and β in the sample, either in the same saccharide or in a mixture of saccharides.

CID after chemical ionization was reported to discern among anomers in a complete set of glucose anomeric pairs of each linkage type, using mass-analyzed ion kinetic energy spectra (MIKES),[45] but this discovery has not been developed into a routine system.

Solving Difficult Problems

Example 1. An extract from a certain tissue may contain a group of 3 to 10 glycosphingolipids with a tetrasaccharide sugar chain, or a glycoprotein may yield as many as 30–60 different N-linked saccharides. Some of these may differ by only one position of linkage, an anomeric configuration or a branch.[40] A fast atom bombardment mass spectrum of such a mixture of several compounds which contain four sugars comprised of galactose, glucosamine, galactosamine, and glucose may be impossible to discern from that of a spectrum of a pure compound. Molecules of the approximate structure Gal-GalNAc-Gal-Glu-ceramide and Gal-GluNAc-Gal-Glu-ceramide may give identical spectra. In fact, they may give identical CID spectra. An example of a mixture consisting of branched and linear compounds includes

$$
\begin{array}{c}
\text{Gal}(\beta1\rightarrow3)\diagdown \\
\qquad\qquad\text{Gal}(\beta1\rightarrow4)\text{Glu}(\beta1\rightarrow1)\text{ceramide} \\
\text{GalNAc}(\beta1\rightarrow4)\diagup \\
\textbf{I}
\end{array}
$$

$$
\begin{array}{c}
\text{Gal}(\beta1\rightarrow4)\diagdown \\
\qquad\qquad\text{Gal}(\beta1\rightarrow4)\text{Glu}(\beta1\rightarrow1)\text{ceramide} \\
\text{GalNAc}(\beta1\rightarrow3)\diagup \\
\textbf{II}
\end{array}
$$

$$
\begin{array}{c}
\text{Gal}(\alpha1\rightarrow3)\diagdown \\
\qquad\qquad\text{Gal}(\beta1\rightarrow4)\text{Glu}(\beta1\rightarrow1)\text{ceramide} \\
\text{GalNAc}(\beta1\rightarrow4)\diagup \\
\textbf{III}
\end{array}
$$

[44] J. Hoffman, B. Lindberg, and S. Svensson, *J. Supramol. Struct. Suppl.* **1**, 31 (1972).
[45] E. G. de Jong, W. Heerma, and D. Dijkstra, *Biomed. Mass Spectrom.* **7**, 127 (1980).

Gal($\beta 1 \rightarrow 3$)GalNAc($\beta 1 \rightarrow 4$)Gal($\beta 1 \rightarrow 4$)Glu($\beta 1 \rightarrow 1$)ceramide

IV

Gal($\beta 1 \rightarrow 4$)GalNAc($\beta 1 \rightarrow 4$)Gal($\beta 1 \rightarrow 4$)Glu($\beta 1 \rightarrow 1$)ceramide

V

A mixture like this is difficult to resolve chromatographically and may give mass spectra from which one could draw one of several conclusions. All the members of this family of compounds have the same mass and relative polarity. In fact, they all have identical sugar compositions. The molecular ion may appear to lose a hexose or an N-acetylhexosamine as the nonreducing terminal moiety, or a mixture of the two losses may be seen. The linear compounds **IV** and **V** will produce fragments from cleavage of the terminal disaccharide while any of the pure compounds **I, II,** or **III** would show no cleavage of a terminal disaccharide, but would show fragments derived from a terminal trisaccharide.

Nilsson's method[42] of periodate, permethylation, and FAB could not distinguish among **I, II,** and **III,** and would have some difficulty with **IV** because the penultimate GalNAc when linked by either its 3- or 4-position is resistant to periodate. His method is a valid third step, however, after determination of sugar composition, FAB-MS, and methylation linkage analysis. The investigator must be extremely vigilant concerning stoichiometry in analysis, and careful about using a number of different purification methods, both with intact and derivatized samples, utilizing several parallel methods before predicting structure. Dell has suggested one of the most powerful combinations in dealing with unknowns[14]: Permethylate the compound; use a part for FAB-MS and a part for methylation linkage analysis.

Example 2. Consider a pure compound of the structure:

Neu5Ac($\alpha 2 \rightarrow 3$)Gal($\beta 1 \rightarrow 4$)GluNAc($\beta 1 \rightarrow 4$)Gal($\beta 1 \rightarrow 4$)Glu-ceramide

VI

The strategy used will be to establish the molecular weight and sugar/lipid composition by FAB-MS, giving a mass which is consistent with the sugar composition listed above and a C_{18}-sphingenine attached to a C_{18} saturated fatty acid.

The FAB spectrum may provide evidence that the sequence of sugar types is as shown above. If the structure were branched, the spectrum may show some differences.[10,14–17] A methylation linkage analysis[4,5] will establish the composition of the linkage types and a partial structure of **VI** can be depicted as below:

Neu5Ac($\alpha 2 \rightarrow$?)Hex?($\beta 1 \rightarrow 4$)GluNAc($\beta 1 \rightarrow$?)Hex?($\beta 1 \rightarrow 4$)Hex?-ceramide

where Hex represents hexose. Obviously a large number of possible compounds still remain to be eliminated to find one structure for **VI**. The questions remain because the FAB ionization mass spectrum cannot be used to distinguish among hexoses within the interior of the molecule, in this case D-galactose and D-glucose, although Richter's results look promising.[21,29] Hellerqvist notes[4] that one should take into account the known pathways of biosynthesis for each type of compound. For example, the most common structures in this class of compound have the glucose attached to the ceramide. Thus, a "likely" sequence would be as follows:

Neu5Ac(α2→?)Gal(β1→4)GluNAc(β1→?)Gal(β1→4)Glu-ceramide

However, it should be remembered that if we are "completely characterizing" molecules, a "likely" structure is not an absolute one. Nilsson's method[42] helps greatly in this situation. The penultimate galactose will survive periodate oxidation, showing that it is the 3-linked galactose, the GlcNAc will survive, but this could mean either a 3- or 4-link. (Prior hydrazinolysis here would give a free amino group at the 2 position of the GlcNAc. The 4-linked GlcNAc would become periodate sensitive, rendering the Nilsson technique more useful.) The next galactose will be cleaved between carbons 2 and 3, showing by a specific set of fragments[42] that this is the 4-linked galactose, etc.

The question marks remaining point to the fact that the molecule under study (**VI**) has both a 3- and a 4-linked D-galactose. Without the Nilsson experiment,[42] their order within the chain is not clear. Since the necessary sugar composition analysis shows 2 galactoses, a sequential enzyme degradation would eventually discriminate among the two possible arrangements of galactoses. Removal of the sialic acid followed by methylation would give a 4-linked and a terminal galactose identifying the original location of the 3-linked galactose. An exciting possibility is that a set of through-the-ring cleavages with CID and MS/MS might lead to the correct answer.[16,17]

Example 3. There are some structures which defy ordinary methods of structural discernment, for example, the lactosamine oligomers shown below.

Gal(β1→4)GluNAc(β1→6)
　　　　　　　　　　　　　　＼
Gal(β1→4)GluNAc(β1→3)Gal(β1→4)GluNAc(β1→3)Gal(β1→4)Glu-
VII

Gal(β1→4)GluNAc(β1→6)
　　　　　　　　　　　　　　＼
Gal(β1→4)GluNAc(β1→3)Gal(β1→4)GluNAc(β1→3)Gal(β1→4)Glu-
VIII

Upon sequential methylation analysis and enzyme degradation, a common simple strategy for complete analysis, the products in both cases are identical. Hence, methylation analysis gives the identical result for both, yielding the following: Terminal galactose, 2 mol; 3-substituted galactose,

1 mol; 3,6-disubstituted galactose, 1 mol; 4-substituted GlcNAc, 3 mol; and 4-substituted glucose, 1 mol. Sequential enzyme degradation with β-galactosidase and β-hexosaminidase yields the following: galactose, 2 mol and GlcNAc, 2 mol. The fragment **IX** results from both structures:

$$Gal(\beta1\rightarrow4)GluNAc(\beta1\rightarrow3)Gal(\beta1\rightarrow4)Glu\text{-ceramide}$$
$$\textbf{IX}$$

Result. The results of example 3 described above lead to the following conclusions:

1. No combination of methylation linkage analysis and exoglycosylhydrolases can resolve structures **VII** and **VIII**.

2. Nilsson's technique cannot resolve structures **VII** and **VIII** because the internal residues are all periodate-resistant.[42]

3. FAB-MS of any of the intact, permethylated or peracetylated compound should give a result which shows that compound **VII** gives a pentasaccharide B fragment[17] while `hot easily yielding a tetrasaccharide B fragment. Compound **VIII** would give a tetrasaccharide B fragment and not a pentasaccharide B fragment.

Some problems with yield of the fragment could occur because of the propensity for charge retention (in positive ion FAB-MS) on the GlcNAc in the cleavage reactions such that compound **VII** could yield a low-abundance pentasaccharide fragment compared with B fragments from cleavages at the GlcNAc glycosidic bonds. A mixture of these two compounds in a sample would be a horrendous analytical enigma.

Treating first with β-galactosidase would give smaller compounds for mass spectrometry which would yield a trisaccharide fragment but not a disaccharide for **VII** while a disaccharide would be easily obtained from **VIII**.[46] CID of the fragment ions formed in either would answer the question.

Solutions to other difficult carbohydrate analytical problems which may be encountered are given below along with pertinent references.

1. Location of substituents on cyclitols was resolved by Hsieh *et al.*[47] where an *myo*-inositol was substituted with both a phosphodiester and a glycosidically linked sugar. The linkages were characterized by a combination of periodate oxidation and analysis of the stereochemically distinct products by gas–liquid chromatography and chemical ionization mass spectrometry.

2. Location of sulfates on heterooligosaccharides was resolved for glycosaminoglycan oligosaccharides by Dell *et al.*[48] using substitution of

[46] K. Watanabe, R. A. Laine, and S.-I. Hakomori, *Biochemistry* **14,** 2725 (1975).

[47] T. C.-Y. Hsieh, K. Kaul, R. A. Laine, and R. L. Lester, *J. Biol. Chem.* **253,** 3575 (1978).

[48] A. Dell, M. E. Rogers, and J. E. Thomas-Oates, *Carbohydr. Res.* **179,** 7 (1988).

the sulfates with acetates after permethylation under conditions which retained the sulfates.

3. Hellerqvist et al.[49] located acyl groups by methyl vinyl ether acetalization[50] of free hydroxyls, an alkali-stable substitution, followed by methyl sulfinyl carbanion and methyl iodide treatment to replace acyl groups with methyl or other alkyl functions.

4. Quantitation of picomole to nanomole amounts of glycosidically bound monosaccharides in complex biological matrices have been attempted by a number of authors. Chemical ionization mass spectrometry was used with gas–liquid chromatography and computer-reconstructed mass chromatograms to analyze N-acetylneuraminic acid directly from erythrocytes by Ashraf et al.[51] Roboz et al.[52] and Sugawara et al.[53] have used similar approaches. Sugiyama et al.[54] have also used GC/MS for estimation of free sialic acid. These methods could be adapted for any monosaccharide or partially methylated sugar.[5]

The problems of purity, stereochemical differences and identical mass and polarity in mixtures of oligosaccharides make mass spectrometry of this class of compounds most challenging. Still, one can see that enormous progress has been made in technology and ingenuity of approach. We can look forward to more sensitive instruments, novel energy-adding methods for the tandem mass spectrometric methods, and computer fingerprinting of data.

[49] C. G. Hellerqvist, B. Lindberg, S. Svennson, and Lindberg, Carbohydr. Res. 8, 43 (1968).

[50] A. N. DeBelder and B. Norrman, Carbohydr. Res. 8, 1 (1968).

[51] J. Ashraf, D. A. Butterfield, J. Jarnefelt, and R. A. Laine, J. Lipid. Res. 21, 1137 (1980).

[52] J. Roboz, R. Suzuki, and J. G. Bekesi, Anal. Biochem. 87, 195 (1978).

[53] Y. Sugawara, M. Iwamori, J. Portaukalian, and Y. Nagai, Anal. Biochem. 132, 147 (1983).

[54] N. Sugiyama, K.-I. Saito, H. Mizu, M. Ito, and Y. Nagai, Anal. Biochem. 170, 140 (1988).

[30] Linkage Analysis Using Lindberg Method

By Carl G. Hellerqvist

Introduction

An ever increasing number of biomedical investigators of the molecular mechanisms underlying biological and pathophysiological phenomena are venturing into the fascinating world of analytical carbohydrate chemistry. Analytical carbohydrate chemistry and the biochemistry of complex carbohydrates had long been ignored in most learning institutions in the United States. Over the last 20 years the number of complex structures elucidated has grown exponentially, as have the number of biological phenomena inferred or shown to involve complex carbohydrate structures. Such growth has gradually convinced many investigators that structural elucidation of the most complex carbohydrate structures is not only possible but essential to understanding biological phenomena.

It helps to overcome the resistance toward attacking structural problems if one considers, the good news, first: if a biological phenomena is governed by the interaction of specific carbohydrate structures with either complimentary protein or carbohydrate structures, then it is safe to assume that the critical carbohydrate structure or epitope is no larger than a hexasaccharide.[1] The bad news is that given the number of different sugars present in biological species, there may be hundreds of millions of hexasaccharide structures possible. At first glance, this thought may be discouraging.

As an example, sugar analysis of a hexasaccharide isolated as a biologically active epitope from a eukaryotic cell might yield D-mannose, D-galactose, L-fucose, and N-acetyl-D-glucosamine, in molar ratios of $2:1:1:2$. The hexoses, mannose and galactose, can each be linked through the oxygens on carbons 2, 3, 4, and 6 if present in the six-membered or pyranose ring form, or through carbons 2, 3, 5, and 6 if in the furanose or five-membered ring form. Thus there are eight different configurations possible for each hexose based on point of attachment. Each hexose can then be linked in either the α or β configuration. It follows that a hexose has 16 distinct possible forms of appearance: furanose, or pyranose, α-linked, or β-linked, and in each of these forms substituted at the oxygens of C-2, C-3, C-4, C-5, or C-6. By the same reasoning, there are

[1] E. A. Kabat, *J. Immunol.* **84,** 82 (1964).

METHODS IN ENZYMOLOGY, VOL. 193

12 distinct structures possible for the fucosyl, and N-acetylglucosaminyl residues which lack linkage positions at O-2 and O-6, respectively.

Based on point or mode of attachment and ring form there are 16 × 16 × 16 × 16 × 12 × 12 = 15,099,494 possible structures for our simple hexasaccharide. Moreover, these possible linkage configurations can be in different orders. With six sugars there are 6! = 720 possibilities. Since mannose and N-acetylglucosamine are present in two copies each we divide with 2! × 2! = 4 for a total of 15,099,494 × 720/4 or 2.72 billion different and distinct structures.

That the seemingly incredible task is an accomplishable one follows from the fact that methylation analysis which as discussed below will reveal mode of attachment and, in most instances, ring form, and thereby, will reduce the number of possible structures to a mere 720/4 for possible orders × 2^6 for different anomeric configurations in the six residues. Thus the 2.72 billion possibilities can be reduced to 11,520 possible structures. Data reduction systems are still obviously of the greatest importance.

Linkage analysis or methylation analysis as practiced today was developed in Professor Bengt Lindberg's laboratories in the late 1960s in Stockholm, Sweden. Two discoveries previously made this a tedious task of linkage analysis applicable to biological samples. The first was the report by Hakomori[2] that glycolipids could be completely methylated (permethylated) in one step. Second was the discovery by Björndal, Svensson, and Lindberg that partially methylated alditol acetates could be separated by gas chromatography[3] and identified by simple fragmentation rules by mass spectrometry.[4] The feasibility of the technology was first demonstrated by Hellerqvist et al. in the structural analysis of the Salmonella typhimurium lipopolysaccharide[5] which, along with the paper by Björndahl et al.,[6] became citation classics. The validity of the methodology was demonstrated by a series of papers from Professor Lindberg's groups which showed correlations between immunological properties and complex carbohydrate fine structures.

The successful partial sequencing of glycolipids by in-beam electron impact ionization discussed by Karlsson and co-workers ([34] in this volume) constituted the initiation of mass spectral methodologies aimed at resolving the remaining structural possibilities. This technology has

[2] S. Hakomori, J. Biochem. (Tokyo) 55, 205 (1964).

[3] H. Björndal, B. Lindberg, and S. Svensson, Acta Chem. Scand. 21, 1801 (1967).

[4] H. Björndal, B. Lindberg, and S. Svensson, Carbohydr. Res. 5, 433 (1967).

[5] C. G. Hellerqvist, B. Lindberg, S. Svensson, T. Holme, and A. A. Lindberg, Carbohydr. Res. 8, 43 (1968) (citation classic).

[6] H. Björndal, C. G. Hellerqvist, B. Lindberg, and S. Svensson, Angew. Chem. 82, 643 (1979) (citation classic).

evolved into the fast atom bombardment (FAB) methodology described by Angel and Nilsson ([32] in this volume) and Pappas et al.[7] To deal with strategy and all the data bases generated for the successful elucidation of a complex poly- or oligosaccharide, which include biosynthetic knowledge, sugar analysis, linkage analysis, anomeric configuration, and sequence (linkage and order), we have generated a computer-based expert system GLYCOSPEC (patent pending)[8] capable of resolving m/z values obtained from mixtures of periodate-oxidized and reduced oligosaccharide alditol acetates[7] into proposed structures.

The strategies and methods relevant to linkage analysis, which is the subject of this chapter, are based on the principle that *they must be applicable to as many types of sugars and linkages as possible.*[8] Furthermore, the experimental approaches have to be based on established carbohydrate properties and chemical reaction mechanisms.[9]

Methylation Procedures

Note: The methods described herein use acid-washed glassware and fresh glass-distilled highest grade solvents. All glassware used has been acid washed and stored no longer than 1 week wrapped in aluminum foil. Deviation from the highest standards may increase the risk of introducing contaminants during analysis by GC/MS and/or quantitation by gas chromatography.

Preparation of Methyl Sulfoxide Carbanion

Methyl sulfoxide carbanion, 2 M, is made up in 5-ml serum vials and stored in the refrigerator. The reactivity of the base make the driest possible environment necessary. To ensure desiccation we subscribe to the following procedure.[8]

1. Rig a high-wall container to be flushed by a gentle stream of dried nitrogen through an inverted funnel from above to protect the reagents from moisture.

2. Inside the container place a small beaker with dry hexane over calcium hydride, a small beaker with ethanol, and three Pasteur pipettes (acid washed-last reinforcement) with latex bulbs.

3. Weigh 500 mg NaH/oil (1 : 1) in a serum vial and place the vial in the high-wall container.

[7] R. S. Pappas, B. Sweetman, S. Ray, and C. G. Hellerqvist, *Carbohydr. Res.* **197,** 1 (1991).

[8] C. G. Hellerqvist and B. Sweetman, "Methods of Biochemical Analysis," Vol. 34: Biomedical Applications of Mass Spectrometry" (C. H. Suelter and D. T. Watson, eds.). Wiley, New York, 1990.

[9] C. Croon, G. Herrström, G. Kull, and B. Lindberg, *Acta Chem. Scand.* **14,** 1338 (1960).

4. Wash the NaH free of oil by repeating the following sequence three times: (1) add 10 ml of hexane and suspend by repetitive pipetting; (2) let the NaH settle; (3) transfer the liquid to the ethanol-containing beaker.

5. Allow the residual NaH to dry under nitrogen.

6. Add 5 ml of freshly prepared vacuum-distilled dimethyl sulfoxide (DMSO) and cap the vial. Remove the vial from the inert atmosphere and apply the metal seal.

7. Apply dry low vacuum with a needle through the septum sufficient to remove any hydrogen gas formed. A peristaltic pump will suffice for this purpose. Sonicate the vial at 50° in a water bath for 2 hr or until the liquid turns green, whichever comes first.

8. To check the potency of the base, withdraw a small aliquot with a syringe and add it onto a triphenylmethane crystal which should turn deep red from the carbanion formed in the proton extraction reaction. The droplet exiting from the needle should solidify on contact with the atmosphere. If done on a balance in a beaker, water can then be added along with an indicator such as phenolphthalein and molarity of the base determined by titration. The range 2–3 M is optimum. Metallic sodium may be used as an alternative to sodium hydride in the preparation of base.[8]

Sample Handling

Before attempting to do a structural analysis, it is essential to obtain all possible information regarding the unknown. Knowledge of biological origin and of principal biosynthetic pathways reveals essential information related to the structure of the material under investigation. Next NMR analysis should be performed to obtain as much information as possible before destructive procedures are applied. Deoxy as well as acyl and alkyl groups can be detected at the high nanomole level with no loss of material by the NMR analysis. Data obtained by these procedures should be part of the data base which governs the total analytical approach. If uronic acid residues are present, the method by Taylor and Conrad[10] for reduction via the carbodiimide derivative is recommended. Use of sodium borodeuteride in the reduction step will isotopically mark the original uronic acid derivative.

As discussed previously, the linkage analysis will generate the most structural information for the amount of material consumed. The number of possible structures will be reduced to a very manageable number once this information is obtained. To obtain a quantitative and qualitative linkage analysis the alkaline-sensitive reducing end, most likely if not always

[10] R. L. Taylor and H. E. Conrad, *Biochemistry* **11,** 1383 (1972).

present in an isolated oligosaccharide but not in a glycolipid or glycopeptide, must be protected.

The reducing end H-2 is acidic and its presence will initiate a β-elimination reaction, which is rapid and continuous if the sugars are linked through O-3 and more slowly if through O-4. This reaction known as "alkaline peeling," is extremely rapid and makes the use of alkaline conditions in any procedure very hazardous. This danger is, however, overcome by a reduction step preceding the alkaline treatment. In addition to saving the material from alkaline peeling, the reduction step converts the reducing end to an alditol readily identifiable by GC/MS as a $(n\text{-}1)\text{-}O$-methyl derivative, where n is the number of OH groups in the reducing end residue. Since 1% peaks can be readily identified, it follows that the degree of polymerization as well as the identity of the reducing end can be determined for fairly large polysaccharides.

The reduced oligosaccharide or polysaccharide sample is subjected to the following general procedure.[8]

1. Transfer sample (1–1000 μg) into a glass Reacti-vial (Pierce Chemical, Rockford, IL) and add 0.3–1.0 ml dry distilled DMSO. Flush with nitrogen and seal.

2. Sonicate the vial in a water bath until the sample is dissolved. Solubility may be increased by lowering the temperature 10° or elevating it 50°, depending on the sample. If sample polysaccharides of very limited solubility are available in large amounts, it may be feasible to sonicate the sample with a sonicator probe in order to partially degrade the material. This procedure is accomplished in water and is continued until the suspension is converted to a solution. The sample can then be reduced with borodeuteride, lyophilized, and dissolved in DMSO.

3. Add a volume of the 2 M base equal to the volume of the DMSO used. Sonicate for 30 min, allow to sit for 6 hr to overnight at room temperature.

4. Withdraw a minute droplet of the reaction mixture and place it on a triphenylmethane crystal. The crystal will turn bright red if base is still present.

5. If base is present, add freshly distilled methyl iodide (equal volume to the base previously added). Methyl iodide is taken up in a dry syringe and added dropwise to the reaction mixture, which is kept chilled at less than 15° in an ice-water bath.

6. Sonicate the solution in the cold room in a sonicator bath until the solution becomes clear. If it will not clear, after 60 min add one-half the volume of DMSO.

7. The methylated carbohydrate can be recovered by either of three

methods: (a) *Dialysis*. Add chloroform to the reaction mixture and transfer to a dialysis bag containing at least 10 ml of water. Wash the vial twice with chloroform and add to the bag. Chloroform will remain in the bag over a 48-hr dialysis period and prevent losses of even relatively low-molecular weight components. (b) *Chloroform Extraction*. Countercurrent chloroform water extraction can be used with low-molecular weight components. (c) *Adsorption on C_{18} Sep-Pak*. Evaporate off excess methyl iodide. Dilute sample 1 : 1 with water and apply it with a syringe to a C_{18} Sep-Pak (Millipore Corp., Milford, PA) which has been washed sequentially with 10 ml each of chloroform, DMSO, methanol, acetonitrile, and water. After sample has been applied, wash with 10 ml of water and 4 ml of acetonitrile : water, 15 : 85. The sample is eluted with 5 ml of chloroform. All phases should be saved.

8. The purified permethylated sample in chloroform is transferred to a small ampule and dried under nitrogen and then subjected to the reaction scheme outlined below.

Preparation of Alditol Acetates

Quantitative Hydrolysis

Several different hydrolytic conditions are used for the hydrolysis of methylated carbohydrates. The following procedure is applicable to the hydrolysis of acid-labile residues such as 3,6-dideoxyhexoses, sialic acids,[11] amino sugars,[11,12] and aminohexuronic acid derivatives.[13] The initial reaction is a hydrolysis but the organic–inorganic anhydride present yields N-acetylation as well as partial O-acetylation[14] which facilitates the complete hydrolysis of amino–glycosidic linkages and serves to protect sensitive sugar residues. In addition, use of sulfuric acid keeps de-O-methylation to a minimum.[9]

1. The permethylated sample (0.1–100 μg) is dissolved in an (acid-washed) glass ampule or Reacti-vial (2 ml) in 0.9 ml of glacial acetic acid. Sulfuric acid, 100 μl of 2 M, is added and the sealed container heated at 100° for 9 hr.
2. Acetic acid is removed by repetitive codistillation with water in a

[11] D. A. Cumming, C. G. Hellerqvist, M. Harris-Braedts, S. W. Michnick, J. P. Carver, and B. Bendiak, *Biochemistry* **28,** 6500 (1989).
[12] C. G. Hellerqvist and A. A. Lindberg, *Carbohydr. Res.* **16,** 39 (1971).
[13] S. V. K. N. Murthy, M. A. Melly, T. M. Harris, C. G. Hellerqvist, and J. H. Hash, *Carbohydr. Res.* **117,** 113 (1983).
[14] C. G. Hellerqvist, unpublished.

rotary evaporator or in a water bath at 40° under nitrogen to minimize losses of volatile components. When the scent of acetic acid is no longer detectable, water is added to reconstitute the original 1-ml volume and the sample is heated for 2 hr at 100° in a sand bath.

3. The sulfuric acid is neutralized by the addition of an *equimolar amount* of $BaCO_3$. Excess precipitate will lower the recovery.

4. The slurry is vacuum-filtered through a sintered-glass funnel (pore size F) into a 15-ml pear-shaped flask, and washed twice with 1 ml of water. Filtrate and washings are combined.

Reduction and Acetylation

1. To the combined filtrate and washings (<4 ml) is added a tenfold molar excess of sodium borodeuteride. The mixture is kept for 1 hr at ambient temperature. If microbubbles cannot be seen forming after 30 min, add more borodeuteride. If in doubt, use a 5-μl capillary tube to extract 2 μl sample, followed by extraction of 2 μl dilute acetic acid.

2. To terminate the reaction, add a small chip of dry ice to consume excess borodeuteride and to precipitate remaining barium ions as $BaCO_3$.

3. The mixture is then filtered through a sintered-glass funnel into a pear-shaped flask. The residue is washed with 2 × 1 ml water and the filtrate and washings combined.

4. Add approximately 3 ml of glacial acetic acid and 5 ml methanol to the mixture. Rotary evaporation removes the volatiles including the methyl borate which is formed. Before the mixture evaporates to dryness, add 5 ml more methanol. Repeat the procedure of addition of methanol and evaporation three more times. Transparent sodium acetate crystals should be seen inside the flask. White rather than transparent crystals are indicative of the presence of sodium borate, which hinders acetylation. If borate is present, it can be removed by dissolving the sample in 1 ml of water, adding 1 ml glacial acetic acid and 5 ml of methanol. Repeat evaporations with methanol alone three times. When the sample is completely free of borate, add 10 ml absolute ethanol and evaporate to dryness.

5. The residue is acetylated by the addition of ca. 0.5 ml of freshly distilled pyridine and ca. 1 ml of acetic anhydride using acid-washed Pasteur pipettes. The flask is stoppered and is heated in a sand bath at 100° for 30 min.

6. Remove the reaction mixture from the bath, and add about 5 ml toluene. Pyridine and acetic anhydride are codistilled with toluene to dryness at 40° using a rotary evaporator. Fresh toluene is added as needed (2–3 times) to maintain efflux of liquid from the sample. Use of the azeo-

trope minimizes loss of volatile derivatives from the sample. Dissolve the sample in 1 ml of ethyl acetate.

7. Prepare a filter unit from a Pasteur pipette fitted with a glass wool plug and 3–4 mm of active charcoal. Wash the filter unit with several milliliters of ethyl acetate. With the charcoal wet, add the residue from step 6 above, dissolved in ethyl acetate. Wash the filter twice with ethyl acetate (0.5 ml) and combine the filtrate and washings, which are then evaporated to dryness under a stream of nitrogen. The filtrate and washings are best collected in a Pasteur pipette which has been flame-sealed just below the tapered portion. The sample is dissolved in ethyl acetate prior to analysis by gas chromatography and/or GC/MS.

Identification of Partially Methylated Alditol Acetates

Partially methylated alditol acetates can be analyzed on a wide variety of capillary gas chromatography columns. Our laboratory has used OV-225 both in the packed column and in the fused-silica capillary configuration (Quadrex Corp., New Haven, CT) as our primary vehicle for quantitation and tentative identification. The tentative identification is based on relative retention times. Commonly the following conditions are used: injector and detector temperature at 240°; head pressure on fused-silica column (25 m) is set at 30 psi. Partially methylated alditol acetates are separated isothermally with the oven temperature at 190° for 20 min. A gradient of 10°/min to 250° and holding for 10 min elutes partially methylated amino sugar derivatives and assures the elution and identification of monomethylated and simple alditol acetates. The former would be a strong indication of undermethylation. Alditol acetates in the presence of fully methylated residues often suggests that the sample had not fully dissolved prior to the addition of base.

In order to draw conclusions about branches and nonreducing terminal residues, the stoichiometry should add up. A reduced carbohydrate found on linkage analysis to contain n nonreducing residues should contain one reducing end and n branch points. Each branch point is represented by one less O-methyl group in a residue than found when the same residue is present as an internal chain derivative.

Determination of Relative Retention Time

A data base of relative retention times is readily prepared by subjecting standard alkyl glycosides to partial methylation. GC/MS analysis of these samples yields a gas chromatogram as illustrated in Fig. 1. The positions

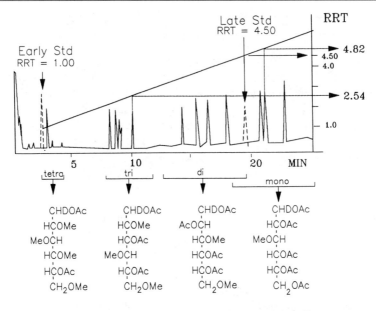

O−Me substitution determined by GC/MS

FIG. 1. The gas chromatogram represents an unknown sample. The relative retention times (RRT) for the partially methylated alditol acetates are calculated relative to an early standard RRT = 1.00 (2,3,4,6-tetra-*O*-methylglucitol) and a late standard RRT = 4.5 (2,3-di-*O*-methylglucitol). Isothermal conditions are derived to give RRT of 1.0 at an elution time of approximately 5 min. Oven temperature is approximately 190°, head pressure is 30 psi on a 25-m fused-silica column. Injector and detector temperatures are at 240°.

of the methyl groups are readily determined from the mass spectra; since all partially methylated components are derived from the same sugar, a retention time table is readily constructed. In order to have reproducible values it is essential to use a late and an early standard. We have elected to relate the retention time of any sugar derivative to 1,5-di-*O*-acetyl-2,3,4,6-tetra-*O*-methylglucitol and 1,4,5,6-tetra-*O*-acetyl-2,3-di-*O*-methylglucitol. A mixture of these components is injected before and after the sample under analysis. The assigned relative retention times and the true retention times in seconds for the late and early standards are entered in a computer program (available from the author) which will then analyze the unknown sample[15] and report relative retention times for all the peaks in the chromatogram as depicted in Fig. 1. In this manner the values are reproducible from packed column to capillary column on the same liquid

[15] J. Gailitt and C. G. Hellerqvist, unpublished.

TABLE I
ISOTHERMAL SEPARATION OF PARTIALLY METHYLATED
ALDITOL ACETATES ON OV-225 COLUMN

Location of methyl group	Relative retention times[a]					
	Xyl	Gal	Glc	Man	Fuc	Rha
2	2.15	—	6.6	5.65	1.43	1.37
3	2.15	—	7.6	6.8	—	1.67
4	2.15	—	8.4	6.8	1.71	1.57
6	—	—	5.0	4.1	—	—
2,3	1.19	4.7	4.50[a]	3.69	—	0.92
2,4	1.06	5.1	4.21	4.51	1.02	0.94
2,5	—	4.65	—	—	—	—
2,6	—	3.14	3.38	2.99	—	—
3,4	1.19	5.5	4.26	4.06[a]	—	0.87
3,5	—	5.1	—	4.51	0.96	—
3,6	—	—	3.73	3.67	—	—
4,6	—	—	3.49	2.92	—	—
2,3,4	0.54	2.89	2.22	2.19	0.58	0.35
2,3,5	—	2.76	—	—	0.47	—
2,3,6	—	2.22	2.32	2.03	—	—
2,4,6	—	2.03	1.82	1.90	—	—
2,5,6	—	1.95	—	—	—	—
3,4,6	—	2.15	1.83	1.82	—	—
2,3,4,6	—	1.19	1.00[a]	1.00	—	—
2,3,5,6	—	1.10	—	—	—	—

[a] 1.00 early standard; 4.50 late standard. Most of the relative retention times were established during the author's tenure at the University of Stockholm.

phase and relatively independent of column condition. Table I shows relative retention times generated by this procedure on OV-225.

Quantitation by Gas Chromatography

Chromatograms using flame ionization detectors are readily quantitated by the area under the peaks. After identification by mass spectrometry, the peaks can be corrected for the number of carbons in the derivative to yield more accurate values. The area representing the amount of a tetra-O-methyl derivative with 14 carbons, for example, would have to be multiplied by 16/14 in order to be compared with the area representing a di-O-methyl derivative. For reasons unknown to us, we need to use a response factor of 1.8 for the partially methylated amino sugar derivatives.

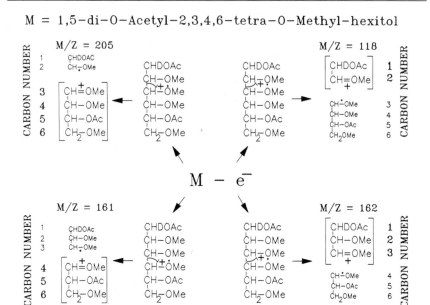

FIG. 2. Fission between carbons in the alditol chain will occur so as to create any one of the depicted possible fragment ions, m/z = 118, 161, 162, and 205 with charge stabilized on a methylated carbon at the site of cleavage (m/z = 45 containing only C-6 is not shown). M, 1,5-Di-O-acetyl-2,3,4,6-tetra-O-methylhexitol.

Fragmentation Rules in Electron Ionization Mass Spectrometry

As mentioned in the introduction, GC/MS became a powerful tool because of the very simple fragmentation patterns seen with partially methylated alditol acetates. The fate of 2,3,4,6-tetra-O-methylhexitol molecules, M, entering an electron ionization (EI) source is illustrated in Fig. 2. Any of the ether oxygens may lose an electron and generate a positive parent ion. The subsequent one-electron transfer results in cleavage of the alditol chain and formation of one radical and one positive ion which is detected and which identifies the number of methoxylated and acetoxylated carbons in the fragment ion. An even number identifies a fragment containing the deuterium on carbon one (C-1); an uneven number shows a fragment ion containing carbon six (C-6). The charge is carried on the methoxylated carbon at the site of cleavage of the alditol chain. The fragmentation rules were derived from the initial studies[6]: *Rule 1.* With the highest preference, fission occurs between an N-methylacetamidylated and a methoxylated carbon with charge retention on the N-methylacetami-

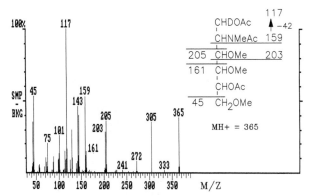

FIG. 3. Fragmentation Rule 1. The inserted structure illustrates the primary fragments formed in EI mass spectrometry of a 2-methylacetamido-2-deoxy-3,4,6-tri-*O*-methylhexitol acetate. Primary fragmentation occurs between C-2 and C-3 with charge retention on C-2 $m/z = 159$. Loss of ketene 42 u yields 117 as the base peak in the spectra. Formation of $m/z = 161$ and $m/z = 203$ follows Rule 2 and formation of $m/z = 45$, Rule 3. The figure shows sample minus background, or SMP − BKG.

dylated carbon. The spectrum in Fig. 3 shows the primary fragment ion m/z 159 with loss of 42 u to yield m/z 117 (base peak) as the signals representing this primary fragment. *Rule 2*. With the next highest preference, fission occurs between two methoxylated carbons with charge retention distributed on either one. The size of the signal is inversely proportional to the size of the fragment. This type of fragmentation is illustrated in Fig. 4. *Rule 3*. With the lowest preference, fission occurs between a methoxylated and an acetoxylated carbon with charge retention preponderantly on the methoxylated carbon (Fig. 5). Exceptions, but of low intensity, may be found in mono-*O*-methyl derivatives.

Identifying Fragment Ions

Aminitol derivatives are readily identified by the fragment ion formed by cleavage on either side of the N-methylacetamidylated carbon, which carries the charge and contributes 85 u to the fragment ion. The remaining primary fragments are formed by addition of methoxylated carbons, 44 u and/or acetoxylated carbons, 72 u (Fig. 3).

Primary fragments derived from a partially methylated [1-^{2}H]hexitol acetate give m/z values which equal ($n \times 44 + m \times 72 + 1$) when C-6 is included and ($n \times 44 + 72 + 2$) when the fragment ion contains C-1. Fragment ions containing a deoxy function have 14 u added. Consequently, the fragment ions containing C-6 in a 6-deoxy sugar give m/z values fitting the equation ($n \times 44 + m \times 72 + 15$).

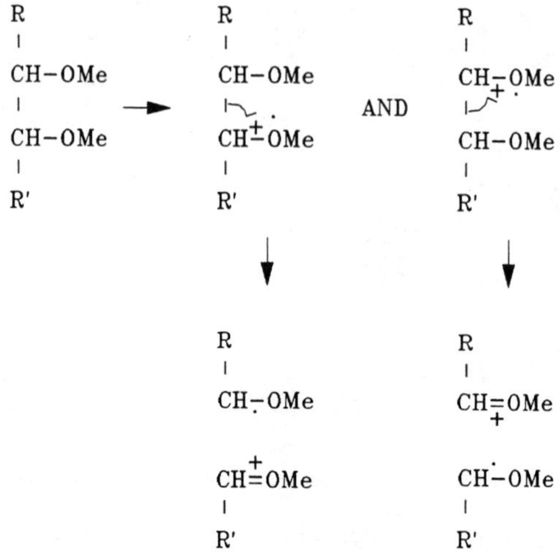

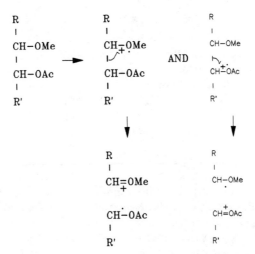

FIG. 4. Fragmentation Rule 2. Fission occurs between two methoxylated carbons with equal chance of charge distribution. Due to mass discrimination, however, the larger the fragment the lower its intensity.

FIG. 5. Fragmentation Rule 3. Fission between a methoxylated and acetoxylated carbon leads generally to the formation of the ion with charge retention on the methylated carbon.

TABLE II
PRIMARY FRAGMENTATION OF PARTIALLY METHYLATED HEXITOL ACETATES[a]

| Position | m/z | | | | | | | | | | | | | |
of O-Me	45	89	117	118	161	162	189	190	205	233	234	261	262	MH
2,3,4,6	■			■	■	■			■					324
2,3,5,6	■	■		■		■			■					324
2,3,4				■		■	■				■			352
2,3,6	■			■		■					■			352
2,3,5			■	■		■					■			352
2,4,6	■			■	■					■				352
2,5,6	■	■		■										352
3,4,6	■			■			■	□		■				352
3,5,6	■	■					■	□						352
2,3				■		■						■		380
2,4				■				■				■		380
2,5			■	■										380
2,6	■			■										380
3,4							■	■		■	■			380
3,5			■				■			■				380
3,6	■						■			■				380
4,6	■				■								■	380
5,6	■	■												380
2				■										408
3								■				■		408
4							■						■	408
5			■											408
6	■		□											408

[a] [MH − 60]⁺ will be significant in ion trap spectra in EI or CI mode; secondary ions may be formed by loss of HOAc (60 u), MeOH (32 u), CH₂CO (42 u), and CH₂O (30 u). □, Possible peak location.

Tables II, III, and IV illustrate the primary fragments anticipated from different substitution patterns of methyl groups in hexitols, 6-deoxyhexitols, and 2-acetamido-2-deoxyhexitols, respectively.

Figures 2, 3, 4, and 5 show that all primary fragments are formed through cleavage of the alditol chain with charge retention on an N-methylacetamidylated or on a methoxylated carbon.

The spectra shown in Figs. 3, 6, 7, 8, 9, and 10 serve to illustrate these principles.

The mass spectrum shown in Fig. 6 represents a 2-acetamido-2-deoxy-hexosidic residue which was 4,6-disubstituted in the carbohydrate component under investigation. The primary fragments are illustrated in the

TABLE III

PRIMARY FRAGMENTATION OF PARTIALLY METHYLATED 6-DEOXYHEXITOL ACETATES[a]

Position of O-Me	m/z											MH
	59	118	131	162	175	190	203	219	234	247	262	
2,3,4		■	■	■	■			■				294
2,3,5	■	■		■	■			■				294
2,3		■		■			■			■		322
2,4		■	■						■	■		322
2,5	■	■								■		322
3,4			■		■	■			■			322
3,5	■				■	■						322
2		■										350
3						■	■					350
4					■						■	350
5	■											350

[a] [MH − 60]+ will be significant in ion trap spectra in EI or CI mode; secondary ions may be formed by loss of HOAc (60 u), MeOH (32 u), CH2CO (42 u), and CH2O (30 u).

TABLE IV

PRIMARY FRAGMENTATION OF PARTIALLY METHYLATED 2-AMINO-2-DEOXYHEXITOLS[a]

Position of O-Me	m/z														MH
	45	89	117	159	161	189	203	205	233	261	318	319	346	374	
N,3,4,6	■		■	■	■		■	■							365
N,3,5,6	■	■	■	■			■	■				■			365
N,3,4			■	■		■	■			■	■				393
N,3,5			■	■			■			■	■				393
N,3,6	■		■	■			■			■	■				393
N,4,6	■		■	■	■						■				393
N,5,6	■	■	■	■							■				393
N,3			■	■			■			■			■		421
N,4			■	■		■							■		421
N,5			■	■									■		421
N,6	■		■	■									■		421
N,0			■	■										■	

[a] [MH − 60]+ will be significant in ion trap spectra in EI or CI mode; secondary ions may be formed by loss of HOAc (60 u), MeOH (32 u), CH2CO (42 u), and CH2O (30 u).

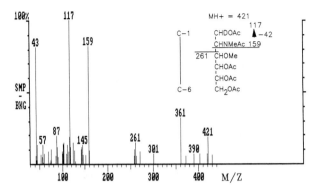

FIG. 6. The mass spectra, ITD in EI mode, illustrates the identification of a 3-*O*-methyl substitution indicating a branched 4,6-disubstituted amino sugar in the original material. The structural insert illustrates the primary fragment ions. $m/z = 346$, a low yield ion generated by fission between C-1 and C-2, may be detected dependent on sample size and instrument condition.

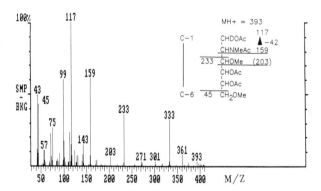

FIG. 7. Primary fragmentation (structural insert) and intensities recorded on the ITD in EI mode of an originally 4-substituted amino sugar.

inserted structure. The spectra was obtained from a fetuin oligosaccharide[11] on a Perkin-Elmer (Newark, NJ) 8500 GC combined with a Finnigan (San Jose, CA) Ion Trap Detector (ITD). The ion trap has a twofold advantage over the quadrupole instruments in linkage analyses where m/z values are less than 600 for most derivatized monomers. First, sensitivity is tenfold higher in the ion trap. Second, due to chemical ionization processes, as illustrated in Figs. 6–8, MH$^+$ and/or [M − 60]$^+$ are detected, which especially with amino sugar derivatives, helps with the correct identification of the number of methoxyl groups.

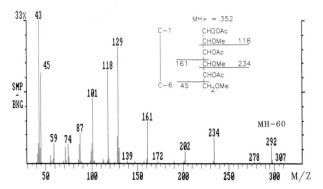

FIG. 8. Mass spectrum, ITD in EI mode, of 2,4,6-tri-*O*-methylhexitol acetate derived from an originally 3-substituted residue.

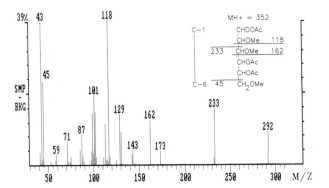

FIG. 9. Mass spectrum, ITD in EI mode, of 2,3,6-tri-*O*-methylhexitol acetate.

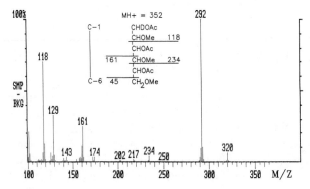

FIG. 10. Mass spectrum, ITD in CI mode, of 2,4,6-tri-*O*-methylhexitol acetate. (From M. Moreland, with permission.)

The mass spectrum derived from an originally 4-substituted 2-acet-amido-2-deoxyhexosidic residue is shown in Fig. 7 and the primary fragments are depicted in the inserted structure. The 6-O-methyl group yields a fragment m/z 45. The next fragment ion formed containing C-6 must contain the methoxylated carbon at C-3 and yield $m/z = 233$ with charge retention on the methoxylated carbon 3; $m/z = 203$ represents cleavage between a methoxylated C-3 and an acetoxylated C-4. This fragment ion, which may or may not be present dependent on the amount of sample being analyzed, contains C-1, C-2, and C-3 with charge retention again on C-3. The di-O-methyl substitution is, however, corroborated in the ion trap spectra by the presence of MH$^+$, $m/z = 393$.

The mass spectra shown in Figs. 8 and 9 represent hexoses originally substituted at O-3 and O-4, respectively. The primary fragment ions are indicated in the structural inserts. As seen, the two substitution patterns are readily distinguishable. These spectra illustrate the benefit of borodeuteride reduction.[12] Were borohydride to have been used, the even-numbered fragment ions 118, 162, and 234 containing C-1 would have been represented by 117, 161, and 233. Thus the two compounds would yield the identical primary fragment ions and could only be distinguished by secondary fragment ions arising from loss of acetic acid (60 u), methanol (32 u), formaldehyde (30 u), and ketene (42 u). The advantage of deuteride reduction is obvious. Use of deuteride makes distinction of mono-O-methyl substitution possible for *any* substitution.[8]

Deuterated methyl iodide is a valuable tool if the sugar analysis revealed naturally occurring O-methyl groups. These methoxylated carbons would contribute 47 u instead of 44 u toward the m/z values obtained from the final derivatives.

Chemical Ionization

Chemical ionization (CI) yields lower intensity fragments than EI, but can still lend itself to the type of identification described herein. CI has been successfully applied to partially methylated alditol acetates in quadrupole instruments for their identification, and for quantitation by single-ion monitoring of [MH − 60]$^+$.[16] The reported spectra are similar to those shown in Figs. 6–10 using an ion trap mass spectrometer in the EI mode except that for the partially methylated alditol acetates characteristic fragment ions are much less intense than [MH − 60]$^+$.

The spectra shown in Figs. 10 and 11 represent those obtained from an ion trap in the CI mode and should be compared to the spectra in

[16] R. A. Laine, *Anal. Biochem.* **116,** 383 (1981).

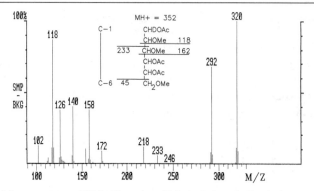

FIG. 11. Mass spectrum, ITD in CI mode, of 2,3,6-tri-*O*-methylhexitol acetate. (From M. Moreland, with permission.)

Figs. 8 and 9, respectively. The *O*-methyl substitution pattern because of deuteride reduction is readily determined. In both Figs. 10 and 11, $m/z = 320$ MH^+ with loss of methanol (32 u), and $m/z = 292$, loss of acetic acid (60 u), reveals a tri-*O*-methylhexitol derivative. The difference in intensities are due to the positions of the *O*-methyl and *O*-acetyl groups to be eliminated, relative to the location of the charge in the protonated parent molecule. The fragment ion $m/z = 118$ identifies a 2-*O*-methyl substitution; $m/z = 161$ identifies a 4,6-di-*O*-methyl substitution; and $m/z = 234$ identifies a 2,4-di-*O*-methyl substitution (Fig. 10). In Fig. 11, $m/z = 233$ identifies a 3,6-di-*O*-methyl substitution; however, $m/z = 162$, which would have corroborated a 2,3-di-*O*-methyl substitution, is missing in the mass spectrum recorded in the CI mode.

In comparing the EI and CI spectra generated on an ion trap, the EI mode has the advantage of giving a higher intensity of the critical primary fragment ions essential for identification of the position of the *O*-methyl groups. As the amount of sample decreases, these essential signals will be the last to vanish. With current software technologies, detection of the sugar derivatives does not depend on the abundance of one ion, rather it can be done by a reconstructed selected-ion chromatogram.

Selected-Ion Monitoring

The most sensitive detectors used in the analysis of sugar derivatives are the mass spectrometers used in selected-ion monitor mode.[16] These outperform the flame ionization detectors by several orders of magnitude. The tables and spectra presented herein illustrate that any partially methylated alditol derivative is identified by a few characteristic primary fragments. The mass spectrometers equipped with software that allows one to scan selected ions and graph composites of selected individual ions allows the investigator to perform linkage analysis on picomole quantities of material.[8]

The mass spectrometer response can be used to quantitate the identified derivatives. Prepare partially methylated derivatives of each individual type of sugar present as described previously. By comparing the flame ionization response to the differently substituted derivatives with the selected-ion composite response for the same derivatives it is possible to generate an adequate quantitative picture from less than microgram quantities of polysaccharides.[8]

The increased interest in complex carbohydrates and their biological functions has stimulated equipment development in this area. Since linkage analysis most often is the most revealing method for generating the fine structure of a complex, biologically active epitope, it is desirable that more sensitive detectors be developed for the small mass range needed for identification of all partially methylated alditol derivatives.

Acknowledgments

The author is indebted to his mentor Professor Bengt Lindberg; to Dr. Margaret Moreland, Glycomed, Inc., Alameda, CA, for submitting Figs. 10 and 11; to Drs. Laine and McCloskey for their comments and patience; to Mr. Edward Byrne for editorial work; and to Mrs. Sandy Rivers for the typing.

[31] Linkage Analysis Using Reductive Cleavage Method

By GARY R. GRAY

Introduction

The relatively recently introduced[1] technique of reductive cleavage allows the simultaneous determination of position(s) of linkage and ring form(s) in polysaccharides. The technique is based on methylation analysis, but departs from it significantly with regard to the types of fragments ultimately analyzed. In standard methylation analysis, the fully methylated polysaccharide is *hydrolyzed*, and the partially methylated sugars so formed are reduced ($NaBH_4$) and acetylated, and then characterized by gas–liquid chromatography-mass spectrometry.[2] Unfortunately, the

---

[1] D. Rolf and G. R. Gray, *J. Am. Chem. Soc.* **104**, 3539 (1982).
[2] C. G. Hellerqvist, this volume [30].

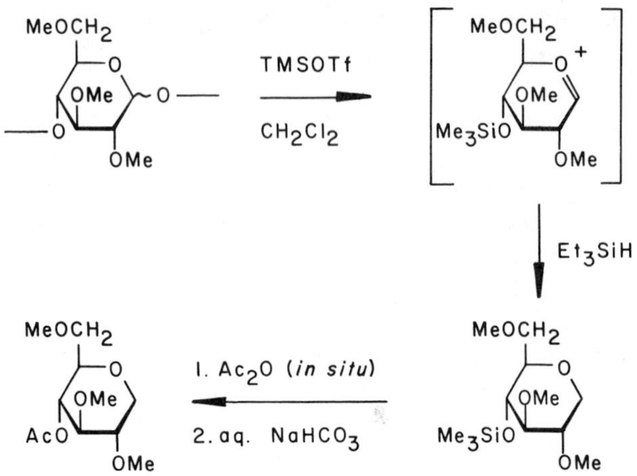

SCHEME 1. Reductive cleavage procedure for 4-linked D-glucopyranosyl residue.

method does not distinguish between 4-linked aldopyranosides and 5-linked aldofuranosides and, moreover, the relative configurations of the resultant partially methylated alditol acetates must be established by comparison of their gas chromatographic retention times with those of authentic standards. Although a procedure based upon standard methylation analysis has been devised[3] that distinguishes between 4-linked aldopyranosyl and 5-linked aldofuranosyl residues, it is quite complex and laborious and thus not well suited for routine analyses.

In contrast, the reductive cleavage procedure, as illustrated in Scheme 1 for a 4-linked D-glucopyranosyl residue, is much easier to perform, and the products, partially methylated *anhydroalditols*, are readily characterized by GC/MS of their acetates or ^{1}H NMR spectroscopy of their benzoates.[4] The latter method of characterization is less sensitive but useful in that it establishes the *identity* of the parent sugar as well as its position(s) of linkage and ring form, whereas the mass spectrometric procedure is more sensitive but does not directly establish the identity of the parent sugar. A comparison of the electron ionization (EI) mass spectra of partially methylated anhydroalditol acetates with those of authentic standards, however, can be used to identify the positions of substitution of methyl and acetyl groups and thus the position of linkage and ring form of the parent sugar residue. Recognizing that D-glucopyranosyl residues and D-mannopyranosyl residues are frequently encountered in complex carbo-

---

[3] A. G. Darvill, M. McNeil, and P. Albersheim, *Carbohydr. Res.* **86**, 309 (1980).
[4] C. K. Lee and G. R. Gray, *J. Am. Chem. Soc.* **110**, 1292 (1988).

hydrates, this chapter provides the mass spectra of the respective partially methylated 1,5-anhydro-D-glucitol (1) and 1,5-anhydro-D-mannitol (2) acetates derived from all terminal (nonreducing), singly, and doubly linked residues.

Experimental Procedures

Reductive Cleavage

Total reductive cleavage is performed with triethylsilane as the reducing agent and with either trimethylsilyl trifluoromethane sulfonate (TMSOTf) or a mixture of trimethylsilylmethane sulfonate (TMSOMs) and boron trifluoride etherate (BF$_3$ · Et$_2$O) as the catalyst. Reductive cleavage with Et$_3$SiH and TMSOTf and subsequent *in situ* acetylation[5] are accomplished as previously described in this series.[6] Reductive cleavage with Et$_3$SiH in the presence of TMSOMs and BF$_3$ · Et$_2$O and separate acetylation are performed as described by Jun and Gray.[7] Briefly, a 5-mg sample of the per-O-methylated polysaccharide and a small stirring bar are added to a Wheaton V-vial (Aldrich, Milwaukee, WI), the inside of which had been previously silanized.[6] The vial and contents are kept under high vacuum for 2 hr, then dichloromethane (0.25 ml, predried with CaH$_2$), Et$_3$SiH (5 Eq/Eq of acetal, Aldrich), TMSOMs (5 Eq/Eq of acetal, prepared as described in Ref. 7), and BF$_3$ · Et$_2$O (1 Eq/Eq of acetal, Aldrich) are sequentially added. The vial is then capped with a Teflon-lined screw top and the contents stirred for 24 hr at room temperature. Methanol (1.0 ml) is then added, stirring is continued for 30 min, and the mixture deionized by passage through a column (0.5 × 5 cm) of Bio-Rad AG501-X8 (Richmond, CA), analytical-grade, mixed-bed resin. Methanol and dichloromethane are removed by careful evaporation under vacuum, and the product acetylated by treating with 5 Eq each of acetic anhydride and 1-methylimidazole in 0.2 ml of dichloromethane for 30 min. Acetylation

[5] D. Rolf, J. A. Bennek, and G. R. Gray, *Carbohydr. Res.* **137**, 183 (1985).
[6] G. R. Gray, this series, Vol. 138, p. 26.
[7] J.-G. Jun and G. R. Gray, *Carbohydr. Res.* **163**, 247 (1987).

is terminated by the addition of saturated, aqueous sodium hydrogen carbonate (0.5 ml), the aqueous and organic layers are separated, and the dichloromethane layer is washed twice with 1-ml portions of water prior to analysis by GLC.

Analysis by Gas Chromatography-Mass Spectrometry

Gas–liquid chromatography is performed with a Hewlett-Packard 5890A gas–liquid chromatograph equipped with a J & W Scientific (Folsom, CA) DB-5 fused-silica capillary column (0.25 mm × 30 m; 0.25 μm film thickness). The column temperature is held at 110° for 2 min and then programmed to 300° at 6°/min. The column is eluted with helium at a flow rate of 1.3 ml/min. Mass spectra are acquired using either a Finnigan 4000 mass spectrometer equipped with a VG Multispec data system or a VG Analytical Ltd. Model VG 7070E-HF high-resolution, double-focusing mass spectrometer, as designated in Figs. 1 and 2. All spectra are acquired at an ionizing energy of 70 eV and at a source temperature of 250°. The VG instrument is scanned from m/z 600 to 35 at 1 sec/decade whereas the Finnigan instrument is scanned from m/z 650 to 20 in 1 sec. The VG instrument is operated at an accelerating voltage of 5 kV.

Authentic Standards

All the compounds whose spectra are reported herein are synthesized by routes involving standard protection/deprotection strategies. Most are available from previous work,[5,8–11] but eight of the compounds are synthesized for the present study. The structures of all compounds are confirmed by ^{1}H NMR spectroscopy.

Results and Discussion

Mass Spectra

Given in Figs. 1 and 2 are the EI mass spectra of methylated/acetylated 1,5-anhydro-D-hexitols having the *gluco* and *manno* configurations, respectively. Included in the figures are all possible products arising from terminal (nonreducing), singly, and doubly linked D-glucopyranosyl and D-mannopyranosyl residues. Inspection of the spectra reveals diagnostic

[8] J. U. Bowie and G. R. Gray, *Carbohydr. Res.* **129,** 87 (1984).
[9] D. Rolf and G. R. Gray, *Carbohydr. Res.* **152,** 343 (1986).
[10] S. A. Vodonik and G. R. Gray, *Carbohydr. Res.* **172,** 255 (1988).
[11] S. A. Vodonik and G. R. Gray, *Carbohydr. Res.* **175,** 93 (1988).

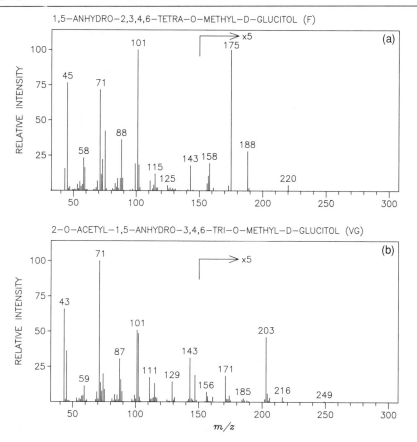

FIG. 1. (a–k) Electron ionization mass spectra of some methylated/acetylated 1,5-anhy-dro-D-glucitol derivatives (1). Spectra denoted with (F) were acquired on a Finnigan 4000 and those denoted with (VG) were acquired on a VG 7070E-HF GC/MS instrument.

differences for the *positional* isomers in terms of the presence or absence of certain ions and their intensity relative to the base peak (usually m/z 43, $CH_3C{\equiv}O^+$). In addition, the molecular ion [or pressure-induced $(M + 1)$ ion] is detected in low intensity in some spectra. The presence of this ion is useful in that it establishes the molecular weight of the component and thus its identity as an anhydrohexitol and its content of O-acetyl and O-methyl groups. The molecular weights of these components are more apparent, however, in chemical ionization (ammonia) mass spectra, wherein $(M + H)^+$ and $(M + NH_4)^+$ ions are detected.

Inspection of Figs. 1 and 2 also reveals small differences in ion intensities and, in a few cases, ion differences, in spectra of positional isomers

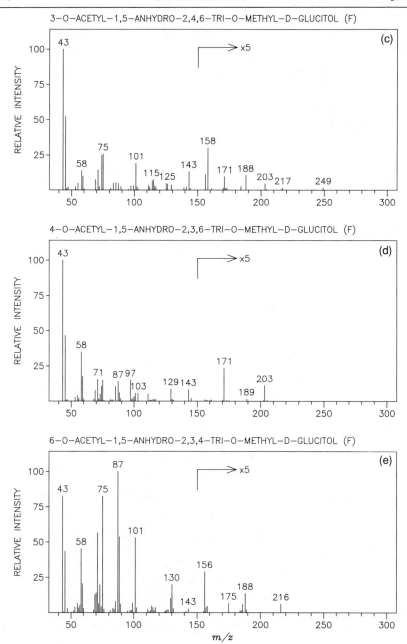

FIG. 1c, d, and e. See legend on p. 577.

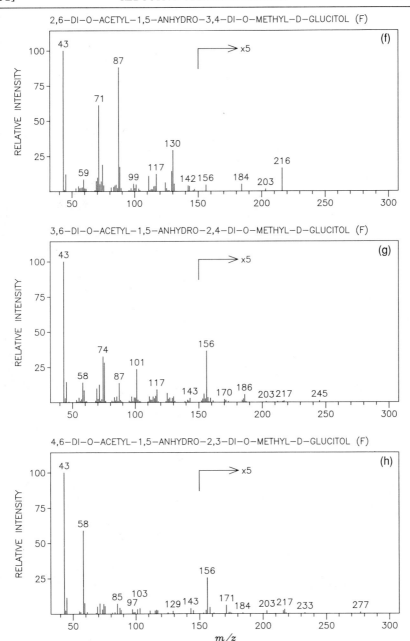

FIG. 1f, g, and h. See legend on p. 577.

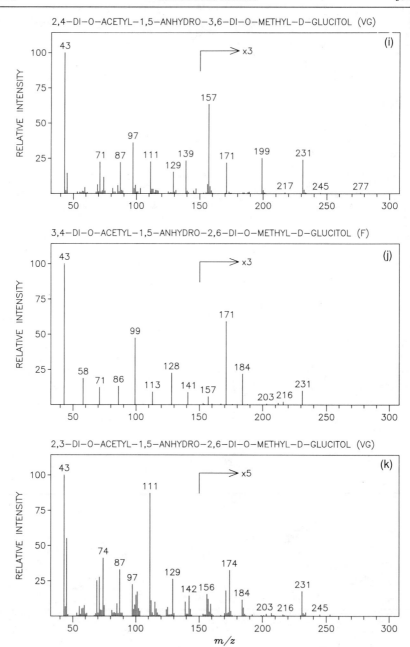

FIG. 1i, j, and k. See legend on p. 577.

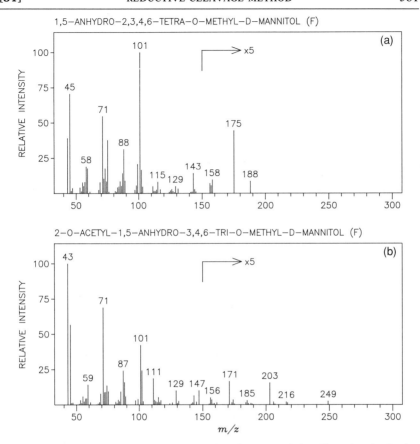

FIG. 2. (a–k) Electron ionization mass spectra of some methylated/acetylated 1,5-anhy-dro-D-mannitol derivatives (2). Spectra denoted with (F) were acquired on a Finnigan 4000 and those denoted with (VG) were acquired on a VG 7070E-HF GC/MS instrument.

having different configurations (*gluco* vs. *manno*). However, the spectra were obtained using different instruments over a period of several years. Since efforts were not made to obtain the spectra under identical conditions, the use of the spectra to assign *configuration* is not warranted. Indeed, even if reproducible differences did exist in the spectra of configurational isomers, the use of such differences for assignment purposes would probably necessitate obtaining a complete set of reference spectra on every instrument used for the analysis. Under these circumstances, a comparison of gas chromatographic retention times would be less laborious.

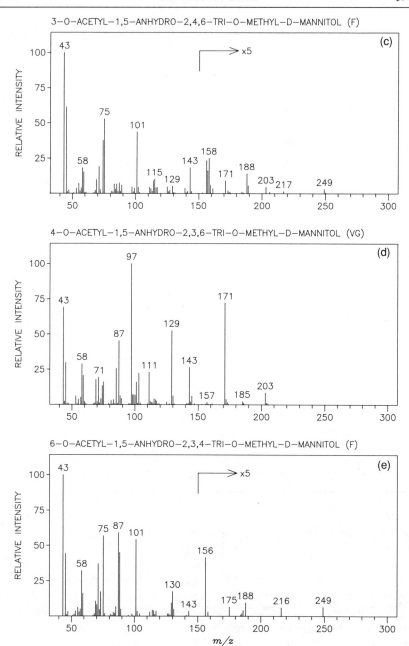

FIG. 2c, d, and e. See legend on p. 581.

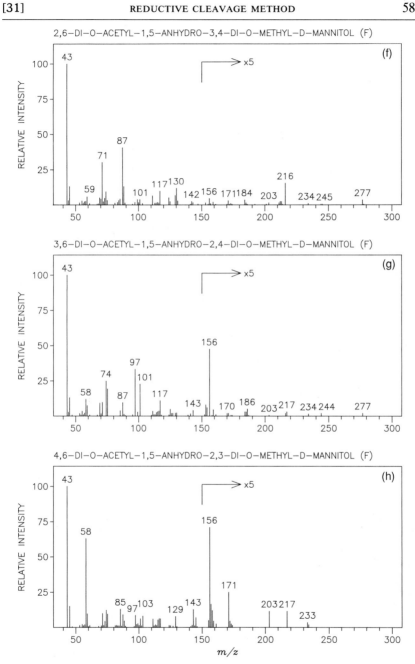

FIG. 2f, g, and h. See legend on p. 581.

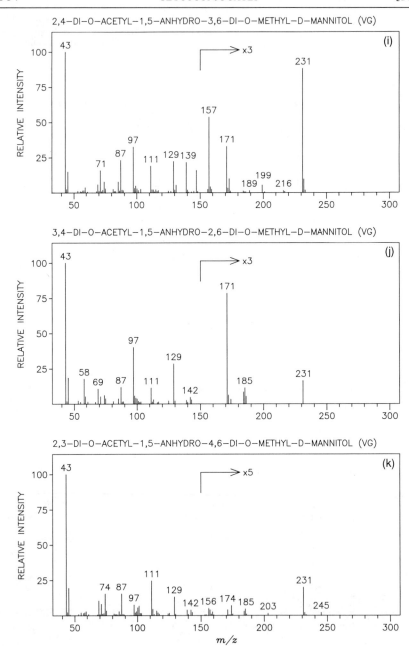

FIG. 2i, j, and k. See legend on p. 581.

TABLE I

INTENSITY RELATIVE TO BASE PEAK OF SELECTED IONS IN MASS SPECTRA OF
METHYLATED/ACETYLATED 1,5-ANHYDRO-D-GLUCITOL DERIVATIVES (1)

Position of O-acetyl	Molecular weight	Relative intensity (%)[a]					
		M − 32	M − 45	M − 60	M − 73	M − 77	M − 105
None	220	5.5	20.8	b	—	17.9	11.9
2	248	0.7	9.3	0.2	0.2	3.7	31.1
3	248	0.1	0.9	2.1	—	1.9	13.1
4	248	—	2.2	—	0.1	4.7	8.1
6	248	1.2	—	2.7	1.3	—	2.7
2,6	276	—	—	3.3	0.3	—	—
3,6	276	—	—	0.1	0.1	—	0.3
4,6	276	—	—	0.4	0.4	—	1.2
2,4	276	—	8.0	—	—	8.4	7.3
3,4	276	—	3.3	0.7	0.2	—	19.7
2,3	276	—	3.5	0.2	0.3	0.1	3.6

[a] Actual intensities vary depending on the instrument used, but observed values are given
to provide an indication of their relative prominence.
[b] Not observed.

Fragmentation Patterns

The fragmentation patterns for derivatives of this type have not been
established, but inspection of the spectra reported herein reveals diagnostic differences based upon the presence or absence of only a few ions in the
spectra. Some key ions are those that arise from elimination of methanol
(M − 32) or acetic acid (M − 60) from the molecular ion, and those that
arise by cleavage between C-5 and C-6 with loss of C-6 and an attached
O-methyl (M − CH$_2$OMe; M − 45) or O-acetyl (M − CH$_2$OAc; M − 73)
group. Fragment ions are also observed at M − 77 and M − 105 due to the
further loss of methanol and acetic acid, respectively, from the M − 45 ion.
The presence of the M − 105 ion can also be explained in terms of loss of
methanol from the M − CH$_2$OAc ion. It should be noted, however, that
there are other ions in the spectra of as yet unexplained origin that are
also diagnostic for identification of a particular positional isomer.

Summarized in Tables I and II are the results obtained from a comparison of the spectra of the *gluco* and *manno* isomers, respectively, for the
relative intensities of the aforementioned fragment ions. Since the spectra
were acquired using different instruments, it is recognized that comparisons based upon the actual intensities are not appropriate. The values
listed do, however, provide an indication of the relative prominence of a
given ion and on this basis they are useful for distinguishing between the

TABLE II

Intensity Relative to Base Peak of Selected Ions in Mass Spectra of
Methylated/Acetylated 1,5-Anhydro-d-mannitol Derivatives (2)

Position of O-acetyl	Molecular weight	Relative intensity (%)[a]					
		M − 32	M − 45	M − 60	M − 73	M − 77	M − 105
None	220	1.7	8.9	[b]	—	14.0	8.1
2	248	0.4	3.1	0.2	0.1	3.3	6.4
3	248	—	0.9	2.8	0.1	1.8	18.4
4	248	—	8.4	—	—	72.2	26.4
6	248	1.2	—	1.9	1.3	—	3.5
2,6	276	0.1	—	3.1	0.3	—	0.6
3,6	276	0.4	—	0.4	0.1	—	0.4
4,6	276	—	—	—	2.3	—	5.0
2,4	276	—	29.4	0.5	—	1.8	11.0
3,4	276	—	5.5	—	—	—	26.1
2,3	276	—	4.0	—	0.3	—	0.8

[a] Actual intensities vary depending on the instrument used, but observed values are given to provide an indication of their relative prominence.
[b] Not observed.

various positional isomers. As is evident in Tables I and II, compounds containing a 6-O-methyl group are readily identified by the presence of an M − 45 ion, which is always much greater in intensity than the M − 73 ion, if indeed the latter is even present. In contrast, the M − 73 ion is always present in the spectra of 6-O-acetyl derivatives and the M − 45 ion is absent. Mono-O-acetyl derivatives (2-, 3-, and 4-O-acetates) that give an M − 45 ion are distinguished by the patterns of elimination of methanol and acetic acid from the molecular ion; i.e., an M − 32 ion at m/z 216 is either weak in intensity or not observed in the spectra of 3-O- and 4-O-acetyl derivatives but is observed in the spectra of 2-O-acetates. The 3-O-acetates are distinguished from the 2- or 4-O-acetates by the presence of an M − 60 ion at m/z 188. The three di-O-acetyl regioisomers (2,4-, 3,4-, and 2,3-diacetates) that give an M − 45 ion are easily distinguished by the intensities of ions at M − 77 and M − 105. In 2,4-di-O-acetyl derivatives, the M − 77 ion at m/z 199 is quite prominent, but this ion is either very weak in intensity or not observed in 2,3- and 3,4-diacetates. Mass spectra of the 3,4-diacetates contain a very intense M − 105 ion at m/z 171, however, distinguishing them from 2,3-diacetates. The three di-O-acetyl regioisomers (2,6-, 3,6-, and 4,6-diacetates) that fail to give an M − 45 ion are distinguished by the intensities of ions at M − 60 and M − 105. The M − 60 ion at m/z 216 is prominent in the spectra of the 2,6-diacetates, thus distinguishing the 2,6-diacetyl from the 3,6- and 4,6-diacetyl regioisomers. The 3,6- and 4,6-diacetyl regioisomers are distinguished, in turn, by the intensity of the $M - 105$ ion at m/z 171 which is prominent only in the spectra of 4,6-diacetates.

Whether the patterns of elimination and fragmentation noted above will be observed for other configurational isomers, such as those of the *galacto* configuration, remains to be established. The similarities noted above for isomers of the *gluco* and *manno* configurations suggest, however, that major differences will not be observed in the spectra of other configurational isomers.

Acknowledgments

The author wishes to thank Samuel Zeller, Angela Ashton, Anello J. D'Ambra, Molly McGlynn, Christine Rozanas, Judith Sherman, and Dinesha Weerasinghe for the synthesis of previously unreported standards, Dr. Edmund Larka for performing GC/MS analyses, and Steven C. Pomerantz (University of Utah) and Kenneth L. Tomer (National Institute for Environmental Health Services) for assistance with plotting of mass spectra. This investigation was supported by PHS grant number GM34710, awarded by the National Institute of General Medical Sciences, DHHS.

[32] Linkage Positions in Glycoconjugates by Periodate Oxidation and Fast Atom Bombardment Mass Spectrometry

By ANNE-SOPHIE ANGEL and BO NILSSON

The introduction of fast atom bombardment mass spectrometry (FAB-MS) started a new era in mass spectrometry. Biological compounds like peptides and underivatized glycoconjugates, which cannot be analyzed by the conventional electron ionization mass spectrometry, could be analyzed using this technique. In contrast to electron ionization, the ionization in FAB is not dependent on volatility, since compounds are ionized in solution by a fast atom beam of argon or xenon. FAB-MS of underivatized glycoconjugates gives information about the molecular weight, but fragment ions formed by cleavage are of low abundance.

Derivatization, for example, peracetylation or permethylation, enhances the fragmentation. For derivatized glycoconjugates the fragment ions, mainly formed by cleavage of the glycosidic bonds, determine the monosaccharide sequence. In FAB-MS, carbon–oxygen bonds are preferentially cleaved in favor of carbon–carbon bonds, due to the low energy of ionization used. Sequence ions containing the reduced terminal, the J series of ions,[1] are therefore usually of low abundance. The monosaccharide sequence in permethylated glycoconjugates is best determined from ions derived from the nonreducing terminal. Binding positions between monosaccharide residues can, in general, not be determined by FAB-MS. An exception is, however, permethylated glycoconjugates containing internal 2-acetamido-2-deoxyhexosyl (HexNAc) residues, where the secondary fragments determine the positions of substitution of the HexNAc residues.[2] Periodate oxidation is a well-known reaction in carbohydrate chemistry.[3] Sugar residues with vicinal hydroxyl groups are cleaved, by periodate, between the carbons resulting in formation of aldehyde groups. This reaction, followed by reduction, permethylation, and FAB-MS, can be used to determine binding positions between monosaccharide residues. The products formed after periodate oxidation depend on position of substitution and the linkages between the monosaccharide residues are deduced from the primary and secondary sequence ions.

Periodate Oxidation

Oligosaccharides, oligosaccharide alditols, and glycosphingolipids (0.1–2.0 mg) are oxidized with periodate in 5 ml of 0.1 M acetate buffer, pH 5.5, containing 8 mM sodium periodate. The oxidation is carried out in the dark at 4° for 48 hr. Excess periodate is destroyed by addition of ethylene glycol (25 μl) and the sample is left at 4° overnight. The pH is adjusted to 7.0 with 0.1 M NaOH; NaBD$_4$ (25 mg) is added and the reduction is carried out overnight at 4°. Acetic acid is added to produce pH 4.5 and, after concentration to dryness, boric acid is removed by evaporations with methanol (3 × 3 ml). The oxidized and reduced product is desalted on a BioGel P-2 column (20 × 1 cm) (Bio-Rad, Richmond, CA) eluted with water. Fractions from all column chromatography are assayed for hexose by the anthrone method.[4]

[1] N. K. Kochetkov and O. S. Chizhov, *Adv. Carbohydr. Chem.* **21**, 39 (1966).
[2] G. Grönberg, P. Lipniunas, T. Lundgren, K. Erlansson, F. Lindh, and B. Nilsson, *Carbohydr. Res.* **191**, 261 (1989).
[3] I. J. Goldstein, G. W. Hay, B. A. Lewis, and F. Smith, *Methods Carbohydr. Chem.* **5**, 361 (1965).
[4] T. A. Scott, Jr., and E. H. Melvin, *Anal. Chem.* **25**, 1656 (1953).

Permethylation and Peracetylation

Periodate oxidized and reduced oligosaccharides and glycosphingolipids are methylated with methyl sulfinylcarbanion and methyl iodide.[5] The permethylated products are purified on a Sep-Pak (Millipore Corp., Milford, MA) C_{18} reversed-phase cartridge,[6] which was previously rinsed with chloroform (20 ml), methanol (10 ml), acetonitrile (4 ml), and water (4 ml). Methylated samples in dimethyl sulfoxide (DMSO) are diluted with water (1 vol) and applied to the cartridge. The following procedure is used for oligosaccharides: washing with water (8 ml), aqueous 15% acetonitrile (4 ml); permethylated oligosaccharides are eluted with acetonitrile (4 ml). For permethylated glycosphingolipids, the procedure involves washing with water (8 ml), acetonitrile (4 ml), and methanol (2 ml) and the permethylated products are eluted with chloroform (4 ml).

Lyophilized samples of periodate oxidized and reduced oligosaccharides are peracetylated in 4 ml of acetic anhydride–pyridine 2 : 1 (v/v) at 100° for 30 min. Ethanol (5 ml) is added and, after concentration to dryness, water (5 ml) is added and the peracetylated oligosaccharides are extracted three times with chloroform (3 × 1 ml). The combined chloroform phases are washed with water (3 × 2 ml) and concentrated to dryness. Residual pyridine is removed by evaporation with toluene (1 ml).

Isolation of Oligosaccharides from Glycoproteins

Fetuin, 1 g, is treated with 100 ml of 1 M NaOH containing 1 M NaBH$_4$ at 100° for 6 hr.[7] The solution is neutralized with 50% aqueous acetic acid, concentrated and chromatographed on a BioGel P-2 column (2.5 × 90 cm), eluted with 0.05 M acetic acid. Pooled fractions are N-acetylated by addition of 10% acetic anhydride in ethanol, in small portions and under constant pH at 7.5, adjusted by addition of 1 M or 0.1 M NaOH.[8] The N-acetylated product is desalted on the same BioGel P-2 column eluted with water. The oligosaccharide alditols are further fractionated on a concanavalin A (Con A)-Sepharose column (2 × 30 cm), eluted at room temperature with 0.05 M Tris-HCl buffer, pH 7.5, containing 0.15 M NaCl.[9] After the first peak, containing triantennary-type structures, the elution is

[5] S. Hakomori, *J. Biochem. (Tokyo)* **55**, 205 (1964).
[6] T. A. Waeghe, A. G. Darvill, M. McNeil, and P. Albersheim, *Carbohydr. Res.* **123**, 281 (1983).
[7] Y. C. Lee and J. R. Scocca, *J. Biol. Chem.* **247**, 5753 (1972).
[8] H. Krotkiewski and E. Lisowska, *Arch. Immunol. Ther. Exp.* **26**, 139 (1978).
[9] H. Krotkiewski, B. Nilsson, and S. Svensson, *Eur. J. Biochem.* **184**, 29 (1989).

continued with 15 mM methyl-α-mannoside in the starting buffer to recover biantennary-type structures. The bi- and triantennary oligosaccharide alditols are separated from low-molecular weight material on the same BioGel P-2 column, eluted with water.

The sialylated bi- and triantennary oligosaccharide alditols are purified from small amounts of contaminating glycopeptides on a HPAE-PAD system consisting of a Dionex BioLC gradient pump and a CarboPac PA-1 column (9 × 250 mm) (Dionex Corp., Sunnyvale, CA). For detection the following pulse potentials and duration times are used: E_1 = 50 mV (t_1 = 360 msec), E_2 = 800 mV (t_2 = 120 msec) and E_3 = -600 mV (t_3 = 420 msec). The separation is accomplished using 200 mM NaOH with a linear gradient of sodium acetate from 50 mM up to 115 mM at 39 min at a flow rate of 5.0 ml/min. Purified fractions are desalted on the same BioGel P-2 column as above. Sialic acid linked to galactose is removed from tetrasialylated triantennary structures, 3 mg, by digestion with 1 unit of neuraminidase from *Clostridium perfringens* Type VI in 0.05 M acetate buffer, pH 4.7, at 37° for 2 hr. The digest is heated at 100° for 2 min and finally purified in water on the above BioGel P-2 column.

Isolation of Oligosaccharides from Glycosphingolipids

Globoside, 50 mg, is dissolved in 20 ml of trifluoroacetic acid–trifluoroacetic anhydride 1 : 50 (v/v) and heated at 100° for 48 hr in a sealed, thick-walled glass tube.[10] (*Caution*: highly corrosive mixture under pressure.) After evaporation of the solvents, 10 ml of methanol is added to remove *O*-trifluoroacetyl groups and, after concentration to dryness, the products are fractionated on a Sephadex G-25 column (2 × 80 cm). The N-trifluoroacetylated globo-*N*-tetraose is N-detrifluoroacetylated in 10 ml of methanol–NH$_4$OH, 4 : 1 (v/v) at room temperature for 24 hr. The compound is acetylated in 5 ml of acetic anhydride–pyridine 2 : 1 (v/v) at 100° for 1 hr. *O*-Acetyl groups are selectively hydrolyzed in 10 ml of methanol–NH$_4$OH 4 : 1 (v/v) at room temperature for 24 hr. The product is finally purified on the same Sephadex G-25 column as above eluted with water. A solution of G$_{D3}$ ganglioside, 50 mg in 30 ml of dry methanol, is treated with ozone at room temperature for 20 min.[11] After concentration to dryness the residue is dissolved in 10 ml of 0.2 M sodium carbonate and left at room temperature for 16 hr. The solution is concentrated and degradation products from the ceramide moiety are removed on a Sep-

[10] B. Nilsson and D. Zopf, this series, Vol. 83 [2].
[11] H. Wiegandt and G. Baschang, *Z. Naturforsch.* **206,** 164 (1965).

Pak C_{18} cartridge eluted with water. The oligosaccharide is finally desalted on a BioGel P-2 column (5 × 90 cm) eluted with water.

Fast Atom Bombardment Mass Spectrometry

FAB mass spectra in positive ion mode are recorded on a VG ZAB SE instrument (VG Analytical, Manchester, UK). Samples are dissolved in thioglycerol (1-thio-2,3-propanediol) and loaded on the stainless steel target. The target is bombarded with xenon atoms with a kinetic energy of 8 keV and an accelerating voltage of 10 kV is used.

Determination of Binding Positions from the Monosaccharide Sequence

In electron ionization mass spectrometry of permethylated glycoconjugates, the sequence ions, from the nonreducing terminal, are formed by ionization of the ring oxygen followed by a homolytic cleavage of the glycosidic bonds.[1] The same mechanism can be assumed to be in effect in FAB-MS. A nonreducing terminal hexosyl residue (Hex) is recognized by the primary and secondary fragments of m/z 219 and m/z 187, a nonreducing terminal HexNAc by m/z 260 and m/z 228, a nonreducing terminal deoxyhexosyl residue (deoxyHex) by m/z 189, and a nonreducing terminal Neu5Ac by m/z 376 and m/z 344. An internal Hex adds 204 mass units to the sequence ions, an internal HexNAc adds 245 mass units, and for an internal Neu5Ac residue 361 mass units are added. Information about binding positions between monosaccharide residues is obtained after applying the following steps: (1) periodate oxidation, (2) borodeuteride reduction, (3) permethylation or peracetylation, and (4) analysis by FAB-MS.

The products obtained after the above steps and the masses for nonreducing terminal and internal hexopyranosyl residues are seen in Table I. For a 2- or 4-substituted Hex residue 208 mass units are added to the sequence ions. A secondary fragment formed by elimination of methanol is seen for a 2-substituted, but not for a 4-substituted, residue.[12] The products formed from the reducing terminal vary with the position of substitution and the ring size. The ions for the modified reducing terminal have, however, too low m/z values to be unambiguously identified. The linkage to a reducing terminal hexopyranose is therefore best determined by addition of 16 mass units (for the glycosidic oxygen) to the sequence ion with the highest m/z value and subtracting the sum from the molecular

---

[12] A.-S. Angel, F. Lindh, and B. Nilsson, *Carbohydr. Res.* **168**, 15 (1987).

TABLE I

STRUCTURE OF PRIMARY SEQUENCE IONS FROM PYRANOSIDES AND PYRANOSES FORMED AFTER
PERIODATE OXIDATION, NaBD$_4$ REDUCTION, AND PERMETHYLATION[a]

Residue	Structure of primary sequence ion	Sequence ion (m/z)	
		Primary	Secondary
Terminal Hex		179	
Terminal 6-deoxy-Hex		149	
Terminal HexNAc		264	232
Terminal Neu5Ac		289	257
Hex 2-Substituted		X + 208	X + 208 − 32
3-Substituted		X + 204	

TABLE I (continued)

Residue	Structure of primary sequence ion	Sequence ion (m/z) Primary	Secondary

4-substituted

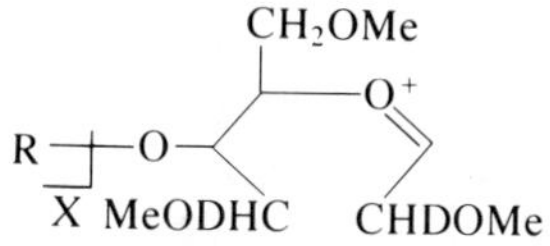

$$CH_2OMe$$
$$R\!-\!\!\!\!|\!-\!O\!-\!CH\!\!\!\!\!\!\!\!\!\!\!\overset{\displaystyle -O^+}{}\!\!\!\!\!\!\!\!\!\!\!\!\!\!\!$$
X MeODHC   CHDOMe

X + 208

6-Substituted

$$R\!-\!\!|\!-\!O\!-\!CH_2$$
X
$$\overset{|}{CH}\!\!\!\!\!\!\overset{-O^+}{}$$
MeODHC
MeODHC

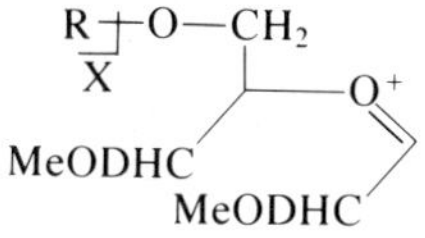

X + 164

Reducing Hex
2-Substituted

CHDOMe
|
$^+$CH
|
CHDOMe

105

3-Substituted

CHDOMe
|
$^+$CH
|
CHOMe
|
CHOMe
|
CH$_2$OMe

192

4-Substituted

CHDOMe
|
$^+$CH
|
CHOMe
|
CH$_2$OMe

148

6-Substituted

CHDOMe
|
CHOMe
|
$^+$CH$_2$

104

(continued)

TABLE I (*continued*)

Residue	Structure of primary sequence ion	Sequence ion (*m/z*)	
		Primary	Secondary
HexNAc-ol			
3-Substituted	CH₂OMe | CHN⟨Me / Ac | ⁺CH | CHDOMe	189	
4-Substituted	CH₂OMe | CHN⟨Me / Ac | CHOMe | ⁺CH | CHDOMe	233	

[a] The secondary fragments are formed from the primary fragments by elimination of methanol. The masses of derivatives obtained from 3- and 4-substituted reduced HexNAc residues are also shown.

weight. The masses of the remainder are the same as discussed above (Table I).

The molecular weight is determined from the protonated molecular ion, $[M + 1]^+$, or the $[M + 23]^+$ ion, which is formed when a sodium salt is present. For ions with $m/z > 1000$ the fractional mass values of the constituent atoms have to be taken into consideration. The annotated masses in the high mass scan correspond to the most abundant ions in the clusters. For compounds with more than 90 carbons, the isotope effect of ^{13}C is significant. The position of substitution of a HexNAc residue is determined from the secondary fragments formed, by elimination, after a primary cleavage of the HexNAc linkage. The primary sequence ion of m/z 424 is characteristic for a product obtained from a sequence of Hex-HexNAc and m/z 558 for the branched sequence of Hex-[6-deoxyHex]-HexNAc. The secondary fragments are preferentially formed by elimination of the substituent in the 3-position of the HexNAc residue. The relative intensities of the primary and secondary ions for 3-substituted, 4-substituted, and 3,4-disubstituted HexNAc residues are shown in Fig. 1.

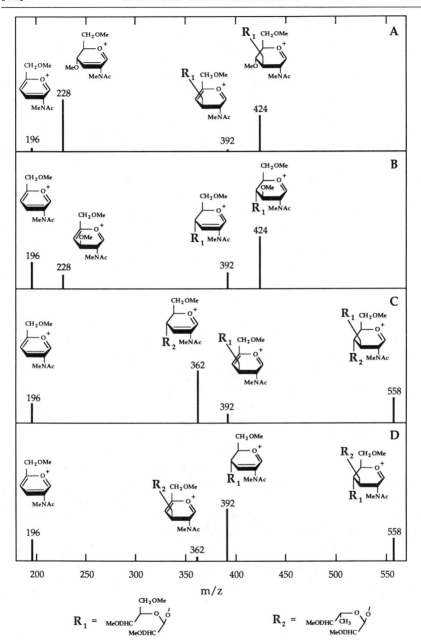

FIG. 1. Primary and secondary fragments formed from 3-substituted (A), 4-substituted (B), and for 3,4-disubstituted (C and D) HexNAc residues. The intensities are normalized to the most abundant ion in each spectrum.

In the FAB mass spectrum of a biantennary glycoprotein fragment, containing 2-substituted Man residues, the primary sequence ion of m/z 632, formed by addition of 208 mass units to m/z 424, gives a secondary fragment of m/z 600 (Fig. 2). The periodate oxidation was carried out on the reducing oligosaccharide which makes the reducing GlcNAc unaffected by periodate. The $[M + 1]^+$ and $[M + 23]^+$ ions of m/z 1748 and m/z 1770, respectively, are consistent with the structure of the product. For a 4-substituted Hex residue, seen in the spectrum of globo-N-tetraose, the primary sequence ion of m/z 676 gives no elimination of methanol (Fig. 3).

In oligosaccharides with a sequence of Neu5Ac-Gal, the linkage is determined by m/z 493 for a 2–3 linkage and for a 2–6 linkage by m/z 378 and m/z 453. These sequences are common in N-linked glycoprotein oligosaccharides. The biantennary oligosaccharides from fetuin[13] were isolated as a mixture of alditols and subjected to periodate oxidation, reduction, permethylation, and FAB-MS with a spectrum shown in Fig. 4. Two branches differing in sialic acid linkages can be recognized. One of the branches is determined by the primary sequence ions of m/z 289, 493, 738, and 946. The secondary fragments of m/z 706 and m/z 914 formed by elimination of methanol from m/z 738 and m/z 946, respectively, establish the linkages in the sequence of Neu5Ac2-3Gal1-4GlcNAc1-2Man. Another sequence of Neu5Ac2-6Gal1-4GlcNAc1-2Man is determined by the primary sequence ions of m/z 289, 378, 453, 698, and 906 and the secondary of m/z 666 and m/z 874, formed from m/z 698 and m/z 906, respectively, by elimination of methanol. Since the periodate oxidation was carried out on the reduced oligosaccharides, the reduced 4-substituted GlcNAc will be degraded to XylNAc-ol as seen from the alditol-containing ions of m/z 233 and m/z 478. The $[M + 23]^+$ ions of m/z 2520 and m/z 2560 show that at least two species of biantennary structures are present, one with both branches of 2–6-linked Neu5Ac and another with one branch each of 2–3- and 2–6-linked Neu5Ac.

A neuraminidase (*Clostridium perfringens*)-digested tetrasialylated triantennary-type structure from fetuin[13] was treated as above (Fig. 5). Certain structural features can be deduced from the spectrum. The primary and secondary sequence ions of m/z 424 and m/z 392 determine a sequence of Gal1-4GlcNAc and a sequence of Gal1-4GlcNAc1-2Man is deduced from ions of m/z 424 and m/z 632. A sialylated branch of Gal1-3[Neu5Ac2-6]GlcNAc is recognized by the ions of m/z 289, 257, and 698. The

[13] E. D. Green, G. Adelt, J. U. Baenziger, S. Wilson, and H. van Halbeek, *J. Biol. Chem.* **263**, 18253 (1988).

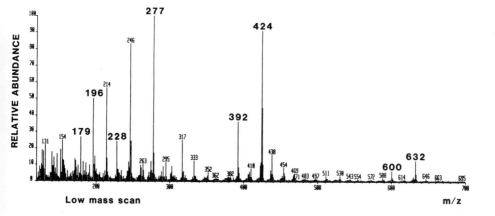

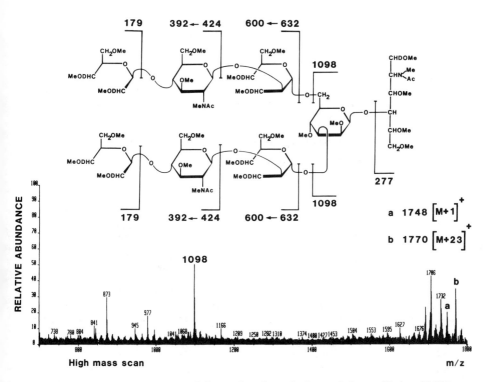

FIG. 2. FAB-mass spectrum of the product formed after periodate oxidation, NaBD₄ reduction, and permethylation of a biantennary glycoprotein fragment. Reprinted from *Carbohydr. Res.* **168,** 15 (1987). Copyright 1989 Elsevier Science Publishers.

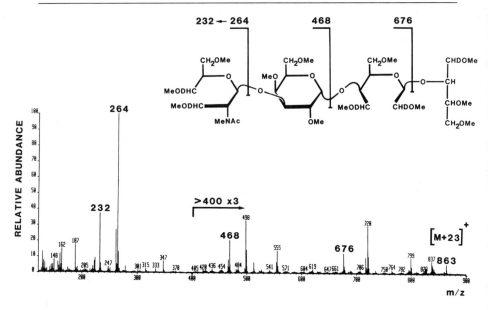

FIG. 3. FAB-mass spectrum of the product obtained after periodate oxidation, NaBD$_4$ reduction, and permethylation of globo-N-tetraose GalNAcβ1-3Galα1-4Galβ1-4Glc. Reprinted from *Carbohydr. Res.* **168**, 15 (1987). Copyright 1989 Elsevier Science Publishers.

[M + 23]$^+$ ion of m/z 2651 is consistent with a triantennary-type structure containing one Neu5Ac residue.

In glycoconjugates containing sialic acid linked to sialic acid, the binding position between these residues is often difficult to determine. The oligosaccharide Neu5Acα2-8Neu5Acα2-3Galβ1-4Glc, isolated from the G$_{D3}$ ganglioside, gave after reduction (NaBD$_4$) and permethylation a FAB spectrum shown in Fig. 6A. The sequence ions of m/z 376, 737, and 941 combined with the [M + 1]$^+$ and [M + 23]$^+$ ions of m/z 1194 and m/z 1216, respectively, determine a sequence of Neu5Ac-Neu5Ac-Gal-Glc-ol-1-d. After periodate oxidation, reduction, and permethylation a FAB spectrum was obtained shown in Fig. 6B. The corresponding sequence ions are m/z 289, 650, and 854 with an [M + 1]$^+$ ion of m/z 1019 and an [M + 23]$^+$ ion of m/z 1041. The linkage between the Neu5Ac residues must be, therefore, 2–8, since any other linkage should give oxidation of the internal Neu5Ac residue. The linkage to the reducing Glc is calculated by 1018 − 870 = 148, where 1018 is the molecular weight and 870 is the sum of the sequence ion with the highest m/z value (m/z 854) and 16 mass units for the glycosidic oxygen. The remainder of 148 shows that the reducing Glc is 4-substituted (Table I).

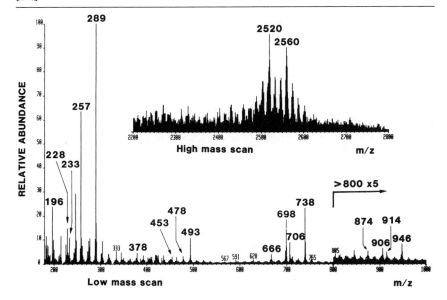

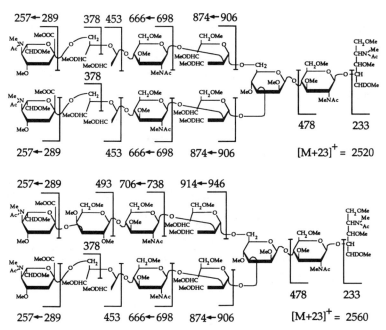

FIG. 4. FAB-mass spectrum of a mixture of biantennary oligosaccharide alditols after periodate oxidation, NaBD₄ reduction, and permethylation.

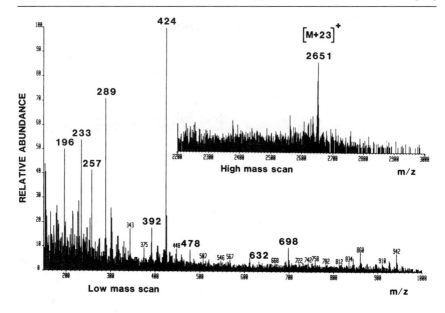

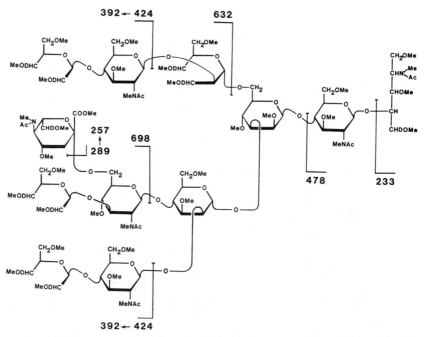

FIG. 5. FAB-mass spectrum of a monosialylated triantennary oligosaccharide alditol after periodate oxidation, NaBD₄ reduction, and permethylation.

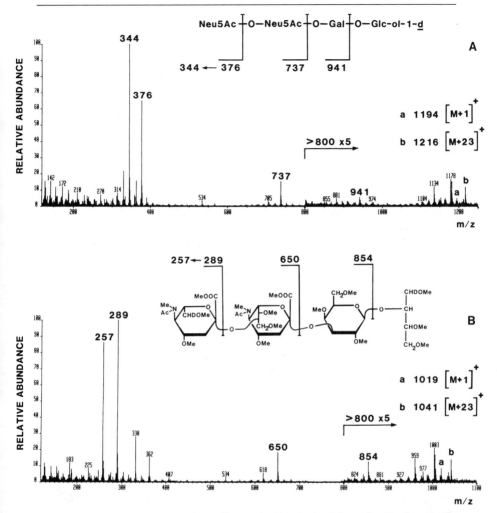

FIG. 6. FAB-mass spectra of an oligosaccharide obtained from the G_{D3} ganglioside Neu5Acα2-8Neu5Acα2-3Galβ1-4Glc as (A) permethylated alditol and (B) after periodate oxidation, NaBD$_4$ reduction, and permethylation. Reprinted from *Carbohydr. Res.* **168**, 15 (1987). Copyright 1989 Elsevier Science Publishers.

Glycoproteins containing O-linked oligosaccharides have, in most cases, GalNAc linked to serine or threonine in an alkaline-labile linkage. O-Linked oligosaccharides released, with alkaline borohydride degradation, as alditols often have a 3,6-disubstituted GalNAc-ol. This GalNAc-ol will be cleaved, between carbons 4 and 5, by periodate to form oligosaccharides with either N-acetylthreosaminitol or ethylene glycol as aglycons.[14] A branched oligosaccharide alditol with a structure of Galβ1-3GlcNAcβ1-3Galβ1-3[Galβ1-4GlcNAcβ1-6]GalNAc-ol, isolated from human blood group substance,[15] gave after permethylation a FAB spectrum shown in Fig. 7A. The two branches linked to the GalNAc-ol residue are recognized by the ions of m/z 228, 432, 464, and 668. Since no fragment ions are formed by cleavage within the alditol, the positions of substitution of the GalNAc-ol cannot be deduced from the spectrum. After periodate oxidation, reduction, and acetylation, a FAB spectrum was obtained shown in Fig. 7B. The primary sequence ions of m/z 263 and m/z 550 combined with the [M + 1]$^+$ ion of m/z 1087 show that the Gal-GlcNAc-Gal sequence is linked to a N-acetylthreosaminitol and, therefore, to the 3-position of the GalNAc-ol in the native compound. An [M + 1]$^+$ ion of m/z 655 together with m/z 263 and m/z 550 determine a structure of Gal-GlcNAc-ethylene glycol, showing that the native compound contains a Gal-GlcNAc sequence linked to the 6-position of the GalNAc-ol.

Another application of this method on O-linked glycoprotein oligosaccharides is given by Galβ1-3GlcNAcβ1-3Galβ1-4GlcNAcβ1-6GalNAc-ol, obtained from human blood group substance.[15] The FAB mass spectrum of the permethylated compound is shown in Fig. 8A. The primary and secondary sequence ions of m/z 464 and m/z 228 show a 3-substituted GlcNAc and m/z 914 together with m/z 882, a 4-substituted GlcNAc residue. A monosubstituted GalNAc-ol, is evident from the alditol containing ions of m/z 276, 521, and 971. The positions of substitution of the Gal and GalNAc-ol residues are determined from the spectrum obtained after periodate oxidation, reduction, and permethylation (Fig. 8B).

The primary sequence ion of m/z 628 is derived by the addition of 204 mass units to m/z 424, which means, as discussed above, a 3-substituted Gal residue (Table I). The alditol-containing ions of m/z 305 and m/z 754, combined with an [M + 1]$^+$ ion of m/z 951, are in agreement with an aglycon of ethylene glycol, thereby showing a 6-substituted GalNAc-ol in the native compound.

This method can also be applied to glycosphingolipids, which is demonstrated by the analysis of the Forssman antigen, GalNAcα1-3GalNAcβ1-3Galα1-4Galβ1-4Glc-Cer; the FAB mass spectrum obtained is shown in Fig. 9. The primary sequence ions of m/z 264, 509, and 713 show a sequence of GalNAc1-3GalNAc1-3Gal. The secondary fragment, m/z 228, shows that the internal GalNAc residue is 3-substituted. No other se-

[14] H. Krotkiewski, E. Lisowska, A.-S. Angel, and B. Nilsson, *Carbohydr. Res.* **184,** 27 (1988).
[15] A. M. Wu, E. A. Kabat, B. Nilsson, D. A. Zopf, F. G. Gruezo, and J. Liao, *J. Biol. Chem.* **259,** 7178 (1984).

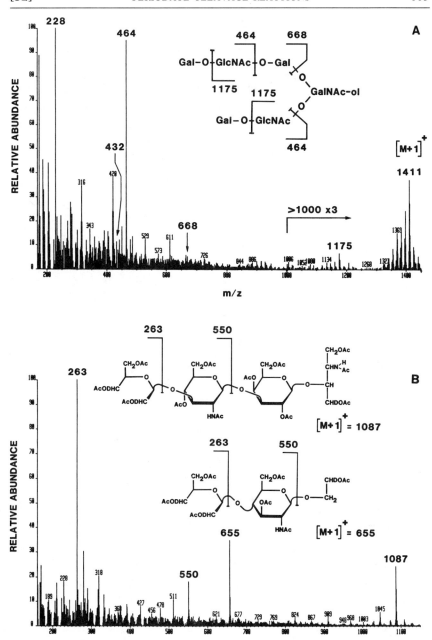

FIG. 7. FAB-mass spectra of a branched oligosaccharide alditol Galβ1-3GlcNAcβ1-3Galβ1-3[Galβ1-4GlcNAcβ1-6]GalNAc-ol, as (A) permethylated derivative and (B) after periodate oxidation, NaBD₄ reduction, and acetylation.

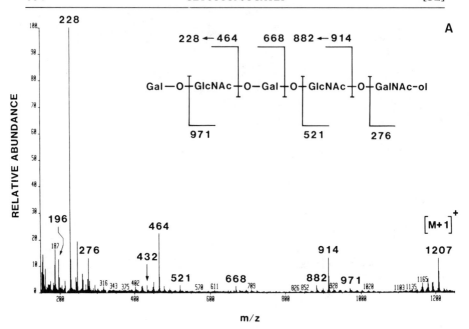

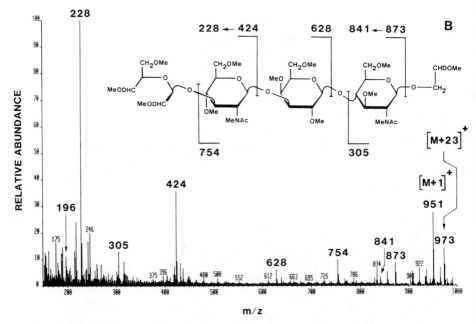

FIG. 8. FAB-mass spectra of a linear oligosaccharide alditol Galβ1-3GlcNAcβ1-3Galβ1-4GlcNAcβ1-6GalNAc-o1, as (A) permethylated derivative and after (B) periodate oxidation, NaBD$_4$ reduction, and permethylation.

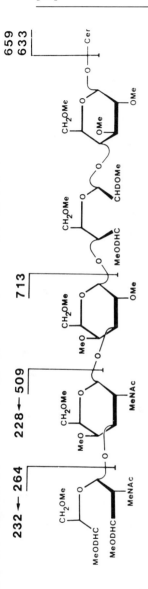

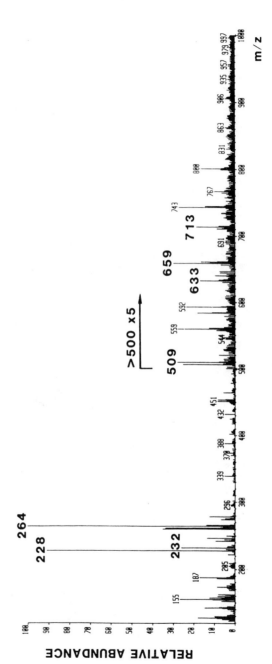

FIG. 9. FAB-mass spectrum obtained after periodate oxidation, NaBD$_4$ reduction, and permethylation of the Forssman antigen GalNAcα1-3GalNAcβ1-3Galα1-4Galβ1-4Glc-Cer. Reprinted from *Carbohydr. Res.* **168**, 15 (1987). Copyright 1989 by Elsevier Science Publishers.

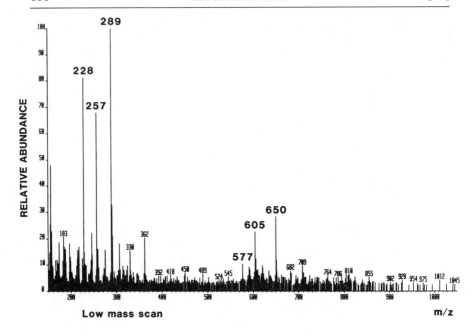

Low mass scan m/z

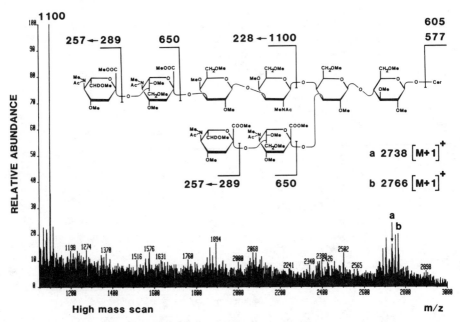

High mass scan m/z

FIG. 10. FAB-mass spectrum obtained after periodate oxidation, NaBD$_4$ reduction, and permethylation of the G$_{Q1b}$ ganglioside, Neu5Acα2-8Neu5Acα2-3Galβ1-3GalNAcβ1-4[Neu5-Acα2-8Neu5Acα2-3]Galβ1-4Glc-Cer.

quence ions are observed, which also was the case for the intact compound. It is also noted that the 4-substituted glucosyl residue, linked to the ceramide moiety, is resistant to periodate oxidation, which also is observed for other glycosphingolipids.[12] The FAB mass spectrum of the product formed from the G_{Q1b} ganglioside is shown in Fig. 10. The primary sequence ions of m/z 289, 650, and 1100 combined with a secondary ion of m/z 228 determine a sequence of Neu5Ac2-8Neu5Ac2-3Gal1-3GalNAc. The linkage in the sequence of Neu5Ac2-8Neu5Ac is deduced as previously discussed (Fig. 6). The $[M + 1]^+$ ions of m/z 2738 and m/z 2766 are in agreement with two species differing in ceramide composition and a carbohydrate structure containing two sequences of Neu5Ac2-8Neu5Ac.

Acknowledgment

We thank Dr. Albert Wu for the generous gift of the oligosaccharides from human blood group substance.

[33] Tandem Mass Spectrometry in Structural Characterization of Oligosaccharide Residues in Glycoconjugates

By WILHELM J. RICHTER, DIETER R. MÜLLER, and BRUNO DOMON

In tandem mass spectrometry (MS/MS), two-stage mass analysis is performed. In the first stage (MS-1), a specific ion of interest is selected according to its mass from the set of ions which constitute the conventional mass spectrum. The second stage (MS-2) generates, from this selected precursor ion, a spectrum of product ions which arise from metastable ion (MI) or collision-induced dissociation (CID).

Applied to structural analysis, MS/MS affords several distinct advantages over customary approaches in that some, or all, of the required tasks are performed with less sample in less time. Fragmentation of the ionized molecule in the ion source of the MS/MS system can take the role of degradation by wet-chemical means. The required separation and isolation of degradation products can be achieved by MS-1 ($<10^{-4}$ sec) rather than chromatography ($>$ min), utilizing ions rather than neutrals that comprise the structural subunit in question. MS-2 provides the necessary product identification by comparison of CID (rather than conventional) spectra of the unknown with an appropriate set of reference data.

The discussion in this chapter is limited to the structural analysis of small carbohydrate subunits (up to 5 sugar units) attached to larger aglycon moieties.

MS/MS Methods

Most of the discussed results were produced using triple-stage quadrupole (QQQ) tandem mass spectrometers (Finnigan MAT models TSQ 45 and 70). To permit GC-MS/MS analyses of sugar derivatives not readily available as pure isomers (α,β-anomers, pyranoside/furanoside mixtures) both systems were coupled to capillary gas chromatographs. As usual, quadrupoles Q_1 (fixed mode) and Q_3 (scanning mode) were operated as MS-1 and MS-2, respectively, while Q_2 (refocusing rf-only mode, with target gas) acted as a high-efficiency collision cell. So-called low-energy CID is the main mode of generating product ions from mass selected precursors in such tandem analyzers (see, e.g., Ref. 1). In the accessible "low-energy" regime, fragmentation depends strongly on collision energy (typically 2–50 eV translational energy of the precursor ion) and on the nature and pressure of target gas (e.g., 0.5 to 5 mtorr argon). Low-energy CID spectra can thus be widely varied and optimized for the required distinction of closely related (stereo)isomers (but care must be taken when spectra from runs with different target gas pressures are matched). Reliability of isomer distinction also profits from high precursor/product ion conversion (sensitivity, "short-time precision") and from unit mass resolution in patterns free from artifact signals.

High-energy CID using collision energies in the 4–10 keV range (cf. Ref. 2), which shows better long-time reproducibility, has been employed only in a few instances (see, e.g., Ref. 3). Use of a four-sector mass spectrometer is necessary to exclude erroneous signals, and also to obtain the desired unit mass resolution both in MS-1 and MS-2. In the VG70-4SE instrument ($E_1B_1E_2B_2$ geometry) used for the measurements, the precursor ion was selected at unit mass resolution by MS-1. Fragmentation was induced by collision in a gas cell located between MS-1 and MS-2 (helium target gas; signal reduction of precursor ion to about 40%). Product ions were analyzed, also at unit mass resolution, in MS-2 (linked scan at constant B_2/E_2 ratio).

[1] R. A. Yost and R. K. Boyd, this volume [7].
[2] M. L. Gross, this volume [6].
[3] B. L. Gillece-Castro and A. L. Burlingame, this volume [37].

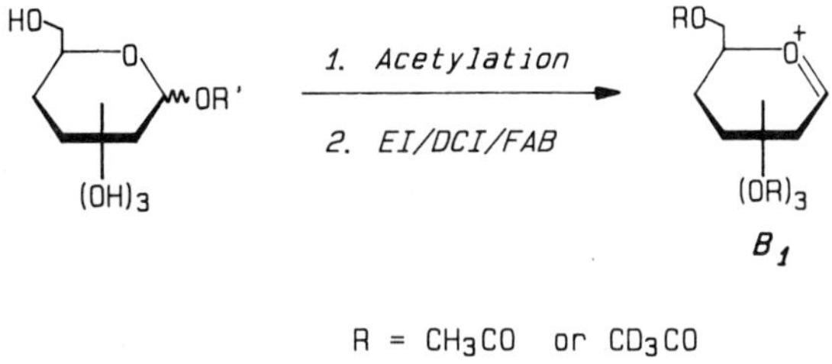

R = CH$_3$CO   or   CD$_3$CO

SCHEME 1. Formation of B_1 ions from monosaccharidic glycosides.

Methods for Comparison of Similar Spectra

The relatively large number of CID spectra to be compared in this application obviously requires objective rather than visual criteria for similarity (or dissimilarity). Derivation of the Euclidian distances (*ED*) between each pair of CID spectra is a simple and valuable means of approaching this objective.[4] The *ED* between two spectra (*i* and *j*) is given by the equation:

$$ED_{i,j} = \sqrt{\sum_{k=1}^{k=N} (I_{i,k} - I_{j,k})^2}$$

where *N* represents the number of relevant signals (*m/z* values), and $I_{i,k}$ and $I_{j,k}$ the intensities (% reconstructed in current) of the *k*th selected *m/z* value in the spectra *i* and *j*, respectively. Very similar (e.g., identical) spectra exhibit *ED* values close to zero, while increasingly larger values reflect increasing dissimilarity. On this basis, spectra can be classified as identical ($ED \le 6$), similar ($6 < ED \le 12$), dubious ($12 < ED \le 18$), and different ($ED > 18$).

Preparation of Monosaccharidic Sugar Ions for MS/MS

Essential for this approach is, of course, the efficient production of suitable precursor ions for the MS/MS experiment. Ideally, these ions comprise the complete carbohydrate moiety of the molecule, yet exclude the otherwise interfering aglycon portion. B_n-type ions[5-7] (B_1 in Scheme 1,

[4] D. L. Massart and L. Kaufman, "Interpretation of Analytical Chemical Data by the Use of Cluster Analysis." Wiley, New York, 1983.

[5] K. Biemann, D. C. DeJongh, and H. K. Schnoes, *J. Am. Chem. Soc.* **85,** 1763 (1963).

[6] T. Radford and D. C. DeJongh, in "Biochemical Applications of Mass Spectrometry" (G. R. Waller, ed.), p. 313. Wiley, New York, 1972.

[7] T. Radford and D. C. DeJongh, in "Biochemical Applications of Mass Spectrometry" (G. R. Waller and O. C. Dermer, eds.), 1st Supplementary Vol., p. 256. Wiley, New York, 1980.

n = number of sugar units, for nomenclature see Ref. 8), which formally represent C-1 carbenium ions, would thus appear to be the candidates of choice. The loss of structural information associated with bond cleavage would be limited to that of the anomeric configuration of C-1. When the sugar portion is properly derivatized, B_n ions are often formed in good yield due to the facile rupture of the glycosidic bond under a variety of ionizing conditions, such as electron impact ionization (EI), desorption chemical ionization (DCI), or fast atom bombardment (FAB). Especially valuable in this respect is acetylation (R = $COCH_3$) and, even more so, trideuterioacetylation (R = $COCD_3$) for reasons to be discussed later. For B_1 ions, trimethylsilylation [R = $(CH_3)_3Si$] and methylation (R = CH_3) is less advantageous. In this chapter only trideuterioacetates will be considered in greater detail.

For an efficient preparation of structurally defined B_n ions, glycosidic substrates with small aglycon residues R′ have proved useful. Methyl glycosides or free oligosaccharides were thus the preferred reference substrates for conversion into per(trideuterioacetyl) derivatives. Aside from the efficiency of preparation, two more criteria have to be met in order that the B_n-type ions qualify for structural assignment: (i) stereochemical integrity must be retained to a sufficient extent (irrespective of actual ion structure), and (ii) the long-lived ions amenable to MS/MS must fragment under significant stereochemical control.

Fulfillment of these requirements for B_1 ions is indicated by the fact that stereoisomeric reference ions give rise to pronounced spectral differences that are independent of the chosen mode of ionization (EI, DCI, or FAB). Occasionally, B_1 ions of test substrates with complex R′ residues (functionalized larger aglycons) give evidence of impaired stereochemistry, although largely to tolerable degrees.[9] In cases of poor match of unknowns with reference B_1 ions, such aglycon residues may be removed as a precaution, and replaced by small substituents (e.g., R′ = CH_3, methanolysis) as are used in the reference substrates.

Assignment of B_1 Ions Derived from Monosaccharidic Glycosides

Spectra of B_1 ions derived from acetylated and trideuterioacetylated methyl gluco- and galactoside (Fig. 1, upper and lower trace, respectively) show immediately that distinction criteria for the labeled derivatives are improved. While spectral differences are quite pronounced in the two unlabeled stereoisomers (e.g., intensity ratios of m/z 169 and m/z 127 ions,

[8] B. Domon and C. E. Costello, *Glycoconjugate J.* **5**, 397 (1988).
[9] D. R. Müller, B. Domon, and W. J. Richter, *Spectrosc. Int. J.* **7**, 11 (1989).

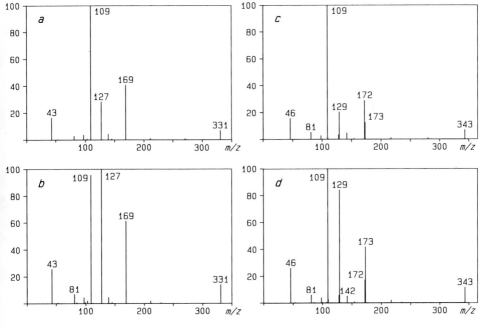

FIG. 1. Low-energy CID spectra of monosaccharidic B_1 ions (m/z 331 and 343, respectively) prepared from methyl-D-gluco- and -D-galactopyranoside tetra-O-acetyl (a and b) and tetra-O-(trideuterioacetyl) derivatives (c and d). Conditions: FAB (thioglycerol), 25 eV collision energy, 1 mtorr argon target gas pressure.

spectra a and b), they are substantially enhanced in the deuterioacetates (spectra c and d) both in number of features and range of variation. These two signals are each shifted and split into two components (m/z 169 → 172/173 and m/z 127 → 128/129). The m/z 172/173 ratio, being about 2:1 for Glc- and 1:2 for Gal-derived B_1, is an especially convenient and sensitive stereochemical indicator which is unlikely to fail when spectral quality in the case of an unknown is low. Thus, assignment may still be feasible in cases of small amounts, moderate enrichment of sample, low B_1 yield, or "foreign" contribution at the selected mass of B_1.

Extension of this approach to the complete set of eight stereoisomeric B_1 ions with hexopyranoside structure (see Table I) shows that three distinct types of behavior can be readily discerned: a glucose type (ratio ca. 2:1, idose and glucose), a mannose type (ratio ca. 1:1, altrose and mannose), and a galactose type (ratio ca. 1:2, talose, gulose, galactose, and allose). Distinction of isomers within these groups is practically impossible under the chosen (and various other) collision conditions.

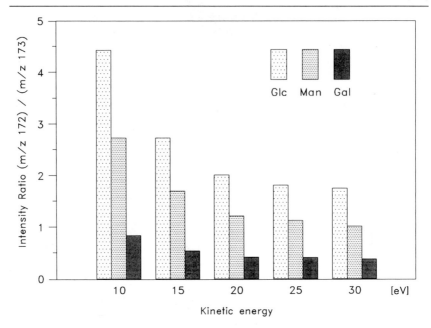

FIG. 2. Dependence of m/z 172/173 ratios in CID spectra of glucose-, mannose-, and galactose-derived B_1 ions on collision energy (trideuterioacetyl derivatives, CH_4-DCI, 1 mtorr argon target gas pressure).

A study of the energy dependence of the CID processes[10] (pseudo-breakdown curves) reveals another attractive feature consisting in the near-constancy of these m/z 172/173 ratios over the entire energy range above ca. 15 eV. In view of this finding, the need for reference spectra taken under unchanged or carefully reproduced collision conditions becomes clearly less stringent. Figure 2 summarizes the 10–30 eV breakdown behavior of the B_1 ions for the three most important aldohexopyranoses (Glc, Man, Gal) in the m/z 172/173 distinction window.

It is interesting to note that further increase of collision energy up to values used in high-energy CID (4 or 8 keV), shows little effect on these ratios.[9] This is compatible with the parallel operation of two (major) fragmentation channels with similar overall activation energy requirements, which contribute with different "weight" depending on the stereochemistry of B_1. In accordance with CID of isotopically labeled analogs (unpub-

[10] D. R. Müller, B. M. Domon, W. Blum, F. Raschdorf, and W. J. Richter, *Biomed. Environ. Mass Spectrom.* **15**, 441 (1988).

TABLE I
CID SPECTRA OF MONOSACCHARIDE IONS B_1[a]

Ion	m/z														
	46	81	98	109	110	128	129	142	154	172	173	217	235	280	343
Me-Ido-*p*	17	4	5	100	1	4	27	6	0	31	15	1	0	1	2
Me-Glc-*p*	18	5	5	100	1	4	28	6	0	26	15	1	0	0	2
Me-Alt-*p*	19	5	5	100	2	5	42	6	1	25	22	1	0	0	4
Me-Man-*p*	21	6	5	100	2	6	45	6	1	25	24	1	0	0	5
Me-Tal-*p*	24	7	5	100	3	7	77	6	1	22	38	1	1	0	7
Me-Gul-*p*	26	8	6	100	3	8	78	7	1	21	39	1	1	0	7
Me-Gal-*p*	25	8	6	100	3	7	91	6	1	20	47	1	1	0	8
Me-All-*p*	29	8	5	100	3	8	93	6	1	19	48	1	1	0	10
Me-Alt-*f*	16	4	3	46	3	4	100	3	0	8	58	3	1	0	8
Me-Tal-*f*	18	5	4	46	3	7	100	3	0	8	49	1	1	0	8
Ph-Glc-*p*	17	3	4	100	1	2	28	4	0	29	15	0	0	0	1
Ph-Gal-*p*	25	3	4	100	0	2	88	4	0	19	50	0	1	0	7

[a] Deuterioacetyl derivatives, CH_4-CI, 25 eV, 1.1 mtorr argon. All, D-Allose; Alt, D-altrose; Gal, D-galactose; Glc, D-glucose; Gul, D-glucose; Man, D-mannose; Ido, D-idose; Tal, D-talose; *p*, pyranosyl; *f*, furanosyl; Me, methyl; Ph, phenyl.

lished results from this laboratory) the two routes proceed via (i) losses of CD_3COOH (m/z 343 → 280), $(CD_3CO)_2O$ (m/z 280 → 172), and CD_3COOH (m/z 172 → 109); and (ii) losses of 2 CD_3COOH (m/z 343 → 217), $CD_2=C=O$ (m/z 217 → 173), and $CD_2=C=O$ (m/z 173 → 129).

Obviously, without deuterium labeling these two routes coincide at the m/z 169 level and thus reduce the number of discriminating features in the spectra (Fig. 1a and b). The low values observed for the m/z 172/173 ratios in hexofuranoside ions (cf. altrose- and talose-derived B_1 ions in Table I) may very well indicate that route (ii) prevails in these ring-size isomers. It is tempting to assume that part of the breakdown of the pyranoside structures also proceeds via ring contraction to furanosidic intermediates.

As test cases for identification of B_1 ions produced from other molecules, phenyl-β-D-glucoside and -galactoside (R' = C_6H_5), as well as the flavonoidic glycosides isoquercitrin (**Ia**) and hyperoside (**Ib**) have been subjected to the same procedure. The results (exemplified at the bottom of Table I for the phenylglycosides) indicate that an immediate and unambiguous assignment of the sugar residue in these substrates is possible, as long as only the three commonly encountered hexopyranosides Glc-*p*, Man-*p*, and Gal-*p* are considered. More importantly, the potential of this

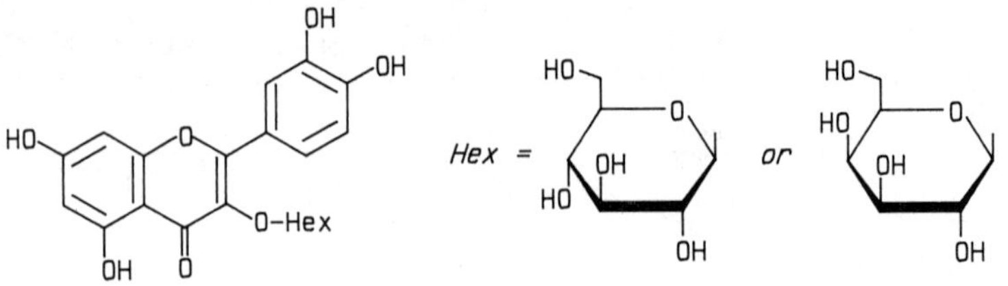

(Ia)   (Ib)

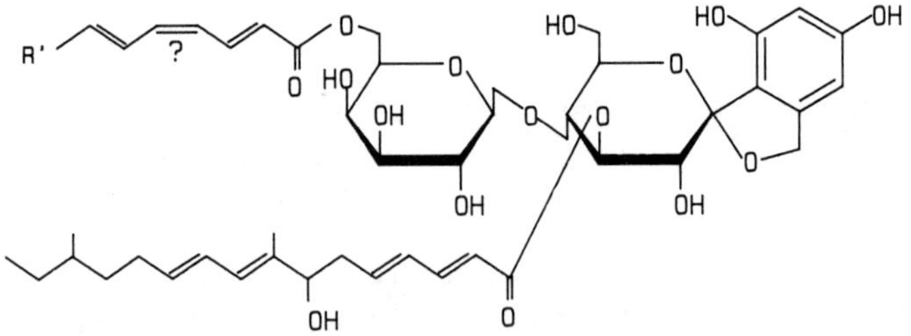

$Hex =$

$R' = CH(OH)CH_2CH_3$ (IIa) $R' = H$ (IIb)

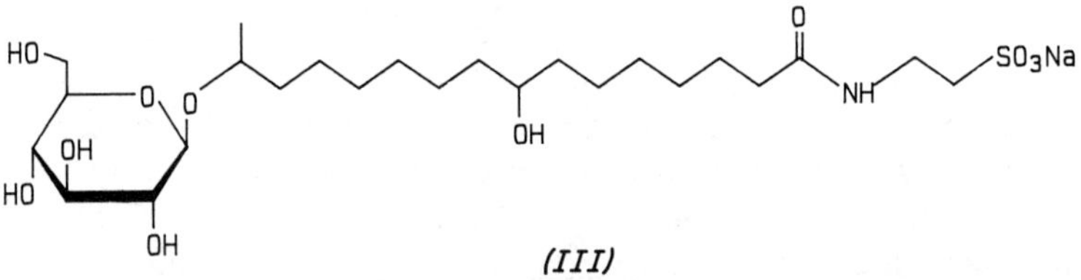

(III)

approach in actual structure elucidation of genuine unknowns has been demonstrated for antibiotics of the papulacandin type (IIa and IIb)[10] and more recently for an oviposition-deterring pheromone ODP (III) produced by the European cherry fruit fly *Rhagoletis cerasi* L.[11] The assignments were straightforward, although in all these cases only limited amounts of (not highly purified) samples were available.

[11] J. Hurter, E. F. Boller, E. Städler, B. Blattmann, H.-R. Buser, N. U. Bosshard, L. Damm, M. W. Kozlowski, R. Schöni, F. Raschdorf, R. Dahinden, E. Schlumpf, H. Fritz, W. J. Richter, and J. Schreiber, *Experientia* **43**, 157 (1988).

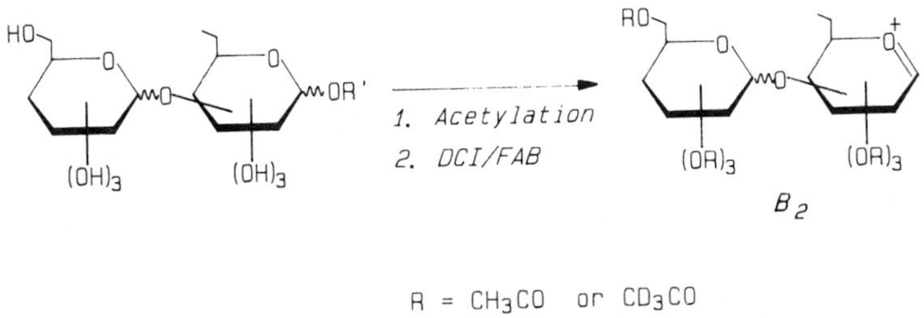

R = CH_3CO or CD_3CO

SCHEME 2. Formation of B_2 ions from disaccharidic glycosides. From Mueller *et al.*[14]

Preparation of Disaccharidic Sugar Ions for MS/MS

For glycoconjugates containing a disaccharidic sugar constituent, the corresponding B_2 ions will be more conveniently produced by FAB or DCI than by EI (Scheme 2). Trideuterioacetylation (R = CD_3CO) can again be used with advantage for improving the fingerprint character of the CID spectra (of which a much greater number are now to be distinguished). In analogy to the B_1 ions, the B_2 homologs are not readily obtained in high yield (and stereochemical integrity) from molecules containing larger aglycones. In practice, these are best removed, e.g., by trifluoroacetolysis,[12] which prevents breaking of the interglycosidic linkage. Yield of B_2 ions will, of course, also suffer from concomitant formation of B_1 due to rupture of the "wrong" glycosidic bond.

Assignment of B_2 Ion Derived from Disaccharidic Glycosides

Compared to B_1, the number of structural parameters defining B_2 ions is markedly increased. Knowledge of the identity of the two sugar units (ring size and configuration), their sequence, as well as interglycosidic linkage (position and stereochemistry) must be acquired. Information of the latter type is not readily obtained on a very small scale by more conventional techniques.

Studies of reference B_2 ions (m/z 640) derived from known disaccharide octa(trideuterioacetates) (R = R' = CD_3CO) or methyl hepta(trideuterioacetates) (R = CD_3CO, R' = CH_3) suggest that MS/MS may fill that gap. The α series of the four diglucoside (1-x) isomers (Fig. 3; x = 2, 3, 4, and 6 for kojibiose, nigerose, maltose, and isomaltose, respectively) illustrates the general trend that linkage positional isomerism is reflected in the CID spectra quite conspicuously. Except for the $(1\rightarrow2)/(1\rightarrow6)$ pair, differentiation of the linkage isomers is possible at a glance, e.g., by considering the

---

[12] B. Nilsson and S. Svensson, *Carbohydr. Res.* **72,** 183 (1978).

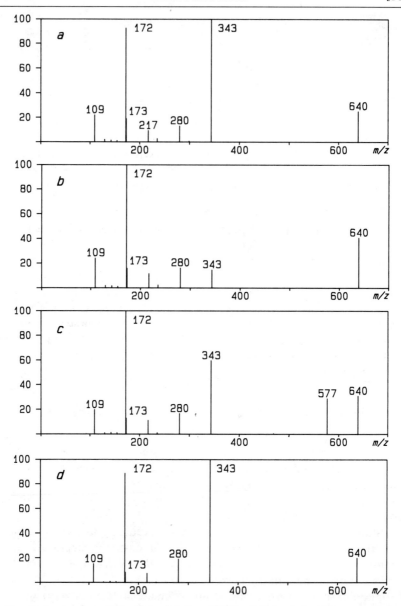

FIG. 3. Low-energy CID spectra of disaccharidic B_2 ions (m/z 640) prepared from ($\alpha 1 \rightarrow x$) diglucoside octa(trideuterioacetyl) derivatives (a, kojibiose; b, nigerose; c, maltose; d, isomaltose). Conditions: FAB (thioglycerol), 25 eV collision energy, 0.7 mtorr argon target gas pressure.

relative abundances of m/z 343. Apart from this, a pronounced loss of CD_3COOH from B_2 provides a specific feature of the 1→4-linkage isomer (maltose). A similar situation is encountered in the corresponding β series (not shown), with the same exception of the (1→2)/(1→6) pair (sophorose/ gentiobiose) which is barely distinguishable.[13]

While m/z 343 relative intensities allow linkage analysis, the m/z 172/ 173 ratios are again useful indicators of the nature of hexose constituents, at least for cases in which they are identical. These ratios vary in both the α and β series of the diglucosides within a range (about 9.0 up to 24.0) characteristic of glucose. Other specific ranges are observed for dimannosides and digalactosides (6.0–8.0 and 2.0–3.0, respectively).[14] The observed Gal-specific (i.e., low) ratios for Gal–Hex-derived ions (Hex = Glc, Man) indicate that both m/z 172 and 173 arise predominantly from the nonreducing sugar unit (B_1 ions). Of course, identification of the latter also defines the sequence of the sugar units in the disaccharides implicitly. Fragmentation of B_2 via B_1 as the exclusive intermediate (absence of "bypass" processes) has, however, been firmly established only for the β1→2-linked diglucosidic B_2 ion [sophorose, analysis of $Glc(OAc)_4$–$Glc(OAc$-$d_3)_3$ derivative].[14] For the 1→3- and 1→4- analogs newer evidence points to contributions from the reducing sugar residues, which may cause the relatively broad variation of the above specific ratios. From the limited data presently available, the validity of this approach cannot be assessed for diglycosides containing nonidentical sugar and sequence will best rely on matching the complete CID spectra with those of reference isomers for identity.

In contrast to positional characterization of interglycosidic bonds, distinction of anomeric configuration of interglycosidic bonds is not readily feasible except for the 1→4 pair. As mentioned before, low-energy CID provides, however, a flexible tool for "tuning" the breakdown pattern in favor of an optimized response to a particular structural parameter. Optimization by adjustment of collision energy (nominally 2 instead of 14 eV) is illustrated by data given for the α and β series of the diglucoside-derived B_2 ions (m/z 640, Table II). Data with similar distinction criteria have also been obtained for unlabeled B_2 analogs (m/z 619).[13]

Aside from differences in precursor ion intensities (e.g., 58 vs. 41% relative intensity for the 1→2 pair), differences are also observed in a "diglycosidic fingerprint" consisting of low (although reproducible) intensity ions still containing both sugar units. Loss of 2 molecules of CD_3COOH (m/z 514) from B_2 is clearly more pronounced for the 1→2 pair in the β-

---

[13] B. Domon, D. R. Müller, and W. J. Richter, *Org. Mass Spectrom.* **24**, 357 (1989).
[14] D. R. Müller, B. Domon, F. Raschdorf, and W. J. Richter, *Adv. Mass Spectrom.* **11B**, 1309 (1989).

TABLE II
CID SPECTRA OF DISACCHARIDIC B_2 IONS[a]

m/z	α1-2	β1-2	α1-3	β1-3	α1-4	β1-4	α1-6	β1-6
109	16.0	12.8	32.4	33.6	12.2	15.1	8.5	9.5
172	36.8	38.0	99.0	100.0	32.7	46.9	27.0	28.5
173	8.5	6.5	11.0	8.3	3.6	4.7	2.3	2.2
217	8.1	8.1	19.6	18.5	6.6	13.3	4.8	4.6
235	1.2	1.1	2.0	1.2	0.9	0.9	0.3	0.4
280	11.4	15.3	44.3	48.3	19.5	25.6	21.1	17.9
343	100.0	100.0	20.9	35.5	46.6	59.5	100.0	100.0
469	0.3	0.2	0.3	0.0	0.3	1.6	0.2	0.8
514	0.8	2.5	0.0	0.0	0.2	0.3	0.2	0.2
577	0.5	0.5	1.9	2.5	40.7	40.5	0.4	1.5
640	58.4	41.1	100.0	89.7	100.0	100.0	55.2	61.2

[a] Deuterioacetyl derivatives, FAB, 2 eV, 1.3 mtorr argon.

isomer, as is loss of 1 molecule of CD_3COOH (m/z 577) in the same isomer for the 1→6 pair. Distinction within the 1→4 pair can finally be made on the basis of loss of $CD_3COOH/(CD_3CO)_2$ (m/z 469). Of course, reliability for such assignment will be improved by considering additional intensity differences in the monoglycosidic fingerprint (e.g., m/z 280 and 217 for the 1→4 isomers). It should be mentioned that differentiaton of 1→2 and 1→6 linkage isomers also becomes feasible at these very low collision energies.

Another parameter of structural complexity of B_2 ions is, of course, due to variation of the sugar constituents. As yet, only B_2 ions have been explored that contain one or two mannoses or galactoses as sugars other than glucose (only hexopyranoses). In this context it will be relevant to evaluate how efficiently low-energy CID can distinguish isomers within an extended set of candidates. One of the simple means that allow an "objective" answer to this question is based on Euclidian distances (ED) (cf. previous section). Other means, involving a more extensive treatment of data, have been discussed elsewhere.[14,15] Similarity assessment of a larger bulk of CID spectra using ED values is illustrated in Table III for a set of eighteen B_2 reference ions.

Table III shows that low-energy CID is, indeed, able to discern most of the isomers by the ED values of the various pairs compared. Amygdalin, which was used as a test substrate, is correctly assigned as a β1→6 diglucoside (ED = 3). It is easily distinguished from all other isomeric possibilities except β1→2 (ED = 4). This "identification" within a reference set, in which (almost) all structural determinants are varied, suggests

[15] B. Domon, D. R. Müller, and W. J. Richter, *Spectros. Int. J.* **7**, 23 (1989).

TABLE III
EUCLIDIAN DISTANCES COMPUTED FOR CID SPECTRA OF DISACCHARIDIC B_2 IONS[a]

Ion derived from	Glc-Glc								Man-Man				Gal-Gal				Gal-Hex	
	α1-2	α1-3	α1-4	α1-6	β1-2	β1-3	β1-4	β1-6	α1-2	α1-3	α1-4	α1-6	α1-3	β1-3	α1-4	β1-6	β1-4[b]	β1-4[c]
Glc(α1-2)Glc	0	44	23	10	8	43	24	8	20	30	19	7	25	36	40	34	24	34
Glc(α1-3)Glc	44	0	25	35	36	4	25	36	42	20	33	49	36	39	83	77	53	54
Glc(α1-4)Glc	23	25	0	17	17	25	5	17	23	15	11	28	22	30	61	55	31	31
Glc(α1-6)Glc	10	35	17	0	4	34	18	3	22	25	17	16	25	36	50	44	30	37
Glc(β1-2)Glc	8	36	17	4	0	35	17	2	19	23	14	14	22	33	48	42	27	34
Glc(β1-3)Glc	43	4	25	34	35	0	25	35	42	22	34	48	37	40	83	77	54	54
Glc(β1-4)Glc	24	25	5	18	17	25	0	18	21	12	10	29	19	27	61	55	29	30
Glc(β1-6)Glc	8	36	17	3	2	35	18	0	20	24	15	14	23	34	48	42	28	35
Man(α1-2)Man	20	42	23	22	19	42	21	20	0	24	14	22	16	24	48	41	18	25
Man(α1-3)Man	30	20	15	25	23	22	12	24	24	0	17	35	17	20	67	60	35	37
Man(α1-4)Man	19	33	11	17	14	34	10	15	14	17	0	22	14	24	53	46	20	24
Man(α1-6)Man	7	49	28	16	14	48	29	14	22	35	22	0	28	39	35	30	23	34
Gal(α1-3)Gal	25	36	22	25	22	37	19	23	16	17	14	28	0	13	55	47	23	29
Gal(β1-3)Gal	36	39	30	36	33	40	27	34	24	20	24	39	13	0	64	56	31	34
Gal(α1-4)Gal	40	83	61	50	48	83	61	48	48	67	53	35	55	64	0	8	40	50
Gal(β1-6)Gal	34	77	55	44	42	77	55	42	41	60	46	30	47	56	8	0	33	44
Gal(β1-4)Glc	24	53	31	30	27	54	29	28	18	35	20	23	23	31	40	33	0	14
Gal(β1-4)Man	34	54	31	37	34	54	30	35	25	37	24	34	29	34	50	44	14	0
Amygdalin	5	39	20	5	4	38	21	3	20	27	17	11	25	36	45	39	27	35

[a] Deuterioacetyl derivatives, CH_4-DCI, 14 eV, 0.8 mtorr argon. Features used for the calculation: m/z 46, 109, 129, 142, 154, 172, 173, 217, 235, 280, 299, 343, 469, 577, 640.

[b] Hex = Glc.

[c] Hex = Man.

IV

that full characterization of a disaccharidic B_2 ion from a genuine unknown is possible.

Application to insulin-like growth factor (IGF)-A_1 (structure **IV**) shows that this is in fact the case.[14] The glycoconjugate, which contains a disaccharide unit attached to a relatively large peptide in the threonine-29 position, has been isolated in small quantities as a congener of its aglycon, the peptide hormone human IGF-I synthesized by recombinant DNA technology with gene expression in yeast. The parent molecule consists of a triply S–S bridged chain of 70 amino acids, whose sequence and folding pattern is known.[16,17] $[M + H]^+$ (FAB-MS, m/z 7972.9) of **IV** indicated the presence of two hexose units (increment of 2×162 Da relative to IGF-I). They were identified (total hydrolysis, GC/MS of trimethylsilyl derivatives) as mannose units, as anticipated for products of posttranslational modification in yeast. Characterization of the dimannosidic subunit had to rely on the currently discussed technique due to lack of quantities necessary for the application of alternative strategies.[18,19]

As outlined before (Scheme II), direct formation of the required B_2 ions from deuterioacetylated **IV** is expected to be impracticable (and actually was). Excellent results were, however, obtained after removal of the large peptide residue by trifluoroacetolysis.[12] Conversion of the result-

[16] E. Rinderknecht and R. E. Humbel, *J. Biol. Chem.* **253**, 2769 (1978).

[17] F. Raschdorf, R. Dahinden, W. Märki, W. J. Richter, and J. P. Merryweather, *Biomed. Environ. Mass Spectrom.* **16**, 3 (1988).

[18] K. Hard, W. Bitter, J. P. Kamerling, and J. F. G. Vliegenthart, *FEBS Lett.* **248**, 111 (1989).

[19] P. Gellerfors, K. Axelsson, A. Hellander, S. Johansson, L. Kenne, S. Lindqvist, B. Pavlu, A. Skottner, and L. Fryklund, *J. Biol. Chem.* **264**, 11444 (1989).

TABLE IV
EUCLIDIAN DISTANCE (ED) "FIT LIST" OF IGF-A$_1$
DISACCHARIDIC B_2 IONS[a]

Number	Fit	ED	Reference disaccharide
1	Identical	6	Man(α1-2)Man
	Similar		
2	Dubious	15	Gal(α1-3)Gal
3		17	Gal(β1-4)Glc
4		18	Man(α1-4)Man
5		22	Gal(β1-3)Gal
7	Different	23	Gal(β1-4)Man
.		.	.
.		.	.
.		.	.
19		47	Gal(α1-4)Gal

[a] Deuterioacetyl derivatives, CH$_4$-DCI, 14 eV, 1 mtorr argon.

ing disaccharide octa(trifluoroacetate) into the trideuterioacetyl analog (deacylation, reacylation) provided no major obstacle, as high yields and the mass selection of B_2 by MS/MS (B_2, m/z 640) rendered the isolation of products at both stages unnecessary. Table IV shows, *inter alia*, the match of the unknown B_2 with dimannoside reference ions (only α series available). Unambiguous assignment in favor of the 1→2 isomer is straight-forward, as in this case all four linkage types are readily discerned. The 1→2 case is especially readily identified by relatively abundant m/z 217 and 235 ions (almost) absent in the other three isomers.[14]

The α configuration of the interglycosidic bone in **IV** suggested by the close fit is in full keeping with the frequent observation of this configuration in O-linked mannosyl residues in glycopeptides and glycoproteins. This assignment, which was only tentative in lack of β-linked B_2 reference ions, has been recently established by NMR.[18,19]

Analysis of B_n Ions Derived from Larger Sugar Moieties

Preliminary results on larger B_n ions (n = 3–5) suggest that this ap-proach may very well allow extension to the analysis of ions comprising more than 2 sugar units. Figure 4 gives an example of a B_4 ion (m/z 1234) produced from perdeuterioacetylcellotetraose (four β1→4-linked glucose units) and subjected to low-energy CID. Comparison with (although a limited number) other B_4 ions shows promise with respect to isomer dis-tinction. For corresponding B_5 ions (m/z 1531), the situation is similar,

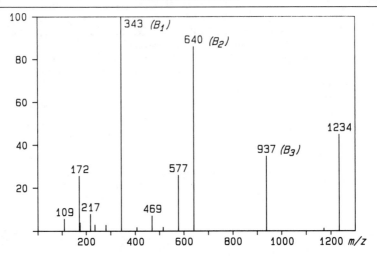

FIG. 4. Low-energy CID spectrum of tetrasaccharidic B_4 ions (m/z 1234) prepared from cellotetraose per(trideuterioacetyl) derivative. Conditions: FAB (thioglycerol), 10 eV collision energy, 0.9 mtorr argon target gas pressure.

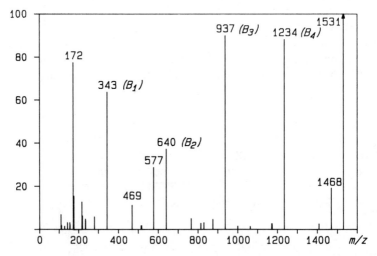

FIG. 5. High-energy CID spectrum of pentasaccharidic B_5 ions (m/z 1531) prepared from cellopentaose per(trideuterioacetyl) derivative. Conditions: FAB (dithiothreitol/dithioerythritol 5 : 1), 8 keV collision energy, helium target gas (precursor ion reduction to 40%).

except that limitations concerning unit mass resolution in precursor ion selection become apparent. Selection of wider mass windows including ^{13}C isotopomers is indispensable for providing the necessary high-mass sensitivity of quadrupole analyzers.

Precursor ion selection at unit mass resolution is, however, attainable for high-energy CID (e.g., 8 keV collision energy) by using four-sector tandem mass spectrometers.[20] Such application is illustrated in Fig. 5 for the B_5 ion derived from perdeuterioacetylcellopentaose as the next higher homolog.

From this example it appears that principally similar information can be obtained from low- and high-energy CID. The latter mode is likely to be increasingly advantageous at higher mass, but a good degree of variability for optimization of the experiment is lost. Use of small substituents (e.g., $R = CH_3$) may also become mandatory in view of the feasibility of MS/MS experiments at high mass.

At present, MS/MS of sugar ions formed from glycoconjugates appears to fulfill some, but not all, of the basic requirements of structure analysis. Within the triad, degradation–separation of products–product identification, MS/MS performs favorably in the latter two steps, in that it substitutes for conventional, less efficient techniques. It is less successful, however, in replacing the wet-chemical degradation step by gas-phase fragmentation aimed at producing an intact sugar subunit in the ion source of the spectrometer.

[20] B. Domon, D. R. Müller, and W. J. Richter, *Biomed. Environ. Mass Spectrom.* **19**, 390 (1990).

[34] Mass Spectrometry of Mixtures of Intact Glycosphingolipids

By BO E. SAMUELSSON, WESTON PIMLOTT, and KARL-ANDERS KARLSSON

Introduction

Glycosphingolipids usually occur in nature in complex mixtures and in small amounts.[1] The number of sugars may range from one to about forty. These substances show distinct cell, tissue, strain, and species specificity in their structures and undergo successive changes during development

[1] J. N. Kanfer and S. Hakomori, "Sphingolipid Biochemistry." Plenum, New York, 1983.

and on tumor transformation. Of particular biomedical importance are the determinants within the major blood group systems (ABO, Lewis, I, P) and the glycolipids which may act as tumor-associated antigens[2] and as receptors for microbes.[3] Structural studies are, therefore, of major importance for the progress of transplantation and tumor biology and for the understanding of infectious diseases. Such studies require sensitive and rapid methods giving information with high structural specificity and with the capability to work with mixtures. Mass spectrometry is not the only method that matches these demands, but taken overall it is perhaps the most valuable tool in this respect. Several chromatographic methods have been tried in direct combination with mass spectrometry, i.e., gas chromatography (GC)/MS, liquid chromatography (LC)/MS, thin-layer chromatography (TLC)/MS, and supercritical fluid chromatography (SFC)/MS, to overcome the complexity of information in mixtures, and some of these techniques are discussed elsewhere in this volume. Our approach has been to take advantage of a temperature gradient in the ion source and distill off different derivatized compounds according to molecular weight, or for glycosphingolipids approximately according to the number of sugars in the carbohydrate chain. The separation is relatively poor but several specific fragments from one species will follow the same distillation curve, different from others, supporting the identity of individual molecular species. An alternate approach, as discussed elsewhere in this volume, is to release the oligosaccharide mixture, which may be preseparated by gas chromatography. However, when this method is used, the information on ceramide structure is lost, and it has been described that this portion of the molecule may influence the biological activity of the carbohydrate, shown both for the interaction with antibodies[4,5] and with bacteria.[3,6]

The electron ionization (EI)/MS distillation technique for the analysis of glycosphingolipid mixtures will be the principal subject of this chapter. A few aspects of TLC/MS are also included.

Instrumentation

Historically the development of the "bleed-off" technique[7] described in this chapter was done using a MS 902 double-focusing instrument (AEI/Kratos Ltd., Manchester, UK) equipped with a standard EI ion source

[2] S. Hakomori, *Chem. Phys. Lipids* **42**, 209 (1986).
[3] K.-A. Karlsson, *Annu. Rev. Biochem.* **58**, 309 (1989).
[4] R. Kannagi, R. Stroup, N. A. Cochran, D. L. Urdal, W. W. Young, and S. Hakomori, *Cancer Res.* **43**, 4997 (1983).
[5] S. J. Crook, J. M. Boogs, A. J. Vistnes, and K. M. Kosky, *Biochemistry* **25**, 7488 (1986).
[6] N. Strömberg, M. Ryd, A. A. Lindberg, and K.-A. Karlsson, *FEBS Lett.* **232**, 193 (1988).
[7] M. E. Breimer, G. C. Hansson, K.-A. Karlsson, H. Leffler, W. Pimlott, and B. E. Samuelsson, *Biomed. Mass Spectrom.* **6**, 231 (1979).

and a heatable direct-insertion probe. The probe was heated by a programmable digital heat controller and samples were loaded into cuvettes (20 mm × 2 mm id) containing glass wool plugs. The cuvette tip (in operation) was connected to the outer wall of the ion source causing gas-phase molecules to traverse areas of hot metal before being ionized, resulting in considerable thermal degradation of the sample. Modification to the in-beam sample technique[8] by which the sample is loaded in a shallow cuvette and brought in close proximity (1–2 mm) to the electron beam diminished thermal degradation and improved high mass fragment intensity and reproducibility.[9]

Later experiments were performed on a high-field ZAB HF 2F instrument equipped with an EI-only ion source and a programmable source heater (VG Analytical, Manchester, UK). Normally samples were run in-beam with a trap current of 500 μA, 30–40 eV ionizing energy, and acceleration voltage of 8 keV. Resolution was adapted to the problem of the moment (usually about 3000 at 10% valley definition). For mixture analysis the source temperature ramp was set to 2°–6°/min in the range of 180° to 360°. Complete or partial spectra were usually recorded with 30 sec/decade scans.

The ZAB HF 2F is now also equipped with a prototype TLC-FAB source (VG Analytical). The probe, which is manually operated, houses a TLC strip 5 × 0.5 cm. By carefully moving the probe forward, different areas along the TLC strip are exposed to the FAB beam.

For positive ion FAB mass spectra, 3-nitrobenzyl alcohol (3NBA), with 1% trifluoroacetic acid is used as matrix; for negative ion FAB, triethanolamine is used. On TLC, the matrix is applied as a thin film only wetting the surface. This is accomplished using a thin glass rod or capillary sliding gently along the surface. Care should be taken to avoid excess matrix fluid which may destroy the separation achieved. A cesium iodide spectrum is used for mass calibration.

The ZAB HF 2F instrument is connected to a data system (VG Analytical 11/250) based on a PDP 11/24 computer with a 1 MB memory and a 96 MB disk system. Fomblin is used up to m/z 3000 as normal primary reference and is introduced at the end of a run when the source has cooled to 250°. A Fomblin mass spectrum is chosen for normal primary calibration but, if considered necessary, a secondary reference calibration file is created which contains mass values of known peaks contained within the current sample. Thus, on mass conversion, these peaks are used as lock

[8] M. E. Breimer, G. C. Hansson, K.-A. Karlsson, G. Larsson, H. Leffler, I. Pascher, W. Pimlott, and B. E. Samuelsson, *Adv. Mass Spectrom.* **8,** 1097 (1980).

[9] M. E. Breimer, G. C. Hansson, K.-A. Karlsson, H. Leffler, W. Pimlott, and B. E. Samuelsson, *FEBS Lett.* **124,** 299 (1981).

points to give improved scan-to-scan reproducibility, so-called accurate mass.

Sample Preparation and Derivatization

The preparation of pure acid and nonacid glycosphingolipid fractions has been described in detail by one of us in this series.[10] In this way, total mixtures are obtained free of nonglycolipid contaminants, which is an essential prerequisite for mass spectrometry.

Substances are analyzed as permethylated, permethylated–reduced, and, in the case of gangliosides, also as permethylated–reduced–trimethylsilylated derivatives. Methylation is performed principally according to Hakomori[11] or by a more recent simplified procedure[12] and reduction is performed in diethyl ether using $LiALH_4$.[13] Trimethylsilylation is performed as has been described.[14] When methylation is used, all hydroxyl and carboxyl groups and amide groups (in hexosamines and ceramide) are methylated. If reduction is utilized, amides are converted to amines which stabilize the adjacent glycosidic linkage, in case of hexosamine,[15] and the fatty acid, in case of ceramide,[15] and methyl esters are converted to alcohols. The production of immonium ions is favored by reduction.

Data Handling of Recorded Spectra

During a sample run, collected data are time-to-mass converted using the primary calibration file. Single selected mass spectra are inspected and intense peaks from known compounds are recognized. These peaks are then incorporated in a secondary reference file and used as "lock masses." Spectra are then time-to-mass adjusted using the secondary reference file. The use of a secondary reference lock mass file is necessary only if accurate masses are needed. Single ion traces are computer reconstructed from the bank of mass spectra of a single run. The accurate mass with normally one decimal is given together with a window. The selection of window size is dependent on selectivity vs. sensitivity but must be carefully selected each time. Standard facilities to average over a number of

---

[10] K.-A. Karlsson, this series, Vol. 138, p. 212.
[11] S. Hakomori, *J. Biochem. (Toyko)* **55**, 205 (1964).
[12] G. Larsson, H. Karlsson, G. C. Hansson, and W. Pimlott, *Carbohydr. Res.* **161**, 282 (1987).
[13] K.-A. Karlsson, *Biochemistry* **13**, 3643 (1974).
[14] K.-A. Karlsson, I. Pascher, and B. E. Samuelsson, *Chem. Phys. Lipids* **12**, 271 (1974).
[15] K.-A. Karlsson, I. Pascher, W. Pimlott, and B. E. Samuelsson, *Biomed. Mass Spectrom.* **1**, 49 (1974).

spectra after a continuous recording are sometimes used when the sample amount is extremely small and diagnostic peaks must be ascertained.

Interpretation of Mass Spectra of Glycosphingolipid Mixtures

EI mass spectra recorded from a mixture of glycolipids after distillation in the ion source may be extremely complex. A prerequisite for their interpretation is a knowledge of the principles of their fragmentation supported by a large number of reference glycolipid spectra.[7] Figures 1 and 2 give the general features of fragmentation for methylated and permethylated–reduced derivatives.

The primary characteristic of a glycosphingolipid species, number and type of carbohydrate units, and spectrum of fatty acids is given by the immonium ions (F) which are specially abundant in the permethylated–reduced derivatives (Fig. 2). A semiquantitation of fatty acid composition is indicated. Sequence information (S fragments in Figs. 1 and 2) is obtained from mass spectra of both derivatives, with characteristic shifts in masses due to reduction of amides (see tabulation in Figs. 1 and 2). A + 1 of Fig. 1 and D + 1 of Fig. 2 are supportive sequence fragments. Information on the lipophilic composition are obtained from C fragments of Fig. 1 and F, D + 1, and M − 1 fragments of Fig. 2. A number of other fragments not shown may also be informative but the most suitable peaks to select for single ion traces are the ones shown.

Selected-Ion Curves from Mass Spectrometry of Permethylated– Reduced Mixtures of Glycosphingolipids

Figure 3 shows selected-ion curves obtained from mass spectrometry of permethylated–reduced glycosphingolipid mixtures of small intestinal epithelial cells of two strains of rat.[16] Selected-ion curves for immonium ions, F fragments, for some of the glycosphingolipid species found in the two strains of rat are shown as a function of temperature. Figure 3 illustrates the partial separation obtained according to number of sugar residues in the carbohydrate chain. In addition, it shows some of the qualitative differences in glycosphingolipid content between the two strains. The black and white strain contained two glycosphingolipids with a terminal hexosamine (m/z 1125 and 1560) which are missing in the white strain. These were blood group A-active glycolipids showing strain differences.

Figure 4 shows similar selected-ion curves from glycolipids of mouse

[16] M. E. Breimer, G. C. Hansson, K.-A. Karlsson, and H. Leffler, *FEBS Lett.* **114,** 51 (1980).

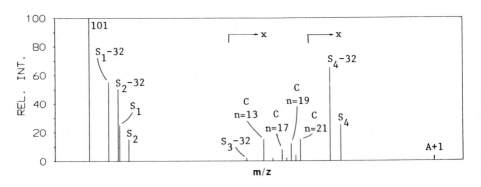

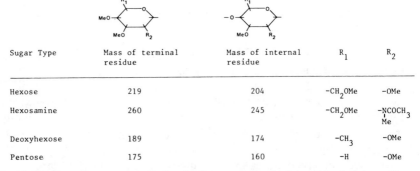

Sugar Type	Mass of terminal residue	Mass of internal residue	R_1	R_2
Hexose	219	204	$-CH_2OMe$	$-OMe$
Hexosamine	260	245	$-CH_2OMe$	$-NCOCH_3$ Me
Deoxyhexose	189	174	$-CH_3$	$-OMe$
Pentose	175	160	$-H$	$-OMe$

FIG. 1. Idealized EI mass spectrum showing fragments of major interest for the interpretation of permethylated glycosphingolipid spectra. Masses of terminal and internal carbohydrate residues are shown below the spectrum. (Reproduced from Ref. 7, with permission of the publisher.)

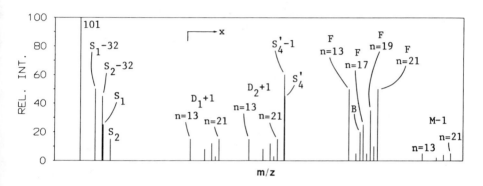

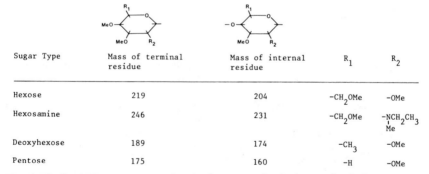

Sugar Type	Mass of terminal residue	Mass of internal residue	R_1	R_2
Hexose	219	204	$-CH_2OMe$	$-OMe$
Hexosamine	246	231	$-CH_2OMe$	$-NCH_2CH_3$ Me
Deoxyhexose	189	174	$-CH_3$	$-OMe$
Pentose	175	160	$-H$	$-OMe$

FIG. 2. Idealized EI mass spectrum showing fragments of major interest for the interpretation of permethylated–reduced glycosphingolipid spectra. Masses of terminal and internal carbohydrate residues are shown below the spectrum. Specific mass changes due to the reduction are evident by a comparison of Figs. 1 and 2. (Reproduced from Ref. 7, with permission of the publisher.)

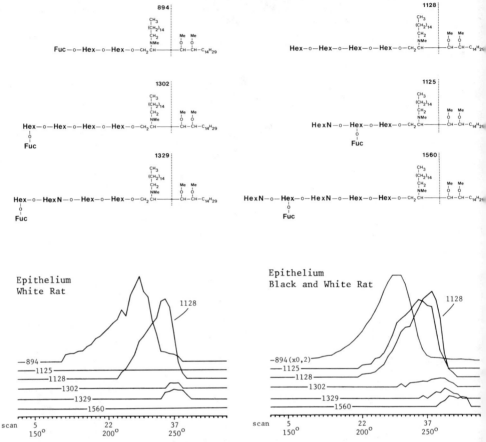

FIG. 3. Selected-ion curves from mass spectrometry of permethylated–reduced glycolip-ids. Total nonacid glycolipids were prepared from epithelial cells of small intestine of a white (left) and a black–white rat strain. Curves for the major glycolipids with one and three hexoses, respectively, are not shown. The sample amounts used were 200 μg, the electron energy was 38 eV, accelerating voltage 4 kV, trap current 500 μA, and ion source temperature 290°. The probe was heated 5°/min and spectra were recorded every 38 sec. The instrument used was an AEI MS 902 mass spectrometer equipped with a computer system (Instem-Kratos Ltd., Manchester, UK). Hex, Hexose; HexN, N-acetylhexosamine; Fuc, fucose. (Reproduced from Ref. 16, with permission of the publisher.)

small intestine.[17] The dominating monohexosylceramide is nicely sepa-rated from the more complex glycolipids. A partial mass spectrum from scan 48 is shown to illustrate the primary data behind the ion curves. The combined use of ion curves and detailed interpretation of selected primary spectra is necessary for valid conclusions. The codistillation of several

[17] M. E. Breimer, G. C. Hansson, K.-A. Karlsson, and H. Leffler, J. Biochem. (Toyko) **90**, 589 (1981).

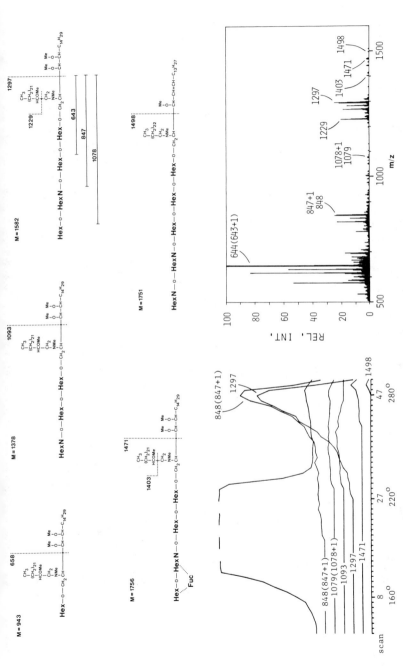

FIG. 4. Mass spectrum and selected-ion curves obtained from glycolipids of mouse small intestine together with simplified formulas for interpretation. A sample of 100 μg of the permethylated–reduced mixture was used. The curves represent relative intensities of selected ions (m/z marked on each curve) as a function of probe temperature. The mass spectrum was reproduced from scan 48. The cuvette was heated 5°/min and mass spectra were recorded every 38 sec. The electron energy was 50 eV, source temperature 300°, accelerating voltage 4 kV, and trap current 500 μA. For abbreviations, see legend to Fig. 3. It could not be decided whether the binding of Fuc was to Hex or HexN of the glycolipid represented by formula on the lower left-hand side. (Reproduced from Ref. 17, with permission of the publisher.)

peaks with temperature strengthens their diagnostic value for a certain molecular species. Inspection of primary spectra minimizes the risk of disregarding unexplained peaks or missing new structures.

Separation by Distillation

As an overall generalization, glycosphingolipids distill off the probe tip according to the number of sugar residues. Although this separation from a chromatographic point of view may appear crude and simple, remarkable results have been achieved. Figure 5 shows the almost complete separation of two components with 4 and 5 sugars in the carbohydrate chain, respectively, in an analysis of permethylated–reduced glycolipids from guinea pig small intestine.[17] Figure 6 shows partial mass spectra from an analysis of permethylated–reduced glycolipids of cat small intestine,[17,18] recorded at 301° and 309°. The spectrum at 301° shows the presence of a Forssman-like pentaglycosylceramide as indicated by, e.g., m/z 1498. The spectrum recorded at 309° almost lack this component. Instead, the presence of m/z 1474 indicates the dominance of a pentahexosylceramide and m/z 1702 and 1675 also the presence of two different blood group A hexaglycosylceramides.

Variation in the lipophilic part, including hydrocarbon chain length, unsaturation, presence of different numbers of hydroxyl groups, and methyl branches, is also a characteristic of glycosphingolipids and will influence the distillation. The immonium ions (F fragments) are thus sensitive both to carbohydrate chain length and the fatty acid. The effect of fatty acid hydrocarbon chain length and hydroxylation versus carbohydrate chain length is illustrated in Fig. 7.[19] Knowledge of these effects is necessary for correct interpretation of selected F fragments from a mixture with heterogeneous and different fatty acids in separate glycolipids.

Selected-Ion Curves from Mass Spectrometry of Permethylated Mixtures of Glycosphingolipids; Carbohydrate Sequence Information

Figure 8 shows selected-ion curves from mass spectrometry of total nonacid glycosphingolipid mixture of cat small intestine.[18] From the mass spectrometric analyses of permethylated–reduced derivatives and the thin-layer chromatographic appearance of the fraction, a number of tenta-

[18] M. E. Breimer, G. C. Hansson, K.-A. Karlsson, H. Leffler, W. Pimlott, and B. E. Samuelsson, *FEBS Lett.* **89**, 42 (1978).

[19] B. E. Samuelsson, *Adv. Mass Spectrom.* **10**, 225 (1986).

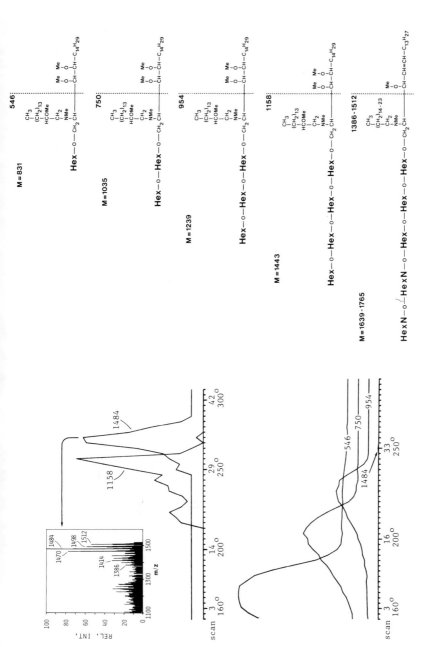

FIG. 5. Mass spectrum and selected-ion curves of permethylated–reduced glycolipids of guinea pig small intestine. For the lower diagram, 140 μg of sample was used at a multiplier setting (detector response) of 6.4. For the upper diagram, 160 μg was used at a multiplier setting of 8.0 giving about a threefold increase of sensitivity. The mass spectrum was recorded during scan 35. The cuvette was heated 5°/min. Mass spectra were recorded every 38 sec; electron energy 50 eV, ion source temperature 300°, accelerating voltage 4 kV, and trap current 500 μA. For abbreviations, see legend to Fig. 3. (Reproduced from Ref. 17, with permission of the publisher.)

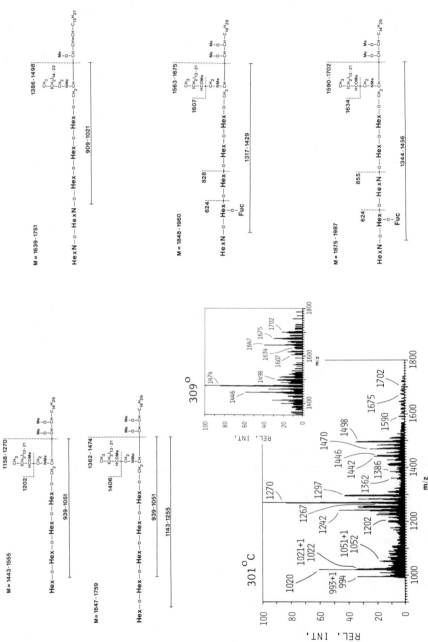

Fig. 6. Mass spectra recorded from 200 μg of permethylated–reduced glycolipids of cat small intestine together with simplified formulas for interpretation. Selected-ion curves of this experiment have been published.[18] The mass spectrum on the lower left-hand side was recorded at 301° and the upper right-hand side at 309°, demonstrating a distinct change in pattern. The electron energy was 48 eV, source temperature 290°, accelerating voltage 4 kV, and trap current 500 μA. (Reproduced from Ref. 17, with permission of the publisher.)

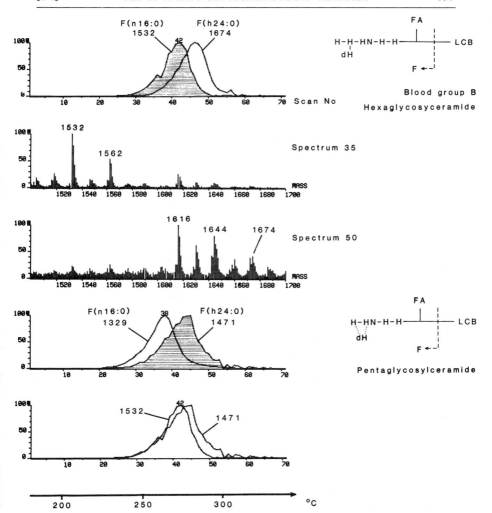

FIG. 7. Effect of hydrocarbon chain length on distillation. In the upper half of the figure, single-ion traces (F fragments) from a blood group B-active hexaglycosylceramide in its permethylated–reduced form and with nonhydroxy C_{16} saturated fatty acid, F(n16:0), and 2-hydroxy C_{24} saturated fatty acid F(h24:0), were reproduced in the same diagram to illustrate fatty acid chain length (and hydroxy group) effect on distillation. Two partial mass spectra (m/z 1500–1700) are also reproduced to show the overall change in fatty acid profile on distillation. Single-ion trances for F(n16:0), F(h24:0) are also reproduced in a similar way for a pentaglycosylceramide. In the lowest diagram, F(n16:0), at m/z 1532 of the hexaglycosylceramide, and F(h24:0), at m/z 1471 of the pentaglycosylceramide, are shown to overlap due to the opposing effect of fatty acid and saccharide. (Reproduced from Ref. 19, with permission of the publisher.)

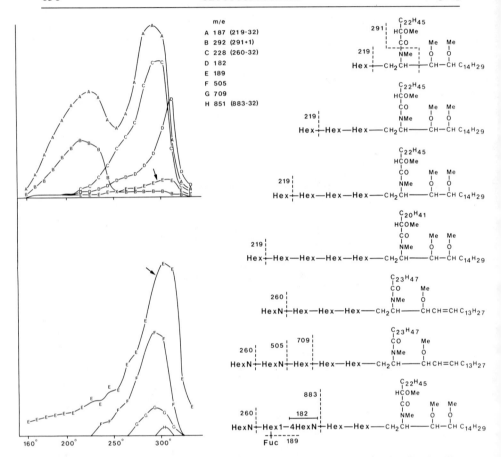

FIG. 8. Selected-ion curves and simplified formulas for interpretation of permethylated total nonacid glycolipids of cat small intestine (30 μg). The curves were recorded during a continuous rise in probe temperature of 5°/min. The arrows indicate the same curve in two reproductions with different intensity scales. Electron energy 44 eV, trap current 500 μA, ion source temperature 290°, accelerating voltage 6 kV. (Reproduced from Ref. 18, with permission of the publisher.)

tive structures were suggested (see formulas). Supportive evidence for partial sequences was then achieved from analysis of the nonreduced permethylated derivatives.

The ion at m/z 187 (219 − 32) showed two intensity maxima. However, m/z 292 (291 + 1) specific for monoglycosylceramide showed only one early maximum. A partial separation was seen between a Forssman-similar pentaglycosylceramide, shown by m/z 505 and 709, and a blood group A

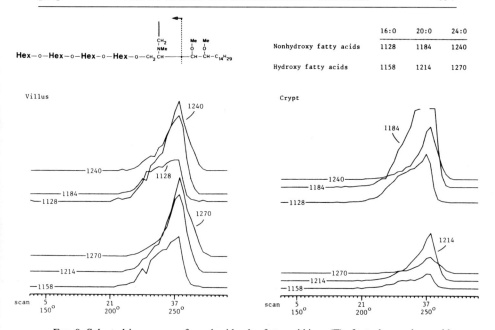

FIG. 9. Selected-ion curves of saccharide plus fatty acid ions (F) of tetrahexosylceramide of villus and crypt epithelial cells of an inbred rat strain. Data were retrieved from 200 μg of the permethylated–reduced mixtures of total glycolipids, evaporated at 5°/min. Spectra were recorded each 38 sec, electron energy was 34 eV, accelerating voltage 4 kV, trap current 500 μA, and ion source temperature 280°. (Reproduced from Ref. 20, with permission of the publisher.)

hexaglycosylceramide, shown by m/z 182, 851 (883 − 32), and 189. The ion at m/z 228 (260 − 32), which is common to both structures, gave a somewhat broader peak. There are a number of other examples within the papers of the reference list showing diagnostic sequence information from individual species in mixtures of permethylated glycolipids (e.g., Refs. 7 and 17). The final conclusion is based on the combined interpretation of spectra of both (or, in case of gangliosides, all three) derivatives.

Information on Ceramide Composition

Advanced information on the lipophilic components of glycosphingolipids in a mixture may be obtained from both derivatives. Figure 9 shows the saccharide plus fatty acid ions (F) of tetrahexosylceramide of villus and

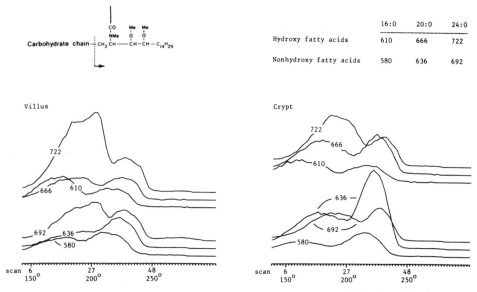

Fig. 10. Selected-ion curves of villus and crypt epithelial glycolipids of an inbred rat strain. Relative abundance of ceramide ions was reproduced. A total of 100 μg each of the permethylated mixture was evaporated by a temperature rise of 5°/min, and spectra were recorded each 38 sec. Electron energy was 49 eV, acceleration voltage 4 kV, trap current 500 μA, and ion source temperature 290°. (Reproduced from Ref. 20, with permission of the publisher.)

crypt epithelial cells of black and white rat small intestine.[20] A hydrocarbon chain lengthening and an increased fatty acid hydroxylation from crypt to villus (successive maturation of a single cell layer) could easily be deduced. Similar analyses could be done for all glycolipids present.

Figure 10 shows single ion traces from permethylated derivatives of the same fractions as for Fig. 9.[20] The information from these ions are more complicated as they are not specific for separate glycolipid species. However the two-peak appearance is explained by the fact that the major components are mono- and trihexosylceramides (as seen from TLC analyses). With this knowledge one may interpret from the curves a relative lengthening of the fatty acid and an increased ceramide hydroxylation from crypt to villus. This is thus in agreement with the above data.

[20] M. E. Breimer, G. C. Hansson, K.-A. Karlsson, and H. Leffler, in "Cell Surface Glycolipids" (C. C. Sweeley, ed.), p. 79. ACS Symp. Ser. No. 128. American Chemical Society, Washington, D.C., 1980.

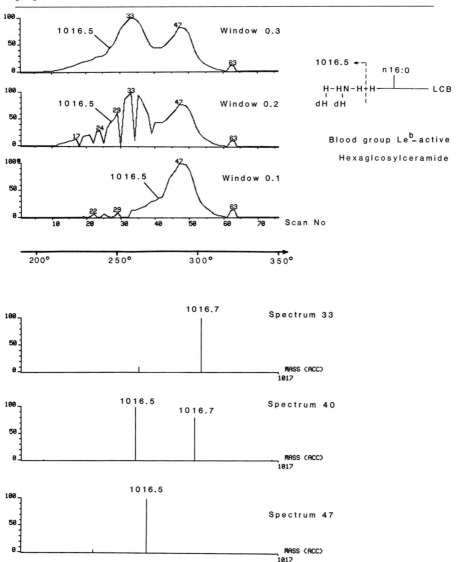

FIG. 11. Selected-ion curve of *m/z* 1016.5 from mass spectrometry of a permethylated mixture of total nonacid glycosphingolipids from plasma of a blood group B Le(a−b+) secretor. About 10 μg of sample was evaporated by a temperature rise of 6°/min and spectra were recorded every 20 sec. Single-ion traces using three different windows are shown. Peaks at *m/z* 1016–1017 were recorded at and in between the two peak maxima of the 0.3 window. Mass marking was done using Fomblin together with sample-derived internal lock masses (instrument resolution 5000). (Reproduced from Ref. 19, with permission of the publisher.)

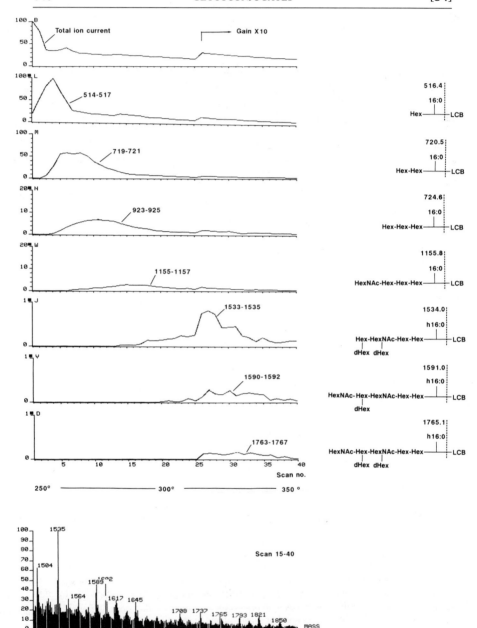

FIG. 12. Selected-ion curves from mass spectrometry of permethylated–reduced total nonacid glycosphingolipids from 5 ml of blood plasma of a human individual with a rare blood group phenotype (*H* gene-deficient blood group A₁ Le(a−b+) secretor from the Reunion

Mass Spectrometer Resolution, Single-Ion Curve Window, and Specificity of Fragments

After a sample run, selected-ion curves may be constructed from collected time-to-mass converted mass spectra. As calibration may vary between different spectra in a consecutive series, especially if the intensity of the peak is low, the computer needs a window to find the requested peak. EI spectra of glycosphingolipids are complex and the complexity increases when analyzing mixtures. The spectrum of an individual component may, as shown above, contain different types of fragments, such as only carbohydrate, only hydrocarbon, or carbohydrate plus hydrocarbon. These fragments may sometimes be superimposed in the spectra of an individual component as well as in spectra of different components in a mixture. The fragments differ however in their relative content of hydrogen atoms and thus in mass, which is the basis for the data of Fig. 11 taken for analysis. By the use of increased instrument resolution, optimized mass calibration by lock masses, and a minimized window when constructing single-ion curves, superimposed fragments may be discriminated. Figure 11 shows an example from the analyses of human blood plasma glycosphingolipids.[19]

Techniques to Improve Sensitivity

A general phenomenon in EI mass spectrometry is a concomitant decrease of sensitivity with mass increase. An accompanying problem is mass assignment due to influence of ion statistics. One way to overcome this problem is to use large windows (± 1–2 mass units) when calculating the single-ion curves. As discussed above, a wide window may corroborate specificity. The correctness of mass assignment can, however, be ascertained by averaging over a large number of consecutive spectra. The mass assignment after such a procedure is often unequivocal.

Figure 12 shows an example from distillation of a permethylated–reduced total nonacid glycosphingolipid sample from 5 ml of human plasma

---

Islands, East Africa). All curves were normalized against the curve for monoglycosylceramide (m/z 514–517). The ion source temperature was raised linearly 10°/min from 250° to 350°. Accelerating voltage was 8 kV and the electron energy 38 eV. Spectra were recorded at a resolution of about 2000. The amount of the seven-sugar compound was estimated to be 10 ng. Note that the detector gain was increased $\times 10$ from spectrum 26. The mass spectrum shown below was computer averaged over scans 15–40. This procedure increased the signal-to-noise ratio and unequivocally gave diagnostic peaks.

obtained from an individual with an extremely rare blood group.[21] The amount of the seven-sugar glycolipid detected was calculated to be 10 ng. The spectrum reproduced below the single-ion curves was obtained by averaging spectra 15–40 and shows clearly the presence of the diagnostic F ions. By the use of this technique, we have been able to detect a glycolipid with as many as eleven sugars in the carbohydrate chain.[22]

Future Developments in Analysis of Mixtures of Intact
  Glycosphingolipids; Possibilities for Thin-Layer
  Chromatography-Mass Spectrometry

The thin-layer plate has recently evolved into an extraordinary assay interface where the separated glycosphingolipids, adsorbed onto the plate, may be exposed to a number of carbohydrate-binding ligands like antibodies, lectins, toxins, microorganisms, and cells (for references, see Ref. 3). One way of using this facility is to characterize glycosphingolipids with ligands of known binding specificity. Another possibility is to screen for possible receptors of a ligand using mixtures of various origins separated on the exposed plate. However, the thin-layer plate in itself is only a chromatographic technique and lacks structural specificity. A direct combination of the separative power of the thin-layer plate and its advanced information on ligand binding with the structural specificity of mass spectrometry may, in practice, be considered a breakthrough in analytical biochemistry, if optimized.

Kushi and Handa[23] were the first to describe this possibility. Only a few papers have so far been published on the application on different kinds of lipids including glycolipids.[23-26] We have used a prototype source from VG Analytical for a ZAB HF 2F instrument with a manually operated probe.

Figure 13 shows the FAB analysis of a standard mixture of globoside and a blood group Le[b]-active hexaglycosylceramide in their permethylated form. As can be seen, the spectra obtained are reasonably good and informative, and the separation of the thin-layer plate is retained after matrix overlay.

[21] R. Mollicone, A. M. Dalix, A. Jacobsson, B. E. Samuelsson, G. Gerard, K. Crainic, T. Caillard, J. LePendu, and R. Oriol, *Glycoconjugate J.* **5,** 499 (1988).
[22] J. Holgersson, W. Pimlott, B. E. Samuelsson, and M. E. Breimer, *Rapid Commun. Mass Spectrom.* **3,** 400 (1989).
[23] Y. Kushi and S. Handa, *J. Biochem.* (*Toyko*) **98,** 265 (1985).
[24] J. Tamura, S. Sakamoto, and E. Kubota, *Jeol News* **23A,** 20 (1987).
[25] H. Karlsson, K.-A. Karlsson, S.-O. Olofsson, W. Pimlott, and B. E. Samuelsson, *Proc. Jpn. Soc. Biomed. Mass Spectrom.* **13,** 35 (1988).
[26] Y. Kushi, C. Rokukawa, and S. Handa, *Anal. Biochem.* **175,** 167 (1988).

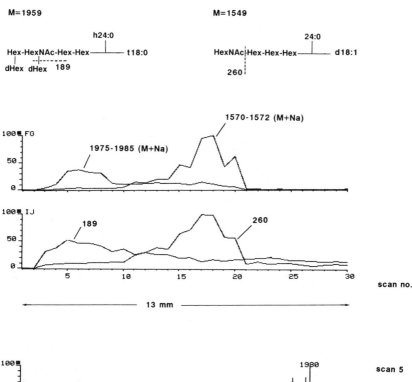

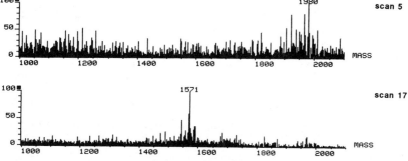

FIG. 13. TLC/MS of globoside and a Le^b-active hexaglycosylceramide separated on a thin-layer plate as permethylated derivatives. HPTLC plates on aluminum sheets (Merck, Darmstadt, Germany) were used. The solvent was chloroform : methanol, 92 : 8 (v/v). About 30 μg of the glycolipid mixture was spotted on the thin-layer plate. The separated spots were located by iodine vapors which were evaporated before introduction into the ion source. The thin-layer plate was moved stepwise within the source by 1 mm. Two scans were recorded for each step. Nitrobenzyl alcohol with 5% trifluoroacetic acid was used as matrix. Spectra were recorded in positive ion FAB mode. Xenon was used for the FAB gun. Acceleration voltage was 8 kV and the resolution was set to about 2000. Single-ion curves are given above and scans 5 and 17 are shown below to illustrate the [M + Na]$^+$ region recorded for the two compounds. The analysis covered about 13 mm of the thin-layer strip.

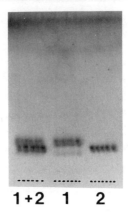

1 + 2   1   2

FIG. 14. Thin-layer chromatogram of 4 µg of a blood-group H-active pentaglycosylcera-mide, lane 1, 4 µg of a blood group A-active hexaglycosylceramide, lane 2, and 8 µg of a 1 : 1 mixture of both, lane 1 + 2. The glycosphingolipids, both with a type 1 carbohydrate chain, were prepared and characterized from pig small intestine (Holgersson, Samuelsson, Breimer, to be published). The thin-layer plate (HPTLC plates on aluminum sheets, Merck, Darmstadt) was developed in chloroform : methanol : water, 60 : 35 : 8 (v/v/v) and the spots revealed by the anisaldehyde reagent. Lane 1 + 2 of a duplicate thin-layer plate, not treated with the anisaldehyde reagent but exposed to iodine vapors to temporarily visualize the spots, was used in the TLC/MS experiment in Fig. 15.

Figure 15 shows another example with a blood group H pentaglycosyl-ceramide and a blood group A-active hexaglycosylceramide analyzed in the negative ion FAB mode and in nonderivatized form. As can be seen from Figure 14, these two substances did not separate completely on the thin-layer plate. However, by using single-ion traces specific for glycolipid species with a certain lipophilic composition, a separation was detected which was not seen on the thin-layer plate by visual inspection.

The method is still in its developmental stage and at least in our hands requires relatively large amounts (several micrograms for each component) of sample.

General Conclusion and Perspective

EI/MS from probe distillation of permethylated, LiAlH$_4$-reduced, and, in case of gangliosides, also LiAlH$_4$-reduced permethylated derivatives where sialyl alcohols have been trimethylsilylated, acetylated, or methyl-ated, appears in our hands to be structurally the most informative mass spectrometric approach when mixtures of biologically active glycolipids are concerned. Dominating simpler species are often completely separated

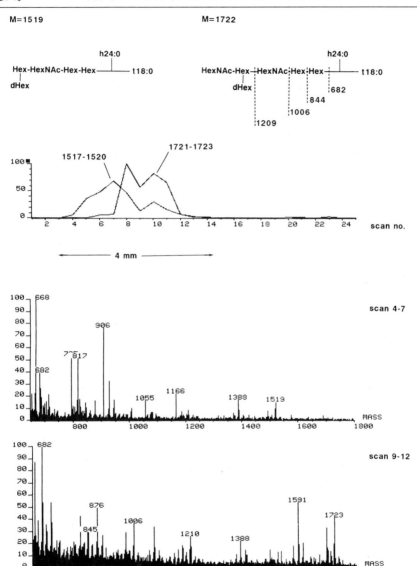

FIG. 15. TLC/MS of a blood group H-active pentaglycosylceramide and a blood group A-active hexaglycosylceramide both with type 1 carbohydrate chains and isolated from pig intestine. The glycolipids were separated on a thin-layer plate as nonderivatized substances (see lane 1 + 2 in Fig. 14). Further conditions were as described in Fig. 13. Spectra were recorded in negative ion FAB mode with triethanolamine as matrix. Ion curves shown represent molecular ions of the major molecular species separated. Mass spectra below were computer averaged over scans 4–7 and scans 9–12, respectively, originating from each of the two compounds separated from each other.

from minor more complex species and information on both ceramide components and saccharide sequences may be extracted by a combined interpretation of the separate derivatives recorded along the evaporation curve. Available methods for the preparation of total glycolipid fractions, even from tissues like blood plasma with very small amounts present, are essential prerequisites for this approach. A fruitful comparison of glycolipids separated on a thin-layer plate and overlayed with various biological reagents, with the same mixture derivatized and analyzed by mass spectrometry is important for preliminary assignment of a receptor-active sequence before the laborious and necessary isolation of a pure species takes place.

In addition to efforts to optimize, e.g., LC/MS and SFC/MS for preseparation purposes, TLC/MS should be seriously developed as an analytical method. The possibility of a direct correlation of overlay binding data with protein reagents (antibodies, lectins, toxins) and even living microbes and cells, and direct scanning of mass spectra over the same mixture, is very attractive for the biochemist and may dramatically improve essential information within important fields like transplantation biology, cancer, and infections.

Several contributions to this volume deal with MS/MS. An interesting supplement would be the use of TLC/MS/MS. But even without TLC, the distillation approach with EI reviewed here may be well suited for MS/MS, especially through the surprisingly abundant F fragments obtained from permethylated–reduced derivatives which carry the complete saccharide chain and the fatty acid.

Acknowledgments

The work presented has been supported by grants from the Swedish Medical Research Council Nos. 3967 and 6521 and the Swedish Board for Technical Development 87-01842P. VG Analytical is gratefully acknowledged for putting the prototype FAB-TLC/MS ion source at our disposal.

[35] Preparation and Desorption Mass Spectrometry of Permethyl and Peracetyl Derivatives of Oligosaccharides

By ANNE DELL

Introduction

The methods described in this chapter are applicable to all types of oligosaccharide and oligosaccharide conjugates including glycopeptides and glycolipids. Although many of these substances can be analyzed by desorption mass spectrometry without prior derivatization, the permethyl and peracetyl derivatives offer the following advantages[1]: (i) higher sensitivity, (ii) applicability to impure samples, e.g., those contaminated with salts and proteins, (iii) reliable and well-defined fragmentation allowing unambiguous sequencing, and (iv) fragmentation data can frequently be obtained from high-molecular weight samples (5000–20,000) which are difficult or impossible to analyze as native compounds.

Permethylation

Principle. Two procedures are suitable for permethylating oligosaccharides and glycoconjugates. The first, which was originally described by Hakomori,[2] uses the anion of dimethyl sulfoxide (DMSO$^-$, usually referred to as the dimsyl anion) to remove protons from the sample prior to their replacement with methyl groups. The second procedure,[3] introduced in 1984, uses solid sodium hydroxide as the base. The latter procedure is experimentally easier and yields a cleaner product, and is therefore preferred for many applications with the important exception of sulfated samples.[4]

Hakomori Permethylation

Reagents

Anhydrous DMSO
Calcium hydride as drying agent

---

[1] A. Dell, *Adv. Carbohydr. Chem. Biochem.* **45,** 19 (1987).
[2] S.-I. Hakomori, *J. Biochem. (Tokyo)* **55,** 205 (1964).
[3] I. Ciucanu and F. Kerek, *Carbohydr. Res.* **131,** 209 (1984).
[4] A. Dell, M. E. Rogers, J. E. Thomas-Oates, T. N. Huckerby, P. N. Sanderson, and I. A. Nieduszynski, *Carbohydr. Res.* **179,** 7 (1988).

Oil-free sodium hydride (NaH), cleaned using sodium-dried ether

Methyl iodide (*Caution:* Methyl iodide is toxic; all manipulations involving this reagent must be performed in a fumehood.)

Procedure. The protocol described here is that used at Imperial College
to permethylate all types of carbohydrate-containing biopolymers and the
conditions are sufficiently mild to preserve the integrity of sulfate and
phosphate esters, while giving excellent methylation. It should be noted
that the term Hakomori permethylation refers to the specific reagents used
and not to a precise protocol. There is wide variation between laboratories
with regard to detailed experimental conditions; some of the more rigorous
procedures reported in the literature may degrade labile bonds, such as
sulfate esters. The method described here was originally introduced by
Morris *et al.*[5] for permethylation of glycopeptides for subsequent analysis
by electron ionization (EI) mass spectrometry. It is a slightly more rigorous
version of the mild permethylation conditions which Morris found to be
suitable for permethylating peptides containing amino acid residues, such
as histidine and methionine, that were prone to undesirable overmethy-
lation.[6]

PURIFICATION OF REAGENTS. Anhydrous DMSO is prepared by standing commercial DMSO over calcium hydride for 2 days and then distilling
from calcium hydride under reduced pressure (oil pump). Although precise
pressure is not important it should be sufficiently low to allow distillation
below about 50°, since decomposition of DMSO can occur at higher temperature. The purified DMSO is stored in a Quickfit flask over fresh
calcium hydride at room temperature. The DMSO will remain dry for
several months provided the stopper is kept tight and is immediately
replaced after DMSO has been dispensed.

Oil-free NaH is prepared from the commercial oil stabilized preparation. Commercial oil-free reagent is not usually active enough to afford
good dimsyl bases. The oil is removed as follows (*Caution:* all these
operations must be carried out well away from sources of water because
sodium and NaH react violently with water): the oil/NaH suspension (200
g) is placed in a large flask in a fumehood and washed 10× with 100 ml of
sodium-dried ether [to ensure that the ether is dry, sodium wire, prepared
using a sodium press, should be carefully added to a previously unused
bottle (2.5 liter) of ether (BDH, Analar grade) which is then left to stand
overnight or until any traces of hydrogen gas evolution have ceased].
Washing is achieved by swirling the flask, allowing the NaH to settle, and

[5] H. R. Morris, M. R. Thompson, D. T. Osuga, A. I. Ahmed, S. M. Chan, J. R. Vandenheede, and R. E. Feeney, *J. Biol. Chem.* **253**, 5155 (1978).
[6] H. R. Morris, *FEBS Lett.* **22**, 257 (1972).

then decanting the ether into a beaker. After the final decanting step, the residual ether is removed under reduced pressure by placing the flask in a large desiccator and attaching it to a vacuum pump. The clean NaH is transferred to a dry screw-capped polypropylene jar and is stored at room temperature in a dry atmosphere, e.g., inside a large (empty) coffee jar covered with dry cotton wool, with the lid screwed tight. NaH prepared in this way should remain stable for months, even years, provided moisture contamination is minimized. The NaH is active if it is dark gray. It decomposes to NaOH which turns the powder white. (*Note*: Ether washings can be safely disposed of by gentle evaporation under a stream of nitrogen in the fumehood followed by dropwise addition of ethanol to destroy the residual NaH.)

PREPARATION OF DIMSYL BASE. Two small microspatula scoops of NaH are placed in the bottom of a Quickfit tube (10 cm × 1 cm) and approx. 3 ml of dry DMSO are added using a Pasteur pipette. After shaking to ensure an even suspension, the tube (without its stopper) is placed in an oven at 90° for about 10 min at which time the mixture is removed, stoppered, and shaken to ensure that the NaH is well mixed. At this point the suspension should be light brown in color. If it is still gray and very little NaH remains on the bottom, an additional microspatula of NaH should be added. The unstoppered tube is returned to the oven for 5–10 min, the exact time being dependent on the depth of color of the suspension; the final mixture should be honey brown in color. The tube containing the base is removed from the oven, stoppered, allowed to cool for a few minutes, and then centrifuged using a bench-top centrifuge at 3000 rpm for 10 min. After centrifugation the DMSO solution containing the dimsyl anion should be crystal clear and honey colored. To check that it is active, a drop of the base is added to a few crystals of triphenylmethane, which will turn scarlet in the presence of active base.

PERMETHYLATION WITH DIMSYL BASE. The sample to be permethylated is dried down in a small glass tube (either Quickfit or Teflon-lined screw-capped) and is dissolved in 2 drops of DMSO (Pasteur pipettes are used for the addition of all liquid reagents in the permethylation reaction). Five drops of the freshly prepared dimsyl base are added followed by 3 drops of methyl iodide. The stoppered tube is left to stand at room temperature for 5 min and then 20 drops of the dimsyl base are added. A drop of the reaction mixture is added to a few triphenylmethane crystals to check for excess base. If the crystals turn scarlet, 0.5 ml of methyl iodide is added to the reaction mixture and the resulting solution left to stand for 10 to 20 min. If excess base is not observed (the triphenylmethane turns yellowish) more base is added until the test shows it to be in excess. Methyl iodide is then added as described above. The reaction is stopped after 20 min by the addition of 1 ml of water. The mixture is shaken vigorously and then left to stand for a few minutes. The methyl iodide settles to the bottom of the tube, and is removed by bubbling nitrogen gas into the mixture using a Pasteur pipette. The DMSO/water solution containing the permethylated sample, together with salts and methylation by-products, is transferred directly to a Sep-Pak (Waters) for sample cleanup using a procedure similar to that originally described by Waeghe *et al.*[7]

SEP-PAK CLEAN-UP. The Sep-Pak is first conditioned by sequential washing with water (5 ml), acetonitrile (5 ml), ethanol (5 ml), and water (10 ml). The permethylation solution is loaded onto the conditioned Sep-Pak which is eluted with water (5 ml), 15% aqueous acetonitrile (2 ml), 30% aqueous acetonitrile (2 ml), 50% aqueous acetonitrile (2 ml), 75% aqueous acetonitrile (2 ml), and acetonitrile (2 ml). If glycolipids are present, two additional eluents, methanol (2 ml) and chloroform (2 ml), are included. Most permethylated oligosaccharides and glycopeptides elute in the 50 and 75% acetonitrile fractions; sulfated samples normally elute in the 15 and 30% acetonitrile fractions; glycolipids are eluted by the chloroform wash. The fractions are dried using a Speedivap (Savant) and, with the exception of the water wash, each is redissolved in 5–10 μl of methanol for analysis by mass spectrometry.

HPLC CLEAN-UP. In some instances, e.g., for complex mixtures of oligosaccharides, a reversed-phase high-performance liquid chromatography (HPLC) step prior to mass spectrometric analysis is desirable. A wide range of column types and elution conditions have been reported for permethylated samples. The following protocol has been successfully used in our laboratory for permethylated oligosaccharides derived from sources as diverse as plant cell walls, bacterial antigens, and mammalian glycoproteins. The column used is a 25 × 0.46 cm Spherisorb S5 ODS2. Solvent A is Milli-Q water and solvent B is acetonitrile (both degassed prior to use). The sample is loaded in 0.5 ml of 36% aqueous acetonitrile and eluted using a 1 ml/min linear gradient from A : B (90 : 10) to A : B (40 : 60) over 100 min followed by a linear gradient from A : B (40 : 60) to A : B (0 : 100) over 10 min. The time of the second gradient is lengthened if high-molecular weight oligosaccharides are expected to be present. The eluent is monitored at 214 nm which detects the amide bond of acetamido residues.

NaOH Permethylation

This procedure can be used to permethylate free oligosaccharides and glycoconjugates and is also useful for screening O-linked glycoproteins for compositions of *O*-glycans. In the last case, the intact glycoprotein is

---

[7] T. J. Waeghe, A. G. Darvill, M. McNeil, and P. Albersheim, *Carbohydr. Res.* **123**, 281 (1983).

permethylated as described below and the NaOH serves both to release the *O*-glycan by β elimination and to deprotonate it for subsequent methylation.[8]

Reagents

NaOH pellets
Dry DMSO
Methyl iodide
Chloroform

Procedure: (*Note*: all manipulations involving methyl iodide or chloroform must be carried out in a fumehood.) About seven pellets of NaOH are placed in a dry mortar and approximately 3 ml of dry DMSO (see above Hakomori permethylation procedure for preparation of dry DMSO) is added using a Pasteur pipette. A pestle is used to grind the NaOH until a slurry is formed with the DMSO. Approximately 1 ml of this slurry is squirted onto the sample which has been previously dried down in the bottom of a Quickfit or Teflon-lined screw-capped tube. Approximately 1 ml of methyl iodide is added using a Pasteur pipette, the mixture is manually shaken, and then placed on an automatic shaker for 10 min at room temperature. The reaction is then quenched by the careful addition of about 1 ml of water which is added in four aliquots with shaking between additions to lessen the effects of the exothermic reaction that occurs. As soon as the vigorous reaction subsides, chloroform (2 ml) is added, and the mixture shaken and allowed to settle. The upper (water) layer is removed with a Pasteur pipette and the lower (chloroform) layer is washed several times with water until the water being removed is completely clear. The chloroform layer is dried under a stream of nitrogen and the residue containing the permethylated sample is redissolved in 5–10 μl of methanol for analysis by mass spectrometry.

Methanolysis

Principle. Partial methanolysis of a permethylated sample using fast atom bombardment (FAB) MS to monitor the reaction provides valuable sequence information including branching patterns.[9] During partial methanolysis the permethylated sample is gradually degraded by hydrolysis of the glycosidic linkages. Methylglycosides (or perdeuteromethylglycosides if deuterated methanol is used in the reaction) are produced at released reducing termini, while each released glycosidic oxygen becomes a free

---

[8] J. E. Thomas-Oates and A. Dell, *Biochem. Soc. Trans.* **17,** 243 (1989).
[9] A. Dell, N. H. Carman, P. R. Tiller, and J. E. Thomas-Oates, *Biomed. Environ. Mass Spectrom.* **16,** 19 (1988).

hydroxyl group. The number of free hydroxyl groups produced is used to define sequence and branching. For example, hydrolytic removal of one or more residues without the generation of a free hydroxyl group indicates that those residues are at the reducing end of the intact oligosaccharide. Similarly, the production of two free hydroxyl groups indicates the simultaneous hydrolysis of two different branches. The analysis is immediate since aliquots of the degradation mixture are loaded directly onto the FAB target with no work-up.

Reagents

Methanol or [²H]methanol
HCl gas

Procedure. Dry HCl gas is bubbled into 0.5–1.0 ml of methanol or [²H]methanol until the solution becomes noticeably warm. After cooling, a 10-μl aliquot of this methanolysis reagent is added to the permethylated sample which has been dried down in a small Reacti-vial (Pierce Chemical Co., Rockford, IL). The sample is incubated at 40° in a heating block and 1-μl aliquots are removed after 2, 20, and 40 min and analyzed immediately by FAB-MS. The temperature is then increased to 60° and 1-μl aliquots are removed and analyzed at 5-, 30-, 60-, and 120-min intervals. During the course of the experiment, additional reagent is added as necessary to prevent drying out. The exact time course can be altered during the reaction if the mass spectra indicate that the reaction is proceeding either very slowly or very rapidly.

Peracetylation

Principle. Peracetylation under acidic conditions[10] preserves base-labile substituents, e.g., acyl,[11] on oligosaccharides and is the preferred method for subsequent FAB-MS analysis. The reagent is prepared by mixing equimolar amounts of trifluoroacetic anhydride and acetic acid (the acetic acid can be replaced with ²H-labeled acetic acid or propionic acid if natural acetates are present in the sample and their presence and location needs to be defined[11]). The resulting mixed anhydride CF_3CO—O—$COCH_3$ can, in principle, transfer either trifluoroacetyl or acetyl to a nucleophilic acceptor. In practice, hydroxyl groups are almost exclusively acetylated. The α-amino group of peptides is trifluoroacetylated with this reagent but amino groups with higher pK values are usually acetylated.

Peracetylation is an excellent method for recovering oligosaccharides

---

[10] E. J. Bourne, M. Stacey, J. C. Tatlow, and J. M. Tedder, *J. Chem. Soc.* p. 2976 (1949).

[11] A. Dell and P. R. Tiller, *Biochem. Biophys. Res. Commun.* **135**, 1126 (1986).

from impure samples, e.g., urine extracts, salty samples that have been initially purified by ion exchange using salt gradients, mixtures of proteins, peptides, and oligosaccharides obtained after digestion procedures.[12] Once the peracetylated sample has been separated from water-soluble contaminants it can be analyzed directly or it can be converted to the permethyl derivative by either Hakomori or NaOH permethylation which releases all the base-labile acetyl groups and replaces them with methyl groups.

Reagents

Trifluoroacetic anhydride
Acetic acid
Chloroform (*Caution*: The toxicity of chloroform requires the use of a fumehood.)
Procedure. The reagent is prepared by mixing trifluoroacetic anhydride and acetic acid (2 : 1, v : v) at room temperature. The mixture is left to stand for about 30 min until the heat from the exothermic reaction has dissipated. Approximately 100 μl of this reagent is added to the dried sample and the reaction is left at room temperature for 10 min; the reagents are then removed under a stream of nitrogen. The residue is taken up in 0.5 ml of chloroform, washed with water (3 × 2 ml), and dried under a stream of nitrogen. The product is dissolved in 5–10 μl of methanol for analysis by mass spectrometry. If the sample is known to be free from impurities the chloroform step may be omitted.

Acetolysis

Principle. Careful acetolysis of peracetylated high mannose structures selectively cleaves the 1→6 linkages.[13] Analysis of the products by FAB-MS[14] or other procedures can therefore reveal branching patterns of these molecules. Acetolysis can also be used as a general hydrolytic procedure affording peracetylated oligosaccharide fragments from glycoproteins or polysaccharides. FAB-MS of these products provides useful information pertaining to the nature of the carbohydrate chains in the intact molecule.[15]

Reagents

Acetic anhydride
Acetic acid
Concentrated sulfuric acid

[12] A. Dell, J. E. Thomas-Oates, M. E. Rogers, and P. R. Tiller, *Biochimie* **70**, 1435 (1988).
[13] L. Rosenfeld and C. E. Ballou, *Carbohydr. Res.* **32**, 287 (1974).
[14] P.-K. Tsai, A. Dell, and C. E. Ballou, *Proc. Natl. Acad. Sci. U.S.A.* **82**, 4119 (1986).
[15] S. Naik, J. E. Oates, A. Dell, G. W. Taylor, P. M. Dey, and J. B. Pridham, *Biochem. Biophys. Res. Commun.* **132**, 1 (1985).

Procedure. (i) SPECIFIC HYDROLYSIS OF 1→6 LINKAGES. The peracetylated sample (prepared as described above) is dissolved in 0.5 ml of a
10 : 10 : 1 (v : v : v) mixture of acetic acid/acetic anhydride/concentrated
sulfuric acid and kept at 40° for 90 min. Aliquots (100 μl) are removed at
10, 30, 60, and 90 min. Each aliquot is quenched with water (1 ml) and
extracted with chloroform (1 ml). The organic layer is washed with water
(3 × 2 ml), evaporated under a stream of nitorgen, and redissolved in 5–10
μl of methanol for analysis by mass spectrometry.

(ii) GENERAL SCREENING PROCEDURE FOR GLYCOPROTEINS AND POLY-
SACCHARIDES. The native sample is dissolved in 1 ml of the acetolysis
reagent (acetic acid/acetic anhydride/concentrated sulfuric acid, 10 : 10 : 1,
v : v : v) and incubated at 60° for 18 hr. Aliquots (200 μl) are removed at
30 min, 2 hr, 6 hr, and 18 hr and are worked up as described in procedure
(i) above.

Desorption Mass Spectrometry

In this section desorption mass spectrometry refers specifically to the
analysis of samples from a liquid matrix using either atoms (FAB-MS) or
ions (LSIMS) in the ionization/desorption process. These two procedures
afford very similar data with the caveat that the quality of data and,
therefore, the presence of some fragment ions is dependent on the energy
of the bombarding species—the greater the energy, the more intense the
spectrum.

Choice of Matrix. For the majority of permethylated and peracetylated
samples, 1-thioglycerol (TG) is the matrix of choice. This can be used
alone or as a 1 : 1 mixture with glycerol. The latter has a longer lifetime
due to the lower volatility of glycerol. 3-Nitrobenzyl alcohol (NBA) is a
matrix which promotes sodium cationization particularly of peracetylated
compounds and may give sequence information additional to that present
in spectra obtained using TG. For peracetyl (but not permethyl) samples,
the NBA matrix appears to be less sensitive to salt contamination than
TG.[16] In the negative ion mode, which for permethylated or peracetylated
oligosaccharides is only appropriate for samples which retain a negative
charge after derivatization, triethanolamine or related basic matrices are
frequently used but TG remains the most versatile one.

Running Samples. Smear 1–2 μl of glycerol/1-thioglycerol (1 : 1, v : v)
onto the metal target on the end of the FAB or SIMS probe. Using a
graduated 1–5 μl micropipette, carefully blow 1 μl of the sample, which

[16] A. Dell, P. Azadi, P. Tiller, J. Thomas-Oates, H. J. Jennings, M. Beurret, and F. Michon,
Carbohydr. Res. **200**, 59 (1990).

TABLE I

MASS INCREMENTS[a] FOR METHYLATED AND ACETYLATED DERIVATIVES OF COMMON
MONOSACCHARIDE RESIDUES

Residue	Methyl	Deuteromethyl	Acetyl	Deuteroacetyl
Hex	204	213	288	297
HexNAc	245	254	287	293
NeuAc	361	376	417	426
NeuGc	391	409	475	487
DeoxyHex	174	180	230	236
HexA	218	227	260	266
Pent	160	166	216	222
O-Me-Hex	204	210	260	266
Hept	248	260	360	372
KDO	276	288	346	355

[a] See Fig. 2.

has been predissolved in 5–10 μl of methanol, onto the surface of the matrix. Introduce the probe into the high vacuum of the source housing for about 30 sec in order to pump off most of the solvent. Do not exhaustively pump down because evaporation of the last traces of solvent during subsequent atom or ion bombardment helps to replenish surface sample. Remove the probe to check that the volume of matrix has not significantly been decreased by coevaporation of methanol and TG. If necessary, carefully replenish the matrix with a small drop of TG. Fully insert the probe into the ion source until it is about 1 cm from its operating position. Switch on the voltages and, while observing the repetitive scanning of a selected region of the spectrum on the visual display unit, push the probe fully into place. The mass range selected for observation will depend on the extent of knowledge of the sample structure. If possible, select the expected molecular ion region, or an area where key fragment ions are expected. For complete unknowns, it is usually appropriate to choose a region of the spectrum near m/z 1500, as the intensities of background signals at this mass will reflect the likelihood of sample signals being present at higher masses. The appearance of the spectrum produced in the first few seconds of atom or ion bombardment is an important indicator of the likelihood of obtaining a good spectrum. If the observed signals are reasonably intense, a full spectrum is recorded immediately. If the visual display shows poor quality data it may be necessary to retune the source rapidly or to remove the probe to add a little more matrix or matrix additive. Suitable additives to try are: (i) 1 μl of 0.1 M HCl, which helps to protonate in the positive mode and is particularly helpful for permethylated sulfated samples in both positive and negative modes, (ii) 1 μl of a 0.1% solution

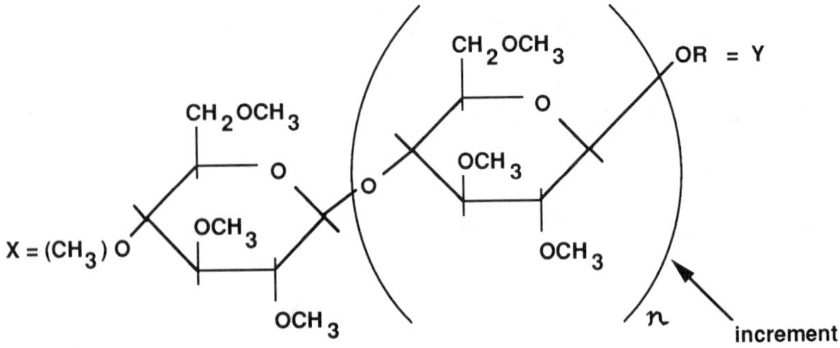

FIG. 1. The terms "increment," "X", and "Y" used in Tables I, III, and IV are illustrated by this schematic representation of a permethylated derivative.

of sodium acetate in methanol, which will enhance the sodiated molecular ions, and (iii) 1 μl of a 0.1 M solution of ammonium thiocyanate, which gives ammonium cationized molecular ions for components which lack amino sugar residues.

Assigning Mass Spectra. Permethylated and peracetylated samples normally afford abundant molecular ions, from which monomeric sugar compositions and substituents can be assigned, together with predictable fragmentation patterns. High-molecular weight samples, e.g., permethylated lactosaminoglycan-containing glycopeptides up to greater than 20,000 Da, may not give observable molecular ions but do, nevertheless, still yield good quality spectra containing nonreducing end fragment ions. The fragmentation behavior of permethylated and peracetylated derivatives has been well documented in reviews.[1,17-19] The purpose of this section is to provide a convenient source of mass data for assisting spectral interpretation.

The molecular weight of a derivatized oligosaccharide is the sum of the mass increments for each residue and for each natural substituent plus the mass of one protecting group (X) at the nonreducing end plus the reducing end increment (Y) (see Fig. 1). Table I gives the mass increments for the common monosaccharides and Table II gives mass increments for their substituents. Table III gives values of X and Table IV gives values of Y for both nonreduced and reduced reducing ends (oligosaccharides are frequently isolated as reduced derivatives). The major fragmentation

[17] V. N. Reinhold and S. A. Carr, *Mass Spectrom. Rev.* **2**, 153 (1983).

[18] A. Dell and J. E. Thomas-Oates, *in* "Analysis of Carbohydrates by GLC and MS" (C. J. Biermann and G. D. McGinnis, eds.), p. 217. CRC Press, Boca Raton, FL, 1989.

[19] H. Egge and J. Peter-Katalinić, *Mass Spectrom. Rev.* **6**, 331 (1987).

TABLE II
MASS INCREMENTS FOR COMMON SUBSTITUENTS[a] PRESENT ON METHYLATED AND
ACETYLATED OLIGOSACCHARIDES

Compound	Methyl	Deuteromethyl	Acetyl	Deuteroacetyl
Pyruvate	56	53	−14	−20
Acetate	NA[b]	NA[b]	0	−3
Phosphate	94	97	38[c]	35[c]
Sulfate	66	63	NA[d]	NA[d]
Inositol	218	230	330	342

[a] Negative values occur if the mass of the substituent is less than that of the derivatizing group.
[b] Acetates are lost during permethylation.
[c] This corresponds to underivatized phosphate; some acetylation of phosphate may occur to give a signal 42 (acetyl) or 45 (deuteroacetyl) mass units higher.
[d] Sulfates are lost during peracetylation.

TABLE III
VALUES OF X (NONREDUCING INCREMENT)[a] FOR DERIVATIZED OLIGOSACCHARIDES

X	Methyl	Deuteromethyl	Acetyl	Deuteroacetyl
Nonreducing	15	18	43	46

[a] See Fig. 2.

TABLE IV
VALUES OF Y (SUBSTITUENT ON REDUCING-END SUGAR)[a] FOR NONREDUCED AND
REDUCED DERIVATIZED OLIGOSACCHARIDES

Y Reducing end	Methyl	Deuteromethyl	Acetyl	Deuteroacetyl
Free	31	34	59	62
Reduced	47	53	103	109

[a] See Fig. 2.

pathways for derivatized oligosaccharides are shown in Fig. 2. By far the most dominant pathway is the A-type cleavage (Fig. 2a). The masses of A-type cleavage ions are calculated by adding X to the sum of the sugar residue and substituent increments. If A-type cleavage is accompanied by β-elimination (Fig. 2b) the resulting ion occurs 32 (permethyl derivative) or 60 (peracetyl derivative) mass units lower. The masses of β-cleavage ions (Fig. 2c) which retain the reducing end of the molecule are calculated

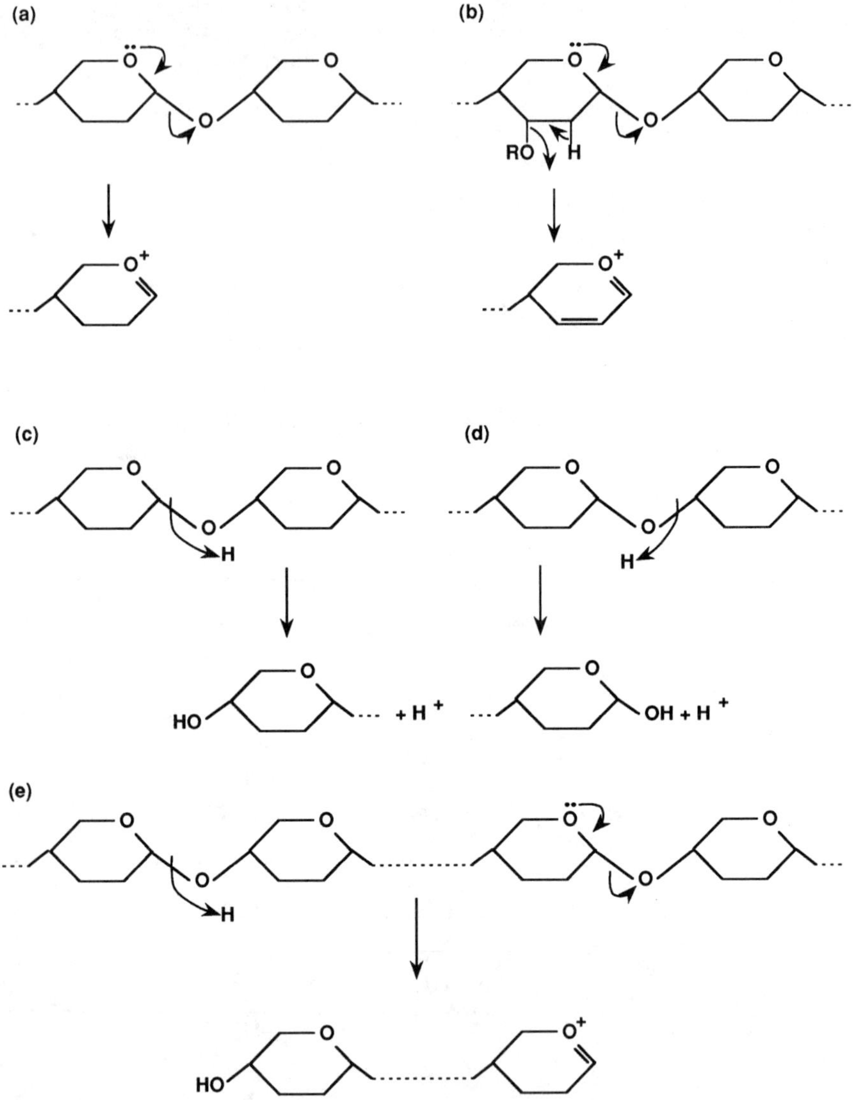

FIG. 2. Major fragmentation pathways for permethylated and peracetylated oligosaccharides. (a) A-type cleavage; this is always the primary mode of fragmentation and occurs most readily at amino sugar residues; (b) A-type cleavage accompanied by elimination of the substituent at position three; (c) β cleavage accompanied by a hydrogen transfer; charge retention on the reducing end; frequently accompanied by a signal 28 mass units higher resulting from ring cleavage to leave a formyl group on the glycosidic oxygen; (d) β cleavage accompanied by hydrogen transfer; charge retention on the nonreducing end; (e) β and A-type cleavage resulting in a double-cleavage ion.

TABLE V

CALCULATED MASSES OF A-TYPE FRAGMENT IONS FORMED BY HexNAc CLEAVAGE OF PERMETHYLATED OLIGOSACCHARIDES OR GLYCOCONJUGATES CONTAINING UP TO NINE Hex OR HexNAc RESIDUES

Fragment	$(HexNAc)_1$	$(HexNAc)_2$	$(HexNAc)_3$	$(HexNAc)_4$	$(HexNAc)_5$	$(HexNAc)_6$	$(HexNAc)_7$	$(HexNAc)_8$	$(HexNAc)_9$
Hex_1	464	709	954	1199	1444	1689	1934	2179	2424
Hex_2	668	913	1158	1403	1648	1893	2138	2383	2628
Hex_3	872	1117	1362	1607	1852	2097	2342	2587	2832
Hex_4	1076	1321	1566	1811	2056	2301	2546	2791	3036
Hex_5	1280	1525	1770	2015	2260	2505	2750	2995	3240
Hex_6	1484	1729	1974	2219	2464	2709	2954	3199	3444
Hex_7	1688	1933	2178	2423	2668	2913	3158	3403	3648
Hex_8	1892	2137	2382	2627	2872	3117	3362	3607	3852
Hex_9	2096	2341	2586	2831	3076	3321	3566	3811	4056

by adding 2 mass units (one for the hydrogen transferred and one for protonation) and Y to the sum of the sugar residue and substituent increments. The masses of β-cleavage ions which retain the nonreducing ends of the molecule (Fig. 2d) are calculated by adding X plus 18 to the sum of the sugar residue and substituent increments. The masses of double-cleavage ions formed by a combination of one β cleavage and one A-type cleavage (Fig. 2e) are obtained by adding 1 mass unit to the sum of the sugar residue and substituent increments. β Cleavage may sometimes be accompanied by a transfer of methyl (giving a 14 u increase in mass) or acetyl (giving a 42 u increase in mass) to the glycosidic oxygen. This is normally, however, a minor pathway resulting in low abundance ions.

Mapping Nonreducing Structures in Glycoproteins. Oligosaccharides and glycoconjugates containing amino sugar residues, e.g., all N- and O-glycans from glycoproteins, show a very specific set of fragment ions when analyzed as their permethyl derivatives. These are A-type ions formed by cleavage at each amino sugar residue and each glycan type affords a predictable and reproducible map of ions containing information on all nonreducing structures. For example, high mannose structures cleave at the chitobiose core to yield fragment ions whose masses define the number of mannosyl residues present, while complex structures afford fragment ions from each of the branches, in addition to chitobiose cleavage ions. The mapping procedure for identifying nonreducing structures can be applied to cell surface screening and is particularly useful for characterizing lactosaminoglycan structures.[20]

PROCEDURE. Cells are extracted with 10 times the volume of chloroform/methanol (2 : 1, v/v) and chloroform/methanol (1 : 2, v/v) to remove lipids. The residue is subjected to extensive pronase digestion, centrifuged, and the soluble material chromatographed on Sephadex G-25. The void-volume fraction containing the glycopeptides is permethylated using the Hakomori procedure described earlier and analyzed by FAB-MS. Table V gives the masses of the A-type ions afforded by glycans containing up to nine Hex or HexNAc residues. High mannose chitobiose cleavage ions are to be found in the first vertical column; characteristic lactosaminoglycan repeats, i.e., $(\text{HexHexNAc})_x$, are on the diagonal. For fucosylated and/or sialylated glycans, increments of 174 (Fuc), 361 (NeuAc), or 391 (NeuGc) are added to each of the numbers in Table V. For example, the chitobiose cleavage ion for a biantennary, disialylated glycan occurs at m/z 2492 ($\text{NeuAc}_2\text{Hex}_5\text{HexNAc}_3^+ = 361 + 361 + 1770$).

[20] M. N. Fukuda, A. Dell, J. E. Oates, and M. Fukuda, *J. Biol. Chem.* **260**, 6623 (1985).

[36] Desorption Mass Spectrometry of Oligosaccharides Coupled with Hydrophobic Chromophores

By LINDA POULTER and A. L. BURLINGAME

Introduction

Early strategies involving mass spectrometric investigation of carbohydrates required chemical derivatization such as permethylation of all labile hydrogens to permit sample vaporization without thermal decomposition for use of electron impact and chemical ionization methods and their GC/MS analogs.[1] Naturally occurring partial methylation and complete chemical methylation was also found to facilitate studies of high mass oligosaccharides using field desorption.[2,3] Karlsson and co-workers exploited a variety of derivatization strategies in order to permit vaporization and also to direct fragmentation in their pioneering systematic studies of high-mass glycolipids.[4,5] These methods and issues have been reviewed elsewhere.[6,7]

Since 1980, liquid matrix sputtering/ionization methods have revolutionized the utility of mass spectrometric-based strategies for the structural characterization of biopolymers.[8] Virtually all classes of labile, polar biological substances may now be investigated throughout an extended mass range (<10,000 Da). This includes a wide variety of structural classes of oligosaccharides occurring as components of natural products, glycolipids, glycoproteins, and proteoglycans.

Although early studies using this new desorption ionization method were carried out on free oligosaccharides[9] or the corresponding oligogly-

[1] T. Radford and D. C. de Jongh, in "Biochemical Applications of Mass Spectrometry (G. R. Waller and O. S. Dermer, eds.), 1st Suppl. Vol., p. 256. Wiley (Interscience), New York, 1980.

[2] M. Linsheid, J. D'Angona, A. L. Burlingame, A. Dell, and C. E. Ballou, *Proc. Natl. Acad. Sci. U.S.A.* **78,** 1471 (1981).

[3] A. L. Burlingame, C. E. Ballou, and A. Dell, *Proc. 29th Annu. Conf. Mass Spectrom. Allied Topics, Minneapolis, MN,* p. 528 (1981).

[4] K.-A. Karlsson, *Progr. Chem. Fats Other Lipids* **16,** 207 (1977).

[5] B. E. Samuelsson, W. Pimlott, and K.-A. Karlsson, this volume [34].

[6] V. N. Reinhold and S. A. Carr, *Mass Spectrom. Rev.* **2,** 153 (1983).

[7] A. Dell and G. N. Taylor, *Mass Spectrom. Rev.* **3,** 357 (1984).

[8] A. L. Burlingame and J. A. McCloskey (Eds.), "Biological Mass Spectrometry." Elsevier, Amsterdam, 1990.

[9] J. P. Kamerling, W. Heerma, J. F. G. Vliegenthart, B. N. Green, I. A. S. Lewis, G. Strecker, and G. Spik, *Biomed. Mass Spectrom.* **10,** 420 (1983).

METHODS IN ENZYMOLOGY, VOL. 193

coalditols,[6,9-11] it has gradually been recognized that such particularly hydrophilic substances present virtually a worst-case scenario for the optimal exploitation of energetic particle-induced sputtering/ionization from the surface layer of viscous hydrophilic liquid matrices such as glycerol, thioglycerol, or triethanolamine. This situation is due both to the lack of surface activity of free carbohydrates in these liquids and to the difficulties in complete separation of such compounds from biological matrices and salts. Hence early recipes advocated the careful addition of alkali metal salts when a sample did not work to form alkali metal attachment ions,[12] which provided a charge and improved the surface activity compared with the corresponding free oligosaccharide. Thus, [M + alkali metal ion]$^+$ species are observed in the positive ion mode and the corresponding [M + halide ion]$^-$ species in the negative ion mode.[13] These studies typically required 1–50 nmol of sample per component.[9,11]

Subsequently, in order to ease the problems of isolation and salt removal as well as increase the sensitivity and promote fragmentation at certain preferred sites, a considerable amount of attention has been directed toward the use of both permethylation and peracetylation combined with sputtering ion soruces, namely, cesium ion liquid secondary ion mass spectrometry (LSIMS) and xenon neutral fast atom bombardment (FAB).[7,11,14-17] These derivatives usually permit the investigator to obtain sufficient mass spectral data to determine some features of the structural class and degree of homogeneity of the sample. However, when such derivatives give poor quality signals, sodium acetate or ammonium thiocyanate is often added to the liquid matrix.[17] While it is clear that permethyl and peracetyl derivatization eases the difficulties of isolation and salt removal, these derivatives are of little use in subsequent separation of components of mixtures. Also, little mention is usually made of the fact that both derivatization procedures yield underderivatization.[15-19] This, of

[10] A. L. Burlingame, in "Secondary Ion Mass Spectrometry, SIMS IV" (A. Benninghoven, J. Okano, R. Shimizu, and H. W. Werner, eds.), p. 399. Springer-Verlag, Berlin, 1984.

[11] J. Peter-Katalinic and H. Egge, Mass Spectrom Rev. 6, 331 (1987).

[12] C. Bosso, J. Defaye, A. Heyraud, and J. Ulrich, Carbohydr. Res. 125, 309 (1984).

[13] D. Promé, J. C. Promé, G. Puzo, and H. Aurelle, Carbohydr. Res. 140, 121 (1985).

[14] S. A. Carr, G. D. Roberts, and M. E. Hemling, in "Mass Spectrometry of Biological Materials," (C. N. McEwen and B. S. Larsen, eds.), p. 87. Dekker, New York, 1990.

[15] A. Dell, Adv. Carbohyd. Chem. Biochem. 45, 19 (1987).

[16] A. Dell and J. E. Thomas-Oakes, in "Analysis of Carbohydrates by GLC and MS" (C. J. Biermann, and G. D. McGinnis, eds.), p. 217. CRC Press, Boca Raton, FL, 1988.

[17] A. Dell, this volume [35].

[18] G. J. Gerwig, P. de Waard, J. P. Kammerling, J. F. G. Vliegenthart, E. Morgenstern, R. Lamed, and E. A. Bayer, J. Biol. Chem. 264, 1027 (1989).

course, distributes the available total ion current among several pseudo-molecular ionic species of the particular component being analyzed[18] and, in effect, compromises the potential gain in overall sensitivity expected from the increased hydrophobicity conveyed by derivatization. In addition, such derivatives increase the molecular weight significantly which results in lowering sensitivity as well.[15,16,20-22]

While use of permethylation and peracetylation provides a convenient strategy to obtain structural class information on carbohydrates attached to glycoproteins and glycolipids without the need for rigorous sample purification,[14,15,21,23] relatively little attention has been devoted to development of chemically and instrumentally integrated strategies including isolation, separation, and derivatization, which would also provide inherently optimized mass spectral quality and sensitivity. While bringing about such an integrated approach is a formidable interdisciplinary challenge, there is still a great need for methodology which could provide rigorous structural characterization of individual carbohydrate isomers at the subnanomole level. This is the subject of the remainder of this chapter.

In this laboratory we have focused attention on the systematic exploration of the use of reductive amination derivatization strategies to effect coupling a hydrophobic chromophore with the free reducing terminus of the chitobiose core upon liberation of asparagine-linked glycosylation by enzymatic treatment of glycoproteins with endoglycosidase H (Endo H) and peptide-N^4-(N-acetyl-β-glucosaminyl) asparagine amidase F (PNGase F). Our initial efforts were devoted to studying the mass spectral fragmentation behavior of the oligoglycoaminodeoxyalditols formed using the *p*-aminobenzoic acid ethyl ester (ABEE) in both positive and negative ion mode and to establish the characteristics of the spectrum–structure correlations.[24] We discovered that the negative ion mode of fragmentation elaborates an unusually abundant, structurally informative ion series. The relative abundances of the fragments observed were shown to correlate with oligosaccharide residue branching patterns. This important finding permitted us to identify the N-linked glycosylation on human hepatitis B

[19] G. O. Aspinall, *in* "The Polysaccharides," (G. O. Aspinall, ed.), Vol. 1, p. 49. Academic Press, New York, 1982.

[20] A. Dell, *Biomed. Environ. Mass Spectrom.* **16**, 19 (1988).

[21] A. Dell, *Biochimie* **70**, 1435 (1988).

[22] H. Sasaki, N. Ochi, A. Dell, and M. Fukuda, *Biochemistry* **27**, 8618 (1988).

[23] A. L. Burlingame, D. Maltby, D. H. Russell, and P. T. Holland, *Anal. Chem.* **60**, 294R (1988).

[24] J. W. Webb, K. Jiang, B. L. Gillece-Castro, A. L. Tarentino, T. H. Plummer, J. C. Byrd, S. J. Fisher, and A. L. Burlingame, *Anal. Biochem* **169**, 337 (1988).

major glycoprotein surface antigen[25] and the high mannose, hybrid, and complex components of glycosylation attached to subunits of nicotinic acetylcholine receptor.[26,27] In addition, the use of this strategy together with the benefit of the knowledge obtained from the systematic investigation of this class of derivative discussed herein has led recently to assignment of a new *Saccharomyces cerevisiae mnn* mutant N-linked oligosaccharide structure in which the $Man_8GlcNAc_2$-core oligosaccharide is enlarged by addition of the outer chain to the $\alpha1\rightarrow3$-linked mannose in the side chain that is attached to the $\beta1\rightarrow4$-linked mannose rather than by its addition to the terminal $\alpha1\rightarrow6$-linked mannose.[28]

We have reported further results of a systematic study of increasing alkyl ester chain length and its concomitant enhancement of relative hydrophobicity yielding up to a factor of 40 increase in LSIMS sensitivity.[29,30] The importance of significantly increased sensitivity coupled with the previously noted advantages of a derivative class containing a chromophore permitting HPLC detection and effective isolation and desalting is further demonstrated in this work by presentation of results *vide infra* obtained from further investigation of the glycosylation of four related glycoproteins which comprise the nicotinic acetylcholine receptor from *Torpedo californica*. This was carried out employing the particular derivative optimized for coupling efficiency and LSIMS sensitivity using the wide diversity of structures expressed at low molar levels on a 250-μg (1 nmol) sample of intact acetylcholine receptor.

Methods

Synthesis of N-Alkyl p-Aminobenzoates

N-Hexyl, *n*-octyl, and *n*-decyl *p*-aminobenzoates are prepared with slight variations according to Kadaba *et al.*[31] The appropriate alcohol (0.1–0.15 mol) is added to *p*-aminobenzoic acid (0.01 mol) in the presence

[25] B. Gillece-Castro, S. J. Fisher, A. L. Tarentino, D. L. Peterson, and A. L. Burlingame, *Arch. Biochem. Biophys.* **256,** 194 (1987).
[26] L. Poulter, J. P. Earnest, R. M. Stroud, and A. L. Burlingame, *Biomed. Environ. Mass Spectrom.* **16,** 25 (1988).
[27] L. Poulter, J. P. Earnest, R. M. Stroud, and A. L. Burlingame, *Proc. Natl. Acad. Sci. U.S.A.* **86,** 6645 (1989).
[28] L. M. Hernandez, L. Ballou, E. Alvarado, B. L. Gillece-Castro, A. L. Burlingame, and C. E. Ballou, *J. Biol. Chem.* **264,** 11849 (1989).
[29] L. Poulter, R. Karrer, K. Jiang, B. L. Gillece-Castro, and A. L. Burlingame, *Proc. 36th Annu. Conf. Mass Spectrom. Allied Topics, San Francisco, CA,* p. 921 (1988).
[30] L. Poulter, R. Karrer, and A. L. Burlingame, *Anal. Biochem.,* submitted.
[31] P. K. Kadaba, M. Carr, M. Tribo, J. Triplett, and A. C. Glasser, *J. Pharmacol. Chem. Sci* **58,** 1422 (1969).

of boron trifluoride ethyl etherate (0.015 mol, 48%). The resulting suspensions are refluxed for between 10 and 24 hr until all of the acid is consumed. The excess alcohol is removed by distillation under reduced pressure (10–100 μm Hg). The products are dissolved in 25–40 ml of ether, washed with sodium carbonate solution (80 ml, 5%), and washed with distilled water (2 × 80 ml). The organic phase is then treated with hexane (300–500 ml) and crystallized overnight at −20°. Recrystallization from hexane provides the desired esters (yield 61–85%).

N-Tetradecyl p-aminobenzoate is prepared according to the description of Flynn *et al.*,[32] by reacting p-nitrobenzoyl chloride (0.05 mol) with n-tetradecanol (0.05 mol) and pyridine (1 ml) in ethanol (100 ml) under reflux for 8 hr. The solvent is removed under reduced pressure and the crude ester is dissolved in ether and washed with sodium carbonate solution (80 ml, 10%) until free from p-nitrobenzoic acid. The ethereal solution is then washed with water (2 × 80 ml) and the ether evaporated. The resulting ester is then reduced under a hydrogen atmosphere in the presence of Pearlman's catalyst (palladium hydroxide on carbon, 0.05 g) at 60°–70° over 24 h to give an overall yield of 42% after recrystallization from hexane.

Preparation of Oligosaccharides for Derivatization

This method is intended to be incorporated as an integral part of our analytical strategy when only very limited quantities of glycoprotein are available for analysis. Oligosaccharides are released from N-linked glycosylation sites on the glycoprotein (1–2 mol) by digestion with peptide-N-glycosidase F (PNGase F).[33] The amount of enzyme needed varies considerably with the nature of the glycoprotein substrate. For example, in our experience 1 mU (milliunit) enzyme/30 μg protein is capable of digesting ribonuclease B, whereas 1 mU enzyme/5 μg protein is required for digestion of α_1-acid glycoprotein and the nicotinic acetylcholine receptor from *Torpedo californica*.[27,30] Digestion is carried out in ammonium bicarbonate buffer (100 mM, pH 8.6) with 0.1% sodium dodecyl sulfate (SDS) and 0.6% Nonidet P-40 (NP-40) at 37° for 18 hr. Analytical SDS–PAGE, according to Laemmli,[34] on an aliquot of the digestion mixture with Coomassie staining (or silver staining for most economic use of sample) is used to show that digestion is complete. SDS present in the digestion mixture is then precipitated by adding guanidine-HCl (1 M)

[32] G. L. Flynn and S. H. Yalkowski, *J. Pharmacol. Chem. Sci.* **61**, 838 (1972).
[33] A. L. Tarentino, C. M. Gomez, and T. H. Plummer, Jr., *Biochemistry* **24**, 4665 (1985).
[34] U. K. Laemmli, *Nature (London)* **227**, 680 (1970).

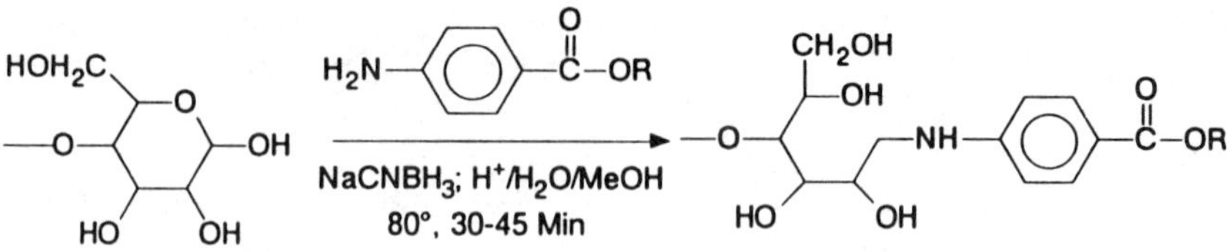

R = Me, Et, Bu, Hex, Oct, Dec, Tdec

SCHEME 1. Oligosaccharide derivatization.

dropwise,[35] and after refrigerating for 2 to 4 hr to ensure complete precipitation, the suspension is centrifuged and the supernatant carefully removed with a micropipette and applied to a C_{18} reversed phase Sep-Pak cartridge (Millipore–Waters, Bedford, MA). Oligosaccharides and salt are eluted by washing with water, leaving the protein and residual detergent (NP-40) absorbed on the cartridge packing. The aqueous sample is then lyophilized for subsequent derivatization. In our experience with subunits of the nicotinic acetylcholine receptor, glycoproteins which have been purified by preparative SDS–PAGE and isolated by electroelution are amenable to the above approach with subsequent derivatization of oligosaccharides for mass spectrometry; but to ensure success; the electroeluted protein should first be subjected to ethanol or acetone precipitation[36] or extensive microdialysis to remove aqueous contaminants also concentrated by the electroelution process.

Care too should be taken not to overload the C_{18} Sep-Pak cartridge, otherwise the removal of detergent and protein will not be complete. Cartridges can be used in series if more than microgram quantities of glycoprotein are digested, and in all cases the protein may be eluted separately with, for example, 80% acetonitrite/0.1% TFA (v/w) for subsequent analysis.

Oligosaccharide Derivatization

Oligosaccharide (1–20 μg) is dissolved in water (40 μl) in a glass Reactivial (Pierce Chemicals, Rockford, IL) (5-ml capacity) that has previously been treated with silylation reagent (Pierce Chemical). n-Alkyl p-aminobenzoate (0.1 mmol), sodium cyanoborohydride (35 mg), glacial acetic acid (41 μl), and methanol (350 μl) are mixed separately in a glass vial to form the reagent mixture (see Scheme 1). Warming of the vial will aid

[35] J. E. Shively, in "Methods of Protein Microcharacterization" (J. E. Shively, ed.), p. 41. Humana Press, Clifton, NJ, 1986.
[36] W. H. Konigsberg and L. Henderson, this series, Vol. 91, p. 254.

solubility of n-hexyl and n-octyl p-aminobenzoates, while the substitution of methanol (120 μl) by an equal volume of chloroform is recommended to solubilize n-decyl and n-tetradecyl p-aminobenzoates. Reagent mixture (40 μl) is added to the solution of oligosaccharide in water and the total volume of the reaction mixture is then made up to 200 μl with methanol. A chloroform/methanol (1 : 2, v/v) should be used in place of methanol when n-decyl and n-tetradecyl p-aminobenzoates are used. The Reactivial is sealed with a Teflon-lined cap, vortexed, and heated at 80°. After 45 min the vial is cooled and distilled water (1 ml) is added. Chloroform (1 ml) is then added when ethyl, n-butyl, n-hexyl or n-octyl p-aminobenzoates are used as derivatizing agents. Hexane (1 ml) is recommended when n-decyl and n-tetradecyl esters are used for derivatization. The mixture is vortexed and then centrifuged to efficiently separate the two phases. The aqueous phase containing the derivatized oligosaccharide is removed and the organic phase is reextracted with water (1 ml). Aqueous phases are combined and lyophilized. The absolute yield of the ethyl ester reductive amination with maltoheptaose (M_7) was measured by GC/MS of the trimethylsilyl methylglucose formed from methanolysis and found to be approximately 60%. Recent studies in this laboratory have lead to a modified protocol for carrying out this reaction using a chitobiose reducing terminal oligosaccharide with a variety of hydrophobic chromophores in similar high yield.[37]

Reversed-Phase Chromatography of Derivatized Oligosaccharides

Derivatized oligosaccharides are both separated and effectively desalted by reversed-phase chromatography using a C_{18} or C_4 column (Vydac, 25 cm × 4.6 mm or 25 cm × 2.1 mm are used by the authors) and a water/acetonitrile solvent system with a gradient of 0 to 60% acetonitrile in 60 min and a flow rate of 1 ml/min or 300 μl/min, depending on the column i.d. A C_{18} column is recommended for highly acidic oligosaccharides, while a C_4 column is best suited for neutral oligosaccharides derivatized with the more hydrophobic n-alkyl p-aminobenzoates. The oligosaccharides are monitored by their absorbance at 304 nm, which is conferred by the p-aminobenzoate group. Peaks may be integrated to provide an estimate of the relative amounts of oligosaccharide present since only one chromophore is present per molecule of oligosaccharide. Eluted samples are either lyophilized or reduced in volume by vacuum centrifugation prior to mass spectrometric analysis.

[37] S. Kaur, W. Liang, R. R. Townsend, and A. L. Burlingame, Anal. Biochem., in preparation.

Liquid Secondary Ion Mass Spectrometry (LSIMS) of Derivatized Oligosaccharides

Samples of derivatized oligosaccharides, typically submicrogram quantities, are applied to a matrix of glycerol/thioglycerol, 2 : 1 (v/v) (1 μl) on the probe tip and excess solvent pumped away after initial evacuation in the mass spectrometer. Trifluoroacetic acid (1%) may be added to the matrix to aid protonation when positive ion spectra are required. Positive-ion LSIMS yields predominantly molecular weight data only, while negative ion spectra show extensive fragmentation. Higher sensitivities can be obtained by using a smaller volume of matrix,[38] but the duration of the ion signal is shortened significantly, thereby making it unsuitable for the recording of a complete spectrum of a large oligosaccharide unless multichannel array detection is available on MS-1. However, if the approximate size of the oligosaccharide is known and only molecular weight data is required, positive ion spectra may be recorded over the molecular ion region successfully using this approach with a postacceleration detector only. Spectra are recorded with the magnet in field control mode, scanning at either 100 or 300 sec/decade and at a mass resolution of 2000–3000. Mass calibration is performed using a mixture of Ultramark 1621 and 443 (PCR Research Chemicals Inc., Gainesville, FL).

Results

Maltoheptaose

In studies aimed at optimizing sensitivity in the analysis of oligosaccharides by LSIMS, we have investigated the relative sputtering efficiencies of a series of alkyl esters of p-aminobenzoic acid by reductive amination of maltoheptaose. Those chosen include ethyl p-aminobenzoate, the derivatizing agent used in many of our earlier studies. Six additional esters were synthesized, including the methyl, n-butyl, n-hexyl, n-octyl, n-decyl, and n-tetradecyl, and coupled with the heptasaccharide maltoheptaose.

Figures 1a and 1b show the variation in the protonated molecular ion abundances of the various derivatives of maltoheptaose as a function of the time spent by the sample in the primary cesium ion beam, using 1.0 and 0.1 μg of derivative, respectively, dissolved in 1 μl of matrix (thioglycerol : glycerol, 1 : 1). It can be seen that there is a steady increase in the molecular ion abundances with increasing alkyl chain length, the effect being most pronounced as the concentration of derivatized oligosaccharide in the liquid matrix is decreased. For example, after 50 sec in the

[38] H. Kambara, in "Mass Spectrometry in the Health and Life Sciences" (A. L. Burlingame and N. Castagnoli, Jr., eds.), p. 65. Elsevier, Amsterdam, 1985.

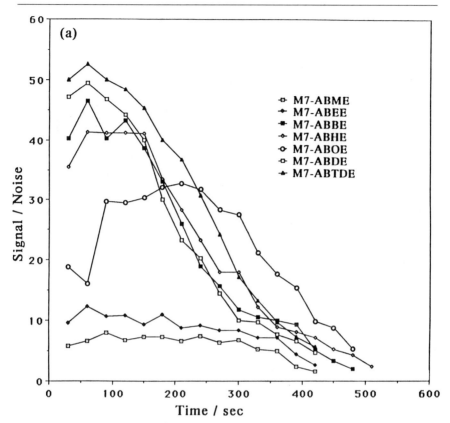

FIG. 1. Protonated molecular ion abundances of *n*-alkyl *p*-aminobenzoates with respect to the chemical noise as a function of time spent in the primary ion beam for (a) 1.0 μg and (b) 0.1 μg of derivatized M7 in 1.0 μl of thioglycerol : glycerol (1 : 1). Plotted points represent the mean of three independent sets of readings taken at fixed time intervals.

primary beam, it can be seen that 0.1 μg of M7 methyl *p*-aminobenzoate is not discernible above the level of the chemical noise, whereas M7 *n*-tetradecyl *p*-aminobenzoate provides a maximum signal-to-noise ratio of 40 : 1. Relative molecular ion abundances were similar in the negative ion mode (data not shown), 0.1 μg of the methyl ester showing once again a signal barely above the level of the noise, while the tetradecyl ester gave an excellent molecular ion as well as distinct fragment ions. Dilution experiments showed that molecular ions for *n*-decyl *p*-aminobenzoate and *n*-tetradecyl *p*-aminobenzoate could be obtained at concentrations of 0.01 μg/μl matrix, the latter having a maximum signal-to-noise of 7 : 1, while even at a concentration as low as 0.001 μg/μl a weak molecular ion for the

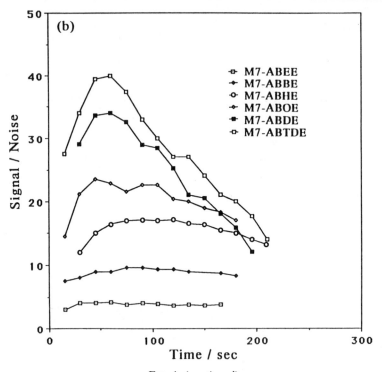

Fig. 1. (*continued*)

tetradecyl ester was observed. Although the sensitivity achieved with *n*-tetradecyl *p*-aminobenzoate is impressive, unfortunately, the coupling yields are low,[30] especially with more hydrophilic oligosaccharides (e.g., large N-linked structures and acidic species), which are not particularly soluble in methanol. In practice, it has been found that *n*-octyl *p*-aminobenzoate provides the optimum alkyl chain length to enhance mass spectral sensitivity while maximizing the yield in preparation of the derivative. The negative ion mass spectrum of 2.0 μg of M7 *n*-octyl *p*-aminobenzoate is shown in Fig. 2 to illustrate mass spectral fragmentation characteristics. The fragment ion nomenclature used is according to Domon and Costello[39] and is detailed in Schemes 2a and 2b. It should be noted that a relatively abundant smooth profile of ions (*Y*) are observed from sequential elimination of anhydro moieties from the nonreducing terminus. In addition, a series of nonreducing terminal charge-bearing ions (²,⁴*A*-type) are present.

[39] B. Domon and C. E. Costello, *Glycoconjugate J.* **5**, 397 (1988).

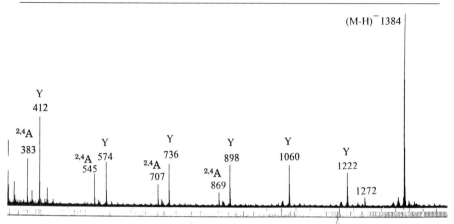

FIG. 2. Negative ion LSIMS mass spectrum of 2.0 μg of *n*-octyl *p*-aminobenzoate in 1.0 μl of thioglycerol : glycerol (1 : 1). Fragment ions of type *Y* arise from glycosidic bond cleavage with hydrogen transfer and charge retention at the derivatized reducing terminus (see Scheme 2a), while those designated $^{2,4}A$ show ring cleavage with charge retention at the nonreducing terminus (see Scheme 2b). The fragment ion at $m/z = 1272$ arises from loss of 1-octene from the *n*-octyl *p*-aminobenzoate group. (Reproduced with permission from Ref. 30.)

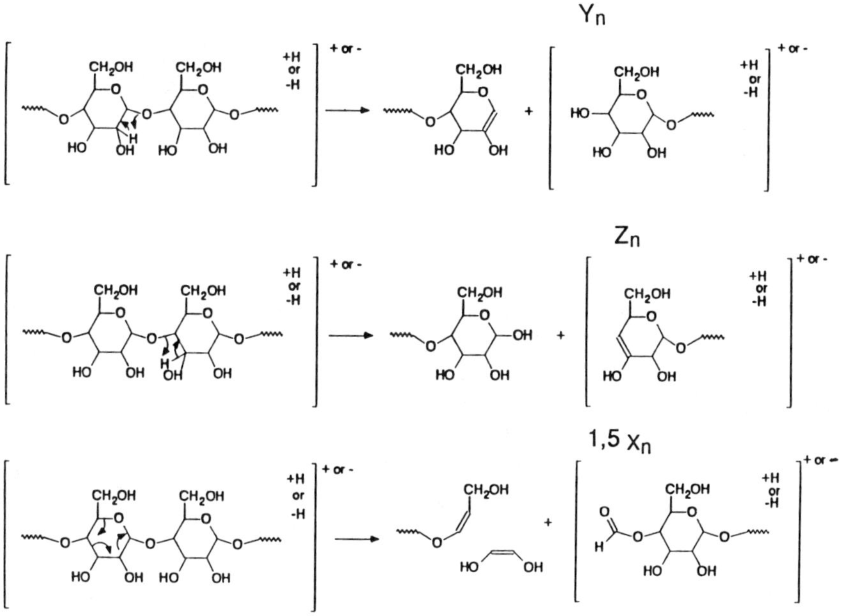

SCHEME 2a. Reducing terminal ion series.

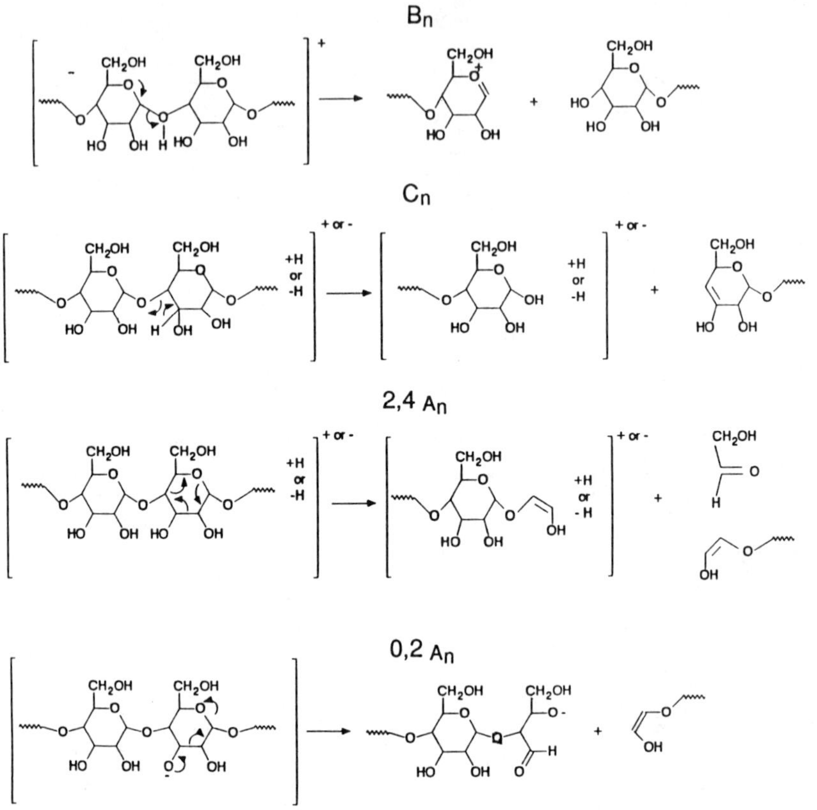

SCHEME 2b. Nonreducing terminal ion series.

Only one significant ion results from fragmentation of the derivatizing moiety itself, m/z 1272, from elimination of octene.

Asparagine-Linked Oligosaccharides

The preponderance of work thus far has centered around structural elucidation of a variety of asparagine-linked oligosaccharides, which may be liberated from natural and recombinant glycoproteins by treatment with Endo H or PNGaseF to yield glycosylation which is structurally homogeneous at their free reducing terminii. Thus, by reductive amination, microheterogeneity is immediately obvious from inspection of the LSIMS spectra in the positive ion mode in which little fragmentation usually occurs. Information regarding the class and type of antennae as well as

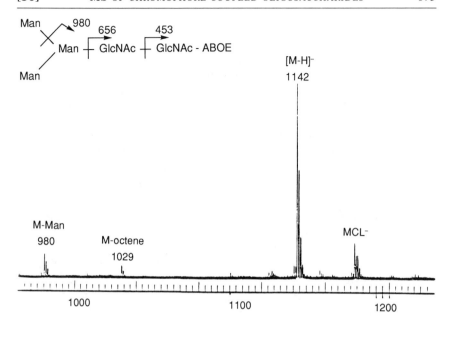

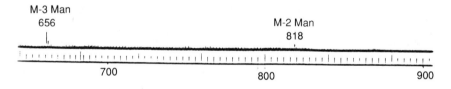

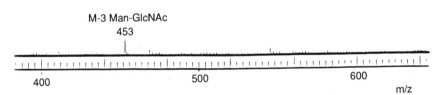

FIG. 3. Negative ion LSIMS spectrum of Man₃GlcNAc₂-ABOE.

information on branching of isobaric isomers may be obtained from the
LSIMS spectra recorded in the negative ion mode.

The negative ion LSIMS mass spectrum of the simplest branched
member of this class is shown in Fig. 3.[37] This trimannosyl structure with
its chitobiose core intact was prepared as the octyl ester, and serves to

illustrate the features of these spectra in the negative ion mode. Depending on the rigor with which the derivatized sample is desalted by washing with water on the reversed-phase HPLC, prior to initiation of the gradient, varying amounts of chloride attachment ions may be observed, such as the isotope cluster at m/z 1178, 1179, and 1180. Such chloride adduct ions in the negative ion mode or sodium adduct ions in the positive ion mode are relatively stable as compared with the corresponding molecular anions occurring for this oligosaccharide at m/z 1142. It can be readily seen that the most abundant fragments occurring in this spectrum are due to loss of a single anhydro mannosyl moiety at m/z 980, a triple mannosyl anhydro moiety at m/z 656, and cleavage of the chitobiose linkage with loss of Man$_3$GlcNAc$_1$ anhydro moiety at m/z 453. In addition, there is a relatively minor loss corresponding to 2 residues of anhydromannose, m/z 818, thought to arise from double-cleavage processes which would be expected to be less abundant.[24]

The negative ion mass spectrum of a Man$_6$GlcNAc$_1$ isomer, prepared as the octyl ester and isolated from a yeast *mnn* mutant, is shown in Fig. 4a. It can readily be seen from the pattern of Y ions representing charge retention at the reducing terminal amino benzoate function that there is abundant loss of moieties resulting from simple elimination reactions corresponding to a monohexosyl, dihexosyl, a trihexosyl moiety, and, finally, a hexahexosyl moiety. Nonreducing A ions for Hex$_2$ and Hex$_3$ m/z 383 and 585, respectively, are also present. The occurrence of these fragments taken together with the absence of significant ion current due to elimination of an analogous 4-mer or 5-mer establishes the branching pattern shown in structure **I**. For comparison, the mass spectrum of another Man$_6$GlcNAc$_1$ carbohydrate, prepared as the ethyl ester, is shown

$$M \rightarrow {}^6M \rightarrow {}^6M \rightarrow {}^4GNAc$$
$$\uparrow^3 \quad \uparrow^3$$
$$M \quad M^6 \leftarrow M$$

I

in Fig. 4b. This oligosaccharide was obtained from a gel filtration column fraction isolated from IgM, which was purified from the plasma of a patient with Waldenstrom's macroglobulinemia.[24] This fraction (d) was reductively aminated as the ethyl ester, and then subjected to reversed-phase HPLC using an aminopropyl-bonded silica column as shown in Fig. 5. The major component d-3 was used to record the mass spectrum shown in Fig. 4b. From the general features of this spectrum, it can be seen that the pattern is similar to that shown in Fig. 4a, except for the shift of 84 mass units due to the mass difference between the octyl and ethyl ester derivatives. Except for the presence of additional, less abundant fragments at mass m/z 707 (Y_4) and m/z 693 (A_4), this spectrum supports structure

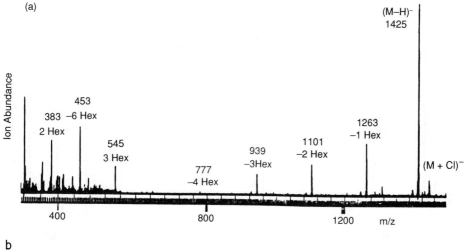

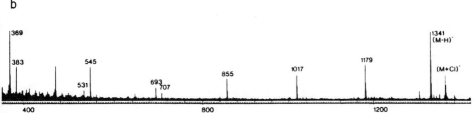

FIG. 4. (a) LSIMS of Man₆GlcNAc₁ obtained from yeast *mnn* mutant. (b) The spectrum obtained from fraction d-3. (Reproduced with permission from (a) Ref. 28 and (b) Ref. 24.)

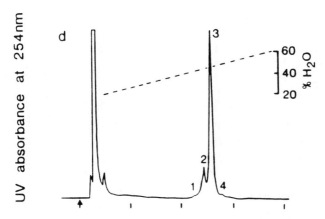

FIG. 5. HPLC separation of IgM-ABEE oligosaccharides: column fraction d. (Reproduced with permission from Ref. 24.)

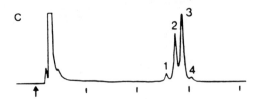

FIG. 6. HPLC separation of IgM-ABEE oligosaccharides: column fraction c. (Reproduced with permission from Ref. 24.)

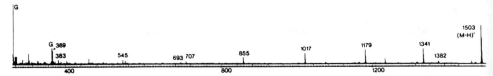

FIG. 7. The spectrum obtained from fraction c-3. (G), Ions that arise from glycerol cluster ions. (Reproduced with permission from Ref. 24.)

II. The presence of these additional fragments indicates the presence of an additional branched isobar structure **III**. The linkage differences

Man
  \
   $^6_3$Man
  /   \
Man ^{6}Man—₄GlcNAc—ABEE
  /
Man—₂Man

II

Man
  \
   $^6_3$Man
  /   \
Man—₂Man $^6_3$Man—₄GlcNAc—ABEE
             /
          Man

III

between structures **I** and **II** have been established from consideration of other data presented in the original literature.[24,28] Whether information on linkage differences may be deduced from the mass spectra or CID spectra, in some cases, remains an open question.

A further interesting example is that of a Man₇GlcNAc₁ component, also from the immunoglobulin, which was treated in an analogous fashion, including HPLC separation, as shown in Fig. 6. The mass spectrum of HPLC fraction c-3 is shown in Fig. 7. From analogous interpretation of the negative ion mass spectrum, it can be seen that the major component is consistent with structure **IV**, but it is clear that there is at least two other possible branched isobars present consistent with structures **V** and **VI**.

These results illustrate the quality of mass spectral information that

Man
  $\diagdown$
    $^6_3$Man
      $\diagdown$
Man$\diagup$ $^6_3$Man—$_4$GlcNAc—ABEE
Man—$_2$Man—$_2$Man$\diagup$

IV

Man—$_2$Man
        $\diagdown$
          $^6_3$Man
            $\diagdown$
Man$\diagup$ $^6_3$Man—$_4$GlcNAc—ABEE
Man—$_2$Man$\diagup$

V

Man
  $\diagdown$
    $^6_3$Man
      $\diagdown$
Man—$_2$Man$\diagup$ $^6_3$Man—$_4$GlcNAc—ABEE
Man—$_2$Man$\diagup$

VI

can be obtained using this derivative in the negative ion mode in addition to the power of this approach in detecting the presence and partial structural nature of isobaric isomeric structures. While, in the published literature these components have not been separated, from recent work in this laboratory using higher resolution chromatographic separation methodology, these branched isomers were separated into individual components such that the mass spectra of isomerically pure substances could be recorded.[37]

To conclude this discussion of spectrum–structure correlations for the high mannose series, the fragmentation patterns of two Man$_9$GlcNAc$_1$ isomers are presented in Figs. 8a and 8b to illustrate their qualitative differences. The mammalian Man$_9$GlcNAc$_1$ shown in Fig. 8a was also obtained from IgM and corresponds to structure **VII** while the Man$_9$ isomer shown in structure **VIII** was from an isolated *mnn* yeast mutant.

Man—$_2$Man
(9)   (6)
        $\diagdown$
          $^6_3$Man
           (4)
            $\diagdown$
Man—$_2$Man$\diagup$ $^6_3$Man—$_4$GlcNAc—ABEE
(10)   (7) $\diagup$
Man—$_2$Man—$_2$Man$\diagup$
(11)   (8)   (5)

VII

M→^{6}M→^{6}M→^{4}GNAc
  ↑^{2} ↑^{3} ↑^{3}
  M   M   M^6←M
                 ↑^{2}
                 M
                 ↑^{2}
                 M

VIII

In Fig. 8a it can be seen that in the Y series, fragments are observed for losses of 1-, 2-, 3-, 5-, and 9-mer moieties, whereas in Fig. 8b the most abundant Y series correspond to loss of 1-, 2-, 4-, and 9-mer, respectively.

A final example will indicate the ease of dealing with oligosaccharides of other classes, using an example from the complex class involving identification of the nature of oligosaccharides attached to recombinant hepatitis B pre-S$_2$ surface antigen. Reversed-phase HPLC of the oligosaccharides liberated by PNGase F were derivatized as the octyl ester as shown in

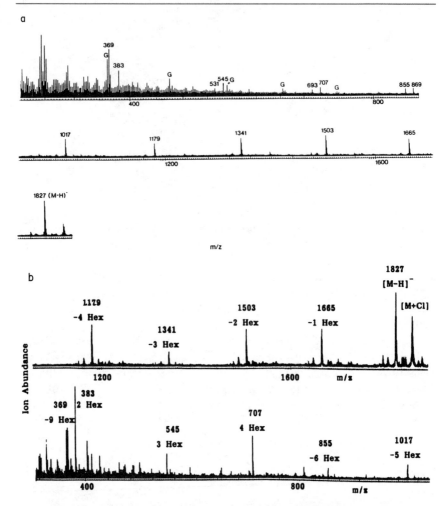

Fig. 8. (a) Negative ion LSIMS of fraction a-4 of the IgM-ABEE oligosaccharides. The spectrum showed an intense molecular anion at 1827. Series Y ions for losses of one, two, three or five mannoses were evident, as well as a less intense ion corresponding to the loss of four mannose residues (m/z 1179). Series $^{2,4}A$ ions for fragments containing one, two, three, or five mannose residues were also observed. These data suggested that the branched residues in the molecule suppressed the formation of fragments consisting of six or seven mannose residues, evidence that the major component is formed from a triantennary core. (Reproduced with permission from Ref. 24.) (b) LSIMS of *mnn*1 *mnn*2 *mnn*10 oligosaccharide Man$_9$GlcNAc. (Reproduced with permission from Ref. 28.)

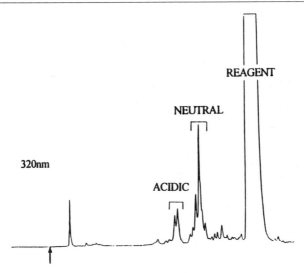

FIG. 9. Reversed-phase HPLC hepatitis B surface antigen pre-S$_2$ oligosaccharide fractionation primarily according to net charge.

Fig. 9. Separation of fractions according to net charge is obtained. The fraction eluting as neutral oligosaccharides are shown in Fig. 10. The major component is a fucosylated biantennary structure **IX**, $(M - H)^-$ m/z 2019.

$$
\begin{array}{c}
\text{Gal---GlcNAc---Man} \overset{\displaystyle 1857 \quad 1653}{} \\
\end{array}
$$

Gal—GlcNAc—Man^{1857} ^{1653} ^{1492} ^{803} ^{600} $(M - H)^-$ 2109

Man—GlcNAc—GlcNAc—ABOE

Gal—GlcNAc—Man ^{1873} Fuc

IX

The presence of higher molecular weight additional neutral components are indicated by peaks at m/z 2384, representing a triantennary structure, and m/z 2165, representing a bifucosylated biantennary structure. Further, peaks at m/z 1857, 1653, and 1492 provide evidence for truncated structures present in this sample.[40]

Oligosaccharides Isolated from Nicotinic Acetylcholine Receptor

Having established that *n*-alkyl *p*-aminobenzoates with ester groups greater than ethyl could be analyzed with higher sensitivity by LSIMS, we incorporated this improvement into a simple, but effective, procedure

[40] B. L. Gillece-Castro and A. L. Burlingame, *Proc. 36th Annu. Conf. Mass Spectrom. Allied Topics, San Francisco, CA,* p. 923 (1988).

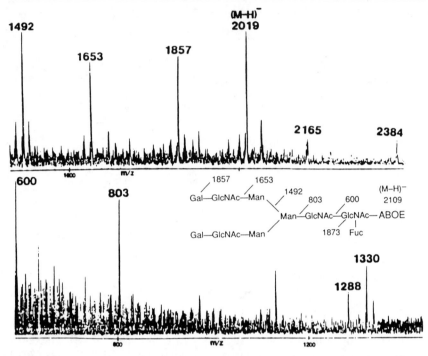

FIG. 10. Negative ion LSIMS of cloned hepatitis B pre-S$_2$ antigen oligosaccharides as the ABOE derivative.

for examining N-linked oligosaccharides released from microgram levels of several glycoproteins. We will present results obtained from studies of the nicotinic acetylcholine receptor (nAcChoR) as an example. In this case, the task was complicated by the size (270 kDa) of this multisubunit protein and its resistance to enzymatic digestion. In particular, the high levels of detergent, needed both to solubilize the protein and to aid digestion by PNGase F, had to be rigorously removed before any derivatization of the oligosaccharide or mass spectrometry could be attempted.

Figure 11 shows the C$_{18}$ reversed-phase HPLC profile of the total oligosaccharide released from nAcChoR (250 μg, 1 nmol) and derivatized with *n*-hexyl *p*-aminobenzoate (no octyl ester of sufficient purity was available at the time of this study). As indicated above, reversed-phase chromatography separates these derivatives largely on the basis of the net charge carried, although shallower water/acetonitrile gradients can produce better resolution of species that carry the same charge but that differ in size. The profile obtained for the oligosaccharides from nAcChoR

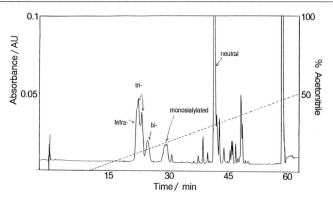

FIG. 11. C_{18} reversed-phase HPLC profile of derivatized oligosaccharides from nAcChoR. (Reproduced with permission from Ref. 30.)

suggested the presence of small amounts of many acidic oligosaccharides, shown to be mono-, bi-, tri-, and tetrasialylated structures by mass spectrometry, as well as a large neutral fraction.

LSIMS of the neutral fraction (20% of total) in both positive and negative ion modes afforded the spectra shown in Fig. 12a and 12b, respectively, revealing the components as high mannose oligosaccharides of composition $Man_8GlcNAc_2$ and $Man_9GlcNAc_2$. No molecular ions suggesting the presence of other high mannose species, hybrid or complex glycans were observed in this fraction. In addition to the singly and doubly charged molecular ions seen in the positive ion spectrum, three series of reducing terminal fragment ions, although of low abundance, were also apparent (see Scheme IIa,b for the nomenclature used). The intensity of the Y series of fragment ions was much greater in the negative ion mode (see Fig. 12b), clearly showing the losses of up to 4 and 5 hexose (mannose) units from $Man_8GlcNAc_2$ and $Man_9GlcNAc_2$, respectively. Further purification of this fraction by HPLC using an amine-bonded column allowed LSIMS data to be obtained on each component. These spectra (data not shown) illustrated that the fragment ion at $m/z = 1438$ resulted from the loss of 3 hexose (mannose) units from $Man_8GlcNAc_2$. This ion was not found in the spectrum of $Man_9GlcNAc_2$ indicating that this oligosaccharide cannot lose 4 hexose (mannose) units by the cleavage of only one bond. The Y ion at $m/z = 1114$ in the negative ion LSIMS spectrum of the high mannose fraction is barely detected above the level of the chemical noise. This may be explained by the presence of a minor isomeric component, or may arise from the cleavage of two bonds of a major component structure.[41] The structures derived from the mass spectral data were consistent

[41] A. Dell and C. E. Ballou, *Carbohydr. Res.* **120**, 95 (1983).

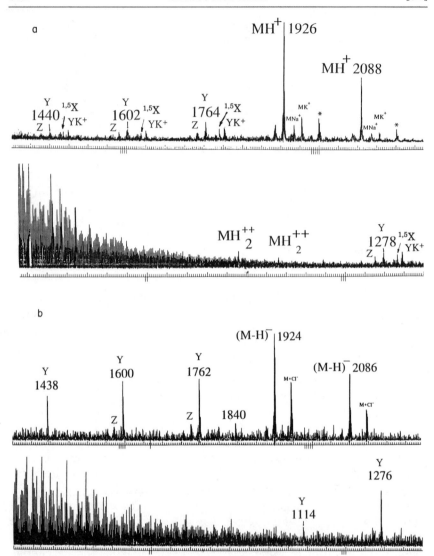

FIG. 12. (a) Positive ion LSIMS mass Spectrum of Man$_8$GlcNAc$_2$-ABHE and Man$_9$Glc-NAc$_2$-ABHE from nAcChoR. Very weak reducing terminal fragment ions are observed (see Scheme 2a for nomenclature). Ions marked with an asterisk are matrix adducts involving the addition of a dehydrated glycerol molecule. (b) Negative ion LSIMS mass spectrum of Man$_8$GlcNAc$_2$-ABHE and Man$_9$GlcNAc$_2$-ABHE from nAcChoR. An abundant series of reducing terminal fragment ions (type Y) are observed (see Scheme 2a for nomenclature). The fragment ion at m/z = 1840 results from the loss of 1-hexene from the n-hexyl p-aminobenzoate group. (Reproduced with permission from Ref. 30.)

with those obtained by the ^{1}H NMR work of Nomoto *et al.*[42] and are shown as structures **X** and **XI**.

$(M - H)^-$ 1924

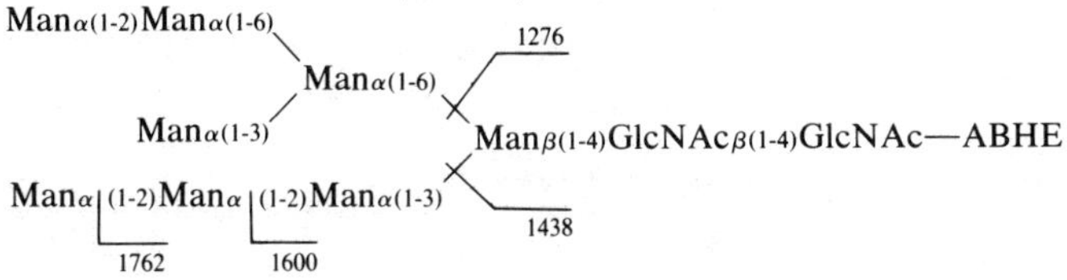

$Man\alpha(1\text{-}2)Man\alpha(1\text{-}6)$
$\qquad\qquad\qquad$ 1276
$\qquad\qquad Man\alpha(1\text{-}6)$
$Man\alpha(1\text{-}3)$
$\qquad\qquad\qquad Man\beta(1\text{-}4)GlcNAc\beta(1\text{-}4)GlcNAc—ABHE$
$Man\alpha\,|\,(1\text{-}2)Man\alpha\,|\,(1\text{-}2)Man\alpha(1\text{-}3)$
$\qquad\qquad\qquad\qquad\qquad 1438$
$\quad\lfloor\quad\quad\lfloor$
$\;\;1762\quad\;\;1600$

X

$(M - H)^-$ 2086

$Man\alpha(1\text{-}2)Man\alpha(1\text{-}6)$
$\qquad\qquad\qquad$ 1276
$\qquad\qquad Man\alpha(1\text{-}6)$
$Man\alpha(1\text{-}2)Man\alpha(1\text{-}3)$
$\qquad\qquad\qquad Man\beta(1\text{-}4)GlcNAc\beta(1\text{-}4)GlcNAc—ABHE$
$Man\alpha\,|\,(1\text{-}2)Man\alpha\,|\,(1\text{-}2)Man\alpha(1\text{-}3)$
$\qquad\qquad\qquad\qquad\qquad 1600$
$\quad\lfloor\quad\quad\lfloor$
$\;\;1924\quad\;\;1762$

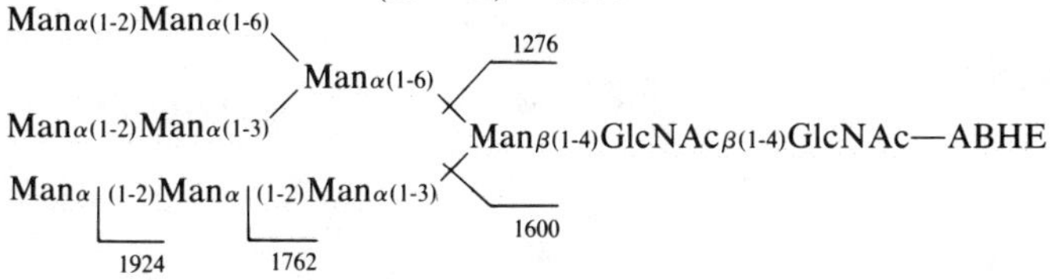

XI

In the case of the whole nAcChoR, however, the total amount of complex oligosaccharide present is low (<30% of all oligosaccharides) and is also extremely heterogeneous, thereby increasing the need for sensitive analysis by mass spectrometry. Figure 13 shows the negative ion spectrum

$(M - H)^-$ 2094   core + 3Hex + HexNAc + NANA

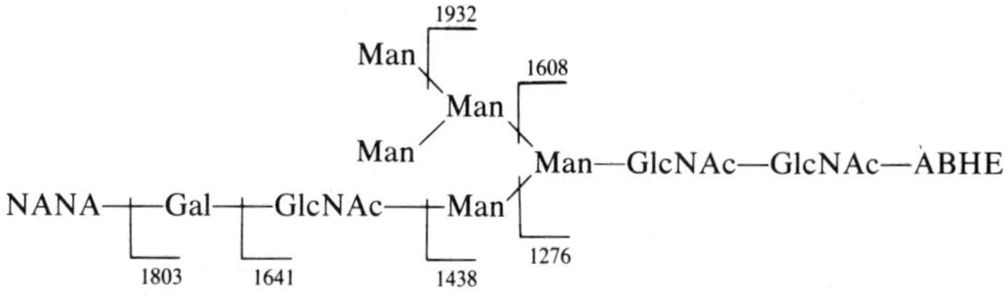

$\qquad\qquad\qquad\qquad$ 1932
$\qquad\qquad\qquad Man$
$\qquad\qquad\qquad\qquad\qquad$ 1608
$\qquad\qquad\qquad\qquad Man$
$\qquad\qquad\qquad Man$
$\qquad\qquad\qquad\qquad\qquad Man—GlcNAc—GlcNAc—ABHE$
$NANA—Gal—GlcNAc—Man$
$\qquad\qquad\qquad\qquad\qquad\qquad\qquad 1276$
$\quad\;\;1803\quad\;1641\quad\;\;1438$

XII

$(M - H)^-$ 1932   core + 2Hex + HexNAc + NANA

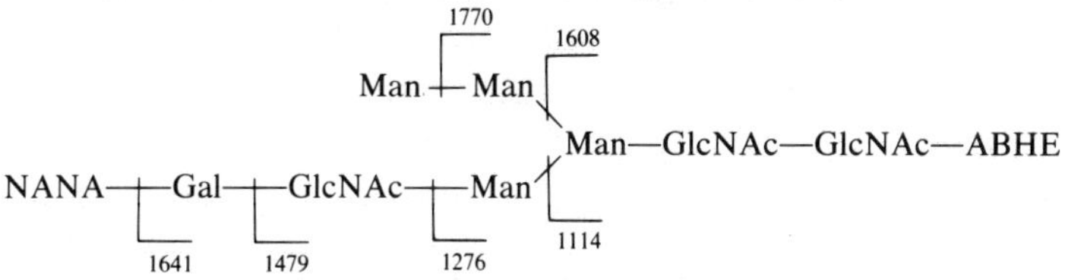

$\qquad\qquad\qquad\qquad$ 1770
$\qquad\qquad\qquad\qquad\qquad$ 1608
$\qquad\qquad\qquad Man—Man$
$\qquad\qquad\qquad\qquad\qquad Man—GlcNAc—GlcNAc—ABHE$
$NANA—Gal—GlcNAc—Man$
$\qquad\qquad\qquad\qquad\qquad\qquad\qquad 1114$
$\quad\;\;1641\quad\;1479\quad\;\;1276$

XIII

[42] H. Nomoto, N. Takahashi, Y. Nagaki, S. Endo, Y. Avata, and K. Hayashi, *Eur. J. Biochem.* **157,** 233 (1986).

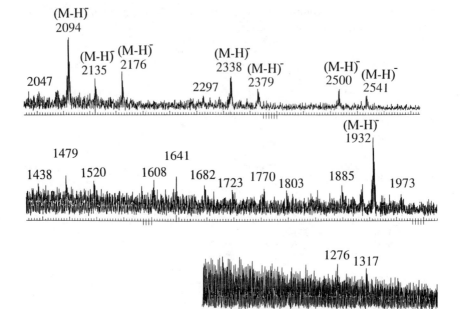

FIG. 13. Negative ion LSIMS spectrum of monosialylated oligosaccharides from nAc-ChoR. All fragment ions labeled are of type Y (see Scheme 2a). (Reproduced with permission from Ref. 30.)

obtained on a fraction of acidic oligosaccharides from nAcChoR, derivatized with n-hexyl p-aminobenzoate. This HPLC fraction contains no less than eight molecular ions, all of which carry one residue of N-acetylneuraminic acid. The major species $(M - H)^- = 2094$ and $(M - H)^- = 1932$ contain more hexose, as judged by their molecular weights, than one would expect for complex structures, suggesting instead hybrid oligosaccharides such as those detailed in structures **XII** and **XIII**. The molecular ions $(M - H)^- = 2135$ and $(M - H)^- = 2500$ possess ratios of hexose to hexosamine that agree with complete monosialylated bi- and triantennary oligosaccharides, respectively, while the remainder of the species present all contain an additional one or two hexosamine residues, which may occur as part of incomplete antennae and/or as a bisecting residue. The fragment ions (predominantly Y ions) are only weakly visible above the level of the chemical noise, but can be used to support or negate structural features that might be suggested by the monosaccharide composition (as calculated from the molecular weight) and a knowledge of common N-linked struc-

tures and biosynthetic pathways.[43,44] It is interesting to note that no fucosylated structures were found in this fraction, or in any other fraction from
nAcChoR (L. Poulter and A. L. Burlingame, unpublished results), in
contrast to the results of Nomoto et al.[42] which stated that fucosylated
bi-, tri-, and tetraantennary oligosaccharides are present. Their evidence,
however, is based predominantly on comparing HPLC retention times
with those of known standards, which clearly is a less rigorous approach
than that offered by mass spectrometry.

Discussion

The experiments performed on the n-alkyl p-aminobenzoate derivatives of maltoheptaose have demonstrated that an increase in the length
of the n-alkyl chain promotes surface activity within the glycerol : thioglycerol matrix, thereby increasing the sensitivity of the mass spectral analysis.
As the amount of oligosaccharide in a given quantity of liquid material is
decreased, the benefits of using the longer chain esters as derivatizing
agents to enhance mass spectral sensitivity become more apparent. Evidence supporting an increase in LSIMS sensitivity with an increase in
the hydrophobicity of the compound analyzed is apparent in the recent
literature.[8,45-48] In particular, Ligon and co-workers[47] have demonstrated
that acyl chain elongation correlates with increasing surface excess concentration and mass spectral sensitivity for a series of acylcarnitines in
glycerol. As previously experienced with ethyl p-aminobenzoate, the
longer chain esters introduce a chromophoric group permitting sensitive
HPLC detection, provide a strong molecular ion in the positive ion mode,
and exhibit extensive fragmentation in negative ion LSIMS. Current derivatization conditions result in lower yields of n-decyl and especially n-
tetradecyl p-aminobenzoate derivatives compared to n-octyl and lower
homologs, which limits their use in the analyses of biological samples.
However, it is clear that if derivatizing conditions were optimized, these
higher homologs would be the reagents of choice. Further work, investigat-

[43] H. Schachter, Biochem. Cell Biol. 64, 163 (1986).
[44] R. Kornfeld and S. Kornfeld, Annu. Rev. Biochem. 54, 631 (1985).
[45] S. Naylor, A. F. Findeis, B. W. Gibson, and D. H. Williams, J. Am. Chem. Soc. 108, 6359 (1986).
[46] M. R. Clench, G. V. Garner, D. B. Gordon, and M. Barber, Biomed. Environ. Mass Spectrom. 12, 355 (1985).
[47] W. V. Ligon and S. B. Dorn, Int. J. Mass Spectrom. Ion Proc. 78, 99 (1987).
[48] A. M. Falick and D. Maltby, Anal. Biochem. 182, 165 (1989).

ing solvent conditions and phase transfer catalysis, will determine whether this is possible.

Taking advantage of the enhanced sensitivity obtainable using the higher esters, e.g., octyl, of *p*-aminobenzoic acid as derivatizing agents, we have described a method particularly suited to working with very limited quantities of glycoprotein. Using this strategy, we have succeeded in characterizing the neutral and acidic oligosaccharides expressed on 250 μg (1 nmol) of a multisubunit integral membrane protein. Clearly, this strategy can be used as a general screening method, providing an "oligosaccharide map" of the gross oligosaccharide structures present on any serum or membrane-bound glycoprotein possessing N-linked oligosaccharides and only available at the picomole to low nanomole level.

The mass spectra of oligosaccharides derivatized with alkyl *p*-aminobenzoates provide monosaccharide composition (on the basis of hexose, deoxyhexose, hexosamine, sialic acid, etc.) and sequence and branching data through the fragment ions observed in the negative ion mode. The mass spectra of these derivatized oligosaccharides do not show extensive ring cleavages as has been demonstrated in the LSIMS,[4,49] Laser desorption-time-of-flight (LD-TOF),[50] and laser desorption-Fourier transform mass spectrometry (LD-FTMS)[51–53] spectra of underivatized oligosaccharides. Whether the analysis of underivatized oligosaccharides by such ionization methods will provide complementary information of general structural utility remains to be established from systematic investigation of related structural members. However, significantly more material is required to obtain spectra of underivatized samples. Moreover, derivatization is an integral part of our analytical strategy, which is crucial for the isolation and purification as well as the mass spectrometry of such small quantities of oligosaccharide. MS/MS methods may prove to be useful for obtaining more extensive ring cleavages of oligosaccharides,[54] but again the sensitivity of the technique, particularly for larger structures, may be a limiting factor (*vide infra*).

Permethylated and peracetylated oligosaccharides, as mentioned ear-

[49] V. N. Reinhold, *in* "Mass Spectrometry in Biomedical Research" (S. J. Gaskell, ed.), chap. 11, p. 181. Wiley, New York, 1986.

[50] M. F. Bean, W. B. Martin, T. J. Raley, Jr., C. M. Murphy, and R. J. Cotter, R. J. *Proc. 36th ASMS Conf. Mass Spectrom. Allied Topics, San Francisco, CA*, p. 1279 (1988).

[51] A. Bjarnason, M. B. Comisarow, G. G. S. Dutton, Z. Lam, and D. A. Weil, *Proc. 36th ASMS Conf. Mass Spectrom. Allied Topics, San Francisco, CA*, p. 1338 (1988).

[52] Z. Lam, M. B. Comisarow, G. G. S. Dutton, D. A. Weil, and A. Bjarnason, *Rapid Commun. Mass Spectrom.* **1**, 5, 83 (1985).

[53] M. L. Coates and C. L. Wilkins, *Anal. Chem.* **59**, 197 (1987).

[54] B. L. Gillece-Castro and A. L. Burlingame, this volume [37].

lier, can be analyzed more sensitively than underivatized samples and this form of derivatization is also of use in the isolation and desalting of oligosaccharides.[14,15,16] LSIMS spectra of permethylated oligosaccharides can also provide data on the substituent at the O-3 position of HexNAc residues within the parent structure. However, spectra of permethylated and peracetylated oligosaccharides commonly show ions corresponding to under- and overderivatized species,[18,19] the presence of which may greatly complicate spectra and also distribute the total ion current over a number of components. Further steps may be incorporated when derivatization is incomplete, such as the use of diazomethane to fully permethylate sialic acid groups,[22] but they can contribute to extensive sample loss when very small quantities of oligosaccharide are available for derivatization. If the sample to be examined is limited in quantity, heterogeneous, and contains many minor components, such as structures **XII** and **XIII** in the monosialylated fraction, derivatization with a long-chain alkyl *p*-aminobenzoate is the better analytical strategy. It should also be noted that permethylation and peracetylation also increase the mass of the parent molecular ion significantly, often allowing only nonreducing terminal fragments to be seen when larger oligosaccharides are examined.[55] In certain cases, this may provide sufficient data; for example, the fragments may characterize nonreducing terminal antigenic determinants.[56,57] However, permethylation or peracetylation is not the most sensitive method for obtaining a "map" of the intact oligosaccharide structures present. Monosaccharide composition and details on the structure of the oligosaccharide core (which may possess antigenic determinants)[58] are also available through our approach.

The extension of our method to provide linkage data and possibly anomeric configuration, using submicrogram quantities of oligosaccharides, should be possible by employing exoglycosidase digestion and examining the truncated structures by mass spectrometry. A similar, although less precise approach employing exoglycosidase digestion of oligosaccharides derivatized with 2-aminopyridine and subsequent HPLC analysis of the truncated species, is already in evidence, and is successful in providing data on as little as 10 pmol of an oligosaccharide structure.[59] The free

[55] J. E. Oates, A. Dell, M. Fukuda, and M. N. Fukuda, *Carbohydr. Res.* **141,** 149 (1985).
[56] J. Peter-Katalinic, H. Egge, F. G. Hanisch, and G. Uhlenbruck, *Adv. Mass Spectrom.* **11B,** 1418 (1989).
[57] J. Peter-Katalinic and H. Egge, *in* "Biological Mass Spectrometry," (A. L. Burlingame and J. A. McCloskey, eds.), p. 397. Elsevier, Amsterdam, 1990.
[58] T. Feizi and R. A. Childs, *Biochem. J.* **245,** 1 (1987).
[59] N. Tomiya, M. Kurono, H. Ishihara, S. Tejima, S. Endo, Y. Arata, and N. Takahashi, *Anal. Biochem.* **163,** 489 (1987).

nonreducing terminal moieties, possessed by oligosaccharides derivatized with alkyl p-aminobenzoates, are directly amenable to exoglycosidase digestion, in contrast to those of permethylated or peracetylated derivatives. Hence sequential enzymatic digestions of the parent derivative can be achieved while minimizing the amount of sample handling needed. When dealing with picomoles of derivatized oligosaccharide, in situ digestion may be the best way to proceed. For example, the sample could be adsorbed to the nitrocellulose foil in the case of plasma desorption mass spectrometry (PDMS) analysis or to a silica plate for LSIMS.[60,61] Pure oligosaccharides or mixtures could be examined in this way, and such data would clearly be complementary to that provided by monoclonal or polyclonal antibody recognition and lectin binding.

Conclusion and Outlook

This work has established that the negative ion mass spectral characteristics of oligoglycoaminodeoxyalditols can be used to determine the isomeric homogeneity of isobaric oligosaccharides and to elucidate the branched structure of individual components. However, further work is required on several interrelated aspects of this problem to effect an overall optimized, integrated strategy for high sensitivity, and unambiguous oligosaccharide structure determination. From a mass spectrometric point of view, this includes increased sensitivity in the LSIMS source (the subject of this chapter), increased mass spectrometer instrument sensitivity for normal LSIMS analyses using multichannel array detection systems,[26,27,62] improved methods for vibronic activation of low internal energy molecular ions, and concomitant increased signal-to-chemical noise using CID and tandem methods when necessary.[54] In addition, there is urgent need to couple higher resolution chromatographic methods[63] with these high-quality mass spectrometric methods to eventually be able to address structure–function or structure–protein recognition issues involving natural glycoproteins available only in limited amounts, such as the nature of

[60] Y. Nakagawa and K. Iwatani, in "Proceedings of the Second Japan-China Symposium on Mass Spectrometry" (H. Matsuda and L. Xiao-tian, eds.), p. 29. Bando Press, Osaka, 1987.

[61] A. M. Lawson, W. Chai, M. Stoll, E. F. Hounsell, T. Feizi, R. H. Bateman, and H. J. Major, Proc. 37th Annu. Conf. Mass Spectrom. Allied Topics, Miami Beach, FL, p. 790 (1989).

[62] B. W. Gibson, J. W. Webb, R. Yamasaki, S. J. Fisher, A. L. Burlingame, R. E. Mandrell, H. Schneider, and J. M. Griffiss, Proc. Natl. Acad. Sci. U.S.A. 86, 17 (1989).

[63] M. R. Hardy and R. R. Townsend, Carbohydr. Res. 188, 107 (1989).

binding between natural human CD_4 and HIV-1 gp120,[64] EGF receptor,[65] and so on.

Acknowledgments

We thank Dr. A. L. Tarentino for generous provision of PNGase F. This work was supported by the NIH, Division of Research Resources, RR01614 to A. L. Burlingame. Linda Poulter is recipient of a NATO postdoctoral fellowship from the Science and Engineering Council, U.K.

[64] D. H. Smith, R. A. Byrn, S. A. Marsters, T. Gregory, J. E. Groopman, and D. J. Capon, *Science* **238**, 1704 (1987).
[65] J. Downward, Y. Yarden, E. Mayes, G. Scrace, N. Totty, P. Stockwell, A. Ullrich, J. Schlessinger, and M. D. Waterfield, *Nature (London)* **307**, 521 (1984).

[37] Oligosaccharide Characterization with High-Energy Collision-Induced Dissociation Mass Spectrometry

By BETH L. GILLECE-CASTRO and A. L. BURLINGAME

The diverse suite of established mass spectrometric methods have been used for over two decades to characterize oligosaccharides and glycoconjugates by identifying molecular weight, sequence, and isomeric structures.[1-16] Typically, glycans[3] and glycoconjugates[2-4] are derivatized

[1] T. Radford and D. C. de Jongh, in "Biochemical Applications of Mass Spectrometry" (G. R. Waller, ed.), p. 313. Wiley (Interscience), New York, 1972.
[2] K.-A. Karlsson, *Progr. Chem. Fats Other Lipids* **16**, 207 (1978).
[3] T. Radford and D. C. de Jongh, in "Biochemical Applications of Mass Spectrometry" (G. R. Waller and O. S. Dermer, eds.), 1st Suppl. Vol., p. 255. Wiley (Interscience), New York, 1980.
[4] G. W. Wood, in "Biochemical Applications of Mass Spectrometry "(G. R. Waller and O. S. Dermer, eds.), p. 173. Wiley (Interscience), New York, 1980.
[5] A. L. Burlingame, T. A. Baillie, P. T. Derrick, and O. S. Chizhov, *Anal. Chem.* **52**, 214R (1980).
[6] A. L. Burlingame, A. Dell, and D. H. Russell, *Anal. Chem.* **54**, 363R (1982).
[7] V. N. Reinhold and S. A. Carr, *Mass Spectrom. Rev.* **2**, 153 (1983).
[8] A. Dell and G. W. Taylor, *Mass Spectrom. Rev.* **3**, 357 (1984).
[9] A. L. Burlingame, J. O. Whitney, and D. H. Russell, *Anal. Chem.* **56**, 417R (1984).
[10] C. C. Sweeley and H. A. Nunez, *Annu. Rev. Biochem.* **54**, 765 (1985).
[11] A. L. Burlingame, T. A. Baillie, and P. J. Derrick, *Anal. Chem.* **58**, 165R (1986).
[12] V. N. Reinhold, in "Mass Spectrometry in Biomedical Research" (S. Gaskell, ed.), 181. Wiley, New York, 1986.
[13] A. Dell, *Adv. Carbohyd. Chem. Biochem.* **45**, 19 (1987).

to increase their volatility and thermal stability in order to permit determination of their molecular weights and structures by electron impact and chemical ionization.[3,7] This is accomplished by either alkylation of all hydroxyl and acetamido functions with methyl groups[17] or esterification with acetyl groups (for detailed procedures see Dell[18]). Structure determination of the permethylated or peracetylated small glycans has been accomplished by electron impact ionization.[1,2]

More recently, the discovery and development of new ionization techniques including field desorption (FD),[19] fast atom bombardment (FAB)[20,21] and liquid secondary ion mass spectrometry (LSIMS)[23] have made possible the analysis of intact oligosaccharides without prior derivatization of all labile hydrogen atoms. While field desorption played an important initial role,[22] it has been overshadowed by the ease of maintaining a stable molecular ion beam with FAB and LSIMS for use with sector instruments. LSIMS is carried out using an energetic primary beam composed of xenon atoms[21] (FAB) or focused cesium ions[23] to eject/ionize the polar, labile analyte from the surface of a polar liquid matrix with ambient vibronic energy content. Oligosaccharides can be sputtered from such viscous liquid matrices in the form of protonated or deprotonated molecular ions despite their extreme hydrophilicity. However, the desorption yields for such hydrophilic neutral oligosaccharides are quite low (1–10 nmol of oligosaccharide required),[14] particularly for positive ions, and unambiguous interpretation of sequence from these spectra can be problematic.[13,16] If negative charge-bearing groups are present (carboxyl, phosphate, sulfate, etc.), the negative ion spectra can be significantly improved.[14] Significant increases in sputtering efficiency may be accomplished through chemical derivatization of the hydrophilic sugar to a more surface-active

[14] H. Egge and J. Peter-Katalinic, *Mass Spectrom. Rev.* **6**, 331 (1987).

[15] A. L. Burlingame, D. Maltby, D. H. Russell, and P. T. Holland, *Anal. Chem.* **60**, 312R (1988).

[16] A. Dell and J. E. Thomas-Oates, in "Analysis of Carbohydrates by GLC and MS" (C. J. Biermann and G. D. McGinnis, eds.), p. 217. CRC Press, Boca Raton, FL, 1989.

[17] S. I. Hakamori, *J. Biochem.* (*Tokyo*) **55**, 205 (1964).

[18] A. Dell, this volume [35].

[19] H. D. Beckey, in "Recent Developments in Mass Spectroscopy" (K. Ogata and T. Hayakawa, eds.), p. 1154. U. Park Press, Baltimore, MD, 1970.

[20] D. J. Surman and J. C. Vickerman, *J. Chem. Res.,* p. 170 (1981).

[21] M. Barber, R. S. Bordoli, R. D. Sedgwick, and A. N. Tyler, *J. Chem. Soc. Chem. Commun.,* p. 325 (1981).

[22] M. Linscheid, J. D'Angona, A. L. Burlingame, A. Dell, and C. E. Ballou, *Proc. Natl. Acad. Sci. U.S.A.* **78**, 1471 (1981).

[23] W. Aberth, K. M. Straub, F. C. Walls, and A. L. Burlingame, *Anal. Chem.* **54**, 2029 (1982).

hydrophobic sugar derivative and selection of an appropriate matrix[16,18,24] since the primary ion or atom beam only sputters the surface layer of the liquid matrix. LSIMS exploits the fact that the surface excess concentration of the analyte determines the sample ion yield.[25,26] For example, increasing the hydrophobicity of glycans by suitable derivatization improves sputtering yields (25 pmol–1 nmol of oligosaccharide), and may also be used to direct the fragmentation by providing preferred sites of protonation or deprotonation, i.e., charge localization. Permethylation or peracetylation,[13,16,18] have been shown to increase sensitivity by approximately a factor of ten. Alternatively, oligosaccharides possessing free reducing termini such as those liberated from N-linked glycoproteins by peptide-N^4-(N-acetyl-β-glucosaminyl) asparagine amidase F (PNGase F) or endoglycosidase H (Endo H), may be coupled to a lipidlike chromophore by reductive amination with concomitant major enhancement of surface activity and sensitivity.[24] The behavior of reducing terminal amine derivatives is significantly different in the positive and negative ion ionization modes. These protonated amines show a preponderance of MH$^+$ with little analytically useful fragmentation[27,28]; however, the corresponding deprotonated species yield a relatively abundant suite of ions which may be used to establish the sequence and branching of the oligosaccharide.[24,29] Of course, advantage may be taken of the fact that these protonated amine derivatives yield primarily molecular ions[29] to establish sample purity or compositional heterogeneity. However, collisional vibronic activation must be carried out in order to induce structurally informative positive ion fragmentation patterns.[28] This conversion of the natural preponderance of low vibronic energy molecular ions into mass spectra containing predictable and analytically useful fragmentation patterns is the subject of this chapter.

Collision-induced dissociation (CID) is a process whereby a small portion of the kinetic energy of the mass-selected incident ion is transformed into vibronic energy upon collision with helium atoms for the purpose of promoting unimolecular decompositions. There are a variety of ways to

[24] L. Poulter and A. L. Burlingame, this volume [36].

[25] W. V. Lignon, Jr. and S. B. Dorn, *Int. J. Mass Spectrom. Ion Proc.* **78,** 99 (1986).

[26] W. Lignon, *in* "Biological Mass Spectrometry" (A. L. Burlingame and J. A. McCloskey, eds.), p. 61. Elsevier, Amsterdam, 1990.

[27] W. T. Wang, N. C. Le Donne, B. Ackerman, and C. C. Sweeley, *Anal. Biochem.* **141,** 366 (1984).

[28] S. A. Carr, V. N. Reinhold, B. N. Green, and J. R. Hass, *Biomed. Mass Spectrom.* **12,** 288 (1985).

[29] J. W. Webb, K. Jiang, B. L. Gillece-Castro, A. L. Tarentino, T. H. Plummer, J. C. Byrd, S. J. Fisher, and A. L. Burlingame, *Anal. Biochem.* **169,** 337 (1988).

carry out the CID experiment using a variety of differing ion-optical instrument configurations and ion sources.[11,15,30,31] Differing instrumental systems can and do yield CID spectra which differ both qualitatively, e.g., the types of fragmentation processes observed, and quantitatively from one another. Low-incident kinetic energy considerations are presented elsewhere[32] and in this volume.[33] High-incident kinetic energy collisional activation[34] is the subject of this chapter. MS/MS methodology is the subject of intense current investigation in many laboratories and a variety of new strategies may be expected to result in higher quality MS/MS information in the near future. A full discussion of the advantages and disadvantages is beyond the scope of this chapter,[15,35] but it should be noted that two generic characteristics of a high-quality MS/MS experiment should be kept in mind. These are the instrumental ability of MS-1 to permit selection of the ^{12}C monoisotopic molecular ion with high efficiency, and the instrumental ability to control the deposition of relatively high quantities of vibronic energy. Thus, if the accuracy of mass measurement of MS-2 is better than 0.2 Da, the CID spectrum will contain both direct cleavage and rearrangement ions without the possibility of mass assignment errors arising from stable isotopic clusters or other components arising from use of a wider mass acceptance window in the ion source. Even under these circumstances, the coejected chemical noise from matrices employed may be observed at the highest sensitivity in the most advanced ion optical systems presently available.[36]

Early work employing MS/MS strategies included use of FD and B/E-linked scanning on a double-focusing instrument to study CID spectra of linkage isomers for a series of disaccharides.[37] Similarly, electron ionization (EI) of small permethyl oligosaccharides has been coupled with collisional activation (CA) to determine the positions of acetyl groups[38] and

[30] F. W. McLafferty, "Tandem Mass Spectrometry." Wiley, New York, 1983.

[31] C. Bruneé, *Int. J. Mass Spectrom. Ion Proc.* **76,** 121 (1987).

[32] L. Poulter and L. C. E. Taylor, *Int. J. Mass Spectrom. Ion Proc.* **91,** 183 (1989).

[33] D. R. Müller, B. Domon, and W. J. Richter, this volume [33].

[34] M. M. Sheil and P. J. Derrick, *in* "Mass Spectrometry in Biomedical Research" (S. Gaskell, ed.), p. 251, Wiley, New York, 1986.

[35] A. L. Burlingame, D. Norwood, D. H. Russell, and D. Millington, *Anal. Chem.* **62,** 268R (1990).

[36] F. C. Walls et al., *in* "Biological Mass Spectrometry" (A. L. Burlingame and J. A. McCloskey, eds.), p. 197. Elsevier, Amsterdam, 1990.

[37] H. Kambara and A. L. Burlingame, *Proc. 27th Annu. Conf. Mass Spectrom. Allied Topics, Seattle, WA,* paper FAMOB13 (1979).

[38] V. Kovácik, E. Petráková, V. Mihálov, I. Tvaroska, and W. Heerma, *Biomed. Mass Spectrom.* **12,** 49 (1985).

oligosaccharide linkage position.[39] Since the introduction of desorption ionization methods, glycans and glycosides have been studied without chemical modification.[14,16,40,41] In addition, studies have been carried out involving branched isomeric structures.

In this chapter the primary structures of neutral oligosaccharides of importance to posttranslational modification of glycoproteins will be considered. These have been studied in positive and negative ion modes, and include oligosaccharides modified by reductive coupling to a hydrophobic chromophoric moiety and by subsequent permethylation and peracetylation. The effect of each type of derivatization on sensitivity and the nature of observed fragmentation will be discussed, and the aspects of structure which can be determined from interpretation of CID mass spectra will be described.

Experimental Procedures

Reagents and Oligosaccharides. All chemicals were obtained from Sigma Chemical Co. (St. Louis, MO) unless otherwise noted. Solvents, including water, were glass distilled by Burdick and Jackson, Baxter Scientific Products, McGaw Park, IL. Glycerol, monothioglycerol, and tetraethylene glycol were products of Aldrich Chemical, Milwaukee, WI. Octyl *p*-aminobenzoate was synthesized in this laboratory.[42] The Manα1→3-(Manα1→6)Man-OMe (Man$_3$-methylglycoside) and Fucα1→2Galβ1→3-(Fucα1→4)GlcNAcβ1→4Glc were products of BioCarb, Lund, Sweden. High mannose oligosaccharides from the IgM of a patient with Waldenstrom's macroglobulinemia were isolated from the Fabμ fragment as the Endo H digestion products as previously described.[29] The eukaryotic N-linked oligosaccharide trimannosyl core was purchased from Oxford Glyco Systems, Abington, U.K. The core N-linked oligosaccharide from *Saccharomyces cerevisiae mnn1mnn2mnn10* mutant was prepared as previously described.[43]

Derivatization of Oligosaccharides. The free reducing termini of the sugars were reductively aminated with a hydrophobic chromophore as described elsewhere.[24,27,29] Coupling with *p*-aminobenzoic acid as either the octyl, butyl, or ethyl ester (ABOE, ABBE or ABEE, respectively) was carried out, and the resulting derivative was isolated by reversed-

[39] E. G. de Jong, W. Heerma, and G. Dijkstra, *Biomed. Mass Spectrom.* **7**, 127 (1980).

[40] D. Promé, J.-C. Promé, G. Puzo, and H. Aurelle, *Carbohydr. Res.* **140**, 121 (1985).

[41] F. W. Crow, K. B. Tomer, J. H. Looker, and M. L. Gross, *Anal. Biochem.* **155**, 286 (1986).

[42] R. Karrer and A. L. Burlingame, unpublished results, this laboratory, April 1988.

[43] L. M. Hernandez, L. Ballou, E. Alvarado, B. L. Gillece-Castro, A. L. Burlingame, and C. L. Ballou, *J. Biol. Chem.* **264**, 11849 (1989).

phase high-performance liquid chromatography (HPLC) on a Vydac (Separations Group, Hesperia, CA) C_4 column using a gradient of aqueous acetonitrile (0 to 60% in 40 min) while monitoring the absorbance at 320 or 306 nm. Permethylation was performed as previously described[17] using sodium hydride in dimethyl sulfoxide (DMSO) as the base, while peracetylation was accomplished using pyridine/acetic anhydride (1:1). The O-linked glycopeptide obtained by pronase digestion of bovine fetuin was prepared as the N-(tert-butoxycarbonyl)-L-tyrosine derivatives and was isolated as previously described.[44]

Mass Spectrometry. MS/MS spectra were obtained on a Kratos Concept II HH four-sector instrument operating in the *EBEB* configuration with a cesium ion LSIMS ion source[45] and an electrooptical microchannel array detector as described elsewhere.[36] Protonated or deprotonated sample (secondary) ions were accelerated to 8 keV with the mass resolution of MS-1 set to approximately 1500. For the CID experiments, the helium pressure was increased until the selected-precursor ion peak was decreased to 30% of its original intensity. This degree of attenuation corresponds to an average of one collision with helium for a given ion.[34] The collision cell was floated at 2 keV to produce 6 keV energy collisions and permit recording of low B/E values. Each magnet of the mass spectrometer was calibrated using a mixture of alkali metal iodide salts (Cs^+, Rb^+, Na^+, and Li^+). Nominal molecular weights are reported for all masses.

As a liquid matrix, 1 μl of glycerol/monothioglycerol (1:1, v/v), tetraethylene glycol, or monothioglycerol was added to the sample probe. Reducing terminal derivatives and permethylated glycans were typically analyzed in mixed monothioglycerol/glycerol or monothioglycerol alone, while underivatized oligosaccharides sputter more efficiently from a tetraethylene glycol solution. For positive ion spectra, 0.5 μl 1 N HCl was also added to the matrix. A cooled probe[46] was used to reduce the matrix evaporation rate. Using a multichannel electrooptical array detection system each simultaneous recording window consisted of 4% of the mass range, and CID data was accumulated at each window for 1 sec.[36] The total mass segment stepping time was 3 to 5 min depending on the molecular weight of the sample.

Nomenclature. We have adopted the nomenclature describing putative fragmentation processes shown in Fig. 1 recently proposed by Domon and Costello.[47] Briefly, fragments with charge retention on the nonreducing

[44] R. R. Townsend, M. R. Hardy, T. C. Wong, and Y. C. Lee, *Biochemistry* **25,** 5716 (1986).
[45] A. M. Falick, G. H. Wang, and F. C. Walls, *Anal. Chem.* **28,** 1308 (1986).
[46] A. M. Falick, F. C. Walls, and R. A. Laine, *Anal. Biochem.* **159,** 132 (1986).
[47] B. Domon and C. E. Costello, *Glycoconjugate J.* **5,** 397 (1988).

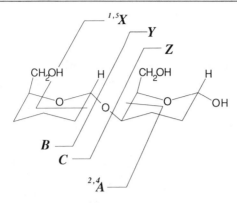

FIG. 1. Oligosaccharide fragmentation nomenclature. (From Ref. 47.)

terminus (drawn left) are designated *A*, *B*, and *C*, while the reducing terminal fragments (drawn right) are similarly designated *X*, *Y*, and *Z*. Ring cleavage fragments *A* and *X* contain superscript numbers designating the two bonds which have been broken. Subscript numbers identify the residue number, and the branches are labeled α, β, etc., where α applies to the branch of highest mass.

CID Spectra of Underivatized Oligosaccharides

Maltoheptaose. The neutral oligosaccharide maltoheptaose was ana- lyzed by MS/MS both without derivatization and after reductive amina- tion. A negative ion MS/MS spectrum was produced by collision of malto- heptaose $[M - H]^-$ m/z 1151 with He at 6 keV (Fig. 2). To 1 μg (0.86 nmol) of an aqueous solution of maltoheptaose was added 1 μl of tetraethylene glycol on the probe tip. Abundant product ions were observed at m/z 221, 383, 545, 707, 869, and 1031 which arise from ring cleavages either with charge retention at the nonreducing ($^{2,4}A$) or reducing ($^{1,5}X$) terminus. Since these ions have been observed in the negative ion spectra of reduc- tively aminated derivatives where the mass of the reducing terminus is shifted, this suggests that some or all of these peaks are ($^{2,4}A$) fragments. It should be noted that the relative abundance of the ($^{2,4}A$) fragments for this linear oligosaccharide form a smooth envelope decreasing in relative peak intensity with decreasing mass. Previously, this laboratory has shown that branched oligosaccharides have an uneven distribution of intensity reflecting features of their structures.[29] In addition, the underivatized mal- toheptaose produces B_1 and $^{2,5}A_1$ ions at m/z 161 and 119 and numerous

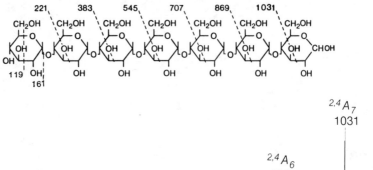

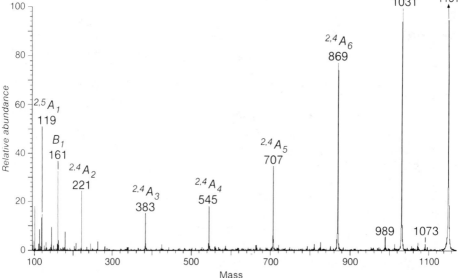

FIG. 2. Negative ion CID spectrum of maltoheptaose. The deprotonated molecule formed by liquid SIMS ionization (m/z 1151) was selected for analysis. The abundant fragments arise from an $^{2,4}A$ (or $^{1,5}X$) ring cleavage process.

low mass fragments of the glucose residues (data not shown below m/z 100).

Complex N-*linked Core Oligosaccharide.* The core oligosaccharide, $Man_3GlcNAc_2$ corresponding to all N-linked oligosaccharides was similarly analyzed. The deprotonated molecule (m/z 909) was the precursor ion selected to produce the spectrum shown in Fig. 3. In this case, charge retention is not exclusive for one end of the molecule as demonstrated by the abundance of X and Y as well as A and C fragment ions. Also, many of the product ions formed require more than one bond to rupture to produce the mass observed. These include X and A ring cleavage products, and ions formed by cleavages at different sites of the molecule. For example, the ion at m/z 586 is formed by two separate Y rearrangements. Most

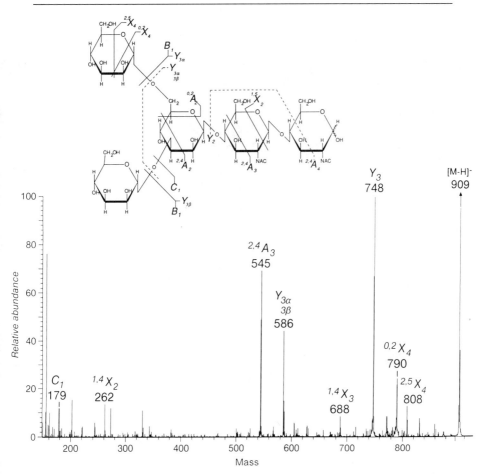

FIG. 3. Negative ion CID spectrum of the free core heptasaccharide obtained from N-linked glycans. Glycosidic bond cleavages (Y and B ions) as well as ring cleavages (A and X) occur in this molecule. Charge is not preferentially retained on either terminus as demonstrated by the formation of both A/B and X/Y product ions. In addition, more than one bond appears to cleave in the same molecule to form ions such as the one at m/z 586, which involves two Y rearrangement losses of both mannose termini.

significantly, the structure of the branched mannose nonreducing terminus can be deduced from the very abundant fragment at m/z 545 for three hexose residues, as well as the Y_3 ion. A linear oligosaccharide would contain a smooth envelope of fragment abundances for losses of each residue (see Fig. 2).

Methylglycoside. If the reducing terminus of an oligosaccharide is

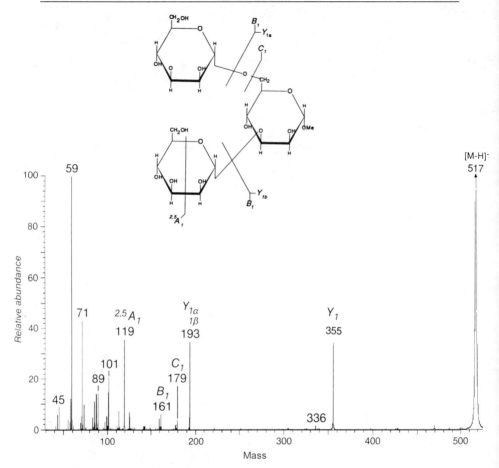

FIG. 4. Negative ion CID spectrum methylglycoside of a branched mannose trisaccharide. Glycosidic bond rearrangements (Y) for the loss of one or both mannose termini form ions at m/z 355 and 193. Charge is also retained on the nonreducing terminus to form A, B, and C ions.

modified by forming the methylglycoside, then charge is retained on the derivatized terminus and Y rearrangements dominate the negative ion spectra. For example, the negative ion CID spectrum (Fig. 4) of the methylglycoside of Manα1→3(Manα1→6)Man (m/z 517) contained abundant Y_1 and $Y_{1\alpha,1\beta}$ ions (m/z 355 and 193) as well as $^{2,5}A_2$, B_1, and C_1 ions (m/z 119, 161, 179). Loss of water and methoxy groups from the deprotonated molecule were also detected.

Lactosamine Hexasaccharide Alditols. Two hexasaccharide branch

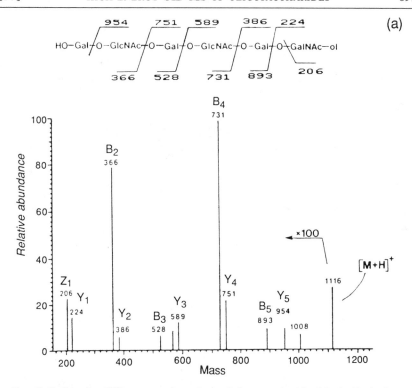

FIG. 5. Positive ion CID spectra of two isomeric hexasaccharide alditols. Both glycans' spectra contain Y and B ions. The spectrum shown in (a) is from a linear N-acetyllactosamine-containing glycan, while (b) is the corresponding branched structure. Reprinted with permission from *Biomed. Mass Spectrom.* **12**, 288 (1985). Copyright 1985 Wiley and Heyden.

isomers were analyzed by positive ion CID (Fig. 5).[28] The protonated molecular ions (m/z 1116) were shown to produce intense nonreducing terminal B ions with preferential cleavage at N-acetylhexosamine residues. The stabilization of hexosamine oxonium ions has been observed in FAB spectra of permethyl derivatives.[13,14,16] Identification of the branching configuration of lactosamine-containing oligosaccharides may be deduced by the relative abundance of the B ions (m/z 366 and 731). For example, in the spectrum of the linear saccharide (Fig. 5a) the B_2 and B_4 (m/z 366 and 731) ions are approximately equal in abundance. In contrast, Fig. 5b shows a spectrum where the B_2 ion predicted from the structure (m/z 366) is approximately twentyfold more abundant than the ion at m/z 731. The presence of this ion even at low abundance must result from either a multiple bond cleavage process[28] or the presence of an isomeric isobar. The complementary Y ions are also present in both spectra.

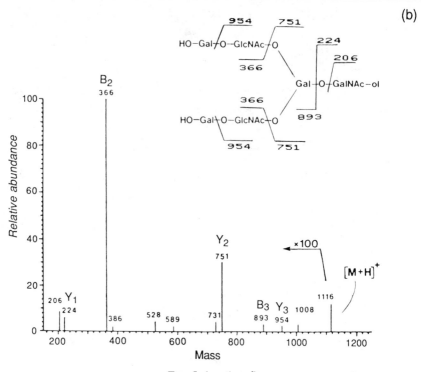

FIG. 5. (continued)

CID Spectra of Reducing Terminal Derivatives

Maltoheptaose. Since the reducing termini of many glycans are free they may be used for coupling to UV-absorbing molecules which provides a means of detecting the oligosaccharides during purification. If the glycan is reductively aminated, a preferred charge site is introduced for ionization, and if the amine contains a lipidlike tail detection by desorption ionization from a liquid matrix is also enhanced.[24] Interestingly, the CID spectra of maltoheptaose, when reductively coupled to esters of *p*-aminobenzoic acid (Fig. 6), contain abundant *Y* ions along with $^{2,4}A$ ions when analyzed in the negative ion mode. This is similar to the spectra observed in the liquid SIMS analysis.[29] The positive ion CID spectrum contains only reducing terminal ions, in this case *Y* ions (Fig. 7). It should be noted that the *Y* fragmentation path involves glycosidic bond cleavage with a hydrogen rearrangement onto the reducing terminal neutral product ion (see Scheme 2a, this volume [36]).

The LSIMS spectra obtained from ABEE derivatives of maltoheptaose have been described.[24,27,29] Importantly, molecular ions are abundant in

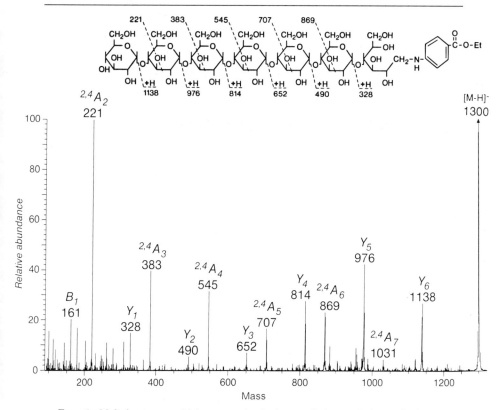

FIG. 6. Maltoheptaose which was reductively coupled to ethyl *p*-aminobenzoate, was analyzed by negative ion CID. Two product ion series dominate the spectrum, the *Y* and $^{2,4}A$.

both positive and negative modes. Although fragments formed by unimolecular decomposition of the protonated maltoheptaose-ABEE are of the $^{1,5}X$, Y, and Z type ions, their relative abundances are too low to be of analytical usefulness. The only abundant fragment in the positive ion CID spectrum was the Y_1 ion containing the reducing terminal hexose with the attached ethyl *p*-aminobenzoate moiety. This is qualitatively dramatically different from the negative ion spectrum which displays a regular array of *Y* fragments (Fig. 6). In positive mode, however, the *Y* series was found to be only slightly more abundant than the $^{1,5}X$ and *Z* ions in LSIMS.

High Mannose Oligosaccharides. Recent studies involving a series of $Man_nGlcNAc_{1-2}$-ABEE oligosaccharides by liquid SIMS have demonstrated the importance of mass spectral studies in the identification of branching isomers.[29] As noted previously, the negative ion liquid SIMS

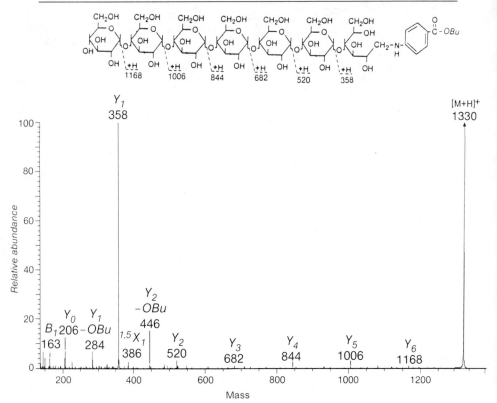

Fig. 7. The positive ion CID spectrum of reductively aminated maltoheptaose. The abundant Y_1 ion at m/z 358 dominates the spectrum demonstrating the charge retention on the aniline moiety.

spectra of the p-aminobenzoates of high mannose oligosaccharides are dominated by the Y and $^{2,4}A$ ion series. Depending on the degree of matrix-associated chemical noise, particularly at lower mass, the presence of $^{2,4}A$ ions may be obscured. These matrix-associated chemical noise peaks occurring at every mass are eliminated in the MS/MS experiment by selecting only the ^{12}C monoisotopic $(M - H)^-$ or $(M + H)^+$ ion formed by LSIMS and recording its CID product ions. The ion types and qualitative pattern observed in LSIMS[29] are also the dominant processes in the negative ion CID spectra, but the matrix-associated noise is absent such that structural use may be made of the lower mass $^{2,4}A$ ion series.

Abundant Y fragments have been observed in both the CID and one-dimensional LSIMS spectra of the major $Man_nGlcNAc_{1-2}$ oligosaccharides isolated from immunoglobulin M (IgM).[29] The CID spectra were very clean

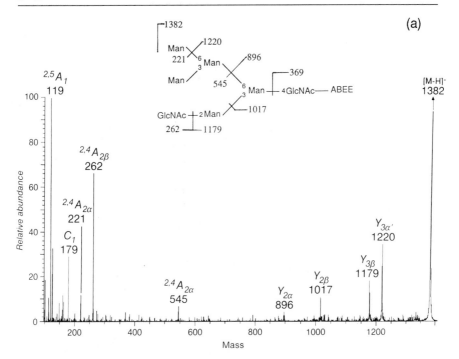

FIG. 8. The CID spectrum of Man$_5$GlcNAc$_2$-ABEE in negative ion mode (a) shows nonreducing terminal $^{2,4}A$ ions and Y glycosidic bond rearrangements with charge retained on the reducing terminal derivative. The positive ion CID spectrum of the same oligosaccharide (b) contains abundant B_1 ions for the nonreducing terminal GlcNAc, in addition to the Y ion series. The Y ions, which require more than one bond cleavage (m/z 533 and 695), are detected only in the positive ion mode suggesting that the negative ion mode is preferred for structural identification.

and low mass fragments such as $^{2,4}A$ at m/z 221 in negative ion mode (Fig. 8a) or B_1 at m/z 163 (Hex), 204 (HexNAc) in positive ion mode (Fig. 8b) were readily observed. These complementary nonreducing terminal fragments are particularly important when questions of multiple Y rearrangements arise. It is suggested by the intensity of the peaks at m/z 533 and 695 in the positive ion spectrum (Fig. 8b), which require the loss of more than one branch from the same molecule, that multiple Y rearrangements occur more frequently from the collisionally activated $[M + H]^+$ ion than the corresponding $[M - H]^-$. It is clear from the intensity of the 204 peak (B_1, GlcNAc) in the positive ion spectrum of the Man$_5$GlcNAc$_2$-ABEE derivative that the nonreducing terminal GlcNAc forms a very stable positive ion of the type characteristic of permethylated lactosamines.[12,16] Nonreducing hexoses also form B ions, although they are much

(b)

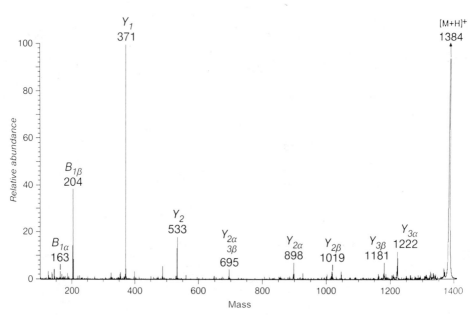

FIG. 8. (*continued*)

less abundant. Recently Johnson, Costello, and Reinhold[48] have isolated two branched isomers of $Man_7GlcNAc_1$ and determined their negative ion CID spectra as the ABEE derivatives. Their fragmentation confirmed previous results from this laboratory.[29]

In similar structural studies of the yeast *mnn1mnn9* mutant high mannose oligosaccharides, it became clear that the $\alpha1\rightarrow6$ elongation in the previously reported structure **I** was inconsistent with the fragmentation

$$*\alpha M \rightarrow {}^6\alpha M \rightarrow {}^6\alpha M \rightarrow {}^6\beta M \rightarrow {}^4\alpha\beta GNAc$$

$$\uparrow^2 \quad \uparrow^2 \quad \uparrow^3 \quad \uparrow^3$$

$$\alpha M \quad \alpha M \quad \alpha M \quad \alpha M$$

$$\uparrow^2$$

$$\alpha M$$

$$\uparrow^2$$

$$\alpha M$$

I

[48] R. S. Johnson, C. E. Costello, and V. N. Reinhold, *Proc. 37th Annu. Conf. Mass Spectrom. Allied Topics, Miami, FL*, p. 1192 (1989).

patterns observed for a series of these oligosaccharides and the sequence-branching correlations previously established.[29] This has led to reinvestigation of the experimental data on which the original structure was based, and to the revision of the structure as **II**.[43]

$$\alpha M \rightarrow {}^6\alpha M \rightarrow {}^6\beta M \rightarrow {}^4\alpha\beta GNAc$$
$$\uparrow^2 \qquad \uparrow^3 \qquad \uparrow^3$$
$$\alpha M \qquad \alpha M \qquad \alpha M^6 \leftarrow \alpha M^*$$
$$\uparrow^2 \qquad \uparrow^2$$
$$\alpha M \qquad \alpha M$$
$$\uparrow^2$$
$$\alpha M$$

II

More recently, MS/MS spectra of the Man_{10} member of this series provided the most dramatic support for the revised Man_{10} structure as shown in the CID spectrum of the octyl ester in negative ion mode in Fig. 9. If structure **I** were considered, a product ion at m/z 1587 would be expected from the loss of three mannose residues. In addition, the abundant ion $Y_{2\alpha}$ observed at m/z 1263 for loss of five hexose residues is inconsistent with structure **I**. However, as is shown in Fig. 9, structure **II** is consistent with the CID spectrum. The Y ions were also detected in the liquid SIMS spectrum, but unique to the CID spectrum are the abundant and unobscured A ions.

Complex Oligosaccharides. A complex mixture of N-linked oligosaccharides isolated from one salivary glycoprotein was reductively coupled to octyl p-aminobenzoate. The neutral fraction was isolated by reversed-phase HPLC and analyzed by negative ion LSIMS (Fig. 10a). Although a multitude of molecular ions were observed from the mixture of oligosaccharides (the possible structure of one component is shown in Fig. 10b), the most abundant ions in the spectrum were found at low mass and represent the common core structure for the components of the mixture. Figure 10b contains the CID spectrum of one fragment ion (m/z 802) formed from the LSIMS ionization of these complex glycans. The selected ion was further fragmented to produce the ion m/z 599 ($Y_{1\alpha}$) which was also observed in the LSIMS spectrum (Fig. 10a). This fragment confirms the location of the fucose residue to be on the reducing terminal GlcNAc.

Permethyl and Peracetyl Oligosaccharides

Permethylated Oligosaccharides. An oligosaccharide Fuc$\alpha1\rightarrow$2Gal$\beta1\rightarrow$3(Fuc$\alpha1\rightarrow$4)GlcNAc$\beta1\rightarrow$4Glc provides an excellent model for a moderate-sized glycan containing N-acetylhexosamine residues. It was

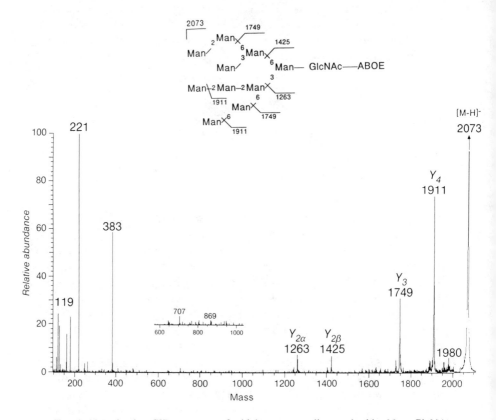

FIG. 9. Negative ion CID spectrum of a high mannose oligosaccharide, Man$_{10}$GlcNAc$_1$-ABOE, isolated from *Saccharomyces cerevisiae*. Inset shows a multiplication ($\times 10$) of one region of the spectrum. The relative abundance of the Y fragment ions suggest that the mannose residues are connected in such a way that single-bond cleavages produce losses of one, two, four, or five hexose residues confirming the revised structure shown.

analyzed as the permethylated derivative with the protonated ^{12}C molecular ion (m/z 1497) selected for collision (see Fig. 11). The most abundant product ion was detected at m/z 260, and represents the $B_{1\alpha}$ ion formed from the nonreducing terminal GalNAc. This is comparable to the 204 ion found in the non-methyated hybrid oligosaccharide (Fig. 8b) in positive

FIG. 10. The negative ion CID spectrum of a complex oligosaccharide fragment ion formed in the liquid SIMS mass spectrum (a) is shown in (b). A mixture of complex oligosaccharides was analyzed as the ABOE derivatives. At low mass intense fragment peaks (asterisks) were selected for CID. The product ions formed from m/z 802 are shown in (b).

(a)

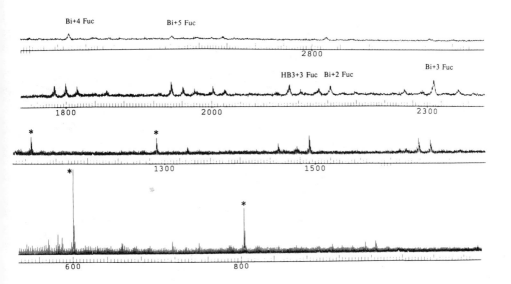

(b)

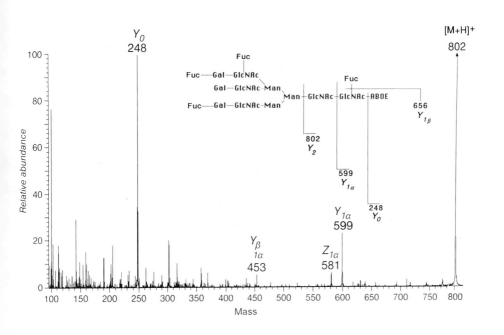

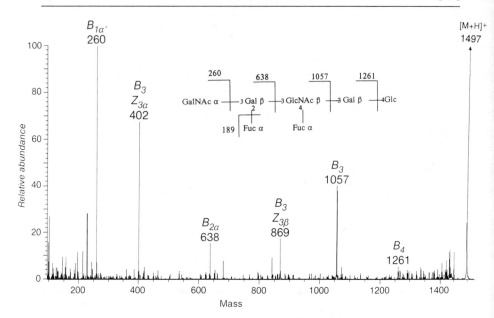

FIG. 11. The positive ion CID spectrum of the permethylated oligosaccharide Fucα1→
2Galβ1→3(Fucα1→4)GlcNAcβ1→4Glc is shown. Nonreducing terminal fragments, *B*, at *N*-
acetylhexosamine residues, *m/z* 260 and 1057, dominate the spectrum. Also, relevant losses
from the 1057 ion (*m/z* 402) reflect the 3-linkage position of the Gal substituent as this branch
is preferentially lost.

ion mode analysis of the ABEE derivative. The B_3 ion at *m/z* 1057 is
characteristic of permethylated complex oligosaccharides when analyzed
by desorption ionization since the positive charge is stabilized by the
hexosamine moiety.[13,14,16] This ion has been observed to lose the residues
substituted in the 3-linkage position in a selective fragmentation.[13,14,16] In
the example shown, the ion at *m/z* 402 is formed by the loss of the
GalNAc1→3 (Fuc1→2) Gal substituent. In addition, a deoxyhexose (Fuc)
is lost from the 1057 ion by a *Y*-type process to form *m/z* 869.

Peracetyl Reductively Coupled Oligosaccharides. The reductive amination may be combined with peracetylation to further increase the hydrophobicity of the oligosaccharide, to direct fragmentation to the nonreducing terminus, and to carry out chemical degradation by partial acetolysis.
For example, the ethyl *p*-aminobenzoate derivative of a high mannose
oligosaccharide was subjected to partial acetolysis to selectively cleave
1→6 linkages. The positive ion CID spectrum of a $Man_6GlcNAc_1$ cleavage
product (*m/z* 2310) is shown in Fig. 12. The *B* fragments (oxonium ions)
for 1, 2, or 3 hexose residues were the only abundant fragments in the

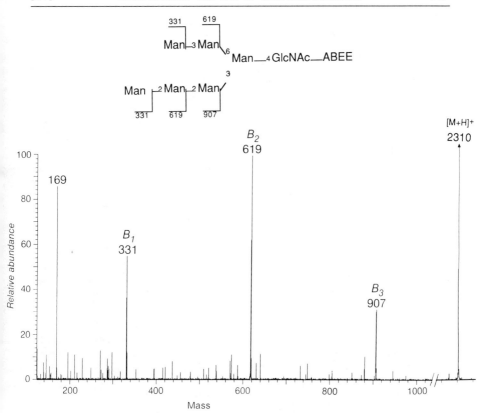

FIG. 12. A high mannose oligosaccharide, $Man_8GlcNAc_1$, isolated from IgM, was reductively coupled to ABEE, and peracetylated. In order to confirm its branching structure, it was subjected to partial acetolysis, and the CID spectrum from one product, a $Man_6GlcNAc$-ABEE, is shown. The peracetylation causes charge to be retained on the nonreducing terminus, and the fragments formed suggest that the structure is branched, and that the longest branch contains three hexose residues. This is consistent with the proposed structure.

spectrum demonstrating the directing effect of peralkyl and peracylations. The fragments are consistent with predicted longest hexose sequence containing three mannose residues.

Glycopeptide with O-Linked Saccharides

The positive ion CID spectrum of the *tert*-butoxycarbonyl-tyrosyl glycopeptide is shown in Fig. 13. The high-energy tandem mass spectrum of the protonated pseudomolecular ion (m/z 1522) was obtained from 2 nmol

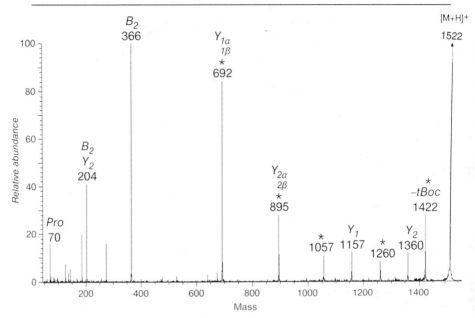

FIG. 13. The positive ion CID spectrum of an O-linked glycopeptide. This glycopeptide which was isolated from the glycoprotein fetuin by exhaustive pronase digestion was N-terminally modified with *t*-Boc-tyrosine. The spectrum contains *Y*-type fragments demonstrating that charge is retained on the peptide portion of the molecule. Multiple bond rearrangements occur to form the ions which lose the *t*-Boc as well as carbohydrate moieties. The structure suggested contains two Hex-HexNAc disaccharides attached to the same peptide.

of the glycopeptide. One striking feature of the spectrum is the appearance of pairs of peaks 100 Da apart. These are due to rearrangement losses of the *t*-Boc moiety. Also, abundant fragment ions due to the loss of carbohydrate residues by glycosidic bond rearrangements (*Y*) are evident by ions at m/z 1360, 1157, 895, and 692 which suggest sequential losses of Hex-HexNAc-Hex-HexNAc. These data are consistent with the presence of four sugar residues attached to the peptide. The ion present at m/z 692 apparently has lost all of the carbohydrate as well as the *t*-Boc moiety and thus represents the gas-phase deglycosylated peptide moiety.

The signal-to-noise ratio is improved compared to the LSIMS spectrum, particularly at low mass, so that nonreducing terminal carbohydrate oxonium ions (*B*) are observed. The most abundant fragment ion, m/z 366, is the Hex + HexNAc (*B*$_2$) ion. From the lack of any larger carbohydrate fragments, it was deduced that two identical Hex-HexNAc glycans are attached to the peptide. Finally, the structure, **III,** of the glycopeptide is shown below, and includes two identical hexosylacetamidohexosamine chains attached to the hydroxyl amino acids serine and threonine in this heptapeptide.

```
                      Hexose          Hexose
                        |               |
                      HexNAc          HexNAc
                        |               |
  t-Boc——Tyr——Gly——Pro——Thr——Pro——Ser——Ala
```

III

Conclusions

Product ions formed by CID of protonated or deprotonated oligosaccharides are shown to be useful in establishing the sequence and branching of various types of glycans. The ions formed by liquid SIMS include molecular ions and some fragment ions produced from species which initially had sufficient excess internal energy to rearrange or cleave bonds. In contrast to peptides, there are cases in which extensive fragmentation occurs during desorption ionization, and therefore sequence can be deduced without the additional vibronic energy deposition supplied by CID. However, underivatized oligosaccharides do not sputter or fragment efficiently by LSIMS, and therefore should be analyzed as a suitable derivative by CID, or by both methods. The CID experiment introduces more internal energy by collisionally exciting the selected precursor ion to induce structurally significant fragmentation. In this study it is shown that both molecular and fragment ions may be selected for collisional activation.

There are three significant observations which have been made from studies of these tandem mass spectra. First, a distinct advantage of the LSIMS plus CID experiment is that the chemical noise arising from the sputtering of the matrix solvent is effectively eliminated producing spectra solely of the precursor ion selected in the first mass spectrometer. This is particularly important at low mass, where nonreducing terminal fragments are present, allowing ready interpretation. Second, when smaller ions are selected, a greater diversity of fragments is observed. Two implications are derived from this observation: (a) the CID spectra of smaller ions ($m/z < 1000$) are more complicated and therefore more difficult to interpret; however (b) the diversity of fragments is a great advantage since more subtle aspects of the structure such as linkage can be determined because the spectral information content can be high. For larger ions ($m/z > 1000$) analyzed by desorption ionization, we have found that the energy is more likely to produce glycosidic bond cleavages (Y, Z, or B ions). In fact, even when more energy is added to larger ions by CID, the glycosidic bond cleavages still predominate. However, as seen in the permethylated oligosaccharide Fucα1→2Galβ1→3(Fucα1→4)GlcNAcβ1→4Glc, the CID spectrum (Fig. 11) is more similar to EI mass spectra in that many cleavages occur which are not observed by LSIMS. Third, it is important to choose the appropriate derivative for particular ionization modes, positive or negative, to provide the type of fragmentation processes and structural information desired. Reducing terminal derivatives have been analyzed by direct chemical ionization,[7] LSIMS,[24,27,28,29,43] and low-energy CID.[32] In all cases the charge is retained preferentially on the reducing terminal moiety. However, the 2-aminopyridine derivative[27,28] while desorbing better than underivatized oligosaccharides, forms only positive ions which fragment poorly under LSIMS ionization conditions. With CID of the molecular ions, however, this derivative gives comparable fragments to the ABXE derivatives, but without the advantage of being able to form negative ions. Under LSIMS ionization, the ABXE derivatives give rich negative ion fragmentation patterns, but the low mass sample ions are difficult to detect due to the matrix solvent chemical noise.

Glycopeptides also provide examples of the charge being retained by the peptide attached to the reducing terminus. In contrast, permethyl and peracetyl glycans favor the formation of nonreducing terminal fragments of the oxonium type (B).[13,14,16] Since sputtering efficiency (and thus sensitivity) is enhanced by all of the derivatization methods, they are strongly recommended prior to CID analysis.

In order to elucidate the structures of glycans and conjugates we have enhanced the information content available from picomole amounts of oligosaccharides by enhancing the sputtering yield and fragmentation of glycans. This is accomplished by derivatization[24] and tandem mass spectrometry.

Acknowledgments

National Institutes of Health Grant RR01614, National Science Foundation Biological Instrumentation Program DIR8700766. We thank K. Jiang for performing the partial acetolysis of Man$_8$GlcNAc, L. Reinders for permethylation of the glycolipid, and F. C. Walls and B. W. Gibson for their excellent technical assistance. The glycolipid was the kind gift of H. Leffler, and the glycopeptide from bovine fetuin was isolated by M. R. Hardy and R. R. Townsend.

[38] Desorption Mass Spectrometry of Glycosphingolipids

By Jasna Peter-Katalinić and Heinz Egge

Introduction

Glycosphingolipids, by definition glycosides of the ceramides, occur ubiquitously on the outer leaflet of cell surface membranes of vertebrates and invertebrates. Their carbohydrate moiety may be composed of one up to more than forty sugar residues.[1] The ceramide portion (Cer), composed of a sphingosine base and a fatty acid moiety linked to each other by an amide bond, can vary in type. More than sixty different sphingosine bases have been described so far.[2] The alkyl chains of the fatty acid moiety vary from C_{14} to C_{26} and may contain double bonds and/or hydroxy groups. The nomenclature of glycosphingolipids (GSLs) is based on the carbohydrate moiety. A glycosphingolipid thus defined may, however, vary in its lipid portion, depending on species, organ, and cell type[3] (Fig. 1). Glucose (Glc), galactose (Gal), N-acetylglucosamine (GlcNAc), N-acetylgalactosamine (GalNAc), fucose (Fuc), sialic acids, and glucuronic acid (GlcA) are the most common sugar components in the mammalian GSLs, whereas mannose (Man), xylose (Xyl), and arabinose (Ara) may occur in GSLs of invertebrates. It is the carbohydrate moiety of a GSL that determines its biological specificity as an antigen or receptor whereas the lipid portion may influence its membrane topology and the physicochemical properties of these membranes.

The complete structural elucidation of a glycosphingolipid implies the determination of the following parameters: (1) the molecular weight; (2) the number and the type of sugar components; (3) their sequence and pattern of branching; (4) the sites of linkages; (5) anomeric configuration; (6) the presence of additional functional groups, e.g., phosphates and sulfates and their site of linkage; (7) the sphingosine base; (8) the fatty acid moiety; and (9) the conformation of the GSL molecule.

The analysis of the structural parameters (1–3) and (6–8) of GSLs has been successfully probed by mass spectrometry for 20 years, generally using thermal evaporation and electron ionization of appropriate derivatives.[4] This methodology, still of general value for the determination of

[1] B. A. Macher and C. C. Sweeley, this series, Vol. 50, p. 236.
[2] K.-A. Karlsson, *Lipids* **5**, 878 (1970).
[3] K. Nakamura, Y. Hashimoto, T. Yamakawa, and A. Suzuki, *J. Biochem. (Tokyo)* **103**, 201 (1988).
[4] H. Egge, *Chem. Phys. Lipids* **21**, 349 (1978).

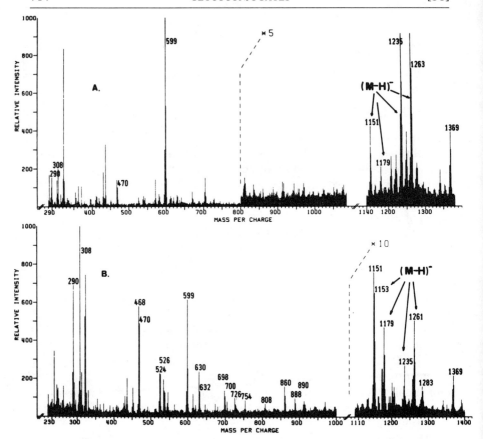

FIG. 1. Negative ion FAB mass spectrum of the native G_{M3} fraction from (A) human liver, (B) human melanoma, and (C) bovine brain. Shown in (D) is the molecular ion region of the permethylated G_{M3} fraction from human melanoma depicted in (B) recorded in positive ion mode. The ganglioside G_{M3} fractions show heterogeneity due to variations in the sphingosine moiety, documented in the molecular ion region by $(M - H)^-$ ions which appear separated by 28 mass units. The structurally unrelated impurity at m/z 599 [in (A) and (B)] is probably of biological origin and is destroyed during the permethylation.

the parameters (2–4) and (7–8), is, however, limited by thermal stability and the absolute necessity to derivatize under conditions that eliminate alkali-labile substituents.

The soft ionization methods especially fast atom bombardment (FAB)[5] and liquid SIMS (secondary ion mass spectrometry),[6] in combination with

[5] M. Barber, R. Bardoli, R. D. Sedgwick, and A. N. Tyler, *J. Chem. Soc. Chem. Commun.* p. 325 (1981).

[6] W. Aberth, K. M. Straub, and A. L. Burlingame, *Anal. Chem.* **54**, 2029 (1982).

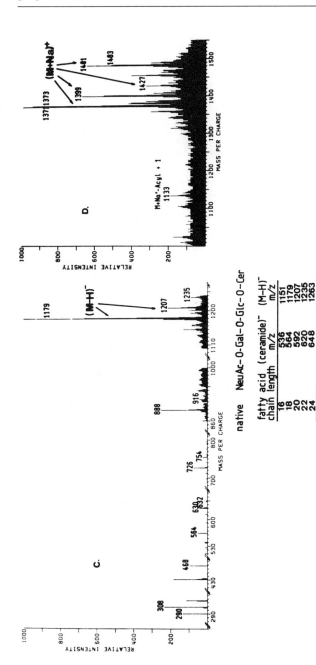

native   NeuAc–0–Gal–0–Glc–0–Cer

fatty acid chain length	(ceramide)⁻ m/z	(M–H)⁻ m/z
16	536	1151
18	564	1179
20	592	1207
22	620	1235
24	648	1263

permethylated

fatty acid chain length	(M+Na)⁺ m/z
16	1371
18	1399
20	1427
22	1455
24	1483

Fig. 1. (*continued*)

the development of high-field instruments, have brought substantial progress. It is now possible to analyze native or derivatized GSL structures within several minutes and to obtain intense molecular and sequence ions on molecules comprising more than 25 sugar residues.[7]

Core Structures in Naturally Occurring Glycosphingolipids

The following five major oligosaccharide structures have been observed in naturally occurring GSLs linked to $Glc(\beta1\rightarrow1)Cer$:

Gal(β1→3)GalNAc(β1→4)Gal(β1→4)-	ganglio core
Gal(β1→3)GlcNAc(β1→3)Gal(β1→4)-	lacto-N core
Gal(β1→4)GlcNAc(β1→3)Gal(β1→4)-	neolacto-N core
GalNAc(β1→3)Gal(α1→4)Gal(β1→4)-	globo core
GalNAc(β1→4)GlcNAc(β1→3)Man(β1→4)-	arthro core

The sugar constituents and sequences of neutral oligosaccharides derived from these cores can be determined by desorption mass spectrometry from the fragment ions in terms of hexoses and N-acetylhexosamines in native or derivatized samples.

Gangliosides (by definition GSLs, which contain sialic acids), can be derived from the ganglio, lacto-N, and/or neolacto-N series. The major representative of the sialic acids is 5-N-acetylneuraminic acid (Neu5NAc). It can be modified by additional hydroxy, deoxy, acetyl, methyl, lactyl, phosphate, and/or sulfate groups. Presently, thirty-two different sialic acids are known.[8] Beside sialic acids, other charged groups like carboxylate, sulfate, phosphate, phosphonate, and/or phosphoethanolamine may also be present.

Such structural features play an important role in the proper choice of the most efficient method for analysis by mass spectrometry. Stereochemical factors may also be involved in the formation of parent as well as daughter ions in the gas phase, but no general rules have been established so far.

Fragmentation Patterns

Molecular ions, $(M - H)^-$, and $(M + H)^+$ or $(M + cation)^+$, provide direct information on molecular weight, whereas information on type and number of sugar components can be deduced indirectly from these values. Primary or parent ions, formed after the cleavage of glycosidic bonds

---

[7] H. Egge and J. Peter-Katalinić, *Mass Spectrom. Rev.* **6**, 331 (1987).
[8] R. Schauer, this series, Vol. 138, p. 132.

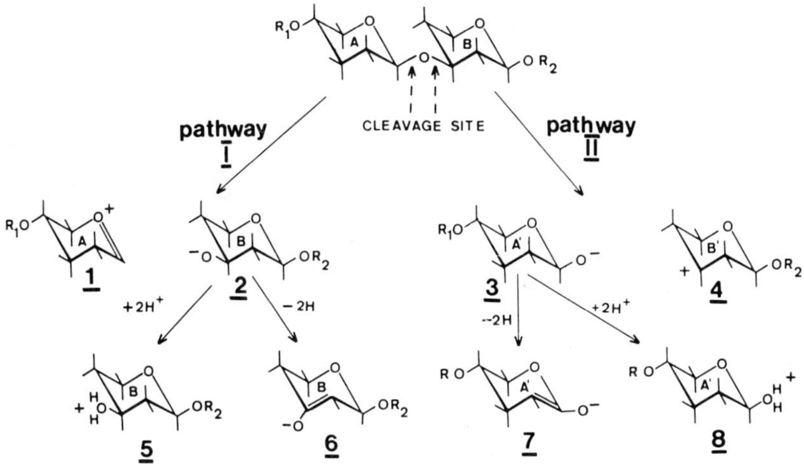

SCHEME 1.

from both possible sites (pathway I and pathway II in Scheme 1), furnish information about sequence and branching pattern.

The cations of types **1** and **4** should be observed in experiments performed in the positive ion mode and the anions of types **2** and **3** in the negative ion mode, if both pathways would randomly take place. In fact, the fragmentation pattern for native GSL in positive ion mode favors the pathway I, but beside the primary ion **1**, the daughter ions **5** and **8**, arising from the protonation of **2** and **3**, respectively, can be observed. Therefore, an exact statement concerning the homogeneity of the molecular species with the same core structures and the branching patterns from these experiments is not possible. In addition, the parent ions are approximately of the same intensity as the daughter ions; consequently, a distribution of the total ion intensity over a number of peaks takes place.

In the positive ion desorption mass spectra of derivatized (permethylated, peracetylated or perbenzoylated) GSL, the oxonium ion of type **1** is so much favored that the sugar sequence from the nonreducing terminus and the branching pattern, in addition to the molecular homogeneity,can easily be deduced directly from the spectrum. The cleavage of sugar C—C bonds, observed in direct chemical ionization (DCI)[9] or fast atom bombardment (FAB)-MS/MS,[10] does not take place.

[9] V. N. Reinhold, this series, Vol. 138, p. 59.
[10] B. Domon and C. Costello, *Biochemistry* **27,** 1534 (1988).

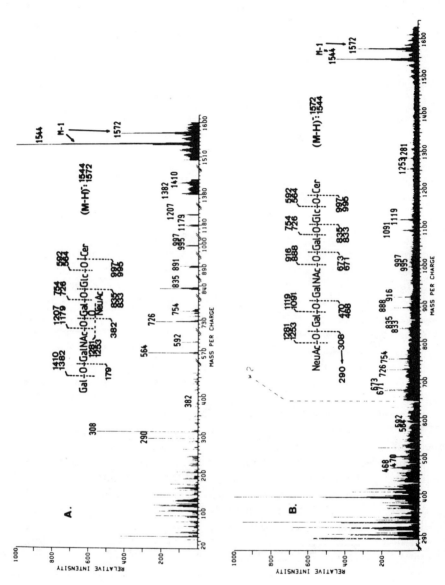

Fig. 2. Negative ion FAB mass spectrum of the gangliosides G_{M1a} (A) and G_{M1b} (B) with their respective fragmentation schemes. The Neu5NAc attachment site is deducible from the different pattern of the sequence ions.

In positive ion desorption mass spectra of permethylated acidic and neutral GSL, the preferred cleavage site in the sugar chain is that of N-methyl-N-acetylamino sugar of type **9** where R is CH_3 or sugar residue. Oxonium ions of type **11** derived from mannose, are more abundant than **10**, derived from glucose or galactose, as found in GSL of the arthro series.[11] The daughter ions, diagnostically important for the discrimination between 1,3 and 1,4/6-substitution at 2-deoxy-2-acetamido sugars, are formed from the corresponding parent ions of type **9** by elimination of the substituent on C-3.[12]

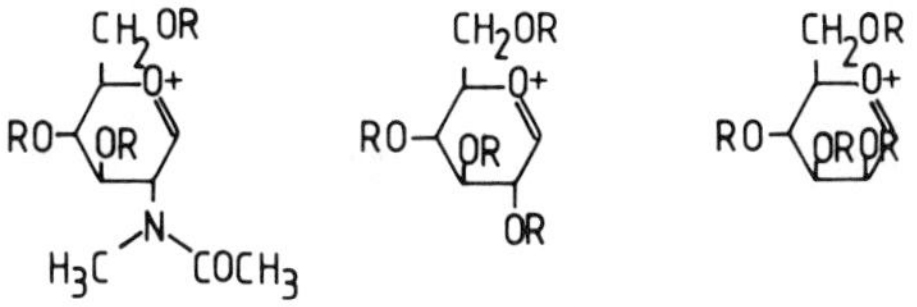

9   **10**   **11**

In negative ion desorption mass spectrometry, performed on the native GSL, generally the formation of alcoholate type of ion **2** (pathway I) is favored. The native GSLs carrying negatively charged substituents, like gangliosides and uronic acid-containing GSLs, however, also exhibit characteristic abundant sequence ions of type **3** (pathway II). By combining the information obtained from the fragmentation patterns via pathways I and II it is possible to obtain the complete sequence information from one single experiment. The sequence of two isomeric gangliosides, G_{Mla} and G_{Mlb}, with different attachment sites of the Neu5NAc, is depicted with their negative ion FAB mass spectra in Fig. 2.

The structure of GSLs containing neutral sugars and alkali-labile substituents can be elucidated preferentially from native samples in the negative ion mode or peracetylated samples in positive or negative ion mode desorption mass spectra. The carbohydrate sequence can be deduced from appropriate abundant primary ions, as discussed previously.

For the structure determination of the ceramide residue, four types of ions can be used: the ceramide cation of the type **12** and anion **13**, in addition to a sphingosine-derived cation **14** and a cation of type **15** arising after the loss of the acyl group from the molecular ion and subsequent protonation:

[11] R. D. Dennis, R. Geyer, H. Egge, J. Peter-Katalinić, S.-C. Li, S. Stirm, and H. Wiegandt, *J. Biol. Chem.* **260**, 5370 (1985).
[12] N. K. Kochetkov and O. S. Chizhov, *Adv. Carbohydr. Chem.* **21**, 39 (1966).

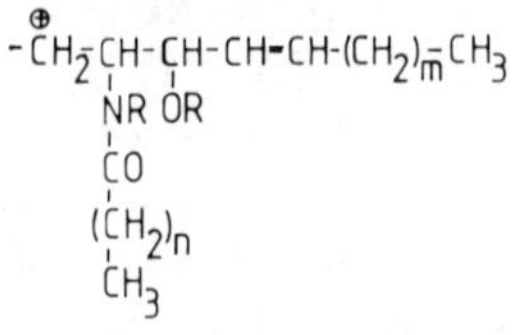

$$\overset{\oplus}{-CH_2}-CH-CH-CH=CH-(CH_2)_{\overline{m}}CH_3$$
$$\quad\quad\;\; NR\;\;OR$$
$$\quad\quad\;\; CO$$
$$\quad\quad\;\; (CH_2)_n$$
$$\quad\quad\;\; CH_3$$

12

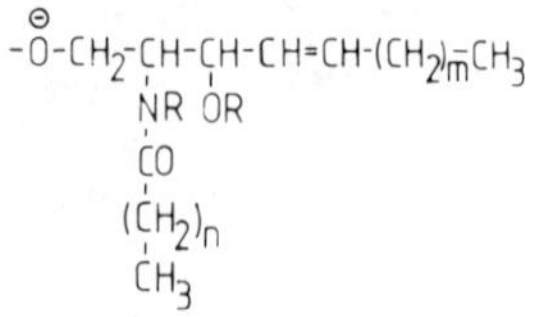

$$\overset{\ominus}{-O}-CH_2-CH-CH-CH=CH-(CH_2)_{\overline{m}}CH_3$$
$$\quad\quad\;\; NR\;\;OR$$
$$\quad\quad\;\; CO$$
$$\quad\quad\;\; (CH_2)_n$$
$$\quad\quad\;\; CH_3$$

13

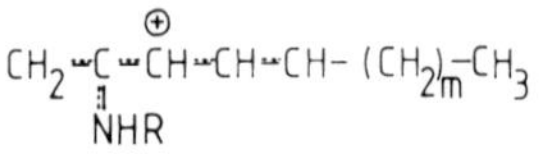

$$CH_2 \text{\small ---} \overset{\oplus}{C} \text{\small ---} CH \text{\small ---} CH \text{\small ---} CH-(CH_2)_{\overline{m}}-CH_3$$
$$\quad\quad NHR$$

14

$$\left[Sugar-O-CH_2-\overset{H}{\underset{NHR}{C}}-\overset{OR'}{CH}-CH=CH-(CH_2)_{\overline{m}}-CH_3 \right] + H^+ \text{ or } Cation^+$$

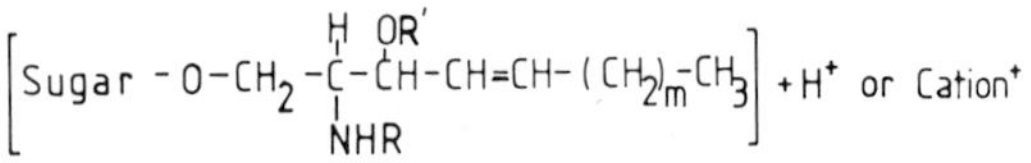

15

where R is H, CH_3; R' is H, CH_3, $COCH_3$. In the positive ion spectra from the permethylated GSLs, the ceramide moiety is generally represented by the cations **12**, **14**, and **15**; in the peracetylated samples however, only **14** is of high abundance. The cation **14** is also abundant in the positive ion desorption mass spectra of the native GSLs.

For the structure determination of the ceramide moiety in sulfatide, the positive ion FAB-MS of peracetylated sulfatide and the negative ion FAB-MS of lysosulfatide were used (Fig. 3).

Design of Appropriate Experimental Protocol for Analysis by Mass Spectrometry

A rational protocol for the structural elucidation of a small amount ($\leq 100\ \mu g$) of a GSL should include the following set of experiments:

1. The negative ion mass spectrum of the native sample provides $(M - H)^-$, $(Cer)^-$, and two series of ions, of type **2** and type **3** (Scheme 1). With these data a structural proposal for the sugar moiety and the ceramide, as a whole, can be made.

2. The positive ion mass spectrum of the native sample furnishes $(M + H)^+$ or $(M + cation)^+$, Cer^+ **12**, and sphingosine-derived ion **14**.

3. After peracetylation the positive ion mass spectrum provides $(M + H)^+$ or $(M + cation)^+$ beside oxonium sugar sequence ions of type **1** and a sphingosine base-derived ion **14**.

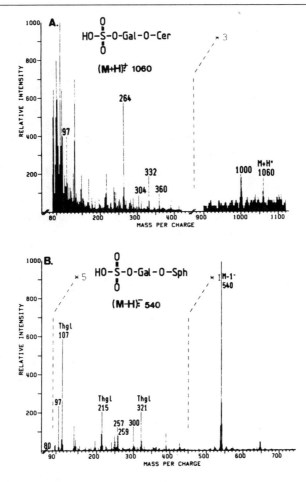

FIG. 3. FAB mass spectra of the sulfatide: HSO₃-Gal-Cer. (A) Positive ion spectrum of peracetylated sulfatide showing (M + H)⁺ at m/z 1060. The Cer⁺ fragment ion 12 at m/z 674 is not present, but the sphingosine ion 14 at m/z 264 indicates that the fatty acid is a lignoceric acid. (B) Negative ion FAB mass spectrum of lysosulfatide. The sugar moiety is unaffected by the change in the ceramide moiety, as documented by the fragment ions at m/z 97 (HSO₄)⁻, m/z 259/257 (HSO₃— Gal)⁻, and m/z 300 (HSO₃—Gal—O—CH=CH—NH₂)⁻.

4. After permethylation using NaOH/methyl iodide in dimethyl sulfoxide (DMSO)[13] the positive ion mass spectrum provides (M + H)⁺ or (M + cation)⁺, oxonium sequence ions **1**, their respective daughter ions, Cer⁺ **12**, sphingosine base-derived ion **14**, and (MH − acyl + 1)⁺ ion **15**.

[13] I. Ciucanu and F. Kerek, *Carbohydr. Res.* **131**, 209 (1984).

Sample Preparation

The isolation of GSLs from biological material is usually achieved by extraction, followed by several separation steps using adsorption and ion-exchange chromatography adapted to the type of GSL under investigation.[14,15] Frequently, as a final purification step, semipreparative chromatography on high-performance TLC plates is included. During the extraction from the plate, considerable amounts of salt, silica, and binder may be coeluted which may subsequently interfere with the ion desorption process. Here the use of high-performance TLC plates with gypsum as binder[16] or an additional clean-up step using reversed-phase C_{18} material is recommended. Alternatively, such contaminants can be cleaned out from the GSL sample by use of 5-cm Iatrobeads (silica gel beads of defined size, supplied by Macherey, Nagel and Co., D-5160 Düren, FRG) columns eluted by solvent mixtures (chloroform/methanol/water) of increasing polarity. Problems rising from minor salt contamination can be resolved by adding 1 μl of 1 N aqueous HCl directly to the sample on the target. In some cases, the samples can be more efficiently purified if derivatized by permethylation, peracetylation, perbenzoylation, or by several derivatization steps, based on the differences in chemical reactivity between functional groups present (see [35] in this volume). The purified and thoroughly desalted native GSLs can be analyzed by positive or negative ion mode desorption.

Types of Ionization for Glycosphingolipids

Field Desorption

Field desorption (FD) ionization, developed by Beckey and Schulten,[17] was used for structural analysis of a few native neutral and acidic GSLs with 1 to 6 sugar residues. The FD mass spectra were dominated by molecular ions $(M + H)^+$ or $(M + cation)^+$, but the fragments of type **1** and **4** were rather weak. In the permethylated samples, the fragments of type **1** became more abundant.[18]

[14] J. N. Kanfer and S. Hakomori, "Sphingolipid Biochemistry." Plenum, New York, 1983.
[15] R. Kannagi, K. Watanabe, and S. Hakomori, this series, Vol. 138, p. 3.
[16] P. Hanfland, *Eur. J. Biochem.* **87**, 161 (1978).
[17] D. Beckey and H. R. Schulten, *Angew. Chem.* **87**, 425 (1975).
[18] Y. Kushi, S. Handa, H. Kambara, and K. Shizukuishi, *J. Biochem.* (*Tokyo*) **94**, 1841 (1983).

Chemical Ionization

Chemical ionization (CI) has been applied to the analysis of GSLs using ammonia as a reagent gas for native and for derivatized neutral[9] and acidic[19,20] GSLs. The oligosaccharides derived from gangliosides after oxidative cleavage of the lipid moiety by ozone and subsequent reduction were analyzed by CI using ammonia and isobutane as reagent gases.[20]

The underivatized GSLs limited to three to five sugar units were analyzed by DCI. The useful results by CI for larger molecules up to the mass of 3000 u were obtained using different sets of derivatization procedures, like permethylation or, alternatively, permethylation with subsequent reduction by $LiAlH_4$, or permethylation, reduction, and trimethylsilylation. The fragmentation patterns are rather complicated because of potential alternate addition of water and/or reagent gas to the parent ions.

Liquid Secondary Ion Mass Spectrometry

In liquid secondary ion mass spectrometry (LSIMS) experiments with a primary beam of xenon ions, native and derivatized GSLs have been analyzed in the positive and in the negative ion mode. With positive ion LSIMS, the molecular ions $(M + H)^+$ and/or $(M + cation)^+$ appeared, in general, to be of lower intensity than the fragment ions. For acidic GSLs, like sulfoglycolipids, the negative ion LSIMS of underivatized samples appeared to give, similar to the FAB ionization, more structural information than with the positive ion mode.[21] Abundant molecular ions $(M - 1)^-$ and/or $(MNa^+ - 2)^-$ were accompanied by significant fragment ions of type **2** and **3**, useful for the determination of the sugar sequence and the site of sulfate substitution.

Thus far, the desorption behavior and fragmentation patterns for LSIMS seem similar to those for FAB.

Fast Atom Bombardment

The majority of fast atom bombardment (FAB) mass spectra of GSLs have been acquired on VG ZAB-1F or HF, VG 70-250, and VG 7070E (VG Analytical, UK), and JMS DX-300, JMS-DX 303, JMS HX-100, and JEOL HX110/HX110 (JEOL, Ltd., Japan) instruments. Ion guns operating with argon or xenon as bombarding gas produced an atom beam

[19] S. A. Carr and V. N. Reinhold, *Biomed. Mass Spectrom.* **11**, 633 (1984).
[20] T. Ariga, R. K. Yu, M. Suzuki, S. Ando, and T. Miyatake, *J. Lipid Res.* **23**, 437 (1982); Y. Tanaka, R. K. Yu, S. Ando, T. Ariga, and T. Itoh, *Carbohydr. Res.* **126**, 1 (1984).
[21] Y. Kushi, S. Handa, and I. Ishizuka, *J. Biochem. (Tokyo)* **97**, 419 (1985).

of an energy equivalent to 8–10 keV and a current of 5×10^{-8} A. The acceleration voltage used varied from 10 to 3 kV, depending on the type of instrument and the extension of the mass range required. On the VG ZAB-HF instrument, an acceleration voltage of 4 kV was necessary to extend the mass range in order to obtain the molecular $(M + Na)^+$ ion at m/z 6181 (nominal mass) of permethylated ceramide pentacosasaccharide from rabbit erythrocyte membranes.[22]

As in the case of liquid SIMS, the solubility of the sample in the liquid matrix is the prerequisite. Matrices like glycerol, thioglycerol, a 1 : 1 mixture of glycerol/thioglycerol, triethanolamine, 1,1,3,3-tetramethylurea, a 5 : 1 mixture of triethanolamine/1,1,3,3-tetramethylurea, and a 4 : 1 mixture of 1,4-dithioerythritol/1,4-dithiothreitol have been successfully applied to native and derivatized neutral and acidic GSLs. The final choice of the appropriate matrix must include considerations about polarity and intrinsic chemical properties of the molecule(s) to be analyzed. In our hands one of the most useful matrices is thioglycerol, having the single disadvantage of rapid evaporization in high vacuum.

The sample of neutral GSL can be dissolved to a clear solution in methanol, chloroform/methanol, 1 : 1 (v/v), or chloroform/methanol/water, 75 : 25 : 4 (v/v/v) containing 1–10 μg/μl. Methanol/concentrated acetic acid (1 : 1, v/v) is considered to be the appropriate solvent for acidic GSLs. The sample solution is applied by a microsyringe to the stainless steel target, coated with about 2 μl of matrix. After the evaporation of the solvent by air or hair dryer, the probe with the sample is introduced to the vacuum and subjected to FAB. Due to the evaporation of the matrix during the acquisition of spectra, it is necessary to run in both up-scan and down-scan mode in order to obtain intense fragments as well as molecular ions. The addition of 1 μl of 1% solution of sodium acetate in methanol to the matrix is followed by an increase of the molecular ion and decrease in the intensity of fragment ions. The typical scan speed for instruments without a data system is approximately 1 sec/decade, depending on mass range and amplification. Use of a data system for multiple acquisitions or signal averaging, can increase the speed of a single scan to about 0.3 sec/decade, thus typically accumulating 10–20 scans. FAB mass spectrometry was applied in positive and negative mode to a large variety of native and derivatized GSLs in order to determine the primary structure of substances involved in recognition processes, pathological states or receptor phenomena. The potency of this method in obtaining several sets of clear-cut data

[22] P. Hanfland, M. Kordowicz, J. Peter-Katalinić, H. Egge, J. Dabrowski, and U. Dabrowski, *Carbohydr. Res.* **178**, 1 (1988).

and the simplicity of sample handling are the reasons for its wide and rapid acceptance.[7,23–25]

Combined Techniques

TLC-LSIMS and TLC-FAB-MS

Matrix-assisted SIMS and FAB-MS can be applied to the analysis of GSLs on TLC plates without prior elution. After TLC separation on an aluminum- or plastic-backed silica gel, the spots are visualized by iodine vapor or Coomassie blue staining. The area of interest on the plate is cut out (max. size 5 mm × 20 mm) and attached with double-face masking tape to a SIMS or FAB probe tip. After adding 1–2 μl of solvent and 2–5 μl matrix to the TLC strip, the probe tip is bombarded by Xe$^+$ ions or Xe atoms under standard conditions.[26] Changes in ganglioside expression have been evaluated in rat fibroblasts after transfection with sarcoma virus DNA[27] using this method. Similarly, the glycolipid mapping of spleen and liver tissues from patients affected by metabolic glycolipid disorders caused by lysosomal enzyme deficiencies[28] and the structure of neoglycolipids after the reductive amination of human milk oligosaccharides with phosphoethanolamine-containing glycerides[29] have been established. Results from other experiments will be needed to evaluate the potency of this method on the higher GSLs.

HPLC-CI and HPLC-FAB

Positive ion CI mass spectra were obtained on HPLC separated neutral mono- to tetraglycosylceramides on-line using ammonia as the reagent gas. The observed fragmentation patterns were quite complex, even for ceramide mono- and disaccharides. As a general feature of CI desorption,[30] the interpretation of these spectra is difficult, because primary ions tend to add or to eliminate water and/or ammonia.

[23] A. Dell, *Adv. Carbohydr. Chem. Biochem.* **45,** 20 (1987).
[24] M. Arita, M. Iwamori, T. Higuchi, and Y. Nagai, *J. Biochem. (Tokyo)* **94,** 249 (1983); Y. Ohashi, M. Iwamori, T. Ogawa, and Y. Nagai, *Biochemistry* **26,** 3990 (1987).
[25] M. Suzuki, K. Nakamura, Y. Hashimoto, A. Suzuki, and T. Yamakawa, *Carbohydr. Res.* **151,** 213 (1986).
[26] Y. Kushi and S. Handa, *J. Biochem. (Tokyo)* **98,** 265 (1985).
[27] H. Nakaishi, Y. Sanai, M. Shibuya, M. Iwamori, and Y. Nagai, *Cancer. Res.* **48,** 1753 (1988).
[28] Y. Kushi, C. Rokukawa, and S. Handa, *Anal. Biochem.* **175,** 167 (1988).
[29] P. W. Tang and T. Feizi, *Carbohydr. Res.* **161,** 133 (1987).
[30] J. E. Evans and R. H. McCluer, *Biomed. Environ. Mass Spectrom.* **14,** 149 (1987).

SFC-CI and SFC-FAB

In supercritical fluid chromatography (SFC), a mobile phase consisting of a highly compressed gas near or above its critical temperature and pressure is used as a mobile phase. This separation method is more rapid than conventional HPLC and provides a higher number of effective plates at low temperature. The separation of GSLs by SFC requires permethylation in order to decrease their polarity and can be combined with CI or FAB mass spectrometry.[31]

Analysis of Mixtures

The complete separation of complex GSL mixtures into individual compounds can seldom be achieved even with the sophisticated methods available at the present time.[14,15] The physical characteristics of individual molecular species of GSL arise from the structural variations of building blocks present in such a multifunctional molecule. The overlapping can occur due to the presence of (a) sugar chain homologs; (b) isomeric sugar sequences; (c) different sites of glycosidic linkages; (d) linearity vs. branching of the core structures; (e) the presence of other functional groups like esters, lactones, sulfates, and phosphates; (f) the modifications of the basic structures by oxidation state (carboxy groups, deoxy sugars); (g) alkyl homologs arising from sphingosine base and/or fatty acid; (h) modifications in the lipid portion, such as in lysocompounds and in neoglycolipids; (i) molecules structurally completely unrelated to GSLs; (j) stereochemistry on the anomeric center of the glycosidic bond.

A general discussion of analysis of mixtures is presented elsewhere in this volume [34] by Samuelsson et al. Only specific issues concerning desorption techniques will be discussed here.

In Fig. 4, the methodological approach of handling a mixture by FAB mass spectrometry is shown. About 30 μg of a ganglioside fraction was obtained after several chromatographic separation steps from a T-lymphoma cell line.[32] Whereas negative ion mode FAB did not give unequivocal results, the positive ion FAB spectrum obtained after permethylation revealed the presence of three molecular species. Their structures were definitely assigned by matching of fragment mass increments with molecular ions present in the mass spectrum of the mixture.

If the analyte represents a mixture of analogs with the same core structure differing only in the absence of the terminal sugar unit, the

[31] J. Kuei, G. R. Her, and V. N. Reinhold, Anal. Biochem. 172, 228 (1988).
[32] J. Müthing, J. Peter-Katalinić, H. Egge, U. Neumann, B. Kniep, A. Loyter, and P. F. Mühlradt, Proc. Xth Int. Symp. Glycoconjugates, Jerusalem, September (1989).

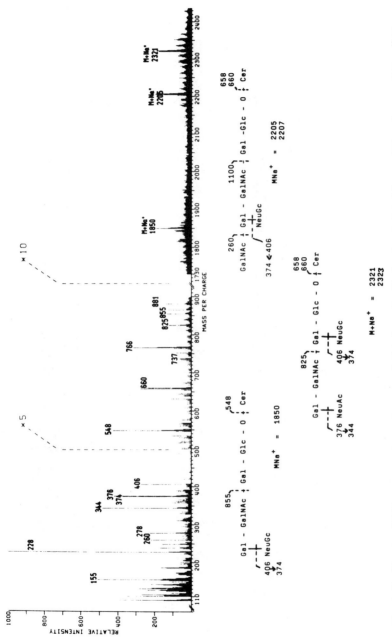

FIG 4. Positive ion FAB mass spectrum of a permethylated ganglioside fraction isolated from a T-lymphoma cell line.[32] Three ganglioside species with the molecular ions (MNa⁺) at m/z 1850 (NeuGcHex₃HexNAcCer₁₆:₀), m/z 2205 (NeuGcHex₃HexNAcCer₂₄:₁), and m/z 2321 (NeuAcNeuGcHex₃HexNAcCer₂₄:₁) are present in the mixture (16:0 and 24:1 indicate the fatty acid moiety of the ceramide). It is possible to deduce the complete sugar sequence and ceramide moiety of each component in the mixture by matching the appropriate fragment elements, as shown in the corresponding fragmentation schemes.

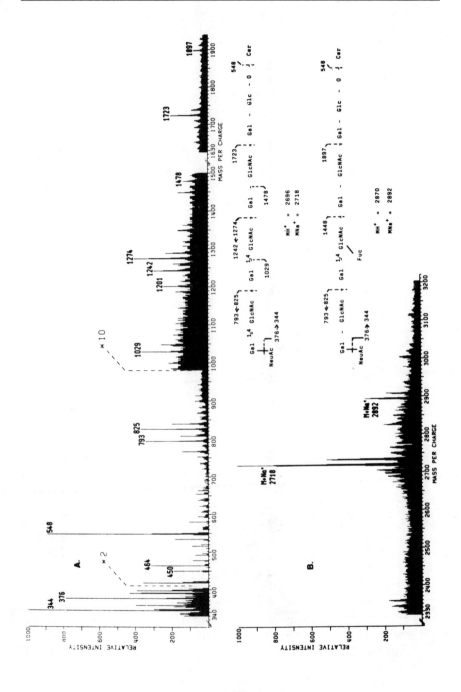

negative ion spectra will not differentiate between ions arising from the cleavage of the glycosidic bond and the molecular ion of the lower homolog. This differentiation is, however, possible in positive ion desorption spectra of the derivatized samples due to pronounced expression of oxonium ions of type 1 and their specific mass increments. In Fig. 5 the positive ion FAB-MS of a permethylated ganglioside fraction, showing immunological activity toward VIM-2 antibody, is depicted. The only difference between the two components is the presence of an additional fucose.[33]

The presence of isomeric structures with the same molecular ion but different sugar sequences can be postulated if more than one set of sequence ions is detected both in negative ion desorption spectra of the native sample and in the positive ion mode of the derivatized sample.

The presence of positional isomers can be detected to some extent if two daughter ions of different specificity occur in the same spectrum of the permethylated GSL.[34] This may be of particular importance for correlation of structural data of specific carbohydrate epitopes with immunological methods, e.g., immunostaining.[35]

The aspect of linearity vs. branching of sugar cores in GSLs can successfully be analyzed in native and in derivatized samples.[7,23,36] It is an important factor in studying the biosynthetic specificity of glycosyltransferases.

Naturally occurring lactones have been found in G_{M1} and G_{D1b} gangliosides as demonstrated in the negative ion desorption spectra of native

---

[33] B. Kniep, J. Peter-Katalinić, H. Egge, W. Knapp, and P. F. Mühlradt, in preparation.

[34] H. Egge, J. Peter-Katalinić, M. Hergersberg, F.-G. Hanisch, R. Bruntz, and G. Uhlenbruck, Proc. XIIIth Int. Carbohydr. Symp. Ithaca, NY, August (1986).

[35] J. Magnani, S. L. Spitalnik, and V. Ginsburg, this series, Vol. 138, p. 195.

[36] H. Egge, M. Kordowicz, J. Peter-Katalinić, and P. Hanfland, J. Biol. Chem. 260, 4927 (1985).

FIG. 5. Positive ion FAB mass spectrum of the permethylated ganglioside fraction isolated from human spleen, carrying the VIM-2 antigen epitope. (A) Lower mass region scanned in thioglycerol without addition of sodium acetate. (B) Molecular ion region scanned after addition of sodium acetate to the matrix. Two gangliosides, NeuAcHex₅HexNAc₃Cer (MNa⁺ = 2718) and dHexNeuAcHex₅HexNAc₃Cer (MNa⁺ = 2892) show the same sugar core structure as documented by common fragment ions at m/z 376–344 (NeuAc⁺) and m/z 825–793 (NeuAcHexHexNAc⁺). The ions at m/z 1274–1242 (NeuAcHex₂HexNAc₂⁺) versus m/z 1448 (dHexNeuAcHex₂HexNAc₂⁺) and m/z 1723 (NeuAcHex₃HexNAc₃⁺) versus m/z 1897 (dHexNeuAcHex₃HexNAc₃⁺) reflect the structural variation on the second lactosamine unit. Both gangliosides have their respective carbohydrate portion bound to the same type of ceramide, carrying palmitic acid.

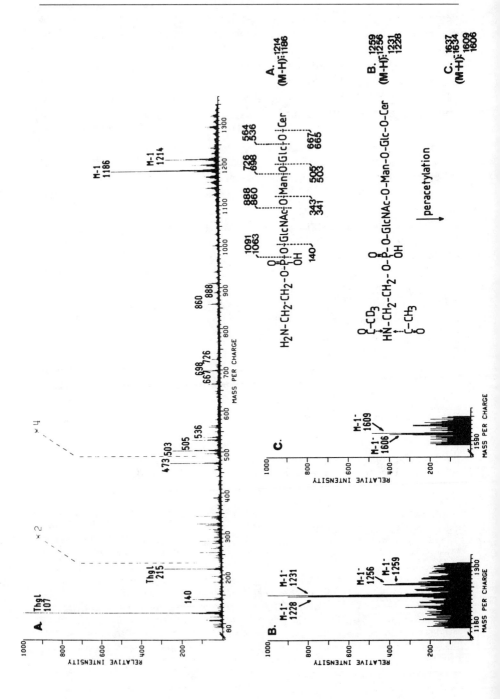

samples.[37] The lactonization is a reversible reaction, which can be induced chemically by adding mineral acids to the samples; lactones will be easily hydrolyzed under basic conditions. Other base- and/or acid-labile esters are the O-acetates,[8] O-sulfates,[38,39] and O-phosphoethanolamines. They can be localized in their desorption mass spectra by altered sugar mass increments and molecular ions. In Fig. 6 the negative ion FAB mass spectrum of a GSL isolated from *Calliphora vicina* is presented. To the sugar core of an arthro type,[11] a phosphoethanolamine substituent is attached. Its free amino group was acetylated by 1:1 mixture of $(CH_3CO)_2O/(CD_3CO)_2O$ and then peracetylated with $(CH_3CO)_2O$.[40] The isotopic label in their respective molecular ions was visible in the negative mode FAB mass spectrum (Fig. 6B and C).

Comparison with Other Analytical Methods

The topo- and stereochemistry of the glycosidic linkages are the dominating factors that control the shape of the terminal epitopes of the carbohydrate antigens of GSLs. Information concerning glycosidic linkages can be obtained from experiments using nuclear magnetic resonance (NMR) and from combined GC/MS techniques; desorption mass spectrometric analysis has also been utilized. Immunostaining methods, based on antigen–antibody complex formation, have become available for searching for specific sugar–antigen epitopes in mixtures of GSLs,[35] particularly tumor-associated antigens.[41] If possible, such techniques should be used in concert with desorption mass spectrometry in order to characterize a yet unknown antigen or to define the specificity of a new antibody. The optimal approach is to permethylate the GSL antigen and to observe its specific parent and respective daughter ions in positive ion mode desorption mass spectra. The use of linked scan techniques[42] should also be considered.

[37] S. Sonnino, G. Kirschner, G. Fronza, H. Egge, R. Ghidoni, D. Acquotti, and G. Tettamanti, *Glycoconjugate J.* **2**, 343 (1985).
[38] H. Leffler, G. C. Hansson, and N. Strömberg, *J. Biol. Chem.* **261**, 1440 (1986).
[39] T. Ariga, T. Kohriyama, and L. Freddo, *et al., J. Biol. Chem.* **262**, 848 (1987).
[40] F. Helling, R. D. Dennis, M. Keller, J. Peter-Katalinić, H. Egge, and H. Wiegandt, in preparation.
[41] S. Hakomori, *Cancer Res.* **45**, 2405 (1985).
[42] A. G. Brenton and J. H. Beynon, *Eur. Spectrosc. News* **29** (1980).

FIG. 6. The negative ion FAB mass spectrum of the GSL from *Calliphora vicina* pupae carrying a phosphoethanolamine substituent.[40] (A) The native sample; (B) the molecular ion region of the same sample after N-acetylation with $(CH_3CO)_2O/(CD_3CO)_2O$; (C) the molecular ion region of the sample depicted in (B) after peracetylation.

Monitoring of Chemical and Enzymatic Reactions

The structural changes induced by chemical reactions of GSLs as polyfunctional molecules include derivatizations ([35] in this volume), oxidative and reductive cleavages of one or more building units ([30]–[32] in this volume), oxidations, and reductions. FAB mass spectrometry has been used to monitor the topospecifity of metaperiodate oxidation in GSLs from rabbit erythrocytes[43] or for acid-catalyzed hydrolysis of neuraminic acid.[44] The synthesis of lyso compounds and their derivatives from GSLs by removal of the fatty acid moiety under basic conditions,[45] leaves the carbohydrate moiety unaffected, as shown by negative ion mode FAB spectra (Fig. 3B). The removal of the lipid moiety from a GSL by the enzyme ceramidase (isolated from leach or from microorganisms)[46] opens the possibility to study the liberated oligosaccharides by desorption mass spectrometry and by comparison with already known sugar structures. On the other side, reductive elimination of the terminal aldehyde group of oligosaccharides with nonpolar amines improves their desorption properties and increases the sensitivity of the measurement.[29,47,48]

Exoglycosidase degradation of the branched GSLs from rabbit erythrocytes by α-galactosidase,[22,36] or a ganglioside from mouse spleen showing choleragenoid-binding activity from mouse spleen by β-galactosidase[49] were monitored by FAB mass spectrometry. It was also shown by FAB methods, that the removal of the neuraminic acid by neuraminidase from *Vibrio cholerae* on the TLC plate is complete only in G_{D1a}, but not in $G_{D1\alpha}$ or G_{D1b}.[50]

Summary

Desorption mass spectrometry has become an important tool for sequencing and mapping of glycosphingolipids of natural, synthetic, or semisynthetic origin. The appropriate combination of different desorption mass

[43] H. Egge and J. Peter-Katalinić, *in* "Mass Spectrometry in the Health and Life Sciences" (A. Burlingame and N. Castagnoli, eds.). Elsevier, Amsterdam, 1985.
[44] J.-E. Mansson, H. Mo, H. Egge, and L. Svennerholm, *FEBS Lett.* **196**, 259 (1986).
[45] G. Schwarzmann and K. Sandhoff, this series, Vol. 138, p. 319.
[46] S.-C. Li, R. DeGasperi, J. E. Muldrey, and Y.-T. Li, *Biochem. Biophys. Res. Commun.* **141**, 346 (1986); M. Ito and T. Yamagata, *J. Biol. Chem.* **261**, 14278 (1986).
[47] S. Dreyer, H. Egge, and J. Peter-Katalinić, in preparation.
[48] L. Poulter and A. L. Burlingame, this volume [36].
[49] K. Nakamura, M. Suzuki, F. Inagaki, T. Yamagawa, and A. Suzuki, *J. Biochem.* (*Tokyo*) **101**, 825 (1987).
[50] J. Müthing, B. Schwinzer, J. Peter-Katalinić, H. Egge, and P. F. Mühlradt, *Biochemistry* **28**, 2923 (1989).

spectrometric techniques with other spectroscopic, enzymatic, chemical, and/or immunological methods represents the most direct and efficient way to establish frequent, yet unknown, molecular structure–function relationships.

Acknowledgments

We thank Dr. A. Vogel and Prof. K. Sandhoff, Bonn, for the sample of lysosulfatide. This work was generously supported by the Deutsche Forschungsgemeinschaft.

[39] High-Mass Gas Chromatography–Mass Spectrometry of Permethylated Oligosaccharides

By GUNNAR C. HANSSON and HASSE KARLSSON

Capillary gas chromatography–mass spectrometry (GC/MS) is a fast, sensitive, and high-resolving method for the characterization of biomolecules. It has been used in glycoconjugate research for analyzing the partially methylated alditol acetates after degradation of the glycoconjugates[1] and, in a few cases, for the analysis of oligosaccharides after permethylation[2–4] or trifluoroacetolysis.[5] The advent of high-temperature capillary GC using thin-film thermostable bonded stationary phases now allows the analysis of large permethylated oligosaccharides,[6–8] an adaptation that should also be useful in other fields of biomedical research.

Preparation of Oligosaccharides

Glycosphingolipids

Glycan is cleaved[7] from the ceramide portion of glycosphingolipids by endoglycoceramidase from *Rhodococcus*[9] (Genzyme Corp., Boston, MA)

[1] This series, Vol. 50 [1].
[2] P. Hallgren and A. Lundblad, *J. Biol. Chem.* **252**, 1014 (1977).
[3] H. van Halbeek, L. Dorland, J. Haverkamp, G. A. Veldink, J. F. G. Vliegenthardt, B. Fournet, G. Ricart, J. Montreuil, W. D. Gathmann, and D. Aminoff, *Eur. J. Biochem.* **118**, 487 (1981).
[4] T. Tsuji and T. Osawa, *Carbohydr. Res.* **151**, 391 (1986).
[5] B. Nilsson and D. Zopf, *Arch. Biochem. Biophys.* **222**, 628 (1983).
[6] H. Karlsson, I. Carlstedt, and G. C. Hansson, *FEBS Lett.* **226**, 23 (1987).
[7] G. C. Hansson, Y.-T. Li, and H. Karlsson, *Biochemistry* **28**, 6672 (1989).
[8] H. Karlsson, I. Carlstedt, and G. C. Hansson, *Anal. Biochem.* **182**, 438 (1989).
[9] M. Ito and T. Yamagata, *J. Biol. Chem.* **261**, 14278 (1986).

or ceramide-glycanase from the leach *Macrobdella decora*[10] (Boehringer Mannheim, Mannheim, FRG). The released oligosaccharides are obtained in the water phase after partitioning with chloroform:methanol: water, 8:4:3 (by vol) and residual detergents trapped by Sep-Pak[7] (Waters, Milford, MA). The ceramides and monoglycosylceramides are obtained in the chloroform phase.[7]

O-Linked Oligosaccharides

The mucins or other glycopeptides (1 mg/ml) are treated with 0.05 M KOH and 1.0 M NaBH$_4$ at 45° for 16 or 45 hr.[6,8,11] Acetic acid (2 M) is added, the sample passed over an ion-exchange column (AG 50W-X8, H$^+$, Bio-Rad, Richmond, CA; 0.5 g resin/ml sample) eluted with water (10 ml/g resin), and evaporated four times after the addition of methanol with a few drops of acetic acid. The neutral oligosaccharides are separated from the acidic ones by ion-exchange chromatography on DEAE-Sephadex.[6,8]

Permethylation

The oligosaccharides are dried in a stream of nitrogen in tubes with a Teflon-faced screw cap, dried under vacuum for 30 min, and permethylated[8,12,13] by the addition of 0.5 or 1.0 ml (less than 0.2 mg and 0.3–2 mg of sample, respectively) of dimethyl sulfoxide (DMSO), 0.1 or 0.2 ml of iodomethane (0.5 or 1.0 ml for oligosaccharides with a Glc-alditol[8]), and 25 or 50 mg of NaOH powder. The samples are stirred with a magnetic bar for 10 min and the reaction stopped by the addition of 2 or 4 ml of water and 1 or 2 ml of chloroform. The water phase is removed after centrifugation and the chloroform phase washed ≥4 times (until neutral) with 2 or 4 ml of water. The chloroform phase is moved to a new tube and evaporated. Oligosaccharides with a reducing end, as obtained from glycosphingolipids, can be reduced with 0.5 ml 1 M NaBH$_4$ overnight, evaporated with nitrogen, and the last step repeated 5 times after the addition of methanol and a drop of acetic acid. The reduced oligosaccharides are dried under vacuum and permethylated as above, but with increased amounts of iodomethane (0.5 or 1.0 ml).[8]

[10] S.-C. Li, R. DeGasperi, J. E. Muldrey, and Y.-T. Li, *Biochem. Biophys. Res. Commun.* **141,** 346 (1986).
[11] D. M. Carlsson, *J. Biol. Chem.* **243,** 616 (1968).
[12] I. Ciucanu and F. Kerek, *Carbohydr. Res.* **131,** 209 (1984).
[13] G. Larson, H. Karlsson, G. C. Hansson, and W. Pimlott, *Carbohydr. Res.* **161,** 281 (1987).

Preparation of Capillary Columns

Capillary columns are prepared from uncoated and deactivated, fused-silica capillary tubing (0.25 mm id) with an outside coating of HT-polyimide, intended for use up to 430° (Chrompack, Middelburg, The Netherlands). Columns (5 or 10 m) are statically coated (0.03–0.05 μm) with freshly made stationary-phase solutions of SE-54 (Alltech, IL) in dichloromethane/pentane, 1 : 1 (by vol) containing 0.8% (w/w stationary phase) dicumyl peroxide (Merck, Darmstadt, FRG). After coating, the capillaries are flushed with dry nitrogen for 30 min at room temperature and then evacuated by attaching both ends to vacuum. The columns are then sealed with a microflame and cross-linked[14] by heating to 145° for 30 min. Columns are then rinsed with 5 ml of dichloromethane and conditioned at 380° for 4 hr using hydrogen as carrier gas.

Conditions for Gas Chromatography–Mass Spectrometry

A Carlo Erba 4160 gas chromatograph was connected to a magnetic sector instrument (VG ZAB-HF, data system VG 11-250) with a home-made interface allowing temperatures up to 400°. The fused-silica columns (10 or 5 m) were introduced directly into the ion source and the tip positioned 1–2 mm from the electron beam. Helium was used as carrier gas with a head pressure of 0.15 bar and a linear gas velocity of 90 cm/sec at 87° for a 10-m column. A high-capacity gas purifier and an OMI-1 indicating purifier (both Supelco, Bellefonte, PA) were used in the carrier gas line. Samples dissolved in ethyl acetate were injected (1 μl; 1–100 ng/component) by the on-column technique. Injections were done at 87° or 200° and after a 1-min isothermal period the temperature was increased 10°/min up to 370°–400° and held for 5 min. The following conditions were used for the mass spectrometer: interface and ion source, 365°–400°; electron energy, 70 eV; trap current 500 μA; accelerating voltage, 8 kV; mass range scanned, m/z 1600–160; scan speed, 2 sec/decade; total cycle time, 3.5 sec; resolution 1400; pressure in the ion source region, 10^{-5} mbar. Fomblin was used as calibrating substance using the direct inlet system. All spectra were background subtracted.

Examples and Comments

Short capillary columns for GC have allowed the analysis of large permethylated oligosaccharides with up to 11 sugar residues and molecular masses up to 2300.[6–8] Due to the special coupling problems (cold and hot

[14] L. Blomberg, J. Buijten, K. Markides, and T. Wännman, *J. Chromatogr.* **239**, 51 (1982).

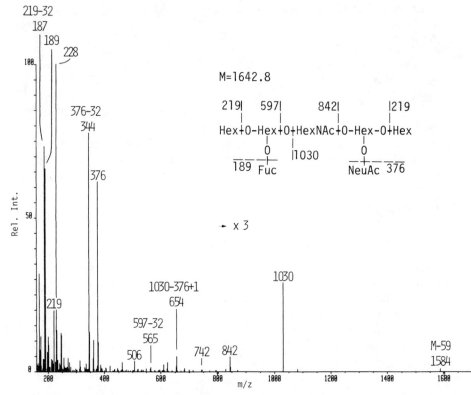

FIG. 1. Mass spectrum from GC/MS of a permethylated oligosaccharide released by ceramide-glycanase from a blood group B-active ganglioside based on G_{M1} obtained from rat stomach.[15] A 10-m column coated with 0.05 μm stationary phase was used, injection at 200°, temperature raised 10°/min up to 390° where the mass spectrum reproduced was recorded.

spots, length of interface, problems with short columns, etc.), it has not yet been possible to reach the equivalent size using GC/MS (the limit at present is eight to nine sugars). A longer column with a length of 10 m can be used for analyses up to seven to eight sugar residues and gives good resolution. A shorter column (5 m) can be used for larger saccharides using faster temperature programs to enhance the resolution and minimize the elution time. Evidence for thermal degradation of the permethylated oligosaccharides has been observed with longer elution times and temperatures higher than 400°.

Oligosaccharides released from neutral glycosphingolipids and gangliosides have been successfully analyzed,[7] as well as similar oligosaccharides found in free form in, for example, milk. Figure 1 shows a mass spectrum

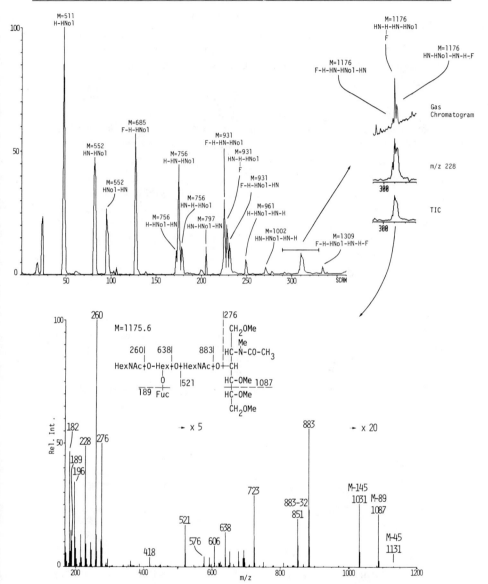

FIG. 2. GC/MS of permethylated neutral oligosaccharides released from porcine small intestinal mucin glycopeptides. The partial total ion current is shown on top with the major oligosaccharide structures marked. A portion of the chromatogram is enlarged (on the right-hand side) which shows a mass chromatogram of m/z 228 (260–32) and part of a gas chromatogram of the same sample. H, Hexose; HN, N-acetylhexosamine; F, fucose; HNol, for N-acetylhexosaminitol (from the GalNAc linked to the peptide). The sequence to the left of HNol is linked to C-3 and that to the right to the C-6 of the HNol. The mass spectrum of scan 311, the middle peak in the cluster, is shown at the bottom of the figure.

from GC/MS of a seven-sugar oligosaccharide containing sialic acid obtained from a purified blood group B ganglioside.[15] This eluted at 390° on a 10-m column. The endoglycoceramidase or ceramide-glycanase enzymes do not split monoglycosylceramides, but these small intact glycosphingolipids can be analyzed by GC/MS.[7]

Oligosaccharides released from mucins and other glycoproteins by alkaline-NaBH$_4$ treatment can be analyzed and the high resolution achieved is necessary for separating the complex mixtures found in nature.[6,8] An example from neutral oligosaccharides of porcine small intestinal mucins is shown in Fig. 2, where three isomeric pentasaccharides are resolved and characterized.

Permethylation using solid NaOH gives good yields for neutral oligosaccharides with a GalNAc-alditol, and neutral and sialic acid-containing oligosaccharides with a Glc-aldose. Oligosaccharides with a Glc-alditol give lower yields by permethylation, although these can be increased by modifying the permethylation procedure[8] (see above). Sialic acid-containing oligosaccharides with a GalNAc-alditol give poor yields.

Acknowledgments

This work was supported by grants from the Swedish Medical Research Council (No. 7461), the Swedish Board for Technical Development, and Gabrielsson's research fund. The mass spectrometer was supported by the Swedish Medical Research Council (No. 3967).

[15] J.-F. Bouhours, D. Bouhours, and G. C. Hansson, J. Biol. Chem. 262, 16370 (1987).

[40] Tandem Mass Spectrometry of Glycolipids

By CATHERINE E. COSTELLO and JAMES E. VATH

Introduction

The structural determination of glycolipids has been considerably advanced through the implementation of mass spectral methods, because of their sensitivity and the wealth of information they convey. There are, however, practical limitations to the utilization of the usual mass spectrometry-based approaches for the analysis of very small quantities of material, particularly when the sample consists of a mixture of closely related compounds. Tandem mass spectrometry (MS/MS) offers a means to address these difficulties, because it reduces the contribution from background and allows the selective determinations of individual components within a complex mixture. In addition, MS/MS spectra, particularly those of derivatives designed to favor selected fragmentation pathways, may include structurally diagnostic ions not present in the normal mass spectrum.

This chapter describes a systematic nomenclature for fragments observed in the mass spectra of glycolipids, the use of a high-performance tandem mass spectrometer (high mass range, with unit resolution or better available in both MS-1 and MS-2), and targeted derivatization procedures for glycolipid structural determinations. Investigation of the MS/MS spectra of underivatized glycolipids in the positive and negative ion modes has demonstrated the usefulness of the technique for structure elucidation at the microgram level. For smaller sample amounts and for compounds with poor fast atom bombardment (FAB) ionization or fragmentation characteristics, microscale derivatization procedures have been developed and the mass spectrometric properties of the derivatives investigated. The high-energy CID (5–10 keV) MS/MS properties of both the native and derivatized glycolipids are discussed and summaries are provided as guides for the analyst.

Low-energy collisions have been demonstrated to be informative about carbohydrate structure during investigations of the behavior of mono- and disaccharides.[1-3] These conditions may furnish additional information about the types of compounds discussed here, but no systematic exploration of the low-energy collision spectra of glycolipids has yet been conducted. Some of the discussion is also relevant to interpretation of the two-sector FAB, field desorption (FD), and direct chemical ionization (DCI) spectra of these compounds, those obtained by normal magnetic scans,[4-7] by linked scans of the electric and magnetic

[1] R. Guevremont and J. L. C. Wright, *Rapid Commun. Mass Spectrom.* **1**, 12 (1987).
[2] R. A. Laine, K. M. Pamidimukkala, A. D. French, R. W. Hall, S. A. Abbas, R. K. Jain, and K. L. Matta, *J. Am. Chem. Soc.* **171**, 6931 (1988).
[3] D. R. Mueller, B. Domon, W. Blum, and W. Richter, *Biomed. Environ. Mass Spectrom.* **15**, 441 (1988); see also [33] in this volume.
[4] C. E. Costello, B. W. Wilson, K. Biemann, and V. N. Reinhold, *in* "Cell Surface Glycolipids" (C. C. Sweeley, ed.), p. 35. ACS Symp. Ser. No. 128, American Chemical Society, Washington, D.C., 1980.
[5] T. H. Wang, T. F. Chen, and D. F. Barofsky, *Biomed. Environ. Mass Spectrom.* **16**, 335 (1988).
[6] M. E. Hemling, R. K. Yu, R. D. Sedgwick, and K. L. Rinehart, *Biochemistry* **23**, 5706 (1984).
[7] J. Kuei, G. R. Her, and V. N. Reinhold, *Anal. Biochem.* **172**, 228 (1988).

fields of a double-focusing instrument,[8-10] by mass-analyzed ion kinetic energy (MIKES) analysis,[11-12] or by use of triple quadrupole and hybrid instruments.

For the analysis of glycolipid mixtures, particularly when sample size is very limited, it is especially desirable to eliminate background interferences by using MS/MS. Unit mass resolution for the selection of the precursor ion (MS-1) in collision experiments is also important, because components that vary in the degree of unsaturation, the presence of amine or hydroxyl functions, or other modifications resulting in mass shifts of only a few daltons are often present as mixtures even in extensively purified extracts. The determination of base and fatty acid chain lengths of individual components, and localization of the site(s) of unsaturation or functional groups may be important for the understanding of biological activity. In addition, product ions may be present at adjacent m/z values even in the collision spectrum of a unit-resolved precursor, since multiple fragmentation pathways related to the various functional groups occur for these compounds. Therefore, while much of the information in this chapter is pertinent to the interpretation of spectra recorded under other instrumental conditions, limited mass resolution in either precursor selection or product ion spectra may result in some residual ambiguities and some of the fragmentations described may not be observed on low-energy collisions.

Nomenclature

Because the nomenclature systems previously employed for small carbohydrates were found to be unwieldy or inadequate when it became necessary to describe the range of fragments observed in the MS/MS spectra of large glycolipids, a nomenclature system for the designation of fragment ions in mass spectra and tandem mass spectra of glycolipids, oligosaccharides, and their derivatives has been developed[13-16] and will be

[8] H. Egge and J. Peter-Katalinić, *Mass Spectrom. Rev.* **6**, 331 (1987); also [38] in this volume.

[9] Y. Ohashi, M. Iwamori, T. Ogawa, and Y. Nagai, *Biochemistry* **26**, 3990 (1987).

[10] S. Ladisch, C. C. Sweeley, H. Becker, and D. Gage, *J. Biol. Chem.* **264**, 12097 (1989).

[11] R. B. Trimble, P. H. Atkinson, A. L. Tarentino, T. H. Plummer, Jr., F. Maley, and K. B. Tomer, *J. Biol. Chem.* **261**, 12000 (1986).

[12] G. Puzo, J.-J. Fournie, and J.-C. Prome, *Anal. Chem.* **57**, 892 (1985).

[13] B. Domon and C. E. Costello, *Biochemistry* **27**, 1534 (1988).

[14] B. Domon and C. E. Costello, *Glycoconjugate J.* **5**, 397 (1988).

[15] B. Domon, J. E. Vath, and C. E. Costello, *Anal. Biochem.* **184**, 151 (1990).

[16] J. E. Vath, B. Domon, and C. E. Costello, *Proc. 37th ASMS Conf. Mass Spectrom. Allied Topics, Miami Beach, FL* p. 770 (1989).

used throughout this discussion. The core of this system will be presented first, and supplementary designations added in the subsequent discussion sections. This nomenclature system is conceptually similar to that in use for the description of FAB mass spectra and tandem mass spectra of peptides[17,18] in that alphabetical symbols and numbers are used to designate bond cleavages along the backbone of the molecule. For glycoconjugates and oligosaccharides, A_n, B_n, C_n are used for fragments that retain the charge on the nonreducing end and X_n, Y_n, Z_n are used for fragments that retain the charge on the reducing (lipid, peptide) end. Superscripts preceding the symbol are used to define cleavages within a carbohydrate ring, as discussed below. Lipid fragments are designated with other letters from the latter part of the alphabet. When lower case symbols are used for the peptide moiety, the system accommodates glycopeptides. The occurrences of these ion types are discussed in the individual sections that describe the spectral characteristics of underivatized and derivatized species. The same letters are used to designate positive and negative ions of similar origin. Hydrogen transfers may result in mass differences between ions of the same letter but opposite charge. When the number of hydrogen transfers that accompany bond-breaking has been found to be constant, the definition of the symbol assumes the transfer, e.g., $B_i^+ = [B_i]^+$ but $B_i^- = [B_i - 2H]^-$. If the number of hydrogens transferred is a variable or differs from the definition, then the number needs to be specified when the symbol is utilized, e.g., $^{3,5}A_4 - 3H$.

Ceramide Portion

Scheme 1 indicates the bond fragmentations and hydrogen transfers included in the designations for product ions of ceramides and the ceramide portion of glycosphingolipids. Scheme 1A shows the product ions observed in the positive ion collision-induced desorption (CID) mass spectra of native and reduced compounds; Scheme 1B shows the product ions observed in negative ion CID mass spectra of native compounds, and includes representations for two important ions that arise via multiple bond cleavages.

Carbohydrate Portion

Because structural variations and, therefore, fragmentation pathways (branching, ring cleavages) are more complex in carbohydrates than in peptides, a more elaborate nomenclature system is necessary. An addi-

---

[17] P. Roepstorff and J. Fohlman, *Biomed. Environ. Mass Spectrom.* **11,** 601 (1984).
[18] Appendix 5, at the end of this volume.

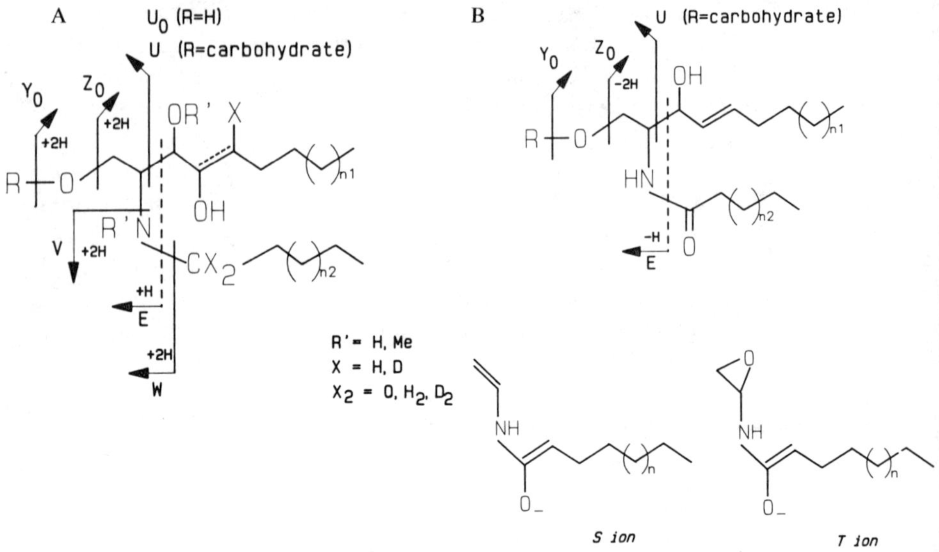

SCHEME 1. Designations for product ions that arise from fragmentations within the ceramide portion of glycolipids. (A) Positive ion fragmentations; (B) negative ion fragmentations. Designations of ion types include the hydrogen transfers indicated.

tional subscript (αn, βn, etc.) has been added to indicate branching (αn is the fragment originating from cleavage within the longest branch). Products of sequential cleavages are designated with a slash in the subscript, e.g., $\alpha n_1/n_2$. Left superscripts indicate the bonds broken during ring cleavages, as shown in Scheme 2.

Materials and Methods

Glycolipids

Lactosylceramide [Gal($\beta 1 \rightarrow 4$)Glc($\beta 1 \rightarrow 1$)Cer, $n16:0/18:0$], galactosylceramide [Gal($\beta 1 \rightarrow 1$)Cer, $n16:0/18:1$], N-palmitoyl-4E-sphingenine, N-oleoyl-4E-sphingenine, G_{M1} {Gal($\beta 1 \rightarrow 3$)GalNAc($\beta 1 \rightarrow 4$)[NeuAc-($\alpha 2 \rightarrow 3$)]Gal($\beta 1 \rightarrow 4$)Glc($\beta 1 \rightarrow 1$)Cer}, and G_{T1b} {NeuAc($\alpha 2 \rightarrow 3$)-Gal($\beta 1 \rightarrow 3$)GalNAc($\beta 1 \rightarrow 4$)[NeuAc($\alpha 2 \rightarrow 8$)NeuAc($\alpha 2 \rightarrow 3$)]Gal($\beta 1 \rightarrow 4$)-Glc$\beta 1 \rightarrow 1$Cer} were obtained from Sigma Chemical Company, St. Louis, MO. The phospholipid samples were a gift from B. N. Singh, SUNY, Syracuse, NY. Sources of the reagents used for derivatizations are noted as the procedures are described.

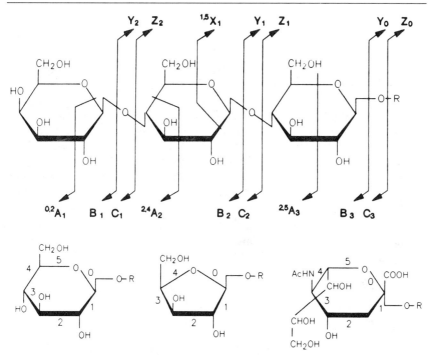

SCHEME 2. Designations for product ions that arise from fragmentations within the carbohydrate portion of glycolipids. Because the number of hydrogen transfers during ring cleavages (A_n, X_n) can vary, these are not included in the designations, but must be indicated in each case. Fragments that arise via glycosidic cleavages include predictable numbers of hydrogen transfers and the designations are thus defined: $B_i^+ = [B_i]$, $C_i^+ = [C_i + 2H]$, $Y_j^+ = [Y_j + 2H]$, $Z_j^+ = [Z_j]$; $B_i^- = [B_i - 2H]$, $C_i^- = [C_i]$, $Y_j^- = [Y_j]$, $Z_j^- = [Z_j - 2]^-$.[14]

Mass Spectrometry

All spectra shown were recorded with a JEOL HX110/HX110 ($E_1B_1 - E_2B_2$) tandem mass spectrometer, operated at ±10 kV accelerating voltage, with ±18 kV postacceleration at the detector. High-energy collisions (5 to 10 keV) with helium in a collision cell (that may be electrically floated at a selected potential) located between MS-1 and MS-2, at a pressure sufficient to reduce the precursor ion abundance to about 20% of its initial value, were employed. The primary ion beam was either 6 keV xenon neutrals or 6–8 keV Cs$^+$ ions. (The choice of primary beam had little effect on the spectra, and was made on the basis of compatibility with other ongoing experiments.) Unless otherwise noted, the spectra were recorded with an electron multiplier during linked scans of MS-2 at constant B/E. The JEOL DA5000 data system provided instrument control and data acquisition and processing. MS-1 was operated at (or above) unit resolution. MS-2 was usually operated at 1 : 1000 resolution. Scan time was 1 min 30 sec to 2 min 30 sec, with filter (30–300 Hz) chosen so as to preserve mass resolution. Sample solutions (1–5 μg/μl) of underivatized compounds were mixed with matrix (1 : 1) and 0.3–0.5 μl of this solution was applied to a stainless steel FAB probe tip (3-mm diameter). For derivatized compounds, sample handling was modified and less sample was required, as noted in the discussions below. In order to allow for postmeasurement selection of parameters such as peak detection threshold and centroid vs. peak maxima mass assignments, and for examination of peak shapes and signal-to-noise (S/N) ratio, all spectra were recorded (and are presented) in the profile mode. Spectra shown are single scans, or the sum of a few scans added together after the measurement. Mass assignment accuracy is normally within 0.3 u of the calculated values.[19] In the figures, the calculated mass values for fragments are indicated on the structures and the experimentally observed values are indicated on the spectra and used in the text. Fractional mass values are included on spectra of compounds above M_r 1000, which have fragments whose fractional mass approaches or exceeds 1 dalton (Da), in order to remove any ambiguity about the assignments.

MS/MS of Underivatized Compounds

Positive Ion MS/MS Spectra

Information about the structure of the aglycon is most readily available from the positive ion MS/MS spectra of the $(M + H)^+$, $(M + cation)^+$, or aglycon-containing fragment ions. The behavior of $(M + H)^+$ ion is easiest to anticipate because prediction of its charge localization is more straightforward. Rules for the interpretation of the CID spectra have therefore been formulated primarily for $(M + H)^+$ ions.[13] Gangliosides and other negatively charged glycolipids are very prone to cationization, forming $(M + Na)^+$ and $(M + K)^+$, and usually do not yield sufficient $(M + H)^+$ ion abundance for MS/MS analysis of the underivatized compounds.

  Ceramides and Neutral Glycosphingolipids. FAB sensitivity for underivatized ceramides, even in concentrated solutions, is poor and derivatization is therefore recommended (see below). For glycosphingolipids that contain one or two sugars, the situation is much improved and 0.5–5 μg of sample provides spectra of excellent quality. The CID spectrum of the

[19] K. Sato, T. Asada, F. Kunihiro, Y. Kammei, E. Kubota, C. E. Costello, S. A. Martin, H. A. Scoble, and K. Biemann, *Anal. Chem.* **59**, 1652 (1987).

$(M + H)^+$ ion, m/z 864.6 of lactosyl-N-palmitoylsphinganine is shown in Fig. 1A. We have obtained the field desorption (FD)-CID MS/MS spectrum of lactosyl-N-palmitoylsphinganine $(M + H)^+$ ion and confirmed that the high-energy CID spectrum is quite independent of the ionization mode, i.e., the FAB-MS/MS and FD-MS/MS spectra are virtually identical. Glycosidic bond fragmentations within the carbohydrate give rise to low abundance ion(s) (m/z 702.6). The ion corresponding to the protonated aglycon (Y_0, m/z 540.5) and its fragments generally have much higher abundances. In the CID spectra of compounds that contain sphinganine, $(M + H - H_2O)^+$ has only low abundance. Both the W and W' ions (m/z 302.3 and m/z 284.3) are present and indicate the length of the base. An abundant V ion (m/z 256.3) indicates the size of the fatty acyl group, and is accompanied by a V' ion if an hydroxyl group is present in the fatty acid. Remote site cleavages[20] along the hydrocarbon chains lead to low abundance ions observed at masses below the aglycon fragment. When the base is (4E)-sphingenine (henceforth called sphingenine), water loss from the molecular ion or the aglycon is extremely facile, and results in the most abundant product ions. For these compounds, the W ion is not observed, but W' and W'' are present and indicate the chain length of the base. The abundances of these ions are much greater than that of the V ion. The weight of the fatty acyl group can still easily be determined, however, by subtraction of the mass of the base from that of the aglycon.

If modifications within the carbohydrate portion of the glycosphingolipid favor fragmentation pathways involving the carbohydrate moiety, thereby diminishing the relative abundance of lipid-related ions in the MS/MS spectrum of $(M + H)^+$, CID spectra of the Y_0 or Y_0' ions can still yield the desired information about the lipid structure. These ions have CID spectra similar to those of the corresponding ceramides, with some variation in relative abundances, due to the preferential formation of one isomer when multiple structures of the fragment are possible. For example, in the case of ceramides and simple glycosphingolipids, the Y' ion may be formed by loss of either hydroxyl group, but loss of the 3-hydroxyl group seems to be preferred. Figure 2A shows the CID spectrum of the Y_0' ion (m/z 520.5) from galactosyl-N-palmitoylsphingenine ($n16:0/18:1$). This spectrum is dominated by the W' (m/z 282.3) and W'' (m/z 264.3) ions originating from the Y_0' ion which has undergone dehydration through loss of the 3-hydroxyl group. The ions at m/z 250.3 and 252.3 are derived by loss of methanol and formaldehyde, respectively, from the W' ion. The low abundance S ion at m/z 280.3 would arise by loss of HOCH=CH-$(CH_2)_{13}CH_3$ from the $Y_0' = Z_0$ ion. When the carbohydrate moiety is

[20] L. J. Deterding and M. L. Gross, *Org. Mass Spectrom.* **23**, 169 (1988).

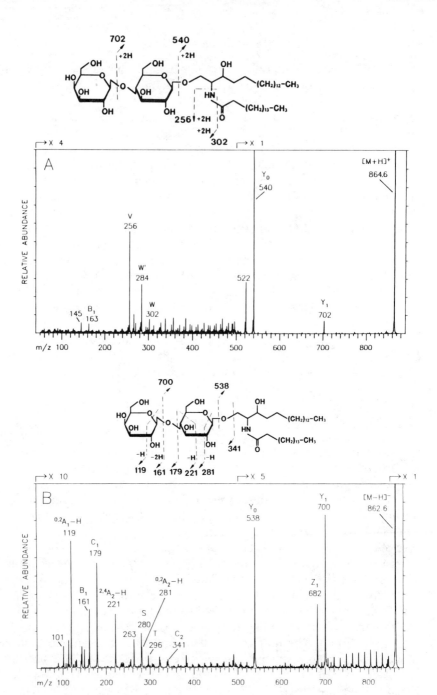

FIG. 1. FAB-CID-MS/MS spectra of lactosyl-*N*-palmitoylsphinganine in DMSO. (A) (M + H)⁺, *m/z* 864.6, glycerol matrix. (B) (M − H)⁻ *m/z* 862.6, triethanolamine matrix. Collision energy in both spectra is 10 keV.

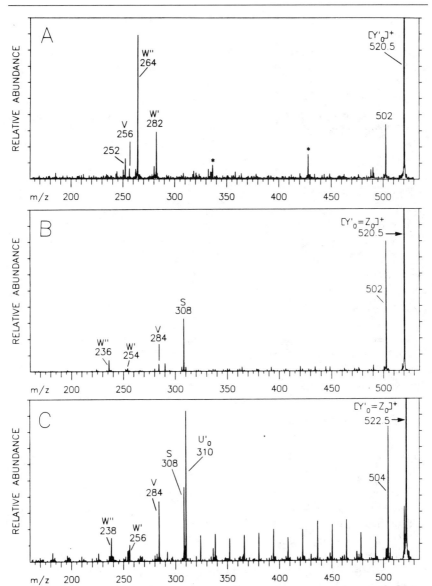

FIG. 2. CID-MS/MS spectra of Y_0' fragment ions in the positive ion FAB mass spectra of glycosphingolipids. (A) m/z 520.5 in the spectrum of galactosyl-N-palmitoylsphingenine ($n16:0/18:1$), M_r 699.6 in DMSO/glycerol; (B) m/z 520.5 in the spectrum of phosphoinositol N-stearoylhexadecasphingenine ($n18:0/16:1$), M_r 779.5, in chloroform/methanol/glycerol, $3:1:4$[21]; (C) m/z 522.5 in the spectrum of phosphoinositol N-stearoylhexadecasphinganine ($n18:0/16:0$), M_r 781.5 in chloroform/methanol/glycerol, $3:1:4$.[21] Collision energy in all three spectra, 10 keV.

attached through a phosphate diester linkage, the apparent Y_0' ion is the most abundant fragment and is formed preferentially by direct loss of glycophosphoric acid, i.e., the carbohydrate including the glycosidic oxygen (equivalent to the Z-type cleavage), as indicated by the appearance of the abundant S fragment in the CID spectrum of Y_0'. (The S designation has previously been used for the negative ion mode, and is used here to designate a positive ion that has the same composition.) Figure 2B shows the CID spectrum of the Y_0' ion (m/z 520.5) from phosphoinositol N-stearoylhexadecasphingenine ($n18:0/16:1$)[21] and Fig. 2C shows the CID spectrum of the Y_0' ion (m/z 522.5) from phosphoinositol N-stearoylhexadecasphinganine ($n18:0/16:0$).[21] Mechanisms for formation of the W', S, and U_0' ions are shown in Scheme 3.

Other Glycolipids. Steroid and terpenoid glycosides yield MS/MS spectra whose general features are similar to the spectra of glycosphingolipids. The base peak in both the normal positive ion FAB mass spectrum and the CID-MS/MS spectrum of the $(M + H)^+$ ion is usually due to the aglycon, but most of the other product ions from $(M + H)^+$ originate from glycosidic cleavages. CID of the aglycon peak produces an extensive fragmentation pattern much like that observed upon electron ionization (EI). For CID spectra of salt-cationized species, the distribution of ion current depends on the charge localization.[22] Likewise, when amino sugars are present, cleavage at the adjacent glycosidic linkage is favored. The CID spectra of glycerophosphoinositols contain abundant ions related to the structures of the acyl and alkyl substituents.[21]

Negative Ion MS/MS Spectra

The negative ion collision spectrum is especially rich in fragments that provide information about the carbohydrate portion of glycolipids. Negative ion spectra are less subject to suppression by endogenous cations such as Na^+ and K^+ and are, therefore, more likely to be useful for the analysis of underivatized gangliosides and of glycolipids that contain sulfate and phosphate.

Neutral Glycosphingolipids. The CID-MS/MS spectrum of the $(M - H)^-$ ion, m/z 862.6, from lactosyl-N-palmitoylsphinganine, shown in Fig. 1B, has features typical for glycolipids that contain neutral sugars, including amino sugars. Ions of the Y series dominate the high-mass portion of the spectrum. At low mass, the prominent ions arise by cleavages within

[21] B. N. Singh, C. E. Costello, D. H. Beach, and G. G. Holz, *Biochem. Biophys. Res. Commun.* **157**, 1239 (1988).

[22] C. E. Costello, B. Domon, C.-H. Zeng, S. Du, F. Qian, and M. Li, *Proc. 35th ASMS Conf. Mass Spectrom. Allied Topics, Denver, CO* p. 1 (1987).

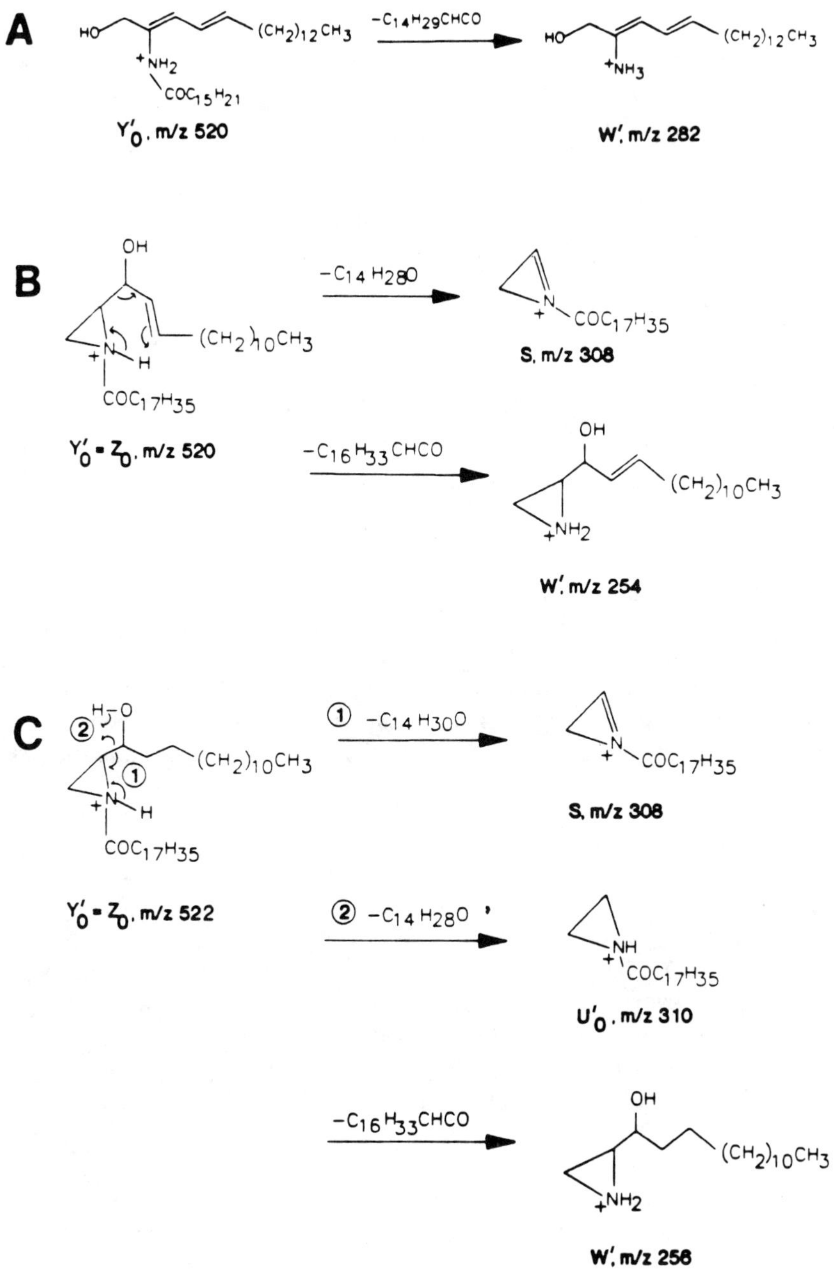

SCHEME 3. Formation of W, S, and U_0' ions from Y_0' precursors.

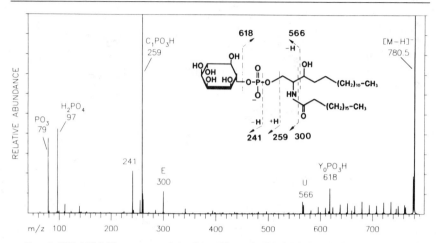

FIG. 3. CID-MS/MS spectrum of the (M − H)⁻, m/z 780.5, of phosphoinositol *N*-stearoyl-hexadecasphinganine (*n*18 : 0/16 : 0) isolated from *Leishmania*,[21] in chloroform/methanol/triethanolamine, 3 : 1 : 4 (v/v/v). Collision energy, 7 keV.

the carbohydrate rings as well as by glycosidic cleavages. Assignments of these ions are indicated on the spectrum. In the spectra of ceramides and cerebrosides that have small carbohydrate moieties, low abundance ions (designated *S* and *T*) characterize the lipid portion of the molecule. These ions do not have useful abundances in the spectra of glycosphingolipids that contain four or more sugars, but CID-MS/MS spectra of the abundant Y_n series fragments ($n = 0 - 2$) of the normal FAB mass spectrum may be used to determine the ceramide structures of such compounds. In the negative ion CID spectra, the series of remote site cleavage ions begins at the molecular ion, rather than from the aglycon, as was described above for positive ion spectra.

Ions related to phosphoric acid (m/z 97.0, $H_2PO_4^-$ and m/z 79.0, PO_3^- and the $(B_n + 80)^-$ and $(C_n + 80)^-$, where *n* equals the number of sugar rings, dominate the negative ion MS/MS spectra of phosphoglycolipids. Figure 3 shows the CID-MS/MS spectrum of the (M − H)⁻ ion, m/z 780.5, of phosphoinositol *N*-stearoylsphinganine (*n*18 : 0/16 : 0), one of several closely related glycolipids from *Leishmania*.[21] Cleavage adjacent to the phosphate at the glycosidic bond results in an abundant Y_0PO_3H ion, m/z 618.5, and cleavages within the ceramide yield U_0^- (m/z 566.3) and *E*-type (m/z 300.0) products, all retaining the phosphate group.

Gangliosides. Localization of the negative charge on the carboxyl group of sialic acid results in a drastic change in the spectra of gangliosides compared to those of cerebrosides, so that the MS/MS spectrum of the

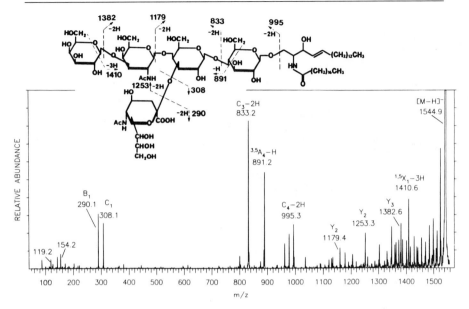

FIG. 4. FAB-CID-MS/MS spectrum of the $(M - H)^-$ ion of the lower molecular weight G_{M1} homolog, m/z 1544.9. Collision energy, 7 keV.

$(M - H)^-$ ion is dominated by ions related to the sialic acid moiety.[13] Figure 4 shows the CID-MS/MS spectrum of the $(M - H)^-$ ion, m/z 1544.9, recorded for the lower homolog of G_{M1}. The most abundant product ions are those that arise by $(C - 2H)$-type glycosidic cleavage beyond the carbohydrate ring that carries the sialic acid substituent (m/z 833.2, m/z 995.3) and by $^{3,5}X$-type cleavage of the carbohydrate ring adjacent to the one substituted (m/z 891.2). Y- and Z-type fragments are also observed only beyond the sialic acid attachment site (m/z 1382.6 and 1179.4, m/z 1161.4). Loss of sialic acid results in a fragment at m/z 1253.3. Abundant ions at m/z 290.1 and m/z 308.1 represent $(M - H)^-$ and $(M - H - H_2O)^-$ for sialic acid, and lower mass product ions are mostly due to fragmentation within the sialic acid moiety.

When more than one sialic acid group is present, the CID-MS/MS spectra of the $(M - H)^-$ or $(M - 2H + Na)^-$ ions allow differentiation of the isomeric possibilities.[13] Information about details of the ceramide structure (other than its total weight) is lacking in the normal FAB mass spectra and the $(M - H)^-$ collision spectra of gangliosides but, once again, may be obtained from CID spectra of ions in the Y_n series ($n = 0 - 2$) in the FAB mass spectrum. Figure 5 shows the CID-MS/MS spectrum of the Y_0 ion, m/z 564.5, of the lower homolog of G_{M1}. This feature has recently

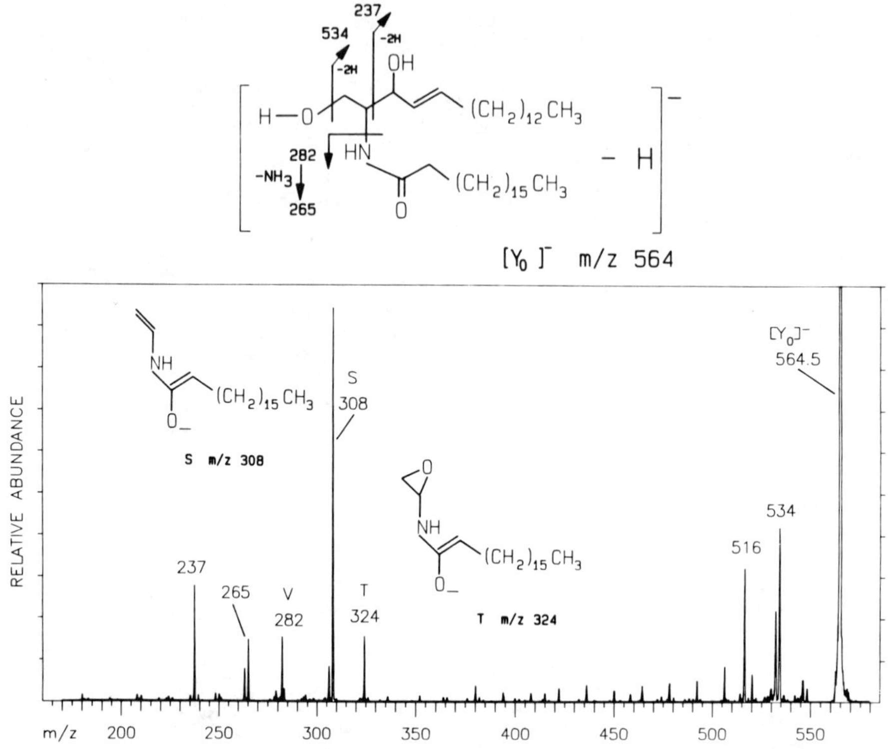

FIG. 5. FAB-CID-MS/MS spectrum of the Y_0 fragment ion, m/z 564.5, due to ceramide portion of the lower homolog, in the FAB mass spectrum of G_{M1}. Collision energy, 7 keV.

been exploited by Sweeley and co-workers for the analysis of gangliosides in tumor cells.[10] The sensitivity for underivatized gangliosides is much lower than that observed for glycolipids which contain only a few sugars and the possibility for adverse effects from the presence of endogenous salts makes CID-MS/MS analysis more problematic for gangliosides. Derivatization overcomes many of these difficulties, improves sensitivity, and introduces control over fragmentation pathways, as described below.

Other Glycolipids. Fragmentation within the carbohydrate portion of other types of glycolipids follows the general patterns described for glycosphingolipids. Unless the lipid has a very favorable site for charge localization, the product ion abundance will generally be distributed primarily over fragments that arise as a result of bond cleavages within the carbohydrate moiety. Some minor fragmentation may also occur within the lipid portion, and may provide unique information about its structure.[22] The most abundant ions in the CID spectra of glycerophosphoinositol compounds are those related to the phospho sugar, and, therefore, resemble the spectrum shown in Fig. 3, but also having abundant ions that correspond to the carboxylate anions of fatty acid substituents.[21]

MS/MS of Reduced and Permethylated Compounds

Derivatization Methods

In many instances it is desirable to derivatize glycolipids prior to MS/ MS analysis. Derivatization can be used to improve secondary ion yield and to direct fragmentation. The types of derivatization that will be described here are permethylation of amide and hydroxyl groups (to improve sensitivity for gangliosides in the positive ion mode), borane reduction of amides to amines (to increase sensitivity and improve MS/MS fragmentation characteristics of both ceramides and glycolipids), hydroboration of olefins (to locate their positions in the lipid chains), and the combined use of permethylation and borane reduction.

Borane reduction of N-acyl groups is performed with condensed reagent vapor and the sample requirement for this reaction is thus decreased to the picomole level. Reactions performed in this manner are able to efficiently convert very small sample amounts into a derivative in a form that permits complete transfer of the sample to the mass spectrometer for GC, EI, and FAB. Contamination is minimized since the sample comes in contact only with minute quantities of the volatile components of the reagent(s). Losses due to surface adsorption and sample transfers are minimized because the sample remains exposed to only a small area on the inside of a capillary tube support during derivatization. The sample is transferred only once, to the instrument for measurement. The additional reactivity of borane with carbon–carbon double bonds in lipids to yield an alkylborane permits the location of unsaturation in the acid and/or sphingoid base of sphingolipids following its conversion to an alcohol using a reagent vapor oxidation step. The reaction conditions employed in the following derivatization reactions and their effect on all of the common functional groups found in glycosphingolipids are summarized in Table I.

Permethylation. The formation of permethylated derivatives for the analysis of glycolipids by positive ion FAB-MS is a well-established strategy.[8,23] The increased sensitivity upon permethylation (about 20- to 50-fold) allows the detection of submicrogram quantities of materials. For MS/MS, increased secondary ion yield is important because the quality of the product ion spectrum obtained upon CID is directly linked to the

---

[23] A. Dell, *Adv. Carbohydr. Chem. Biochem.* **45,** 19 (1987); see also [35] in this volume.

TABLE I
REACTIVITY OF GLYCOLIPID FUNCTIONAL GROUPS UNDER VARIOUS CONDITIONS FOR PERMETHYLATION AND "REDUCTION/HYDROBORATION"[a]

Functional group	Permethylation CH$_3$I/NaOH in DMSO 1 hr room temp.	Borane reactions	
		1 N BX$_3$ in THF[a] 10–30 min 45°	1 N BX$_3$ in THF[a] 2 hr 45°
—CO—NH— Ceramide	—CO—N(CH$_3$)—	—CX$_2$—NR—[b]	—CX$_2$—NR—
CH$_3$—CO—NH— N-acetylhexosamine, sialic acid	CH$_3$—CO—NR— (nr)[c]	CH$_3$—CO—NR—	CH$_3$—CX$_2$—NR—
—CH=CH— Ceramide	—CH=CH— (nr)	—CHOH—CHX— —CHX—CHOH—	—CHOH—CHX— —CHX—CHOH—
—CHOH—CH=CH— Ceramide	—CH(OCH$_3$)—CH=CH—	—CHOR—CHOH—CHX—	—CHOR—CHOH—CHX—
—COOH Sialic acid	—COOCH$_3$	—COOR (nr)	—COOR (nr)
—OH Ceramide, carbohydrate	—OCH$_3$	—OR (nr)	—OR (nr)

[a] X, H or D.
[b] R, CH$_3$ after permethylation; R, H if permethylation step is omitted.
[c] (nr), no reaction.

amount of precursor ion available, among other variables. The goal of many of the derivatization strategies described here was to develop methods where sample waste is minimized. Generally, the techniques developed start with one to two times the sample amount required for a single MS/MS experiment and consume all of the derivatized sample in the measurements (FAB-MS and MS/MS). The permethylation technique best suited for this approach is essentially that of Ciucanu and Kerek[24] as modified by Gunnarsson.[25] The sample (1–10 μg) is dried in the bottom of a screw-top culture tube and subsequently dissolved in 200–300 μl of anhydrous dimethyl sulfoxide (DMSO) (Pierce Chemical, Rockford, IL) which contains about 10 mg of finely powdered NaOH. The DMSO/NaOH solution is kept as a stock solution prepared by finely pulverizing NaOH pellets with a mortar and pestle and combining with dry DMSO in a dry culture tube under a nitrogen atmosphere to form a saturated solution. The stock solution is vortexed and the 200–300-μl aliquot for the reaction is drawn from the area of cloudy suspension above the solid. Such a stock solution can be used reliably for several weeks, if the solution is discarded when discoloration or odor develop. The sample in the DMSO/NaOH solution is then sonicated for 5 min at room temperature. Methyl iodide (Aldrich Chemicals, Milwaukee, WI) (10–20 μl) is added and the solution sonicated at room temperature for an additional hour. Water (1 ml) containing 1–5% acetic acid (enough to neutralize the NaOH) is added to the reaction mixture. The neutralization of the NaOH is important to avoid methyl ester hydrolysis in the NeuAc residues of gangliosides. The samples are purified on Sep-Pak cartridge (Millipore Corp., Milford, MA) which has been successively conditioned with 20–50 ml chloroform, 20 ml methanol, 5 ml acetonitrile, and 5 ml water. Adequate washing of the cartridge with chloroform is important for small sample amounts in order to minimize silica contamination from the cartridge which can be detrimental to FAB-MS analysis. The entire sample is applied to the conditioned cartridge which is then washed with 5 ml water, 5 ml acetonitrile, and 1 ml methanol. The sample is eluted with 1 ml of chloroform and this fraction is concentrated to about 50 μl by evaporation with a nitrogen stream. The solution is either transferred to a melting point capillary tube for further reagent vapor derivatization (described below) or 1–2 μl of 3-nitrobenzyl alcohol is added and the sample concentrated *in vacuo* for FAB-MS and MS/MS analysis.

Borane Reduction. Permethylated, reduced derivatives of glycosphingolipids have made an important contribution to extending the upper limit

[24] I. Ciucanu and F. Kerek, *Carbohydr. Res.* **131,** 209 (1984).
[25] A. Gunnarsson, *Glycoconjugate J.* **4,** 239 (1987).

of the mass range and improving the information contained in the EI mass spectra of these compounds.[26-27] Similar reduced and permethylated–reduced derivatives of sphingolipids have also proved useful for FAB-MS/MS. For this application, the borane reduction and associated reactions (solvolysis and oxidation) are conducted with condensed reagent vapor derivatization.[28] With this technique the sample is deposited as a solid on a small area of the inner wall of a capillary tube. A 1- to 20-μl aliquot of sphingolipid solution [either native or permethylated in 2 : 1 (v/v) chloroform/methanol solution] is placed into the bottom of a melting point capillary (100 mm long, 1.5 mm id) using a 10-μl fused-silica capillary syringe (J&W Scientific, Folsom, CA) as shown in Fig. 6A, left-hand side. The tube is placed in a vacuum centrifuge and the solvent evaporated *in vacuo* leaving a thin film of solid on the walls at the bottom of the capillary tube (Fig. 6A, right-hand side). The sample tube is placed into an apparatus for conducting reagent vapor reactions (shown in Fig. 6B). The apparatus has a side arm well which contains about 200 μl of the appropriate reagent or reagent mixture. The reagent well is cooled with either dry ice or liquid nitrogen and the apparatus is evacuated (<50 μm). The apparatus is then sealed at the stopcock, the coolant is removed from the reagent well, and the entire apparatus placed into an oven at the appropriate temperature to effect the reaction. After suitable time to complete the reaction, the apparatus is removed from the oven and the reagent well is again cooled. The apparatus is then evacuated to remove any residual reagents or volatile by-products of the reaction. The sample in the tube can either be prepared for FAB-MS as described below or transferred to another apparatus loaded with the next set of reagents for a subsequent derivatization step.

The reduction step is carried out by reacting the sample on the capillary tube with the vapor from a 1 M BH$_3$ or BD$_3$ solution in tetrahydrofuran (Alpha Products, Beverly, MA) at 45° for 10 min to 2 hr, depending upon the sample type and the degree of amide reduction desired (see Table I). The boron complexes of amine and hydroxyl groups which result from the above reaction are solvolyzed with the vapor from a 0.5 N HCl solution in methanol (prepared by bubbling HCl through dry methanol or available commercially from Supelco, Bellefonte, PA) for 3 hr at room temperature. For samples which contain olefinic bonds, an alkaline oxidation step is required to convert the alkyl borane functions to the corresponding alcohol. This is accomplished by reaction with the vapor from a solution of

[26] K.-A. Karlsson, *Biochemistry* **13,** 3643 (1974); see also [34] in this volume.
[27] G. Larson, H. Karlsson, G. C. Hansson, and W. Pimlott, *Carbohydr. Res.* **161,** 281 (1987).
[28] J. E. Vath, M. Zollinger, and K. Biemann, *Fresenius Z. Anal. Chem.* **331,** 248 (1988).

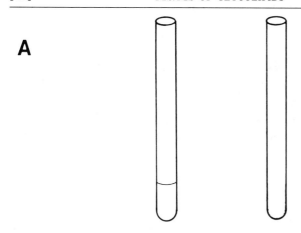

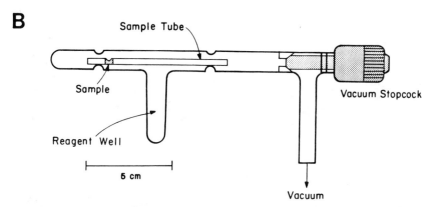

FIG. 6. Apparatus for microscale derivatizations using vapor-phase reagents. (A) Placement of sample within capillaries; (B) reaction vessel. (Reprinted with permission from Ref. 28.)

1 : 1 30% aqueous H_2O_2/3 N aqueous ammonia for 1 hr at 45° (for ceramides and small neutral glycosphingolipids) or 6 hr at 45° for gangliosides.

Following the reagent vapor reactions, the samples are prepared for FAB-MS by placing 5–10 μl of a 10% solution of 3-nitrobenzyl alcohol (Aldrich, Milwaukee, WI) in 3 : 1 (v/v) chloroform/MeOH in the sample tube using a fused-silica syringe. The tube is then placed into a vacuum centrifuge and the sample is concentrated into the FAB matrix *in vacuo*. A 0.3- to 0.5-μl aliquot of the sample solution in matrix is retrieved from

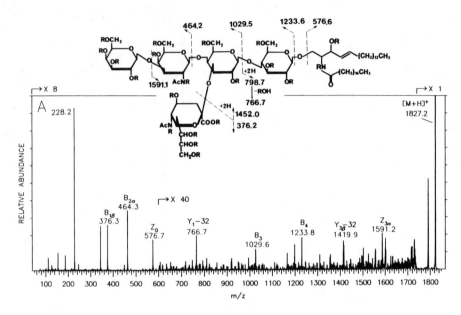

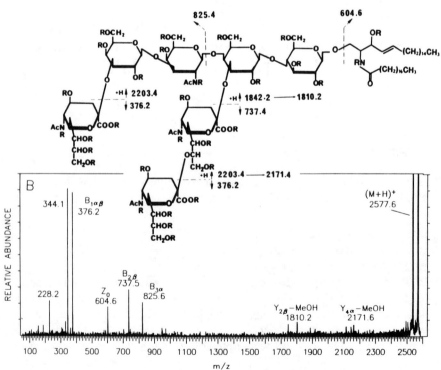

the capillary tube with the fused-silica syringe and placed onto the FAB probe which has been previously coated with a very thin layer of glycerol (0.1 μl) in order to increase the longevity of the sample signal. In this manner the reagent vapor derivatized sample can either be used entirely or in portions for FAB-MS and MS/MS, which are often performed at different times.

MS/MS Spectra of Derivatized Glycosphingolipids

Permethylated Gangliosides. As previously stated, the MS/MS analysis of low microgram or submicrogram quantities of gangliosides in the positive ion mode usually requires permethylation. An example of a FAB-MS/MS spectrum of such a derivative is given for the lower homolog of G_{M1}, $(M + H)^+$ m/z 1827.2, in Fig. 7A. The product ion spectrum is primarily composed of fragment ions resulting from cleavages of the interglycosidic linkages (*B* and *Y* ions) with minimal fragmentation in the ceramide portion of the molecule. The *B* ions occurring adjacent to the amino sugars ($B_{2\alpha}$ at m/z 464.3 next to GalNAc and $B_{1\beta}$ at m/z 376.2 next to NeuAc) are particularly abundant fragment ions that indicate to which residue sialic acid is attached. Even though the spectrum shown in Fig. 7A is informative, it is not ideal for low-level structural characterization by MS/MS. The center and high mass portion of the spectrum, containing much of the sugar sequence information, is very weak in this derivative and with slightly less material, or with larger molecules, this region would disappear. This effect can be observed in the FAB-MS/MS spectrum of permethylated G_{T1b}, $(M + H)^+$ m/z 2577.6, in Fig. 7B. For this spectrum, the *B* ions are once again very informative for the sialic acid attachment. In addition to the B_1 ion at m/z 376.2, the serial attachment of two sialic acids beyond the GalNAc residue is indicated by the $B_{2\beta}$ ion at m/z 737.5 and the fact that only one of the sialic acids present is attached at the nonreducing end of the carbohydrate is apparent from the $B_{3\alpha}$ ion at m/z 825.6. The middle and high mass regions of this spectrum are extremely weak considering the amount of material used to obtain this spectrum is the same as that in Fig. 7A for permethylated G_{M1} (5 μg). The only structurally informative ions that appear above the noise in this region are $Y_{2\beta}$–methanol and $Y_{4\beta}$–methanol at m/z 1810.2 and 2171.6, respectively. Despite the increased signal size over the native compound, the fragmenta-

---

FIG. 7. FAB-CID-MS/MS spectra of permethylated gangliosides at 7 keV collision energy. (A) $(M + H)^+$ m/z 1827.2 of the lower molecular weight homolog of permethylated G_{M1}. (B) $(M + H)^+$ m/z 2577.6 of the lower molecular weight homolog of permethylated G_{T1b}.

tion characteristics of the permethylated gangliosides in MS/MS limits the amount of sequence information for small sample amounts and those with extensive carbohydrate moieties. The combined use of amide reduction with permethylation, described below, is able to overcome some of these problems.

Amide-Reduced Glycolipids. For ceramides and small neutral glycosphingolipids there is a considerable sensitivity enhancement upon reduction of the amide group in the sphingolipid to an amine. Borane reduction, as opposed to reduction with lithium aluminum hydride, offers additional reactivity with alkenes as well as the ability to conduct the derivatization at much lower levels. Figures 8A and 8B compare the FAB-MS spectrum of 1 μg of N-palmitoylsphingenine [$(M + H)^+$ m/z 538.5] with that of 10 ng of the BH_3-derivatized sample [$(M + H)^+$ m/z 542.5], respectively. The most striking difference between the spectra of these two samples is the enhancement in ion yield upon derivatization. For ceramides, the increase in sensitivity upon reduction is about 10^3 while for monoglycosyl glycosphingolipids it is about 10^2. The ion at m/z 520.5 in the spectrum of the underivatized compound is the result of dehydration of the $(M + H)^+$ species while the ion at m/z 526.5 in the spectrum of the derivatized compound is due to an elimination–rehydroboration by-product of derivatization that is typical for the hydroboration of allylic alcohols[29] such as that present in most sphingolipids. This by-product is most noticeable in derivatized ceramides where it represents about 20% of the desired product (its yield is much lower during derivatization of cerebrosides and gangliosides). The sensitivity enhancement of the major product over that observed for the underivatized compound is such that the formation of the by-product is only a minor inconvenience. The major product of hydroboration and oxidation of the allylic alcohol is a vicinal diol by virtue of the directing influence of the hydroxyl group in the 3-position of the sphingoid base. This product can be differentiated from the naturally occurring 4-hydroxysphinganine (phytosphingosine) when the reduction is performed with BD_3 because a deuterium will be placed at the nonhydroxylated carbon atom (C-5).

As opposed to allylic carbon–carbon double bonds, hydroboration and oxidation of an isolated olefinic bond results in a mixture of two isomeric alcohols with the hydroxyl group attached to either carbon atom of the original double bond. A deuterium atom is introduced at the nonhydroxylated carbon atom upon BD_3 reduction. This reactivity permits the location of double bonds present in either the acyl or sphingoid base chains. Figure

[29] G. M. L. Cragg, "Organoboranes in Organic Synthesis." Dekker, New York, 1973.

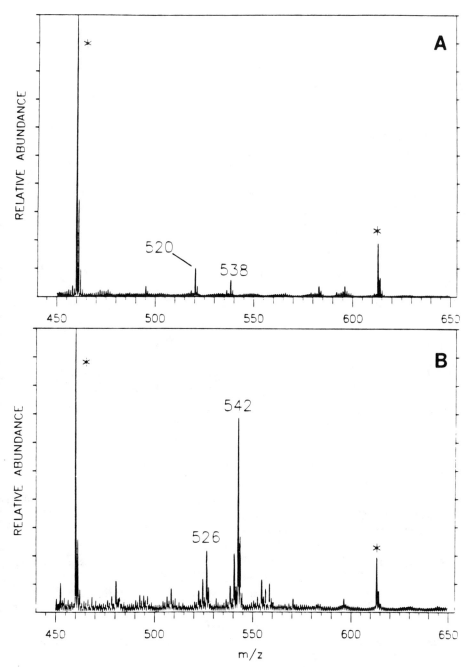

FIG. 8. FAB mass spectra of *N*-palmitoylsphingenine (A) before and (B) after reduction/ hydroboration. Asterisks (∗) indicate matrix-related cluster ions.

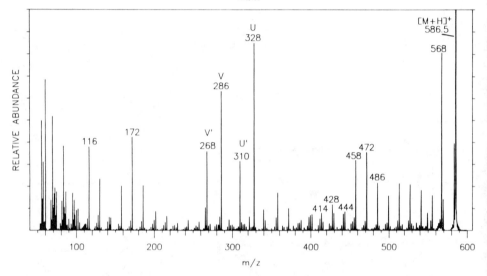

FIG. 9. FAB-CID-MS/MS spectrum of the (M + H)⁺ ion, *m/z* 586.5, obtained for *N*-oleoylsphingenine, after reduction/hydroboration, showing product ions that indicate the location of the double bond(s) in the starting material. Collision energy, 7 keV.

9 shows the CID mass spectrum of the (M + H)⁺ *m/z* 586.5 ion produced by reagent vapor hydroboration and oxidation of *N*-oleoylsphingenine. The major fragment ions for this derivative, the *U* and *V* ions at *m/z* 328.3 and 286.3, respectively, are typical for amide-reduced ceramides. These ions confirm the location of the alkene in the acyl chain of the native molecule. Also, additional ions due to loss of water from the *U* and *V* ions are present (labeled *U′* and *V′* at *m/z* 310.3 and 268.3, respectively). These ions are indicative of the presence of hydroxyl groups in the acyl chain and were used to provide extended information with regard to the characterization of hydroxyacyl and hydroxysphinganine-containing cerebrosides from cestodes.[15,30] The new hydroxyl group introduced in the derivatization shifts the mass at the carbon atoms of the former double bond and

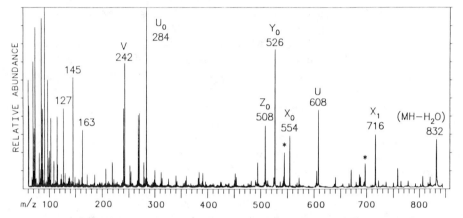

FIG. 10. FAB-CID-MS/MS spectrum obtained for the (M + H)$^+$ m/z 850.7 of the product obtained by reduction of 1 ng (1.2 pmol) lactosyl-*N*-palmitoylsphingenine. Collision energy, 5 keV. The spectrum was recorded with a photodiode array detector on MS-2, 10 × 0.3 sec integrations per segment.[31,32] Asterisks (*) indicate matrix-related fragments.

generally stabilizes elimination of alkene and H_2 to the new C–OH bond in charge remote site fragmentation. The increased intensity of the charge remote site fragment ions at m/z 472.4 and 458.4 allow almost instant visual identification of the former double bond location between C-9 and C-10 of the acyl chain. This technique was used to locate the position of the carbon–carbon double bond in the sphingolipid base of a ceramide isolated from soft coral.[15]

As an example of amide-reduced neutral glycosphingolipids, the FAB-MS/MS spectrum of reduced lactosylsphinganine (M + H)$^+$ m/z 850.7 is shown in Fig. 10. A 1.0 ng sample was used for the borane reduction and acquisition of the CID spectrum. In addition, the spectrum was recorded

[30] B. N. Singh, C. E. Costello, S. B. Levery, R. W. Walenga, D. H. Beach, J. F. Mueller, and G. G. Holz, *Mol. Biochem. Parasitol.* **26**, 99 (1987).

with a multichannel array detector[31,32] which lends an additional factor of 50 to 100 increase in sensitivity above that of derivatization. As was the case for reduced ceramides, the U (m/z 608.3) and V (m/z 242.3) ions are present in this spectrum. Now there is a U_0 ion at m/z 284.3 which represents the same bond cleavage in the sphingoid base as the U ion, but without the carbohydrate unit attached. The Y_0 ion at m/z 526.5, an ion encompassing the entire aglycon, is also prominent in the normal FAB mass spectrum of the derivative and can be used to thoroughly characterize the ceramide (i.e., locate double bonds) by obtaining its MS/MS spectrum analogous to the case shown in Fig. 5. The fragment ions for the carbohydrate moiety of the derivative are quite different than those of the native compound.[13,15] Instead of the familiar cleavages of the interglycosidic linkage seen in the spectra of the native and permethylated compounds, the major fragmentations in the sugars occur within the ring yielding $^{1,5}X_0$ (m/z 554.1) and $^{1,5}X_1$ (m/z 716.5) ions. It shall be shown in a later series of examples for G_{M1} (Figs. 12A and 13) that this type of fragmentation is due to specific charge location in the ceramide portion of the derivative at the site of high proton affinity (the amino group resulting from reduction). This results in charge remote site-induced fragmentation in the carbohydrate rather than the site-induced protonation and cleavage which is responsible for the B and Y ions in compounds which lack such a directing influence.

A very similar spectrum is observed for the amide-reduced derivative of the ganglioside G_{M3}, (M + H)$^+$ m/z 1171.8 shown in Fig. 11. The Y_0, U, X_0, and X_1 ions are found at m/z 570.6, 913.5, 598.6, and 760.5, respectively, although, in this case, the U ion abundance is decreased significantly. The difference that now exists in the derivative of this compound compared to Lac-Cer is the presence of the amine in the carbohydrate from the reduced amide in the NeuAc residue. This residue provides a strong B_1 ion at m/z 278.2 and an additional Y_2 ion at m/z 894.5. Overall the amine present in the lipid still seems to dictate the fragmentation of this derivative.

Permethylated and Amide-Reduced Gangliosides. Reduction of amides to amines in gangliosides can create one or more sites of enhanced basicity in the molecule, which can protonate preferentially in the formation of the (M + H)$^+$ ion and provide discrete charge locations which are able to channel the fragmentation toward specific pathways. Combined with an additional 2- to 5-fold increase in sensitivity over that of the

[31] J. A. Hill, S. A. Martin, J. E. Biller, and K. Biemann, *Biomed. Environ. Mass Spectrom.* **17,** 147 (1988).

[32] J. A. Hill, S. A. Martin, J. E. Biller, and K. Biemann, *Int. J. Mass Spectrom. Ion Proc.* **92,** 211 (1989).

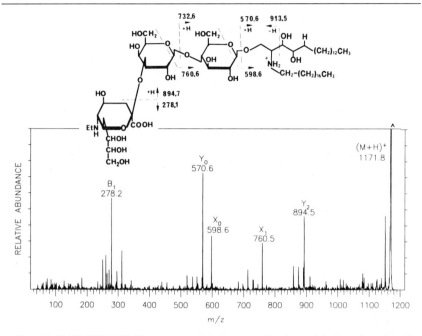

FIG. 11. FAB-CID-MS/MS spectrum of the lower molecular weight homolog of amide-reduced G_{M3} $(M + H)^+$ 1171.9. Collision energy, 7 keV.

permethylated alone, the improved fragmentation of permethylated and reduced gangliosides significantly lowers the limit for obtaining CID mass spectra of these compounds. Figure 12A is the FAB-CID mass spectra of the permethylated and reduced C_{14}-sphingoid homolog of G_{M1} at $(M + H)^+$ m/z 1803.1. In contrast to that of the permethylated-only compound in Fig. 7A, the spectrum contains many more significant ions, even though it was acquired with less than one-half the material (2 μg). Cleavages still occur adjacent to the reduced amino sugars, $B_{2\alpha}$ m/z 450.4 and $B_{1\beta}$ m/z 362.3, but now more fragmentation is observed near the reducing end of the carbohydrate portion giving rise to C and A ions. The C_4–methanol ion at m/z 1191.8 is very abundant and, in fact, the C_n–methanol ion (where $n =$ the number of sugars in the longest chain) is the base peak in the spectra of most permethylated, amide-reduced gangliosides. This ion defines the size of the carbohydrate unit. Loss of the substituent in the 3-position of the terminal residue of C ions is common and has been observed in the FAB mass spectra of permethylated glycoconjugates.[8] Thus, the C_3 ion also loses sialic acid (SA) to yield an abundant ion at m/z 640.6. Instead of the normal Y_n ion, such as in the spectrum of permethyl derivatives,

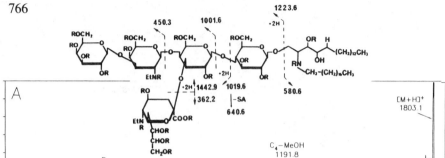

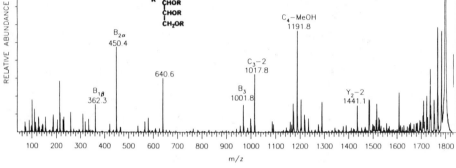

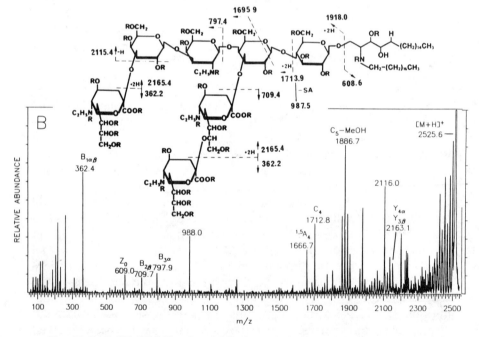

FIG. 12. FAB-CID-MS/MS spectra of the lower molecular weight homologs of ganglio-sides obtained after permethylation and reduction with BH_3. (A) Permethylated G_{M1} fully amide-reduced, $(M + H)^+$ 1803.1. (B) Permethylated G_{T1b} fully amide-reduced, $(M + H)^+$ 2525.6. Collision energy, 7 keV.

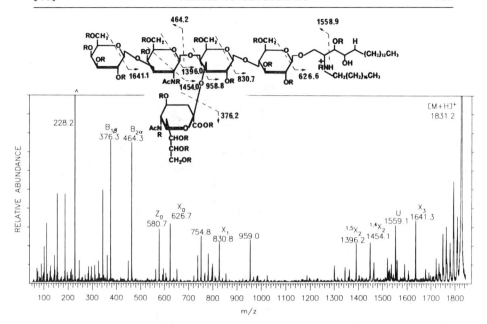

FIG. 13. FAB-CID-MS/MS spectrum of the lower molecular weight homolog of permethylated, partially reduced G_{M1} $(M + H)^+$ 1831.2. Collision energy, 7 keV.

which include an hydroxyl group at the site of cleavage as well as a proton due to ionization at the same site or elsewhere in the ion, a Y_2-2 ion is observed at m/z 1441.1. This ion would appear to result from a charge remote site process. Virtually all of the analogous ions are present in the spectrum of permethylated, amide-reduced GT_{1b} $(M + H)^+$ m/z 2525.6 shown in Fig. 12B. Note that the C_n–methanol ion again reveals the size of the carbohydrate (C_5–methanol m/z 1886.7). The attachment of a single sialic acid on the Gal residue of the nonreducing end of GT_{1b} is diagnosed from the $B_{3\alpha}$ ion at m/z 797.9 as was the case in the permethylated-only derivative. The fact that the other two sialic acids are consecutively linked is again indicated by the $B_{2\beta}$ ion at m/z 709.7.

The reactivity difference between amides in ceramides versus those in N-acetyl sugars in gangliosides with respect to hydroboration allows for the generation of a ceramide-only reduced ganglioside derivative, having the strong directing influence of a single amino group in the aglycon. The same G_{M1} homolog reacted with a shorter reduction time (10 min) is shown in Fig. 13. A complete series of X_n ions (m/z 626.7, 830.8, 1396.2, and 1641.3) and a U ion at m/z 1559.1 are present. The C_4–methanol ion, which was the most abundant in the permethylated, totally amide-reduced case, is now absent as well as most other ions with charge retention on the nonreducing end due to the absence of amine in the carbohydrate portion. The B ion adjacent to the GalNAc residue ($B_{2\alpha}$ m/z 464.3) remains, which is useful for confirming the position of sialic acid residues. The spectrum also contains some ions resulting from the elimination of two neighboring substituents on a particular sugar residue as noted on the structure in Fig. 13. This occurs mostly with the substituents in positions 3 and 4 of Glc and Gal residues (m/z 754.8 and 959.0, respectively). The MS/MS spectrum of this ceramide-only reduced species is clearly an improvement over that of the permethylated or permethylated, amide-reduced derivatives. The sequence of the sugars is more efficiently determined since there is a complete series of a single ion type. The problem with this derivative lies in the fact that the reduction time required for a suitable yield of ceramide-reduced derivative varies depending on the ganglioside (i.e., shorter reaction times are required for G_{M1} than G_{D1}). Also with these short reaction times, variables such as temperature regulation and strength of the borane reagent (which changes with age) can make a substantial difference in yield of the derivative. Consequently, the reaction has to be optimized individually for each compound type. This is best done by running a range of reactions from 5 to 30 min, and determining by FAB-MS the reaction time most advantageous for MS/MS analysis.

Conclusion

Tandem mass spectrometry of glycolipids represents a powerful new approach to the determination of structural details, especially when only small quantities of the compound of interest are available and may be present as components of complex mixtures of closely related species. Derivatization can improve sensitivity and can direct the fragmentation along pathways that maximize spectral information content. This approach is still in its early stages of development, and should increase in importance as more experimental data becomes available for a broader range of native compounds, for sets of rationally designed derivatives, and for a variety of collision regimes.

Acknowledgments

The authors wish to acknowledge extensive contributions of B. Domon to development of the methodology and the nomenclature system described in this chapter, helpful discussions with K. Biemann, and the recording of the array detector spectrum by S. A. Martin, J. A. Hill, and J. E. Biller. The MIT Mass Spectrometry Facility is supported by Grant No. RR00317 from the NIH Center for Research Resources.

Section IV

Nucleic Acid Constituents

[41] Constituents of Nucleic Acids: Overview and Strategy

By JAMES A. McCLOSKEY

Mass spectrometry has played a significant role in the characterization of bases, nucleosides, and their analogs, both natural and synthetic, while recent advances in methods of ionization[1-3] hold promise for routine extension to polynucleotides. This overview outlines the experimental approaches which are applicable in six primary areas, with emphasis on the methods described in chapters [42]–[47] of this volume. Although mass spectrometry has been effectively utilized in work in related areas, such as nucleoside antibiotics, cyclic nucleotides, cytokinins, and nucleosides in physiological fluids, the present coverage is limited primarily to nucleic acid constituents. The reader is also referred to earlier detailed reviews of a general nature in this field,[4-9] and to specialized reviews dealing with nucleic acid photoproducts,[10] xenobiotic modification,[11] desorption ionization,[6,12] the characterization of nucleosides,[13] and liquid chromatography-mass spectrometry (LC/MS) and tandem mass spectrometry (MS/MS).[14]

When compared with its successes for other classes of biopolymers, the applications of mass spectrometry to nucleic acids have long been restricted due to experimental difficulties associated with the intrinsic high

[1] A. G. Harrison and R. J. Cotter, this volume [1].

[2] F. Hillenkamp and M. Karas, this volume [12].

[3] C. G. Edmonds and R. D. Smith, this volume [22].

[4] J. A. McCloskey, *in* "Basic Principles in Nucleic Acid Chemistry" (P. O. P. Ts'o, ed.), Vol. I, p. 209. Academic Press, New York, 1974.

[5] C. Hignite, *in* "Biochemical Applications of Mass Spectrometry" (G. R. Waller and O. C. Dermer, eds.), 1st suppl. vol. p. 527. Wiley (Interscience), New York, 1980.

[6] J. A. McCloskey, *in* "Mass Spectrometry in the Health and Life Sciences" (A. L. Burlingame and N. Castagnoli, Jr., eds.), p. 521. Elsevier, Amsterdam, 1985.

[7] K. H. Schram, *in* "Mass Spectrometry" (A. M. Lawson, ed.), p. 507. de Gruyter, New York, 1989.

[8] P. F. Crain, *Mass Spectrom. Rev.* **9**, 505 (1990).

[9] A. L. Burlingame, D. Millington, D. Norwood, and D. H. Russell, *Anal. Chem.* **62**, 268R (1990).

[10] C. Fenselau, *in* "Photochemistry and Photobiology of Nucleic Acids" (S. Y. Wang, ed.), p. 419. Academic Press, New York, 1976.

[11] A. L. Burlingame, K. Straub, and T. Baille, *Mass Spectrom. Rev.* **2**, 331 (1978).

[12] D. L. Slowikowski and K. H. Schram, *Nucleosides Nucleotides* **4**, 309 (1985).

[13] J. A. McCloskey, *in* "Mass Spectrometry in Biomedical Research" (S. J. Gaskell, ed.), p. 75. Wiley, New York, 1986.

[14] C. C. Nelson and J. A. McCloskey, *Adv. Mass Spectrom.* **11B**, 1296 (1989).

METHODS IN ENZYMOLOGY, VOL. 193

polarity of nucleotides and nucleosides. Thus, chemical derivatization[15-17] and methods based on desorption ionization,[1] including direct coupling of mass spectrometers and liquid chromatographs,[18,19] have had considerable impact on certain of the major areas described below. In each of these areas, sample quantity (often dictated by sample polarity and ionization yield) and purity are major determinants of the experimental approach which is most likely to be effective. In general, if the component of interest is present at the submicrogram level and in the presence of other constituents of similar structure, chromatographic separation, either alone or in combination with mass spectrometry, will usually be required. Although procedures based on tandem mass spectrometry have been developed which are, in principle, applicable to mixtures of nucleic acid products,[8,14] relatively few actual applications have been reported, (e.g., Refs. 20, 21), and so the potential of MS/MS has yet to be realized.

The following are six principal areas in which mass spectrometry can be applied to nucleic acid-related problems, based on experimental methods which have reached the stage of routine application. The first of these is discussed in detail, because it represents the area in which mass spectrometry has been employed most effectively. Most of the approaches referred to in this section are relevant to the other five areas and reflect common experimental issues, such as that of sample quantity and purity. Additional references to examples of specific applications can be found in chapters [42]–[47] in this volume and Ref. 8.

Structure Determination of New Natural Nucleosides

Posttranscriptional processing of RNA and DNA produces a wide structural variety of modifications, principally in the base, and now known to number approximately 77 nucleosides in RNA and 15 in DNA. In tRNA, where the greatest number have been discovered, approximately 74 are known, of which 52 have been placed in specific locations in the 455

[15] D. R. Knapp, this volume [15].

[16] K. H. Schram, this volume [43].

[17] K. H. Schram and J. A. McCloskey, in "GLC and HPLC Determination of Therapeutics Agents, Part 3" (K. Tsuji, ed.), Chap. 3. Dekker, New York, 1979.

[18] M. L. Vestal, this volume [5].

[19] S. C. Pomerantz and J. A. McCloskey, this volume [44].

[20] For example, S. E. Unger, A. E. Schoen, and R. G. Cooks, *J. Org. Chem.* **46**, 4767 (1981).

[21] P. F. Crain, T. Hashizume, C. C. Nelson, S. C. Pomerantz, and J. A. McCloskey, in "Biological Mass Spectrometry" (A. L. Burlingame and J. A. McCloskey, eds.), p. 509. Elsevier, New York, 1990.

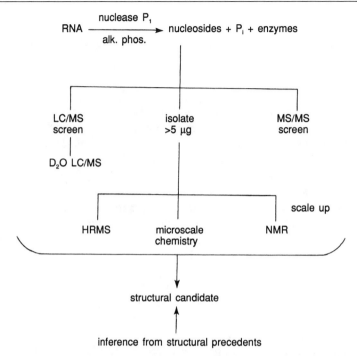

FIG. 1. Principal elements of microscale characterization of nucleosides from RNA.

known sequences.[22] Mass spectrometry has played a significant role in the characterization of nearly all of the new nucleosides discovered in nucleic acids in the past 20 years, primarily due to the sensitivity of the method, and the relative importance of molecular weight and elemental composition data in a structural system which is relatively conserved.

Because tRNA is a heterogeneous mixture, individual modified nucleosides usually constitute between about 0.001 and 0.4% of the total nucleoside population, so that isolation of microgram-level quantities, for example, for proton NMR, may be a formidable task. An overall experimental scheme[21] used in the author's laboratory is shown in Fig. 1. Because the nucleoside is the smallest structural unit in which nucleic acid modification is expressed, it is the preferable level at which to work, although nucleoside monophosphates may be preferred in special instances, such as to avoid alkaline pH during dephosphorylation by alkaline phosphatase, or for negative ion desorption experiments. In any event, acid or base chemi-

[22] M. Sprinzl, T. Hartmann, J. Weber, J. Blank, and R. Zeidler, *Nucleic Acids Res.* **17,** r1 (1989).

cal hydrolysis of RNA and DNA should be avoided in the determination of new structures to preclude the possibility of chemical degradation or rearrangement. The enzymatic hydrolysis protocols discussed elsewhere in this volume[23] are generally applicable to procedures involving analysis by mass spectrometry.

Screening of hydrolytic mixtures by gas chromatography-mass spectrometry (GC/MS) of trimethylsilylated digests is possible in restricted cases[24,25] when it is known that the component of interest will produce a sufficiently volatile derivative.[26] However, such is not the case for many complex nucleosides, such as members of the t⁶A and queuosine families (abbreviations are defined in Table II of [44]), so that GC/MS is much less well suited than LC/MS as a general screening method.

Depending on the complexity of the hydrolyzate, initial decisions must be made as to whether to isolate the component of interest, or to derive structural information directly from the mixture. If isolation is to be carried out, LC/MS provides an effective means for locating the component of interest in a high-performance liquid chromatography (HPLC) effluent, as described in this volume.[19] In the case of isolations below 1 μg per component, care must be taken to avoid adsorptive losses during further sample handling, even though, in principle, most measurements using mass spectrometry can be carried out on quantities below 100 ng. In favorable cases, tentative structure assignments can be made from LC/ MS and D$_2$O-LC/MS experiments,[19] permitting further testing by other means, or by synthesis.

Isolation of individual nucleoside components, preferably from HPLC systems that employ volatile buffers,[27] is obviously preferable if circumstances permit. It is emphasized that high-resolution mass spectrometry[28] (see Fig. 1), preferably in the form of a complete electron ionization (EI) mass spectrum, remains as one of the single most effective methods for determination of nucleoside structure. Although high-resolution measurements can be made directly on enzymatic digests,[29] measurements made on isolated nucleosides are much more practicable.

[23] P. F. Crain, this volume [42].

[24] M. Dizdaroglu, this volume [46].

[25] For example, M. Buck, J. A. McCloskey, B. Basile, and B. N. Ames, *Nucleic Acids Res.* **10,** 5649 (1982).

[26] For leading references to the gas chromatography of volatile nucleoside derivatives, see Ref. 14.

[27] M. Buck, M. Connick, and B. N. Ames, *Anal. Biochem.* **129,** 1 (1983).

[28] K. Biemann, this volume [13].

[29] H. Pang, D. L. Smith, K. Yamaizumi, S. Nishimura, and J. A. McCloskey, *Eur. J. Biochem.* **127,** 459 (1982).

When exact mass (high-resolution) data are acquired, derivation of the correct elemental composition can be significantly aided by the application of four general rules which are applicable to ribonucleoside structure[30]:

1. Total number of rings-and-double bonds (RDB), calculated by standard means[31] for each computer-generated composition candidate, is highly unlikely to fall outside the range 4–12.

2. Nitrogen rule: even number of N atoms is required by an even molecular weight; odd number of N atoms is required by an odd molecular weight.

3. Nitrogen atom content must be ≥2.

4. Oxygen atom content must be ≥4 (≥3 for deoxynucleosides).

Of the various derivatization alternatives available, trimethylsilylation[16,24,32] and alternatively, permethylation,[33] constitutes the most effective overall approach in terms of structural information to be gained. Mass spectrometric measurements using thermal vaporization of free nucleosides entails significant risk of pyrolytic degradation in the case of most of the hypermodified nucleosides from RNA and DNA, while desorption ionization,[1] even with collision-induced dissociation (CID), tends to provide somewhat limited structural information compared with EI spectra of volatile derivatives.

The isolated material may be further subjected to microscale chemical reactions which test for selected structural features. A partial list of such reactions used in the author's laboratory is given in Table I, most of which are applicable, in some circumstances, to nucleoside mixtures.

The deduction of a structural candidate (see Fig. 1) may rely on information other than mass spectrometry, such as sequence location. In addition, unlike DNA damage products or nucleoside antibiotics, which exhibit greater structural diversity, the types of structural modifications in natural nucleic acids are relatively conserved. For example, methylation of ribose in RNA is only known at O-2', and glycosidic bonds are of β configuration. As a result, some structural inferences can be made from a list of guidelines[21] which greatly narrow the choice of structural candidates.

Final proof of structure should preferably be made by EI mass spec-

[30] P. F. Crain, H. Yamamoto, J. A. McCloskey, Z. Yamaizumi, S. Nishimura, K. Limburg, M. Raba, and H. J. Gross, *Adv. Mass Spectrom.* **8,** 1135 (1980).

[31] RDB = $x - \frac{1}{2}y + \frac{1}{2}z + 1$ for the neutral molecule $C_xH_yN_zO_n$. For protonated molecules (MH$^+$), the calculated value will end in 0.5. For discussion see F. W. McLafferty, "Interpretation of Mass Spectra," p. 23. Univ. Sci. Books, Mill Valley, CA, 1980.

[32] J. A. McCloskey, this volume [45].

[33] D. L. von Minden and J. A. McCloskey, *J. Am. Chem. Soc.* **95,** 7480 (1973).

TABLE I
MICROSCALE REACTIONS FOR STRUCTURAL TESTS OF NUCLEOSIDES
BY MASS SPECTROMETRY

Reaction	Comment	Ref.
H–D exchange of active H	Determine number of active H; distinguish isomers	a
H–D exchange of slightly acidic H	Test for presence of H-8 in purines	b, c
O-Isopropylidine formation	Test for *cis*-diol moiety in side chains	b, d
Oxidation of base during trimethylsilylation	Test for 7-methylpurines	e
Formation of -Si(CD$_3$)$_3$ or per-CD$_3$ derivatives	Assist interpretation of mass spectrum; establish M_r	f, g
Formation of per-CH$_3$ or per-CD$_3$ derivatives	EI spectrum: loss of methylene imine as test for adenosine N^6-methylation	h, i

[a] J. A. McCloskey, this volume [16].

[b] H. Kasai, Z. Ohashi, F. Harada, S. Nishimura, N. J. Oppenheimer, P. F. Crain, J. G. Liehr, D. L. von Minden, and J. A. McCloskey, *Biochemistry* **14**, 4198 (1975).

[c] Z. Yamaizumi, S. Nishimura, K. Limburg, M. Raba, H. J. Gross, P. F. Crain, and J. A. McCloskey, *J. Am. Chem. Soc.* **101**, 2224 (1979).

[d] D. W. Phillipson, C. G. Edmonds, P. F. Crain, D. L. Smith, D. R. Davis, and J. A. McCloskey, *J. Biol. Chem.* **262**, 3462 (1987).

[e] D. L. von Minden, R. N. Stillwell, W. A. Koenig, K. J. Lyman, and J. A. McCloskey, *Anal. Biochem.* **50**, 110 (1972).

[f] J. A. McCloskey, R. N. Stillwell, and A. M. Lawson, *Anal. Chem.* **40**, 233 (1968).

[g] H. Pang, K. H. Schram, D. L. Smith, S. P. Gupta, L. B. Townsend, and J. A. McCloskey, *J. Org. Chem.* **47**, 3923 (1982).

[h] D. L. von Minden, J. G. Liehr, M. H. Wilson, and J. A. McCloskey, *J. Org. Chem.* **39**, 285 (1974).

[i] P. F. Crain, H. Yamamoto, J. A. McCloskey, Z. Yamaizumi, S. Nishimura, K. Limburg, M. Raba, and H. J. Gross, *Adv. Mass Spectrom.* **8**, 1135 (1980).

trometry, by comparison with the mass spectrum of an authentic nucleoside. The use of chromatography as a sole means of comparison is discouraged because of the potential ambiguity associated with use of retention time for characterization. Obviously, the use of UV absorbance, as from a photodiode array or multichannel HPLC UV detector, greatly increases the reliability of identification based primarily on chromatography. When applicable, capillary GC/EI-MS offers the most secure test of structural identity in a single method,[34] because of the accuracy and precision with which retention times of unknown and reference nucleosides can be compared, and the great detail inherent in EI mass spectra. If LC/MS is

[34] For example, J. A. McCloskey, P. F. Crain, C. G. Edmonds, R. Gupta, T. Hashizume, D. W. Phillipson, and K. O. Stetter, *Nucleic Acids Res.* **15**, 683 (1987).

available, the combination of retention times, mass, and UV absorbance (as discussed elsewhere in this volume[19]) is also a sensitive and effective test of identity, which is advantageous in not requiring purification of the unknown nucleoside.

Qualitative Analysis of Structurally Known Nucleosides in Nucleic Acid Hydrolysates

The recognition of modified nucleosides in digests of nucleic acids is an analytical problem of importance in several areas, including studies of structure–function relationships and biosynthetic modification pathways,[35] and in conjunction with sequence determination.[36] If hydrolysates are compositionally simple, then standard chromatographic procedures alone may suffice for identification.[27,37] Otherwise, thermospray LC/MS is an effective method at the nanogram level, which is advantageous because chromatographic overlap is generally not a serious problem.[19] Although identification of modified nucleosides from RNA sequence measurements is more sensitive if ^{32}P postlabeling methods[38] are used, the analysis of RNA digests by LC/MS is considerably faster, and of equal or greater objectivity with respect to structure assignment.

Tandem mass spectrometry[39,40] offers a chromatography-independent means of effectively identifying *preselected* nucleosides (or those having preselected structure features), at sensitivity levels approximately the same as for thermospray LC/MS. However, caution must be exercised in the event of occurrence of isobaric compounds (those having the same nominal molecular weight). Examples of such occurrences are 4-thiouridine and 5-hydroxyuridine, N^4-methyl-2'-deoxycytidine and 5-methyl-2'-deoxycytidine, and numerous monomethylpurine nucleosides. Clearly, if only one isobar is present, MS/MS can be used for isomer differentiation,[41–43] but if multiple compounds are represented in the MS-1 selected

[35] G. R. Björk, J. U. Ericsson, C. E. D. Gustafsson, T. G. Hagervall, Y. H. Jönsson, and P. M. Wikstrom, *Annu. Rev. Biochem.* **56**, 263 (1987).
[36] F. Harada and S. Nishimura, *Biochemistry* **11**, 301 (1972).
[37] C. W. Gehrke and K. C. Kuo, *J. Chromatogr.* **471**, 3 (1989).
[38] Y. Kuchino, N. Hanyu, and S. Nishimura, this series, Vol. 155, p. 379.
[39] M. L. Gross, this volume [6].
[40] R. K. Boyd and R. A. Yost, this volume [7].
[41] S. E. Unger, A. E. Schoen, R. G. Cooks, D. J. Ashworth, J. D. Gomes, and C.-J. Chang, *J. Org. Chem.* **46**, 4765 (1981).
[42] R. L. Hettich, *Biomed. Environ. Mass Spectrom.* **18**, 265 (1989).
[43] T. Hashizume, C. C. Nelson, S. C. Pomerantz, and J. A. McCloskey, *Nucleosides Nucleotides* **9**, 355 (1990).

ion beam, overlapping CID spectra will result which are easily misinterpreted.

High-Sensitivity Quantitative Analysis of Methylated Bases in DNA

Gas chromatography-mass spectrometry has long been a major analytical method for high sensitivity analysis of complex mixtures. Using selected-ion monitoring[44] and stable isotope dilution, detection limits in the picogram range are routinely possible, but with the relevant provision that the sample be sufficiently volatile for gas chromatography. Numerous approaches to quantitative measurements of nucleic acid bases using mass spectrometry have been tried,[8] but have generally not enjoyed the combination of sensitivity and accuracy, both being necessary criteria because of the availability of less direct but simpler methods, such as use of restriction endonucleases. For quantitative measurement of trace levels of material, the vagaries of derivatization yield and adsorptive losses are such that use of stable isotope-labeled internal standards is highly desirable. The isotope dilution method (described elsewhere[45]) exemplifies the considerable potential of trace analysis of nucleic acid bases using GC/MS. As discussed[45,46] the use of internal standards of this type greatly reduces the usual experimental problems of microscale derivatization and analysis of polar compounds at the picogram level, but with the significant requirement that a suitable standard be available. Similar measurements can be made using other directly combined chromatography-mass spectrometry approaches in which derivatization is not required, but undoubtedly without the ultimate sensitivity offered by GC/MS.

Molecular Weight or Sequence Determination of Synthetic Oligonucleotides

The evolution of desorption methods of ionization[1] over the past decade has provided a means for direct measurement of molecular weights of polynucleotides, and in some cases, sequence information.[8] From the standpoint of experimental applicability, these methods as a whole are advantageous in being relatively rapid (for example, compared with polyacrylamide gel electrophoresis), but compare poorly in terms of sensitivity with desorption ionization yields of other biopolymers (peptides, glyco-

[44] J. T. Watson, this volume [4].
[45] P. F. Crain, this volume [47].
[46] L. Siekmann, in "Mass Spectrometry" (A. M. Lawson, ed.), p. 647. de Gruyter, New York, 1989.

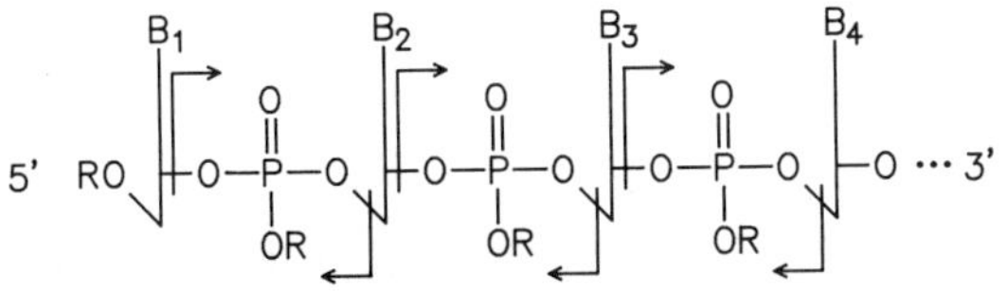

$$\text{5'} \quad RO \left[-O-\overset{\overset{\displaystyle O}{\|}}{\underset{\underset{\displaystyle OR}{|}}{P}}-O \right] \left[-O-\overset{\overset{\displaystyle O}{\|}}{\underset{\underset{\displaystyle OR}{|}}{P}}-O \right] \left[-O-\overset{\overset{\displaystyle O}{\|}}{\underset{\underset{\displaystyle OR}{|}}{P}}-O \right] -O \cdots \text{3'}$$

SCHEME 1. Sequence-determining cleavages in the mass spectra of oligonucleotides. R, H or protecting group; B, base.

conjugates) or with numerous conventional biochemical methods that employ ^{32}P labeling. For this reason, routine applications of mass spectrometry to oligonucleotides are, in general, restricted to synthetic materials for which sample quantities are of less importance. For example, conventional fast atom bombardment (FAB) mass spectra of deoxy octamers and decamers are reported to require 10 nmol of material,[47] and quantitative measurement of ion yields of short oligonucleotides ($\leq$8-mer) by continuous-flow-FAB shows 50- to 100-fold less sensitivity than for peptides of similar molecular weight.[48] A promising method for enhancing the surface concentration of ATP in the FAB matrix has been demonstrated,[49] but has not been tested with oligonucleotides.

The principal sequence determining ions in the negative ion desorption mass spectra of oligonucleotides, either from spontaneous decomposition[50] or collision-induced dissociation[51] (CID), are as shown by the arrows in Scheme 1. Analogous cleavages (with hydrogen transfer) are obtained in positive ion spectra, but insufficient evidence presently exists to generalize the advantages of one ionization mode over the other.

Detailed studies of CID of small oligonucleotides show that processes which compete with the cleavages shown in Scheme I, such as elimination of a neutral base, become increasingly prominent with increasing chain length and seriously impede sequence determination.[52] Indeed, it has been concluded that spectra from spontaneous decompositions are more useful for sequence determination than those from CID.[8]

In the case of chemically blocked oligonucleotides or their analogs, molecular weight determination by mass spectrometry may serve to indicate the correctness of the structure, since the synthetic route to the

[47] L. Grotjahn, H. Blöcker, and R. Frank, *Biomed. Mass Spectrom.* **12,** 514 (1985).

[48] M. M. Sheil, J. A. Kowalak, and J. A. McCloskey, *Proc. 37th ASMS Conf. Mass Spectrom. Allied Topics, Miami Beach, FL* p. 728 (1989).

[49] W. V. Ligon and S. B. Dorn, *Fresenius Z. Anal. Chem.* **325,** 626 (1986).

[50] C. J. McNeal, S. A. Narang, R. D. Macfarlane, H. M. Hsiung, and R. Brousseau, *Proc. Natl. Acad. Sci. U.S.A.* **77,** 735 (1980).

[51] M. Linsheid and A. L. Burlingame, *Org. Mass Spectrom.* **18,** 245 (1983).

[52] R. L. Cerny, K. B. Tomer, M. L. Gross, and L. Grotjahn, *Anal. Biochem.* **165,** 175 (1987).

$$5'-dGpdTp-O \overset{\overset{\displaystyle H}{\overset{\displaystyle |}{\underset{\displaystyle O}{\wedge}}}}{\diagdown} O-pdCpdG-3'$$

1

$$
\begin{array}{c}
\text{pa bz bz bz \quad bz bz pa} \\
\text{| \ | \ | \ | \quad | \ | \ |} \\
5'-\text{mmt}-\text{NH}-\text{d}(\text{GcCcCcAcTcAcCcG})-\text{OH}-3'
\end{array}
$$

2

Abbreviations in **2** are as follows: mmT, monomethoxytrityl; pa, phenylacetyl; bz, benzoyl; c, carbamoyl.

product will generally be known. Verification of sequence location of a modified residue or internucleotide linkage can be made in favorable cases from ions analogous to those shown in Scheme I. For examples of the applications of negative ion FAB mass spectrometry to the characterization of oligonucleotide analogs, the reader is referred to studies of compound **1** in which one residue in the nucleotide chain is a 1,4-anhydro-2-deoxyribityl moiety,[53] and **2** in which the backbone phosphodiester bonds have been replaced by uncharged carbamate linkages.[54]

Characterization of in Vitro Reaction Products of Nucleic Acids and of Model Subunits

The characterization of chemical, metabolic, or radiolytic reaction products from nucleic acids by mass spectrometry, in general, follows the experimental protocols and considerations discussed in the first section, with two principal differences. First, the products of such reactions may be unstable or highly reactive, so that it may be desirable to minimize sample handling procedures unless the products of interest are known to be stable under conditions of chromatography or derivatization. Second, the products of in vitro reactions potentially exhibit broader structural diversity and less predictable structures than in the case of natural modifications. As shown elsewhere in this volume[24] electron ionization (EI) GC/MS, although applicable only to products which can be made sufficiently

---

[53] C. R. Iden and R. A. Rieger, *Biomed. Environ. Mass Spectrom.* **18**, 617 (1989).

[54] D. Griffin, J. Laramee, M. Deinzer, E. Stirchak, and D. Weller, *Biomed. Environ. Mass Spectrom.* **17**, 105 (1988).

volatile for gas chromatography, offers the structural detail characteristic of EI, as well as the significant advantages of capillary gas chromatography for dealing with complex mixtures. In addition, experimental approaches based on tandem mass spectrometry are relevant to this field (reviewed in Ref. 14), but have been generally underutilized in terms of applications.

Measurement of Biologically Incorporated Stable Isotopes

The analysis of biologically incorporated stable isotopes (^2H, ^{13}C, ^{15}N, ^{18}O) in intact molecules by mass spectrometry offers several general advantages and characteristics which are applicable to the biosynthesis of nucleic acids. (1) Measurements are generally rapid because they can be directly applied to mixtures without isolation of the component of interest when directly combined chromatography-mass spectrometry is used. (2) The mass spectrum will, in general, provide the percentage of each labeled species present (cf. [16]) rather than a weighted average as in the case of radioisotopes. (3) In favorable cases, the extent of incorporation can be separately determined for the same isotope present in several positions in the molecule, from a single mass spectrum. (4) Nanogram levels of material are generally required, which is favorable compared with other methods for determination of stable isotopes, such as NMR. The principal disadvantage is a requirement for minimum incorporation of approximately 1 mol% of isotope (depending on experimental circumstances) in order to be distinguished from naturally occurring heavy isotopes.

The degree of structural specificity in determining or in verifying the location of an isotopic label depends on both the availability of suitable fragment ions in the mass spectrum, and the reliability of peak assignments in the spectrum. The reader is directed to two earlier reports for typical examples of applications: measurement of ^{18}O at specific sites in the ribose skeleton of nucleosides from *Escherichia coli* RNA in studies of ribose biosynthesis[55]; and study of the mechanism of methylation of the uracil moiety by 5,10-methylene tetrahydrofolate to produce thymine riboside in tRNA.[56] In both cases, reliance was placed on earlier work involving detailed studies of EI fragmentation reactions.

Acknowledgments

Methods described from the author's laboratory were developed with support from the National Institutes of Health (GM 21584, CA 18024).

[55] R. Caprioli and D. Rittenberg, *Biochemistry* **8,** 3375 (1969).
[56] A. S. Delk, D. P. Nagle, Jr., J. C. Rabinowitz, and K. M. Straub, *Biochem. Biophys. Res. Commun.* **86,** 244 (1979).

[42] Preparation and Enzymatic Hydrolysis of DNA and RNA for Mass Spectrometry

By PAMELA F. CRAIN

Introduction

The utilization of mass spectrometry for direct analysis of DNA and RNA enzymatic digests places certain restrictions on the types of buffers and salts tolerated in the hydrolysate. While thermospray liquid chromatography-mass spectrometry (LC/MS) is relatively insensitive to their presence, when a fast atom bombardment (FAB)-based method is used,[1] there is already substantial background from the matrix alone, and reduction of any additional interferences from the digest can only simplify the analysis. For an electron ionization (EI)-based method, in which the dried digest is trimethylsilylated, the presence of ammonium sulfate, Mg^{2+}, Tris-Cl, etc., utilized in typical digestion protocols will attenuate yields of derivatized nucleosides.[2] Thymidine and cytidine are most severely affected, and removal of salts following digestion is recommended. The operational equivalent of desalting is to conduct the hydrolysis using volatile buffers, and this approach is successful for examining digests at the nucleoside[3] and nucleotide[4] level, because the hydrolysate may simply be lyophilized without clean-up and the concomitant potential for losses of trace constituents.

Both DNA and RNA can be digested to nucleoside 5'-phosphates with nuclease P_1, a phosphodiesterase which is qualitatively insensitive to the nature of the sugar and base.[5,6] Later studies by Gehrke and colleagues showed a slight resistance to digestion in the case of pseudouridine and the 2'-O-methyl derivatives of guanosine and uridine, and established detailed protocols for complete digestion of tRNA[7] and DNA[8] to nucleo-

---

[1] C. C. Nelson and J. A. McCloskey, *Adv. Mass Spectrom.* **11A,** 260 (1989).

[2] A. B. Patel and C. W. Gehrke, *J. Chromatogr.* **130,** 115 (1977).

[3] H. Pang, D. L. Smith, P. F. Crain, K. Yamaizumi, S. Nishimura, and J.A. McCloskey, *Eur. J. Biochem.* **127,** 459 (1982).

[4] D. Swinton, S. Hattman, P. F. Crain, C. S. Cheng, D. L. Smith, and J. A. McCloskey, *Proc. Natl. Acad. Sci. U.S.A.* **80,** 7400 (1983).

[5] M. Fujimoto, A. Kuninaka, and H. Yoshino, *Agric. Biol. Chem.* **38,** 1555 (1974).

[6] M. Fujimoto, A. Kuninaka, and H. Yoshino, *Agric. Biol. Chem.* **38,** 785 (1974).

[7] C. W. Gehrke, K. C. Kuo, R. A. McCune, K. O. Gerhardt, and P. Agris, *J. Chromatogr.* **230,** 297 (1982).

[8] C. W. Gehrke, R. A. McCune, M. A. Gama-Sosa, M. Ehrlich, and K. C. Kuo, *J. Chromatogr.* **301,** 199 (1984).

METHODS IN ENZYMOLOGY, VOL. 193

sides in less than 4 hr using nuclease P_1 and alkaline phosphatase in a sodium acetate, zinc sulfate, and Tris solution. Overnight incubation with the same enzymes in ammonium acetate solutions was found convenient and was used in early studies of tRNA modification using LC/MS.[9–11] During the course of these studies, the author observed that several highly modified nucleosides, notably N^2,N^2-O-2′-trimethylguanosine (m$_2^2$Gm)[10] and N^4-acetyl-O-2′-methylcytidine (ac^{4}Cm),[10] each containing modifications in both the sugar and base residues, and the tricyclic guanosine analog mimG[11] (for chemical names of abbreviations used, see Table II in [44]) could not be completely recovered using any of the nuclease P_1-only procedures.[7,9] The latest version of the nuclease P_1-only protocol for digestion of tRNA also leaves the highly modified wyeosine residue of yeast tRNA^{Phe} substantially unreleased after overnight incubation.[12]

During studies which established the structure of the highly modified tRNA constituent ms^{2}t^{6}A, partial digestion with nuclease P_1 was used to recover the dinucleotide ms^{2}t^{6}ApA, from which the new constituent was released with venom phosphodiesterase.[13] This result suggests that the two enzymes have different tolerances for substrate structural features, and prompted the development of a universally applicable two-enzyme protocol for digestion of nucleic aids which contain hypermodified constituents. Minimization of exposure to extremes of pH, and brevity of incubation times, have been emphasized in order to minimize degradation[12] or modification[14] of labile nucleosides.

Procedures

All solutions for dissolving nucleic acids and enzymes should be made using autoclaved distilled water and the highest available grade of salt(s), and should be filter-sterilized with 0.20 μm filters and dispensed into sterile containers; disposable presterilized 15-ml capped culture tubes are convenient and readily available from laboratory supply houses. Human skin has been shown to contain nucleases,[15] and gloves are generally

[9] C. G. Edmonds, M. L. Vestal, and J. A. McCloskey, *Nucleic Acids Res.* **13**, 8197 (1985).

[10] C. G. Edmonds, P. F. Crain, T. Hashizume, R. Gupta, K. O. Stetter, and J. A. McCloskey, *J. Chem. Soc. Chem. Commun.* p. 909 (1987).

[11] J. A. McCloskey, P. F. Crain, C. G. Edmonds, R. Gupta, T. Hashizume, D. W. Phillipson, and K. O. Stetter, *Nucleic Acids Res.* **15**, 683 (1987).

[12] C. W. Gehrke and K. C. Kuo, *J. Chromatogr.* **471**, 3 (1989).

[13] Z. Yamaizumi, S. Nishimura, K. Limburg, M. Raba, H. J. Gross, P. F. Crain, and J. A. McCloskey, *J. Am. Chem. Soc.* **101**, 2224 (1979).

[14] H. Kasai, P. F. Crain, Y. Kuchino, S. Nishimura, A. Ootsuyama, and H. Tanooka, *Carcinogenesis* **7**, 1849 (1986).

[15] R. W. Holley, J. Apgar, and S. H. Merrill, *J. Biol. Chem.* **236**, PC42 (1961).

recommended when preparing nucleic acids to avoid contamination of tubes and pipettes and subsequent degradation of the nucleic acid. Solutions should be dispensed with sterile disposable micropipettes.

Preparation of DNA and RNA

Isolated DNA or RNA typically contains varying amounts of the other nucleic acid, and while traces of RNA in DNA (and vice versa) generally do not confound the biological experiment, they will be readily apparent in the mass spectrometric analysis, and have the potential to interfere in the analysis. Their elimination or reduction to minimal levels is recommended. Enzymatic removal of the contaminating nucleic acid with the appropriate nuclease and removal of the enzyme and digested nucleic acid by precipitation, or microscale batch chromatography,[16] is the simplest approach and is suitable for treatment of the small amounts of nucleic acid required for mass spectrometry. Elimination of additional sample handling such as dialysis, required if alternative methodologies such as ultracentrifugation or electrophoresis are used for this purpose, also assists in maintaining sample integrity.

Removal of RNA from DNA. The usual procedure for elimination of RNA from DNA during isolation is digestion of the isolate with DNase-free RNase A.[17] If the amount of RNA remaining is unacceptably great, the author prefers RNase T_2 (EC 3.1.27.1) for a second treatment because, at this stage, the contaminating ribonucleotide oligomers will be purine-rich, and therefore more efficiently digested. RNase T_2 suitably free of contaminating DNase activities is widely available (for example, Life Technologies, Inc., Gaithersburg, MD; Boehringer-Mannheim, Indianapolis, IN).

Removal of DNA from RNA. Treatment of the isolate with RNase-free DNase[18] is the simplest option in this case, and suitable enzyme preparations are readily available from the sources listed above.

Purification and Storage. A simple procedure has been described for recovering DNA from mixtures containing protein and nucleotides,[19] which is also suitable for RNA.[20] After the unwanted nucleic acid is digested, the RNase or DNase may be precipitated by adding $\frac{1}{2}$ vol of 7.5

[16] P. F. Crain, this volume [47].

[17] T. Maniatis, E. F. Fritsch, and J. Sambrook, "Molecular Cloning: A Laboratory Manual," p. 280. Cold Spring Harbor Laboratories, Cold Spring Harbor, NY, 1982.

[18] T. Maniatis, E. F. Fritsch, and J. Sambrook, "Molecular Cloning: A Laboratory Manual," p. 192. Cold Spring Harbor Laboratories, Cold Spring Harbor, NY, 1982.

[19] J. Crouse and D. Amorese, *Focus* **9**(2), 3 (1987).

[20] J. Crouse, personal communication, 1989.

M ammonium acetate to the solution and centrifuging the mixture for 15 min at 12,000 rpm at room temperature. Remove the supernatant, add 2.5 volumes of 95% ethanol, and then centrifuge the suspension 15 min at 12,000 rpm to precipitate the nucleic acid; nucleotides from the digested nucleic acid remain in solution. Suspension of the nucleic acid pellet in 2.5 M ammonium acetate and reprecipitation with ethanol should remove essentially all of the low-molecular weight contaminants. Storage of nucleic acids as ethanol suspensions at $-20°$ is recommended to minimize degradation.[21]

The amount of purified nucleic acid in solution may be estimated spectrophotometrically: One A_{260} unit is 50 μg/ml of DNA, 40 μg/ml of RNA, and 20 μg/ml of oligonucleotides, when measured at 260 nm through a 1-cm path length. The appropriate volume of solution may then be removed, and the remaining nucleic acid precipitated for storage by addition of 2.5 volumes of 95% ethanol.

Preparation of Enzyme Solutions

General Comments. Enzymes should always be of the highest purity and specific activity available, and should be stored as recommended by the supplier. For crystalline enzymes, formulation of working solutions by weighing and dissolving a suitable amount is recommended; if the entire preparation is dissolved the solution should be subdivided and stored at $-20°$ to minimize repetitive freeze–thaw cycles. Once a portion has been thawed for use, it should be stored at 4°. Each individual enzyme working solution should be evaluated for the presence of contaminants before use by the mass spectrometric method to be used for analyzing the nucleic acid digest. The enzyme sources listed below are those which have been successfully used by the author; other sources may provide preparations of equally suitable quality. The user should note, however, that alternative suppliers may use different assays to define a unit of activity and the amounts required for complete digestion may differ from those described below.

Nuclease P_1 (EC 3.1.30.1). Lyophilized enzyme (e.g., Sigma N8630, St. Louis, MO; Boehringer-Mannheim 236 225) is dissolved in 0.05 M ammonium acetate, pH 5.3, to a concentration of 2 units per microliter.

Phosphodiesterase I (from Snake Venom) (EC 3.1.4.1). Lyophilized enzyme in a buffer salts matrix (Sigma P6903) is dissolved in distilled water to a concentration of 0.001 unit per microliter.

Alkaline Phosphatase (EC 3.1.3.1). The bacterial enzyme is preferred (e.g., Sigma P4252) because it is thermally stable and can be readily

[21] J. Apgar, R. W. Holley, and S. H. Merrill, *J. Biol. Chem.* **237**, 796 (1962).

purified from any contaminating activities by heating (see below). The protein suspension in 2.5 M ammonium sulfate can be used directly if the digest is to be examined solely by LC/MS. If the hydrolyzate is to be examined as a mixture with a desorption ionization method such as FAB,[1] or if chemical derivatization is required,[3,4,22] desalting of the enzyme is recommended because the amount of ammonium sulfate (330 $\mu g/\mu l$) introduced will be in substantial excess of the analyte and may interfere with the analysis. Alkaline phosphatase may contain deaminases[7,8] which can transform adenosine (A) and deoxyadenosine (dA) to inosine (I) and deoxyinosine (dI), respectively; if present in excessive amount, they can be removed.

REMOVAL OF EXCESS AMMONIUM SULFATE. A simple procedure for desalting has been described[23] which yields a phosphatase preparation stable for several months at 4°. Centrifuge the suspension at 12,000 rpm at 4° for 1 min. Remove the supernatant and add distilled water to the precipitated enzyme to yield the desired enzymatic activity in units per microliter.

ASSAY FOR ADENOSINE AND DEOXYADENOSINE DEAMINASES. A rapid HPLC-based assay for deaminase activity may be conducted as follows. Prepare solutions of A and dA in distilled water at concentrations of 1 to 2 nmol/μl. Incubate 20 nmol of each with 1 unit of alkaline phosphatase in 50 μl of 0.05 M ammonium bicarbonate for 1 hr at 37°. Inject 10 μl onto a C_{18} reversed-phase HPLC column (250 × 4.6 mm LC-18S, Supelco, Bellefonte, PA) eluted with 8% acetonitrile (40% aq.) in 0.25 M ammonium acetate, pH 6, at a flow rate of 2 ml/min. The following retention times will be observed: I (4.20 min), dI (5.07 min), A (10.95 min), and dA (13.27 min).

REMOVAL OF DEAMINASES. If deaminase activity is present, it can be removed by the following procedure.[12,24] Centrifuge the alkaline phosphatase suspension 5 min at 12,000 rpm. Remove the supernatant and dissolve the precipitated enzyme in 0.05 M Tris-Cl, pH 7.8. Heat the solution at 95° for 10 min, then centrifuge the solution 5 min at 12,000 rpm to precipitate contaminating enzyme(s). Transfer the phosphatase-containing supernatant to a separate vial for assay and storage.

ASSAY FOR PHOSPHATASE ACTIVITY. One unit (assayed as described by the supplier using p-nitrophenylphosphate as a substrate) of alkaline phosphatase will convert 100 nmol of pA in 100 μl of 0.05 M ammonium

[22] M. Dizdaroglu, *J. Chromatogr.* **367,** 357 (1986).

[23] T. Maniatis, E. F. Fritsch, and J. Sambrook, "Molecular Cloning: A Laboratory Manual," p. 134. Cold Spring Harbor Laboratories, Cold Spring Harbor, NY, 1982.

[24] H. O. Smith and M. L. Birnstiel, *Nucleic Acids Res.* **3,** 2387 (1976).

bicarbonate to A in less than 5 min at 37°. Using the HPLC system described above, the retention time for pA is 3.50 min. The phosphatase preparation is usable if 50% or greater activity is retained.[8]

Protocol for Enzymatic Digestion of DNA and RNA

For addition of enzymes, the author prefers 5 μl Wiretrol micropipettes (Drummond Scientific Co., Broomall, PA) calibrated in 1-μl increments. To avoid contamination of the stock solutions, a separate set is reserved for dispensing each enzyme and the enzyme solution is deposited onto the side of the sample tube just above the solution surface. If the plunger is withdrawn about 1 cm, the micropipette can be used to combine the two solutions by gentle stirring; any solution drawn into the pipette by capillary action can readily be redispensed without contacting the plunger.

Procedure. The nucleic acid should be dissolved in water or 0.001 M Tris-Cl, pH 7.4, at a concentration of about 0.5 μg/μl for DNA and 1–3 μg/μl for RNA. Denature the nucleic acid by heating for 3 min at 100°, then rapidly chill the solution in an ice-water slush. Add $\frac{1}{10}$ vol of 0.1 M ammonium acetate, pH 5.3, and mix well. The amounts of enzymes specified are for each 0.5 A_{260} unit of nucleic acid. Add 2 units of nuclease P_1, and incubate the solution at 45° for 2 hr. Add $\frac{1}{10}$ vol of 1 M ammonium bicarbonate,[25] and 0.002 units of venom phosphodiesterase and continue incubation for an additional 2 hr at 37°.[26] Last, add 0.5 units of alkaline phosphatase and incubate for 1 hr at 37°.[27] The solution is then ready for immediate analysis by LC/MS.[28]

General Comments. If other types of mass spectrometric experiments are to be performed, verification of completeness of digestion by HPLC[29] prior to analysis is strongly recommended because unexpected failures may be encountered due to decomposition of enzyme(s) during storage and repetitive handling, or unanticipated excess salt content of the starting nucleic acid. Examples of "failed" digests using DNA (from herring

---

[25] The pH of freshly made 1 M reagent-grade ammonium bicarbonate is 7.8–7.9, and the solution may be used directly. The pH will increase with time, and should be checked before use.

[26] If mixture analysis at the nucleotide level is desired, the alkaline phosphatase treatment is omitted.

[27] Although venom phosphodiesterase and alkaline phosphatase are used at the same pH, the enzymes should be added consecutively and not concurrently; otherwise, attenuation of phosphodiesterase activity will occur.

[28] S. C. Pomerantz and J. A. McCloskey, this volume [44].

[29] M. Buck, M. Connick, and B. N. Ames, *Anal. Biochem.* **129**, 1 (1983).

sperm; Sigma D1632) are shown in Fig. 1. Figure 1 shows completely digested material containing a typically small amount of RNA contamination. 5′-Deoxyribonucleotides resulting from incomplete dephosphorylation are shown in Fig. 1b; these peaks are characteristically broad and elute earlier than the corresponding nucleosides. (See Table III in [44][28] for retention times of 5′-ribonucleotides from incompletely dephosphorylated RNA digests.) Figure 1c illustrates incomplete P_1 digestion resulting in accumulation of small oligomers, identified by LC/MS.[28] If the digest is successful and is to be stored frozen for repetitive use, the solution should be neutralized with acetic acid. In addition, if thiolated nucleosides are present, the mixture should be made 0.001 M in dithiothreitol,[30] otherwise loss of sulfur will occur with time. If analysis of the mixture is to be performed using FAB,[1] or if the mixture is to be derivatized for EI-MS,[3,4,22] it may be directly dried in a suitable vial; removal of enzymes is not necessary in either case.

Discussion

Figure 2 shows chromatograms of mixed tRNA from the thermophilic archaebacterium *Sulfolobus solfataricus* digested by two different protocols. Figure 2a shows the digest from a 2-hr nuclease P_1-only protocol[7] [the same solution incubated an additional 12 hr (overnight)[12] is essentially unchanged and is shown and interpreted in [44]). Two prominent late-eluting peaks are dinucleotides determined by LC/MS to consist of m_2^2Gm[10] and C, and mimG[11] and A. The chromatogram of the digest in Fig. 2b, conducted as described in this chapter, reveals that these extremely nuclease-resistant dinucleotides have been digested to their constituent nucleosides by a brief additional treatment with venom phosphodiesterase. Note an increase in the amount of m^6A (generated from base-catalyzed rearrangement of m^1A) as a consequence of the additional 2-hr incubation at pH 7.9.

The following comments can serve as a guide for adaptation of this two-enzyme protocol if sufficient nucleic acid is available for testing. The original short protocols for digestion of tRNA[7] and DNA[8] using only nuclease P_1 and alkaline phosphatase, substituting the enzyme solutions, and volatile buffers described here, may prove entirely suitable. If this is not the case, evidenced by the accumulation of dinucleotides (Figs. 1b and 2a), then it is likely that extension of the digestion time to 12 hr[12] will still

---

[30] For analysis of seleno nucleosides, which are readily oxidized, the nuclease digestion buffer contains 1 mM dithiothreitol [W. M. Ching, *Arch. Biochem. Biophys.* **244,** 137 (1986)].

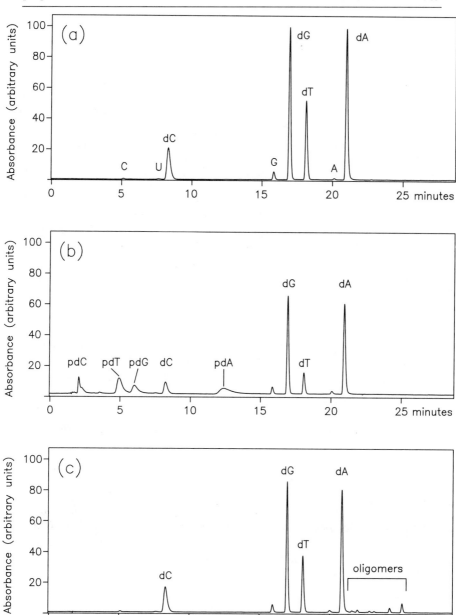

FIG. 1. HPLC chromatograms of deoxynucleosides from digested herring sperm DNA. HPLC conditions are as described elsewhere in this volume.[28] (a) Completely digested DNA; small amounts of RNA nucleosides are apparent. (b) Incompletely dephosphorylated digest from an insufficient amount of alkaline phosphatase. (c) Incompletely digested DNA from an insufficient amount of nuclease P_1.

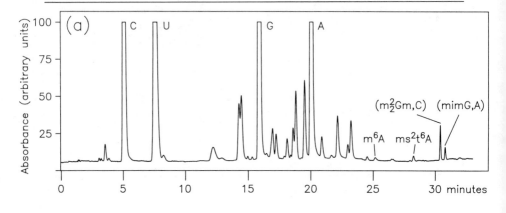

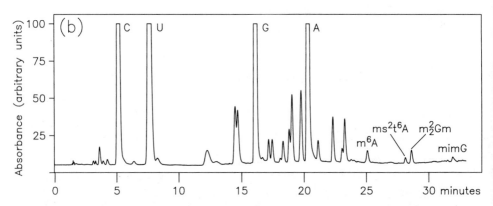

FIG. 2. HPLC chromatograms of nucleosides from *Sulfolobus solfataricus* tRNA. (a) tRNA digested using only nuclease P_1 and alkaline phosphatase.[7] (b) tRNA digested using the protocol described in this chapter. HPLC conditions as described elsewhere in this volume.[28]

leave highly modified constituents incompletely released, and only lead to loss of acid-labile species through prolonged incubation at pH 5.3. In any case, as a general principle, limiting exposure to extremes of pH can only improve the recovery of labile constituents.

Acknowledgments

The author is indebted to S. C. Pomerantz for preparation of figures. A portion of the methods described were not previously published. Their development was supported by National Institutes of Health grant GM 21584.

[43] Preparation of Trimethylsilyl Derivatives of Nucleic
Acid Components for Analysis by Mass Spectrometry

By KARL H. SCHRAM

Introduction

Analysis of nucleic acid bases, nucleosides, and nucleotides by electron ionization (EI) or chemical ionization (CI) mass spectrometry is best performed following conversion to their trimethylsilyl (TMS) derviatives.[1] In terms of gas chromatography-mass spectrometry (GC/MS) characteristics, and relative to other derivatives, the TMS analogs (1) provide good gas chromatographic peak shape with reproducible retention times[2] and narrow solvent peaks, (2) afford a wealth of structural information, with the fragmentation of the TMS derivatives of bases,[3] nucleosides,[4] and nucleotides[5] being, in general, well understood (see [46]), and (3) are easily prepared, in a quantitative manner, on a microscale. Reagents are also commercially available for the preparation of the deuterium-labeled TMS derivative, the mass spectrum of which can be used to confirm tentative ion structures assigned on the basis of the spectrum of the unlabeled TMS analog.[6] The degree of success which may be expected from the GC/MS separation and analysis of nucleic acid components depends on the complexity and polarity of the compound(s) being examined. Generally, good success may be expected from the GC/MS analysis of the TMS derivatives of nucleic acid bases and nucleosides using either packed[7,8] or capillary GC columns. More complex nucleosides containing polar modifications in the carbohydrate or aglycon may decompose during chromatography and analysis by direct-probe sample introduction will be required. Higher order samples, i.e., the nucleotide monophosphates, are also best analyzed by direct probe since, even as their TMS derivatives,

---

[1] J. A. McCloskey, A. M. Lawson, K. Tsuboyama, P. M. Krueger, and R. N. Stillwell, *J. Am. Chem. Soc.* **90,** 4182 (1968).

[2] S. E. Hattox and J. A. McCloskey, *Anal. Chem.* **46,** 1378 (1974).

[3] E. White, P. M. Krueger, and J. A. McCloskey, *J. Org. Chem.* **37,** 430 (1972).

[4] H. Pang, K. H. Schram, D. L. Smith, S. P. Gupta, L. B. Townsend, and J. A. McCloskey, *J. Org. Chem.* **47,** 3923 (1982).

[5] A. M. Lawson, R. N. Stillwell, M. M. Tacker, K. Tsuboyama, and J. A. McCloskey, *J. Am. Chem. Soc.* **93,** 1014 (1971).

[6] J. A. McCloskey, R. N. Stillwell, and A. M. Lawson, *Anal. Chem.* **40,** 233 (1968).

[7] C. W. Gehrke and A. B. Patel, *J. Chromatogr.* **123,** 335 (1976).

[8] C. W. Gehrke and A. B. Patel, *J. Chromatogr.* **130,** 103 (1977).

the nucleotides are not sufficiently volatile for GC separation. Finally, analysis of the TMS derivatives of nucleosides and nucleotides by fast atom bombardment (FAB) ionization provides greater sensitivity[9] and considerably more fragmentation[10] compared to analysis of the free compounds.

Preparation of TMS Derivatives

Reagents

N,O-Bis(trimethylsilyl)trifluoroacetamide (BSTFA) containing 1% trimethylchlorosilane (TMCS) and N,O-bis(trimethylsilyl)acetamide (BSA) is available from a number of suppliers including Pierce (Rockford, IL), Regis Chemical Co. (Morton Grove, IL), Alltech Associates, Inc. (Deerfield, IL), and Supelco, Inc. (Bellefonte, PA). $^2H_9$-Labeled reagents may be purchased from Regis and MSD Isotopes (Montreal, Canada). Derivatization-grade solvents, e.g., pyridine and acetonitrile, are available from the above mentioned sources.

General Method for Bases, Nucleosides, and Nucleotides

A general method for the preparation of TMS derivatives of nucleic acid bases, nucleosides, and nucleotides has been described[11] and is performed as follows. The sample is thoroughly dried overnight above P_2O_5 in a vacuum desiccator to ensure the removal of all water. An estimated 50 μg of the sample is transferred to a suitable reaction vessel, BSTFA + 1% TMCS (40 μl) and pyridine (10 μl) are added, the reaction tube tightly closed, and the mixture heated at 100° for 1 hr. After cooling to room temperature, 0.5–1 μl of the reaction mixture is injected onto the GC column or analyzed by direct-probe introduction of the sample. TMS derivatives prepared in the above manner have been reported to remain stable for days or weeks if tightly capped and refrigerated.[11]

Smaller scale derivatization reactions may be performed using a melting point capillary tube as the reaction vessel. A solution containing the sample is transferred to the tube and the sample thoroughly dried. An appropriate quantity of the derivatization reagents is added to afford a

[9] Q. M. Weng, W. M. Hammargren, D. Slowikowski, K. H. Schram, C. Borysko, L. L. Wotring, and L. B. Townsend, *Anal. Biochem.* **178,** 102 (1989).

[10] D. Slowikowski and K. H. Schram, *Biomed. Environ. Mass Spectrom.* **13,** 263 (1986).

[11] K. H. Schram and J. A. McCloskey, *in* "GLC and HPLC Determination of Therapeutic Agents" (K. Tsuji, ed.), p. 1149. Dekker, New York, 1979.

concentration in the range of 100 ng to 1 $\mu g/\mu l$; the tube is sealed and heated as described above. The dilution factor used is dependent on the sensitivity and operating mode of the mass spectrometer, with significantly smaller quantities of sample being required if the selected-ion monitoring technique is used. Additional details of derivative formation may be found in elsewhere in this volume [14].

Preparation of TMS Derivatives of Nucleic Acid Bases. The general procedure is applicable to mixtures of nucleic acid bases and either BSTFA + 1% TMCS or BSA + 1% TMCS may be used with heating at 100° for 1 to 2 hr. Alternatively, the reaction may be performed at room temperature if a mixture of BSA/acetonitrile (3 : 1, v/v) is used and the sample is occasionally shaken.[3] Other solvents may also be used, but side reactions initiated by the solvent may occur, as has been noted in the use of BSTFA in dimethylformamide (DMF).[12]

The general procedure is also applicable to the preparation of the TMS derivatives of DNA samples which have been acid hydrolyzed to the component nucleic acid bases.

Preparation of TMS Derivatives of Nucleosides. The general procedure may also be applied to pure samples or mixtures of nucleosides of synthetic or biological origin. However, in some cases, for example "hypermodified" nucleosides in RNA hydrolysates[13] and complex nucleoside antibiotics,[14] derivatization may be incomplete or the sample may not be sufficiently volatile to permit analysis by GC/MS. In such cases, the sample will have to introduced into the mass spectrometer using the direct probe. Although not a problem with purified samples, identification or quantitation of individual components in a mixture from the direct probe may be difficult and successful experiments will probably require the use of alternative methods, e.g., FAB combined with MS/MS.[15-17]

Another problem with the use of TMS derivatives of nucleosides is the formation of undesired side reaction products with some nucleosides, a particular problem when the structure elucidation of an unknown nucleo-

[12] S. K. Sethi, P. F. Crain, and J. A. McCloskey, *J. Chromatogr.* **254**, 109 (1983).

[13] H. Kasai, K. Nakanishi, R. D. Macfarlane, D. F. Torgerson, Z. Ohashi, J. A. McCloskey, H. J. Gross, and S. Nishimura, *J. Am. Chem. Soc.* **98**, 5044 (1976).

[14] J. A. McCloskey, *in* "Mass Spectrometry in Biomedical Research" (S. J. Gaskell, ed.), p. 75. Wiley, New York, 1986.

[15] F. W. Crow, K. B. Tomer, M. L. Gross, J. A. McCloskey, and D. E. Bergstrom, *Anal. Biochem.* **139**, 243 (1984).

[16] K. H. Schram, *Trends Anal. Chem.* **7**, 28 (1988).

[17] K. H. Schram, *in* "Clinical Biochemistry: Principles, Methods, Applications, Vol. 1: Mass Spectrometry" (A. M. Lawson, ed.), p. 507. de Gruyter, New York, 1989.

side is being attempted. Examples include incorporation of an oxygen at the C-8 position of 7-methylguanosine[18] and the dehydration of reduced pyrimidine nucleosides to their analogous aromatic analogs.[19] Finally, special problems exist with the chromatographic behavior of cytidine and some of its analogs and preparation of mixed derivatives may be required for the GC/MS analysis of this class of nucleosides.[20,21]

TMS Derivatives of RNA Hydrolysates.[22] The hydrolysate from 40 μg of tRNA is thoroughly dried over P_2O_5 under vacuum. Dry pyridine (10 μl) is added and the solution redried to remove all traces of water. To the dried mixture, contained in the same tube as was used for hydrolysis, is added BSA or BSTFA, TMCS, and pyridine (100 : 1 : 10, v/v/v; total volume, 20 μl) and the vial tightly sealed and heated at 100° for 1 hr (see this volume [42]); this same procedure is applicable to the silylation of mononucleotides produced by enzymatic hydrolysis of DNA.[23] An aliquot of the derivatized mixture is then used for analysis by mass spectrometry.

TMS Derivatives of the Nucleoside Fraction of Human Urine. The nucleoside fraction from human urine, obtained using solid-phase extraction of 3 ml of urine,[24] is thoroughly dried in a suitable reaction vial. BSTFA + 1% TMCS (80 μl) and pyridine (20 μl) are added, the mixture tightly sealed, and heated at 100° for 1 hr. A 0.5- to 1-μl portion of reaction mixture is injected onto the capillary column for GC/MS analysis. Nucleosides present in larger quantities, e.g., pseudouridine, may require dilution with fresh BSTFA/TMCS to prevent saturation of the detector. Combined chromatographic methods may be used for the isolation of larger amounts of sample.[25]

TMS Derivatives of Nucleotides. Nucleotides, as either the free compound or in the salt form, may require a higher temperature for complete derivatization, e.g., 150° compared to 100° for the bases and nucleosides,

[18] D. L. von Minden, R. N. Stillwell, W. A. Koenig, K. J. Lyman, and J. A. McCloskey, *Anal. Biochem.* **50,** 110 (1972).

[19] J. A. Kelley, M. M. Abbasi, and J. A. Beisler, *Anal. Biochem.* **103,** 203 (1980).

[20] K. H. Schram, T. Taniguchi, and J. A. McCloskey, *J. Chromatogr.* **155,** 355 (1978).

[21] U. I. Krahmer, J. G. Liehr, K. J. Lyman, E. A. Orr, R. N. Stillwell, and J. A. McCloskey, *Anal. Biochem.* **82,** 217 (1977).

[22] H. Pang, D. L. Smith, P. F. Crain, K. Yamaizumi, S. Nishimura, and J. A. McCloskey, *Eur. J. Biochem.* **127,** 459 (1982).

[23] D. Swinton, S. Hattman, P. F. Crain, C.-S. Cheng, D. L. Smith, and J. A. McCloskey, *Proc. Natl. Acad. Sci. U.S.A.* **80,** 7400 (1983).

[24] V. V. Mykytyn and K. H. Schram, *Proc. 37th ASMS Conf. Mass Spectrom. Allied Topics,* Miami, FL p. 734 (1989).

[25] G. B. Chheda, H. B. Patrzyc, H. A. Tworek, and S. P. Dutta, *in* "Chromatography and Modification of Nucleosides" (C. W. Gherke and K. C. Kuo, eds.), pp. 185–230. Elsevier, Amsterdam, 1990.

and shorter reaction time (30 min); room temperatures can also be used with reaction times of from 30 min to overnight, but lower derivatives may form.[5] If the analysis of cytidine 5'-monophosphate (CMP) or adenosine 3',5'-cyclic monophosphate (cAMP) is required, different conditions are used for preparation of the TMS derivatives.

CYTIDINE 5'-MONOPHOSPHATE.[5] Approximately 200 μg of the thoroughly dried nucleotide or its salt is dissolved in BSA (35 μl), TMCS (5 μl), and acetonitrile (16 μl) and heated at 70° for 60 to 90 min. Full derivative formation, i.e., addition of 5 TMS groups, is only 20–30% that observed for the other major nucleotides.

ADENOSINE 3',5'-CYCLIC MONOPHOSPHATE.[5] A 50-μg sample of the dried acid or its salt is dissolved in a mixture of BSTFA + 1% TMCS and pyridine (10:1, v/v) and heated at 100° for 3 hr.

Gas Chromatographic and Mass Spectrometric Conditions

Gas Chromatography

Development of a temperature program which affords the necessary resolution of components is empirical and dependent on the complexity of the nucleoside mixture. General conditions used in this laboratory[26] for the analysis of complex mixtures of urinary nucleosides are as follows. A 0.5- to 1-μl sample of the TMS reaction mixture is introduced via splitless injection onto a fused-silica DB-5 capillary column (30 m × 0.25 mm; 0.25 μm film thickness; J&W Scientific, Folsom, CA) directly coupled to the mass spectrometer. The initial column temperature is 150° with a program rate of 6°/min to 300° with a final hold time of 5 min. Helium is used as the carrier gas at a head pressure of 10 psi with the injection port and transfer lines at 250°. Estimates of the elution order of a number of nucleosides can be made based on previously published data using packed gas chromatographic columns.[2,11]

Mass Spectrometry

Conditions used for the acquisition of full-scan, EI data on complex mixtures of TMS-derivatized nuclosides are as follows[26]: source temperature, 250°; mass range scanned, 70–1000 Da; scan rate, 0.4 sec/decade; resolution, 1000 (10% valley definition); and ionizing energy, 70 eV.

If the derivatized sample is introduced into the mass spectrometer using the direct-insertion probe, care should be taken that the fore-vacuum

[26] M. L. J. Reimer, T. D. McClure, and K. H. Schram, *Biomed. Environ. Mass Spectrom.* **17**, 533 (1989).

line is opened slowly during evaporation of reagents and solvent to prevent "bumping" the sample from the crucible.

Operation in the chemical ionization (CI) mode is performed by adjusting the reagent gas pressure until the following ion ratios are obtained: (a) m/z 17/15 approximately 3 : 1 for methane, (b) m/z 57/43 approximately 30 : 1 for isobutane, and (c) m/z 18/35 approximately 25 : 1 for ammonia.[27] Other chromatographic and mass spectrometer operating conditions are as described above.

Analysis of nucleoside and nucleotide TMS derivatives in the FAB mode is performed as follows[9,10]: 1 μl of the derivatization mixture is added to 1 μl of an appropriate matrix, e.g., diglyme (tetraethyleneglycol dimethyl ether), previously applied to the tip of the FAB probe. The probe is introduced into the mass spectrometer and the mass range of interest scanned at 11 sec/decade. Samples are ionized using an Ion Tech FAB 11N saddle field atom gun (Ion Tech Ltd., Teddington, England) operating at 8 kV and 1 mA with argon as the bombarding gas. The major advantages[9,10] of analysis of the TMS-derivatized nucleosides in the FAB mode include (1) the presence of an intense MH$^+$ ion which allows easy assignment of molecular weight, (2) the presence of a large number of structurally relevant ions related to the intact molecule, the sugar moiety, and the aglycon with portions of the sugar residue attached, (3) considerably better sensitivity relative to the FAB analysis of the free sample, and (4) the acquisition of the mass spectrum at ambient temperature, which decreases the possibility of thermal degradation.

[27] M. S. Wilson and J. A. McCloskey, *J. Am. Chem. Soc.* **97**, 3436 (1975).

[44] Analysis of RNA Hydrolyzates by Liquid Chromatography–Mass Spectrometry

By STEVEN C. POMERANTZ and JAMES A. MCCLOSKEY

Reversed-phase high-performance liquid chromatography (HPLC) is an experimentally effective technique for the separation of nucleosides,[1] particularly for the analysis of enzymatic digests of RNA[2,3] and DNA.[4,5]

[1] C. W. Gehrke and K. C. Kuo, *J. Chromatogr.* **188**, 129 (1980).
[2] C. W. Gehrke and K. C. Kuo, *J. Chromatogr.* **471**, 3 (1989).
[3] M. Buck, M. Connick, and B. N. Ames, *Anal. Biochem.* **129**, 1 (1983).
[4] K. C. Kuo, R. A. McCune, and C. W. Gehrke, *Nucleic Acids Res.* **8**, 4763 (1980).
[5] C. W. Gehrke, R. A. McCune, M. A. Gama-Sosa, M. Ehrlich, and K. C. Kuo, *J. Chromatogr.* **301**, 199 (1984).

Depending on the problem at hand, identification of nucleoside constituents can often be made on the basis of retention times and UV absorbance characteristics. As is common, in general, when chromatographic methods are used for purposes of identification, as opposed to simply for separation, the reliability of the method suffers as the complexity of the mixture increases, or when components of unknown or unexpected identity are encountered. The development of directly combined HPLC-mass spectrometry [liquid chromatography (LC)/MS] based on the thermospray interface[6,7] provides a method which can be effectively applied to the analysis of nucleosides in nucleic acid digests,[8] and is a powerful extension of the capabilities of either technique alone.

Identification of structurally known nucleosides present in mixtures using LC/MS is based on: (a) relative HPLC retention times, which are generally reproducible to ±0.15 min from run-to-run and ±0.50 min from day-to-day, or better if precautions such as rigorous temperature control are taken[1]; (b) characteristic UV absorbance,[2,9,10] for example, recorded using a dual wavelength detector or photodiode array detector; (c) complete mass spectra, recorded every 2–3 sec over the duration of a chromatographic run, or in the form of selected-ion recordings,[11] made several times per second. In the case of structurally unknown components, the thermospray mass spectrum will establish whether a given constituent is a nucleoside or nucleotide, as opposed to a UV-absorbing impurity. For unknown nucleosides, the mass spectrum will, in general, establish the masses of the molecule and base moiety, and whether ribose is methylated or otherwise substituted. In some cases, LC can be carried out using deuterated mobile phase, thus exchanging all heteroatom-bound hydrogen atoms by deuterium.[12] The resulting mass spectrum can be used to differentiate isomers having different numbers of active hydrogens without isolation of the component of interest, and provides constraints in the characterization of structural unknowns. In the case of the unknown nucleosides, LC/MS data may, in favorable cases, provide sufficient leads to permit structural candidates to be proposed, and then tested by synthesis or other means.

[6] C. R. Blakley, J. J. Carmody, and M. L. Vestal, *Anal. Chem.* **52**, 1636 (1980).
[7] M. L. Vestal, this volume [5].
[8] C. G. Edmonds, M. L. Vestal, and J. A. McCloskey, *Nucleic Acids Res.* **13**, 8197 (1985).
[9] R. H. Hall, "The Modified Nucleosides in Nucleic Acids," Chap. 2. Columbia Univ. Press, New York, 1971.
[10] H. Ishikura, K. Watanabe, and T. Ohishima, *in* "Handbook of Biochemistry" (Japanese Biochemical Society, ed.), Vol. I, p. 1032. Tokyo Kagaku Dozin, Tokyo, 1979.
[11] J. T. Watson, this volume [4].
[12] C. G. Edmonds, S. C. Pomerantz, F. F. Hsu, and J. A. McCloskey, *Anal. Chem.* **60**, 2314 (1988).

In any event, LC/MS provides a rapid means of screening RNA or DNA digests so that specific components of interest can be rigorously identified,[13-17] or structural unknowns can be targeted for isolation and further characterization.[18,19] In favorable situations in which modifications are relatively simple, preliminary structure assignments to unknown nucleosides can be made, without isolation of components, directly from digests of unfractionated tRNA.[20,21] The thermospray LC/MS technique can also be utilized in conjunction with RNA sequencing studies by analysis of oligonucleotides produced by selective cleavage by ribonuclease T_1,[17] and is effective for rapid screening of synthetic nucleoside reaction mixtures,[22] and hydrolyzates of DNA *in vitro* reaction products.[23]

The use of a mass-specific HPLC detector (the mass spectrometer) introduces a high degree of structural selectivity, such that minor components of interest can generally be detected without chromatographic resolution from other constituents. On the other hand, thermospray mass spectra are less reproducible in terms of ion yields and relative abundances than ions produced by other means (in particular, electron ionization) and so are less well suited for quantitative analysis. In common with other ionization methods which produce parent molecular species by protonation,[24] and in distinct contrast to electron ionization (EI),[25] thermospray generates relatively few ions and thus less structural detail, although the

[13] T. G. Hagervall, C. G. Edmonds, J. A. McCloskey, and G. R. Bjork, *J. Biol. Chem.* **262**, 8488 (1987).

[14] G. M. Kirtland, T. D. Morris, P. H. Moore, J. J. O'Brian, C. G. Edmonds, J. A. McCloskey, and J. R. Katze, *J. Bacteriol.* **170**, 5633 (1988).

[15] T. G. Hagervall, Y. H. Jonsson, C. G. Edmonds, J. A. McCloskey, and G. R. Bjork, *J. Bacteriol.* **172**, 252 (1990).

[16] R. P. Martin, A.-P. Sibler, C. W. Gehrke, K. Kuo, C. G. Edmonds, J. A. McCloskey, and G. Dirheimer, *Biochemistry* **29**, 956 (1990).

[17] P. F. Crain, T. Hashizume, C. C. Nelson, S. C. Pomerantz, and J. A. McCloskey, in "Biological Mass Spectrometry" (A. L. Burlingame and J. A. McCloskey, eds.), p. 509. Elsevier, Amsterdam, 1990.

[18] D. W. Phillipson, C. G. Edmonds, P. F. Crain, D. L. Smith, D. R. Davis, and J. A. McCloskey, *J. Biol. Chem.* **262**, 3462 (1987).

[19] J. A. McCloskey, P. F. Crain, C. G. Edmonds, R. Gupta, T. Hashizume, D. W. Phillipson, and K. O. Stetter, *Nucleic Acids Res.* **15**, 683 (1987).

[20] C. G. Edmonds, P. F. Crain, T. Hashizume, R. Gupta, K. O. Stetter, and J. A. McCloskey, *J. Chem. Soc. Chem. Commun.* p. 909 (1987).

[21] J. A. McCloskey, C. G. Edmonds, R. Gupta, T. Hashizume, C. H. Hocart, K. O. Stetter, *Nucleic Acids Res. Symp. Ser.* **20**, 45 (1988).

[22] T. Hashizume, C. C. Nelson, S. C. Pomerantz, and J. A. McCloskey, *Nucleosides Nucleotides* **9**, 355 (1990).

[23] S. M. Musser, S.-S. Pan, and P. S. Callery, *J. Chromatogr.* **474**, 197 (1989).

[24] A. G. Harrison and R. J. Cotter, this volume [1].

[25] J. A. McCloskey, this volume [45].

principal side chains common to posttranscriptionally modified RNA can usually be recognized[26] (see below).

Materials

Reagents

HPLC-grade water
Deuterium oxide, 98 atom % D or higher
Ammonium acetate
Acetonitrile
Acetic acid
Trifluoroacetic anhydride
Reagents should be the highest purity obtainable (HPLC-grade).

Water. Water quality is of paramount importance for both chromatography and mass spectrometry. Optimum thermospray performance for molecular ion production requires low concentrations of alkali cations in the mobile phase. Water that is free of organic impurities and has a resistivity of at least 16 $M\Omega$-cm^{-1} is required. In this laboratory, water is obtained from the house-distilled water feed stock, which is subsequently treated by passage through an ion-exchange bed, an activated-carbon filter, a Nanopure II ion-exchange system (Barnstead, Boston, MA), a second activated-carbon filter, and a final particulate filter (0.2 μm).

Deuterium Oxide. Previous experience with commercially available D_2O indicated a purity problem, causing thermospray vaporizers to become occluded by deposits, probably siliceous in nature, within 30 to 60 min of operation. This problem can be avoided if D_2O is purchased that has been stored and shipped in non-glass containers. Low conductivity, 99.8% minimum isotopic purity D_2O from Merck Isotopes (St. Louis, MO) is satisfactory.

Ammonium Acetate Buffer. The buffer (A) is prepared by adding 77.08 g of deliquescent ammonium acetate crystals (HPLC-grade, J. T. Baker, Phillipsburg, NJ) to 4 liters of water. While stirring vigorously, the pH of the solution is adjusted to 6.0 with glacial acetic acid. The buffer is vacuum degassed and filtered through a 0.2 μm Nylon-66 filter in a single step with a solvent filter/degasser (EM Science, Cherry Hill, NJ). Aqueous buffer should be stored in polyethylene containers to prevent leaching of cations from glass vessels, and should be used within 2 weeks of preparation.

---

[26] C. G. Edmonds, T. C. McKee, and J. A. McCloskey, *Proc. 33rd ASMS Conf. Mass Spectrom. Allied Topics, San Diego, CA* p. 514 (1985).

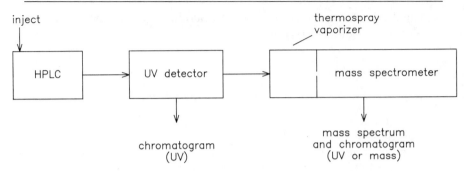

FIG. 1. Configuration of instrument components for thermospray LC/MS; see text for discussion.

Organic Modifier. Organic modifier (B) is prepared by adding 2.4 liters of water to 1.6 liters of acetonitrile (HPLC-grade, American Burdick and Jackson, Muskegon, MI), and degassed by stirring vigorously under vacuum for 20 min. Alternatively, a filter/degasser, ultrasonication, or helium sparging may be employed to degas the organic modifier.

Deuterated Buffer. Deuterated ammonium acetate (ND$_4$OAc) is prepared by three successive evaporations from D$_2$O solution. The deuterated ammonium acetate crystals are stored under dry nitrogen to prevent back-exchange with atmospheric water. The buffer solution is made immediately prior to use, with pD adjustment to 7.25 by deuterated trifluoroacetic acid (prepared by deuterolysis of trifluoroacetic anhydride). The pD of the buffer is measured with a standard pH electrode, and corrected by the relation pD = pH + 0.4 units.[27] Neat acetonitrile is used for organic modifier in both isocratic and gradient elutions to maximize deuterium exchange levels.

Instrumentation

The configuration of instrument components is shown schematically in Fig. 1.

Liquid Chromatograph. The chromatographic system employed for analyses described here is a Beckman Instruments (Fullerton, CA) Model 342. This system consists of a Model 114M single-piston pump for delivery of aqueous buffer, and a Model 100A dual-piston pump for delivery of

---

[27] P. K. Glasoe and F. A. Long, *J. Phys. Chem.* **64,** 188 (1960).

organic modifier. These pumps are connected to a Model 400 dual-chamber high-pressure mixer with a 2.8 ml mixing volume. Because high-pressure mixing and gradient formation is generally more reproducible than low-pressure mixing, two pumps and a high-pressure mixer are used. In general the pumps should be the most pulse-free available, and capable of operation at 5000 psi back-pressure. The injector is a Beckman Model 210 fitted with a 100 μl loop. Microparticulates are prevented from entering the injector by the insertion of a 0.5 μm frit assembly (Scientific Systems, State College, PA) between the mixer and injector. Chromatographic separations are performed on a 4.6 × 250 mm LC-18S analytical column (Supelco, Bellefonte, PA) preceded by a 4.6 × 30 mm RP-18 guard column (Applied Biosystems, Foster City, CA). Reproducibility of retention times is increased by maintaining the analytical and guard columns at 30° in a thermostatically controlled column heater (Model TCM, Waters Associates, Milford, MA). UV absorbance detection at 254 and 280 nm is performed using a Model 440 detector (Waters Associates) serially connected between the column and the mass spectrometer, as shown in Fig. 1. In principle, any UV detector can be used, but the detector flow cell must be capable of withstanding a pressure differential of approximately 1500 psi. If possible, the absorbance detector output should be digitized by the mass spectrometry data system to facilitate time alignment of the UV chromatogram and mass spectra, as discussed in a later section. All interconnecting tubing is 0.01-inch id, and lengths are minimized to the greatest extent possible.

Mass Spectrometer. All spectra presented here were obtained on a noncommercial quadrupole mass spectrometer and data system previously described.[8] The quadrupoles are maintained at 150° to prevent vapor condensation in the analyzer section. Mass spectra are scanned every 1.7 sec from m/z 100 to m/z 360. The UV absorbance signal is digitized and recorded by the mass spectrometer data system each time a mass spectral scan is made. A UV chromatogram is, therefore, produced by the data system for accurate time alignment with mass channels. In the resulting chromatograms, the mass channels time scale lags the UV time scale by approximately 6 sec, due to the transit time between UV and mass detectors.

HPLC Gradient. The gradient profile used for nucleoside separations is a slight modification of the gradient developed by Buck *et al.*[3] and is listed in Table I. If only a narrow range of elution times is of interest, or for simple mixtures, isocratic conditions or a modified version of the gradient shown can be used to reduce the analysis time.

Thermospray Vaporizer Controller and Operating Conditions. The vaporizer controller used is from Vestec Corp. (Houston, TX). Most mass

TABLE I
GRADIENT PROFILE FOR HPLC OF NUCLEOSIDES

Interval time (min)		Composition (%B)	
Start	End	Start	End
0.0	3.0	0.0	0.0
3.0	4.4	0.0	0.2
4.4	5.8	0.2	0.8
5.8	7.2	0.8	1.8
7.2	8.6	1.8	3.2
8.6	10.0	3.2	5.0
10.0	25.0	5.0	25.0
25.0	30.0	25.0	50.0
30.0	34.0	50.0	75.0
34.0	37.0	75.0	75.0
37.0	45.0	75.0	100.0
45.0	48.0	100.0	100.0
48.0	53.0	100.0	0.0

spectrometer manufacturers offer thermospray interface options, either as part of a new instrument or for retrofitting to an existing instrument. The controller should provide complete temperature control of the vaporizer probe, and should ideally include provision for automatic temperature compensation for the variable heat capacity of the gradient eluant. The operational parameters of the vaporizer for nucleoside analysis are in general accord with most other classes of compounds. Sensitivity is maximized when the tip temperature (T_2) is maintained within 2° to 5° of the "burnout" temperature, i.e., the temperature at which 100% of the effluent is vaporized within the vaporizer. The determination of this operating point from the T_1/T_2 curve has been previously described.[7] Operation at this point for vaporizers of average orifice size (75 μm diameter) yields tip temperatures of approximately 240°–270°, and control temperatures (T_1) in the range of 125°–140°. The source block should be maintained at approximately 350° to provide sufficient heat to complete the vaporization process. The resultant vapor temperature is approximately 270°–300°. In thermospray ion sources equipped with vaporizer tip heaters, the tip should be maintained at approximately 300°. Optimization of the operating conditions can be accomplished by pumping 2 ml/min of a solution of ammonium acetate buffer containing 1 ppm adenosine and 5 ppm guanosine directly into the ion source, bypassing the column. A broadly useful operating temperature is obtained by maximizing the response of the guanosine molecular ion (MH$^+$, m/z 284), without significant sacrifice of the adenosine response (MH$^+$, m/z 268). The intensity ratio is a moderately strong function of temperature, and is highly vaporizer-dependent. Generally, the best achievable response ratio is approximately 10 : 1 (268 : 284), although in many cases, maximum performance will only yield a response ratio of 20 : 1 or less.

Enzymatic Hydrolysis of RNA. Procedures for nuclease digestion of RNA to ribonucleotides, followed by dephosphorylation using alkaline phosphatase, are followed which, in general, permit direct injection of the crude digest directly into the HPLC. Detailed protocols and discussion are given elsewhere in this volume.[28]

Interpretation of Results

HPLC Retention Times and UV Absorbance

The elution positions in the reversed-phase HPLC system of Buck *et al.*[3] of ribonucleosides from RNA are listed in Table II; some common contaminants which may occur in enzymatic digests of RNA are given in Table III. These retention times data were compiled from this laboratory and the earlier listing of Buck *et al.*[3] Because individual retention time values were, in general, determined from multiple experiments and not necessarily from the same analysis, their values in relation to each other may exhibit minor variance. In some cases, elution order has been estimated from the standard reversed-phase retention data published by Gehrke,[2] based on a phosphate buffer system. In general, retention times can be referenced to the elution times of known nucleosides within ±0.50 min on a day-to-day basis. The major ribonucleosides (cytidine, uridine, guanosine, adenosine) can be used as internal reference points, although peak centers are often difficult to accurately define if large quantities have been injected. Depending on the source of RNA, common modified nucleosides present in lower amounts, such as pseudouridine, N^2-methylguanosine, or N^6,N^6-dimethyladenosine, can also be used as convenient reference points.

Common contaminants such as the four major 2'-deoxynucleosides from traces of DNA can usually be recognized from their retention times (Table III), in particular, for deoxycytidine and deoxyadenosine, which are generally well resolved in the presence of ribonucleosides. Mononucleotides resulting from incomplete dephosphorylation (see also Ref. 28) constitute a more difficult problem because the elution times for most of the monophosphates of the nucleosides listed in Table II are not known.

---

[28] P. F. Crain, this volume [42].

TABLE II
THERMOSPRAY LC/MS DATA FOR RNA NUCLEOSIDES

Nucleoside	Symbol	Retention time (min)[a]	Mass spectrum, m/z^b		
			MH$^+$	BH$_2^+$	S − H·NH$_4^+$
5,6-Dihydrouridine	D	3.2	247	115	150
Pseudouridine	Ψ	3.5	245	—	—
5-Carboxymethylaminomethyluridine	cmnm^{5}U	4.7	332	202	150
Cytidine	C	5.2	244	112	150
5-Hydroxyuridine	ho^{5}U	5.2	261	129	150
5-Carboxymethoxyuridine	cmo^{5}U	5.6	319	187	150
3-(3-Amino-3-carboxypropyl)uridine	acp^{3}U	6.5^c	346	214	150
Uridine	U	7.4	245	113	150
1-Methylpseudouridine	m^{1}Ψ	8.6	259	—	—
2-Thiocytidine	s^{2}C	8.7	260	128	150
2′-O-Methylpseudouridine	Ψm	11.9	259	—	—
3-Methylcytidine	m^{3}C	12.0	258	126	150
N^4-Methylcytidine	m^{4}C	12.0	258	126	150
5-Methylcytidine	m^{5}C	12.4	258	126	150
5-Carboxymethylaminomethyl-2-thiouridine	cmnm^{5}s^{2}U	12.7	349	216	150
1-Methyladenosine	m^{1}A	13.9	282	150	150
5-Methylaminomethyl-2-thiouridine	mnm^{5}s^{2}U	14.1	304	172	150
2′-O-Methylcytidine	Cm	14.3	258	112	164
Inosine	I	15.2	269	137	150
5-Methoxyuridine	mo^{5}U	15.2	275	143	150
5-Methyluridine (ribosylthymine)	m^{5}U (T)	15.4	259	127	150
2-Thiouridine	s^{2}U	15.5	261	129	150
Guanosine	G	15.7	284	152	150
7-Methylguanosine	m^{7}G	16.0	298	166	150
2′-O-Methyluridine	Um	17.0	259	113	164
4-Thiouridine	s^{4}U	17.0	261	129	150
3-Methyluridine	m^{3}U	17.1	259	127	150
4,2′-O-Dimethylcytidine	m^{4}Cm	17.8	272	126	164
5,2′-O-Dimethylcytidine	m^{5}Cm	18.0	272	126	164
Epoxyqueuosine	oQ	18.1	426	294	150
1-Methylinosine	m^{1}I	18.2	283	151	150
Queuosine	Q	18.3	410	278	150
1-Methylguanosine	m^{1}G	18.7	298	166	150
2′-O-Methylguanosine	Gm	18.7	298	152	164
Lysidine	acp^{2}C	19.0	372	240	150
N^4-Acetylcytidine	ac^{4}C	19.4	286	154	150
N^2-Methylguanosine	m^{2}G	19.5	298	166	150
Adenosine	A	19.9	268	136	150
5-Methyl-2-thiouridine	m^{5}s^{2}U (s^{2}T)	20.2	275	143	150
5-Methoxycarbonylmethoxyuridine	mcmo^{5}U	20.2	333	201	150
2′-O-Ribosyladenosine	Ar	21.6	400	136	282
1,2′-O-Dimethylinosine	m^{1}Im	21.9	297	151	164

TABLE II (continued)

Nucleoside	Symbol	Retention time (min)[a]	Mass spectrum, m/z[b]		
			MH⁺	BH₂⁺	S − H · NH₄⁺
N^2,N^2-Dimethylguanosine	m₂²G	22.0	312	180	150
2′-O-Methyladenosine	Am	22.8	282	136	164
N^6-Threonylcarbamoyladenosine	t⁶A	23.0	413	281	150
5-Methoxycarbonylmethyl-2-thiouridine	mcm⁵s²U	23.2[c]	333	201	150
N^4-Acetyl-2′-O-methylcytidine	ac⁴Cm	23.9	300	154	164
2-Methyladenosine	m²A	24.4	282	150	150
N^6-Methyladenosine	m⁶A	25.0	282	150	150
2-Thio-2′-O-methyluridine	s²Um	25.3	275	129	164
N^6-Methyl-N^6-threonylcarba-moyladenosine	mt⁶A	27.0[c]	427	295	150
N^6-Threonylcarbamoyl-2-methyl-thioadenosine	ms²t⁶A	27.8	459	327	150
2-Methylthioadenosine	ms²A	28.2	314	182	150
$N^6,2′$-O-Dimethyladenosine	m⁶Am	28.5[c]	296	150	164
$N^2,N^2,2′$-O-Trimethylguanosine	m₂²Gm	28.7	326	180	164
N^6,N^6-Dimethyladenosine	m₂⁶A	30.6	296	164	150
3-(β-D-Ribofuranosyl)-4,9-dihydro-4,6,7-trimethyl-9-oximidazo[1,2-a]purine	mimG	31.5	350	218	150
N^6-(trans-4-Hydroxy-3-methyl-2-butenyl)adenosine (zeatin riboside)	io⁶A	32.5[c]	352	220	150
N^6-(cis-4-Hydroxy-3-methyl-2-butenyl)adenosine (zeatin riboside)	io⁶A	32.5[c]	352	220	150
Wyobutosine	yW	34.5[c]	509	377	150
N^6-(3-Methyl-2-butenyl)adenosine	i⁶A	36.0	336	204	150
N^6-(cis-4-Hydroxy-3-methyl-2-butenyl)-2-methylthio-adenosine	ms²io⁶A	36.2[c]	398	266	150
N^6-(3-Methyl-2-butenyl)-2-methylthioadenosine	ms²i⁶A	45.0[c]	382	250	150

[a] Nucleosides listed with the same retention times are given in the order of elution.

[b] Values given are calculated, and do not imply that the ion will be observed under typical operating conditions.

[c] Time taken from M. Buck, M. Connick, and B. N. Ames, *Anal. Biochem.* **129,** 1 (1983).

TABLE III
THERMOSPRAY LC/MS DATA FOR COMMON CONTAMINANTS IN RNA HYDROLYZATES

Compound	Symbol	Retention time (min)	Mass spectrum, m/z^a		
			MH$^+$	BH$_2^+$	S $-$ H · NH$_4^+$
Cytidine 5'-monophosphate	pC	2.0	324	112	150^b
Uridine 5'-monophosphate	pU	2.3	325	113	150^b
Guanosine 5'-monophosphate	pG	3.4	364	152	150^b
Adenosine 5'-monophosphate	pA	7.4	348	136	150^b
2'-Deoxycytidine	dC	8.3	228	112	134
5-Methyl-2'-deoxycytidine	m^{5}dC	16.2	242	126	134
2'-Deoxyguanosine	dG	16.4	268	152	134
2'-Deoxyinosine	dI	16.8	253	137	134
Thymidine	dT	17.8	243	127	134
2'-Deoxyadenosine	dA	20.8	252	136	134
N^6-Methyl-2'-deoxyadenosine	m^{6}dA	26.2	266	150	134

^{a} Values given are calculated, and do not imply that the ion will be observed under typical operating conditions.

^{b} In the mass spectra of mononucleotides, the value of S is defined as that of the corresponding nucleoside. See text for discussion.

In the case of very complex nucleoside mixtures, such as digests of unfractionated tRNA, identification of minor components based solely on retention time is unreliable (see discussion elsewhere in this volume[29]), due not only to the large number of known natural nucleosides, but also to the possible presence of small oligonucleotides or to UV-absorbing impurities.

UV absorbance characteristics,[2,9,10] for example, in the form of 254/280 nm ratios, can be useful in limiting cases for confirmation of structure or as an aid in recognizing mononucleotides through the characteristic absorbance of the corresponding nucleoside. However, chromatographic overlap from minor components or impurities, which may be unrecognized, seriously limits the reliability of UV absorbance, particularly in the most crowded region of the chromatogram between about 17 and 23 min. With the exception of 5,6-dihydrouridine, all natural nucleosides from RNA and DNA absorb strongly in the UV, and are readily detected by UV absorbance at levels down to about 1 ng. Dihydrouridine, the earliest eluting nucleoside (Table II), is easily detected by mass spectrometry, as discussed below.

In principle, a UV detector is not required for detection, and mass spectrometer total ion current can be used to monitor elution of nucleosides. In practice the ion current signal from thermospray is sufficiently

[29] J. A. McCloskey, this volume [41].

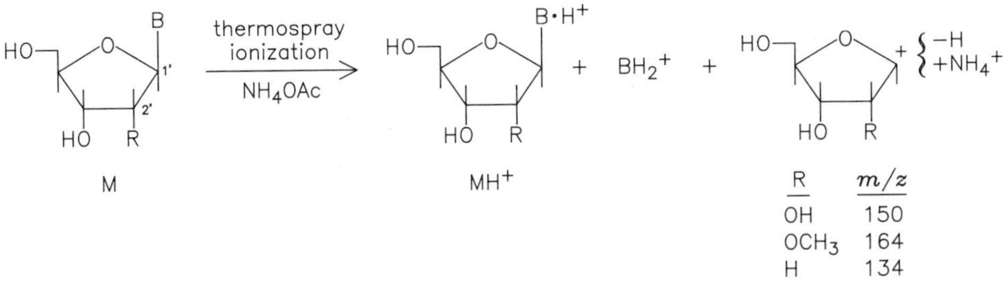

SCHEME 1. Principal products from thermospray ionization of nucleosides.

noisy that fine detail of chromatographic resolution is lost and occasional interferences are encountered from nonnucleoside (non-UV-absorbing) impurities which contribute to the total ion current signal. The use of a UV detector placed in series between the chromatograph and mass spectrometer as shown in Fig. 1 is, therefore, strongly recommended.

Thermospray Mass Spectra of Nucleosides

The principal products of thermospray ionization of nucleosides in the presence of NH_4^+, are schematically shown in Scheme 1,[8] and are similar to those resulting from NH_3 chemical ionization.[30] These are the protonated molecule (MH^+), the protonated free base (BH_2^+), and a sugar ion $[(S - H) \cdot NH_4^+]$ corresponding to loss of one hydroxyl hydrogen from the sugar fragment and formation of an adduct with NH_4^+ from the mobile-phase buffer. Thermospray mass spectra of the four major ribonucleosides from RNA are shown in Fig. 2. Molecular ion (MH^+) abundances vary widely, and are influenced by vapor temperature and other experimental and physical conditions which affect droplet size and rate of evaporation of the spray, such as flow rate and wear at the tip of the vaporizer nozzle. Among the common nucleosides, guanosine and 7-methylguanosine produce extensive dissociation of MH^+ to form BH_2^+, a tendency exhibited by some other derivatives of guanine. MH^+ abundance is not strictly related to sample size, and for large quantities (>1 μg), MH^+ peaks may be observed at higher relative intensity values than is the case in the range below $\sim$50 ng. The principal sugar ions of all nucleosides are usually of low abundance and assume one of the three values shown in Scheme 1, characteristic of the two principal sugars known in natural RNA (ribose or 2'-O-methylribose), or DNA (2'-deoxyribose).

---

[30] M. S. Wilson and J. A. McCloskey, *J. Am. Chem. Soc.* **97**, 3436 (1975).

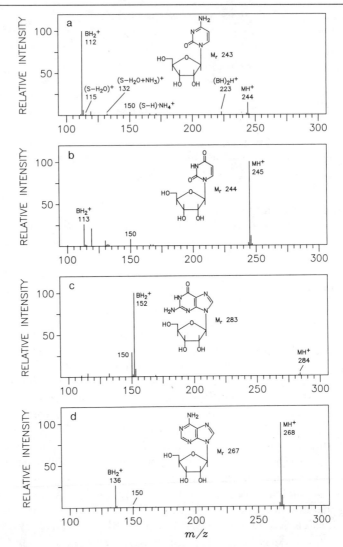

FIG. 2. Thermospray mass spectra of (a) cytidine, (b) uridine, (c) guanosine, (d) adenosine. Some minor sugar ions are denoted in panel (a). See text for experimental details.

Minor ions, which vary in abundance and depend in part on operating conditions, and are not always observed, include the sugar moiety (S) ions[31] as marked in Fig. 2a, and adduct ions of the molecule and base

[31] F. F. Hsu, Ph.D. Dissertation, University of Utah, Salt Lake City, Utah, 1986.

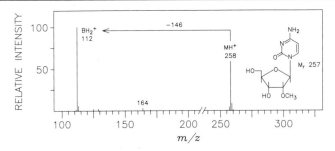

FIG. 3. Thermospray mass spectrum of 2'-O-methylcytidine, showing 146 u difference characteristic of methylribose.

with NH_4^+, Na^+ or protonated acetonitrile. In some cases, particularly derivatives of cytidine, proton-bound base–base dimers may be formed (see Fig. 2a, m/z 223) and recognized by their exact chromatographic correspondence with the corresponding BH_2^+ or MH^+ ions.

Simple Modification. Methylation and other simple forms of modification are most easily recognized by mass shifts: in the base by differences between the observed value for BH_2^+ and the corresponding values of the four major nucleosides, and in the sugar by differences between MH^+ and BH_2^+ in the same spectrum (e.g., 146 mass units for 2'-O-methylation) as seen in Fig. 3. Note that hydroxylation and thiation (replacement of O by S) in the base each result in a 16 mass unit shift, so isomers such as 5-hydroxyuridine and 2- or 4-thiouridine must be differentiated by other means such as relative retention time (Table II).

Pseudouridine and its derivatives are distinguished by the absence of BH_2^+ ions (Fig. 4), a characteristic of C-nucleosides resulting from increased glycosidic bond strength.[32] Fragmentation energy is, therefore, channeled into other dissociation routes which follow analogous pathways reported in the NH_3-chemical ionization spectrum[30] involving expulsion of H_2O (m/z 227, 209) and cleavage through the ribose ring (m/z 155, 185) with corresponding ammonia complexes (m/z 172, 202).

Complex Modification of the Base. Posttranscriptional processing of RNA produces more than 50 forms of base modification, mostly in tRNA and many with side chains that produce characteristic fragment ions. Thermospray mass spectra of some of the major classes of modified nucleosides from tRNA are shown in Fig. 5, and serve as examples of the influence of side chains on the dissociation of MH^+.[26] In this respect, two processes are represented: one is protonation of basic sites in the side chain which strongly directs side chain fragmentation; the other involves

[32] J. M. Rice and G. O. Dudek, *Biochem. Biophys. Res. Commun.* **35**, 383 (1969).

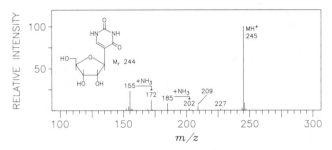

FIG. 4. Thermospray mass spectrum of pseudouridine, showing the absence of $BH_2{}^+$ ions characteristic of C—C glycosidic bonds.

thermal effects, which contributes to the absence of MH^+ ions in the spectra of some compounds having complex side chains (Fig. 5c,d).

N-Acetylation of cytidine, a modification found in many tRNAs, leads to elimination of ketene (42 u) from both the molecular and base ions (Fig. 5a), a reaction which is broadly characteristic of O- and N-acetyl derivatives in the case of other ionization modes. In 5-methylaminomethyl-uridine (Fig. 5b), loss of CH_3 from the base produces a stable protonated 5-methylamino ion, m/z 158. This ion is characteristic of several related 5-substituted uridines and, although of relatively low abundance, can be effectively monitored for detection in hydrolyzates of unfractionated tRNA by LC/MS.[13]

Very polar nucleosides dissociate completely in the heated spray and [in contrast to fast atom bombardment (FAB)] molecular species are of low abundance or are absent, although this disadvantage is, in part, compensated for by the occurrence of characteristic and structurally informative fragment ions. Two leading examples are represented by the highly modified nucleosides N-[(9-β-D-ribofuranosyl-9H-purine-6-yl)carbamoyl]-threonine (t^6A) (Fig. 5c) and queuosine (Q) (Fig. 5d), common constituents of tRNA. In both cases, side chain ions constitute the most abundant fragments formed as shown in Scheme 2: m/z 120 and 116, respectively. Formation of protonated adenosine (m/z 268) and adenine (m/z 136) from t^6A (Fig. 5c) clearly marks the spectrum as that of an adenosine derivative. With the exception of the minor ion m/z 294, structural detail of the threonyl side chain is absent, in contrast to the EI mass spectrum of the volatile trimethylsilyl derivative.[25,33]

[33] H. Kasai, K. Murao, S. Nishimura, J. G. Liehr, and P. F. Crain, *Eur. J. Biochem.* **69,** 444 (1976). (Microfilm Suppl. AO-553.)

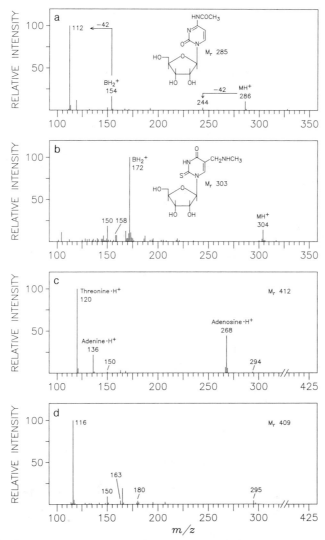

FIG. 5. Thermospray mass spectra of side chain containing nucleosides from RNA. (a) N^4-Acetylcytidine; (b) 5-methylaminomethyl-2-thiouridine; (c) t^{6}A; (d) queuosine. See also Scheme 2.

Although the thermospray mass spectrum of queuosine was shown earlier to exhibit an MH$^+$ peak,[18] the abundance of the MH$^+$ ion is dependent on vaporization conditions and the spectrum in Fig. 5d is more representative of that usually obtained. The most important ions, shown in Scheme 2, result from initial protonation of the side chain nitrogen, in

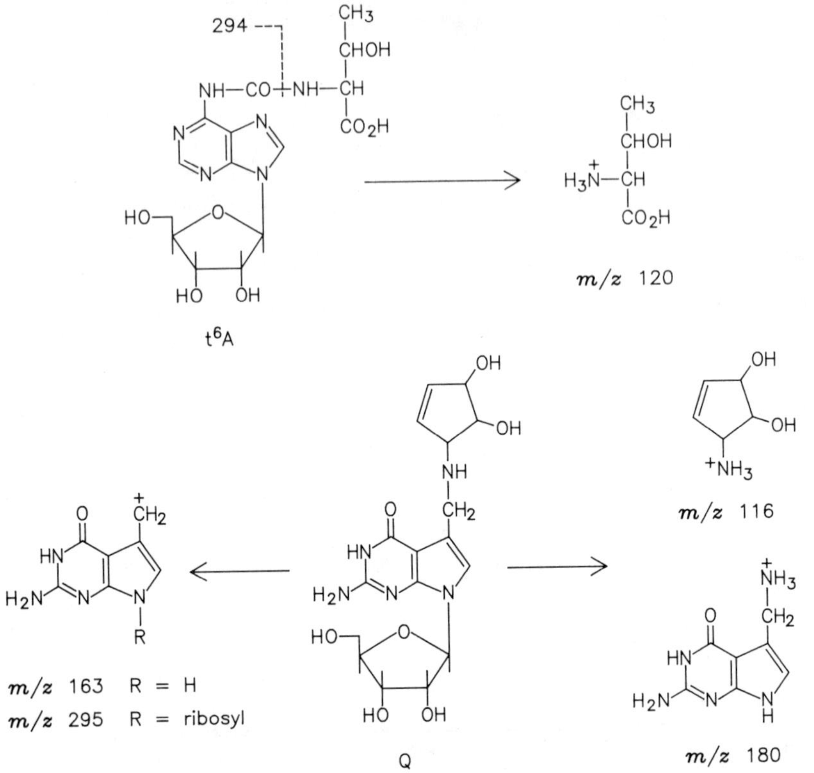

SCHEME 2. Ions from thermospray ionization of hypermodified nucleosides t⁶A and Q.

accord with the solution-phase properties of queuosine.[34] The abundant side-chain ion m/z 116 played a major role in the discovery of the epoxide derivative (oQ), in which the analogous ion occurs at m/z 132.[18]

Thermospray Mass Spectra Acquired in D₂O

If LC/MS is carried out in D_2O or similarly labeled mobile phases, active hydrogen will be rapidly exchanged for deuterium, leading to very high exchange levels of approximately 97%. Details of column equilibration and other experimental procedures are given by Edmonds *et al.*[12] The exchange procedure is useful under two circumstances: (1) to distinguish

[34] H. Kasai, Z. Ohashi, F. Harada, S. Nishimura, N. J. Oppenheimer, P. F. Crain, J. G. Liehr, D. L. von Minden, and J. A. McCloskey, *Biochemistry* **14**, 4198 (1975).

isomers which differ in the number of exchangeable hydrogen atoms, a circumstance most likely to occur when one or both reference isomers are unavailable to establish HPLC elution times, or when chromatographic separation is not possible; and (2) for characterization of structural unknowns, as an aid in developing plausible structures which can then be tested by other means. In either case, caution should be exercised that some exchange of "nonactive" carbon-bound protium is not effected. Such exchange occurs in the heated thermospray jet[12,35] and is characteristically quantitative at C-8 of guanosine derivatives, and to the extent of 45 to 59% at C-8 in adenosine derivatives.[36]

The D$_2$O-LC/MS method is particularly useful because exchange data can readily be obtained, on nanogram-level quantities of material in complex mixtures, without necessity of isolation of components of interest. An example of the method is given by the differentiation of the isomers N^4,2'-O-dimethylcytidine and 5,2'-O-dimethylcytidine in digests of unfractionated *Pyrodictium occultum* tRNA (Scheme 3). The isomers, which could not be reliably chromatographically distinguished or resolved from another nucleoside, both would exhibit the same MH$^+$ value. Upon LC/MS analysis in D$_2$O, the molecular species shifted very clearly from m/z 272 to 277, as shown in Scheme 3, indicating four exchangeable hydrogens in the neutral molecule, in favor of the 5,2'-O-dimethyl isomer. The same approach applies to other nucleoside or base isomers which differ by N or O vs. C substitution.

Analysis of RNA Hydrolyzates

Because of the added dimension of mass as a HPLC detection parameter, both reliability of identification and increased dynamic range of analysis are markedly enhanced using LC/MS (as configured in Fig. 1) compared with UV detection alone. If the analysis is oriented toward detection of structurally known or expected nucleosides, a direct combination of retention time and mass (usually MH$^+$ and BH$_2$$^+$ ions) for direct comparison with reference data as in Table II, will usually suffice.

The HPLC chromatogram shown in Fig. 6 is typical of a complex RNA hydrolyzate, with the exception that enzymatic digestion was carried out so as to produce several dinucleotides (for discussion in a subsequent section). The assignments marked were made by comparison of mass and retention time data with those in Table II, and following the guidelines for

[35] M. M. Siegel, *Anal. Chem.* **60**, 2090 (1988).
[36] C. G. Edmonds, S. C. Pomerantz, F. F. Hsu, and J. A. McCloskey, *Proc. 36th ASMS Conf. Mass Spectrom. Allied Topics, San Francisco, CA* p. 1256 (1988).

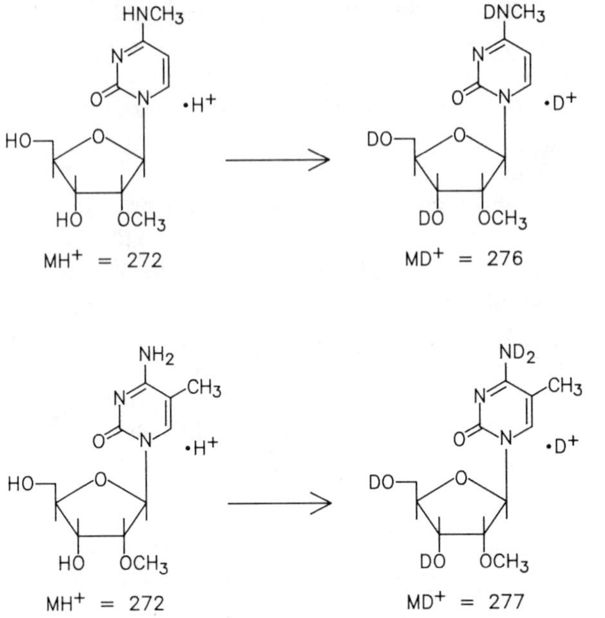

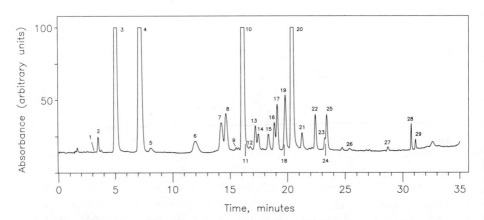

SCHEME 3. Mass shifts from D_2O-LC/MS used to distinguish N- and C-methyl nucleoside isomers.

FIG. 6. Chromatographic separation of nucleosides from LC/MS analysis of an enzymatic hydrolyzate of 18 μg of unfractionated *Sulfolobus solfataricus* tRNA. UV detection at 254 nm. Peak identities: 1, D; 2, Ψ; 3, C; 4, U; 5, dC; 6, m⁵C; 7, m¹A; 8, Cm; 9, I; 10, G; 11, s²U; 12, m⁷G; 13, dG; 14, Um; 15, dT; 16, m¹I; 17, m¹G; 18, ac⁴C; 19, m²G; 20, A; 21, dA; 22, m²₂G; 23, t⁶A; 24, Am; 25, unknown nucleoside; 26, m⁶A; 27, ms²t⁶A; 28, [m²₂Gm,C]; 29, [mimG,A]. Unnumbered peaks were shown by their mass spectra not to be nucleosides. For nomenclature, see Tables II and III. Data shown in Figs. 7–13 were taken from this experiment.

interpretation discussed below. If HPLC operating conditions are such that retention times shift slightly from those listed in reference tables, the components of interest can be internally referenced to other nucleosides in the same chromatogram. As an alternative means of correcting the retention time scale, analysis of the hydrolyzate can be immediately followed by analysis of an appropriate reference mixture of nucleosides or of a model RNA hydrolyzate. In the latter case, *Escherichia coli* tRNA^{Tyr} (Subriden RNA, Rolling Bay, WA) is a useful standard, containing six modified nucleosides (pseudouridine, 4-thiouridine, queuosine, 2'-*O*-methylguanosine, thymine riboside, and 2-methylthio-N^6-isopentenyladenosine),[37] spanning a range of retention times from 3.5 to 45 min.

If nucleosides of unknown or unexpected structure are encountered, the relatedness of ions in an individual mass spectrum should be verified through their chromatographic correspondence by replotting their abundances as time-based chromatograms. If mass spectra have been recorded every 2 sec, then the retention time of each eluant can be determined to within ±2 sec. For mass-based chromatographic resolution of components which very nearly coelute, a separate analysis using selected-ion monitoring may provide more useful ion current profiles of higher quality and resolution.

As indicated earlier, the quality of total ion signals from thermospray is insufficient to retain detailed chromatographic features, so that use of an intermediate UV detector (Fig. 1) is desirable. In the case of complex mixtures it is important to maintain a rigorous time scale correspondence between UV and mass signals, so that the mass spectra and ion current profiles can be reliably related to detailed features of the UV chromatogram.

As with other forms of combined chromatography-mass spectrometry, most notably GC/MS, the change of ion current with time (as a chromatogram) is an effective means of distinguishing ions from the component of interest from unrelated ions in the same mass spectral scan. In some instances, the major ribonucleosides, when present in relatively large quantity, will exhibit slight tailing as represented by the base ions (BH_2^+) for several minutes after elution. This effect, which is most pronounced for guanosine, is due to adsorption of traces of the base formed on the vaporizer tip and will not be evident in the UV chromatogram. For ions at all abundance levels, it is recommended that ion current profiles from all peaks of potential interest in the mass spectrum be displayed as an integral part of the interpretation of the data. In this manner, characteristic

[37] H. M. Goodman, J. N. Abelson, A. Landy, S. Zadrazil, and J. D. Smith, *Eur. J. Biochem.* **13**, 461 (1970).

and structurally informative ions can often be recognized that are otherwise indistinguishable from background or from signals due to other components. In many instances, systematic interrogation of low abundance ions in this fashion will uncover the presence of chromatographically hidden minor components, and will provide leads in the assignment of structure to unknown constituents.

The major ions produced from the presently described HPLC solvents are m/z 100, 101, 105, 118, 119, 124, 182, 184, and 227 and, in general, represent mass channels that cannot be used effectively for analysis at low levels. Because the changing composition of the gradient (Table I) causes changes in the abundances of most minor background ions with time, a conventional background subtraction of mass spectra at appropriate points in the chromatogram is usually desirable (see below).

Sensitivity of Detection. Scanned mass spectra (m/z 100–360, 1.7 sec) of satisfactory quality can usually be acquired from 10 to 30 ng or more of each component, or from 1 to 5 ng using selected-ion monitoring. Requirements for total sample size submitted to analysis will, therefore, depend on the levels of minor constituents of interest, and whether the UV or mass signals of major components are permitted to be recorded at saturation levels. In the case of isoaccepting tRNAs, 3–5 μg (0.06–0.1 A_{260} units) will produce a hydrolyzate containing ~30–50 ng of each modified nucleoside which occurs once per tRNA molecule, leading to mass spectra of good quality. Of more practical relevance, the analysis of unfractionated tRNA for components at or above the 0.05% level requires $\geq$30 μg of tRNA (which yields 20 μg of nucleosides). Obviously, the amount of nucleic acid required depends to some extent on experimental circumstances, and whether the components of interest occur principally in minor tRNA species. Figure 7a shows the mass spectrum recorded at 28.8 min. in Fig. 6 (the apex of peak 27), estimated to correspond to 18 ng. For comparison, the same data after background subtraction are shown in Fig. 7b from which m/z 101, 120, 182, and 314 can be tentatively concluded to arise from the HPLC peak in question. The chromatographic alignment of these four ions were plotted as shown in Fig. 8, and therefore confirms the relatedness of all but m/z 101. The 46 mass unit difference exhibited by m/z 182 and 314 compared with the corresponding peaks (m/z 136 and 268 in Fig. 5c) reflects replacement of H by CH_3S in the base of t^6A, in agreement with the assignment 2-methylthio-t^6A (ms^2t^6A) (see Table II).

Detection of Nucleosides Not Apparent in UV Chromatogram. The most direct means of searching for specific nucleosides is simply to generate the appropriate ion current profiles in the required retention time region of the chromatogram, usually within ±0.5 min of the expected *relative* retention time. Figure 9 shows the result of a test for presence of the

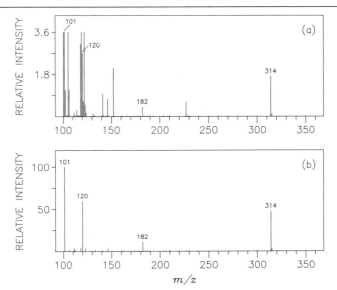

FIG. 7. Mass spectrum recorded at 28.8 min in Fig. 6. (a) Single scan, raw data; (b) after subtraction of background at 28.4 min.

uridine derivative s^2U, which is predicted (from Table II), if present, to coelute with a very large amount of guanosine. The m/z 261 chromatogram clearly shows the elution of s^2U approximately 0.25 min after guanosine, but well within the elution profile of guanosine and not detectable in the UV chromatogram. That the component represented by the m/z 261 signal is not the isomer s^4U is established by relative retention times (s^4U elutes 1.3 min after guanosine).

The interrogation of mass spectral files as described above is also a useful method to locate components of interest when using new or unfamiliar chromatographic procedures. By prediction of prominent ions for nucleosides which are either unknown or unavailable, hydrolyzates can be rapidly screened for specific structural features, as in the case of certain side chains (e.g., Fig. 5c,d).

It must be kept in mind that selected-ion data presentations and experiments may be misleading if not considered in the context of other ions or data from the chromatogram. A typical example of the problem is shown in Fig. 10 by the selected-ion profiles from m/z 154 and 286 which are characteristic of the cytidine derivative ac^4C (see Fig. 5a). The m/z 154 response at 20.4 min (corresponding to 20.3 min in the UV chromatogram) suggests the presence of an unexpected nucleoside not catalogued in Table II, but is instead due to a very minor ion associated with the elution of the

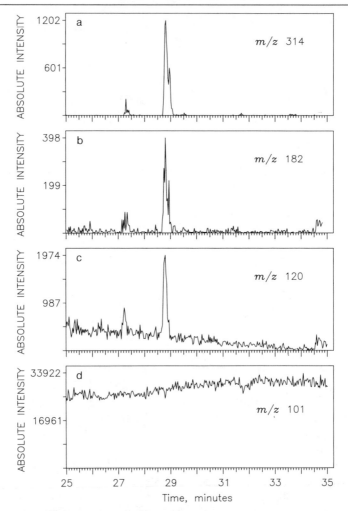

FIG. 8. Mass chromatograms from the four ions which survived background subtraction in Fig. 7b.

major component adenosine. Unlike the example shown in Fig. 9, the later m/z 154 signal exhibits an identical retention time to m/z 268, suggesting they are related to the same component.

Effect of Non-UV-Absorbing Constituents. The presence of non-UV-absorbing materials in the hydrolyzate will not be recognized in the UV chromatogram, but may give discrete chromatographic peaks or a tailing background continuum when monitored mass spectrometrically. With one

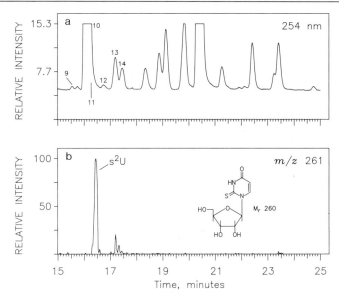

FIG. 9. Test for presence of 2-thiouridine (s^2U) by interrogation of LC/MS data recorded in Fig. 6. (a) UV detection; (b) detection using m/z 261, the MH^+ ion value for s^2U.

exception (discussed below) such components are of no interest, although they may interfere with other types of measurements if the HPLC fraction (in a separate experiment) is collected for further analysis. The direct correspondence between profiles from UV absorbance and ion current is the best means for determining which peaks in the mass spectrum are associated with non-UV-absorbing contaminants.

Dihydrouridine is the only natural nucleoside from RNA or DNA which has no appreciable absorbance in the UV, which is due to absence of the α,β-unsaturated ketone moiety. The earliest eluting natural nucleoside in the gradient system of Buck et al.,[3] it is readily detected in hydrolytic mixtures,[8] even in trace quantities as demonstrated in Fig. 11. The MH^+ ion, m/z 247 is much more abundant than that of the base BH_2^+ and is preferable for detection or monitoring. Typically, additional responses from the isotope peaks of pseudouridine, cytidine, and uridine will be observed in the m/z 247 channel, at appropriate retention times following dihydrouridine, as shown.

UV-Absorbing Components Which Are Not Ribonucleosides. The most common HPLC eluants in this category are small amounts of 2′-deoxynucleosides arising from contamination by DNA during the isolation of RNA. Although some modified deoxynucleosides may be present (nota-

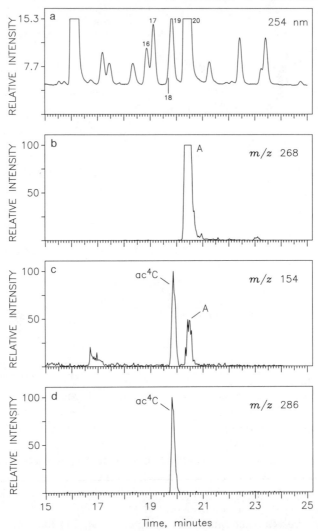

FIG. 10. Effect of minor ions associated with major nucleoside components. (a) UV detection; (b) m/z 268 channel, marking the elution position of adenosine; (c) m/z 154, characteristic of N^4-acetylcytidine (ac⁴C), showing "false positive" at 20.4 min; (d) m/z 286 channel, characteristic of ac⁴C.

bly 5-methyl-2′-deoxycytidine, Table III) under the present circumstance they will usually be at undetectable levels, and the only problem from DNA contamination will be the four major deoxynucleosides, at predictable elution times (Table III). As shown by the spectrum of thymidine in Fig. 12, deoxynucleosides are readily recognized by the 116 mass unit

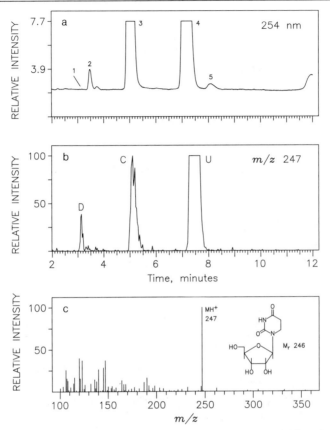

FIG. 11. Detection of the non-UV-absorbing nucleoside dihydrouridine (D). (a) UV detection; (b) m/z 247 channel (MH$^+$ for D); (c) mass spectrum recorded at 3.1 min.

difference between MH$^+$ and BH$_2$$^+$ ions, compared with 132 or 146 units for ribonucleosides (Scheme 1).

Mono- and dinucleotides may result from incomplete enzymatic digestion (cf. elsewhere in this volume[28]), and may be difficult to recognize under some circumstances due to the absence or very low abundance of molecular species. The monophosphates of the major nucleosides—those most prevalent as a result of incomplete hydrolysis—will elute earlier than the corresponding nucleosides (compare data in Tables II and III), but will exhibit essentially the mass spectrum of the nucleoside, which is undoubtedly formed in the heated spray.[31] A more difficult problem arises in the (less likely) event that monophosphates of modified nucleosides are present at detectable levels, because very few reference elution times are

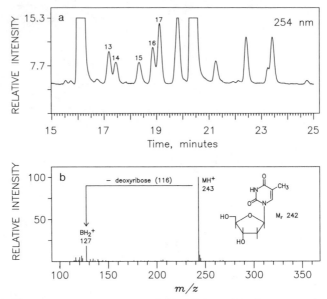

Fig. 12. Thermospray mass spectrum of thymidine resulting from contamination of the *S. solfataricus* tRNA isolate by DNA. (a) UV detection; (b) mass spectrum recorded at 18.4 min, showing 116 u difference characteristic of deoxynucleosides.

known. Here also, tentative assignment as a nucleotide can be made by finding ions corresponding to the nucleoside, but at the wrong elution time.

Dinucleotides may be formed principally due to marked resistance of some modified residues to nuclease digestion.[28] Among these modifications is ribose methylation, which renders the 3'-phosphate group to slowed digestion by nuclease P_1 so as to form nucleotides of the type NmpN. In general, dinucleotides can be characterized by the observation of ions corresponding to both constituent nucleosides and their bases, having identical elution time profiles, but at a retention time corresponding to neither nucleoside. A typical result (from intentional underdigestion) is shown in Fig. 13a, in which is plotted the mass spectrum of the 30.7-min HPLC peak from Fig. 6, from which ions were chosen for profiling. The matched elution profiles for cytosine (m/z 112), the MH^+ ion of a trimethylated guanosine (m/z 326), and the base dimethylguanine (m/z 180) are experimentally indistinguishable, leading to the probable partial structure m_2GmpC. In some instances such as this, one of the four diagnostic ions is absent (MH^+ of cytidine in the present case), an effect probably associated with the chemistry of thermally assisted dissociation in the spray, for which there is presently insufficient information to permit pre-

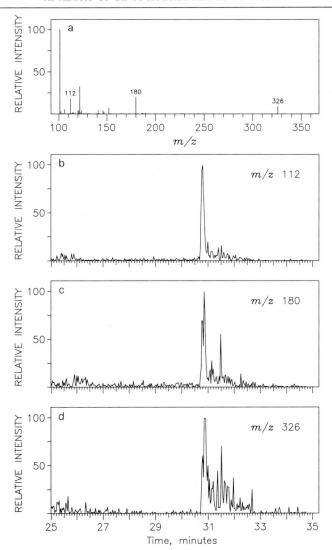

FIG. 13. Recognition of dinucleotides as impurities in RNA digests. (a) Mass spectrum recorded at 30.7 min (Fig. 6, peak 28). Mass chromatogram from ions shown in panel (a) (data from channels not exhibiting chromatographic correspondence not shown); (b) *m/z* 112; (c) *m/z* 180; (d) *m/z* 326.

diction of occurrence. From the evidence in hand, there is no direct means to place the two methyl groups in guanine, although a parallel approach in such cases is to collect the HPLC peak, and digest it completely to nucleosides for subsequent LC/MS analysis.[38] In the present example, the guanosine derivative is probably $N^2,N^2,2'-O$-trimethylguanosine (m_2^2Gm), based on the occurrence of this unusual nucleoside in *S. solfataricus* tRNA[20] and limited sequence data.[39] Other responses in the m/z 180 and 326 channels around 31.7 min suggest the presence of related nucleotides, but at levels too low for further characterization.

If dinucleotides are formed which do not contain ribose methylation, the possibility of two sequence isomers results, which cannot be distinguished on the basis of the data in hand. For example, the HPLC peak at 31.2 min in Fig. 6 exhibits ion current profiles (not shown) clearly demonstrating presence of the elements of adenosine and the tricyclic guanine derivative mimG,[19] but the dinucleotide isomers mimGpA and ApmimG cannot be distinguished.

Acknowledgments

The authors acknowledge numerous contributions of C. G. Edmonds, F. F. Hsu, and M. L. Vestal in early stages of development of the methods described, under support by the National Institute of General Medical Sciences (GM 21584). Hydrolyzates for the presently described experiments were prepared by P. F. Crain. Samples of lysidine and 2'-*O*-ribosyl-adenosine were generously provided by S. Yokoyama and G. Keith, respectively.

---

[38] N. Takeda, unpublished experiments, 1989.
[39] Y. Kuchino, M. Ihara, Y. Yabusaki, and S. Nishimura, *Nature (London)* **298**, 684 (1982).

[45] Electron Ionization Mass Spectra of Trimethylsilyl Derivatives of Nucleosides

By JAMES A. McCLOSKEY

Introduction

Trimethylsilyl (TMS) derivatives of nucleosides[1] have found extensive use for structural characterization by electron ionization (EI) mass spectrometry. Examples of applications include nucleosides from RNA,[2-7] physiological fluids,[8-11] nucleoside antibiotics,[12-15] and nucleoside adduct and DNA damage products.[16-19]

[1] J. A. McCloskey, A. M. Lawson, K. Tsuboyama, P. M. Krueger, and R. N. Stillwell, *J. Am. Chem. Soc.* **90,** 4182 (1968).

[2] H. Kasai, Z. Ohashi, F. Harada, S. Nishimura, N. J. Oppenheimer, P. F. Crain, J. G. Liehr, D. L. von Minden, and J. A. McCloskey, *Biochemistry* **14,** 4198 (1975).

[3] Z. Yamaizumi, S. Nishimura, K. Limburg, M. Raba, H. J. Gross, P. F. Crain, and J. A. McCloskey, *J. Am. Chem. Soc.* **101,** 2224 (1979).

[4] H. Pang, M. Ihara, Y. Kuchino, S. Nishimura, R. Gupta, C. R. Woese, and J. A. McCloskey, *J. Biol. Chem.* **257,** 3589 (1982).

[5] J. A. McCloskey, P. F. Crain, C. G. Edmonds, R. Gupta, T. Hashizume, D. W. Phillipson, and K. O. Stetter, *Nucleic Acids Res.* **15,** 683 (1987).

[6] T. Muramatsu, S. Yokoyama, N. Horie, A. Matsuda, T. Ueda, Z. Yamaizumi, Y. Kuchino, S. Nishimura, and T. Miyazawa, *J. Biol. Chem.* **263,** 9261 (1988).

[7] J. Desgres, G. Keith, K. C. Kuo, and C. W. Gehrke, *Nucleic Acids Res.* **17,** 865 (1989).

[8] G. B. Chheda, S. P. Dutta, A. Mittleman, J. A. Montgomery, S. K. Sethi, J. A. McCloskey, and H. B. Patrzyc, *Cancer Res.* **45,** 5958 (1985).

[9] G. B. Chheda, H. B. Patrzyc, A. K. Bhargava, P. F. Crain, S. K. Sethi, J. A. McCloskey, and S. P. Dutta, *Nucleosides Nucleotides* **6,** 597 (1987).

[10] G. B. Chheda, H. A. Tworek, A. K. Bhargava, E. M. Rachlin, S. P. Dutta, and H. B. Patrzyc, *Nucleosides Nucleotides* **7,** 417 (1988).

[11] K. H. Schram, *in* "Mass Spectrometry" (A. M. Lawson, ed.), p. 507. de Gruyter, New York, 1989.

[12] G. O. Morton, J. E. Lancaster, G. E. Van Lear, W. Fulmor, and W. E. Meyer, *J. Am. Chem. Soc.* **91,** 1535 (1969).

[13] K. Isono, P. F. Crain, and J. A. McCloskey, *J. Am. Chem. Soc.* **97,** 943 (1975).

[14] M. Uramoto, K. Kobinata, K. Isono, T. Higashijima, T. Miyazawa, E. E. Jenkins, and J. A. McCloskey, *Tetrahedron* **38,** 1599 (1982).

[15] K. Isono, P. F. Crain, T. J. Odiorne, J. A. McCloskey, and R. J. Suhadolnik, *J. Am. Chem. Soc.* **95,** 5788 (1973).

[16] D. B. Ludlum, B. S. Kramer, J. Wang, and C. Fenselau, *Biochemistry* **14,** 5480 (1975).

[17] N. K. Scribner, J. D. Scribner, D. L. Smith, K. H. Schram, and J. A. McCloskey, *Chem. Biol. Interact.* **26,** 27 (1979).

[18] D. Kanne, K. Straub, J. E. Hearst, and H. Raporport, *J. Am. Chem. Soc.* **104,** 6754 (1982).

[19] M. Dizdaroglu, this volume [46].

The present chapter principally describes steps and procedures for the interpretation of mass spectra of trimethylsilylated nucleosides. Most examples presented deal with ribonucleosides because they have had the greatest attention in terms of both model studies and applications, but the procedures described are generally relevant to deoxynucleosides,[20] various nucleoside analogs, and mononucleotides.[21] Relatively little discussion of the details of reaction pathways and mechanisms of fragmentation is presented, although in certain cases such considerations can be useful in making structure assignments, and, in particular, in the determination of sites of chemically or biologically incorporated stable isotopes. For details of mechanisms and ion structures, and citations to other work in the field, the reader is referred to the paper by Pang et al.[22] Procedures for hydrolysis of nucleic acids to nucleosides prior to derivatization are described elsewhere in this volume,[23] as are microscale trimethylsilylation procedures.[24]

Compared with other approaches based on mass spectrometry,[25-27] TMS derivatives offer the following advantages: (1) their EI mass spectra produce great structural detail, and models and isotopically labeled analogs have been extensively studied; (2) in terms of increased mass resulting from derivatization, TMS groups produce greater mass dispersion and therefore less chance of overlap when working with mixtures than in the case of the unblocked nucleosides, and result in molecular weights ($\sim$500–900) which fall in a generally background-free mass region; (3) the derivatives are easily prepared on a small scale and are sufficiently volatile and thermally stable, so that thermal decomposition during direct-probe vaporization is very unusual; most of the simpler nucleosides are amendable to gas chromatography[28] and thus GC/MS; (4) silylation reagents which contain fully deuterated methyl groups [$-Si(CD_3)_3$] are commer-

---

[20] J. A. McCloskey, in "Basic Principles in Nucleic Acid Chemistry" (P. O. P. Ts'o, ed.), Vol. I, p. 209. Academic Press, New York, 1974.

[21] A. M. Lawson, R. N. Stillwell, M. M. Tacker, K. Tsuboyama, and J. A. McCloskey, J. Am. Chem. Soc. 93, 1014 (1971).

[22] H. Pang, K. H. Schram, D. L. Smith, S. P. Gupta, L. B. Townsend, and J. A. McCloskey, J. Org. Chem. 47, 3923 (1982). (Microfilm Suppl. 53 pp.)

[23] P. F. Crain, this volume [42].

[24] K. H. Schram, this volume [43].

[25] J. A. McCloskey, in "Mass Spectrometry in the Health and Life Sciences" (A. L. Burlingame and N. Castagnoli, Jr., eds.), p. 521. Elsevier, Amsterdam, 1985.

[26] J. A. McCloskey, in "Mass Spectrometry in Biomedical Research" (S. J. Gaskell, ed.), p. 75. Wiley, New York, 1986.

[27] P. F. Crain, Mass Spectrom. Rev. 9, 505 (1990).

[28] For example, S. E. Hattox and J. A. McCloskey, Anal. Chem. 46, 1378 (1974).

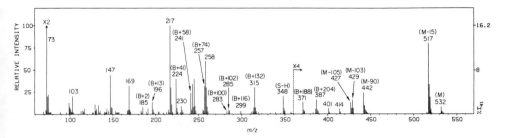

FIG. 1. Mass spectrum of uridine-(TMS)$_4$ with ions of the molecular ion, base, and sugar series indicated. B = 183 u. [From H. Pang, K. H. Schram, D. L. Smith, S. P. Gupta, L. B. Townsend, and J. A. McCloskey, *J. Org. Chem.* **47**, 3923 (1982), with permission.)

cially available (Regis Chemical Co., Morton Grove, IL; MSD Isotopes, Montreal), so that labeled derivatives can be routinely prepared.

The disadvantages of TMS derivatives are: (1) TMS groups (particularly N-TMS) are subject to hydrolysis in the presence of traces of moisture, frequently giving small but variable amounts of lower derivatives; (2) molecular ion abundances are often low, particularly in the case of EI spectra of the pyrimidine deoxynucleosides; (3) in some cases, notably pyrimidine deoxynucleosides, a large portion of the total ion current is consumed by silyl ions which have little or no structural utility (e.g., m/z 73, 147, 217); (4) when nucleosides from biological isolates are derivatized, sample sizes in the range below ca. 0.5 μg may be subject to the influence of extraneous materials, such as salts, resulting in inexplicably very low derivatization yields.

The overall potential advantages of trimethylsilylation must be weighed in the light of other factors—in particular, sample quantity and purity. Further discussion of these issues with respect to alternative methods is given elsewhere in this volume.[29]

Interpretation of Mass Spectra

The basic features of EI mass spectra of trimethylsilylated nucleosides are shown for the major ribonucleosides from RNA (Figs. 1 and 2). Detailed assignments are given for uridine-(TMS)$_4$, as provided in Fig. 1.

General Procedure

The order and relative importance of individual steps may vary according to the problem at hand. The following procedures assume little or no prior structural information. The plotting of ion current profiles can be

[29] J. A. McCloskey, this volume [41].

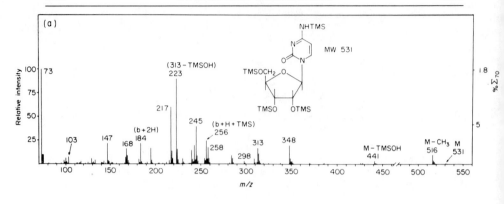

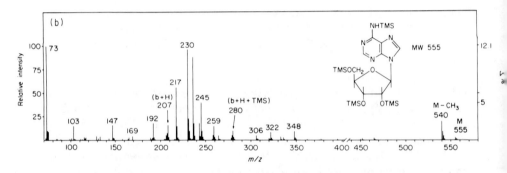

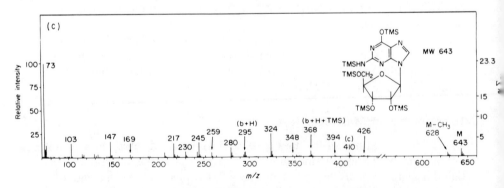

Fig. 2. Mass spectra of (a) cytidine-$(TMS)_4$ (B = 182 u); (b) adenosine-$(TMS)_4$ (B = 206 u); (c) guanosine-$(TMS)_5$ (B = 294 u). For detailed assignments, see Tables I and II to calculate specific values of M and B ions series, and Table III.

TABLE I
MOLECULAR ION SERIES FROM
TRIMETHYLSILYLATED NUCLEOSIDES[a]

Ion	Composition	Mass
M − 15	M − CH$_3$	M − 15.0235
M − 90	M − C$_3$H$_{10}$OSi	M − 90.0501
M − 103	M − C$_4$H$_{11}$OSi	M − 103.0579
M − 105	M − C$_4$H$_{13}$OSi	M − 105.0736
M − 118	M − C$_4$H$_{10}$O$_2$Si	M − 118.0450
M − 131	M − C$_5$H$_{11}$O$_2$Si	M − 131.0528
M − 180	M − C$_6$H$_{20}$O$_2$Si$_2$	M − 180.1002
M − 195	M − C$_7$H$_{23}$O$_2$Si$_2$	M − 195.1237

[a] From H. Pang, K. H. Schram, D. L. Smith, S. P. Gupta, L. B. Townsend, and J. A. McCloskey, *J. Org. Chem.* **47**, 3923 (1982).

made at any stage of the interpretation procedure as an aid to establishing the relatedness of ions to the same component. The details of structural assignments are given in later sections.

1. Assign the molecular ion and other members of the molecular ion series (listed in the following section). From the upper mass range of the spectrum, estimate the extent of contributions from other components, including derivatives which have incorporated more or fewer silyl groups than the principal derivative. Ion series resulting from multiple derivatives are marked by differences of 72 u, the replacement value of H vs. TMS. In the event that the molecular ion (M) is not observed and the value of M is not made clear by subsequent assignments of the base ion series, deuterium labeling in the TMS group will readily distinguish between M and M − CH$_3$ ions; for incorporation of n TMS groups, M species must shift in exact multiples of 9 n [i.e., 36 u in the case of uridine-(TMS)$_4$], while M − CH$_3$ ions will shift 6 + 9(n − 1) u compared with the spectrum of the unlabeled derivative, due to loss of one CD$_3$.

2. Determine the value of B (the base fragment, 183 u in Fig. 1) by subtracting the most likely value of the sugar fragment (S). For nucleosides from RNA the values of S, with very rare exceptions, are 349 [ribosyl-(TMS)$_3$] or 291 [2′-*O*-methylribosyl-(TMS)$_2$], and from DNA the value is 261 [2′-deoxyribosyl-(TMS)$_2$]. Next calculate and assign other members of the base ion series.

3. Assign ions of the sugar series; check for ions characteristic of phosphate in the event the component is unexpectedly a nucleotide. The value of S for mononucleotides, which can be used in step 2, is 501 [ribosylphosphate-(TMS)$_4$], and 413 for monodeoxynucleotides [deoxyribosylphosphate-(TMS)$_3$].

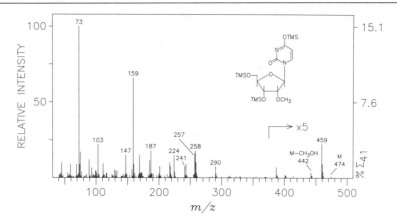

FIG. 3. Mass spectrum of 2'-O-methyluridine-(TMS)$_3$. Ions which contain the 2'-OCH$_3$ group are shifted 58 u lower compared with Fig. 1.

4. Seek evidence for side-chain substituents, either from ions which cannot be assigned by the above procedures, or by characteristic mass differences associated with the common nucleic acid side chains.

5. Consider the information to be gained from additional experiments (see below), in particular, use of deuterium-labeled TMS groups, exact mass measurement, or alternative ionization methods to test or verify the molecular weight.

Molecular Ion (M) Series of Ions

Ions resulting from simple losses from the molecular ion are listed in Table I. Not all members of the series are usually observed, and M − 15 and M − 90 ions are the most abundant and thus useful for identification of M. The abundances of M vary widely and are characteristically high for nucleosides of guanine and 1-methylpurines. The characteristic formation of M − 15 provides a useful pattern with M to permit visual recognition of M in the spectrum. The ion abundance ratio M/(M − 15) is usually <1 with the general exception of guanosine (see Fig. 2) and its analogs, an effect due primarily to the 2-aminopurine moiety.[22]

Other members of the M series are due to losses of TMSOH (M − 90), and TMSOH in combination with CH$_3$ (M − 105), CO (M − 118), and a second TMSOH (M − 180), and loss of 5'-CH$_2$OTMS (M − 103). Because the elimination of TMSOH is highly selective for the O-2' ether function, methylation at O-2' leads to loss of CH$_3$OH from M in lieu of M − 90, as shown by the mass spectrum of the 2'-O-methyluridine derivative in Fig. 3.

TABLE II
BASE ION SERIES FROM
TRIMETHYLSILYLATED NUCLEOSIDES[a]

Ion	Composition	Mass
B + 204	B + $C_8H_{20}O_2Si_2$	B + 204.1002[b]
B + 188	B + $C_8H_{20}OSi_2$	B + 188.1053[b]
B + 132[c]	B + $C_5H_{12}O_2Si$	B + 132.0607[b]
B + 128	B + $C_6H_{12}OSi$	B + 128.0657
B + 116	B + $C_5H_{12}OSi$	B + 116.0658[b]
B + 102	B + $C_4H_{10}OSi$	B + 102.0501
B + 100	B + C_4H_8OSi	B + 100.0344[b]
B + 74	B + $C_3H_{10}Si$	B + 74.0552
B + 58	B + C_2H_6Si	B + 58.0239
B + 41	B + C_2HO	B + 41.0027
B + 30	B + CH_2O	B + 30.0106
B + 13	B + CH	B + 13.0078
B + 2	B + H_2	B + 2.0157
B + 1	B + H	B + 1.0078
B	B	B
B − 14	B − CH_2	B − 14.0157

[a] From H. Pang, K. H. Schram, D. L. Smith,
S. P. Gupta, L. B. Townsend, and J. A.
McCloskey, *J. Org. Chem.* **47**, 3923 (1982).
[b] 88.03444 u lower in spectra of 2′-deoxynucleo-
side derivatives.
[c] B + 131 in some mass spectra.

Base (B) Series of Ions

Ions containing the base plus portions of the sugar constitute the base series (Table II). In the case of nucleosides derived from nucleic acids, this series is of considerable importance because the base bears the most frequently found sites of posttranscriptional modification, and can be used in establishing the exact mass and elemental composition of the base. Fifteen base series ions are identified in the spectrum of uridine (Fig. 1), although six to ten is a more typical value for nucleosides not having side chains which compete for fragmentation routes. If exact mass data are available, for example, from a full high-resolution spectrum, the exact mass interrelationships represented in Table II can be used to assign the base series in mixtures when more than one value of B is represented.[30]

Most ions listed in Table II bear hydrogen rearranged from that portion of the sugar which is lost as a neutral species, and which is derived from the ribose skeleton rather than TMS methyl groups. In some cases, an

[30] H. Pang, D. L. Smith, P. F. Crain, K. Yamaizumi, S. Nishimura, and J. A. McCloskey, *Eur. J. Biochem.* **127**, 459 (1982).

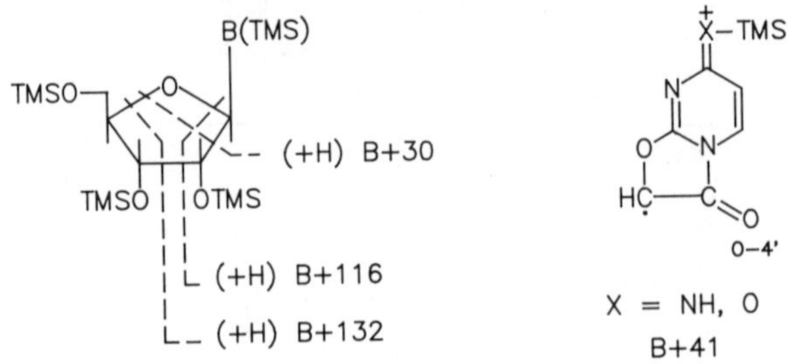

SCHEME 1. Selected ion assignments from trimethylsilylated nucleosides.

analogous ion is also formed by transfer of TMS instead of H, therefore appearing 72 u higher than the H-transfer species. Such pairs of ions are $B + 132/204$, $B + 116/188$, $B + 30/102$, and $B + 2/74$. Care must be taken not to confuse such 72 u differences with the presence of higher or lower derivatives, which can usually be recognized from peaks in the molecular ion series.

The most important members of the base series for ribonucleosides are usually $B + 30$, $B + 116$, and $B + 132$, as schematically depicted in Scheme 1. These ions are useful to indicate sites of substitution in the sugar, for example, methylation at $O - 2'$. As shown by comparison of Figs. 1 and 3, the ions $B + 132$ and $B + 116$ in Fig. 1 shift 58 u lower (replacement value of TMS by CH_3) to m/z 257 and 241 in Fig. 3. The ions $B + 74$, $B + 30$, $B + 116$, and $B + 132$ also constitute a useful set of usually abundant ions for which exact mass measurements can be made to establish an accurate mean value for B in order to determine elemental composition, a method demonstrated earlier using permethyl derivatives.[31] Losses of either CH_4 or CH_3 result in other product ions listed in Table II: $B + 1 \rightarrow B - 14$; $B + 74 \rightarrow B + 58$; $B + 116 \rightarrow B + 100$.

Two of the B series ions are characteristic of pyrimidines (Figs. 1, 2a). $B + 13$ is a complex ion which is compositionally, but not structurally, equivalent to $B + CH$. In the case of uridine (but not 2-thiopyrimidines[22]), this ion has lost oxygen at $C - 2$ of the base, but has retained $C - 1'$, $H - 1'$, and $O - 4'$ (m/z 223, Fig. 2a). The $B + 41$ ion (Scheme 1) is most abundant in derivatives of cytidine but is also significant in derivatives of uridine (m/z 224, Fig. 1).[22,32]

In the case of pseudouridine and other C-nucleosides, which contain the stronger C–C glycosidic bonds, ions which require breakage of the

[31] P. F. Crain, H. Yamamoto, J. A. McCloskey, Z. Yamaizumi, S. Nishimura, K. Limburg, M. Raba, and H. J. Gross, Adv. Mass Spectrom. 8, 1135 (1980).

[32] J. G. Liehr, D. L. von Minden, S. E. Hattox, and J. A. McCloskey, Biomed. Mass Spectrom. 1, 281 (1974).

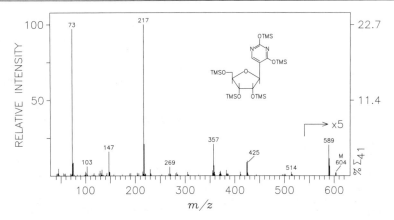

FIG. 4. Mass spectrum of pseudouridine-(TMS)$_5$, showing absence of ions resulting from cleavage of the glycosidic bond. B = 255 u.

glycosidic bond are essentially absent,[33,34] as demonstrated by pseudouridine-(TMS)$_5$[20] in Fig. 4. This striking characteristic is useful in assigning the C–C glycosidic linkage in nucleosides of unknown structure[4] and is manifested in the absence of certain ions in both the base (B + 1, B + 2, B + 74, B + 58) and sugar (following section) series.

Sugar (S) Series of Ions

Sugar-derived ions for simple pentose-containing nucleosides, as from RNA and DNA, are listed in Table III. As seen for the ribonucleosides in Figs. 1–3, such ions are easily assigned, and except for questions of sugar methylation, are invariant in mass and convey relatively little structural information. As shown in Fig. 3, ribose methylation is easily detected by the S series ions m/z 290 and 159 (see Table III). Deoxynucleosides (as shown in Fig. 5) exhibit many of the same common ions (m/z 73, 103, 217) with the important exception of those closely related to the intact sugar fragment, which contain the C − 2′ grouping and that are thus shifted 88 u lower compared with the corresponding ribonucleoside ions as indicated in Table III. Ions of the sugar series tend to be much less abundant in the case of deoxynucleosides (compare Figs. 2b and 5), reflecting absence of O − 2′, which is otherwise influential in directing fragmentation.[22] The

[33] J. M. Rice and G. O. Dudek, *Biochem. Biophys. Res. Commun.* **35**, 383 (1969).
[34] S. M. Hecht, A. S. Gupta, and N. J. Leonard, *Anal. Biochem.* **30**, 249 (1969).

TABLE III
SUGAR ION SERIES FROM TRIMETHYLSILYLATED NUCLEOSIDES

	Mass (m/z)		
Ion	Pentose	2'-O-Methylpentose	2'-Deoxypentose
S	349.1687	291.1448	261.1343
S − H	348.1609	290.1370	260.1265
S − CH₃OH	—	259.1186	—
S − Me₃SiOH	259.1186	—	171.0842
S − H − Me₃SiOH	258.1108	200.0869	170.0764
S − H − CH₂OSiMe₃	245.1030	187.0791	157.0686
S − CH₄ − Me₃SiOH	243.0873	185.0634	155.0529
C₄H₄O₂(Me₃Si)₂	230.1159	230.1159	—
C₅H₇O₂Me₃Si	—	172.0920	—
C₃H₃O₂(Me₃Si)₂	217.1081	217.1081	217.1081
C₄H₆O₂Me₃Si	—	159.0842	—
S − 2 Me₃SiOH	169.0685	—	81.0341
CH₂OSiMe₃	103.0580	103.0580	103.0580

most direct indication of the presence of a deoxy sugar is simply from the mass differences between the B and M ion series, as seen in Fig. 5.

The mechanism of formation of the common sugar ion S − H has been studied in detail in ribonucleosides, and its relative abundance has been shown to depend on the steric accessibility of H − 2' to the base. This is a consequence of its mechanism of formation,[22] in which H − 2' is selectively extracted by the base to eliminate the neutral base (BH) molecule concomitant with formation of the odd-electron S − H ion. As a result, the S − H ion abundance is markedly lower in the mass spectra of α-anomer derivatives, and may be useful in assigning anomeric configuration when both the α and β anomers are available for comparison.[22] The S − H ion abundance is also reduced in the case of purine nucleosides, and in the absence of the C − 2' heteroatom as noted above, with both effects due to differences in charge localization.

Also worthy of note in the sugar ion series is m/z 103, due primarily, but not exclusively, to the 5'-silyloxy group, TMSO⁺=CH₂. This ion is formed in small amounts in the absence of 5'-OTMS, such as 5'-deoxy, 5'-phosphoryl or ribopyranosyl nucleosides.[22] Caution should therefore be exercised in correlating presence of m/z 103 solely with the 5'-silyloxy group, which might be of interest if isotopic substitution or modification at C − 5' were an issue.

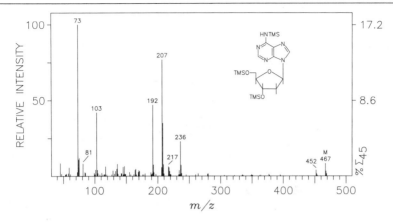

FIG. 5. Mass spectrum of 2-deoxyadenosine-(TMS)$_3$. B = 206 u.

Effects of Multiple Derivatives and Other Mixtures

Even if GC/MS is used for sample introduction, more than one component may be represented in a given mass spectrum, leading to potential problems in assigning the sources of fragment ions. The presence of mixtures may be initially recognized from the pattern of ions in the molecular ion region. Whenever possible the relationship of ions to a common molecular precursor should be tested by data system-assisted construction of ion current profiles, even if spectra are acquired from direct-probe sample introduction. In the case of nucleosides differing only in the number of TMS groups present, the lower derivative (having an additional unblocked polar site) will, in general, vaporize at a slightly higher temperature. Ions used to monitor individual nucleoside components in this fashion should be chosen for their uniqueness, thus excluding most ions from the sugar series. On the other hand, if there is interest in simply distinguishing nucleosides from impurities, then sugar series ions (in particular m/z 103, 245, 348) are suitable for profiling, as class-selective markers.

The fragmentation patterns resulting from derivatives having one less or more TMS blocking group may be quite different in appearance, as shown by comparison of Figs. 1 and 6. As observed, absence of the TMS group in the base generally lowers the abundances of base series ions. In particular, the overall abundance and abundance ratios of M/(M − 15) may be quite different, so that accurate estimates of molar ratios of higher and lower derivatives based on ion abundances may be subject to considerable error. In any event, properly executed derivatization reactions[24] should lead to no more than 5–15% of the minor derivatized species.

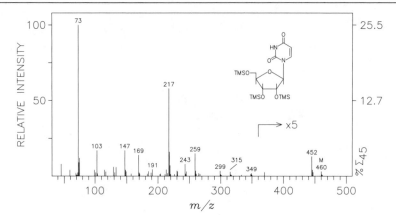

FIG. 6. Mass spectrum of uridine-(TMS)$_3$, for comparison with Fig. 1. B = 111 u.

Recognition of Side-Chain Substitution

Examples of side-chain-containing nucleosides from RNA and DNA are shown in Scheme 2 as the TMS derivatives. The presence of side chain or heteroatom substitutions in the base can be recognized in two ways. First, mass differences between common side chains and observed molecular mass can be calculated to determine whether the residual value corresponds to one of the four major nucleosides. Such differences represent simple shifts (14 u for CH_3, 28 for $(CH_3)_2$, 16 for OH or for replacement of oxygen by sulfur, 68 for isopentenyl, etc.) but the calculation must reflect whether the putative substituent is replacing H (as when substituted on carbon) or TMS (as when substituted on N). Guidelines for the expected sites of silylation are discussed elsewhere.[24] For example, dimethyl substitution in adenosine increases the molecular mass by 28 u but such substitution at N^6 blocks silylation at that site (see Scheme 2, structure **1**), and so the net mass expected for derivative **1** is 44 u below the observed value for adenosine (see Fig. 2b). Because there may be experimental uncertainty as to the anticipated extent of silylation (e.g., at the amide nitrogen in **6**) several allowable mass values must be considered, differing by 72 u.

The second and usually more useful method, especially for complex side chains, is the appearance of characteristic fragment ions resulting from side-chain cleavage. Some leading examples are listed in Table IV, corresponding to derivatives shown in Scheme 2. The observed ions may be due to loss of side chain atoms with charge retention on the nucleoside portion (as in **5**), or less commonly will be due to ionized portions of the side chain (as in **7**). In either case, side chain-directed fragmentation processes may compete so favorably with the generic nucleoside fragmen-

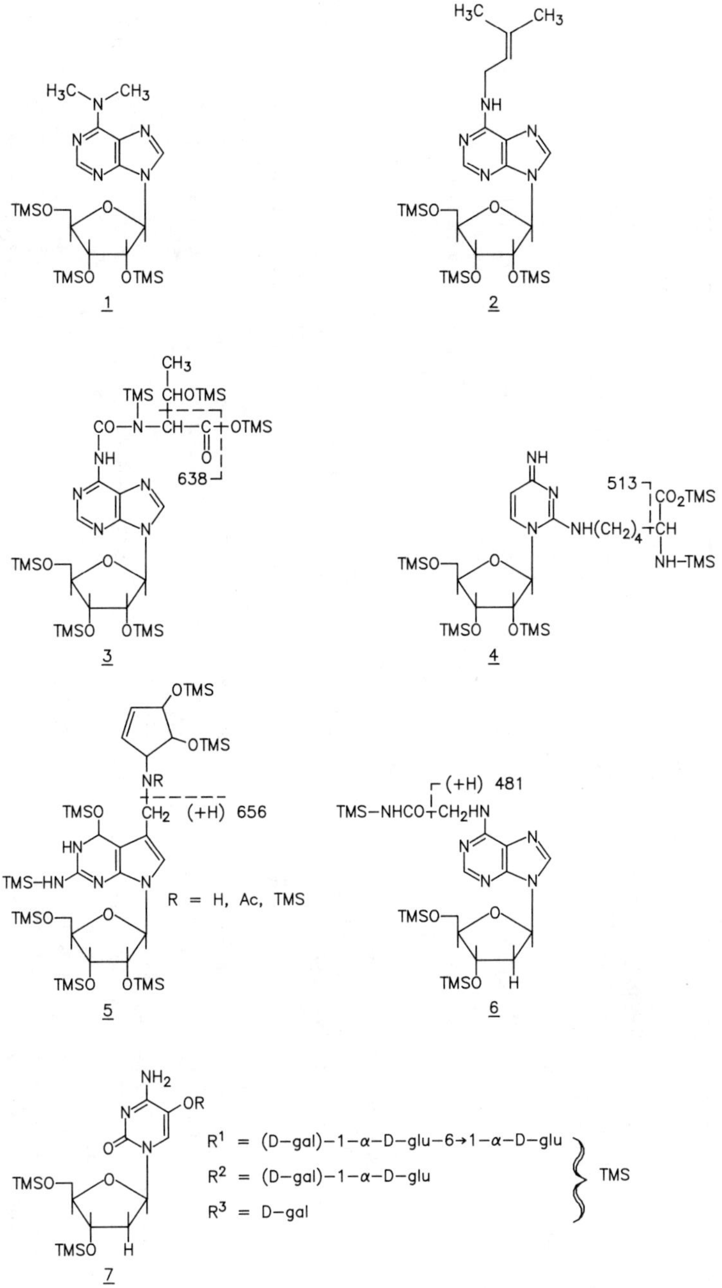

SCHEME 2. Structures (1)–(7) of trimethylsilylated nucleosides from RNA and DNA which produce characteristic side-chain fragment ions.

TABLE IV

CHARACTERISTIC FRAGMENTATION REACTIONS RESULTING FROM SIDE-CHAIN SUBSTITUENTS IN RNA AND DNA

Nucleoside (source)	Structure of derivative (Scheme 2)	m/z (% relative intensity)	Assignment	Reference
N^6,N^6-Dimethyladenosine (rRNA)	1	134 (10)	$(B + 1) - CH_3N$	a
N^6-Isopentenyladenosine (tRNA)	2	160 (18)	$(B + 1) - C_3H_7$	b
		508 (2.6)	$M - C_3H_7$	
N-[(9-β-D-Ribofuranosyl-9H-purin-6-yl)-carbamoyl]threonine (tRNA)	3	290 (68)	$(B + 1) - C_2H_4(OTMS)_2$	c
		638 (29)	$M - C_2H_4(OTMS)_2$	
Lysidine (tRNA)	4	513 (65)[d]	$M - CH(NHTMS)CO_2TMS$	e
Queuosine (tRNA)	5	656 (57)[f]	See Scheme 2	g
		307 (37)[f]	655-Ribosyl	
α-N-(9-β-D-2'-deoxyribofuranosylpurin-6-yl)-glycinamide (DNA)	6	481 (19)	$M - CONHTMS$	h
		221 (79)	481 – Deoxyribosyl	
Glycosylated derivatives of deoxycytosine (DNA)	7	451 (3)	Hexosyl(TMS)$_4$	i
		361 (35)	451 – TMSOH	

[a] For leading references see D. L. von Minden, J. G. Liehr, M. H. Wilson, and J. A. McCloskey, J. Org. Chem. 39, 285 (1974).

[b] For leading references see T. Hashizume, J. A. McCloskey, and J. G. Liehr, Biomed. Mass Spectrom. 3, 177 (1976).

[c] Microfilm supplement to H. Kasai, K. Murao, J. G. Liehr, P. F. Liehr, P. F. Crain, J. A. McCloskey, and S. Nishimura, Eur. J. Biochem. 69, 435 (1976).

[d] Base peak reported as off-scale.

[e] T. Muramatsu, S. Yokoyama, N. Horie, A. Matsuda, T. Ueda, Z. Yamaizumi, Y. Kuchino, S. Nishimura, and T. Miyazawa, J. Biol Chem. 263, 9261 (1988).

[f] Relative intensity value is for R = acetyl.

[g] H. Kasai, Z. Ohashi, F. Harada, S. Nishimura, N. J. Oppenheimer, P. F. Crain, J. G. Liehr, D. L. von Minden, and J. A. McCloskey, Proc. Natl. Acad. Sci. U.S.A. 80, 7400 (1983).

[h] D. Swinton, S. Hattman, P. F. Crain, C. S. Cheng, D. L. Smith, and J. A. McCloskey, Biochemistry 14, 4198 (1975).

[i] F. F. Hsu, P. F. Crain, J. A. McCloskey, Adv. Mass Spectrom. 11B, 1450 (1989).

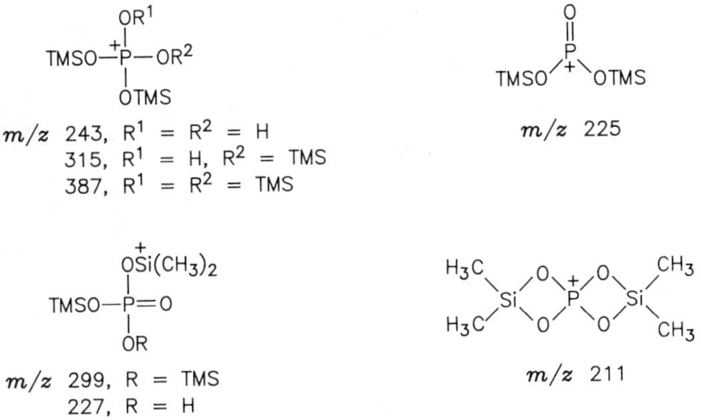

SCHEME 3. Phosphate-containing ions characteristic of trimethylsilylated nucleotides.

tation shown in Tables I–III that the usual base or other ion series is not observed or is of low abundance; such is the case with the nucleosides of type **7** and to a lesser extent with **5** and the family of compounds it represents. In such instances, low abundance ions of the base series can sometimes be recognized by measurement and testing of exact mass interrelationships (Table II), including the required difference between M and putative values of B. The use of exact mass measurements in this fashion is particularly effective when the sample is impure and the base-containing ions are not evident in the low-resolution mass spectrum.

A typical example of the influence of a biologically common and relatively complex side chain on the overall fragmentation pattern is shown in Fig. 7 (structure **3**). Because of the influence of the threonyl side chain, this spectrum is different in appearance from that of the parent nucleoside (Fig. 2b), and the assignment of M (m/z 844) relies to a large extent on the M series ions m/z 754 (M − 90) and 739 (M − 105). Sugar and general silyl ions are readily assigned (m/z 73, 75, 103, 147, 169, 217, 230, 243, 245, 259, 348; see Table III), but from a putative value for B (844 − 349 = 495) no ions of the base series are evident. However, confirmation of the spectrum as that of a ribonucleoside is made through a series of "ribose loss" differences of 348 u: 739 → 391, 710 → 362, 668 → 320, 638 → 290. The reactions represented by these differences are analogous to the formation of the B + 1 ion from M in which ribose is lost with retention of one H by the base, and such reactions are often reflected in mass spectra of complex nucleosides which exhibit ions resulting from side-chain cleavages. The only ion evident from the side chain itself is m/z 117, shown by high-resolution measurements to be due to both possible portions of the

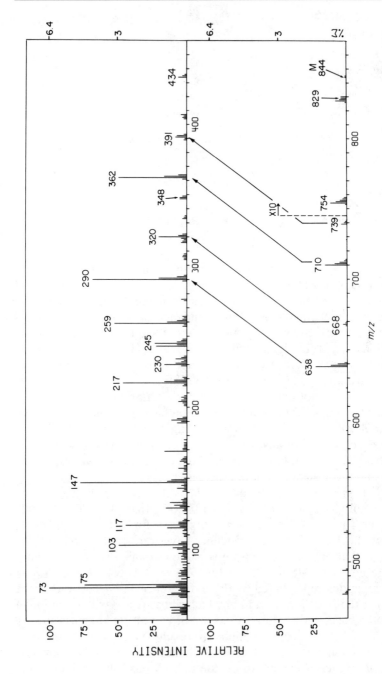

Fig. 7. Mass spectrum of the TMS derivative of N-[(9-β-D-ribofuranosyl-9H-purin-6-yl)carbamoyl]threonine. [From microfilm supplement to: H. Kasai, K. Murao, J. G. Liehr, P. F. Crain, J. A. McCloskey, and S. Nishimura, *Eur. J. Biochem.* **69**, 435 (1976), with permission.] Additional assignments: m/z 710 (754-44), 668 (710-42); see text for discussion.

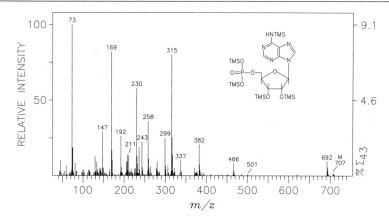

FIG. 8. Mass spectrum of the (TMS)₅ derivative of adenosine 5'-monophosphate (5'-AMP). [Adapted from A. M. Lawson, R. N. Stillwell, M. M. Tacker, K. Tsuboyama, and J. A. McCloskey, *J. Am. Chem. Soc.* **93**, 1014 (1971), with permission.]

threonyl moiety: $TMSCO_2^+$ and $CH_3^+CHOTMS$ (1 : 2). For additional discussion the reader is referred to Ref. 35.

Recognition of Nucleotides

Mononucleotides, which may be inadvertently formed due to incomplete enzymatic digestion,[23] produce mass spectra which exhibit M and S series of ions shifted 156 u higher than the corresponding nucleoside. A typical example is given by the silyl derivative of 5'-AMP in Fig. 8. The presence of phosphate is usually marked by a highly characteristic set of ions as indicated in Scheme 3, some of which are related by loss of CH_4 (315 → 299, 243 → 227 → 211). In most nucleotide spectra, the ion pair *m/z* 315–299 is the most abundant of the phosphate ions and therefore most readily recognized. In the event that inorganic phosphate is present in the silylation mixture (for example, from partial phosphatase degradation), this pair of ions will instead appear at 314 [M for *tris*-(TMS) phosphate] and 299, resulting from the loss of CH_3. In the case of isolated samples, both *m/z* 314 and 315 may, therefore, be present.

For further discussion of the mass spectra of mono-[21] and dinucleotide TMS derivatives,[36] the earlier literature should be consulted.[20]

[35] H. Kasai, K. Murao, J. G. Liehr, P. F. Crain, J. A. McCloskey, and S. Nishimura, *Eur. J. Biochem.*, **69**, 435 (1976).

[36] C. Hignite, *in* "Biomedical Applications of Mass Spectrometry" (G. R. Waller, ed.), p. 429. Wiley, New York, 1972.

Acknowledgments

The author is indebted to many colleagues who contributed to this work over a period of time, including K. J. Lyman, B. Basile, H. Pang, K. H. Schram, and S. C. Pomerantz. The methods described were developed and supported by NIH grants GM 21584 and CA 18024.

[46] Gas Chromatography–Mass Spectrometry of Free Radical-Induced Products of Pyrimidines and Purines in DNA

By MIRAL DIZDAROGLU

Oxygen-derived species such as superoxide radical ($O_2{}^-$) and H_2O_2 are produced in mammalian cells as a result of aerobic metabolism.[1,2] Excess generation of these species *in vivo* by endogenous sources (e.g., oxidant enzymes, phagocytic cells) or exogenous sources (e.g., redox-cyclic drugs) can result in damage to cellular DNA. Thus, oxygen-derived species are mutagenic and may act as promoters of carcinogenesis.[3,4] However, much of the toxicity of the superoxide radical and H_2O_2 *in vivo* is thought to arise from their metal ion-dependent conversion into highly reactive hydroxyl radical (·OH) and iron ions appear to be the most likely catalysts of such conversion (iron-catalyzed Haber–Weiss reaction).[2] As an exogenous source, ionizing radiation can also cause formation of ·OH among other radical species (H atoms, hydrated electrons) in cells by interaction with cellular water.[5] In the past, radiation chemists have extensively studied the reactions of free radicals with DNA and chemically characterized the resulting products.[5] Free radicals, especially ·OH, produce a large number of sugar and base products in DNA,[5] and DNA–protein crosslinks in nucleoprotein *in vivo* and *in vitro*.[6–9] For an understanding of the

[1] I. Fridovich, *Science* **209**, 875 (1978).

[2] B. Halliwell and J. M. C. Gutteridge, *Mol. Aspects Med.* **8**, 89 (1985).

[3] M. Larramendy, A. C. Melho-Filho, E. A. Leme Martins, and R. Meneghini, *Mutat. Res.* **178**, 57 (1983).

[4] P. A. Cerutti, *Science* **227**, 375 (1985).

[5] C. von Sonntag, "The Chemical Basis of Radiation Biology," Taylor and Francis, London, 1987.

[6] L. K. Mee and S. J. Adelstein, *Proc. Natl. Acad. Sci. U.S.A.* **78**, 2194 (1981).

[7] N. L. Oleinick, S. Chiu, N. Ramakrishnan, and L. Xue, *Br. J. Cancer* **55**, Suppl. VIII, 135 (1987).

[8] E. Gajewski, A. F. Fuciarelli, and M. Dizdaroglu, *Int. J. Radiat. Biol.* **54**, 445 (1988).

[9] M. Dizdaroglu, E. Gajewski, P. Reddy, and S. A. Margolis, *Biochemistry* **28**, 3625 (1989).

biological consequences of free radical-induced lesions in DNA and in nucleoprotein, it is essential to chemically characterize such lesions and measure their quantities. A number of analytical techniques has been used in the past for this purpose. This chapter describes the use of the technique of gas chromatography-mass spectrometry (GC/MS) for chemical characterization and quantitative measurement of free radical-induced products of pyrimidines and purines in DNA and DNA–protein cross-links in nucleoprotein.

Materials and Methods

Reagents. Reagents and some reference compounds used in the methodology described are available commercially.

Synthesis of Reference Compounds. 6-Amino-8-hydroxypurine (8-hydroxyadenine) is synthesized from 8-bromoadenine by treatment with formic acid, and is purified by recrystallization from water.[10] 2,6-Diamino-4-hydroxy-5-formamidopyrimidine is synthesized by treatment of 2,5,6-triamino-4-hydroxypyrimidine with formic acid, and is recrystallized from water.[11] Thymine glycol is synthesized by treatment of thymine with osmium tetroxide.[12]

Hydrolysis

Prior to analysis by GC/MS, DNA and nucleoprotein must be hydrolyzed. Acidic hydrolysis cleaves the glycosidic bonds between bases and sugar moieties in DNA and thus frees intact and modified bases. Enzymatic hydrolysis is used to hydrolyze DNA to nucleosides. In the case of DNA–protein cross-links, the simplest way for hydrolysis of nucleoprotein appears to be the standard method of protein hydrolysis, i.e., hydrolysis with 6 M HCl, which cleaves peptide bonds in proteins as well as glycosidic bonds in DNA to free base–amino acid cross-links.[8,9] Prior to hydrolysis, DNA and nucleoprotein samples should be extensively dialyzed against water and subsequently lyophilized.

Acidic Hydrolysis. Of the various acids, formic acid appears to be the most suitable for hydrolysis of DNA.[13] The following procedure has been used in the studies reviewed here. An aliquot of DNA (usually 0.1–1 mg) is treated with 1 ml of formic acid (88%) in evacuated and sealed tubes at 150° for 30 to 40 min. The sample is then lyophilized. This procedure of

[10] M. Dizdaroglu and D. S. Bergtold, *Anal. Biochem.* **156,** 182 (1986).
[11] L. F. Cavalieri and A. Bendich, *J. Am. Chem. Soc.* **72,** 2587 (1950).
[12] M. Dizdaroglu, E. Holwitt, M. P. Hagan, and W. F. Blakely, *Biochem. J.* **235,** 531 (1986).
[13] G. R. Wyatt and S. S. Cohen, *Biochem. J.* **55,** 774 (1953).

hydrolysis has been found to be optimal for quantitative removal of free radical-induced products of bases from DNA.[14] Furthermore, this procedure does not alter the products of thymine, adenine, and guanine in DNA.[15] However, the products of free radical-damaged cytosine in DNA can be modified by deamination and/or dehydration as follows: cytosine glycol yields a mixture of 5-hydroxycytosine and 5-hydroxyuracil, the former by dehydration and the latter by deamination and dehydration. 5-Hydroxy-6-hydrocytosine, 5,6-dihydroxycytosine, and 4-amino-5-hydroxy-2-imidazolidinon-3-ene deaminate to give 5-hydroxy-6-hydrouracil, 5,6-dihydroxyuracil, and 5-hydroxyhydantoin, respectively. These modifications are evidently quantitative, as indicated by studies on the release of the products as a function of the time of acidic hydrolysis as well as by the measurement of the dose–yield relationships.[14]

Nucleoprotein (approximately 1 mg) is hydrolyzed with 1 ml of 6 M HCl in evacuated and sealed tubes at 120° for 18 hr, and then lyophilized.

Enzymatic Hydrolysis. This type of hydrolysis of DNA is discussed in detail elsewhere in this volume.[16] The following procedure has been used in the studies reviewed here. An aliquot (0.5–1 mg) of lyophilized DNA samples is incubated in 0.5 ml of 10 mM Tris-HCl buffer, pH 8.5 (containing 2 mM MgCl$_2$) with deoxyribonuclease I (100 units), spleen exonuclease (0.01 unit), snake venom exonuclease (0.5 units), and alkaline phosphatase (10 units) at 37° for 24 hr. After hydrolysis, the sample is lyophilized.

The drawback of enzymatic hydrolysis is that the products of the cytosine moiety in DNA cannot readily be analyzed by gas chromatography as their nucleosides, and only one product of the thymine moiety is observed as its nucleoside.[17] On the other hand, the enzymatic hydrolysis permits the analysis of 8,5′-cyclopurine-2′-deoxynucleoside residues, which are not released from DNA by acidic hydrolysis.[17-19] In a recent work, the deamination of 2′-deoxyadenosine and its products has been observed during the course of enzymatic hydrolysis.[17] This was presumably due to the presence of deaminase activity in one of the enzymes used. This can be avoided by using enzymes which are free of deaminase activity.

[14] A. F. Fuciarelli, B. J. Wegher, E. Gajewski, M. Dizdaroglu, and W. F. Blakely, *Radiat. Res.* **119**, 219 (1989).
[15] M. Dizdaroglu, *Anal. Biochem.* **144**, 593 (1985).
[16] P. F. Crain, this volume [42].
[17] M. Dizdaroglu, *J. Chromatogr.* **367**, 357 (1986).
[18] M. Dizdaroglu, *Biochem. J.* **238**, 247 (1986).
[19] M.-L. Dirksen, W. F. Blakely, E. Holwitt, and M. Dizdaroglu, *Int. J. Radiat. Biol.* **54**, 195 (1988).

Enzymatic hydrolysis of nucleoprotein to release DNA base–amino acid cross-links for GC/MS analysis has not been reported.

Derivatization

The GC/MS technique is applicable only to compounds that are volatile, or can be made sufficiently volatile. Bases, nucleosides, and DNA base–amino acid cross-links are not sufficiently volatile for gas chromatography, and thus must be converted into volatile derivatives. Generally, trimethylsilylation is the most widely used mode of derivatization. The use of *tert*-butyldimethylsilylation has been also reported, but so far only for qualitative analysis of free radical-induced products of bases in DNA.[15,20,21] Trimethylsilylation is discussed in detail elsewhere in this volume.[22] The following procedure has been used for trimethylsilylation in the studies reviewed here. Lyophilized hydrolysates of DNA or nucleoprotein (0.5–1 mg) are trimethylsilylated with 0.25 ml of a mixture of bis(trimethylsilyl)trifluoroacetamide (containing 1% trimethylchlorosilane) and acetonitrile (1.5 : 1, by vol) at 140° for 30 min in polytetrafluoroethylene-capped vials (sealed under nitrogen). The amount of the reagents can be modified according to the amount of DNA or nucleoprotein used. After derivatization, samples are allowed to cool to room temperature, and then directly injected into the injection port of the gas chromatograph without any further treatment.

Apparatus

Any conventional GC/MS instrument equipped with a capillary inlet system and a computer work station can be used for this purpose. The present methodology utilizes fused-silica capillary columns, which are commercially available. These types of columns provide high inertness and separation efficiency, and permit measurements of high sensitivity. Columns coated with cross-linked 5% phenyl methylsilicone gum phase appear to be best for the purpose. Column length varies depending on the type of analysis. Helium (ultra high purity) is used as the carrier gas. Generally, the split mode of injection is used, and the split ratio (i.e., ratio of carrier gas flow through the splitter vent to flow through the column) is adjusted according to the concentration of the analyte(s) in a given mixture. The injection port of the gas chromatograph, the GC/MS interface, and the ion source of the mass spectrometer are kept at 250°. The glass

---

[20] M. Dizdaroglu, *J. Chromatogr.* **295,** 103 (1984).
[21] M. Dizdaroglu, *BioTechniques* **4,** 536 (1986).
[22] K. H. Schram, this volume [43].

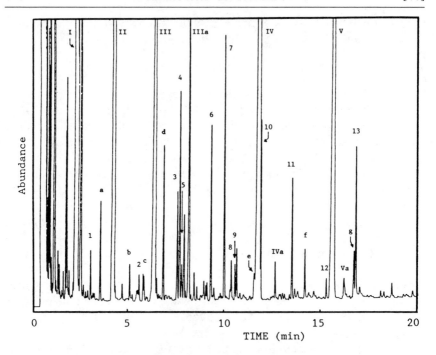

Fig. 1. Gas chromatogram of a DNA sample, which was γ-irradiated in aqueous solution [dose, 330 Gy (joule/kg)] followed by hydrolysis with formic acid and trimethylsilylation. Column: fused-silica capillary (12 m; 0.2 mm id) coated with cross-linked 5% phenyl methyl-silicone gum phase (film thickness, 0.11 μm). Temperature program: 100° to 250° at 7°/min after 3 min at 100°. Peaks: I, phosphoric acid; 1, uracil; II, thymine; 2, 5,6-dihydrothymine; III, cytosine; d, 5-methylcytosine; 3, 5-hydroxy-6-hydrothymine; 4, 5-hydroxyuracil; 5, 5-hydroxy-6-hydrouracil; IIIa, cytosine; 6, 5-hydroxycytosine; 7 and 8, cis- and trans-thymine glycol; 9, 5,6-dihydroxyuracil; IV, adenine; 10, 4,6-diamino-5-formamidopyrimidine; IVa, adenine; 11, 8-hydroxyadenine; 12, 2,6-diamino-4-hydroxy-5-formamidopyrimidine; V, gua-nine; Va, guanine; 13, 8-hydroxyguanine. Compounds represented by peaks a–g were not defined, and were also present in control samples (all compounds as their Me₃Si derivatives). (From Ref. 15 with permission.)

liner in the injection port is filled with silanized glass wool. This permits the homogeneous vaporization of injected samples in the injection port, and avoids peak tailing during analysis. Mass spectra are recorded in the electron ionization (EI) mode at 70 eV.

Gas Chromatography–Mass Spectrometry of Pyrimidine and Purine Products in DNA

Free Bases. Figure 1 illustrates a typical gas chromatographic separa-tion of trimethylsilyl (Me₃Si) derivatives of intact and modified bases

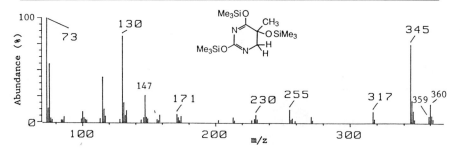

FIG. 2. Mass spectrum of 5-hydroxy-6-hydrothymine-(Me₃Si)₃.

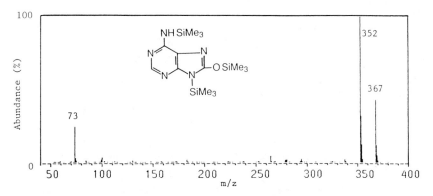

FIG. 3. Mass spectrum of 8-hydroxyadenine-(Me₃Si)₃. (From Ref. 15 with permission.)

released by formic acid hydrolysis from γ-irradiated DNA (peak identification is given in the legend). As Fig. 1 clearly shows, the products of all four bases are well separated from one another and from the four intact bases. Mass spectra of the trimethylsilyl (Me₃Si) derivatives of the modified bases are dominated generally by an intense molecular ion (M⁺ ion) and an intense (M − 15)⁺ ion,[15,20] as are those of the four intact bases.[23,24] The latter ion is due to characteristic loss of a methyl radical [15 mass units (u)] from M⁺ ion.[23,24] In some instances, (M − 1)⁺ ion, which results from loss of an H atom from M⁺ ion, also appears as an intense ion in the mass spectra.[15,20] Figures 2–4 illustrate some typical mass spectra. In its mass spectrum (Fig. 2), 5-hydroxy-6-hydrothymine-(Me₃Si)₃ (M_r 360) provides an intense (M − 15)⁺ ion at m/z 345 and an intense M⁺ ion at m/z 360. The (M − 1)⁺ ion (m/z 359) is also produced. Loss of CO from

[23] E. White, P. M. Krueger, and J. A. McCloskey, *J. Org. Chem.* **37**, 430 (1972).
[24] J. A. McCloskey, *in* "Basic Principles in Nucleic Acid Chemistry" (P. O. P. Ts'o, ed.), Vol. I, p. 209. Academic Press, New York, 1974.

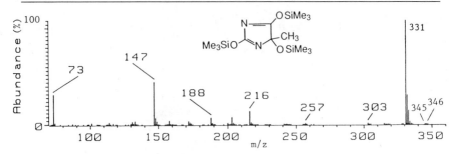

FIG. 4. Mass spectrum of 5-hydroxy-5-methylhydantoin-(Me$_3$Si)$_3$.

$(M - 15)^+$ ion accounts for m/z 317. The m/z 271 and 255 ions result from the loss of Me$_3$SiO· (89 u) from M‡ ion and by loss of Me$_3$SiOH (90 u) from $(M - 15)^+$ ion, respectively. Elimination of Me$_3$SiOCN (115 u) from $(M - 15)^+$ ion, which is typical of trimethylsilylated pyrimidines and purines,[23] presumably accounts for m/z 230. Ions at m/z 73 and 147 are common fragmentation products of Me$_3$Si derivatives and serve no diagnostic purpose.[23] The abundant ion at m/z 130 is a companion ion of m/z 147 and is diagnostically unimportant. The mass spectrum of 8-hydroxyadenine-(Me$_3$Si)$_3$ (M_r 367) is dominated by an abundant $(M - 15)^+$ ion and an abundant M‡ ion at m/z 352 and 367, respectively (Fig. 3). The high abundance of these ions reflects the aromatic character of the molecule. Figure 4 illustrates the mass spectrum of 5-hydroxy-5-methylhydantoin-(Me$_3$Si)$_3$ (M_r 346). This compound is a major ·OH-induced product of thymine and is formed in the presence of oxygen.[5] Thus, this product is not represented in Fig. 1. The $(M - 15)^+$ ion appears as the most abundant ion in the mass spectrum, whereas M‡ and $(M - 1)^+$ ions are of low intensity. Loss of CO from $(M - 15)^+$ ion accounts for m/z 303. The $(M - 89)^+$ ion is present at m/z 257. The abundant ion at m/z 216 is produced by elimination of Me$_3$SiOCN from the $(M - 15)^+$ ion.

The separation by capillary gas chromatography of the *tert*-butyldimethylsilyl derivatives of modified bases and four intact bases released from γ-irradiated DNA by acidic hydrolysis has been recently reported.[21] The mass spectra of these compounds contain an intense $(M - 57)^+$ ion, which is due to typical loss of the *tert*-butyl radical (57 u) from M‡ ion.[21,25] In some instances, M‡ and $(M - 15)^+$ ions are also observed.[21]

Nucleosides. The Me$_3$Si derivatives of free radical-induced products of 2'-deoxynucleosides generally follow the same fragmentation patterns

[25] P. F. Crain and J. A. McCloskey, *Anal. Biochem.* **132**, 124 (1983).

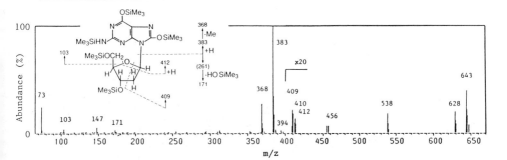

Fig. 5. Mass spectrum of 8-hydroxy-2'-deoxyguanosine-(Me₃Si)₅. (From Ref. 29 with permission.)

as those of other nucleosides previously studied.[26,27] EI mass spectra of trimethylsilylated nucleosides are discussed in detail elsewhere in this volume.[28] As an example of free radical-induced products of nucleosides, the mass spectrum of 8-hydroxy-2'-deoxyguanosine-(Me₃Si)₅ (M_r 643) is illustrated in Fig. 5. The M⁺ and (M − 15)⁺ ions are present at m/z 643 and 628, respectively. The most intense ion at m/z 383 and the ion at m/z 368 represent the characteristic (base + H)⁺ ion [(B + 1)⁺ ion] and the (base + H − Me)⁺ ion, respectively.[26,27] The high intensity of the (B + 1)⁺ ion (m/z 383) is caused by stabilization through an electron-donating substituent at the C-8 atom of the purine ring.[27,29] The ion at m/z 538 results from M⁺ ion by typical loss of Me₃SiOH (90 u).[27] The ion at m/z 456 arises from addition of HMe₃Si to the base [(B + 74)⁺ ion]. Other ions are formed as illustrated in the insert in Fig. 5.

8,5'-Cyclopurine nucleosides represent a concomitant damage to the purine base and sugar moieties of the same nucleoside. Free radical-induced formation of these compounds in DNA *in vitro* and *in vivo* has been demonstrated recently.[18,19,30] As expected, Me₃Si derivatives of these compounds provide partly different fragmentation patterns from those of other nucleosides.[18,19] As an example, the mass spectrum of (5'R)-8,5'-

[26] J. A. McCloskey, A. M. Lawson, D. Tsuboyama, P. M. Krueger, and R. N. Stillwell, *J. Am. Chem. Soc.* **90,** 4182 (1968).

[27] H. Pang, K. H. Schram, D. L. Smith, S. P. Gupta, L. B. Towsend, and J. A. McCloskey, *J. Org. Chem.* **47,** 3923 (1982).

[28] J. A. McCloskey, this volume [45].

[29] M. Dizdaroglu, *Biochemistry* **24,** 4476 (1985).

[30] M. Dizdaroglu, M.-L. Dirksen, H. Jiang, and J. H. Robbins, *Biochem. J.* **241,** 929 (1987).

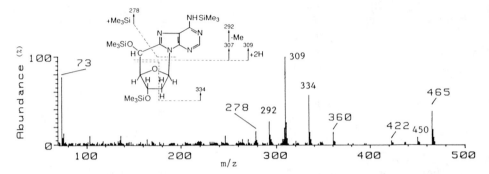

FIG. 6. Mass spectrum of (5'R)-8,5'-cyclo-2'-deoxyadenosine-(Me$_3$Si)$_3$. (From Ref. 19 with permission.)

cyclo-2'-deoxyadenosine-(Me$_3$Si)$_3$ (M_r 465) is illustrated in Fig. 6. The insert shows the structure of this compound and its fragmentation patterns. The (5'S)-diastereomer of this compound gives an essentially identical mass spectrum.[19] Unlike other trimethylsilylated nucleosides, 8,5'-cyclopurine nucleosides provide an abundant M$^+$ ion in their mass spectra,[18,19] most likely due to stabilization of M$^+$ ion by the increased number of rings in the molecule.[31] In Fig. 6, typical (M − 15)$^+$ and (M − 15 − 90)$^+$ ions are present at m/z 450 and 360, respectively. In contrast to other trimethylsilylated nucleosides, the mass spectra of these 8,5'-cyclopurine nucleosides are characterized by the high abundance of ions containing the base plus portions of the sugar moiety.[18,19] One of those ions appears as the most abundant ion (m/z 309) in the mass spectrum in Fig. 6. Another difference is the absence of the (B + 1)$^+$ ion. Instead, the (B + Me$_3$Si)$^+$ ion appears as an abundant peak at m/z 278. The fragmentation patterns shown in the insert in Fig. 6, which are typical of trimethylsilylated 8,5'-cyclopurine nucleosides, have been ascertained by measurement of the exact masses of the ions by high-resolution mass spectrometry.[18]

DNA Base–Amino Acid Cross-Links. In the past several years, the GC/MS technique has been used extensively for the study of free radical-induced DNA base–amino acid cross-links in model systems consisting of an aqueous mixture of a DNA base and an amino acid.[8,20,32,33] The goal of these studies was the understanding of gas chromatographic and mass spectrometric properties of cross-links and development of methodologies

[31] F. W. McLafferty, "Interpretation of Mass Spectra." Univ. Sci. Books, Mill Valley, CA, 1980.

[32] S. A. Margolis, B. Coxon, E. Gajewski, and M. Dizdaroglu, *Biochemistry* **27**, 6353 (1988).

[33] M. Dizdaroglu and M. G. Simic, *Int. J. Radiat. Biol.* **47**, 63 (1985).

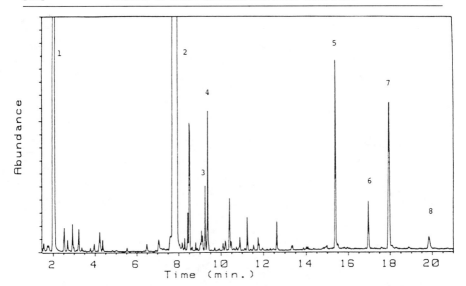

Fig. 7. Total-ion chromatogram obtained from a γ-irradiated mixture of Thy and Tyr after treatment with 6 M HCl followed by trimethylsilylation. Column: fused-silica capillary (15 m; 0.25 mm id) coated with cross-linked 5% phenyl methylsilicone gum phase (film thickness, 0.25 μm). Temperature program: 140° to 270° at 10°/min after 2 min at 140°. Peaks: 1, thymine; 2, tyrosine; 3, 2-hydroxytyrosine; 4, DOPA; 5, Thy–Tyr cross-link; 6–8, dimeric products of tyrosine (all compounds as their Me_3Si derivatives). (From Ref. 32 with permission.)

for identification of free radical-induced DNA–protein cross-links in nucleoprotein. Model systems have been exposed to ionizing radiation after N_2O saturation. Under those conditions, mainly ·OH was produced as a radical species and base–amino acid cross-links were formed as a result of reactions of this radical. The total-ion chromatogram of a typical model system consisting of thymine (Thy) and tyrosine (Tyr) is illustrated in Fig. 7. Under the conditions used, one Thy–Tyr cross-link was observed (peak 5). The exact structure of this compound has been elucidated by the combined use of high-performance liquid chromatography (HPLC), GC/MS, high-resolution MS, and ^1H NMR and ^{13}C NMR spectroscopy.[32] The mass spectrum of the Me_3Si derivative of the Thy–Tyr cross-link is illustrated in Fig. 8. The insert shows the structure and fragmentation patterns of the molecule. The M^+ and $(M - 15)^+$ ions are present at m/z 665 and 650, respectively. The cleavage of the bond between α- and β-carbons of the Tyr moiety accompanied by an H atom transfer gives rise to the most abundant ion at m/z 448 $[(M - 218 + 1)^+$ ion]. The high abundance of this ion is most likely due to its resonance stabilization through the aromatic ring. The $(M - 218)^+$ ion (m/z 447) arising from the

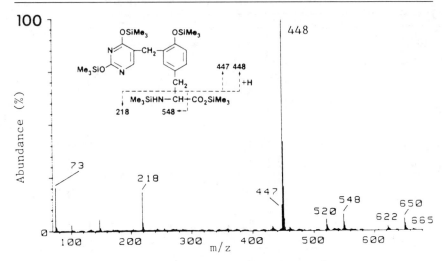

FIG. 8. Mass spectrum of the Me$_3$Si derivative of Thy–Tyr cross-link.

same cleavage without the H atom transfer is also present. This cleavage, which is typical of Me$_3$Si derivatives of amino acids,[34] also accounts for the m/z 218 ion when charge is retained on the α-carbon. Another characteristic fragmentation is the loss of ·CO$_2$SiMe$_3$ from M‡ ion, which leads to the ion at m/z 548 [(M − 117)$^+$ ion]. Ions at m/z 520 and 622 result by typical loss of CO from m/z 548 and 650, respectively.[34] The fragmentation patterns illustrated in Fig. 8 have been ascertained by measurement of the exact masses of the ions by high-resolution MS.[32] The fragmentation of the Me$_3$Si derivative of the Thy-Phe cross-link also follows the same patterns giving rise to an intense (M − 218 + 1)$^+$ ion among other ions.[33] In the case of cross-links involving aliphatic amino acids, the same fragmentations occur. However, the (M − 117)$^+$ ion generally appears as one of the most prominent ions in the mass spectra, whereas the abundance of the (M − 218)$^+$ and (M − 218 + 1)$^+$ ions varies substantially depending on the aliphatic amino acid residue.[8,35]

Gas Chromatography–Mass Spectrometry with Selected-Ion Monitoring

Identification at low concentrations (e.g., femtomole to picomole) of components of a complex mixture of organic compounds is generally carried out using GC/MS with selected-ion monitoring (SIM).[36] When

[34] K. R. Leimer, R. H. Rice, and C. W. Gehrke, *J. Chromatogr.* **141**, 355 (1977).
[35] M. Dizdaroglu and E. Gajewski, *Cancer Res.* **49**, 3463 (1989).
[36] J. T. Watson, this volume [4].

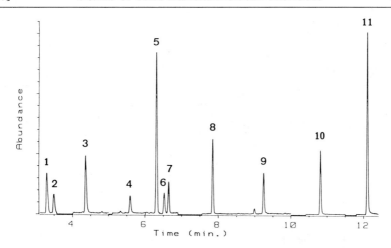

FIG. 9. Ion-current profiles obtained during GC/MS-SIM analysis of a DNA sample, which was treated with H_2O_2/Fe^{3+}–NTA followed by formic acid hydrolysis and trimethylsilylation. Column as in Fig. 1 except for film thickness (0.33 μm). Temperature program: 150° to 250° at 8°/min after 2 min at 150°. Peaks [ion monitored in the expected retention time region]: 1, 5-hydroxy-5-methylhydantoin-(Me$_3$Si)$_3$ (m/z 331); 2, 5-hydroxyhydantoin-(Me$_3$Si)$_3$ (m/z 317); 3, 5-hydroxyuracil-(Me$_3$Si)$_3$ (m/z 329); 4, 5-hydroxycytosine-(Me$_3$Si)$_3$ (m/z 343); 5, cis-thymine glycol-(Me$_3$Si)$_4$ (m/z 259); 6, 5,6-dihydroxyuracil-(Me$_3$Si)$_4$ (m/z 417); 7, trans-thymine glycol-(Me$_3$Si)$_4$ (m/z 259); 8, 4,6-diamino-5-formamidopyrimidine-(Me$_3$Si)$_3$ (m/z 354); 9, 8-hydroxyadenine-(Me$_3$Si)$_3$ (m/z 352); 10, 2,6-diamino-4-hydroxy-5-formamidopyrimidine-(Me$_3$Si)$_4$ (m/z 442); 11, 8-hydroxyguanine-(Me$_3$Si)$_4$ (m/z 440).

using this technique, a mass spectrometer is set to monitor a number of typical ions of an analyte during the time interval during which the analyte elutes from the column. The analyte is then identified when the signals of the monitored ions with typical abundances all line up at its retention time. For this purpose, the knowledge of the mass spectrum and the retention time of the analyte is required. When using capillary gas chromatography, retention times can be measured with great accuracy and precision (1 in 500 or 1 in 1000; ± 2 sec in 20 min, etc.), and thus play an important role in reliable identification, in addition to the simultaneous measurement of mass.

Identification of Modified Bases in DNA by GC/MS-SIM. As a typical example, Fig. 9 illustrates ion-current profiles obtained during GC/MS-SIM analysis of a trimethylsilylated hydrolysate of DNA, which was treated with H_2O_2 in aqueous solution in the presence of Fe^{3+}–chelate of nitrilotriacetic acid (Fe^{3+}–NTA). The H_2O_2/Fe^{3+}–NTA system generates · OH via superoxide radical.[37] As Fig. 9 clearly shows, a large number of

[37] S. Inoue and S. Kawanishi, *Cancer Res.* **47**, 6522 (1987).

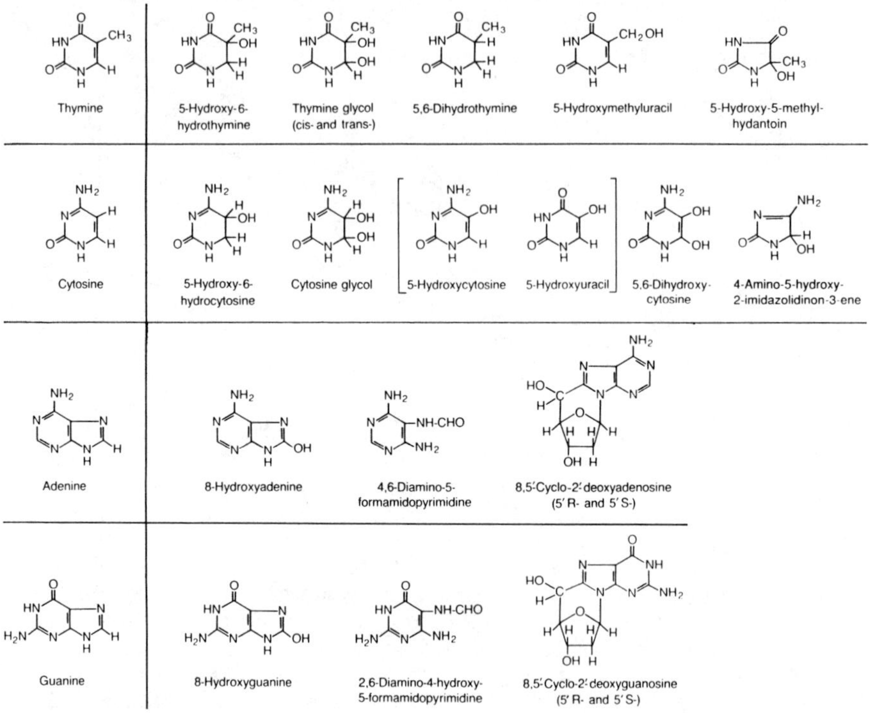

FIG. 10. Structure of free radical-induced products of pyrimidines and purines in DNA, which were identified by the use of GC/MS technique.

products can be monitored in a single run within a short analysis time. Generally, a small amount of DNA (0.4 μg in this case) is required for such an analysis. The amount of DNA, of course, may vary depending on the yields of the products in DNA. In Fig. 9, the current profile of only one ion of each compound is plotted in the expected retention time region. For an unequivocal identification, a number of characteristic ions of a compound are monitored in the same time interval during GC/MS-SIM analysis. Subsequently, a partial mass spectrum is obtained on the basis of the monitored ions and their relative abundances, and compared with that of authentic compound. For this purpose, the mass spectrum of the desired authentic compound should be recorded under the same tuning conditions of the mass spectrometer as are used to monitor actual samples.

Figure 10 illustrates the structure of free radical-induced products of pyrimidines and purines in DNA, which have been identified so far by the use of the GC/MS technique. 5-Hydroxy-6-hydrocytosine, 5,6-dihydroxy-cytosine, and 4-amino-5-hydroxy-2-imidazolidinon-3-ene deaminate during acidic hydrolysis and are identified as their analogs derived from uracil, namely, 5-hydroxy-6-hydrouracil (peak 5 in Fig. 1), 5,6-dihydroxyuracil (peak 9 in Fig. 1 and peak 6 in Fig. 9), and 5-hydroxyhydantoin (peak 2 in Fig. 9), respectively. Cytosine glycol is also modified during acidic hydrolysis and is identified as 5-hydroxyuracil and 5-hydroxycytosine (peaks 4 and 6 in Fig. 1, and peaks 3 and 4 in Fig. 9, respectively).

Identification of DNA–Protein Cross-Links in Nucleoprotein by GC/ MS-SIM. Having obtained gas chromatographic and mass spectrometric properties of ·OH-induced base–amino acid cross-links in model systems as was explained above, the GC/MS-SIM technique has been applied to identification of corresponding DNA–protein cross-links in nucleoprotein γ-irradiated in aqueous solution. Calf thymus nucleohistone has been used as nucleoprotein. As a typical example, Fig. 11 illustrates ion-current profiles of several ions of the Me_3Si derivative of the Thy–Tyr cross-link (for the mass spectrum see Fig. 8), which were obtained during GC/MS-SIM analysis of an HCl-hydrolysate of calf thymus nucleohistone.[9] The signals of the monitored ions with appropriate relative abundances are seen in Fig. 11A obtained with γ-irradiated nucleohistone. Other DNA–protein cross-links identified in nucleohistone using GC/MS-SIM involve Thy, Cyt, Gly, Ala, Val, Leu, Ile, Thr, Lys, and Tyr.[8,35,38]

Quantitative Measurements of Free Radical-Induced Products in DNA and Nucleoprotein by GC/MS-SIM. The quantity of an analyte in a complex mixture can often be accurately measured by GC/MS-SIM.[39] For this purpose, an internal standard is used, which is added to the samples prior to GC/MS-SIM analysis. Ideally, a stable isotope-labeled analog of the analyte is used as an internal standard as discussed elsewhere in this volume.[40] The mass spectrometer is calibrated first, using samples containing known quantities of the analyte and internal standard. This is done by monitoring an intense and characteristic ion of each compound during GC/MS-SIM analysis. The ion-current ratio of the ions monitored is plotted as a function of the ratio of the molar amounts of both compounds. The slope of such a plot is the relative molar response factor, which is used to calculate the quantity of the analyte in the mixture. Relative molar response factors depend upon experimental and instrumental conditions and should be determined in each laboratory. In the case of free radical-induced products in DNA and nucleoprotein, isotope-labeled analogs are not available. Thus, structurally similar compounds must be used as internal standards.[39] The suitability of the GC/MS-SIM technique has been

---

[38] E. Gajewski and M. Dizdaroglu, *Biochemistry* **29,** 977 (1990).
[39] J. T. Watson, "Introduction to Mass Spectrometry." Chap. 3. Raven, New York, 1985.
[40] P. F. Crain, this volume [47].

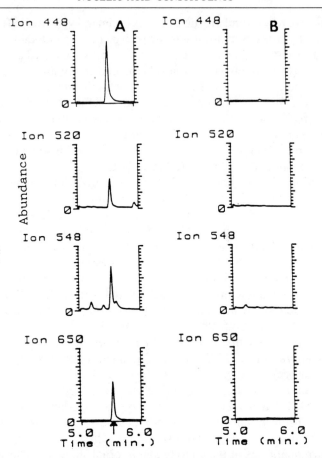

FIG. 11. Ion-current profiles obtained during GC/MS-SIM analysis of a nucleohistone sample, which was hydrolyzed by HCl and trimethylsilylated. (A) γ-Irradiated in aqueous solution (dose, 300 Gy); (B) unirradiated. Column as in Fig. 1 except for internal diameter (0.32 mm) and film thickness (0.17 μm). Temperature program: 200° to 270° at 10°/min after 1 min at 200°. (From Ref. 9 with permission.)

demonstrated recently for quantitative measurement of free radical-induced products of pyrimidines and purines in DNA,[14,19] and of free radical-induced DNA–protein cross-links in nucleohistone.[8,9,35,38]

Conclusions

The GC/MS technique is well suited for analysis of modified pyrimidines and purines, and of DNA–protein cross-links produced by free radical reactions in DNA and nucleoprotein. Derivatives of these compounds possess excellent gas chromatographic properties and provide characteristic mass spectra. The high selectivity and sensitivity of the SIM technique permit the unequivocal identification of a large number of these compounds in complex hydrolysates of DNA and nucleoprotein in one single run at quantities from femtomole to picomole per injection. Their quantitative measurement is also accomplished by GC/MS-SIM using suitable internal standards. The GC/MS technique is expected to greatly contribute to the understanding of the free radical-induced damage to DNA and nucleoprotein *in vitro* and *in vivo*.

[47] Analysis of 5-Methylcytosine in DNA by Isotope Dilution Gas Chromatography–Mass Spectrometry

By Pamela F. Crain

Introduction

5-Methylcytosine is a minor base present in DNA from many organisms, and in higher eukaryotes, there is, in many cases, a general correspondence between hypomethylation and gene activation.[1] For cases where an accurate and unbiased direct chemical measurement of the amount of 5-methylcytosine in DNA is required, a sensitive assay has been developed based on stable isotope dilution gas chromatography–mass spectrometry (GC/MS) with selected-ion monitoring.[2-4] Nanogram amounts of DNA have been analyzed for 5-methylcytosine content from 0.2 to 1.5 mol %, and the method is suited both to the high sensitivity measurement of trace amounts of 5-methylcytosine,[2,3] and to the accurate determination of variation of 5-methylcytosine content with growth stage,[4-6] and similar studies.

The basic structure of the assay consists of the addition of isotopically labeled thymine and 5-methylcytosine to the DNA prior to hydrolysis, to

[1] W. Doerfler, *Annu. Rev. Biochem.* **52,** 93 (1983).

[2] P. F. Crain and J. A. McCloskey, *Anal. Biochem.* **132,** 124 (1983).

[3] J. A. McCloskey, E. M. Rachlin, C. W. Whitehead, and P. F. Crain, *Nucleic Acids Res. Symp. Ser.* **20,** 47 (1988).

[4] P. J. Russell, J. A. Welch, E. M. Rachlin, and J. A. McCloskey, *J. Bacteriol.* **169,** 4393 (1987).

[5] P. J. Russell, K. D. Rodland, J. E. Cutler, E. M. Rachlin, and J. A. McCloskey, "Molecular Genetics of Filamentous Fungi," p. 321. Liss, New York, 1985.

[6] P. J. Russell, K. D. Rodland, E. M. Rachlin, and J. A. McCloskey, *J. Bacteriol.* **169,** 2902 (1987).

serve as internal standards for measurement of the corresponding DNA bases as their *tert*-butyldimethylsilyl (TBDMS) derivatives, and then analysis by capillary column GC/MS with selected-ion monitoring. Microscale hydrolysis and derivatization are performed in the same tube, and the bases are analyzed directly from the reaction solution. No clean-up is required past the hydrolysis step thus minimizing the potential for sample loss. An important function of the labeled analogs is to serve as carriers to prevent absorptive losses of trace constituents during sample manipulation and analysis. The mole percent of 5-methylcytosine is determined from the known thymine content of the DNA, or when the thymine content is not known, from the amount of DNA analyzed, although with lessened accuracy.

General features of the assay steps will be discussed in sufficient detail to permit the user to adapt protocols to available mass spectrometers and to the amounts of DNA available for analysis. No exact quantities can be given because conditions and amounts of DNA required will be instrument-sensitive, and must be individually determined.

Procedures

Preparation of DNA

The presence of traces of RNA in the DNA can potentially interfere with 5-methylcytosine determination because both tRNA[7] and rRNA[8] contain 5-methylcytidine. When the aglycons from acid-digested nucleic acid are analyzed (and not molecular ion-related species of nucleosides from enzymatic digests), the identity of the attached sugar cannot be determined, making attribution of the 5-methylcytosine solely to DNA impossible. It is, therefore, important to remove RNA from DNA, or at least reduce the content to an acceptably low level. If sufficient DNA is available, high-performance liquid chromatography (HPLC) analysis of enzymatic digests will reveal RNA contamination,[9] and is especially recommended where commercially supplied DNAs are to be examined. Suitable methods for RNA removal include CsCl equilibrium centrifugation,[4] or digestion with RNase T_2.[9] Either of these methods, however, introduces low-molecular weight contaminants, which must be removed. In addition, commercially supplied DNAs may be shipped in buffered salt solutions; acid hydrolysis of NaCl-containing DNAs has been reported to result in

[7] M. Sprinzl, T. Hartmann, J. Weber, J. Blank, and R. Zeidler, *Nucleic Acids Res.* **17**, r1 (1989).
[8] H. Noller, *Annu. Rev. Biochem.* **53**, 119 (1984).
[9] P. F. Crain, this volume [42].

deamination of cytosines,[10] and although the internal standards compensate for this occurrence, it is generally undesirable. In addition, Tris salts can attentuate the acid strength during hydrolysis leading to poor recoveries, and in addition consume derivatizing reagent. Accordingly, the DNA to be analyzed should be salt-free.

DNA can conveniently be desalted using NENSORB 20 columns (DuPont/NEN Products, Boston, MA), from which nucleic acid is eluted using aqueous alcohol solutions.[11] Salts, buffers, and nucleotides and enzyme from RNase T_2 treatment are removed with water washes. The capacity of the column is 20 μg of nucleic acid plus protein, so the amount of enzyme added to the DNA to remove RNA must be taken into account. The supplier's protocol may be used; an additional water wash is recommended. The importance of not allowing the column dry out at the top (or at the bottom, as a consequence of removing the solvent transfer syringe from the inlet too quickly, before the pressure has equalized) must be emphasized. If a drop of water remains at the tip of the column when the last wash solution has reached the top of the column bed, remove it with a clean tissue; the first two drops of eluate may then be discarded and the nucleic acid recovered in a minimal volume. The next sixteen drops will contain DNA (or tRNA).[12] Recoveries are excellent, typically 60–90% or higher,[11] but are nonetheless not quantitative, and the amount of DNA recovered should be determined so that the proper amount of labeled standards to be added may be calculated. This is conveniently done by measurement of UV absorbance: for double-stranded DNA, one A_{260} corresponds to 50 μg/ml (40 μg/ml for single-stranded DNA) when measured at 260 nm through a 1-cm path length cell.

Synthesis of Labeled 5-Methylcytosine and Thymine
Internal Standards

2,4-Dimethoxy-5-([$^2H_3$]methyl)pyrimidine.[3]   5-Bromo-2,4-dimethoxypyrimidine (3.46 g; 15.9 m)[13] in 60 ml dry tetrahydrofuran was cooled to $-65°$ under dry bubbling nitrogen. *n*-Butyllithium (7.15 ml; 2.4 M in hexane; Aldrich, Milwaukee, WI) was added with stirring, followed by 22 gm (15.2 mmol) of [$^2H_3$]methyl iodide (99.5% ^2H; MSD Isotopes, St. Louis,

[10] J. P. Ford, M. Coca-Prados, and M. T. Hsu, *J. Biol. Chem.* **255**, 7544 (1980).

[11] M. T. Johnson, B. A. Read, A. M. Monko, G. Pappas, and B. A. Johnson, *BioTechniques* **4**, 64 (1986).

[12] P. F. Crain, unpublished observations, 1989.

[13] Commercially available material may be used, but in any case, it must be free of sodium methoxide from which it is commonly synthesized. A convenient means of purification is sublimation onto a cold finger from which the crystalline product may be washed off with ethyl ether.

MO) in 5 ml dry tetrahydrofuran. The reaction was stopped after 2 hr by addition of small chips of dry ice. The solution was poured into 50 ml H_2O and extracted twice with diethyl ether. The combined ether solutions were dried over $MgSO_4$, filtered, and the filtrate concentrated to give a yellow oil. Extraction of this material with hot hexane, and concentration of the extract gives an earlier described dimer[14] in 3.5% yield. Filtration of the hexane solution, followed by concentration and recrystallization five times gives 2,4-dimethoxy-5-([²H₃]methyl)pyrimidine in 12% yield.

¹⁵N⁴-5-([²H₃]methyl)Cytosine and [methyl-²H₃]Thymine.[3] 2,4-Dimethoxy-5-([²H₃]methyl)pyrimidine (100 mg) was mixed with 0.5 ml of dry acetyl chloride and allowed to stand at room temperature for 2 days. After sample work-up,[15] the white solid material (0.06 g) was added to 1.5 ml of ethanol and placed on ice. The solution was saturated with 99.9% [¹⁵N]ammonia (81 mg; Mound Laboratories, Miamisburg, OH), and heated 20 hr at 110° in a sealed tube. The residue after drying was fractionated by anion-exclusion chromatography on Bio-Rad AG 50W-X8 (NH_4^+) (Richmond, CA). The first peak is [*methyl*-²H₃]thymine (formed as a by-product) and the second is ¹⁵N⁴-5-([²H₃]methyl)cytosine (in 62% yield). Contaminating unlabeled products are undetectable under typical assay conditions.[4]

Hydrolysis of DNA

Microscale hydrolysis of DNA is conveniently carried out in small tubes fashioned from 3 mm od Pyrex tubing cut to lengths of approximately 5 cm. After one end is flame-sealed, the tubes should be cleaned with NOCHROMIX (Godax Laboratories, New York), thoroughly rinsed with deionized H_2O, and oven-dried at 180°, preferably overnight. Silanization of the tubes with 5% dimethyldichlorosilane in toluene, followed by overnight baking at 180°, is recommended.

The desired amounts of salt-free DNA and internal standards (see later Discussion) are mixed, transferred to a tube, and dried *in vacuo*. Solutions of labeled and unlabeled thymine and 5-methylcytosine used for constructing the calibration curve, and a blank, should also be prepared and dried. To each tube is added 10 μl 98% formic acid, which is then flame-sealed, spun briefly at 12,000 rpm to disperse any trapped air, and placed at 180° for 1 hr. After hydrolysis, the formic acid is removed *in vacuo*.

[14] T. L. V. Ulbricht, *Tetrahedron* **6**, 225 (1959).
[15] M. Prystasz, *in* "Nucleic Aid Chemistry: Improved and New Synthetic Procedures, Methods and Techniques," Part 1 (L. B. Townsend and R. S. Tipson, eds.), p. 77. Wiley (Interscience), New York, 1978.

Preparation of tert-Butyldimethylsilyl Derivatives

Immediately prior to derivatization, the dried DNA hydrolysate or standard compounds (in the original hydrolysis tubes) are treated with 10 μl of dry pyridine and dried *in vacuo* to remove any traces of water by forming an azeotrope. The derivatizing reagent is a 1 : 1 solution of dry pyridine and N-methyl-N-[(*tert*-butyl)dimethylsilyl]trifluoroacetamide containing 1% *tert*-butyldimethylchlorosilane (Regis Chemical Co., Morton Grove, IL). To the dried sample is added 10 μl of derivatizing reagent and the tube is flame-sealed and heated for 20 min at 120°.

Gas Chromatography–Mass Spectrometry

Gas Chromatography. Fused-silica capillary columns of 15 m × 0.32 mm id can be used;[3] bonded phases are recommended, with the choice of phase not being critical. Methyl silicone (e.g., J & W DB-1, Folsom, CA) and 5% phenyl–95% methyl silicone (e.g., J & W DB-5) phases have been successfully utilized.[3,4] The di-TBDMS derivatives of thymine and 5-methylcytosine elute in the region of *n*-hydrocarbons C_{19}–C_{20}, and gas chromatographic conditions may be optimized on this basis. A submicroliter syringe (SGE, Austin, TX) is convenient for splitless injection of small volumes. A syringe-cleaning apparatus (Hamilton, Reno, NV) is required for thorough cleaning of the syringe between injections.

Mass Spectrometry. Mass spectra of the di-*tert*-butyldimethylsilyl derivatives of thymine and 5-methylcytosine are shown in Fig. 1, and reveal domination of the spectra by the characteristic M − 57 ion arising from loss of the *tert*-butyl radical from the molecular ion species. Doubly charged ions from loss of two TBDMS radicals are analogous to the doubly charged M − 30 species from trimethylsilyl derivatives of bases.[16]

Quantification of thymine and 5-methylcytosine in DNA is accomplished by selected-ion recording of their M − 57 ions, and those of the corresponding isotopically labeled internal standards. The values of these ions are: 5-methylcytosine, m/z 296.16; thymine, m/z 297.14; and [*methyl*-$^2H_3$]thymine and $^{15}N^4$-5-([$^2H_3$]methyl)cytosine, m/z 300.17. [The molecular ion (m/z 285.09) of tris(trifluoromethyl)-*s*-triazine (PCR, Inc., Gainesville, FL), provides a suitable lock mass for selected-ion recording.] The peak height ratios (297/300 for thymine; 296/300 for 5-methylcytosine) are used to derive the amounts of thymine and 5-methylcytosine in the DNA. If the mole percent of thymine in the DNA is known, or can be calculated, then the mole percent of 5-methylcytosine can be determined from the 5-methylcytosine/thymine ratio.

[16] E. White V, P. M. Krueger, and J. A. McCloskey, *J. Org. Chem.* **37**, 420 (1972).

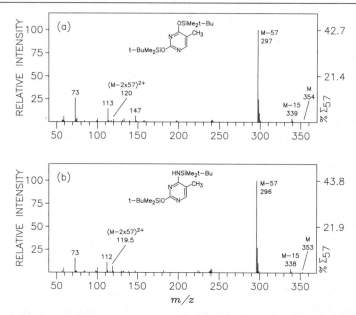

FIG. 1. Electron ionization mass spectra (70 eV) of the di-*tert*-butyldimethylsilyl derivatives of (a) thymine and (b) 5-methylcytosine.

A calibration curve must be constructed from which the amounts of thymine and 5-methylcytosine present can be determined relative to the known amount of internal standard added to the DNA prior to acid digestion. Unlike the situation where the amount of analyte(s) cannot be anticipated and variations in concentration of several orders of magnitude must be allowed for, DNA base analysis presents a more restricted situation. Thymidine, for example, may typically comprise about 20 mol%, but in any case, is unlikely to vary by more than a factor of ±2, and so can be estimated at 10–40 mol%; the amount of thymine may, therefore, be anticipated based on the amount of DNA to be analyzed. Accordingly, the range of 5-methylcytosine content may also be anticipated because it will be generally known whether the 5-methylcytosine analysis is to determine trace amounts or to measure variations in content of a larger amount of base. The calibration curve is then constructed by adding varied amounts of analyte to a predetermined (see below) fixed amount of internal standard.[17] These reference mixtures should also be acid-treated in parallel with the DNA, so that if deamination of 5-methylcytosine should occur

[17] L. Siekmann, *in* "Mass Spectrometry" (A. M. Lawson, ed.), p. 647. de Gruyter, New York, 1989.

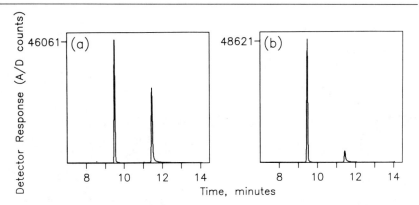

FIG. 2. Selected-ion recordings of the M − 57 ions (m/z 300) of the di-*tert*-butyldimethylsilyl derivatives of [*methyl*-^{2}H$_3$]thymine (eluting at 9.42 min) and $^{15}N^4$-5-([^{2}H$_3$]methyl)cytosine (11.45 min). The amount of solution injected for (a) was halved for (b), but the detector sensitivity for (b) was twice that of (a).

(generally not an issue with salt-free DNA[2, 18]) it can be compensated for by the corresponding process in the isotopically labeled standard.

Discussion

The amounts of DNA and internal standards to be used must be determined for each mass spectrometer and GC column. The thymine content of the DNA will determine the maximum amount of DNA that can be analyzed in a single run. If too large an amount is injected, the detector and/or the GC column capacity may be exceeded, resulting in signal saturation and distortion of the peak shape. Ion source defocusing may also be a consequence of too large a sample amount; it is especially important to keep the amount of [*methyl*-^{2}H$_3$]thymine per injection well less than the maximum tolerable amount because it elutes about 2 sec ahead of unlabeled thymine and if present in excess will distort the thymine signal. The amount of $^{15}N^4$-5-([^{2}H$_3$]methyl)cytosine, on the other hand, must be sufficiently large to serve as a "carrier" to prevent losses of trace amounts of 5-methylcytosine in the DNA during chromatography. Figure 2 illustrates the presence of a "carrier effect," which can be recognized because as increasingly smaller sample amounts are injected, the signal ceases to be proportional to the amount injected. In Fig. 2b, the amount of labeled 5-methylcytosine has decreased relative to the amount in Fig. 2a under conditions when the signals should have remained the same, as

[18] D. Eick, H. J. Fritz, and W. Doerfler, *Anal. Biochem.* **135,** 165 (1983).

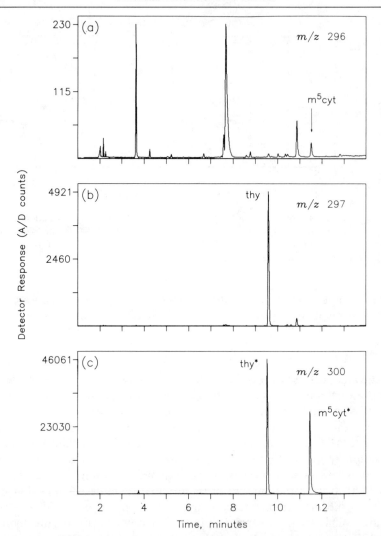

FIG. 3. Selected-ion recordings for determination of 5-methylcytosine in pBR322 DNA. (a) m/z 296, 5-Methylcytosine; (b) m/z 297, thymine; (c) m/z 300, [$methyl$-$^2H_3$]thymine and $^{15}N^4$-5-([$^2H_3$]methyl)cytosine internal standards. DNA (200 ng) was acid-digested, then derivatized in 10 μl of reagent. The amount of solution injected corresponds to 1 ng of DNA, and the amount of 5-methylcytosine determined is 0.5 pg (4 fmol). Conditions: splitless injection of 0.05 μl onto a 15 m × 0.32 mm id Flexibond SE-30 column (Pierce Chemical Co., Rockford, IL) in a Hewlett-Packard 5890 gas chromatograph (injector 250°; column 60° for 1 min, then programmed to 180° at 50°/min, followed by a program of 180° to 250° at 5°/min). The chromatograph is interfaced to a VG 70-SEQ mass spectrometer; conditions were 100 eV ionizing energy, 8 kV accelerating voltage, 100 μA filament current; multiplier gain 1 × 10^5.

observed for thymine. The injected amount of the labeled 5-methylcytosine internal standard is, therefore, too low to serve as a carrier. This amount will also need to be determined for a given instrumental system. It should be noted that the amounts of bases and standards per injection may change with time, depending on the condition of the injection port and column.

As a general observation, the limiting factor for sensitivity of the assay is more likely to result from chemical "noise" than inherent instrumental limitations. A selected-ion recording from the analysis of 1 ng of pBR322 DNA is shown in Fig. 3 and demonstrates the sensitivity which may be achieved for assay of a highly purified reference DNA. Note, however, that at the low picogram sample range for 5-methylcytosine, shown in trace (a), there are a number of species of undetermined origin and identity which produce ions of m/z 296, and which may be present in much larger amount. Their production may be difficult to control or anticipate, so the potential for interference should be recognized when small amounts of 5-methylcytosine or DNA must be analyzed.

Acknowledgments

The author is indebted to S. C. Pomerantz for preparation of figures. A portion of the methods described were not previously published. Their development was supported by NIH grant GM 21584.

Appendixes

Appendix 1. Mass and Abundance Values

Appendix 2. Reference Ions for Exact Mass Measurements

Appendix 3. Mass Spectra of Matrix Materials

Appendix 4. Calculation of Isotopic Abundance Distributions

Appendix 5. Nomenclature for Peptide Fragment Ions

Appendix 6. Mass Values for Amino Acid Residues in Peptides

Appendix 1. Mass and Abundance Values

By JAMES A. MCCLOSKEY

ISOTOPIC MASS AND ABUNDANCE VALUES

Isotope	Mass[a]	Natural abundance (%)[b]
^{1}H	1.007 825 035	99.985
^{2}H	2.014 101 779	0.015
^{12}C	12	98.90
^{13}C	13.003 354 826	1.10
^{14}N	14.003 074 002	99.63
^{15}N	15.000 108 97	0.37
^{16}O	15.994 914 63	99.76
^{17}O	16.999 131 2	0.04
^{18}O	17.999 160 3	0.200
^{19}F	18.998 403 22	100
^{23}Na	22.989 767 7	100
^{28}Si	27.976 927 1	92.23
^{29}Si	28.976 494 9	4.67
^{30}Si	29.973 770 1	3.10
^{31}P	30.973 762 0	100
^{32}S	31.972 070 698	95.02
^{33}S	32.971 458 428	0.75
^{34}S	33.967 866 650	4.21
^{36}S	35.967 080 620	0.02
^{35}Cl	34.968 852 728	75.77
^{37}Cl	36.965 902 619	24.23
^{39}K	38.963 707 4	93.2581
^{40}K	39.963 999 2	0.012
^{41}K	40.961 825 4	6.7302
^{79}Br	78.918 336 1	50.69
^{81}Br	80.916 289	49.31
^{127}I	126.904 473	100

[a] A. H. Wapstra and G. Audi, *Nucl. Phys.* **A432,** 1 (1985). Standard errors are given in this article but are omitted from the present table.

[b] "CRC Handbook of Chemistry and Physics," 70th ed., p. B-227. CRC Press, Boca Raton, FL, 1989.

SELECTED ATOMIC WEIGHTS

Element	Atomic weight[a]
H	1.007 94(7)
C	12.011(1)
N	14.006 74(7)
O	15.999 4(3)
F	18.998 403 2(9)
Na	22.989 768 (6)
Si	28.085 5(3)
P	30.973 762(4)
S	32.066(6)
Cl	35.452 7(9)
K	39.098 3(1)
Br	79.904(1)
I	126.904 47(3)

[a] "Atomic Weights of the Elements 1987," *Pure Appl. Chem.* **60,** 841 (1988). Values in parentheses indicate the level of uncertainty, which is related mainly to natural variations in isotopic abundance. For example, the atomic weight of hydrogen is 1.00794 ± 0.00007.

Appendix 2. Reference Ions for Exact Mass Measurements

By RICHARD M. MILBERG

TRIS(PERFLUOROHEPTYL)-s-TRIAZINE[a]

Mass	Mass	Mass	Mass	Mass
49.99680	151.99976	301.99645	615.97409	927.95493
68.99521	168.98882	325.98391	646.97250	965.95174
75.99988	180.98882	375.98072	670.96622	1015.94854
85.00763	192.98882	420.98220	720.96303	1065.94535
99.99361	218.98563	470.97900	770.95984	1115.94216
118.99202	230.98563	520.97581	815.96132	1127.94216
130.99202	268.98244	565.97729	865.95813	1165.93896
137.99669	280.98244	570.97262	915.95493	1184.93737

[a] See also Fig. 3, this volume [14].

METHODS IN ENZYMOLOGY, VOL. 193

PERFLUOROTRIBUTYLAMINE[a]

Mass	Mass	Mass	Mass	Mass
30.99840	118.99202	175.99249	263.98711	463.97433
49.99379	130.99202	180.98882	313.98391	501.97114
68.99521	149.99042	213.99030	325.98391	537.97114
92.99521	161.99042	218.98563	375.98072	575.96795
99.99361	163.99349	225.99030	413.97753	613.96475
113.99669	168.98882	230.98563	425.97753	

[a] See also Fig. 2, this volume [14].

PFK[a]

Mass	Mass	Mass	Mass	Mass
51.00462	218.98563	392.97605	566.96647	742.95369
68.99521	230.98563	404.97605	580.96327	754.95369
80.99521	242.98563	416.97605	592.96327	766.95369
92.99521	254.98563	430.97285	604.96327	780.95050
99.99361	268.98244	442.97285	616.96327	792.95050
111.99361	280.98244	454.97285	630.96008	804.95050
118.99202	292.98244	466.97285	642.96008	816.95050
130.99202	304.98244	480.96966	654.96008	830.94731
142.99202	318.97924	492.96966	666.96008	842.94731
149.99042	330.97924	504.96966	680.95689	854.94731
154.99202	342.97924	516.96966	692.95689	866.94731
168.98882	354.97924	530.96647	704.95689	880.94411
180.98882	366.97924	542.96647	716.95689	892.94411
192.98882	380.97605	554.96647	730.95369	904.94411
204.98882				

[a] See also Fig. 1, this volume [14].

GLYCEROL[a]

Mass	Mass	Mass	Mass	Mass
57.03404	369.19720	737.38658	1105.57596	1473.76533
75.04460	461.24455	829.43392	1197.62330	1565.81268
93.05517	553.29189	921.48127	1289.67064	1657.86002
185.10251	645.33923	1013.52861	1381.71799	1749.90737
277.14986				

[a] Subtract 2.10565 from each mass for the corresponding negative ion. See also mass spectrum of glycerol in Appendix 3.

ULTRAMARK 2500F

Mass	Mass	Mass	Mass	Mass
30.99840	265.97895	500.95949	1064.90721	1744.84347
46.99332	268.98244	516.95441	1114.90402	1778.84536
49.99681	280.98244	534.96138	1130.89894	1794.84027
68.99521	284.97735	566.95121	1180.89574	1828.84217
80.99521	296.97735	616.94802	1230.89255	1844.83708
96.99012	300.97226	632.94298	1296.88427	1860.83199
99.99361	312.97226	664.94293	1346.88108	1894.83388
118.99202	315.97575	666.94482	1396.87788	1944.83069
130.99202	318.97924	682.93970	1462.86960	1960.82561
134.98693	334.97416	716.94163	1496.87150	1994.82750
146.98693	346.97416	732.93654	1512.86641	2010.82241
149.99042	350.96907	782.93335	1546.86830	2026.81733
168.98882	368.97605	798.92827	1562.86322	2044.82430
180.98882	378.96399	832.93016	1578.85813	2060.81922
184.98374	384.97096	848.92507	1596.86511	2076.81413
196.98374	400.96588	882.92697	1612.86002	2094.82111
212.97865	428.96079	898.92188	1628.85494	2110.81603
218.98563	434.96777	914.91680	1662.85683	2126.81094
230.98563	450.96268	948.91869	1678.85174	2176.80775
234.98054	462.96268	964.91360	1694.84666	2192.80266
246.98054	466.95760	998.91549	1712.85364	2210.80964
262.97546	478.95760	1014.91041	1728.84855	

POLYETHYLENE GLYCOL[a]

n	PEG + H$^+$ Mass	n	PEG + H$^+$ Mass	n	PEG + H$^+$ Mass
1	63.04460	16	723.43783	31	1383.83105
2	107.07082	17	767.46404	32	1427.85726
3	151.09703	18	811.49026	33	1471.88348
4	195.12325	19	855.51647	34	1515.90969
5	239.14946	20	899.54269	35	1559.93591
6	283.17568	21	943.56890	36	1603.96212
7	327.20189	22	987:59511	37	1647.98834
8	371.22811	23	1031.62133	38	1692.01455
9	415.25432	24	1075.64754	39	1736.04077
10	459.28054	25	1119.67376	40	1780.06698
11	503.30675	26	1163.69997	41	1824.09320
12	547.33297	27	1207.72619	42	1868.11941
13	591.35918	28	1251.75240	43	1912.14563
14	635.38540	29	1295.77862	44	1956.17184
15	679.41161	30	1339.80483	45	2000.19805

[a] Cluster ions from desorption ionization. Add 17.02655 to each mass for PEG + NH$_4^+$ and 21.98195 for PEG + Na$^+$.

CESIUM IODIDE (CsI), SODIUM IODIDE (NaI), RUBIDIUM IODIDE (RbI) MIXTURE
POSITIVE IONS[a]

Mass	Abundance (%)	Mass	Abundance (%)
22.98976	5.00	770.35164	1.12
45.97954	5.00	816.34789	0.68
84.911791	5.00	864.34152	1.68
107.90099	3.47	912.33516	4.86
132.90543	>100.00	934.17429	0.65
155.89519	3.07	982.16791	0.49
172.88402	>100.00	1030.16159	0.12
217.81721	4.65	1072.24950	0.47
234.80605	19.04	1124.15140	0.92
265.81087	9.47	1172.14510	2.03
282.79967	23.72	1222.14369	0.31
296.72807	59.45	1335.96770	0.05
322.77826	45.03	1372.03799	0.25
344.72169	45.04	1431.95499	0.34
392.71533	90.08	1521.93219	0.05
432.69392	0.62	1595.77759	0.15
472.67250	7.89	1643.77129	0.43
508.54434	5.70	1691.76489	0.45
542.60959	0.60	1821.72069	0.09
556.53999	2.94	1903.58119	0.05
582.58817	0.31	1971.61499	0.04
604.53162	3.96	2121.50899	0.03
622.56676	3.80	2163.39100	0.08
652.52524	11.42	2211.38469	0.05
722.35800	2.00	2471.19460	0.05

[a] FAB 8 keV Xe. Mixture of equal amounts of 1.4 M solutions of CsI, NaI, and RbI.

POSITIVE ION MS/MS CALIBRATION MIXTURE[a]

Mass	Mass	Mass	Mass	Mass
6.0151	132.9054	472.6720	1431.9550	3510.4343
7.0160	155.8952	510.5316	1691.7649	3770.2442
22.9898	172.8840	604.5316	1951.5748	4030.0541
38.9637	217.8172	652.5252	2211.3846	4289.8640
45.9795	265.8109	722.3580	2471.1946	4549.6739
59.0448	282.7997	816.3479	2731.0045	4809.4838
84.9118	322.7783	912.3352	2990.8144	5069.2937
107.9016	392.7153	1172.1451	3250.6244	

[a] See this volume [14].

Mass	Mass	Mass	Mass	Mass
132.90543	5329.1036	10525.302	15721.500	20917.698
392.71534	5588.9135	10785.112	15981.310	21177.508
652.52525	5848.7235	11044.922	16241.120	21437.318
912.33516	6108.5334	11304.732	16500.930	21697.128
1172.1451	6368.3433	11564.541	16760.740	21956.938
1431.9550	6628.1532	11824.351	17020.550	22216.748
1691.7649	6887.9631	12084.161	17280.359	22476.558
1951.5748	7147.7730	12343.971	17540.169	22736.368
2211.3847	7407.5829	12603.781	17799.979	22996.178
2471.1946	7667.3928	12863.591	18059.789	23255.987
2731.0045	7927.2027	13123.401	18319.599	23515.797
2990.8144	8187.0126	13383.211	18579.409	23775.607
3250.6244	8446.8226	13643.021	18839.219	24035.417
3510.4343	8706.6325	13902.831	19099.029	24295.227
3770.2442	8966.4424	14162.641	19358.839	24555.037
4030.0541	9226.2523	14422.450	19618.649	24814.847
4289.8640	9486.0622	14682.260	19878.459	
4549.6739	9745.8721	14942.070	20138.269	
4809.4838	10005.682	15201.880	20398.078	
5069.2937	10265.492	15461.690	20657.888	

[a] See also Fig. 4, this volume [14].

CESIUM IODIDE NEGATIVE IONS

Mass	Mass	Mass	Mass	Mass
126.90448	5323.1027	10519.301	15715.499	20911.697
386.71439	5582.9126	10779.111	15975.309	21171.507
646.52430	5842.7225	11038.921	16235.119	21431.317
906.33421	6102.5324	11298.731	16494.929	21691.127
1166.1441	6362.3423	11558.541	16754.739	21950.937
1425.9540	6622.1522	11818.350	17014.549	22210.747
1685.7639	6881.9621	12078.160	17274.359	22470.557
1945.5738	7141.7720	12337.970	17534.168	22730.367
2205.3838	7401.5820	12597.780	17793.978	22990.177
2465.1937	7661.3919	12857.590	18053.788	23249.986
2725.0036	7921.2018	13117.400	18313.598	23509.796
2984.8135	8181.0117	13377.210	18573.408	23769.606
3244.6234	8440.8216	13637.020	18833.218	24029.416
3504.4333	8700.6315	13896.830	19093.028	24289.226
3764.2432	8960.4414	14156.640	19352.838	24549.036
4024.0531	9220.2513	14416.450	19612.648	24808.846
4283.8630	9480.0612	14676.259	19872.458	
4543.6729	9739.8711	14936.069	20132.268	
4803.4829	9999.6811	15195.879	20392.077	
5063.2928	10259.491	15455.689	20651.887	

REFERENCE COMPOUNDS FOR HIGH-RESOLUTION FAST ATOM BOMBARDMENT EXACT MASS MEASUREMENT

Sequence	Formula $(M + H)^+$	Mass $(M + H)^+$	Source[a]	Catalog number
1. Gly-Leu	$C_8H_{17}N_2O_3$	189.12392	P	3022
2. Gly-Phe	$C_{11}H_{15}N_2O_3$	223.10827	P	3053
3. Leu-Gly-Gly	$C_{10}H_{20}N_3O_4$	246.14538	P	3025
4. Glu-Glu	$C_{10}H_{17}N_2O_7$	277.10358	P	3080
5. Tyr-Arg · acetate	$C_{15}H_{24}N_5O_4$	338.18283	B	G-2450
6. Z-Gly-Phe	$C_{19}H_{21}N_2O_5$	357.14505	P	3020
7. Gly-Gly-Phe-Leu	$C_{19}H_{29}N_4O_5$	393.21380	B	N-1175
8. Phe-Gly-Leu-Met	$C_{22}H_{36}N_5O_4S$	466.24880	P	7464
9. Arg-Pro-Lys-Pro	$C_{22}H_{41}N_8O_5$	497.31999	B	H-1875
10. Tyr-Gly-Gly-Phe-Leu	$C_{28}H_{38}N_5O_7$	556.27712	B	H-2740
11. Tyr-D-Ala-Gly-Phe-Met	$C_{28}H_{38}N_5O_7S$	588.24920	B	H-2790
12. Tyr-D-Pen-Gly-Phe-Pen	$C_{30}H_{40}N_5O_7S_2$	646.23692	P	8849
13. Arg-Pro-Pro-Gly-Phe-Ser-Pro-Leu	$C_{41}H_{64}O_{10}N_{11}$	870.48376	P	7061
14. Arg-Pro-Pro-Gly-Phe-Ser Pro-Phe · acetate	$C_{44}H_{62}N_{11}O_{10}$	904.46811	B	H-1965
15. Arg-Pro-Pro-Gly-Phe-Ser-Pro-Phe-Arg · acetate	$C_{50}H_{74}N_{15}O_{11}$	1060.56922	B	H-1970
16. Lys-Arg-Pro-Pro-Gly-Phe-Ser-Pro-Phe-Arg	$C_{56}H_{86}N_{17}O_{12}$	1188.66419	B	H-2180

[a] B, BACHEM Bioscience, Inc., 3700 Market Street, Philadelphia, PA 19104; P, Peninsula Laboratories, Inc., 611 Taylor Way, Belmont, CA 94002.

Appendix 3. Mass Spectra of Matrix Materials

By C. E. COSTELLO

Presented on the following pages are the mass spectra of materials commonly used as matrices for fast atom bombardment (FAB) and liquid secondary ionization mass spectrometry (LSIMS), recorded over the mass range m/z 100–1000. The primary beam was 6 or 15 keV Cs^+ or 6 keV Xe^0. Since the observed spectra are fairly independent of the choice of primary beam, the figures are labeled only with respect to the matrix and the mass spectrometer's operation mode (positive or negative). The trivial and chemical names, the elemental composition, and the exact mass value of the $^{12}C^1H^{14}N^{16}O^{32}S$ monomer are given below for each matrix.

Glycerol. 1,2,3-Propanetriol, $C_3H_8O_3$, m/z 92.0473.

1-Thioglycerol. 3-Mercapto-1,2-propanediol, $C_3H_8O_2S$, *m/z* 108.0245. Ammonium chloride is frequently present as an impurity, even in high-quality commercial lots, such as this Fluka >99%(GC) grade, and its presence leads to ion series at $[(108)_n + 18]$ in the positive ion mode and $[(108)_n + 35]$ in the negative ion mode.

5:1 Dithiothreitol/Dithioerythritol. (Magic Bullet, DTT/DTE). 5:1 Mixture of DL-*threo-* and *erythro*-1,4-dimercapto-2,3-butanediol, $C_4H_{10}O_2S_2$, *m/z* 154.0122. Larger clusters contain increasing amounts of the oxidized (2 u lower) species, and these shift spacing of some maxima to 152 u intervals.

3-Nitrobenzyl Alcohol. $C_7H_7NO_3$, *m/z* 153.0416.

2-Nitrophenyl Octyl Ether. $C_{14}H_{21}NO_3$, *m/z* 251.1521.

Sulfolane. Tetramethylene sulfone, $C_4H_8SO_2$, *m/z* 120.0245.

Triethanolamine. 2,2′,2″-Nitrilotriethanol, $C_6H_{15}NO_3$, *m/z* 149.1052.

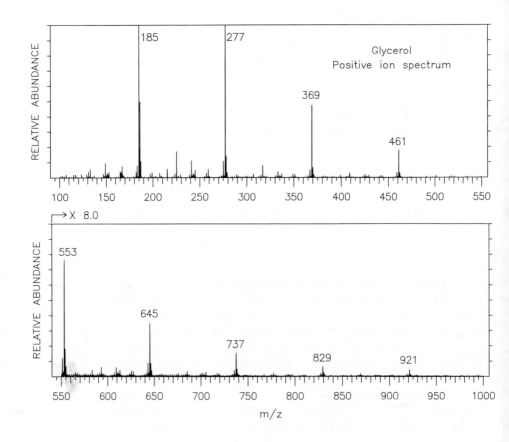

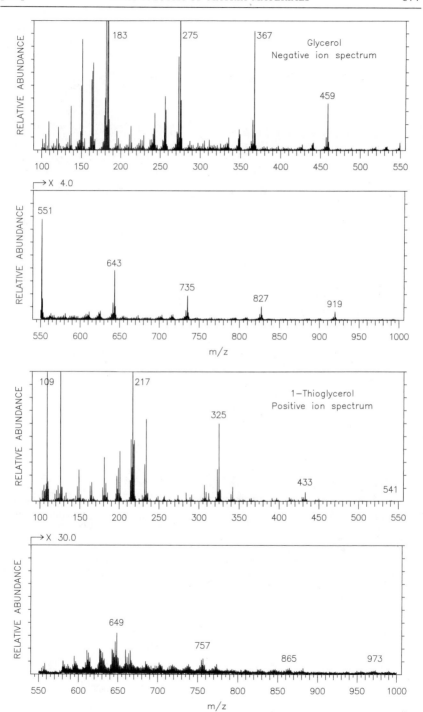

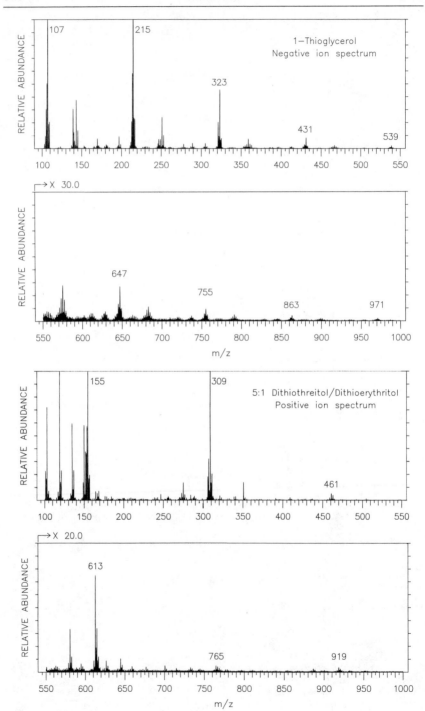

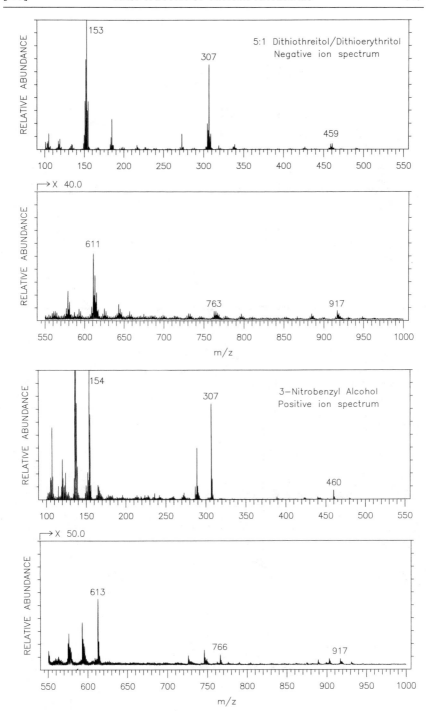

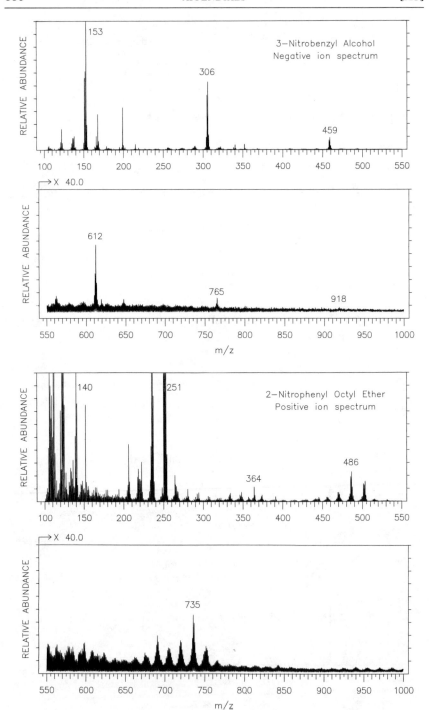

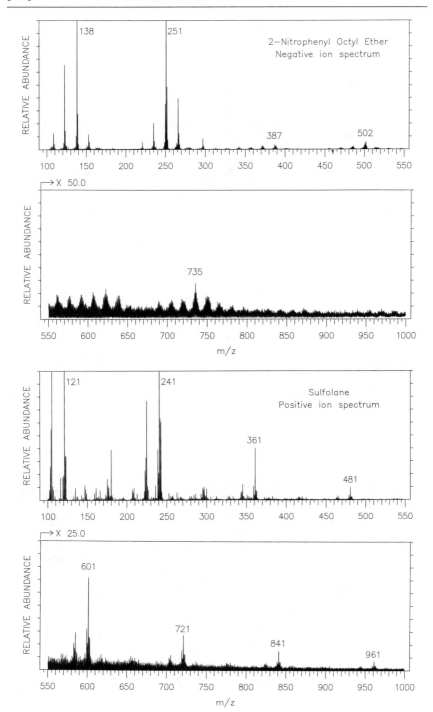

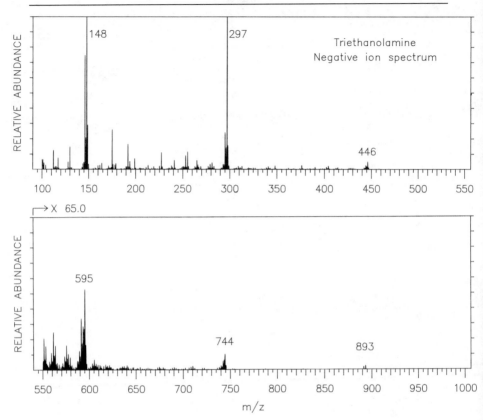

Appendix 4. Calculation of Isotopic Abundance Distributions

By JAMES A. McCLOSKEY

   This appendix contains an overview of equations for the calculation of isotopic abundance distributions. From the examples given below, extensions can be made to larger molecules and to those which contain a greater number of elements. For further discussion and specialized calculations the reader is referred to Refs. 1–5.

[1] J. H. Beynon, "Mass Spectrometry and Its Applications to Organic Chemistry," p. 294. Elsevier, New York, 1960.
[2] H. Yamamoto and J. A. McCloskey, *Anal. Chem.* **49,** 281 (1977).

Generalized Calculation of Mass and Abundance Ratios

The basic relationship which governs isotopic abundance patterns is given as Eq. (1) which shows the probability (P) that no heavy isotopes will occur in a molecule of composition $C_wH_xN_yO_z$, where c, h, n, and o represent the relative isotopic abundances (expressed as percentages) of the heavy isotopes of carbon, hydrogen, nitrogen, and oxygen, respectively.

$$P = \left(\frac{100 - c}{100}\right)^w \left(\frac{100 - h}{100}\right)^x \left(\frac{100 - n}{100}\right)^y \left(\frac{100 - o_1 - o_2}{100}\right)^z \tag{1}$$

A generalized form of Eq. (1) describing the probability of occurrence of any two-isotope element (e.g., hydrogen) can be written as[2]

$$P = \left(\frac{100 - X}{100}\right)^N \sum_{i=0}^{N} \binom{N}{i} \left(\frac{X}{100 - X}\right)^i \tag{2}$$

in which N = number of atoms of the given element; X = percentage isotopic abundance of heavy isotope, $\binom{N}{i} = N!/((N - i)!i!)$; $0! = 1$; and $i = 0$ corresponds to the molecular species containing no heavy isotopes.

Likewise, the probability of occurrence of three-isotope elements (e.g., oxygen) is

$$P = \left(\frac{100 - X - Y}{100}\right)^N \sum_{i=0}^{N} \sum_{k=0}^{i} \binom{N}{i} \binom{i}{k} \left(\frac{X}{100 - X - Y}\right)^i \left(\frac{Y}{X}\right)^k \tag{3}$$

in which X and Y = percentage isotopic abundances of two heavy isotopes; the indexes i and k govern the isotopic combination considered, such that $i + k$ designates the number of combinations greater than the all-light isotope species; e.g., $i = 1$, $k = 1$ corresponds to the third isotopic species, or the "second isotope peak" in a mass spectrum.

From expansions of Eq. (1), the abundance relationships can be calculated between the all-light isotopic species of any of the heavier isotopic species. For example, for the all-light molecular species M the abundance of the first isotope peak (M + 1) can be expressed[1] as

[3] J. A. Yergey, *Int. J. Mass Spectrom. Ion Phys.* **52,** 337 (1983).
[4] C. S. Hsu, *Anal. Chem.* **56,** 2263 (1984).
[5] M. L. Brownawell and J. San Filippo, Jr., *J. Chem. Ed.* **59,** 663 (1982).

$$\frac{P_{M+1}}{P_M} = w\left(\frac{c}{100-c}\right) + x\left(\frac{h}{100-h}\right)$$

$$+ y\left(\frac{n}{100-n}\right) + z\left(\frac{o_1}{100-o_1-o_2}\right) \quad (4)$$

Although seldom justified by experimental circumstances, slight variations in natural isotopic abundances dictate that accurate values of c, h, n, etc., be used in some circumstances. Otherwise the mean values given in Appendix 1 will suffice, in which case values for terms in Eq. (4) become

$$\frac{c}{100-c} = 1.1225 \times 10^{-2} \qquad \frac{h}{100-h} = 1.5002 \times 10^{-4}$$

$$\frac{n}{100-n} = 3.6130 \times 10^{-3} \qquad \frac{o_1}{100-o_1-o_2} = 4.0177 \times 10^{-4}$$

Calculation of the entire isotopic abundance pattern for a given molecule can be carried out using a small computer, and represents the product of individual probabilities for each element. For example, the isotopic distribution for methionine ($C_5H_{11}NO_2S$) is described by Eq. (5), where ΣP represents the sum of probabilities of all isotopic species in the molecule,[2] and s_1 and s_2 represent percentage abundances of ^{33}S and ^{34}S. The sum of values of i and k, from each bracketed statement, represents the number of heavy isotopes under consideration.

$$\Sigma P = \left\{\left(\frac{100-c}{100}\right)^5\left(\frac{100-h}{100}\right)^{11}\right.$$

$$\times \left(\frac{100-n}{100}\right)\left(\frac{100-o_1-o_2}{100}\right)^2\left(\frac{100-s_1-s_2}{100}\right)\right\}$$

$$\times \left\{\sum_{i=0}^{5}\binom{5}{i}\left(\frac{c}{100-c}\right)^i\right\}\left\{\sum_{i=0}^{11}\binom{11}{i}\left(\frac{h}{100-h}\right)^i\right\}$$

$$\times \left\{\sum_{i=0}^{1}\binom{1}{i}\left(\frac{n}{100-n}\right)^i\right\}$$

$$\times \left\{\sum_{i=0}^{2}\sum_{k=0}^{i}\binom{2}{i}\binom{i}{k}\left(\frac{o_1}{100-o_1-o_2}\right)^i\left(\frac{o_2}{o_1}\right)^k\right\}$$

$$\times \left\{\sum_{i=0}^{1}\sum_{k=0}^{i}\binom{1}{i}\binom{i}{k}\left(\frac{s_1}{100-s_1-s_2}\right)^i\left(\frac{s_2}{s_1}\right)^k\right\} \quad (5)$$

By changes in the value of c (or other isotopes), the isotopic patterns for isotopically enriched molecules can be calculated.[2] In cases in which the same element occurs at two different isotopic enrichments in the same molecule, the "enriched" isotope can be defined as a separate element. For example, the tri-$^{13}CH_3$ derivative of methionine prepared using 98 atom % $^{13}CH_3I$ would be represented in an expanded version of Eq. (5) as $C_5{}^{13}C_3H_{17}NO_2S$, in which five carbons are treated at natural isotopic abundance levels, and three carbons are treated with $c = 98$.

Expansions of Eq. (1) can be used to calculate the isotopic patterns for large molecules or ions, e.g., over 10,000 Da. It has been pointed out that in such cases variations in the natural isotopic abundance of carbon may have a significant effect on the average mass value of the isotopic cluster.[6]

Approximation for Calculation of Simple Isotopic Patterns

For measurements involving small organic molecules, in which the qualitative appearance of isotopic patterns is of interest, the height of the first isotope peak (M + 1) can be approximated relative to M = 100% by

$$\% \ (M + 1) = (w \times 1.1) + (y \times 0.4) \tag{6}$$

Contributions from ^2H and ^{17}O are low (see Appendix 1) and so can usually be ignored. Likewise, the elements F, P, and I are monoisotopic (see Appendix 1) and so do not contribute to higher isotope peaks. If s atoms of S or t atoms of Si are present, the terms ($s \times 0.8$) and ($t \times 5$) are added.

The second isotope peak (M + 2) can be approximated by

$$\% \ (M + 2) \simeq \frac{(w \times 1.1)^2}{200} + (z \times 0.2) \tag{7}$$

If s atoms of S or t atoms of Si are present, the terms ($s \times 4$) and ($t \times 3$) are added.

When elements containing large amounts of more than one isotope are present (e.g., Cl, Br), the primary isotope patterns can be estimated by expansion of the binomial $(a + b)^n$ for each element, where a = abundance of the light isotope, b = abundance of the heavy isotope, and n = number of atoms present. For example: for Cl_3, $a \simeq 3$, $b \simeq 1$, $n = 3$. Therefore $(a + b)^3 = a^3 \times 3a^2b + 3ab^2 + b^3$ and abundance ratios $= 27:27:9:1$.

For presence of m atoms of a second such element having isotopes d and e, a product is expanded:

$$(a + b)^n \times (d + e)^m$$

[6] S. C. Pomerantz and J. A. McCloskey, *Org. Mass Spectrom.* **22**, 251 (1987).

For example, for Cl_2Br, $a \simeq 3$, $b \simeq 1$, $n = 2$, $d \simeq 1$, $e \simeq 1$, $m = 1$. Therefore, $(a + b)^2 \times (d + e) = a^2d + 2abd + b^2d + a^2e + 2abe + b^2e$. Combination of the terms in the expansion which corresponds to equivalent nominal mass values (0, 1, 2 . . . occurrences of a heavy isotope), and substitution of the numerical values for a, b, d, and e, yields the approximate abundance ratio $9 : 15 : 7 : 1$.

To include contributions from C, N, O, etc., in such cases, the principal isotope peaks (9, 15, 7, and 1 abundance units in the preceding example) can each be treated as "all-light" isotopic species and contributions from ^{13}C, etc., calculated relative to each of those peaks using Eqs. (6) and (7). For example, for dichlorobromobenzene ($C_6H_3Cl_2Br$) each of the principal halogen isotope peaks (separated by 2 u) would be accompanied by a first isotope peak due to ^{13}C of $(6 \times 1.1)\%$, giving the approximate pattern $9 : 0.60 : 15 : 1.0 : 7 : 0.47 : 1 : 0.07$, with each species separated by 1 u.

Appendix 5. Nomenclature for Peptide Fragment Ions (Positive Ions)*

By KLAUS BIEMANN

For further discussion of peptide ion nomenclature, see Refs. 1 and 2.

N-terminal Ions

a_n: $H—(NH—CHR—CO)_{n-1}—\overset{+}{N}H{=}CHR_n$

or

H^+

$H—(NH—CHR—CO)_{n-1}—NH—\overset{CR_n{}^aR_n{}^b}{\underset{\|}{C}}H$

$a_n + 1$: $H—(NH—CHR—CO)_{n-1}—NH—\overset{R_n}{\underset{|}{C}}H\cdot$ with H^+

b_n: $H—(NH—CHR—CO)_{n-1}—NH—CHR_n—C{\equiv}O^+$

c_n: $H—(NH—CHR—CO)_n—NH_2$ with H^+

---

* R represents the side chains of the amino acids; $R_n{}^a$ and $R_n{}^b$ are the beta substituents of the nth amino acid.
¹ P. Roepstorff and J. Fohlman, *Biomed. Mass Spectrom.* **11**, 601 (1984).
² K. Biemann, *Biomed. Environ. Mass Spectrom.* **16**, 99 (1988).

$$d_n: \quad H\overline{-(NH-CHR-CO)}_{n-1}-NH-\overset{\overset{\displaystyle CR_n^{\,b}}{\|}}{CH} \qquad \text{(with } H^+ \text{ over the bracketed portion)}$$

C-terminal Ion Types

$$v_n: \quad HN=CH-CO\overline{-(NH-CHR-CO)}_{n-1}-OH \qquad (H^+)$$

$$w_n: \quad \overset{\overset{\displaystyle CR_n^{\,b}}{\|}}{CH}-CO\overline{-(NH-CHR-CO)}_{n-1}-OH \qquad (H^+)$$

$$x_n: \quad {}^+O\equiv C-NH-CHR_n-CO-(NH-CHR-CO)_{n-1}-OH$$

or

$$O=C=N\overline{-CHR_n-CO-(NH-CHR-CO)}_{n-1}-OH \qquad (H^+)$$

$$y_n: \quad H\overline{-(NH-CHR-CO)}_n-OH \qquad (H^+)$$

$$y_n - 2: \quad \overline{HN=CR_n-CO-(NH-CHR-CO)}_{n-1}-OH \qquad (H^+)$$

$$z_n: \quad \overset{\overset{\displaystyle CR_n^{\,a}R_n^{\,b}}{\|}}{CH}-CO\overline{-(NH-CHR-CO)}_{n-1}-OH \qquad (H^+)$$

$$z_n + 1: \quad \cdot CHR_n-CO\overline{-(NH-CHR-CO)}_{n-1}-OH \qquad (H^+)$$

Non-Sequence Specific Ions

Internal acyl ions (denoted by single letter codes) (e.g., GA at m/z 129 for R = H, CH$_3$):

$$H_2N-CHR-CO-NH-CHR-C\equiv O^+$$

Internal immonium ions (denoted by single-letter code followed by the notation -28):

$$H_2N-CHR-CO-\overset{+}{NH}=CHR$$

Amino acid immonium ions (denoted by single letter code) (e.g., F at m/z 120 for R = C$_6$H$_5$CH$_2$):

$$H_2\overset{+}{N}=CHR$$

Loss of amino acid side chains from $(M + H)^+$ is denoted as the single letter code of the amino acid involved, preceded by a minus sign (e.g., $-V$ for $[M + H - 43]^+$).

Appendix 6. Mass Values for Amino Acid Residues in Peptides

By KLAUS BIEMANN

		Residue mass[a]	
Amino acid	Single letter code	Monoisotopic[b]	Average[c]
Glycine	G	57.02147	57.052
Alanine	A	71.03712	71.079
Serine	S	87.03203	87.078
Proline	P	97.05277	97.117
Valine	V	99.06842	99.133
Threonine	T	101.04768	101.105
Cysteine	C	103.00919	103.144
Isoleucine	I	113.08407	113.160
Leucine	L	113.08407	113.160
Aspargine	N	114.04293	114.104
Aspartic acid	D	115.02695	115.089
Glutamine	Q	128.05858	128.131
Lysine	K	128.09497	128.174
Glutamic acid	E	129.04260	129.116
Methionine	M	131.04049	131.198
Histidine	H	137.05891	137.142
Phenylalanine	F	147.06842	147.177
Arginine	R	156.10112	156.188
Tyrosine	Y	163.06333	163.17
Tryptophan	W	186.07932	186.213
Homoserine lactone		83.03712	83.090
Homoserine		101.04768	101.105
Pyroglutamic acid		111.03203	111.100
Carbamidomethylcysteine		160.03065	160.197
Carboxymethylcysteine		161.01466	161.181
Pyridylethylcysteine		208.06703	208.284

[a] Defined as the mass of subunit A:

$$\text{---CO} + \text{NH---}\underset{\text{A}}{\overset{\overset{\displaystyle R}{|}}{\text{CH}}}\text{---CO} + \text{NH---}$$

[b] Calculated using all-light isotopic exact mass values in Appendix 1.
[c] Calculated using atomic weight values in Appendix 1.

Author Index

Numbers in parentheses are footnote reference numbers and indicate that an author's work is referred to although the name is not cited in the text.

Subject Index

A

structural elucidation of, 672
Asparagine residues, glycosylation of, 501
AspN, protein digestion, 466
Associative ion-molecular reactions, 172
Asymmetric fission, 276
Atmospheric pressure ionization mass
  spectrometers, 13, 35–36
Atomic weights, selected, 870
Average mass, of molecule, calculation,
  443–444

B

BEB instrument, 135
Beef spleen purple acid phosphatase. *See*
  Violet phosphatase from beef spleen
BE instruments, 42, 134
Bent quadrupole collision cell, 173–174
Beryllium/copper dynode, 65
Biantennary glycoprotein fragment, FAB-
  mass spectrum of product formed
  after periodate oxidation, NaBD₄
  reduction, and permethylation of,
  596–597
Bile acids, ¹⁸O-labeled, synthesis of, 345–
  346
Binary model, for rotational–vibrational
  excitation, 252–253
N,O-Bis(trimethylsilyl)trifluoroacetamide,
  327
Blood group antigens, 624
Box and grid electron multiplier, 66
Bradykinin
  electrospray ionization mass spectra of,
    419–421
  phosphotyrosine from, MS/MS spectra
    of, 494–497
Brønsted acid chemical ionization, 14–16
  formation of cluster ions in, 15
  proton exothermicity, 14–15
  reagents, 14
Brønsted base chemical ionization, 19–20
  for molecular weight determination, 20
  reagent ions, 19–20
BSTFA. *See N,O*-Bis(trimethylsilyl)tri-
  fluoroacetamide
tert-Butoxycarbonyl-tyrosyl glycopeptide,
  positive ion CID spectrum of, 709–711
tert-Butyldimethylsilylation, 318

of DNA, 845
method, 327

C

CAD (collisional-activated dissociation).
  See Collisional activation; Collision-
  induced dissociation
Caerulein, LSIMS molecular ion region of,
  in positive and negative ion modes,
  486–487
Californium-252, 351
  airborne, danger of, 277–278
  asymmetric fission, 276
  fission barrier, 275–276
  neutron multiplicity, 276
  nuclear decay properties, 275–276
  production, 275
  radiation exposure with, 277–278
  radiation from, kinds of, 277
  radiation levels, 277
  radiological aspects of, 277–278
  self-transfer of, 277–278
Californium-252 plasma desorption mass
    spectrometer
  acceleration grid, 267
  ²⁵²Cf source, 267–268
  chemical properties, 275
  cover foil, 278
  dose levels from, 278–279
  license required to procure, 278
  nuclear radiation surrounding, 276
  open, 277
  preparation of, 275
  properly sealed, 278
  sealed, 278
  strength of, 277
  commercial instrument, 264
  configuration of elements of, 267
  constant fraction discriminator, 268
  conversion foil, 267
  electron detector, 267
  fast logic pulses, 268
  flight tube, 266–267
  operation, principles and practice, 266–268
  start detector, 267
  stop detector, 268
  target stick, 266

HV8073.6.F66   3rd floor

QD1.A355 no.40 c.1 } 2nd floor
QD1.A355 no.40 c.2 }

QP601.CM33 vol. 193   2nd floor

Q

S

reagents for, 792
of RNA hydrolyzates, 794
Triperfluoroalkyltriazines, 296
Triple quadrupole mass spectrometers,
   154, 159–160, 200, 353, 608
   low-energy collisional activation carried
      out in, 251–252
Tris(perfluoroheptyl)-s-triazine
   as mass reference standard, 313
   mass values, 870
   positive ion electron ionization mass
      spectrum of, 308
Tris(perfluorononyl)-s-triazine, as mass
   reference standard, 313
Trypanosomes, variant surface glycoprotein, carbohydrate heterogeneity in
   high mannose glycopeptide from, 25–
   26
Trypsin
   bovine, M_r determination, and elec-
      trospray ionization-MS, 424
   chymotryptic behavior, in protein diges-
      tion, 381–383
   cleavage of glycoproteins, 503
   cleavage of nitrocellulose-bound sample
      in situ, 439
   digestion of cytochrome c, for mass
      spectrometry, 364–367
   digestion of protein, 464–465
      to produce disulfide-containing pep-
         tides, 379
   purified, by reversed-phase HPLC, 365–
      366
Tryptic digests, microbore LC/MS with
   CF-FAB for analysis of, 227–233
Tryptic peptides
   CID spectra, interpretation, 465–466
   doubly charged
      produced by electrospray, 178–179
      tandem mass spectrometry of, 426–
         427
   electrospray mass spectrum of, 178–179
   mapping, 389–390
   sequencing, 390
Tumor-associated antigens, 624
Two-dimensional nuclear magnetic reso-
   nance, in carbohydrate analysis, 545
Two-sector instruments, 133–134
Two-sector mass spectrometry, sensitivity,
   522

Tyrosine residues, sulfated, LSIMS analysis, 483, 486
Tyrosyl sulfotransferases, 480

U

Ubiquitin
   bovine red blood cell, source, 443
   calculated molecular weight, vs. molecu-
      lar weight from electrospray ioniza-
      tion-MS, 445
UHQ-water, 438, 441
Ultramark 1621, 307–308
   as mass reference standard, 313
Ultramark 2500F, 307–308
   mass values, 872
Ultramark F series, 307–308
   chemical ionization spectra of, 309
   as mass reference standard, 313
Unimolecular decomposition reactions, 238
Universal interface, 124–130
   block diagram of, 125
   control system, 128–129
   electron ionization system, 126
   flow conditions, 126
   flow velocities through, 126
   ionization techniques with, 129–130
   ion source, 127–128
   liquid chromatographic conditions with,
      129
   mass spectrometer used with, 128
   membrane separator, 127
   momentum separator, 127
   optimizing, for particular applications,
      129–130
   solvent removal, 128–129
   vacuum system, 128
Uridine, ^{18}O incorporation into, by exchange with $H_2^{18}O$, 346
Urine, nucleoside fraction of, trimethylsilyl
   derivatives of, 794
UV laser desorption. See Laser desorption

V

Val¹-gramicidin A, CID mass spectrum of,
   458–459
Vanillic acid, as matrix for LDI-MS, 283
Vaporization, minimum temperatures for,
   for several common solvents, 113–114